Springer Texts in Statistics

Advisors:
George Casella, Stephen Fienberg, Ingram Olkin

Springer Texts in Statistics

(continued after index)

Christian P. Robert

The Bayesian Choice

From Decision-Theoretic Foundations to Computational Implementation

Second Edition

 Springer

Christian P. Robert
CEREMADE
Universite Paris Dauphine
Place du Maréchal de Lattre de Tassigny
75775 Paris cedex 16
France
xian@ceremade.dauphine.fr

Library of Congress Control Number: 2007926596

ISBN 978-0-387-71598-8   e-ISBN 978-0-387-71599-5

Printed on acid-free paper.

9 8 7 6 5 4 3 2 1

springer.com

To my reference prior,
Brigitte,
and to two major updates,
Joachim and Rachel.

Preface to the Paperback Edition

What could not be changed must be endured.

Robert Jordan, *New Spring, Prequel to The Wheel of Time*

THANKS

While this paperback edition is almost identical to the second edition of *The Bayesian Choice*, published in 2001, and thus does not require a specific introduction, it offers me the opportunity to thank several groups of people for their contributions that made this edition possible.

First, the changes, when compared with the second edition, are only made of corrections of typographical and conceptual errors (whose updated list can be found on my Webpage[1]). Almost all errors have been pointed out to me by friends, colleagues, unknown lecturers or anonymous readers who (always kindly and sometimes apologetically) sent me emails asking me to clarify a specific paragraph, a formula or a problem that did not make sense to them. Needless to say, I am very grateful to those numerous contributors for making the book more accurate and I obviously encourage all contributors who think there could be an error in the current edition to contact me because they cannot be wrong! Either there indeed is a mistake that needs to be set right or there is no mistake but the context is ambiguous at best and the corresponding text needs to be rewritten. Thanks, then, to Guido Consonni, Estelle Dauchy, Arnaud Doucet, Pierre Druihlet, Ed Green, Feng Liang, Jean-Michel Marin, M.R.L.N. Panchanana, Fabrice Pautot, and Judith Rousseau.

Second, working with my colleague Jean-Michel Marin on the design of a course for teaching Bayesian Statistics from a practical and computational perspective (a venture now published as *Bayesian Core* by Springer in early 2007) was a very important moment in that I realized that the material in this very book, *The Bayesian Choice*, was essential in communicating the essential relevance and coherence of the Bayesian approach

---

[1] http://www.ceremade.dauphine.fr/~xian/books.html

through its decision-theoretic foundations, while the message contained in the other book and transmitted only through processing datasets is that the Bayesian methodology is a universal and multifaceted tool for data analysis. While introducing wider and less mathematical audiences to the elegance and simplicity of the Bayesian methodology in a shorter and therefore more focussed volume was also necessary, if only because some learn better from examples than from theory, I came to the conclusion that there was no paradox in insisting on those foundations in another book! I am therefore immensely thankful to Jean-Michel Marin for initiating this epiphany (if I may rightly borrow this expression from Joyce!), as well as for several years of intense collaboration. Similarly, the DeGroot Prize committee of the ISBA—International Society for Bayesian Analysis—World meeting of 2004 in Valparaiso, Chile, greatly honored me by attributing to *The Bayesian Choice* this prestigious prize. In doing so, this committee highlighted the relevance of both foundations and implementation for the present and future of Bayesian Statistics, when it stated that the *"book sets a new standard for modern textbooks dealing with Bayesian methods, especially those using MCMC techniques, and that it is a worthy successor to DeGroot's and Berger's earlier texts"*. I am quite indebted to the members of the committee for this wonderful recognition.

Third, it has been more than 18 years since I started working with John Kimmel from Springer New York (on a basic Probability textbook with Arup Bose that never materialized), and I always appreciated the support he provided over the various editions of the books. So, when he presented me with the possibility to publish this paperback edition, I first got some mixed feelings, because he made me feel like a classics author! This caused my kids poking endless fun at me and, in the end, I am quite grateful to John for the opportunity to teach from this book to a wider audience and thus hopefully exposing them to the beauty of Bayesian theory and methodology. Short of embarking upon a translation of *The Bayesian Choice* into Chinese or Arabic, I do not think there is much more he could do to support the book!

In Memoriam

This is a sheer consequence of time moving on, unfortunately, but I lost another dear friend since the last publication of *The Bayesian Choice*. José Sam Lazaro passed away last Spring: a mathematician, a professor and a colleague at the Université de Rouen, a music addict and a movie aficionado that made me discover *Der Tod und das Mädchen* as well as *The Night of The Hunter*, an intense piano player, a memorable tale teller, he was above all a philosopher and a friend. Although he would have made a joke out of it, I would like to dedicate this edition to his memory and wish him well to play this final and endless sonata...

Valencia and Paris   **Christian P. Robert**
February 2007

Preface to the Second Edition

"You can never know everything," Lan said quietly, "and part of what you know is always wrong. Perhaps even the most important part. A portion of wisdom lies in knowing that. A portion of courage lies in going on anyway."

Robert Jordan, *Winter's Heart, Book IX of the Wheel of Time.*

Overview of Changes

Why a second edition? When thinking about it, this is more like a third edition, since the previous edition of *The Bayesian Choice* was the translation of the French version, and already included updates and corrections. The reasons for a new edition of the book are severalfold. The Bayesian community has grown at an incredible pace since 1994. The previous version not only overlooks important areas in the field but misses the significant advances that have taken place in the last seven years.

Firstly, the MCMC[2] revolution has fueled considerable advances in Bayesian modeling, with applications ranging from medical Statistics, to signal processing, to finance. While present in the 1994 edition, these methods were not emphasized enough: for instance, MCMC methods were not presented until the penultimate chapter.

Another significant advance that needed attention is the development of new testing approaches and, more generally, of model choice tools in connection with, and as a result of, MCMC techniques such as reversible jump. Other important expansions include hierarchical and dynamic models, whose processing only began to emerge in the early 1990s.

This second edition is not revolutionary, compared with the 1994 edition. It includes, however, important advances that have taken place since then. The only new chapter deals with model choice (Chapter 7) and is isolated from general testing theory (Chapter 5), since model choice is indeed a different problem and also because it calls for new, mostly computational,

---

[2] MCMC stands for *Markov chain Monte Carlo*, a simulation methodology which was (re)discovered in the early 1990s by the Bayesian community.

tools. For this reason, and also to emphasize the increasing importance of computational techniques, Chapter 6—previously Chapter 9—has been placed earlier in the book, after the presentation of the fundamentals of Bayesian Statistics. Chapter 6 could almost be called a new chapter in that its presentation has been deeply renovated in the light of ten years of MCMC practice. In Chapter 3, the material on noninformative priors has been expanded and includes, in particular, matching priors, since the research activity has been quite intense in this area in the past few years. Chapter 4 still deals with general estimation problems, but I have incorporated a new section on dynamic models, since those are quite central to the development of Bayesian Statistics in applied fields such as signal processing, finance and econometrics. Despite Delampady's criticisms of Chapter 11 in *The Mathematical Reviews*, I have decided to leave this chapter in: it does not hurt, when one is finished reading a book, to take an overall and more philosophical view of the topic because the reader has very likely acquired enough perspective to understand such arguments. (In a strictly textbook implementation, this chapter can be suggested as an additional reading, comparable with the Notes.)

Another noteworthy change from the previous edition is the decreased emphasis on decision-theoretic principles. Although I still believe that statistical procedures must be grounded on such principles, the developments in the previous decade have mainly focused on methodology, including computational methodology, rather than attacking broader and more ambitious decision problems (once again, including computational methodology). The second part of the book (starting with Chapter 6) is therefore less decision-theoretic and, in contrast to others, chapters such as Chapters 8 and 9 have hardly been changed.

At a more typographical level, subsections and separations have been introduced in many sections to improve visibility and reading, and more advanced or more sketchy parts have been relegated to a *Notes* section at the end of each chapter, following the approach adopted in *Monte Carlo Statistical Methods*, written with George Casella. The end of an example is associated with the ‖ symbol, while the end of a proof is indicated by the □□ symbol.

Several books on Bayesian Statistics have appeared in the interim, among them Bernardo and Smith (1994), Carlin and Louis (1996, 2000a), Gelman et al. (1996), O'Hagan (1994), and Schervish (1995). However, these books either emphasize deeper theoretical aspects at a higher mathematical level (Bernardo and Smith (1994), O'Hagan (1996), or Schervish (1996)) and are thus aimed at a more mature audience than this book, or they highlight a different vision of the practice of Bayesian Statistics (Carlin and Louis (2000a) or Gelman et al. (1996)), missing for instance the connection with Decision Theory developed in this book.

Course Schedules

My advice about running a course based on this book has hardly changed. In a first course on Bayesian analysis, the basic chapters (Chapters 1–6) should be covered almost entirely, with the exception of the Notes and Sections 4.5 and 5.4, while a course focusing more on Decision Theory could skip parts of Chapters 1 and 3, and Chapter 4 altogether, to cover Chapters 7–9. For a more advanced curriculum for students already exposed to Bayesian Statistics, my suggestion is first to cover the impropriety issue in Section 1.5, the noninformative priors in Section 3.5, the dynamic models in Section 4.5 and Notes 4.7.3 and 4.7.4. I would also spend time on the testing issues of Chapter 5 (with the possible exception of Sections 5.3 and 5.4). Then, after a thorough coverage of simulation methods via Chapter 6, I would move to the more controversial topic of model choice in Chapter 7, to recent admissibility results as in Section 8.2.5 and Note 8.7.1, and to the hierarchical and empirical modelings of Chapter 10. In this later scenario, the Notes should be most helpful for setting out reading seminars.

Thanks

I have always been of two minds about including a thank-you section in a book: on the one hand, it does not mean anything to most readers, except maybe to bring to light some of the author's idiosyncrasies that might better remain hidden! It may also antagonize some of those concerned because they are not mentioned, or because they are not mentioned according to their expectations, or even because they *are* mentioned! On the other hand, the core of ethical requirements for intellectual works is that sources should be acknowledged. This extends to suggestions that contributed to making the work better, clearer or simply different. And it is a small token of gratitude to the following people for the time spent on the successive drafts of this edition that their efforts should be acknowledged in print for all to behold!

Although this is "only" a revision, the time spent on this edition was mostly stolen from evenings, early mornings and week-ends, that is from Brigitte, Joachim and Rachel's time! I am thus most grateful to them for reading and playing (almost) quietly while I was typing furiously and searching desperately through piles of material for this or that reference, and for listening to Bartoli and Gudjónsson, rather than to Manau or Diana Krall! I cannot swear this book-writing experience will never happen again but, in the meanwhile, I promise there will be more time available for reading *Mister Bear to the Rescue*, and for besieging the Playmobil castle in full scale, for playing chess and for biking on Sunday afternoons!

I am thankful to several people for the improvements in the current edition! First, I got a steady stream of feedback and suggestions from those who taught from the book. This group includes Ed Green, Tatsuya Kubokawa, and Marty Wells. In particular, Judith Rousseau, radical biker and Jordanite as well as Bayesian, definitely was instrumental in the re-organization of Chapter 3. I also got many helpful comments from many people, including the two "Cambridge Frenchies" Christophe Andrieu and Arnaud Doucet (plus a memorable welcome for a retreat week in Cambridge to finish Chapter 6), Jim Berger (for his support in general, and for providing preprints on model choice in particular), Olivier Cappé (who also installed Linux on my laptop, and consequently brought immense freedom for working on the book anywhere, from the sand-box to the subway, and, lately, to CREST, where Unix is now banned!), Maria DeIorio, Jean-Louis Fouley, Malay Ghosh (through his very supportive review in JASA), Jim Hobert (who helped in clarifying Chapters 6 and 10), Ana Justel, Stephen Lauritzen (for pointing out mistakes with Wishart distributions), Anne Philippe, Walter Racugno (who gave me the opportunity to teach an advanced class in model choice in Ca'liari last fall, thus providing the core of Chapter 7), Adrian Raftery, Anne Sullivan Rosen (about the style of this preface), and Jean-Michel Zakoian (for his advice on the new parts on dynamic models). I also take the opportunity to thank other friends and colleagues such as George Casella, Jérôme Dupuis, Merrilee Hurn, Kerrie Mengersen, Eric Moulines, Alain Monfort, and Mike Titterington, since working with them gave me a broader vision of the field, which is hopefully incorporated in this version. In particular, the experience of writing *Monte Carlo Statistical Methods* with George Casella in the past years left its mark on this book, not only through the style file and the inclusion of Notes, but also as a sharper focus on essentials. Manuela Delbois helped very obligingly with the transformation from TEX to LATEX, and with the subsequent additions and indexings. And, last but not least, John Kimmel and Jenny Wolkowicki, from Springer-Verlag, have been very efficient and helpful in pushing me to write this new edition for the former, in keeping the whole schedule under control and in getting the book published on time for the latter. Needless to say, the usual *proviso* applies: all remaining typos, errors, confusions and obscure statements are mine and only mine!

IN MEMORIAM

A most personal word about two people whose *absence* has marked this new edition: in the summer 1997, I lost my friend Costas Goutis in a diving accident in Seattle. By no means am I the only one to feel keenly his absence, but, beyond any doubt, this book would have benefited from his vision, had he been around. Two summers later, in 1999, Bernhard K. Flury died in a mountain accident in the Alps. While the criticisms of our respective books always focussed on the cover colors, even to the extent of sending one another pirated versions in the "right" color, the disappearance of his unique humor has taken a measure of fun out of the world.

Paris, France   Christian P. Robert
March 2001

Preface to the First Edition

From where we stand, the rain seems random.
If we could stand somewhere else, we would see the order in it.

— **T. Hillerman** (1990) *Coyote Waits.*

This book stemmed from a translation of a French version that was written to supplement the gap in the French statistical literature about Bayesian Analysis and Decision Theory. As a result, its scope is wide enough to cover the two years of the French graduate Statistics curriculum and, more generally, most graduate programs. This book builds on very little prerequisites in Statistics and only requires basic skills in calculus, measure theory, and probability. Intended as a preparation of doctoral candidates, this book goes far enough to cover advanced topics and modern developments of Bayesian Statistics (complete class theorems, the Stein effect, hierarchical and empirical modelings, Gibbs sampling, etc.). As usual, what started as a translation eventually ended up as a deeper revision because of the comments of French readers, adjustments to the different needs of American programs, and because my perception of things has changed slightly in the meantime. As a result, this new version is quite adequate for a general graduate audience of an American university.

In terms of level and existing literature, this book starts at a level similar to those of the introductory books of Lee (1989) and Press (1989), but it also goes further and keeps up with most of the recent advances in Bayesian Statistics, while justifying the theoretical appeal of the Bayesian approach on decision-theoretic grounds. Nonetheless, this book differs from the reference book of Berger (1985a) by including the more recent developments of the Bayesian field (the Stein effect for spherically symmetric distributions, multiple shrinkage, loss estimation, decision theory for testing and confidence regions, hierarchical developments, Bayesian computation, mixture estimation, etc.). Moreover, the style is closer to that of a textbook in the sense that the progression is intended to be linear. In fact, the exposition of the advantages of a Bayesian approach and of the existing links with other axiomatic systems (fiducial theory, maximum likelihood, frequentist theory, invariance, etc.) does not prevent an overall unity in the discourse. This should make the book easier to read by students; through the years and on both sides of the blackboard(!), I found most Statistics courses disturbing because a wide scope of methods was presented simultaneously with very little emphasis on ways of discriminating between competing approaches. In particular, students with a strong mathematical background are quite puzzled by this multiplicity of theories since they have not been exposed previously to conflicting systems of axioms. A unitarian presentation that includes other approaches as limiting cases is thus more likely to reassure the students, while giving a broad enough view of Decision Theory and even of parametric Statistics.

The plan[3] of the book is as follows: Chapter 1 is an introduction to statistical models, including the Bayesian model and some connections with the Likelihood Principle. The book then proceeds with Chapter 2 on Decision Theory, considered from a classical point of view, this approach being justified through the axioms of rationality and the need to compare decision rules in a coherent way. It also includes a presentation of usual losses and a discussion of the Stein effect. Chapter 3 gives the corresponding analysis for prior distributions and deals in detail with conjugate priors, mixtures of conjugate priors, and noninformative priors, including a concluding section on prior robustness. Classical statistical models are studied in Chapter 4, paying particular attention to normal models and their relations with linear regression. This chapter also contains a section on sampling models that allows us to include the pedagogical example of capture-recapture models. Tests and confidence regions are considered separately in Chapter 5, since we present the usual construction through $0 - 1$ losses, but also include recent advances in the alternative decision-theoretic evaluations of testing problems. The second part of the book dwells on more advanced topics and can be considered as providing a basis for a more advanced graduate course. Chapter 8 covers complete class results and sufficient/necessary admissibility conditions. Chapter 9 introduces the notion of invariance and its relations with Bayesian Statistics, including a heuristic section on the Hunt–Stein theorem. Hierarchical and empirical extensions of the Bayesian approach, including some developments on the Stein effect, are treated in Chapter 10. Chapter 6 is quite appealing, considering the available literature, as it incorporates in a graduate textbook an introduction to state-of-the-art computational methods (Laplace, Monte Carlo and, mainly, Gibbs sampling). In connection with this chapter, a short appendix provides the usual pseudo-random generators. Chapter 11 is a more personal conclusion on the advantages of Bayesian theory, also mentioning the most common criticisms of the Bayesian approach. French readers may appreciate that

---

[3] The chapter and section numbers have been adapted to the current edition.

a lot of effort has been put into the exercises of each chapter in terms of volume and difficulty. They now range from very easy to difficult, instead of being uniformly difficult! The most difficult exercises are indexed by asterisks and are usually derived from research papers (covering subjects such as *spherically symmetric distributions* (1.1), *the Pitman nearness criticism* (2.57–2.62), *marginalization paradoxes* (3.44–3.50), *multiple shrinkage* (10.38), etc.). They should benefit most readers by pointing out new directions of Bayesian research and providing additional perspectives.

A standard one-semester course should cover the first five chapters (with the possible omission of Note 2.8.2, §2.5.4, §2.6, §3.4, Note 4.7.1, §4.3.3, and §5.4). More advanced (or longer) courses can explore the material presented in Chapters 8, 9, and 10, bearing in mind that a detailed and rigorous treatment of these topics requires additional reading of the literature mentioned in those chapters. In any case, I would advise against entirely forgoing Chapter 6. Even a cursory reading of this chapter may be beneficial to most students, by illustrating the practical difficulties related to the computation of Bayesian procedures and the corresponding answers brought by simulation methods.

This book took many excruciatingly small steps and exacted a heavy toll on evenings, weekends, and vacations... It is thus only a small indication of my gratitude that this book be dedicated to Brigitte (although she might take this as a propitiatory attempt for future books!!!). Many persons are to be thanked for the present version of this book. First and foremost, Jim Berger's "responsibility" can be traced back to 1987 when he invited me to Purdue University for a year and, as a result, considerably broadened my vision of Statistics; he emphasized his case by insisting very vigorously that I translate the French version and urging me along the whole time. My gratitude to Jim goes very deep when I consider his strong influence in my "coming-of-age" as a statistician. Mary-Ellen Bock, Anirban Das Gupta, Edward George, Gene (formerly Jiunn) Hwang, and Marty Wells were also very instrumental in my progression towards the Bayesian choice, although they do not necessarily support this choice. In this regard, George Casella must be singled out for his strong influence through these years of intense collaboration and friendship, even during his most severe (and "unbearable") criticisms of the Bayesian paradigm! I am also quite grateful to Jean-François Angers, Dean Foster, and Giovanni Parmigiani for taking the risk of using a preliminary version of these notes in their courses, as well as for their subsequent comments. Thanks to Teena Seele for guiding my first steps in TeX, as well as some delicate points in this book—never use \def as an abbreviation of definition! I am also grateful to Elsevier North-Holland for granting me permission to use Diaconis and Ylvisaker's (1985) figures in §3.3. Last, and definitely not least, Kerrie Mengersen and Costas Goutis put a lot of time and effort reading through a preliminary version and provided many helpful comments on content, style, and clarity, while adding a

touch of Ausso-Greek accent to the tone. (In addition, Costas Goutis saved
the subject index from utter destruction!) They are thus partly responsible for the improvements over previous versions (but obviously not for the
remaining defects!), and I am most grateful to them for their essential help.

Paris, France   Christian P. Robert
May 1994

Contents

List of Tables

List of Figures

CHAPTER 1

Introduction

"Sometimes the Pattern has a randomness to it—to our eyes, at least—but what chance that you should meet a man who could guide you in this thing, and he one who could follow the guiding?"

Robert Jordan, *The Eye of the World, Book I of the Wheel of Time.*

1.1 Statistical problems and statistical models

The main purpose of statistical theory is to derive from observations of a random phenomenon an *inference* about the probability distribution underlying this phenomenon. That is, it provides either an analysis (description) of a past phenomenon, or some predictions about a future phenomenon of a similar nature. In this book, we insist on the *decision-oriented* aspects of statistical inference because, first, these analysis and predictions are usually motivated by an objective purpose (whether a company should launch a new product, a racing boat should modify its route, a new drug should be put on the market, an individual should sell shares, etc.) having measurable consequences (monetary results, position at the end of the race, recovery rate of patients, benefits, etc.). Second, to propose inferential procedures implies that one should stand by them, i.e., that the statistician thinks they are preferable to alternative procedures. Therefore, there is a need for an evaluative tool that allows for the comparison of different procedures; this is the purpose of *Decision Theory*. As with most formal definitions, this view of Statistics ignores some additional aspects of statistical practice such as those related to *data collection* (surveys, design of experiments, etc.). This book does, as well, although we do not want to diminish the importance of these omitted topics.

We also insist on the fact that Statistics should be considered an *interpretation* of natural phenomena, rather than an *explanation*. In fact, statistical inference is based on a *probabilistic modeling* of the observed phenomenon and implies a necessarily reductive formalization step since without this probabilistic support it cannot provide any useful conclusion (or decision).

Example 1.1.1 Consider the problem of forest fires. They usually appear at random, but some ecological and meteorological factors influence their eruption. Determining the probability p of fire as a *function* of these factors should help in the prevention of forest fires, even though such modeling is obviously unable to lead to the eradication of forest fires and cannot possibly encompass all the factors involved. A more reductive approach is to assume a parametrized shape for the function p, including physical constraints on the influential factors. For instance, denoting by h the humidity rate, t the average temperature, x the degree of management of the forest, a *logistic model*

$$p = \exp(\alpha_1 h + \alpha_2 t + \alpha_3 x) / \left[1 + \exp(\alpha_1 h + \alpha_2 t + \alpha_3 x) \right]$$

could be proposed, the statistical step dealing with the evaluation of the parameters $\alpha_1, \alpha_2, \alpha_3$.   ‖

To impose a probability modeling on unexplained phenomena seems to be overly reductive in some cases because a given phenomenon can be entirely deterministic, although the regulating function of the process is unknown and cannot be recovered from the observations. This is, for instance, the case with *chaotic phenomena*, where a deterministic sequence of observations cannot be distinguished from a sequence of random variables, in the sense of a statistical test (see Bergé, Pommeau, and Vidal (1984) and Gleick (1987) for introductions to chaos theory). *Pseudo-random generators* are actually based on this indeterminacy. While they use iterative deterministic algorithms such as

$$a_{t+1} = f(a_t),$$

they imitate (or *simulate*) rather well the behavior of a sequence of random variables (see Devroye (1985), Gentle (1998), Robert and Casella (2004), and Appendix B for a list of the most common generators).

Although valid on philosophical grounds, this criticism does not hold if we consider the probabilistic modeling from the *interpretation* perspective mentioned above. This modeling simultaneously incorporates the available information about the phenomenon (influential factors, frequency, amplitude, etc.) *and* the uncertainty pertaining to this information. It thus authorizes a *quantitative* discourse on the problem by providing via probability theory a genuine *calculus of uncertainty* going beyond the mere description of deterministic modelings. This is why a probabilistic interpretation is necessary for statistical inference; it provides a framework replacing a singular phenomenon in the globality of a model and thus allows for analysis and generalizations. Far from being a misappropriation of the inferential purposes, the imposition of a probabilistic structure that is only an approximation of reality is essential for the subsequent statistical modeling to induce a deeper and more adequate understanding of the considered phenomenon.

Obviously, probabilistic modeling can only be defended if it provides an adequate representation of the observed phenomenon. A more down-to-earth criticism of probabilistic modeling is therefore that, even when a modeling is appropriate, it is difficult to know exactly the probability distribution underlying the generation of the observations, e.g., to know that it is normal, exponential, binomial, etc., except in special cases.

Example 1.1.2 Consider a radioactive material with unknown half-life H. For a given atom of this material, the time before disintegration follows exactly an exponential distribution[1] with parameter $\log(2)/H$. The observation of several of these particles can then lead to an inference about H.   ‖

Example 1.1.3 In order to determine the number N of buses in a town, a possible inferential strategy goes as follows: observe buses during an entire day and keep track of their identifying number. Then repeat the experiment the next day by paying attention to the number of buses been already observed on the previous day, n. If 20 buses have been observed the first day and 30 the second day, n is distributed as a hypergeometric[1] random variable, $\mathcal{H}(30, N, 20/N)$, and the knowledge of this distribution leads, for instance, to the approximation of N by $20(30/n)$. This method, called *capture-recapture*, has induced numerous and less anecdotal developments in ecology and population dynamics (see Chapter 4).   ‖

We could create many other examples where the distribution of the observations is exactly known, its derivation based upon physical, economical, etc., considerations. In the vast majority of cases, however, statistical modeling is reductive in the sense that it only approximates the reality, losing part of its richness but gaining in efficiency.

Example 1.1.4 Price and salary variations are closely related. One way to represent this dependence is to assume the linear relation

$$\Delta P = a + b\,\Delta S + \epsilon,$$

where ΔP and ΔS are the price and salary variations, a and b are unknown coefficients and ϵ is the error factor. A further, but drastic, reduction can be obtained by assuming that ϵ is normally distributed because, while ϵ is indeed a random variable, there are many factors playing a role in the determination of prices and salaries and it is usually impossible to determine the distribution of ϵ. Nonetheless, besides a justification through the *Central Limit Theorem* (i.e., the additional influence of many small factors of similar magnitude), this advanced modeling also allows for a more thorough statistical analysis, which remains valid even if the distribution of ϵ is not exactly normal. (See also Exercise 1.3.)   ‖

---

[1] See Appendix A for a survey of the most common distributions.

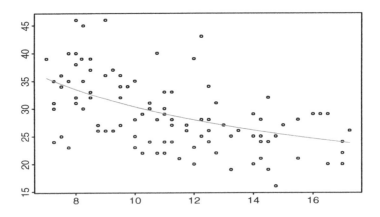

Figure 1.1.1. *Plot of monthly unemployment rate versus number of accidents (in thousands) in Michigan, from 1978 to 1987. (Source: Lenk (1999).)*

Example 1.1.5 Consider the dataset depicted by Figure 1.1, which plots the monthly unemployment rate against the monthly number of accidents (in thousands) in Michigan, from 1978 to 1987. Lenk (1999) argues in favor of a connection between these two variates, in that higher unemployment rates lead to less traffic on the roads and thus fewer accidents. A major step towards reduction is then to postulate a parametric structure in the dependence, such as the *Poisson regression* model

(1.1.1) $$N|\varrho \sim \mathcal{P}(\exp\{\beta_0 + \beta_1 \log(\varrho)\})\,,$$

where N denotes the number of accidents and ϱ the corresponding unemployment rate. Figure 1.1 also depicts the estimated expectation $\mathbb{E}[N|\varrho]$, which tends to confirm the decreasing impact of unemployment upon accidents. But the validity of the modeling (1.1.1) first needs to be assessed, using goodness-of-fit and other model choice techniques (see Chapter 7). ‖

In some cases, the reductive effect is deliberately sought as a positive *smoothing effect* which partly removes unimportant perturbations of the phenomenon and often improves its analysis by highlighting the major factors, as in the following example.

Example 1.1.6 Radiographs can be represented as a 1000×1200 grid of elementary points, called *pixels*, which are grey levels represented by numbers between 0 and 256. For instance, Figure 1.1.6 provides the histogram of the grey levels for a typical chest radiograph. If we consider a pixel to be a discrete random variable taking values in $\{0, 1, \ldots, 256\}$, the histogram gives an approximation of the distribution of this random variable. As shown by the figure, this distribution is rather complex, but approximately bimodal. This particularity is observed on most radiographs and suggests

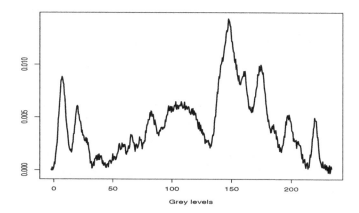

Figure 1.1.2. *Histogram of the grey levels of a chest radiograph and its modeling by a two-component mixture. (Source: Plessis (1989).)*

a modeling of the distribution through a continuous approximation by a *mixture of two normal distributions*, with density

$$(1.1.2) \quad f(x) = \frac{p}{\sqrt{2\pi}\sigma_1} \exp\left[-\frac{(x-\mu_1)^2}{2\sigma_1^2}\right] + \frac{1-p}{\sqrt{2\pi}\sigma_2} \exp\left[-\frac{(x-\mu_2)^2}{2\sigma_2^2}\right].$$

Obviously, this modeling considerably smoothes the histogram (see Figure 1.1.6), but also allows a description of the image through five parameters, with no substantial loss of information. The two important modes of the true distribution have actually been shown to correspond to two regions of the chest, the *lungs* and the *mediastinum*. This smoothing technique is used in an image-processing algorithm called *Parametric Histogram Specification* (see Plessis (1989)). We will consider the Bayesian estimation of mixture distributions in detail in Section 6.4.   ‖

   Given this imperative of reduction of the complexity of the observed phenomenon, two statistical approaches contend. A first approach assumes that statistical inference must incorporate as much as possible of this complexity, and thus aims at estimating the distribution underlying the phenomenon under minimal assumptions, generally using functional estimation (density, regression function, etc.). This approach is called *nonparametric*. Conversely, the *parametric* approach represents the distribution of the observations through a density function $f(x|\theta)$, where only the parameter θ (of finite dimension) is unknown.

   We consider that this second approach is more pragmatic, since it takes into account that a finite number of observations can efficiently estimate only a finite number of parameters. Moreover, a parametric modeling authorizes an evaluation of the inferential tools for *finite sample sizes*, contrary to the more involved nonparametric methods, which are usually justified only asymptotically, therefore strictly only apply when the sample size becomes *infinite* (see, however, Field and Ronchetti (1990), who study the applicability of asymptotic results for finite sample sizes). Of course, some nonparametric approaches, like rank tests (Hajek and Sidàk (1967)), completely evacuate the estimation aspect by devising distribution-free statistics, but their applicability is limited to testing settings.

Both approaches have their interest and we shall not try to justify any further the parametric choice. Quite naturally, there is also an extensive literature on model construction. See Cox (1990) and Lehmann (1990) for references, as well as reflections on the very notion of a statistical model. We will see in Chapter 7 some approaches to the comparison of models, which can be used in the modeling stage, that is, when a model is sought in order to fit the data and several potential models contend.

In this book, we only consider parametric modeling. We assume that the observations supporting the statistical analysis, $x_1, \ldots, x_n$, have been generated from a parametrized probability distribution, i.e., x_i ($1 \leq i \leq n$) has a distribution with density $f_i(x_i|\theta_i, x_1, \ldots, x_{i-1})$ on $\mathbb{R}^p$, such that the parameter θ_i is unknown and the function f_i is known (see Exercise 1.2 about the formal ambiguity of this definition and Note 1.8.2 for indications about the Bayesian approach to nonparametrics). This model can then be represented more simply by

$$x \sim f(x|\theta),$$

where x is the vector of the observations and θ is the set of the parameters, $\theta_1, \ldots, \theta_n$, which may all be equal. This approach is unifying, in the sense that it represents similarly an isolated observation, dependent observations, and repeated independent and identically distributed (*i.i.d.*) observations $x_1, \ldots, x_n$ from a common distribution, $f(x_1|\theta)$. In the latter case, $x = (x_1, \ldots, x_n)$ and

$$f(x|\theta) = \prod_{i=1}^{n} f(x_i|\theta).$$

Notice that densities of discrete and continuous random variables will be denoted identically in this book, the reference measure being generally provided by the setting. Moreover, we use the notation "*x is distributed according to f*" or "*x ~ f*" instead of "*x is an observation from the distribution with density f*" for the sake of conciseness[2]. Most of the time, the sample is reduced to a single observation, for simplification reasons, but

---

[2] This book does not follow the usual probabilistic convention that random variables are represented by capital letters, X say, and their *realization*, that is, their observed value, by the corresponding lower case letter, x, as in $P(X \leq x)$. This is because from a Bayesian point of view, we always condition on the realized value x and, besides, consider the parameter, θ say, as a random variable: the use of capital Greek letters then gets confusing in the extreme since Θ is rather, by convention, the parameter space. This also facilitates the use of conditional expressions, which abound in Bayesian computations. In cases potentially provoking confusion, we will revert to the capital-lower case convention.

also because we are usually dealing with distributions where the sample size does not matter, since they allow for sufficient statistics of constant dimension (see Section 1.3 and Chapter 3).

Definition 1.1.7 *A parametric statistical model consists of the observation of a random variable x, distributed according to $f(x|\theta)$, where only the parameter θ is unknown and belongs to a vector space Θ of finite dimension.*

Once the statistical model is defined, the main purpose of the statistical analysis is to lead to an *inference* on the parameter θ. This means that we use the observation x to improve our knowledge on the parameter θ, so that one can take a decision related with this parameter, i.e., either estimate a function of θ or a future event which distribution depends on θ. The inference can deal with some components of θ, precisely (*"What is the value of θ_1?"*) or not (*"Is θ_2 larger than θ_3?"*). A distinction is often made between *estimation* and *testing* problems, depending on whether the exact value of the parameters (or of some functions of the parameters), or just a hypothesis about these parameters, is of interest. For instance, the two reference books of classical Statistics, Lehmann (1986) and Lehmann and Casella (1998), deal, respectively, with each of these themes. Other authors have proposed a more subtle distinction between *estimation* and *evaluation* of estimation procedures (see, for instance, Casella and Berger (1990)). More generally, inference covers the random phenomenon directed by θ and thus includes *prediction*, that is, the evaluation of the distribution of a future observation y depending on θ (and possibly the current observation x), $y \sim g(y|\theta, x)$. As shown later, these divisions are somehow artificial, since all inferential problems can be expressed as estimation problems when considered from a decision-theoretic perspective.

The choice of the parametric approach made in this book can be criticized, since we cannot always assume that the distribution of the observations is known up to a (finite dimensional) parameter, but we maintain that this reduction allows for deeper developments in the inferential process, even though this may seem a paradoxical statement. Criticisms on the reductive aspects of the statistical approach and, a fortiori, on the parametric choice, are actually seconded by other criticisms about the choice of the evaluation criteria and the whole purpose of Decision Theory, as we will see in Chapter 2. However, we stand by these choices on the ground that these increasingly reductive steps are minimal requirements for a statistical approach to be coherent (that is, self-consistent). Indeed, the ultimate goal of statistical analysis is, in the overwhelming majority of cases, to support a *decision* as being *optimal* (or at least reasonable). It is thus necessary to be able to compare the different inferential procedures at hand. The next section presents the foundations of Bayesian statistical analysis, which seems to us to be the most appropriate approach for this determination of optimal procedures, while also being the most coherent method[3],

---

[3] As reported in Robins and Wasserman (2000), there are several formal definitions of

since it builds up these procedures by starting from required properties, instead of the reverse, namely, verifying the good behavior of procedures selected in an ad-hoc manner. The Bayesian choice, as presented in this book, may appear as an unnecessary reduction of the inferential scope, and it has indeed been criticized by many as being so. But we will see in the following chapters that this reduction is both necessary and beneficial. Chapter 11 summarizes various points of a defense of the Bayesian choice, and can be read in connection with the previous arguments[4].

Notice that there also exists a Bayesian approach to nonparametric Statistics. It usually involves prior distributions on functional spaces, such as Dirichlet processes. See Ferguson (1973, 1974) and Escobar (1989), Escobar and West (1994), Dey et al. (1998), Müller et al. (1999) and Note 1.8.2 for references in this area. Example 1.4.3 provides an illustration of the interest of the Bayesian approach in this setting.

1.2 The Bayesian paradigm as a duality principle

Compared[5] with probabilistic modeling, the purpose of a statistical analysis is fundamentally an *inversion* purpose, since it aims at retrieving the causes—reduced to the parameters of the probabilistic generating mechanism—from the effects—summarized by the observations[6]. In other words, when observing a random phenomenon directed by a parameter θ, statistical methods allow to deduce from these observations an *inference* (that is, a summary, a characterization) about θ, while probabilistic modeling characterizes the behavior of the future observations *conditional* on θ. This inverting aspect of Statistics is obvious in the notion of the *likelihood* function, since, formally, it is just the sample density rewritten in the proper order,

(1.2.1) $$\ell(\theta|x) = f(x|\theta),$$

i.e., as a function of θ, which is *unknown*, depending on the observed value x. Historically, the *fiducial* approach of Fisher (1956) also relies on this inversion (see Note 1.8.1).

A general description of the inversion of probabilities is given by *Bayes's Theorem*: If A and E are events such that $P(E) \neq 0$, $P(A|E)$ and $P(E|A)$ are related by

$$P(A|E) \quad = \quad \frac{P(E|A)P(A)}{P(E|A)P(A) + P(E|A^c)P(A^c)}$$

---

coherence, from Savage (1954) to Heath and Sudderth (1989), which all lead to the conclusion that a procedure is coherent if, and only if, it is Bayesian.

[4] This chapter and Chapter 11 are worth re-reading once the more technical points of the inferential process and the issues at hand have been fully understood.

[5] The word *paradigm*, which is a grammatical term, is used here as an equivalent for *model* or *principles*.

[6] At the time of Bayes and Laplace, i.e., at the end of the eighteenth century, Statistics was often called *Inverse Probability* because of this perspective. See Stigler (1986, Chapter 3).

$$= \frac{P(E|A)P(A)}{P(E)} .$$

In particular,

(1.2.2) $$\frac{P(A|E)}{P(B|E)} = \frac{P(E|A)}{P(E|B)} ,$$

when $P(B) = P(A)$. To derive this result through the machinery of modern axiomatized probability theory is trivial. However, it appears as a major conceptual step in the history of Statistics, being the first *inversion* of probabilities. Equation (1.2.2) expresses the fundamental fact that, for two equiprobable causes, the ratio of their probabilities given a particular effect is the same as the ratio of the probabilities of this effect given the two causes. This theorem also is an actualization principle since it describes the updating of the likelihood of A from $P(A)$ to $P(A|E)$ once E has been observed. Thomas Bayes (1764) actually proved a continuous version of this result, namely, that given two random variables x and y, with conditional distribution[7] $f(x|y)$ and marginal distribution $g(y)$, the conditional distribution of y given x is

$$g(y|x) = \frac{f(x|y)g(y)}{\int f(x|y)g(y)\, dy} .$$

While this inversion theorem is quite natural from a probabilistic point of view, Bayes and Laplace went further and considered that the *uncertainty* on the parameters θ of a model could be modeled through a *probability* distribution π on Θ, called *prior distribution*. The inference is then based on the distribution of θ conditional on x, $\pi(\theta|x)$, called *posterior distribution* and defined by

(1.2.3) $$\pi(\theta|x) = \frac{f(x|\theta)\pi(\theta)}{\int f(x|\theta)\pi(\theta)\, d\theta} .$$

Notice that $\pi(\theta|x)$ is actually proportional to the distribution of x conditional upon θ, i.e., the likelihood, multiplied by the prior distribution of θ. (It seems that the full generality of (1.2.3) was not perceived by Bayes, but by Laplace, who developed it to a greater extent.) The main addition brought by a Bayesian statistical model is thus to consider a probability distribution on the parameters.

Definition 1.2.1 *A Bayesian statistical model is made of a parametric statistical model, $f(x|\theta)$, and a prior distribution on the parameters, $\pi(\theta)$.*

In statistical terms, Bayes's Theorem thus actualizes the information on θ by extracting the information on θ contained in the observation x. Its impact is based on the daring move that puts causes (observations) and

[7] We will often replace *distribution* with *density*, assuming that the later is well defined with respect to a natural dominating measure, like the Lebesgue measure. It is only in advanced settings, such as the Haar measure in Chapter 9, that a finer level of measure theory will be needed.

effects (parameters) on the same conceptual level, since both of them have probability distributions. From a statistical modeling viewpoint, there is thus little difference between observations and parameters, since conditional manipulations allow for an interplay of their respective roles. Notice that, historically, this perspective that parameters directing random phenomena can also be perceived as random variables goes against the atheistic determinism of Laplace[8] as well as the clerical position of Bayes, who was a nonconformist minister. By imposing this fundamental modification to the perception of random phenomena, these two mathematicians created modern statistical analysis and, in particular, Bayesian analysis.

Indeed, the recourse to the prior distribution π on the parameters of a model is truly revolutionary. There is in fact a major step from the notion of an *unknown* parameter to the notion of a *random* parameter, and many statisticians place an absolute boundary between the two concepts, although they accept the probabilistic modeling on the observation(s). They defend this point of view on the ground that, even though in some particular settings the parameter is produced under the simultaneous action of many factors and can thus appear as (partly) random, as for instance, in quantum physics, the parameter to be estimated cannot be perceived as resulting from a random experiment in most cases. A typical setting occurs when estimating physical quantities like the speed of light, c. An answer in this particular setting is that the limited accuracy of the measurement instruments implies that the true value of c will never be known, and thus that it is justified to consider c as being uniformly distributed on $[c_0 - \epsilon, c_0 + \epsilon]$, if ϵ is the maximal precision of the measuring instruments and c_0 the obtained value.

We will consider in Chapter 3 some approaches to the delicate problem of prior distribution determination. However, more fundamentally, we want to stress here that the importance of the prior distribution in a Bayesian statistical analysis is not at all that the parameter of interest θ can (or cannot) be perceived as generated from π or even as a random variable, but rather that the use of a prior distribution is the best way to summarize the available information (or even the lack of information) about this parameter, as well as the residual uncertainty, thus allowing for incorporation of this imperfect information in the decision process. (Similar reasoning led Laplace to develop statistical models, despite his determinism.) A more technical point is that the only way to construct a mathematically justified approach operating conditional upon the observations is to introduce a corresponding distribution on the parameters. See also Lindley (1990, §3) for a detailed axiomatic justification of the use of prior distributions.

Let us conclude this section with the historical examples of Bayes and Laplace.

Example 1.2.2 (Bayes (1764)) A billiard ball W is rolled on a line of

---

[8] *"We must envision the present state of the Universe as the effect of its anterior state and as the cause of the following state"* – Laplace (1795).

length one, with a uniform probability of stopping anywhere. It stops at p.
A second ball O is then rolled n times under the same assumptions and X
denotes the number of times the ball O stopped on the left of W. *Given
X, what inference can we make on p?*

In modern terminology, the problem is then to derive the posterior distribution of p given X, when the prior distribution on p is uniform on $[0, 1]$
and $X \sim \mathcal{B}(n, p)$, the binomial distribution (see Appendix A). Since

$$P(X = x|p) \;=\; \binom{n}{x} p^x (1-p)^{n-x},$$

$$P(a < p < b \text{ and } X = x) \;=\; \int_a^b \binom{n}{x} p^x (1-p)^{n-x} dp$$

and

$$P(X = x) = \int_0^1 \binom{n}{x} p^x (1-p)^{n-x} \, dp,$$

we derive that

$$
\begin{aligned}
P(a < p < b | X = x) \;&=\; \frac{\int_a^b \binom{n}{x} p^x (1-p)^{n-x} \, dp}{\int_0^1 \binom{n}{x} p^x (1-p)^{n-x} \, dp} \\[2mm]
&=\; \frac{\int_a^b p^x (1-p)^{n-x} \, dp}{B(x+1, n-x+1)},
\end{aligned}
$$

i.e., that the distribution of p conditional upon $X = x$ is a beta distribution,
$\mathcal{B}e(x+1, n-x+1)$ (see Appendix A).   ‖

In the same spirit, Laplace introduced a probabilistic modeling of the
parameter space. But his examples are more advanced than Bayes's, in the
sense that the prior distributions Laplace considers are based on abstract
reasoning, instead of the physical basis of Bayes's prior distribution.[9]

Example 1.2.3 (Laplace (1773)) An urn contains a number n of black
and white cards. If the first card drawn out of the urn is white, what is the
probability that the proportion p of white cards is p_0? In his resolution of
the problem, Laplace assumes that all numbers from 2 to $n-1$ are equally
likely values for pn, i.e., that p is uniformly distributed on $\{2/n, \dots, (n-1)/n\}$. The posterior distribution of p can then be derived using Bayes's
Theorem and

$$
\begin{aligned}
P(p = p_0 | \text{data}) \;&=\; \frac{p_0 \times 1/(n-2)}{\sum_{p=2/n}^{(n-1)/n} p \times 1/(n-2)} \\[2mm]
&=\; \frac{n \, p_0}{n(n-1)/2 - 1}.
\end{aligned}
$$
          ‖

---

[9] It is also possible to picture a more Machiavellian Bayes who picked up this particular
example in order to circumvent potential criticisms of the choice of the prior. But it
seems that this was not the case, i.e., that Bayes was actually studying this example
for its own sake. See Stigler (1986) for more details.

The above choice of the prior distribution can obviously be attacked as being partly arbitrary. However, in Laplace's view of probability theory, most events can be decomposed into elementary *equiprobable* events and, therefore, in this particular case, it seems reasonable to consider the events $\{p = i/n\}$ $(2 \le i \le n-1)$ as elementary events. A similar reasoning justifies the following example.

Example 1.2.4 (Laplace (1786)) Considering male and female births in Paris, Laplace wants to test whether the probability x of a male birth is above $1/2$. For $251,527$ male and $241,945$ female births, assuming that x has a uniform prior distribution on $[0,1]$, Laplace obtains

$$P(x \le 1/2|(251,527;241,945)) = 1.15 \times 10^{-42}.$$

(see Stigler (1986, p. 134) and Exercise 1.6). He then deduces that this probability x is more than likely to be above 50%. Still assuming a uniform prior distribution on this probability, he also compares the male births in London and Paris and deduces that the probability of a male birth is significantly higher in England.   ‖

The following example worked out by Laplace is even more interesting because, from a practical point of view, it provides a method of deriving optimal procedures and, from a theoretical point of view, it is the first formal derivation of a Bayes estimator.

Example 1.2.5 In astronomy, one frequently gets several observations of a quantity ξ. These measurements are independently distributed according to a distribution that is supposed to be unimodal and symmetric around ξ. If we put a uniform distribution on the parameter ξ, it should be a "uniform distribution on $(-\infty, +\infty)$", which is not defined as a probability distribution. However, if we agree on this formal extension (see Section 1.5 for a justification), we can work with the Lebesgue measure on $(-\infty, +\infty)$ instead.

Using this *generalized distribution*, Laplace (1773) establishes that the *posterior median* of ξ, i.e. the median for the distribution of ξ conditional on the observations, is an optimal estimator in the sense that it minimizes the average absolute error

(1.2.4) $\mathbb{E}^{\xi}[\,|\xi - \delta|\,]$

in δ, where $\mathbb{E}^{\xi}[\cdot]$ denotes the expectation under the distribution of ξ (see Appendix C for a list of usual notations). This result justifies the use of the posterior median as an estimator of ξ, whatever the distribution of the observation. Although established more than two centuries ago, it is strikingly modern (generality of the distribution, choice of a loss function to evaluate the estimators) and Laplace extended it in 1810 by establishing a similar result for squared error.

Surprisingly, though, Laplace was rather unsatisfied with this result because he still needed the distribution of the observation error to be able to calculate the resulting estimator. He first considered, in 1774, the double exponential distribution

$$(1.2.5) \qquad \varphi_\xi(x) = \frac{\xi}{2} e^{-\xi|x|}, \qquad x \in \mathbb{R}, \ \xi > 0,$$

also called the *Laplace distribution*, which supposedly involved the resolution of a fifteenth degree equation for three observations. (Actually, Laplace made a mistake and the correct equation is cubic, as shown by Stigler (1986).) Then, in 1777, he looked at the even less tractable alternative

$$\varphi_\xi(x) = \frac{1}{2\xi} \log\left(\xi/|x|\right) \mathbb{I}_{|x|\leq\xi}, \qquad \xi > 0,$$

where $\mathbb{I}$ denotes the indicator function. It was only in 1810 when Legendre and Gauss independently exposed the importance of the *normal distribution*, that Laplace was able to compute his (Bayes) estimators explicitly, since he then thought this was the ideal error distribution.   ‖

We will consider again this example, along with other optimality results, in Chapter 2, when we study different loss functions to evaluate estimation procedures and the associated Bayes estimators. Let us stress here that the main consequence of the Bayes and Laplace works has been to introduce the *conditional perspective* in Statistics, i.e., to realize that parameters and observations are fundamentally identical objects, albeit differently perceived.[10] To construct in parallel a probability distribution on the parameter space completes this equivalence and, through Bayes's Theorem, allows a quantitative discourse on the causes, i.e., in our parametric framework an inference on the parameters. As mentioned above, the choice of the prior distribution is delicate, but its determination should be incorporated into the statistical process in parallel with the determination of the distribution of the observation. Indeed, a prior distribution is the best way to include residual information into the model. In addition, Bayesian statistical analysis provides natural tools to incorporate the uncertainty associated with this information in the prior distribution (possibly through a hierarchical modeling, see Chapter 10). Lastly, as pointed out in Lindley (1971), the Bayesian paradigm is intrinsically logical: given a set of required properties represented by the loss function and the prior distribution, the Bayesian approach provides estimators satisfying these requirements, while other approaches evaluate the properties of estimators derived for reasons external to the inferential framework.

1.3 Likelihood Principle and Sufficiency Principle

1.3.1 Sufficiency

Classical Statistics can be envisaged as being directed by principles often justified by "common sense" or additional axioms. On the contrary, the

---

[10] Again, this is why this book indistinctly writes random variables, observations and parameters in lower case.

Bayesian approach naturally incorporates most of these principles with no restraint on the procedures to be considered, and also definitely rejects other principles, such as *unbiasedness*. This notion once was a cornerstone of Classical Statistics and restricted the choice of estimators to those who are on average correct (see Lehmann and Casella (1998)). While intuitively acceptable, it imposes too stringent conditions on the choice of the procedures and often leads to inefficiency in their performances. (See, e.g., the Stein effect in Note 2.8.2.) More importantly, the number of problems which allow for unbiased solutions is a negligible percentage of all estimation problems (Exercise 1.12). Despite these drawbacks, a recent statistical technique like the bootstrap (Efron (1982), Hall (1992)) was introduced as a way to (asymptotically) reduce the bias.

Two fundamental principles are followed by the Bayesian paradigm, namely the Likelihood Principle and the Sufficiency Principle.

Definition 1.3.1 *When $x \sim f(x|\theta)$, a function T of x (also called a statistic) is said to be sufficient if the distribution of x conditional upon $T(x)$ does not depend on θ.*

A sufficient statistic $T(x)$ contains the whole information brought by x about θ. According to the *factorization theorem*, under some measure theoretic regularity conditions (see Lehmann and Casella (1998)), the density of x can then be written as

$$f(x|\theta) = g(T(x)|\theta)h(x|T(x)),$$

if g is the density of $T(x)$. We will see in Chapter 2 that, when an estimator is evaluated under a convex loss, the optimal procedures only depend on sufficient statistics (this is the *Rao–Blackwell Theorem*). In particular, when the model allows for a *minimal sufficient* statistic, i.e., for a sufficient statistic that is a function of all the other sufficient statistics, we only have to consider the procedures depending on this statistic or equivalently the restricted statistical model associated with this statistic. The concept of sufficiency has been developed by Fisher and is associated with the following principle.

Sufficiency Principle *Two observations x and y factorizing through the same value of a sufficient statistic T, that is, such that $T(x) = T(y)$, must lead to the same inference on θ.*

Example 1.3.2 Consider $x_1, \ldots, x_n$ independent observations from a normal distribution $\mathcal{N}(\mu, \sigma^2)$ (see Appendix A). The factorization theorem then implies that the pair $T(x) = (\bar{x}, s^2)$, where

$$\bar{x} = \frac{1}{n}\sum_{i=1}^{n} x_i \quad \text{and} \quad s^2 = \sum_{i=1}^{n}(x_i - \bar{x})^2,$$

is a sufficient statistic for the parameter (μ, σ), with density

$$g(T(x)|\theta) = \sqrt{\frac{n}{2\pi\sigma^2}} e^{-(\bar{x}-\theta)^2 n/2\sigma^2} \frac{(s^2)^{(n-3)/2} e^{-s^2/2\sigma^2}}{\sigma^n \Gamma(n-1/2) 2^{n-1/2}}.$$

Therefore, according to the Sufficiency Principle, inference on μ should only depend on this two-dimensional vector, whatever the sample size n is. We will see in Chapter 3 that the existence of a sufficient statistic of constant dimension is in a sense characteristic of *exponential families*[11].   ‖

Example 1.3.3 Consider $x_1 \sim \mathcal{B}(n_1, p)$, $x_2 \sim \mathcal{B}(n_2, p)$, and $x_3 \sim \mathcal{B}(n_3, p)$, three binomial independent observations when the sample sizes n_1, n_2, and n_3 are known. The likelihood function is then

$$f(x_1, x_2, x_3|p) = \binom{n_1}{x_1}\binom{n_2}{x_2}\binom{n_3}{x_3} p^{x_1+x_2+x_3}(1-p)^{n_1+n_2+n_3-x_1-x_2-x_3}$$

and the statistics

$$T_1(x_1, x_2, x_3) = x_1 + x_2 + x_3 \quad \text{or} \quad T_2(x_1, x_2, x_3) = \frac{x_1 + x_2 + x_3}{n_1 + n_2 + n_3}$$

are sufficient, on the contrary of $x_1/n_1 + x_2/n_2 + x_3/n_3$.   ‖

The Sufficiency Principle is generally accepted by most statisticians, in particular because of the Rao–Blackwell Theorem, which rules out estimators which do not depend only on sufficient statistics. In *model choice* settings, it is sometimes criticized as being too drastically reductive, but note that the Sufficiency Principle is only legitimate when the statistical model is actually the one underlying the generation of the observations. Any uncertainty about the distribution of the observations should be incorporated into the model, a modification which almost certainly leads to a change of sufficient statistics. A similar cautionary remark applies to the Likelihood Principle.

1.3.2 The Likelihood Principle

This second principle is partly a consequence of the Sufficiency Principle. It can be attributed to Fisher (1959) or even to Barnard (1949), but was formalized by Birnbaum (1962). It is strongly defended by Berger and Wolpert (1988) who provide an extended study of the topic. In the following definition, the notion of *information* is to be considered in the general sense of the collection of all possible inferences on θ, and not in the mathematical sense of Fisher information, defined in Chapter 3.

Likelihood Principle *The information brought by an observation x about θ is entirely contained in the likelihood function $\ell(\theta|x)$. Moreover, if x_1 and*

---

[11] For other distributions, sufficiency is not a very interesting concept, since the dimension of a sufficient statistic is then of the order of the dimension of the observation x (or of the corresponding sample), as detailed in Chapter 3.

x_2 are two observations depending on the same parameter θ, such that there exists a constant c satisfying

$$\ell_1(\theta|x_1) = c\ell_2(\theta|x_2)$$

for every θ, they then bring the same information about θ and must lead to identical inferences.

Notice that the Likelihood Principle is only valid when

(i) inference is about the *same* parameter θ; and

(ii) θ includes *every* unknown factor of the model.

The following example provides a (now classical) illustration of this principle.

Example 1.3.4 While working on the audience share of a TV series, $0 \le \theta \le 1$ representing the part of the TV audience, an investigator found nine viewers and three nonviewers. If no additional information is available on the experiment, two probability models at least can be proposed:

(1) the investigator questioned 12 persons, thus observed $x \sim \mathcal{B}(12, \theta)$ with $x = 9$;

(2) the investigator questioned N persons until she obtained 3 nonviewers, with $N \sim \mathcal{N}eg(3, 1 - \theta)$ and $N = 12$.

In other words, the random quantity in the experiment can be either 9 or 12. (Notice that it could also be both.) The important point is that, for both models, the likelihood is proportional to

$$\theta^3(1 - \theta)^9.$$

Therefore, the Likelihood Principle implies that the inference on θ should be identical for both models. As shown in Exercise 1.23, this is not the case for the classical approach.   ‖

Since the Bayesian approach is entirely based on the posterior distribution

$$\pi(\theta|x) = \frac{\ell(\theta|x)\pi(\theta)}{\int \ell(\theta|x)\pi(\theta)d\theta}$$

(see (1.2.3) and Section 1.4), which depends on x only through $\ell(\theta|x)$, the Likelihood Principle is automatically satisfied in a Bayesian setting.

On the contrary, the classical or *frequentist* approach[12] focuses on the *average* behavior properties of procedures and thus justifies the use of an estimator for reasons that can contradict the Likelihood Principle. This perspective is particularly striking in testing theory. For instance, if $x \sim$

---

[12] The theory built up by Wald, Neyman and Pearson in the 1950s is called *frequentist*, because it evaluates statistical procedures according to their long-run performances, that is, on the average (or in *frequency*) rather than focusing on the performance of a procedure for the obtained observation, as a conditional approach would do. The frequentist approach will be considered in detail in Chapters 2 and 5.

$\mathcal{N}(\theta, 1)$ and if the hypothesis to be tested is $H_0 : \theta = 0$, the classical Neyman–Pearson test procedure at level 5% is to reject the hypothesis if $x = 1.96$, on the basis that $P(|x - \theta| \geq 1.96) = 0.05$, thus conditioning on the event $|x| > 1.96$ rather than $x = 1.96$ (which is impossible for the frequentist theory). The frequency argument associated with this procedure is then that, in 5% of the cases when H_0 is true, it rejects wrongly the null hypothesis. Such arguments come to contradict the Likelihood Principle because tail behaviors may vary for similar likelihoods (see Exercises 1.17 and 1.23). The opposition between the frequentist and Bayesian paradigms is stronger in testing theory than in point estimation, where the frequentist approach usually appears as a limiting case of the Bayesian approach (see Chapter 5).

Example 1.3.5   Consider x_1, x_2 i.i.d. $\mathcal{N}(\theta, 1)$. The likelihood function is then

$$\ell(\theta|x_1, x_2) \propto \exp\{-(\bar{x} - \theta)^2\}$$

with $\bar{x} = (x_1 + x_2)/2$. Now, consider the alternative distribution

$$g(x_1, x_2|\theta) = \pi^{-3/2} \frac{e^{-(x_1+x_2-2\theta)^2/4}}{1 + (x_1 - x_2)^2}.$$

This distribution gives a likelihood function proportional to $\ell(\theta|x_1, x_2)$ and therefore should lead to the same inference about θ. However, the distribution g is quite different from $f(x_1, x_2|\theta)$; for instance, the expectation of $(x_1 - x_2)$ is not defined. Therefore, the estimators of θ will have different frequentist properties if they do not depend only on $\bar{x}$. In particular, the confidence regions on θ may differ significantly because of the heavier tails of g.   ∥

Example 1.3.6   Another implication of the Likelihood Principle is the *Stopping Rule Principle* in sequential analysis. A *stopping rule* τ can be defined as follows. If the experiments $\mathcal{E}_i$ lead to observations $x_i \in \mathcal{X}_i$, with $x_i \sim f(x_i|\theta)$, consider a corresponding sequence $\mathcal{A}_i \subset \mathcal{X}_1 \times \ldots \times \mathcal{X}_i$ such that the criterion τ takes the value n if $(x_1, \ldots, x_n) \in \mathcal{A}_n$, i.e., the experiment stops after the nth observation only if the first n observations are in $\mathcal{A}_n$. The likelihood of $(x_1, \ldots, x_n)$ is then

$$\begin{aligned}
\ell(\theta|x_1, \ldots, x_n) &= f(x_1|\theta)f(x_2|x_1, \theta) \\
&\quad \ldots f(x_n|x_1, \ldots, x_{n-1}, \theta) \mathbb{I}_{\mathcal{A}_n}(x_1, \ldots, x_n),
\end{aligned}$$

thus depends only on τ through the sample $x_1, \ldots, x_n$. This implies the following principle.

Stopping Rule Principle   *If a sequence of experiments, $\mathcal{E}_1, \mathcal{E}_2, \ldots$, is directed by a stopping rule, τ, which indicates when the experiments should stop, inference about θ must depend on τ only through the resulting sample.*

Example 1.3.4 illustrates the case where two different stopping rules lead to the same sample: either the sample size is fixed to be 12, or the experiment is stopped when 9 viewers have been interviewed. Another striking (although artificial) example of a stopping rule is to observe $x_i \sim \mathcal{N}(\theta, 1)$ and to take τ as the first integer n such that

$$|\bar{x}_n| = \left|\sum_{i=1}^{n} x_i/n\right| > 1.96/\sqrt{n}.$$

In this case, the stopping rule is obviously incompatible with frequentist modeling since the resulting sample *always* leads to the rejection of the null hypothesis $H_0 : \theta = 0$ at the level 5% (see Chapter 5). On the contrary, a Bayesian approach avoids this difficulty (see Raiffa and Schlaifer (1961) and Berger and Wolpert (1988, p. 81)). ‖

1.3.3 Derivation of the Likelihood Principle

A justification of the Likelihood Principle has been provided by Birnbaum (1962) who established that it is implied by the Sufficiency Principle, conditional upon the acceptance of a second principle.

Conditionality Principle *If two experiments on the parameter θ, $\mathcal{E}_1$ and $\mathcal{E}_2$, are available and if one of these two experiments is selected with probability p, the resulting inference on θ should only depend on the selected experiment.*

This principle seems difficult to reject when the selected experiment is known, as shown by the following example.

Example 1.3.7 (Cox (1958)) In a research laboratory, a physical quantity θ can be measured by a precise but often busy machine, which provides a measurement, $x_1 \sim \mathcal{N}(\theta, 0.1)$, with probability $p = 0.5$, or through a less precise but always available machine, which gives $x_2 \sim \mathcal{N}(\theta, 10)$. The machine being selected at random, depending on the availability of the more precise machine, the inference on θ when it has been selected should not depend on the fact that the alternative machine *could have been selected*. In fact, a classical confidence interval at level 5% taking into account this selection, i.e., averaging over all the possible experiments, is of half-length 5.19, while the interval associated with $\mathcal{E}_1$ is of half-length 0.62 (Exercise 1.20). ‖

The equivalence result of Birnbaum (1962) is then as follows.

Theorem 1.3.8 *The Likelihood Principle is equivalent to the conjunction of the Sufficiency and the Conditionality Principles.*

Proof. We define first the *evidence* associated with an experiment $\mathcal{E}$, $Ev(\mathcal{E}, x)$, as the collection of the possible inferences on the parameter θ directing this experiment. Let $\mathcal{E}^*$ denote the *mixed* experiment starting with the choice of $\mathcal{E}_i$ with probability 0.5 ($i = 1, 2$), thus with result (i, x_i). Under these notations, the Conditionality Principle can be written as

(1.3.1) $$Ev(\mathcal{E}^*, (j, x_j)) = Ev(\mathcal{E}_j, x_j).$$

Consider x_1^0 and x_2^0 such that

(1.3.2) $$\ell(\cdot | x_1^0) = c\ell(\cdot | x_2^0).$$

The Likelihood Principle then implies

(1.3.3) $$Ev(\mathcal{E}_1, x_1^0) = Ev(\mathcal{E}_2, x_2^0).$$

Let us assume that (1.3.2) is satisfied. For the mixed experiment $\mathcal{E}^*$ derived from the two initial experiments, consider the statistic

$$T(j, x_j) = \begin{cases} (1, x_1^0) & \text{if } j = 2, \ x_2 = x_2^0, \\ (j, x_j) & \text{otherwise,} \end{cases}$$

which takes the same value for $(1, x_1^0)$ and for $(2, x_2^0)$. Then, this statistic is sufficient, since, if $t \neq (1, x_1^0)$,

$$P_\theta(X^* = (j, x_j) | T = t) = \mathbb{I}_t(j, x_j)$$

and

$$P_\theta(X^* = (1, x_1^0) | T = (1, x_1^0)) = \frac{c}{1 + c},$$

due to the proportionality of the likelihood functions. The Sufficiency Principle then implies that

(1.3.4) $$Ev(\mathcal{E}^*, (1, x_1)) = Ev(\mathcal{E}^*, (2, x_2))$$

and, combined with (1.3.1), leads to (1.3.3).

The reciprocal of this theorem can be derived for the Conditionality Principle from the fact that the likelihood functions of (j, x_j) and x_j are proportional, and for the Sufficiency Principle from the factorization theorem. $\quad\square\square$

Evans, Fraser and Monette (1986) have shown that the Likelihood Principle can also be derived as a consequence of a stronger version of the Conditionality Principle.

1.3.4 Implementation of the Likelihood Principle

It thus seems quite justified to follow the Likelihood Principle because this principle can be derived from the unassailable Sufficiency and Conditionality Principles. However, this principle is altogether too vague, since it does not lead to the selection of a particular procedure when faced with a given inferential problem. It has been argued that the role of the statistician should stop with the determination of the likelihood function (Box and Tiao (1973)) since it is sufficient for clients to draw their inference, but this extreme view only stands in the most simple cases (or from a Bayesian decisional point of view, if the decision-maker also provides a prior distribution and a loss function). In large (parameter) dimensions, the likelihood function is also difficult to manipulate because of the lack of proper representations tools.

The vagueness of the Likelihood Principle calls for a reinforcement of the axiomatic bases of the inferential process, i.e., for additional structures in the construction of statistical procedures. For instance, an effective implementation of the Likelihood Principle is the *maximum likelihood estimation* method, as briefly described in Section 1.3.5. Similarly, the Bayesian paradigm allows for implementation of the Likelihood Principle in practice, with the additional advantage of including the decision-related requirements of the inferential problem, and even getting optimal procedures from a frequentist point of view (see below).

If we keep in mind the inversion aspect of Statistics presented in Section 1.2, it is tempting to consider the likelihood as a generalized density in θ, whose mode would then be the maximum likelihood estimator, and to work with this density as with a regular distribution. This approach seems to have been advocated by Laplace when he suggested using the uniform prior distribution when no information was available on θ (see Examples 1.2.3–1.2.5). Similarly, Fisher introduced the fiducial approach (see Note 1.8.1) to try to circumvent the determination of a prior distribution while putting into practice the Likelihood Principle, the choice of his distribution being objective (since depending only on the distribution of the observations). However, this approach is at its most defensible when θ is a location parameter (see also Example 1.5.1), since it leads in general to paradoxes and contradictions, the most immediate being that $\ell(\theta|x)$ is not necessarily integrable as a function of θ (Exercise 1.26). The derivation of objective posterior distributions actually calls for a more advanced theory of *noninformative* distributions (see Chapter 3), which shows that the likelihood function cannot always be considered the most natural posterior distribution.

Many approaches have been suggested to implement the Likelihood Principle like, for instance, *penalized likelihood theory* (Akaike (1978, 1983)) or *stochastic complexity theory* (Rissanen (1983, 1990)). See also Bjørnstad (1990) for a survey of non-Bayesian methods derived from the Likelihood Principle in the prediction area. The overall conclusion of this section is nonetheless that, apart from the fact that many of these theories have a Bayesian flavor, a truly Bayesian approach is the most appropriate to take advantage of the Likelihood Principle (see Berger and Wolpert (1988, Chapter 5) for an extensive discussion of this point).

1.3.5 Maximum likelihood estimation

The Likelihood Principle is altogether distinct from the *maximum likelihood estimation* approach, which is only one of several ways to implement the Likelihood Principle. Because we encounter this technique quite often in the next chapters, and also because it can be situated at the fringe of the Bayesian paradigm, we recall briefly some basic facts about the maximum likelihood approach. Extended coverage can be found in Lehmann and Casella (1998).

When $x \sim f(x|\theta)$ is observed, the maximum likelihood approach considers the following estimator of θ

$$(1.3.5) \qquad \hat{\theta} = \arg \sup_{\theta} \ell(\theta|\mathbf{x}),$$

i.e., the value of θ that maximizes the density at x, $f(x|\theta)$, or, informally, the probability of observing the given value of x. The maximization (1.3.5) is not always possible (see, e.g., the case of a mixture of two normal distributions, which is detailed in Chapter 6) or can lead to several (equivalent) global maxima (see, e.g., the case of a Cauchy distribution, $\mathcal{C}(0,1)$, with two well-separated observations). Nevertheless, the maximum likelihood estimator method is widely used, partly because of this intuitive motivation of maximizing the probability of occurrence and partly because of strong asymptotic properties (*consistency* and *efficiency*). An interesting feature of maximum likelihood estimators is also that they are *parameterization-invariant*. That is to say, for any function $h(\theta)$, the maximum likelihood estimator of $h(\theta)$ is $h(\hat{\theta})$ (even when h is not one-to-one). This property is not enjoyed by any other statistical approach (except by Bayes estimators in the special case of *intrinsic losses*. See Section 2.5.4).

The maximum likelihood method also has drawbacks. First, the practical maximization of $\ell(\theta|x)$ can be quite complex, especially in multidimensional and constrained settings. Consider, for instance, the examples of a mixture of normal distributions, of a truncated Weibull distribution

$$\ell(\theta_1, \theta_2|x_1, \ldots, x_n) = (\theta_1 \theta_2)^n (x_1 \ldots x_n)^{\theta_1} \exp\left\{-\theta_2 \sum_{i=1}^{n} x_i^{\theta_1}\right\}$$

(see Exercise 1.29), or of a 10×10 table where $x_{ij} \sim \mathcal{N}(\theta_{ij}, 1)$ when θ_{ij} is increasing in i and j (see Robert and Hwang (1996) and Exercises 1.30 and 1.31). Some numerical procedures, such as the EM algorithm of Dempster et al. (1977) for missing data models or the algorithm of Robertson et al. (1988) for order-restricted parameter spaces, have been tailored to this approach, but unsolved difficulties remain.

Second, a maximization technique is bound to give estimators that lack smoothness, as opposed to integration for instance. This is particularly true when the parameter space is restricted. For example, Saxena and Alam (1982) show that, if $x \sim \chi_p^2(\lambda)$, that is, a noncentral chi-squared distribution with p degrees of freedom[13], the maximum likelihood estimator of λ is equal to 0 for $x < p$. Similarly, maximum likelihood estimators can be quite unstable, i.e., vary widely for small variations of the observations, at least for reduced sample sizes (see Exercise 1.32).

---

[13] This example also exhibits a limitation of the invariance mentioned above: when $y \sim \mathcal{N}_p(\theta, I_p)$, the maximum likelihood estimator of $\lambda = ||\theta||^2$ is $||y||^2 = x \sim \chi_p^2(\lambda)$, which differs from the maximum likelihood estimator based on x (see Exercise 3.55).

A last but important defect of the maximum likelihood approach is that it lacks decision-theoretic and probabilistic supports. In fact, it does not incorporate the requirements of a decision-theoretic analysis and also fails to provide evaluation tools for the estimators it proposes. For instance, tests are not possible in a purely maximum likelihood context: it is necessary to call for frequentist justifications, even if they are based upon a likelihood ratio (see Section 5.3). Similarly, confidence regions of the form $C = \{\theta;\ \ell(\theta)/\ell(\hat{\theta}) \geq c\}$, which are asymptotically shortest, will not depend solely on the likelihood function if the bound c is to be chosen to achieve coverage at a given level α.

1.4 Prior and posterior distributions

Let us assume at this point that, in addition to the sample distribution, $f(x|\theta)$, a prior distribution on θ, $\pi(\theta)$, is available, that is, that we deal with a complete Bayesian model. Chapter 3 considers the preliminary problem of deriving this distribution from the prior information. Given these two distributions, we can construct several distributions, namely:

(a) the *joint distribution* of (θ, x),

$$\varphi(\theta, x) = f(x|\theta)\pi(\theta)\,;$$

(b) the *marginal distribution* of x,

$$
\begin{aligned}
m(x) &= \int \varphi(\theta, x)\, d\theta \\
&= \int f(x|\theta)\pi(\theta)\, d\theta\,;
\end{aligned}
$$

(c) the *posterior distribution* of θ, obtained by Bayes's formula,

$$
\begin{aligned}
\pi(\theta|x) &= \frac{f(x|\theta)\pi(\theta)}{\int f(x|\theta)\pi(\theta)\, d\theta} \\
&= \frac{f(x|\theta)\pi(\theta)}{m(x)}\,;
\end{aligned}
$$

(d) the *predictive distribution* of y, when $y \sim g(y|\theta, x)$, obtained by

$$g(y|x) = \int g(y|\theta, x)\pi(\theta|x)d\theta\,.$$

Example 1.4.1 (Example 1.2.2 continued) If $x \sim \mathcal{B}(n, p)$ and $p \sim \mathcal{B}e(\alpha, \beta)$ (with $\alpha = \beta = 1$ in the particular case of Bayes),

$$
\begin{aligned}
f(x|p) &= \binom{n}{x} p^x (1-p)^{n-x}, \qquad x = 0, 1, ..., n, \\
\pi(p) &= \frac{1}{B(\alpha, \beta)} p^{\alpha-1}(1-p)^{\beta-1}, \quad 0 \leq p \leq 1.
\end{aligned}
$$

The joint distribution of (x, p) is then

$$\varphi(x, p) = \frac{\binom{n}{x}}{B(\alpha, \beta)} p^{\alpha + x - 1}(1 - p)^{n - x + \beta - 1}$$

and the marginal distribution of x is

$$
\begin{aligned}
m(x) &= \frac{\binom{n}{x}}{B(\alpha, \beta)} B(\alpha + x, n - x + \beta) \\
&= \binom{n}{x} \frac{\Gamma(\alpha + \beta)}{\Gamma(\alpha)\Gamma(\beta)} \frac{\Gamma(\alpha + x)\Gamma(n - x + \beta)}{\Gamma(\alpha + \beta + n)},
\end{aligned}
$$

since the posterior distribution of p is

$$\pi(p|x) = \frac{p^{\alpha + x - 1}(1 - p)^{\beta + n - x - 1}}{B(\alpha + x, \beta + n - x)},$$

i.e., a beta distribution $\mathcal{B}e(\alpha + x, \beta + n - x)$.   ∥

Among these distributions, the central concept of the Bayesian paradigm is the *posterior distribution*. In fact, this distribution operates conditional upon the observations, thus operates automatically the *inversion* of probabilities defined in Section 1.2, while incorporating the requirement of the Likelihood Principle. It thus avoids averaging over the unobserved values of x, which is the essence of the frequentist approach. Indeed, the posterior distribution is the updating of the information available on θ, owing to the information contained in $\ell(\theta|x)$, while $\pi(\theta)$ represents the information available a priori, that is, before observing x.

Notice that the Bayesian approach enjoys a specific kind of coherence (we will meet others in the following chapters) in that the order in which i.i.d. observations are collected does not matter (this is a consequence of the Likelihood Principle), but also that updating the prior one observation at a time, or all observations together, does not matter. In other words,

$$
\begin{aligned}
\pi(\theta|x_1, \ldots, x_n) &= \frac{f(x_n|\theta)\pi(\theta|x_1, \ldots, x_{n-1})}{\int f(x_n|\theta)\pi(\theta|x_1, \ldots, x_{n-1})d\theta} \\
&= \frac{f(x_n|\theta)f(x_{n-1}|\theta)\pi(\theta|x_1, \ldots, x_{n-2})}{\int f(x_n|\theta)f(x_{n-1}|\theta)\pi(\theta|x_1, \ldots, x_{n-2})d\theta} \\
&= \cdots \\
&= \frac{f(x_n|\theta)f(x_{n-1}|\theta)\ldots f(x_1|\theta)\pi(\theta)}{\int f(x_n|\theta)f(x_{n-1}|\theta)\ldots f(x_1|\theta)\pi(\theta)d\theta}.
\end{aligned}
$$

(1.4.1)

It may happen that the observations do not modify the distribution of some parameters. This is obviously the case when the distribution of x does not depend on these parameters, as in some nonidentifiable settings.

Example 1.4.2 Consider one observation x from a normal

$$\mathcal{N}\left(\frac{\theta_1 + \theta_2}{2}, 1\right)$$

distribution, with a prior π on (θ_1, θ_2) such that $\pi(\theta_1, \theta_2) = \pi_1(\theta_1 + \theta_2)\pi_2(\theta_1 - \theta_2)$. If we operate the change of variables

$$\xi_1 = \frac{\theta_1 + \theta_2}{2}, \qquad \xi_2 = \frac{\theta_1 - \theta_2}{2},$$

the posterior distribution of ξ_2 is then

$$
\begin{aligned}
\pi(\xi_2) \quad &\propto \quad \int_{\mathbb{R}} \exp\left\{-(x - \xi_1)^2/2\right\} 2\pi_1(2\xi_1) 2\pi_2(2\xi_2) d\xi_1 \\
&\propto \quad \pi_2(2\xi_2) \int_{\mathbb{R}} \exp\left\{-(x - \xi_1)^2/2\right\} \pi_1(2\xi_1) d\xi_1 \\
&\propto \quad \pi_2(2\xi_2)
\end{aligned}
$$

for every observation x. The observation thus brings no information on ξ_2.   ‖

We must warn the reader that *not* every nonidentifiable setting leads to this simple conclusion: depending on the choice of the prior distribution and on the reparameterization of the parameter θ in (θ_1, θ_2), where the distribution of x only depends on θ_1, the marginal posterior distribution of θ_2 may or may not depend on x (Exercise 1.45). An important aspect of the Bayesian paradigm in nonidentifiable settings is, however, that the prior distribution can be used as a tool to *identify* the parts of the parameter that are not covered by the likelihood, even though the choice of prior may have a bearing on the identifiable part.

This invariance from prior distribution to posterior distribution may also occur for some parameters when the number of parameters becomes too large compared with the sample size (Exercise 1.39).

Example 1.4.3 A general setting where this situation occurs is found when the number of parameters is infinite, for instance, when the inference encompasses a whole distribution. Studden (1990) considers n observations $x_1, \ldots, x_n$ from a *mixture of geometric distributions*,

$$x \sim \int_0^1 \theta^x (1 - \theta) \, dG(\theta),$$

x taking its values in $\mathbb{N}$ and the probability distribution G being unknown. In this setting, G can be represented by the sequence of its noncentral moments $c_1, c_2, \ldots$. The likelihood function is then derived from $P(X = k) = c_k - c_{k+1}$. Studden (1990) shows that, although the c_i are constrained by an infinite number of inequalities (starting with $c_1 > c_2 > c_1^2$), it is possible to derive (algebraically) independent functions of the c_i's, $p_1, p_2, \ldots$, taking values in $[0, 1]$ and such that c_i only depends on $p_1, \ldots, p_i$ (see Exercise 1.46 for details). Therefore, if the prior distribution of $p_1, p_2, \ldots$ is

$$\pi(p_1, p_2, \ldots) = \prod_{i=1}^{+\infty} \pi_i(p_i),$$

and if the largest observation in the sample is k, the posterior distribution of $p_{k+2}, p_{k+3}, \ldots$ does not depend on the observations:

$$\pi(p_{k+2}, \ldots | x_1, \ldots, x_n) = \pi(p_{k+2}, \ldots) = \prod_{i=k+2}^{+\infty} \pi_i(p_i).$$

$\|$

Conversely, the marginal distribution does not involve the parameter of interest θ. It is therefore rarely of direct use, except in the *empirical Bayesian approach* (see Chapter 10), since the posterior distribution is much more adapted to inferential purposes. The marginal distribution can, however, be used in the derivation of the prior distribution if the available information has been gathered from different experiments, that is, dealing with different θ's as in *meta-analysis* (see Mosteller and Chalmers (1992), Mengersen and Tweedie (1993), and Givens et al. (1997)).

Given a probability distribution π on θ, the Bayesian inferential scope is much larger than the classical perspective. For instance, not only the mean, mode, or median of $\pi(\theta|x)$ can be computed, but also evaluations of the performances of these estimators (through their variance and higher-order moments) are available. Moreover, the knowledge of the posterior distribution also allows for the derivation of *confidence regions* through highest posterior density (HPD) regions, that is, regions of the form

$$\{\theta; \pi(\theta|x) \geq k\},$$

in both unidimensional and multidimensional cases. Similarly, it is possible to derive quite naturally the probability of a hypothesis H_0, by conditioning on the observations, i.e., $P^\pi(\theta \in H_0|x)$. Let us stress that the Bayesian approach is the only one justifying such an expression because the expression $P(\theta = \theta_0) = 0.95$ is meaningless unless θ is a random variable. From a Bayesian point of view, this expression signifies that we are ready to bet that θ is equal to θ_0 with a 95/5 odds ratio, or, in other words, that the uncertainty about the value of θ is reduced to a 5% zone. Chapters 4 and 5 are devoted to the study of estimation techniques that incorporate the decisional requirements. We just illustrate the simplicity of this derivation by constructing a confidence interval in the following example.

Example 1.4.4 Consider $x \sim \mathcal{N}(\theta, 1)$ and $\theta \sim \mathcal{N}(0, 10)$. Therefore, for a given[14] x,

$$\pi(\theta|x) \quad \propto \quad f(x|\theta)\pi(\theta) \propto \exp\left(-\frac{(x-\theta)^2}{2} - \frac{\theta^2}{20}\right)$$

---

[14] The proportionality symbol $\propto$ is to be taken for functions of θ (not of x). While being entirely rigorous, computations using proportionality signs lead to greater efficiency in the derivation of posterior distributions. In fact, probability densities are uniquely determined by their functional form, and the normalizing constant can be recovered, when necessary, at the end of the computation. This technique will therefore be used extensively in this book. Obviously, it is not always appropriate, for instance when the proportionality constant is 0 or infinity, as seen in Section 1.5.

$$\propto \quad \exp\left(-\frac{11\theta^2}{20} + \theta x\right)$$

$$\propto \quad \exp\left(-\frac{11}{20}\{\theta - (10x/11)\}^2\right)$$

and $\theta|x \sim \mathcal{N}(\frac{10}{11}x, \frac{10}{11})$. A natural confidence region is then

$$
\begin{aligned}
C &= \{\theta; \pi(\theta|x) > k\} \\
&= \left\{\theta; \left|\theta - \frac{10}{11}x\right| > k'\right\} .
\end{aligned}
$$

We can also associate a *confidence level* α with this region in the sense that, if $z_{\alpha/2}$ is the $\alpha/2$ quantile of $\mathcal{N}(0,1)$,

$$C_\alpha = \left[\frac{10}{11}x - z_{\alpha/2}\sqrt{\frac{10}{11}}, \frac{10}{11}x + z_{\alpha/2}\sqrt{\frac{10}{11}}\right]$$

has a posterior probability $(1 - \alpha)$ of containing θ. $\qquad\qquad\qquad \|$

We will see in Chapter 10 that a posterior distribution can sometimes be decomposed into several levels according to a hierarchical structure, the parameters of the first levels being treated as random variables with additional prior distributions. But this decomposition is instrumental and does not modify the fundamental structure of the Bayesian model.

A problem we did not mention above is that, although all posterior quantities are automatically defined from a conceptual point of view as integrals with respect to the posterior distribution, it may be quite difficult to provide a numerical value in practice and, in particular, an explicit form of the posterior distribution cannot always be derived. In fact, the complexity of the posterior distributions increases when the parameters are continuous and when the dimension of Θ is large.

These computational difficulties are studied in Chapter 6, where we provide some general solutions. Still, they should not be considered a major drawback of the Bayesian approach. Indeed, Computational Statistics is currently undergoing such a rapid development that we can clearly reject the notion of a prior distribution chosen for its computational tractability, even though we may still rely on these particular distributions to present simpler and clearer examples in this book. On the contrary, it is stimulating to see that we are getting closer to the goal of providing more powerful and efficient statistical tools because of these new computational techniques, as they allow for the use of more complex prior distributions, which are in turn more representative of the available prior information.

1.5 Improper prior distributions

When the parameter θ can be treated as a random variable with known probability distribution π, we saw in the previous section that Bayes's

Theorem is the basis of Bayesian inference, since it leads to the posterior distribution. In many cases, however, the prior distribution is determined on a subjective or theoretical basis that provides a σ-finite measure on the parameter space Θ instead of a probability measure, that is, a measure π such that

$$\int_{\Theta} \pi(\theta)\, d\theta = +\infty.$$

In such cases, the prior distribution is said to be *improper* (or *generalized*). (An alternative definition of generalized Bayes estimators is considered in Chapter 2.)

When this distribution stems from subjective reasons, that is, when the decision-maker is evaluating the relative likelihoods of different parts of the parameter space Θ (see Chapter 3), it really makes sense that, for large parameter spaces, for instance when Θ is noncountable, the sum of these weights, that is, the measure of Θ, should be infinite.

Example 1.5.1 Consider a distribution $f(x - \theta)$ where the *location parameter* θ is in $\mathbb{R}$ with no restriction. If no prior information is available on the parameter θ, it is quite acceptable to consider that the likelihood of an interval $[a, b]$ is proportional to its length $b - a$, therefore that the prior is proportional to the *Lebesgue measure* on $\mathbb{R}$. This was also the distribution selected by Laplace (see Example 1.2.5). ‖

When such improper prior distributions are derived by automatic methods from the density $f(x|\theta)$ (see Chapter 3), they seem more open to criticism, but let us point out the following points.

(1) These automatic approaches are usually the only way to derive prior distributions in noninformative settings, that is, in cases where the only available (or retained) information is the knowledge of the sample distribution, $f(x|\theta)$. This generalization of the usual Bayesian paradigm thus makes possible a further extension of the scope of Bayesian techniques.

(2) The performances of the estimators derived from these generalized distributions are usually good enough to justify these distributions. Moreover, they often permit recovery of usual estimators like maximum likelihood estimators, thus guaranteeing a closure of the inferential field by presenting alternative approaches at the boundary of the Bayesian paradigm.

(3) The generalized prior distributions often occur as limits of proper distributions (according to various topologies). They can thus be interpreted as extreme cases where the reliability of the prior information has completely disappeared and seem to provide a more *robust* (or more *objective*) answer in terms of a possible *misspecification* of the prior distribution (i.e., a wrong interpretation of the sparse prior information).

(4) Such distributions are generally more acceptable to non-Bayesians, partly for reasons (2) and (3), but also because they may have frequentist justifications, such as:

(i) *minimaxity*, which is related to the usually improper *"least favorable* distributions", defined in Chapter 2);

(ii) *admissibility*, as proper and some improper distributions lead to admissible estimators, while admissible estimators sometimes only correspond to Bayes estimators (see Chapter 8); and

(iii) *invariance*, as the best equivariant estimator is a Bayes estimator for the generally improper *Haar measure* associated with the transformation group (see Chapter 9).

(5) A recent perspective (see, e.g., Berger (2000)) is that improper priors should be preferred to vague proper priors such as, a $\mathcal{N}(0, 100^2)$ distribution say, because the later gives a false sense of safety owing to properness, while lacking robustness in terms of influence on the resulting inference.

These reasons do not convince all Bayesians (see, e.g., Lindley (1965)), but the inclusion of improper distributions in the Bayesian paradigm allows for a closure of the inferential scope (figuratively as well as topologically).

From a more practical perspective, the fact that the prior distribution is improper weakens the above symmetry between the observations and the parameters, but *as long as the posterior distribution is defined*, Bayesian methods apply as well. In fact, the notion of conditional measures is not clearly defined in measure theory, although Hartigan (1983) advocates such an extension, but the convention is to take the posterior distribution $\pi(\theta|x)$ associated with an improper prior π as given by Bayes's formula

$$\pi(\theta|x) = \frac{f(x|\theta)\pi(\theta)}{\int_\Theta f(x|\theta)\pi(\theta)\,d\theta},$$

when the pseudo marginal distribution $\int_\Theta f(x|\theta)\pi(\theta)\,d\theta$ is well defined. This is an imperative condition for using improper priors, which (almost) always hold for proper priors (Exercise 1.47).

Example 1.5.2 (Example 1.5.1 continued) If $f(x - \theta)$ is the density of the normal distribution $\mathcal{N}(\theta, 1)$ and $\pi(\theta) = \varpi$, an arbitrary constant, the pseudo marginal distribution is the measure

$$m(x) = \varpi \int_{-\infty}^{+\infty} \frac{1}{\sqrt{2\pi}} \exp\left\{-(x - \theta)^2/2\right\} d\theta = \varpi$$

and, by Bayes's formula, the posterior distribution of θ is

$$\pi(\theta \mid x) = \frac{1}{\sqrt{2\pi}} \exp\left\{-\frac{(x - \theta)^2}{2}\right\},$$

i.e., corresponds to $\mathcal{N}(x, 1)$. Notice that the constant ϖ does not play a role in the posterior distribution, and that the posterior distribution is actually the likelihood function. Therefore, even though improper priors cannot be normalized, it does not matter because the constant is of no interest for the statistical inference (but see Chapter 5 for an important exception). ‖

According to the Bayesian version of the Likelihood Principle, only posterior distributions are of importance. Therefore, the generalization from proper to improper prior distributions should not cause problems, in the sense that the posterior distribution corresponding to an improper prior can be used similarly to a regular posterior distributions, *when it is defined.* (Obviously, the interpretation of the prior distribution is more delicate.) For instance, in Example 1.5.1, the relative prior weight of any interval is null, but this does not mean that this interval is unlikely a priori. Actually, a misinterpretation of improper priors as regular prior distributions may lead to difficulties like *marginalization paradoxes* (see Chapter 3) because the usual calculus of conditional probability does not apply in this setting. As expressed by Lindley (1990), *the mistake is to think of them [non-informative priors] as representing ignorance.*

It may happen that, for some observations x, the posterior distribution is not defined (Exercises 1.49–1.52). The usual solution is to determine the improper answer as a limit for a sequence of proper distributions (while also checking the justifications of the improper distribution).

Example 1.5.3 Consider a binomial observation, $x \sim \mathcal{B}(n, p)$, as in the original example of Bayes. Some authors (see Novick and Hall (1965) and Villegas (1977)) reject Laplace's choice of the uniform distribution on $[0, 1]$ as automatic prior distribution because it seems to be biased against the extreme values, 0 and 1. They propose to consider instead Haldane's (1931) prior

$$\pi^*(p) \propto [p(1 - p)]^{-1}.$$

In this case, the marginal distribution,

$$m(x) = \int_0^1 [p(1 - p)]^{-1} \binom{n}{x} p^x (1 - p)^{n-x} dp$$
$$= B(x, n - x),$$

is only defined for $x \neq 0, n$. Therefore, $\pi(p|x)$ does not exist for these two extreme values of x, since the product $\pi^*(p) p^x (1 - p)^{n-x}$ cannot be normalized for these two values. For the other values, the posterior distribution is $\mathcal{B}e(x, n-x)$, with posterior mean x/n, which is also the maximum likelihood estimator.

The difficulty in 0 and n can be overcome as follows. The prior measure π^* appears as a limit of unnormalized beta distributions,

$$\pi_{\alpha,\beta}(p) = p^{\alpha-1}(1 - p)^{\beta-1},$$

when α and β go to 0. These distributions $\pi_{\alpha,\beta}$ lead to beta posterior distributions, $\mathcal{B}e(\alpha+x, \beta+n-x)$, notwithstanding the lack of the normalizing factor, since the choice of the constant in the prior distribution is irrelevant. The posterior distribution $\pi_{\alpha,\beta}(p|x)$ has the expectation

$$\delta^\pi_{\alpha,\beta}(x) = \frac{x + \alpha}{\alpha + \beta + n},$$

which goes to x/n when α and β go to 0. If the posterior mean is the quantity of interest, we can then extend the inferential procedure to the cases $x = 0$ and $x = n$ by taking also x/n as a formal Bayes estimator. ‖

Example 1.5.4 Consider $x \sim \mathcal{N}(0, \sigma^2)$. It follows from invariance considerations that an interesting prior distribution on σ is the measure $\pi(\sigma) = 1/\sigma$ (see Chapter 9). It gives the posterior distribution

$$\pi(\sigma^2|x) \propto \frac{e^{-x^2/2\sigma^2}}{\sigma^2},$$

which is not defined for $x = 0$. However, owing to the continuity of the random variable x, this difficulty is of little importance compared with Example 1.5.3.   ‖

Obviously, these limiting arguments are ad-hoc expedients which are not always justified, in particular because the resulting estimator may depend on the choice of the converging sequence. An example of this phenomenon is provided by Richard (1973) (see also Bauwens (1991)) in the case of a normal distribution $\mathcal{N}(\theta, \sigma^2)$, when $\pi(\theta)$ is the Lebesgue measure and σ^{-2} is distributed according to a gamma distribution $\mathcal{G}(\alpha, s_0^2)$, i.e., when

$$\pi(\theta, \sigma^2) \propto \frac{1}{\sigma^{2(\alpha+1)}} e^{-s_0^2/2\sigma^2} ;$$

the estimator of θ then depends on the behavior of the ratio $s_0^2/(\alpha - 1)$ when both numerator and denominator go to 0.

Moreover, when estimating a *discontinuous* function of θ, the estimator for the limiting distribution may differ from the limit of the estimators. This is, for instance, the case in testing theory with the *Jeffreys–Lindley paradox* (see Chapter 5). Finally, there may be settings such that improper prior distributions cannot be used easily, like in mixture estimation (see Exercise 1.57 and Chapter 6) or in testing theory when testing two-sided hypotheses (see Exercises 1.61–1.63 and Chapter 5).

It is thus important to exercise additional caution when dealing with improper distributions in order to avoid ill-defined distributions. In this book, improper distributions will always be used under the implicit assumption that the corresponding posterior distributions are defined, even though there are settings where this condition could be relaxed (see Note 1.8.3).

The practical difficulty is in checking the propriety (or *properness*) condition

$$\int f(x|\theta)\pi(\theta) \, d\theta < \infty$$

in complex settings like hierarchical models (see Exercise 1.67 and Chapter 10), where the use of improper priors on the upper level of the hierarchy is quite common. The problem is even more crucial because new computational tools like MCMC algorithms (Chapter 6) do not require in practice this checking of properness (see Note 1.8.3, and Hobert and Casella (1996, 1998)).

Let us stress again that the main justification for using improper prior distributions is to provide a completion of the Bayesian inferential field for subjective, axiomatic (in relation with *complete class results*, see Chapter 8), and practical reasons. This extension does not modify the complexity of the inference, however, because the posterior distribution is truly a probability distribution.

1.6 The Bayesian choice

To close this introduction, let us impress upon the reader that there is such a thing as a Bayesian choice. Thus, it is always possible to adhere to this choice, or to opt for other options. While we are resolutely advocating for this choice, there is no reason to become overly strident. Most statistical theories, such as those presented in Lehmann and Casella (1998), have a reasonable level of coherence and most often agree when the number of observations gets large compared with the number of parameters (see Note 1.8.4).

If we do not present these options here, it is both for philosophical and practical reasons (exposed in Chapter 11), and also for the purpose of presenting an unified discourse on Statistics, where all procedures logically follow from a given set of axioms. This is indeed for us *the* compelling reason for adhering to the Bayesian choice, namely, the ultimate coherence of the axioms of Bayesian statistical inference. By modeling the unknown parameters of the sampling distribution through a probability structure, i.e., by probabilizing uncertainty, the Bayesian approach authorizes a quantitative discourse on these parameters. It also allows incorporation in the inferential procedure of the prior information *and* of the imprecision of this information. Besides, apart from subjective and axiomatic arguments in favor of the Bayesian approach, which is the only system allowing for conditioning on the observations (and thus for an effective implementation of the Likelihood Principle), Bayes estimators are also quintessential for the frequentist optimality notions of Decision Theory. In fact, they can provide essential tools even to those statisticians who reject prior elicitation and the Bayesian interpretation of reality.

1.7 Exercises

Section[15] **1.1**

1.1 *(Kelker (1970)) A vector $x \in \mathbb{R}^p$ is distributed according to a *spherically symmetric distribution* if $e.x$ has the same distribution than x for every

---

[15] The exercises with stars are more advanced, but offer a broader view of the topics treated in each chapter. They can be be treated as useful complements, or as a guided lecture of relevant papers by most readers.

orthogonal transform e.

a. Show that, when a spherically symmetric distribution has a density, it is a function of $x^t x$ only.

b. Show that, if the density of x is $\varphi(x^t x)$, the density of $r = ||x||$ is proportional to

$$r^{p-1}\,\varphi(r^2),$$

and give the proportionality coefficient.

c. Show that, if $x = (x_1', x_2')'$ with $x_1 \in \mathbb{R}^q$ and $x_2 \in \mathbb{R}^{p-q}$, and $||x||^2 = ||x_1||^2 + ||x_2||^2$, the density of $(r_1, r_2) = (||x_1||, ||x_2||)$ is proportional to

$$r_1^{q-1}\, r_2^{p-q-1}\, \varphi\left(r_1^2 + r_2^2\right).$$

d. Deduce that

$$U = \frac{||x_1||^2}{||x_1||^2 + ||x_2||^2}$$

is distributed according to a beta distribution $\mathcal{B}e(q/2, (p-q)/2)$.

e. Conclude that

$$\frac{p-q}{q}\,\frac{||x_1||^2}{||x_2||^2}$$

is distributed according to the F-distribution $\mathcal{F}_{p-q,q}$ independently of the spherically symmetric distribution of x. Deduce that the F-ratio is a *robust* quantity in the sense that its distribution is constant on a range of spherically symmetric distributions.

1.2 *(Gouriéroux and Monfort (1996)) This exercise points out that the boundary between parametric and nonparametric models is quite difficult to determine. However, in the second case, the parameter cannot be identified.

a. Show that a c.d.f. is characterized by the values it takes at the rational numbers.

b. Deduce that the collection of the c.d.f.'s on $\mathbb{R}$ has the power of continuum (i.e., the cardinal of the set of the parts of $\mathbb{N}$, the set of natural integers) and thus that all probability distributions on $\mathbb{R}$ can be indexed by a real parameter.

1.3 Show that, if $x_1, \ldots, x_n$ are known explanatory variables and $y_1, \ldots, y_n$ are distributed as $\mathbb{E}[y_i] = bx_i$, the *least-squares estimator* of b, solution of

$$\min_b \sum_{i=1}^{n} (y_i - bx_i)^2,$$

is also a maximum likelihood estimator under a normality assumption.

1.4 In Example 1.1.3, give the expectation of n. Does that mean that $20 \times 30/n$ is an unbiased estimator of N?

1.5 In Example 1.1.6, show that the moments of $x \sim f(x)$ can be written as $\mathbb{E}[x^k] = p\mathbb{E}[x_1^k] + (1-p)\mathbb{E}[x_2^k]$. Deduce a moment estimator of $(p, \mu_1, \mu_2, \sigma_1^2, \sigma_2^2)$. [*Note:* Historically, this is the estimate of Pearson (1894).]

Section 1.2

1.6 Derive the probabilities of Example 1.2.4 from the approximation

$$\Phi(-x) \simeq \frac{1}{\sqrt{2\pi}x}e^{-x^2/2},$$

which is valid when x is large.

1.7 An examination has 15 questions, each with 3 possible answers. Assume that 70% of the students taking the examination are prepared and answer correctly each question with probability 0.8; the remaining 30% answer at random.

a. Characterize the distribution of S, score of a student if one point is attributed to each correct answer.

b. Eight correct answers are necessary to pass the examination. Given that a student has passed the examination, what is the probability that she was prepared?

1.8 Prove the discrete and continuous versions of Bayes's Theorem.

1.9 *(Romano and Siegel (1986)) The *Simpson paradox* provides an illustration of the need for a conditional approach in Statistics. Consider two medical treatments, T_1 and T_2, T_1 being applied to 50 patients and T_2 to 50 others. The result of the experiment gives the following survival percentages: 40% for treatment T_1, 32% for treatment T_2. Therefore, treatment T_1 seems better because it leads to an higher survival rate. However, if age is taken into account, dividing the subjects between juniors (50) and seniors (50), the success rates are described in the following table:

	T_1	T_2
junior	40	50
senior	10	35

and T_1 is worse than T_2 in both cases. Explain the paradox in terms of Bayes's Theorem.

1.10 Show that the quantity δ that minimizes (1.2.4) is the median of the distribution of ξ. Give the quantity δ that minimizes the average squared error $\mathbb{E}^\xi[(\xi - \delta)^2]$.

1.11 Find the median of the posterior distribution associated with the sampling distribution (1.2.5) and a flat prior $\pi(\xi) = 1$ on ξ. [*Note:* See Stigler (1986) for a resolution.]

Section 1.3

1.12 Show that, for a sample from a normal $\mathcal{N}(\theta, \sigma^2)$ distribution, there does not exist an unbiased estimator of σ but only of powers of σ^2.

1.13 Consider $x \sim P(\lambda)$. Show that $\delta(x) = \mathbb{I}_0(x)$ is an unbiased estimator of $e^{-\lambda}$ which is null with probability $1 - e^{-\lambda}$.

1.14 *A statistic S is said to be *ancillary* if its distribution does not depend on the parameter θ and it is said to be *complete* if $\mathbb{E}_\theta[g(S)] = 0$ for every θ implies $g(s) \equiv 0$. Show that, if S is complete and minimal sufficient, it is independent of every ancillary statistic. [*Note:* This result is called *Basu's Theorem*. The reverse is false.]

1.15 Consider a sample $x_1, \ldots, x_n$ of i.i.d. variables with c.d.f. F.

a. Give the density of the order statistic.

b. Show that $O = (X_{(1)}, ..., X_{(n)})$ is sufficient. What is the conditional distribution of $(X_1, ..., X_n)$ given O?

c. Consider $X_1, ..., X_n$ i.i.d. with totally unknown density. Show that O is then complete.

1.16 Show that a statistic T is sufficient if and only if

$$\ell(\theta|x) \propto \ell(\theta|T(x)).$$

1.17 (Berger and Wolpert (1988, p. 21)) Consider x with support $\{1, 2, 3\}$ and distribution $f(\cdot \,|\, 0)$ or $f(\cdot \,|\, 1)$, where

	1	2	3
$f(x|0)$	0.9	0.05	0.05
$f(x|1)$	0.1	0.05	0.85

(column group header: x)

Show that the procedure that rejects the hypothesis $H_0 : \theta = 0$ (to accept $H_1 : \theta = 1$) when $x = 2, 3$ has a probability 0.9 to be correct (under H_0 as well as under the alternative). What is the implication of the Likelihood Principle when $x = 2$?

1.18 Show that the Stopping Rule Principle given in Example 1.3.6 is a consequence of the Likelihood Principle for the discrete case. [*Note:* See Berger and Wolpert (1988) for the extension to the continuous case.]

1.19 For Example 1.3.6, show that the stopping rule τ is finite with probability 1. (*Hint:* Use the *law of the iterated logarithm*. See Billingsley (1986).)

1.20 Show that the confidence intervals of Example 1.3.7 are correct: under the mixed experiment, $x \sim 0.5\mathcal{N}(\theta, 0.1) + 0.5\mathcal{N}(\theta, 10)$ and $P(\theta \in [x - 5.19, x + 5.19]) = 0.95$, while, under experiment $\mathcal{E}_1$, $x \sim \mathcal{N}(\theta, 0.1)$ and $P(\theta \in [x - 0.62, x + 0.62]) = 0.95$.

1.21 (Raiffa and Schlaifer (1961)) Show that, if $z \sim f(z|\theta)$ and if $x = t(z)$, x is a sufficient statistic if and only if for every prior π on θ, $\pi(\theta|x) = \pi(\theta|z)$.

1.22 Consider $x_1, \ldots, x_n$ distributed according to $\mathcal{E}xp(\lambda)$. The data is *censored* in the sense that there exist n random variables $y_1, \ldots, y_n$ distributed according to $f(y)$, independent of λ, and $z_1 = x_1 \wedge y_1, \ldots, z_n = x_n \wedge y_n$ are the actual observations.

a. Show that, according to the Likelihood Principle, the inference on λ should not depend on f.

b. Extend this independence to other types of censoring.

1.23 Compare the lengths of the confidence intervals at level 10% in the setting of Example 1.3.7.

1.24 (Berger (1985a)) In the setting of Example 1.3.4, show that, for the UMPU test of $H_0 : p = 1/2$, the null hypothesis will be accepted or rejected at level 5%, depending on the distribution considered. Deduce that the frequentist theory of tests is not compatible with the Likelihood Principle. (*Hint:* See Chapter 5 for definitions.)

1.25 This exercise aims at generalizing Examples 1.3.4 and 1.3.5 in the continuous case by showing that there can also be incompatibility between the frequentist approach and the Likelihood Principle in continuous settings.

a. If $f(x|\theta)$ is a density such that x is a complete statistic, show that there is no other density $g(x|\theta)$ such that the two likelihood functions $\ell_f(\theta|x) = f(x|\theta)$ and $\ell_g(\theta|x) = g(x|\theta)$ are proportional (in θ) for every x.

b. Consider now a sample $x_1, \ldots, x_n$ from $f(x|\theta)$. We assume that there exists a complete sufficient statistic $T(x_1, \ldots, x_n)$ of dimension 1 and an ancillary statistic $S(x_1, \ldots, x_n)$ such that the couple (T, S) is a one-to-one function of $(x_1, \ldots, x_n)$. Show that, if there exists another density $g(x_1, \ldots, x_n|\theta)$ such that the two likelihood functions are proportional,

$$\ell_g(\theta|x_1, \ldots, x_n) = \omega(x_1, \ldots, x_n)\ell_f(\theta|x_1, \ldots, x_n),$$

the proportionality factor ω only depends on $S(x_1, \ldots, x_n)$.

c. In the particular case when $f(x|\theta)$ is the exponential density, $f(x|\theta) = \theta e^{-\theta x}$, give an example of a density $g(x_1, \ldots, x_n|\theta)$ such that the two likelihood functions are proportional. (*Hint:* Find an ancillary statistic S and derive a function $h(x_1, \ldots, x_n)$ depending only on $S(x_1, \ldots, x_n)$ such that $\mathbb{E}_\theta[h(x_1, \ldots, x_n)] = 1$.)

The following exercises (1.27 to 1.36) present some additional aspects of maximum likelihood estimation.

1.26 Show that, if the likelihood function $\ell(\theta|x)$ is used as a density on θ, the resulting inference does not obey the Likelihood Principle (*Hint:* Show that the posterior distribution of $h(\theta)$, when h is a one-to-one transform, is not the transform of $\ell(\theta|x)$ by the Jacobian rule.)

1.27 Consider a Bernoulli random variable $y \sim \mathcal{B}([1 + e^\theta]^{-1})$.

a. If $y = 1$, show that there is no maximum likelihood estimator of θ.

b. Show that the same problem occurs when $y_1, y_2 \sim \mathcal{B}([1 + e^\theta]^{-1})$ and $y_1 = y_2 = 0$ or $y_1 = y_2 = 1$. Give the maximum likelihood estimator in the other cases.

1.28 Consider x_1, x_2 two independent observations from $\mathcal{C}(\theta, 1)$. Show that, when $|x_1 - x_2| > 2$, the likelihood function is bimodal. Find examples of x_1, x_2, x_3 i.i.d. $\mathcal{C}(\theta, 1)$ for which the likelihood function has three modes.

1.29 The *Weibull distribution* $\mathcal{We}(\alpha, c)$ is widely used in engineering and reliability. Its density is given by

$$f(x|\alpha, c) = c\alpha^{-1}(x/\alpha)^{c-1}e^{-(x/\alpha)^c}.$$

a. Show that, when c is known, this model is equivalent to a gamma model.

b. Give the likelihood equations in α and c and show that they do not allow for explicit solutions.

c. Consider an i.i.d. $x_1, \ldots, x_n$ sample from $\mathcal{We}(\alpha, c)$ censored from the right in y_0. Give the corresponding likelihood function when α and c are unknown and show that there is no explicit maximum likelihood estimators in this case either.

1.30 *(Robertson et al. (1988)) For a sample $x_1, \ldots, x_n$, and a function f on $\mathcal{X}$, the isotonic regression of f with weights ω_i is the solution of the minimization in g of

$$\sum_{i=1}^n \omega_i(g(x_i) - f(x_i))^2,$$

under the constraint $g(x_1) \leq \ldots \leq g(x_n)$.

a. Show that a solution to this problem is obtained by the *pool-adjacent-violators* algorithm: if f is not isotonic, find i such that $f(x_{i-1}) > f(x_i)$, replace $f(x_{i-1})$ and $f(x_i)$ by

$$f^*(x_i) = f^*(x_{i-1}) = \frac{\omega_i f(x_i) + \omega_{i-1} f(x_{i-1})}{\omega_i + \omega_{i-1}},$$

and repeat until the constraint is satisfied. Take $g = f^*$.

b. Apply to the case $n = 4$, $f(x_1) = 23$, $f(x_2) = 27$, $f(x_3) = 25$, $f(x_4) = 28$, when the weights are all equal.

1.31 *(**Exercise 1.30 cont.**) The simple *tree-ordering* is obtained when one compares some treatment effects with a control state. The isotonic regression is then obtained under the constraint $g(x_i) \geq g(x_1)$ for $i = 2, \ldots, n$.

a. Show that the following algorithm provides the isotonic regression g^*: if f is not isotonic, assume w.l.o.g. that the $f(x_i)$ are in increasing order ($i \geq 2$). Find the smallest j such that

$$A_j = \frac{\omega_1 f(x_1) + \ldots + \omega_j f(x_j)}{\omega_1 + \ldots \omega_j} < f(x_{j+1})$$

and take $g^*(x_1) = A_j = g^*(x_2) = \ldots = g^*(x_j)$, $g^*(x_{j+1}) = f(x_{j+1})$, $\ldots$.

b. Apply to the case where $n = 5$, $f(x_1) = 18$, $f(x_2) = 17$, $f(x_3) = 12$, $f(x_4) = 21$ and $f(x_5) = 16$, with $\omega_1 = \omega_2 = \omega_5 = 1$ and $\omega_3 = \omega_4 = 3$.

1.32 (Olkin et al. (1981)) Consider n observations $x_1, \ldots, x_n$ from $\mathcal{B}(k, p)$ where both k and p are unknown.

a. Show that the maximum likelihood estimator of k, $\hat{k}$, is such that

$$(\hat{k}(1 - \hat{p}))^n \geq \prod_{i=1}^n (\hat{k} - x_i) \quad \text{and} \quad ((\hat{k} + 1)(1 - \hat{p}))^n < \prod_{i=1}^n (\hat{k} + 1 - x_i),$$

where $\hat{p}$ is the maximum likelihood estimator of p.

b. If the sample is $16, 18, 22, 25, 27$, show that $\hat{k} = 99$.

c. If the sample is $16, 18, 22, 25, 28$, show that $\hat{k} = 190$ and conclude on the stability of the maximum likelihood estimator.

1.33 Give the maximum likelihood estimator of p for Example 1.1.6 if the other parameters are known and if there are two observations. Compare with the mean of the posterior distribution if $p \sim \mathcal{U}_{[0,1]}$.

1.34 (Basu (1988)) An urn contains 1000 tickets; 20 are tagged θ and 980 are tagged 10θ. A ticket is drawn at random with tag x.

a. Give the maximum likelihood estimator of θ, $\delta(x)$, and show that $P(\delta(x) = \theta) = 0.98$.

b. Suppose now there are 20 tickets tagged θ and 980 tagged $a_i\theta$ ($i \leq 980$), such that $a_i \in [10, 10.1]$ and $a_i \neq a_j$ ($i \neq j$). Give the new maximum likelihood estimator, δ', and show that $P(\delta'(x) < 10\theta) = 0.02$. Conclude about the appeal of maximum likelihood estimation in this case.

1.35 (Romano and Siegel (1986)) Given

$$f(x) = \frac{1}{x} \exp\left[-50\left(\frac{1}{x} - 1\right)^2\right] \qquad (x > 0),$$

show that f is integrable and that there exist $a, b > 0$ such that

$$\int_0^b af(x)dx = 1 \quad \text{and} \quad \int_1^b af(x)dx = 0.99.$$

For the distribution with density

$$p(y|\theta) = a\theta^{-1}f(y\theta^{-1})\mathbb{I}_{[0,b\theta]}(y),$$

give the maximum likelihood estimator, $\delta(y)$, and show that $P(\delta(y) > 10\theta) = 0.99$.

1.36 (Romano and Siegel (1986)) Consider x_1, x_2, x_3 i.i.d. $\mathcal{N}(\theta, \sigma^2)$.

a. Give the maximum likelihood estimator of σ^2 if $(x_1, x_2, x_3) = (9, 10, 11)$ or if $(x_1, x_2, x_3) = (29, 30, 31)$.

b. Given three additional observations x_4, x_5, x_6, give the maximum likelihood estimator if $(x_1, \ldots, x_6) = (9, 10, 11, 29, 30, 31)$. Does this result contradict the Likelihood Principle?

Section 1.4

1.37 If $x \sim \mathcal{N}(\theta, \sigma^2)$, $y \sim \mathcal{N}(\varrho x, \sigma^2)$, as in an autoregressive model, with ϱ known, and $\pi(\theta, \sigma^2) = 1/\sigma^2$, give the predictive distribution of y given x.

1.38 If $y \sim \mathcal{B}(n, \theta)$, $x \sim \mathcal{B}(m, \theta)$, and $\theta \sim \mathcal{B}e(\alpha, \beta)$, give the predictive distribution of y given x.

1.39 Given a proper distribution $\pi(\theta)$ and a sampling distribution $f(x|\theta)$, show that the only case such that $\pi(\theta|x)$ and $\pi(\theta)$ are identical occurs when $f(x|\theta)$ does not depend on θ.

1.40 Consider a prior distribution π positive on Θ and $x \sim f(x|\theta)$. Assume that the likelihood $\ell(\theta|x)$ is bounded, continuous, and has a unique maximum $\hat{\theta}(x)$.

a. Show that, when considering a virtual sample $x_n = (x, \ldots, x)$ made of n replications of the original observation x, the posterior distribution $\pi(\theta|x_n)$ converges to a Dirac mass in $\hat{\theta}(x)$.

b. Derive a Bayesian algorithm for computing maximum likelihood estimators.

1.41 *Given a couple (x, y) of random variables, the marginal distributions $f(x)$ and $f(y)$ are not sufficient to characterize the joint distribution of (x, y).

a. Give an example of two different bivariate distributions with the same marginals. (*Hint:* Take these marginals to be uniform $\mathcal{U}([0, 1])$ and find a function from $[0, 1]^2$ to $[0, 1]^2$ which is increasing in both its coefficients).

b. Show that, on the contrary, if the two conditional distributions $f(x|y)$ and $f(y|x)$ are known, the distribution of the couple (x, y) is also uniquely defined.

c. Extend b. to a vector $(x_1, \ldots, x_n)$ such that the full conditionals $f_i(x_i|x_j, j \neq i)$ are known. [*Note:* This result is called the *Hammersley–Clifford* Theorem, see Robert and Casella (2004).]

d. Show that property b. does not necessarily hold if $f(x|y)$ and $f(x)$ are known, i.e., that several distributions $f(y)$ can relate $f(x)$ and $f(x|y)$. (*Hint:* Exhibit a counter-example.)

e. Give some sufficient conditions on $f(x|y)$ for the above property to be true. (*Hint:* Relate this problem to the theory of complete statistics.)

1.42 Consider $x_1, \ldots, x_n$ i.i.d. $\mathcal{P}(\lambda)$. Show that $\sum_{i=1}^{n} x_i$ is a sufficient statistic and give a confidence region as in Example 1.4.4 when $\pi(\lambda)$ is a $\mathcal{G}(\alpha, \beta)$ distribution. For a given α level, compare its length with an equal tail confidence region.

1.43 Give the posterior and the marginal distributions in the following cases:

(i) $x|\sigma \sim \mathcal{N}(0, \sigma^2)$, $1/\sigma^2 \sim \mathcal{G}(1, 2)$;

(ii) $x|\lambda \sim \mathcal{P}(\lambda)$, $\lambda \sim \mathcal{G}(2, 1)$;

(iii) $x|p \sim \mathcal{N}eg(10, p)$, $p \sim \mathcal{B}e(1/2, 1/2)$.

1.44 Show that, for a sample $x_1, \ldots, x_n$ from a distribution with conditional density $f(x_i|\theta, x_{i-1})$, the actualizing decomposition (1.4.1) also applies. [*Note:* The sequence x_i is then a Markov chain.]

1.45 Show that, in the setting of Example 1.4.2, the marginal posterior distribution on ξ_2 is different from the marginal prior distribution if $\pi(\xi_1, \xi_2)$ does not factorize in $\pi_1(\xi_1)\pi_2(\xi_2)$.

1.46 *(Studden (1990)) In the setting of Example 1.4.3, we define the *canonical moments* of a distribution and show that they can be used as a representation of this distribution.

a. Show that the first two moments c_1 and c_2 are related by the two following inequalities:
$$c_1^2 \le c_2 \le c_1$$
and that the sequence (c_k) is monotonically decreasing to 0.

b. Consider a kth degree polynomial
$$P_k(x) = \sum_{i=0}^{k} a_i x^i.$$

Deduce from

(1.7.1) $\int_0^1 P_k^2(x) g(x) \, dx \ge 0$

that

(1.7.2) $a^t C_k a \ge 0, \qquad \forall a \in \mathbb{R}^{k+1},$

where
$$C_k = \begin{pmatrix} 1 & c_1 & c_2 & \cdots & c_k \\ c_1 & c_2 & c_3 & \cdots & c_{k+1} \\ \cdots & \cdots & \cdots & \cdots & \cdots \\ c_k & c_{k+1} & & \cdots & c_{2k} \end{pmatrix}$$
and $a^t = (a_0, a_1, \ldots, a_k)$.

c. Show that for every distribution g, the moments c_k satisfy

(1.7.3) $\begin{vmatrix} 1 & c_1 & c_2 & \cdots & c_k \\ c_1 & c_2 & c_3 & \cdots & c_{k+1} \\ \cdots & \cdots & \cdots & \cdots & \cdots \\ c_k & c_{k+1} & & \cdots & c_{2k} \end{vmatrix} > 0.$

(*Hint:* Interpret (1.7.2) as a property of C_k.)

d. Using inequalities similar to (1.7.1) for the polynomials $t(1-t)P_k^2(t)$, $tP_k^2(t)$, and $(1-t)P_k^2(t)$, derive the following inequalities on the moments of g:

(1.7.4)
$$\begin{vmatrix} c_1 - c_2 & c_2 - c_3 & \cdots & c_{k-1} - c_k \\ c_2 - c_3 & c_3 - c_4 & \cdots & c_k - c_{k+1} \\ \cdots & \cdots & \cdots & \cdots \\ c_{k-1} - c_k & \cdots & \cdots & c_{2k-1} - c_{2k} \end{vmatrix} > 0,$$

(1.7.5)
$$\begin{vmatrix} c_1 & c_2 & \cdots & c_k \\ c_2 & c_3 & \cdots & c_{k+1} \\ \cdots & \cdots & \cdots & \cdots \\ c_k & c_{k+1} & \cdots & c_{2k-1} \end{vmatrix} > 0,$$

(1.7.6)
$$\begin{vmatrix} 1 - c_1 & c_1 - c_2 & \cdots & c_{k-1} - c_k \\ c_1 - c_2 & c_2 - c_3 & \cdots & c_k - c_{k+1} \\ \cdots & \cdots & \cdots & \cdots \\ c_{k-1} - c_k & \cdots & \cdots & c_{2k-2} - c_{2k-1} \end{vmatrix} > 0.$$

e. Show that (1.7.3) (resp. (1.7.4)) induces a lower (resp. upper) bound $\underline{c}_{2k}$ (resp. $\bar{c}_{2k}$) on c_{2k} and that (1.7.5) (resp. (1.7.6)) induces a lower (resp. upper) bound $\underline{c}_{2k-1}$ (resp. $\bar{c}_{2k-1}$) on c_{2k-1}.

f. Defining p_k as
$$p_k = \frac{c_k - \underline{c}_k}{\bar{c}_k - \underline{c}_k},$$
show that the relation between $(p_1, ..., p_n)$ and $(c_1, ..., c_n)$ is one-to-one for every n and that the p_i are independent.

g. Show that the inverse transform is given by the following recursive formulas. Let us define
$$q_i = 1 - p_i, \quad \zeta_1 = p_1, \quad \zeta_i = p_i q_{i-1} \quad (i \geq 2).$$

Then
$$\begin{cases} S_{1,k} = \zeta_1 + \ldots + \zeta_k & (k \geq 1), \\ S_{j,k} = \sum_{i=1}^{k-j+1} \zeta_i S_{j-1,i+j-1} & (j \geq 2), \\ c_n = S_{n,n}. \end{cases}$$

Section 1.5

1.47 The problem with improper priors that the integral $\int_\Theta f(x|\theta)\pi(\theta)\,d\theta$ may not exist does not appear with proper priors.

a. Recall Fubini's theorem and apply to the couple of functions $(f(x|\theta), \pi(\theta))$.

b. Deduce that, if π is a finite positive measure,

(1.7.7)
$$\int_\Theta f(x|\theta)\pi(\theta)\,d\theta < \infty$$

almost everywhere.

c. Show that, if π is improper and $f(x|\theta)$ has a finite support, then $\pi(\theta|x)$ is defined if, and only if, (1.7.7) is finite for every x in the support of $f(x|\theta)$.

1.48 Show that, if π is a positive measure on Θ, the integral (1.7.7) is positive almost everywhere.

1.49 (Fernandez and Steel (1999)) Consider n i.i.d. observations $x_1, \ldots, x_n$ from the mixture
$$p\mathcal{N}(\mu_0, \sigma_0^2) + (1 - p)\mathcal{N}(\mu_0, \sigma_1^2),$$

where p, μ_0 and σ_0 are known. The prior on σ_1 is a beta $\mathcal{Be}(\alpha, \beta)$ distribution. Show that, if $r \geq 1$ observations are equal to μ_0, the posterior distribution is only defined when $\alpha > r$. [*Note:* From a measure theoretical point of view, the set of x_i's equal to μ_0 is of measure zero. If one (or more) observation is exactly equal to μ_0, it means that the continuous mixture model is not appropriate.]

1.50 (Exercise 1.49 cont.) Consider an observation x from a normal distribution $\mathcal{N}(0, \sigma^2)$.

   a. If the prior distribution on σ is an exponential distribution $\mathcal{E}xp(\lambda)$, show that the posterior distribution is not defined for $x = 0$.
   b. If the prior distribution on σ is the improper prior $\pi(\sigma) = \sigma^{-1} \exp(-\alpha\sigma^{-2})$, with $\alpha > 0$, show that the posterior distribution is always defined.

1.51 (Exercise 1.50 cont.) Consider an observation y with $y = x - \lambda$, where x is distributed from Laplace's distribution,

$$f(x|\theta) = \theta^{-1} \exp(-|x|/\theta),$$

and λ is distributed from

$$\pi(\lambda) = |\lambda|^{-1/2} \mathbb{I}_{[-1/2,1/2]}(\lambda).$$

If θ is distributed from a gamma $\mathcal{G}(1/2, a)$ $(a > 0)$, show that, if $y = 0$, the posterior distribution is not defined.

1.52 (Musio and Racugno (1999)) Consider the Poisson $\mathcal{P}(\theta)$ model

$$P_\theta(X = x) = \frac{\theta^x}{x!} e^{-\theta}, \quad x = 0, 1, \ldots, \quad \theta > 0,$$

and the prior distribution $\pi(\theta) = 1/\theta$. Show that for $x = 0$, the posterior distribution is not defined.

1.53 (Raiffa and Schlaifer (1961)) Consider a $\mathcal{Be}(\alpha m, (1 - m)\alpha)$ prior on $p \in [0, 1]$. Show that, if m is held fixed and α approaches 0, the prior distribution converges to a two-point mass distribution with weight m on $p = 1$ and $(1-m)$ on $p = 0$. Discuss the drawbacks of such a setting.

1.54 (Bauwens (1991)) Consider $x_1, \ldots, x_n$ i.i.d. $\mathcal{N}(\theta, \sigma^2)$ and

$$\pi(\theta, \sigma^2) = \sigma^{-2(\alpha+1)} \exp(-s_0^2/2\sigma^2).$$

   a. Compute the posterior distribution $\pi(\theta, \sigma^2 | x_1, \ldots, x_n)$ and show that it only depends on $\bar{x}$ and $s^2 = \sum_{i=1}^n (x_i - \bar{x})^2$.
   b. Derive the posterior expectation $\mathbb{E}^\pi[\theta | x_1, \ldots, x_n]$ and show that its behavior when α and s_0 both converge to 0 depends on the limit of the ratio $s_0^2/\alpha - 1$.

1.55 Show that if the prior $\pi(\theta)$ is improper and the sample space $\mathcal{X}$ is finite, the posterior distribution $\pi(\theta | x)$ is not defined for some values of x.

1.56 Consider $x_1, \ldots, x_n$ distributed according to $\mathcal{N}(\theta_j, 1)$, with $\theta_j \sim \mathcal{N}(\mu, \sigma^2)$ $(1 \leq j \leq n)$ and $\pi(\mu, \sigma^2) = \sigma^{-2}$. Show that the posterior distribution $\pi(\mu, \sigma^2 | x_1, \ldots, x_n)$ is not defined.

1.57 In the setting of Example 1.1.6, that is, for a mixture of two normal distributions,

   a. Show that the maximum likelihood estimator is not defined when all the parameters are unknown.

b. Similarly, show that it is not possible to use an improper prior of the form

$$\pi_1(\mu_1, \sigma_1)\pi_2(\mu_2, \sigma_2)\pi_3(p)$$

to estimate these parameters. (*Hint:* Write the likelihood as a sum of $n + 1$ terms, depending on the number of observations allocated to the first component.)

[*Note:* Mengersen and Robert (1996) show that it is possible to use some improper priors by introducing prior dependence between the components.]

1.58 *(**Exercise 1.57 cont.**) For a mixture of two normal distributions (1.1.2), if the prior distribution on the parameters is of the form

$$\pi_1(\mu_1, \sigma_1)\pi_1(\mu_2, \sigma_2)\pi_3(p)$$

and $\pi_3(p) = \pi_3(1-p)$, show that the marginal posterior distribution of (μ_1, σ_1) is the same as the marginal posterior distribution of (μ_2, σ_2), whatever the sample. Deduce that the posterior mean of (μ_1, σ_1) is equal to the posterior mean of (μ_2, σ_2) and therefore that it is not a pertinent estimator. [*Note:* This problem is a consequence of the nonidentifiability of the component labels in a mixture. Solutions involve imposing identifying constraints such as the ordering $\mu_1 \leq \mu_2$, or using loss functions that are invariant under permutation of the component labels.]

1.59 Construct a limiting argument as in Example 1.5.3 to solve the indeterminacy of Example 1.5.4. Derive the posterior mean.

1.60 Show that, if the prior distribution is improper, the pseudo marginal distribution is also improper.

1.61 *(Hobert and Casella (1998)) Consider a random-effect model,

$$y_{ij} = \beta + u_i + \varepsilon_{ij}, \qquad i = 1, \ldots, I, \ j = 1, \ldots, J,$$

where $u_i \sim \mathcal{N}(0, \sigma^2)$ and $\varepsilon_{ij} \sim \mathcal{N}(0, \tau^2)$. Under the prior

$$\pi(\beta, \sigma^2, \tau^2) = \frac{1}{\sigma^2 \tau^2} ,$$

the posterior does not exist.

a. By integrating out the (unobservable) random-effects u_i, show that the full posterior distribution of $(\beta, \sigma^2, \tau^2)$ is

$$\pi(\beta, \sigma^2, \tau^2 | y) \propto \sigma^{-2-I} \tau^{-2-IJ} \exp\left\{-\frac{1}{2\tau^2} \sum_{i,j} (y_{ij} - \bar{y}_i)^2\right\}$$

$$\times \exp\left\{-\frac{J \sum_i (\bar{y}_i - \beta)^2}{2(\tau^2 + J\sigma^2)}\right\} (J\tau^{-2} + \sigma^{-2})^{-1/2} .$$

b. Integrate out β to get the marginal posterior density

$$\pi(\sigma^2, \tau^2 | y) \propto \frac{\sigma^{-2-I} \tau^{-2-IJ}}{(J\tau^{-2} + \sigma^{-2})^{1/2}} (\tau^2 + J\sigma^2)^{1/2}$$

$$\times \exp\left\{-\frac{1}{2\tau^2} \sum_{i,j} (y_{ij} - \bar{y}_i)^2 - \frac{J}{2(\tau^2 + J\sigma^2)} \sum_i (\bar{y}_i - \bar{y})^2\right\} .$$

c. Show that the full posterior is not integrable. (*Hint:* For $\tau \neq 0$, $\pi(\sigma^2, \tau^2 | y)$ behaves like σ^{-2} in a neighborhood of 0.)

d. Show that the conditional distributions

$$U_i | y, \beta, \sigma^2, \tau^2 \ \sim \ \mathcal{N}\left(\frac{J(\bar{y}_i - \beta)}{J + \tau^2 \sigma^{-2}}, (J\tau^{-2} + \sigma^{-2})^{-1}\right),$$

$$\beta | u, y, \sigma^2, \tau^2 \ \sim \ \mathcal{N}(\bar{y} - \bar{u}, \tau^2/JI),$$

$$\sigma^2 | u, \beta, y, \tau^2 \ \sim \ \mathcal{IG}\left(I/2, (1/2)\sum_i u_i^2\right),$$

$$\tau^2 | u, \beta, y, \sigma^2 \ \sim \ \mathcal{IG}\left(IJ/2, (1/2)\sum_{i,j}(y_{ij} - u_i - \beta)^2\right),$$

are well defined. [*Note:* The consequences of this definition of the full posterior will be clarified in Chapter 6.]

1.62 *Consider a dichotomous probit model, where $(1 \leq i \leq n)$

$$(1.7.8) \qquad\qquad P(d_i = 1) = 1 - P(d_i = 0) = P(z_i \geq 0),$$

with $z_i \sim \mathcal{N}(r_i\beta, \sigma^2)$, $\beta \in \mathbb{R}$, r_i being a covariate. (Note that the z_i's are *not* observed.)

a. Show that the parameter (β, σ) is not identifiable.
b. For the prior distribution $\pi(\beta, \sigma) = 1/\sigma$, show that the posterior distribution is not defined.
c. For the prior distribution

$$\sigma^{-2} \sim \mathcal{G}a(1.5, 1.5), \qquad \beta|\sigma \sim \mathcal{N}(0, 10^2),$$

show that the posterior distribution is well defined.
d. An identifying constraint is to impose $\sigma = 1$ on the model. Give sufficient conditions on the observations (d_i, r_i) for the posterior distribution on β to be defined if $\pi(\beta) = 1$.
e. Same question as d. when the normal distribution on the z_i's is replaced with the logistic function, that is,

$$P(d_i = 1) = 1 - P(d_i = 0) = \frac{\exp(r_i\beta)}{1 + \exp(r_i\beta)},$$

which gives the dichotomous logit model.

1.63 *(Kubokawa and Robert (1994)) In *linear calibration models*, the interest is in determining values of the regressor x from observed responses y, as opposed to standard linear regression. A simplified version of this problem can be put into the framework of observing the independent random variables

$$(1.7.9) \qquad\qquad y \sim \mathcal{N}_p(\beta, \sigma^2 I_p), \ z \sim \mathcal{N}_p(x_0\beta, \sigma^2 I_p), \ s \sim \sigma^2 \chi_q^2,$$

with $x_0 \in \mathbb{R}$, $\beta \in \mathbb{R}^p$. The parameter of interest is x_0.

a. A reference prior on (x_0, β, σ) yields the joint posterior distribution

$$\pi(x_0, \beta, \sigma^2 | y, z, s) \ \propto \ \sigma^{-(3p+q)-\frac{1}{2}} \exp\{-(s + \|y - \beta\|^2 \\ + \|z - x_0\beta\|^2)/2\sigma^2\} (1 + x_0^2)^{-1/2}.$$

Show that this posterior is compatible with the sampling distribution (1.7.9).

b. Show that the marginal posterior distribution of x_0 is

$$\pi(x_0|y, z, s) \propto \frac{(1 + x_0^2)^{(p+q-1)/2}}{\left\{\left(x_0 - \frac{y^t z}{s + \|y\|^2}\right)^2 + \frac{\|z\|^2 + s}{\|y\|^2 + s} - \frac{(y^t z)^2}{(s + \|y\|^2)^2}\right\}^{(2p+q)/2}}.$$

c. Deduce that the posterior distribution of x_0 is well defined.

[*Note:* See Osborne (1991) for an introduction and review of calibration problems. The model (1.7.9) is also equivalent to Fieller's (1954) problem. See, e.g., Lehmann and Casella (1998).]

Note 1.8.2

1.64 *(Diaconis and Kemperman (1996)) Show that the definition of the Dirichlet process $\mathcal{D}(F_0, \alpha)$ given in 1.8.2 is compatible with the following one: given an i.i.d. sequence x_i from F_0 and a sequence of weights ω_i such that

$$\omega_1 \sim \mathcal{Be}(1, \alpha), \quad \omega_1 + \omega_2 \sim \mathcal{Be}(1, \alpha)\mathbb{I}_{[\omega_1, 1]}, \dots$$

the random distribution

$$F = \sum_{i=1}^{\infty} \omega_i \delta_{x_i}$$

is distributed from $\mathcal{D}(F_0, \alpha)$.

1.65 *(**Exercise 1.64 cont.**) If $F \sim \mathcal{D}(F_0, \alpha)$, the quantity $X = \int x F(dx)$ is a random variable.

a. If $\alpha = 1$ and F_0 is a Cauchy distribution, show that X is also distributed as a Cauchy random variable. [*Note:* This relates to the characterizing property of Cauchy distributions that the average of Cauchy random variables is a Cauchy variable with the same parameters.]

b. If $\alpha = 1$ and $F_0 = \varrho\delta_0 + (1 - \varrho)\delta_1$, show that X is distributed as a beta $\mathcal{Be}(\varrho, 1 - \varrho)$ random variable.

c. Show that, if $\alpha = 1$ and F_0 is $\mathcal{U}_{[0,1]}$, X has the density

$$\frac{e}{\pi} \frac{\sin(\pi y)}{(1 - y)^{(1-y)} y^y}.$$

[*Note:* See Diaconis and Kemperman (1996) for the general formula relating F_0 and the density of X.]

1.66 *(Diaconis and Kemperman (1996)) The Dirichlet process prior $\mathcal{D}(F_0, \alpha)$ can also be described via the so-called *Chinese restaurant process*. Consider a restaurant with many large tables and label each table j with a realization y_j from F_0. Then seat arrivals as follows: the first person to arrive sits at the first table. The $(n + 1)$ th person sits at an empty table with probability $\alpha/(\alpha + n)$ and to the right of a seated person with probability $n/(\alpha + n)$.

a. If x_i denotes the label z_j of the table where the i th person sits, show that the sequence $x_1, x_2, \dots$ is exchangeable (that is, that the distribution is invariant under any permutation of the indices).

b. Show that this sequence can be seen as i.i.d. replications from F, with F distributed from $\mathcal{D}(F_0, \alpha)$, using the conditional distribution given in Note 1.8.2.

c. Show that this definition is also compatible with the definition of Exercise 1.64.

Note 1.8.3

1.67 *(Hadjicostas and Berry (1999)) Consider independent observations x_i $(i = 1, \ldots, n)$ from Poisson distributions $\mathcal{P}oi(\lambda_i t_i)$, where the durations t_i are known. The prior on the λ_i's is a gamma distribution $\mathcal{G}(\alpha, \beta)$ with an independence assumption. The model is *hierarchical* because the parameters (α, β) are supposed to be distributed from a prior distribution $\pi(\alpha, \beta)$ such that

(1.7.10) $\pi(\alpha, \beta) \propto \alpha^{k_1} (\alpha + s_1)^{k_2} \beta^{k_3} (\beta + s_2)^{k_4}$,

where the values k_i and $s_j > 0$ are known $(i = 1, \ldots, 4, \ j = 1, 2)$.

a. Show that the prior distribution (1.7.10) is proper if, and only if,

$$k_1 + k_2 + 1 < 0, \quad k_1 + 1 > 0, \quad k_3 + k_4 + 1 < 0, \quad k_3 + 1 > 0.$$

b. By integrating out the λ_i's from the joint distribution of the λ_i's and of (α, β), derive the posterior (marginal) distribution of (α, β).

c. Show that the posterior (marginal) distribution of (α, β) is defined (proper) if, and only if,

$$k_1 + y + 1 > 0, \quad k_3 + r + 1 > 0, \quad k_3 > k_1 + k_2$$

and either $k_3 + k_4 + 1 < 0$ or $k_3 + k_4 + 1 = 0$ and $k_1 + y > 0$, where

$$y = \sum_{i=1}^{n} \mathbb{I}_0(x_i), \qquad r = \sum_{i=1}^{n} x_i \,.$$

d. Verify that the conditions of a. imply the conditions of b. (as they should).

e. Show that the conditions of b. are satisfied when $(k_1, \ldots, k_4) = (-8, 0, -5, 0)$ and $(y, r) = (10, 337)$, while the conditions of a. are not satisfied.

f. Show that the conditions of b. are not satisfied when $(k_1, \ldots, k_4) = (-12, 0, 1, 1)$ and $(y, r) = (10, 337)$.

Note 1.8.4

1.68 *(Robins and Ritov (1997)) Consider i.i.d. observations (x_i, y_i) in $(0, 1)^k \times \mathbb{R}$ from the following model: $x \sim f(x)$, $y|x \sim \mathcal{N}(\theta(x), 1)$, with the mean function θ uniformly bounded on $(0, 1)^k$ and f in the set of densities such that $c < f(x) < 1/c$ uniformly on $(0, 1)^k$, where $c < 1$ is a fixed constant. Assume that the quantity of interest is

$$\varphi = \int_{(0,1)^k} \theta(x) dx \,.$$

a. Show that the space Θ of mean functions θ is infinite dimensional.

b. Give the likelihood $\ell(\theta, f)$ and show that it factorizes in a function of f times a function of θ.

c. When f is known, show that $(x_1, \ldots, x_n)$ is ancillary.

d. When f is unknown, show that $(x_1, \ldots, x_n)$ is θ-ancillary, in the sense that the conditional likelihood given $(x_1, \ldots, x_n)$ is a function of θ only, the marginal distribution of $(x_1, \ldots, x_n)$ is a function of f only, and the parameter space is a product space. (See Cox and Hinkley (1974) and Robins and Wasserman (2000) for details about this notion.)

e. When f is known, show that

$$\frac{1}{n} \sum_{i=1}^{n} \frac{y_i}{f(x_i)}$$

is a consistent estimator of φ. (In fact, it is $\sqrt{n}$ uniformly consistent.)

f. When f is unknown, Robins and Ritov (1997) have shown that there is no uniformly consistent estimator of φ. Deduce that, if the prior distribution on (θ, f) factorizes as $\pi_1(\theta)\pi_2(f)$, the Bayesian inference on θ (and thus φ) is the same whatever the value of f.

g. On the contrary, if the prior distribution on (θ, f) makes θ and f dependent, and if f is known to be equal to f_0, the posterior distribution will depend on f_0. Deduce that this dependence violates the Likelihood Principle.

[*Note:* The previous simplified description of Robins and Ritov (1997) follows from Robins and Wasserman (2000).]

1.8 Notes

1.8.1 A brief history of Bayesian Statistics

Books have been written on the history of Bayesian Statistics, including Stigler (1986), Dale (1991), Lad (1996) and Hald (1998), and we only point out here a few highlights in the development of Bayesian Statistics in the last two centuries.

As detailed in this chapter, the first occurrence of Bayes's formula took place in 1761, in the setting of the binomial example of Section 1.2, exposed by the Reverent Thomas Bayes before the Royal Society, and published posthumously by his friend R. Price in 1763. Pierre Simon Laplace then rediscovered this formula in a greater generality in 1773, apparently ignoring Bayes's previous work. The use of the Bayesian principle then became common in the 19th century, as reported in Stigler (1986), but criticisms started to arise by the end of the 19th century, as for instance in Venn (1886) or Bertrand (1889), focusing on the choice of the uniform prior and the resulting reparameterization paradoxes, as reported by Zabell (1989).

Then, despite further formalizations of the Bayesian paradigm by Edgeworth and Karl Pearson at the turn of the century and Keynes (1921) later, came first Kolmogorov, whose axiomatization of the theory of probabilities in the 1920s seemed to go against the Bayesian paradigm and the notion of subjective probabilities, and second Fisher, who moved away from the Bayesian approach (Fisher (1912)) to the definition of the likelihood function (Fisher (1922)), then to fiducial Statistics (Fisher (1930)), but never revised his opinion on Bayesian Statistics. This is slightly paradoxical, since fiducial Statistics was, in a sense, an attempt to overcome the difficulty of selecting the prior distribution by deriving it from the likelihood function (Seidenfeld (1992)), in the spirit of the *noninformative approaches* of Jeffreys (1939) and Bernardo (1979).

For instance, considering the relation $O = P + \epsilon$ where ϵ is an error term, fiducial Statistics argues that, if P (the cause) is known, O (the effect) is distributed according to the above relation. Conversely, if O is known, $P = O - \epsilon$ is distributed according to the symmetric distribution. In this perspective, observations and parameters play a *symmetric* role, depending on the way the model is analyzed, i.e., depending on what is known and what is unknown. More generally, the fiducial approach consists of renormalizing the likelihood (1.2.1) so that it becomes a density *in θ* when

$$\int_\Theta \ell(\theta|x)\, d\theta < +\infty,$$

thus truly inverting the roles of x and θ. As can be seen in the above example, the argument underlying the causal inversion is totally conditional: conditional upon P, $O = P + \epsilon$ while, conditional upon O, $P = O - \epsilon$. Obviously, this argument does not hold from a probabilistic point of view: if O is a random variable and P is a (constant) parameter, to write $P = O - \epsilon$ does not imply that P becomes a random variable. Moreover, the transformation of $\ell(\theta|x)$ into a density is not always possible. The fiducial approach was progressively abandoned after the exposure of fundamental paradoxes (see Stein (1959), Wilkinson (1977) and the references in Zabell (1992)).

Jeffreys's (1939) book appears as the first modern treatise on Bayesian Statistics: it contains, besides the idea of a noninformative prior, those of predictive distribution, Bayes factors and improper priors. But it came out at the time of Fisher's development of likelihood Statistics and Neyman's (1934) confidence intervals and did not meet with the same success. Alternatives to Bayesian Statistics then became the standard in the 1930s with the introduction of maximum likelihood estimators and the development of a formalized theory of Mathematical Statistics, where prior distributions only appeared as a way of constructing formally optimal estimators as in Wald (1950) or Ibragimov and Has'minskii (1981) (see Chapter 8). Attempts to formalize further the Bayesian approach to Statistics by Gini or de Finetti from the 1930s to the 1970s did not bring it more popularity against the then-dominant Neyman–Pearson paradigm, even though the Bayesian community was growing and produced treatises such as those of Savage (1954) and Lindley (1965, 1971).

It can be argued that it is only recently that Bayesian Statistics got a new impetus, thanks to the development of new computational tools—which have always been central to the Bayesian paradigm—and the rapidly growing interest of practitioners in this approach to statistical modeling, as stressed in Berger's (2000) view of the present and future states of Bayesian Statistics. The vitality of current Bayesian Statistics can be seen through the percentage of Bayesian papers in Statistics journals as well as in other fields. It thus looks as though practitioners in this century will be taking better heed of Bayesian Statistics than their twentieth-century counterparts.

1.8.2 Bayesian nonparametric Statistics

While this book sticks to the parametric approach to Statistics, there is a large literature on Bayesian nonparametric Statistics. First, optimality notions such as minimaxity are central to functional estimation; similar to the parametric case (see Chapter 3), Bayes estimators can be used to determine minimaxity bounds and minimax estimators.

A second and much less formal aspect is to envision Bayesian prior modeling in a infinite dimensional space. This is obviously harder, for mathematical as well as prior construction, reasons. But a first solution is to stand in the grey area between parametric and non-parametric Statistics as in Example 1.4.3: the number of parameters is finite but grows to infinity with the number of observations. This is, for instance, the case with kernel estimation, where a density is approximated by a mixture

$$\frac{1}{n\sigma}\sum_{i=1}^{n}K\left(\frac{x-x_i}{\sigma}\right),$$

K being a density function, and where σ can be estimated in a Bayesian way, with Hermite expansions (Hjort (1996)), or with wavelets bases (Müller and Vidakovic (1999, Chapter 1)), where a function f is decomposed in a functional basis,

$$f(x)=\sum_{i}\sum_{j}\omega_{ij}\Psi\left(\frac{x-\mu_i}{\sigma_j}\right),$$

where Ψ denotes a special function called the *mother wavelet*, like the Haar wavelet

$$\Psi(x)=\mathbb{I}_{[0,1/2)}-\mathbb{I}_{[1/2,1)},$$

where the scale and location parameters μ_i and σ_j are fixed and known, and where the coefficients ω_{ij} can be associated with a prior distribution like (Abramovich et al. (1998))

$$\omega_{ij}\sim\varrho_i\mathcal{N}(0,\tau_i^2)+(1-\varrho_i)\delta_0,$$

where δ_0 denotes the Dirac mass at 0.

A second solution, when estimating a c.d.f. F, is to put a prior distribution on F. The most common choice is to use a Dirichlet distribution $\mathcal{D}(F_0,\alpha)$ on F, F_0 being the prior mean and α the precision, as introduced in Ferguson (1974). This prior distribution enjoys the coherency property that, if $F\sim\mathcal{D}(F_0,\alpha)$, the vector $(F(A_1),\ldots,F(A_p))$ is distributed as a Dirichlet variable in the usual sense $\mathcal{D}_p(\alpha F_0(A_1),\ldots,\alpha F_0(A_p))$ for every partition $(A_1,\ldots,A_p)$. But it also leads to posterior distributions which are partly discrete: if $x_1,\ldots,x_n$ are distributed from F and $F\sim\mathcal{D}(F_0,\alpha)$, the marginal conditional distribution of x_1 given $(x_2,\ldots,x_n)$ is

$$\frac{\alpha}{\alpha+n-1}F_0+\frac{1}{\alpha+n-1}\sum_{i=2}^{n}\delta_{x_i}.$$

(See also Exercises 1.64 and 1.66 for other characterizations.) The approximation of the posterior distribution requires advanced computational tools that will be developed in Chapter 6. (See Note 6.6.7 for more details.) Other proposals have thus appeared in the literature such as the *generalized Dirichlet distribution* (Hjort (1996)), *Pólya tree priors* (Fabius (1964), Lavine (1992)), *beta process prior* (Hjort (1990)), *Lévy process priors* (Phillips and Smith (1996)).

As a concluding note, let us mention that a recent trend in Bayesian statistics has been to study models with varying dimensions, such as mixtures, hidden Markov models and other dynamic models, as well as neural networks, thanks to new computational tools developed by Grenander and Miller (1994), Green (1995), Phillips and Smith (1996), or Stephens (1997) (see Chapter 6). This is, for instance, the case with mixtures

$$\sum_{i=1}^{k}p_{ik}\varphi(x|\theta_{ik})$$

where $\varphi(\cdot|\theta)$ is a parametrized density, the sum of the weights p_{ik} sum up to 1, and the number of components k is unknown. While this is a well-defined parametric problem, it is closer to nonparametric imperatives than to standard parametric estimation (see Richardson and Green (1997) or Stephens (2000)).

1.8.3 Proper posteriors

It must by now be clear from Section 1.5 that an improper prior π can only be used for inference purposes if (1.7.7) holds for the observation x at hand. If this condition is not satisfied, posterior quantities like the posterior mean or posterior median have no meaning, since, for instance, the ratio

$$\frac{\int_{\Theta} f(x|\theta)\pi(\theta)\,d\theta}{\int_{\Theta} \theta f(x|\theta)\pi(\theta)\,d\theta}$$

is not defined. To verify that (1.7.7) is satisfied in complex models can be quite difficult (see Exercises 1.61 and 1.62) or even simply impossible. Unfortunately, because of computational innovations such as the Gibbs sampler (see Chapter 6), it is possible to work directly from the relation $\pi(\theta|x) \propto f(x|\theta)\pi(\theta)$ to simulate values from the posterior $\pi(\theta|x)$, but the simulation output does not always signal that the posterior does not exist (see Hobert and Casella (1998)). There are therefore examples in the literature where data have been analyzed with such undefined posteriors, this lack of definition of the posterior been only discovered years later.

We will see in Note 6.6.4, however, that there are good reasons to work with improper posteriors on extended spaces, that is, through a completion of θ in (α, θ), as long as the properness of the posterior $\pi(\theta|x)$ is satisfied.

1.8.4 Asymptotic properties of Bayes estimators

We do not develop the asymptotic point of view in this book for two main reasons, the first of which being that the Bayesian point of view is intrinsically conditional. When conditioning on the observation x, which may be a sample $(x_1, \ldots, x_n)$, there is no reason to wonder what might happen if n goes to infinity since n is fixed by the sample size. Theorizing on future values of the observations thus leads to a frequentist analysis, opposite to the imperatives of the Bayesian perspective. The second point is that, even though it does not integrate asymptotic requirements, Bayesian procedures perform well in a vast majority of cases under asymptotic criteria. It is not so paradoxical that, most often, the Bayesian perspective, and in particular the choice of the prior distribution, cease to be relevant when the number of observations gets infinitely large compared with the number of parameters. (There are well-known exceptions to this ideal setting, as in the *Neyman–Scott problem* of Example 3.5.10, in Diaconis and Freedman (1986), where the number of parameters increases with the number of observations and leads to inconsistent Bayes estimators, or in Robins and Ritov (1997), as detailed in Exercise 1.68.)

In a general context, Ibragimov and Has'minskii (1981, Chapter 1) show that Bayes estimators are *consistent*, that is, that they almost surely converge to the true value of the parameter when the number of observations goes to infinity. This is, for instance, the case with estimators δ_α ($\alpha \geq 1$) that minimize the posterior loss (see Chapter 2) associated with the loss function

$L(\delta, \theta) = |\theta - \delta|^{\alpha}$, under fairly weak constraints on the prior distribution π and the sampling density $f(x|\theta)$. Ibragimov and Has'minskii (1981, Chapter 3) also establish (under more stringent conditions) the asymptotic efficiency of some Bayes estimates, that is, that the posterior distribution converges towards the true value at the rate $n^{-1/2}$. (See Schervish (1995) for more details.)

Barron et al. (1999) also give general conditions for consistency of a posterior distribution in the following sense: the posterior probability of every Hellinger neighborhood of the true distribution tends to 1 almost surely when the sample size goes to infinity. (The *Hellinger distance* between two densities f_1 and f_2 (or the corresponding distributions) is defined as

$$d(f_1, f_2) = \int \left(f_1(x)^{1/2} - f_2(x)^{1/2} \right)^2 dx \,.$$

We will use it for decision-theoretic purposes in Chapter 2.) The basic assumption on the prior distribution π is that it gives positive mass to every Kullback–Leibler neighborhood of the true distribution. (We will also use the Kullback–Leibler pseudo-distance in Chapter 2.)

We will come back to asymptotics, nonetheless, in Chapter 3 with the definition of noninformative priors via the asymptotic approximation of tail behaviors, and in Chapter 6 with the *Laplace approximation* to posterior integrals.

CHAPTER 2

Decision-Theoretic Foundations

Today would run out according to the Pattern. But over and over he mulled over the decisions he had made since he first entered the Waste. Could he have done something different, something that would have avoided this day, this place? Next time, perhaps.

Robert Jordan, *The Fires of Heaven, Book V of the Wheel of Time.*

2.1 Evaluating estimators

Considering that the overall purpose of most inferential studies is to provide the statistician (or a client) with a *decision*, it seems reasonable to ask for an *evaluation* criterion of decision procedures that assesses the consequences of each decision and depends on the parameters of the model, i.e., the true state of the world (or of Nature). These *decisions* can be of various kinds, ranging from buying equities depending on their future returns θ, to stopping an agricultural experiment on the productivity θ of a new crop species, to estimating the underground economy contribution θ to the U.S. GNP, to deciding whether the number θ of homeless people has increased since the last census. They also include assessing whether a new scientific theory is compatible with the experimental evidence at hand. If no evaluation criterion is available, it is impossible to compare different decision procedures and absurd solutions, such as proposing $\hat{\theta} = 3$ for any real estimation problem or even more dramatically the answer one wants to impose, can only be eliminated by ad-hoc reasoning. To avoid such reasoning implies a reinforced axiomatization of the statistical inferential framework, called *Decision Theory*. This augmented theoretical structure is necessary for Statistics to reach a coherence otherwise unattainable[1].

Although almost everybody agrees on the need for such an evaluation criterion, there is an important controversy running about the choice of this

---

[1] The Bayesian approach is, from our point of view, the ultimate step in this quest for coherence.

evaluation criterion, since the consequences on the decision are not innocuous. This difficulty even led some statisticians to totally reject Decision Theory, on the basis that a practical determination of the decision-maker evaluation criterion is utterly impossible in most cases.

This criterion is usually called *loss* and is defined as follows, where $\mathcal{D}$ denotes the set of possible decisions. $\mathcal{D}$ is called the *decision* space and most theoretical examples focus on the case $\mathcal{D} = \Theta$, which represents the standard estimation setting.

Definition 2.1.1 *A loss function is any function* L *from* $\Theta \times \mathcal{D}$ *in* $[0, +\infty)$.

This loss function is supposed to evaluate the penalty (or error) $L(\theta, d)$ associated with the decision d when the parameter takes the value θ. In a traditional setting of parameter estimation, when $\mathcal{D}$ is Θ or $h(\Theta)$, the loss function $L(\theta, \delta)$ measures the error made in evaluating $h(\theta)$ by δ. Section 2.2 introduces a set of so-called rationality axioms that ensures the existence of such a function in a decision setting.

The actual determination of the loss function is often awkward in practice, in particular because the determination of the consequences of each action for each value of θ is usually impossible when $\mathcal{D}$ or Θ are large sets, for instance when they have an infinite number of elements. Moreover, in qualitative models, it may be delicate to quantify the consequences of each decision. We will see through paradoxes like the *Saint Petersburg paradox* that, even when the loss function seems obvious, for instance when errors can be expressed as monetary losses, the actual loss function can be quite different from its intuitive and linear approximation.

The complexity of determining the subjective loss function of the decision-maker often prompts the statistician to use classical (or *canonical*) losses, selected because of their simplicity and mathematical tractability. Such losses are also necessary for a theoretical treatment of the derivation of optimal procedures, when there is no practical motivation for the choice of a particular loss function. The term *classical* is related to their long history, dating back to Laplace (1773) for the absolute error loss (2.5.3) and Gauss (1810) for the quadratic loss (2.5.1), when *errors* in terms of performance of estimators or consequences of decisions were confused with *errors* in terms of the irreducible variability of random variables (variance). But this attribute should not be taken as a value statement, since an extensive use of these losses does not legitimize them any further. In fact, the recourse to such automatic (or generic) losses, although often justified in practice—it is still better to take a decision in a finite time using an approximate criterion rather that spending an infinite time to determine exactly the proper loss function—has generated many of the criticisms addressed to Decision Theory.

A fundamental basis of Bayesian Decision Theory is that statistical inference should start with the rigorous determination of three factors:

(1) the distribution family for the observations, $f(x|\theta)$;

(2) the prior distribution for the parameters, $\pi(\theta)$;

(3) the loss associated with the decisions, $L(\theta, \delta)$;

the prior and the loss, and even sometimes the sampling distribution, being derived from partly subjective considerations. Classical decision-theoreticians omit the second point. The frequentist criticisms of the Bayesian paradigm often fail to take into account the problem of the construction of the loss function, even though this may be at least as complicated as the derivation of the prior distribution. In addition, to presuppose the existence of a loss function implies that some information about the problem at hand is available. This information could therefore be used more efficiently by building up a prior distribution. Actually, Lindley (1985) states that loss and prior are difficult to separate and should be analyzed simultaneously. We will see in Section 2.4 an example of the *duality* existing between these two factors. We also mention in Section 2.5.4 how classical losses could be replaced by more intrinsic losses (similar to the noninformative priors introduced in Chapter 3), when no information at all is available on the penalty associated with erroneous decisions or even with the parameterization of interest.

In some cases, it is possible to reduce the class of acceptable loss functions by *invariance* considerations, for example when the model is invariant under the action of a group of transformations. Such considerations apply as well to the choice of the prior distribution, as we will see in Chapter 9. It is also interesting to note that these invariance motivations are often used in other decision-theoretic approaches, where a drastic reduction of the class of inferential procedures is necessary to select a best solution.

Example 2.1.2 Consider the problem of estimating the mean θ of a normal vector, $x \sim \mathcal{N}_n(\theta, \Sigma)$, where Σ is a known diagonal matrix with diagonal elements σ_i^2 $(1 \leq i \leq n)$. In this case, $\mathcal{D} = \Theta = \mathbb{R}^p$, and δ stands for an evaluation of θ. If no additional information is available on the model, it seems logical to choose the loss function so that it weights equally the estimation of each component, i.e., to use a loss of the form

$$\sum_{i=1}^{n} L\left(\frac{\delta_i - \theta_i}{\sigma_i}\right),$$

where L takes its minimum at 0. Indeed, for such losses, the components with larger variances do not strongly bias the selection of the resulting estimator. In other words, the components with a larger variance are not overly weighted when the estimation errors $(\delta_i - \theta_i)$ are normalized by σ_i. The usual choice of L is the quadratic loss $L(t) = t^2$, i.e., the global estimation error is the sum of the squared componentwise errors.   ‖

2.2 Existence of a utility function

The notion of utility (defined as the opposite of loss) is used not only in Statistics, but also in Economics and in other fields like Game Theory where it is necessary to *order* consequences of actions or decisions. *Consequences* (or *rewards*) are generic notions which summarize the set of outcomes resulting from the decision-maker's action. In the simplest cases, it may be the monetary profit or loss resulting from the decision. In an estimation setting, it may be a measure of distance between the evaluation and the true value of the parameter, as in Example 2.1.2. The axiomatic foundations of utility are due to Von Neumann and Morgenstern (1947) and led to numerous extensions, in particular in Game Theory. This approach is considered in a statistical framework by Wald (1950) and Ferguson (1967). Extensions and additional comments can be found in DeGroot (1970, Chapter 7) and recent references on utility theory are Fishburn (1988) and Machina (1982, 1987). See also Chamberlain (2000) for a connection with econometrics.

The general framework behind utility theory considers $\mathcal{R}$, space of *rewards*, which is assumed to be completely known. For instance, $\mathcal{R} = \mathbb{R}$. We also suppose that *it is possible to order the rewards*, i.e., that there exists a *total ordering*, denoted $\preceq$, on $\mathcal{R}$ such that, if r_1 and r_2 are in $\mathcal{R}$,

(1) $r_1 \preceq r_2$ or $r_2 \preceq r_1$; and

(2) if $r_1 \preceq r_2$ and $r_2 \preceq r_3$, then $r_1 \preceq r_3$.

These two properties seem to be minimal requirements in a decision-making setting. In particular, *transitivity* (2) is absolutely necessary to allow a comparison of decision procedures. Otherwise, we may end up with cycles such as $r_1 \preceq r_2 \preceq r_3 \preceq r_1$ and be at a loss about selecting the best reward among the three. Section 2.6 presents a criterion which is intransitive (and thus does not pertain to Decision Theory). We denote by $\prec$ and $\sim$ the *strict* order and *equivalence* relations derived from $\preceq$ respectively. Therefore, one and only one of the three following relations is satisfied by any pair (r_1, r_2) in $\mathcal{R}^2$

$$r_1 \prec r_2, \qquad r_2 \prec r_1, \qquad r_1 \sim r_2.$$

To proceed further in the construction of the utility function, it is necessary to extend the reward space from $\mathcal{R}$ to $\mathcal{P}$, the space of probability distributions on $\mathcal{R}$. This also allows the decision-maker to take into account partly randomized decisions; moreover, the extended reward space is convex.

Example 2.2.1 In most real-life settings, the rewards associated with an action are not exactly known when the decision is taken or, equivalently, some decisions involve a gambling step. For instance, in finance, the monetary revenue $r \in \mathcal{R} = \mathbb{R}$ derived from stock market shares is not guaranteed when the shareholder has to decide from which company she should buy shares. In this case, $\mathcal{D} = \{d_1, \ldots, d_n\}$, where d_k represents the action "buy the share from company k." At the time of the decision, the rewards associated with the different shares are random dividends, only known by the end of the year. ‖

The order relation $\preceq$ is also assumed to be available on $\mathcal{P}$. For instance, when the rewards are monetary, the order relation on $\mathcal{P}$ can be derived by comparing the average yields associated with the distributions P. Therefore, it is possible to compare two distributions of probability on $\mathcal{R}$, P_1 and P_2. We thus assume that $\preceq$ satisfies the extensions of the two hypotheses (1) and (2) to $\mathcal{P}$:

(A₁) $P_1 \preceq P_2$ or $P_2 \preceq P_1$; and

(A₂)   if $P_1 \preceq P_2$ and $P_2 \preceq P_3$, then $P_1 \preceq P_3$.

The order relation on $\mathcal{R}$ then appears as a special case of the order on $\mathcal{P}$ by considering the Dirac masses δ_r $(r \in \mathcal{R})$.

The existence of the order $\preceq$ on $\mathcal{P}$ relies on the assumption that there exists an optimal reward, therefore, that there exists at least a partial ordering on the consequences, even when they are random. This is obviously the case when there exists a function U on $\mathcal{R}$ associated with $\preceq$, such that $P_1 \preceq P_2$ is equivalent to

$$\mathbb{E}^{P_1}[U(r)] \leq \mathbb{E}^{P_2}[U(r)],$$

as in the above monetary example. This function U is called the *utility function*. We now present an axiomatic system on $\preceq$ that ensures the existence of the utility function.

For simplicity's sake, we only consider here the set of *bounded* distributions, $\mathcal{P_B}$, corresponding to the distributions with bounded support, for which there exist r_1 and r_2 such that

$$[r_1, r_2] = \{r : r_1 \preceq r \preceq r_2\} \quad \text{and} \quad P([r_1, r_2]) = 1.$$

For P_1, P_2 in $\mathcal{P_B}$, we define the *mixture* $P = \alpha P_1 + (1 - \alpha) P_2$ as the distribution that generates a reward from P_1 with probability α and a reward from P_2 with probability $(1 - \alpha)$. For instance, $\alpha r_1 + (1 - \alpha) r_2$ is the distribution that gives the reward r_1 with probability α and the reward r_2 with probability $(1 - \alpha)$. Two additional assumptions (or axioms) are necessary to derive the existence of a utility function on $\mathcal{R}$. First, there must be *conservation of the ordering under indifferent alternatives*:

(A₃)   if $P_1 \preceq P_2$, $\alpha P_1 + (1 - \alpha)P \preceq \alpha P_2 + (1 - \alpha)P$ for every $P \in \mathcal{P}$.

For example, if the share buyers of Example 2.2.1 can compare two companies with dividend distributions P_1 and P_2, they should be able to keep a ranking of the two companies if there is a chance $(1 - \alpha)$ that both dividends are replaced by state bounds with dividend distribution P. The order relation must also be *connected* (or *closed*):

(A₄)   if $P_1 \preceq P_2 \preceq P_3$, there exist α and $\beta \in (0,1)$ such that
$$\alpha P_1 + (1 - \alpha)P_3 \preceq P_2 \preceq \beta P_1 + (1 - \beta)P_3.$$

The last assumption then implies the following result.

Lemma 2.2.2 *If r_1, r_2, and r are rewards in $\mathcal{R}$ with $r_1 \prec r_2$ and $r_1 \preceq r \preceq r_2$, there exists a unique v $(0 \leq v \leq 1)$ such that $r \sim vr_1 + (1-v)r_2$.*

Lemma 2.2.2 is actually the key to the derivation of the *utility function*, U, on $\mathcal{R}$. Indeed, given r_1 and r_2, two arbitrary rewards such that $r_2 \prec r_1$, we can define U in the following way. For every $r \in \mathcal{R}$, consider

(i) $U(r) = v$ if $r_2 \preceq r \preceq r_1$ and $r \sim vr_1 + (1-v)r_2$;

(ii) $U(r) = \frac{-v}{1-v}$ if $r \preceq r_2$ and $r_2 \sim vr_1 + (1-v)r$; and

(iii) $U(r) = \frac{1}{v}$ if $r_1 \preceq r$ and $r_1 \sim vr + (1-v)r_2$.

In particular, $U(r_1) = 1$ and $U(r_2) = 0$. Moreover, this function U preserves the order relation on $\mathcal{R}$ (see DeGroot (1970, p. 105) for a proof).

Lemma 2.2.3 *If r_1, r_2, and r_3 are three rewards in $\mathcal{R}$ such that $r_2 \sim \alpha r_1 + (1-\alpha)r_3$*

$$U(r_2) = \alpha U(r_1) + (1-\alpha)U(r_3).$$

Actually, the axioms (A_3) and (A_4) can be further reduced while Lemma 2.2.3 still holds. It is indeed sufficient that they are satisfied on $\mathcal{R}$ only. The extension of the definition of the utility function to $\mathcal{P}_\mathcal{B}$ calls for an additional assumption. Given P such that $P([r_1, r_2]) = 1$, define

$$\alpha(r) = \frac{U(r) - U(r_1)}{U(r_2) - U(r_1)}$$

and

$$\beta = \int_{[r_1, r_2]} \alpha(r) \, dP(r).$$

Then the additional axiom

(A_5) $$P \sim \beta \delta_{r_2} + (1-\beta)\delta_{r_1}$$

implies that, if r is equivalent to $\alpha(r)r_1 + (1-\alpha(r))r_2$ for every $r \in [r_1, r_2]$, this equivalence must hold on average. In fact, notice that β is derived from the expected utility,

$$\beta = \frac{\mathbb{E}^P[U(r)] - U(r_1)}{U(r_2) - U(r_1)},$$

and this assumption provides a definition of U on $\mathcal{P}_\mathcal{B}$. As in Lemma 2.2.3 where U is restricted to $\mathcal{R}$, and as shown by the following result, axiom (A_5) indicates that U provides a *linearization* (or a linear parameterization) of the order relation $\preceq$ on $\mathcal{P}_\mathcal{B}$. Although slightly tautological—since it involves in its formulation the utility function we are trying to derive—, (A_5) indeed leads to the following extension of Lemma 2.2.3 to $\mathcal{P}_\mathcal{B}$.

Theorem 2.2.4 *Consider P_1 and P_2 in $\mathcal{P}_\mathcal{B}$. Then,*

$$P_1 \preceq P_2$$

if and only if

$$\mathbb{E}^{P_1}[U(r)] \leq \mathbb{E}^{P_2}[U(r)].$$

Moreover, if U^* is another utility function satisfying the above equivalence relation, there exist $a > 0$ and b such that

$$U^*(r) = aU(r) + b.$$

Proof. Consider r_1 and r_2 such that

$$P_1([r_1, r_2]) = P_2([r_1, r_2]) = 1$$

(with $r_1 \prec r_2$). Since

$$P_1 \sim \frac{\mathbb{E}^{P_1}[U(r)] - U(r_1)}{U(r_2) - U(r_1)} r_2 + \frac{U(r_2) - \mathbb{E}^{P_1}[U(r)]}{U(r_2) - U(r_1)} r_1$$

and

$$P_2 \sim \frac{\mathbb{E}^{P_2}[U(r)] - U(r_1)}{U(r_2) - U(r_1)} r_2 + \frac{U(r_2) - \mathbb{E}^{P_2}[U(r)]}{U(r_2) - U(r_1)} r_1,$$

$P_1 \preceq P_2$ is truly equivalent to

$$\frac{\mathbb{E}^{P_1}[U(r)] - U(r_1)}{U(r_2) - U(r_1)} \leq \frac{\mathbb{E}^{P_2}[U(r)] - U(r_1)}{U(r_2) - U(r_1)},$$

i.e., $\mathbb{E}^{P_1}[U(r)] \leq \mathbb{E}^{P_2}[U(r)]$. Moreover, for any other utility function U^*, there exist a and b such that $U^*(r_1) = aU(r_1) + b$, $U^*(r_2) = aU(r_2) + b$. The extension of this relation to every $r \in \mathcal{R}$ follows from Lemma 2.2.3. □□

Notice that the above derivation does not involve any restriction on the function U. Therefore, it does not need to be bounded, although this condition is often mentioned in textbooks. It can be argued that this generality is artificial and formal, since subjective utility functions are always bounded. For instance, when considering monetary rewards, there is a psychological threshold, say \$100,000,000, above which (most) individuals have an almost constant utility function.

However, this upper bound varies from individual to individual, and even more so from individuals to companies or states. It is also important to incorporate unacceptable rewards, although the assumption (A$_4$) prevents rewards with utility equal to $-\infty$. (This restriction implies that the death of a patient in a pharmaceutical study or a major accident in a nuclear plant have a finite utility.) Moreover, most theoretical losses are not bounded. A counterpart of this generality is that the above results have only been established for $\mathcal{P}_\mathcal{B}$. Actually, they can be extended to $\mathcal{P}_\mathcal{E}$, the set of distributions P in $\mathcal{P}$ such that $\mathbb{E}^P[U(r)]$ is finite, under the assumption that (A$_1$)–(A$_5$) and two additional hypotheses are satisfied for $\mathcal{P}_\mathcal{E}$ (see Exercise 2.3).

Theorem 2.2.5 *Consider P and Q, two distributions in $\mathcal{P}_\mathcal{E}$. Then, $P \preceq Q$ if and only if*

$$\mathbb{E}^P[U(r)] \leq \mathbb{E}^Q[U(r)].$$

Of course, Theorem 2.2.5 fails to deal with infinite utility distributions. If such distributions exist, they must be compared between themselves and a separate utility function constructed on this restricted class, since they are in a sense the only distributions of interest. However, the loss functions considered in the sequel are bounded from below, usually by 0. Therefore, the corresponding utility functions, opposites of the loss functions, are always bounded from above and infinite reward paradoxes can be avoided. (Rubin (1984) and Fishburn (1987) provide reduced axiomatic systems ensuring the existence of a utility function.)

Many criticisms have been addressed on theoretical or psychological grounds against the notion of *rationality of decision-makers* and the associated axioms (A_1)–(A_4). First, it seems illusory to assume that individuals can compare all rewards, that is, that they can provide a total ordering of $\mathcal{P}$ (or even of $\mathcal{R}$) because their discriminating abilities are necessarily limited, especially about contiguous or extreme alternatives. The *transitivity* assumption is also too strong, since examples in sports or politics show that real-life orderings of preferences often lead to nontransitivity, as illustrated by *Condorcet and Simpson* paradoxes (see Casella and Wells (1993) and Exercises 1.9 and 2.2). More fundamentally, the assumption that the ordering can be extended from $\mathcal{R}$ to $\mathcal{P}$ has been strongly attacked because it implies that a social ordering can be derived from a set of individual orderings and this is not possible in general (see Arrow (1951) or Blyth (1993)). However, while recognizing this fact, Rubin (1987) notes that this impossibility just implies that utility and prior are not separable, not that an optimal (Bayesian) decision cannot be obtained, and he gives a restricted set of axioms pertaining to this purpose. In general, the criticisms above are obviously valuable, but cannot stand against the absolute need of an axiomatic framework validating decision-making under uncertainty. As already mentioned in Chapter 1, statistical modeling *is and must be* reductive; although necessarily missing part of the complexity of the world, the simplified representation it gives of this world allows statisticians and others to reach decisions. Decision Theory thus describes an idealized setting, under an ultimate rationality real decision-makers fail to attain, but aim at nonetheless[2].

From a more practical point of view, the above derivation of the utility function can be criticized as being unrealistic. Berger (1985a) provides a few examples based on DeGroot (1970), deriving the utility function from successive partitions of the reward space (see also Raiffa and Schlaifer (1961)). However, if $\mathcal{R}$ is large (e.g., noncountable), U cannot be evaluated for each reward r, even though the linearity exhibited by Lemma 2.2.3 allows for approximations when $\mathcal{R} \subset \mathbb{R}$. In a multidimensional setting, linear approximations are no longer possible unless one uses a linear combination of

---

[2] To borrow from Smith (1984), to criticize the idealized structures of Decision Theory because of human limitations is somehow akin to attacking integration because some integrals can only be solved numerically.

componentwise utilities, i.e.,

$$U(r_1, r_2, \ldots, r_n) = \sum_{i=1}^{n} \alpha_i U_i(r_i)$$

(see Raiffa (1968), Keeney and Raifa (1976) and Smith (1988) for a discussion). In general, practical utility functions will thus only approximate the true utility functions.

Even when the reward is purely monetary there is a necessity of rigorous determination of the utility function because U may be far from linear, especially for large rewards. This means that a gain of \$3000 with probability $1/2$ may not be equivalent to earning \$1500 with certainty. To solve this paradox, Laplace (1795) introduced the notion of *moral expectation*, derived from the relative value of an increase of wealth, *"absolute value divided by the total wealth of the involved person."* Laplace deduces that the moral expectation *"coincides with the mathematical expectation when the wealth becomes infinite compared with the variations due to uncertainty,"* meaning the utility is indeed linear only around 0. Otherwise, *risk aversion* attitudes slow down the utility curve, which is typically concave for large values of rewards and bounded above. (Persons with a convex utility function are called *risk lovers* because they prefer a random gain to the expectation of this gain. Notice that this attitude is quite understandable in a neighborhood of 0.) To construct the money utility function is obviously more cumbersome than to use a linear utility, but this derivation gives a more accurate representation of reality and can even prevent paradoxes such as the following one.

Example 2.2.6 (Saint Petersburg Paradox) Consider a game where a coin is thrown until a *head* appears. When this event occurs at the nth throw, the player gain is 3^n, leading to an average gain of

$$\sum_{n=1}^{+\infty} 3^n \frac{1}{2^n} = +\infty.$$

Every player should then be ready to pay an arbitrarily high entrance fee to play this game, even though there is less than a 0.05 chance to go beyond the fifth throw! This modeling does not take into account that the fortune of a player is necessarily bounded and that he or she can only play a limited number of games. A solution to this paradox is to substitute for the linear utility function a bounded utility function, such as

$$U(r) = \frac{r}{\delta + r} \qquad (\delta > 0, \ r > -\delta),$$

and $U(r) = -\infty$ otherwise. This construction is quite similar to Laplace's moral expectation. An acceptable entrance fee e will then be such that the expected utility of the game is larger than the utility of doing nothing, i.e.,

$$\mathbb{E}[U(r - e)] \geq U(0) = 0.$$

Consider now a modification to the game such that the player can leave the game at any time n and take the gain 3^n if a head has not yet appeared. The average gain at time n is then

$$\frac{3^n}{\delta + 3^n}\, 2^{-n},$$

which can provide an optimal leaving time n_0, depending on the utility parameter δ, which in its turn somehow characterizes the *risk aversion* of the player (see Smith (1988) for a more thorough description). For instance, δ may represent the fortune of the player, since $U(\tau)$ goes to $-\infty$ when τ goes to $-\delta$. The particular choice of U can obviously be criticized, but a more accurate representation of the utility function requires a detailed analysis of the motivations of the player (see also Exercise 2.9).   ‖

See also Bernardo and Smith (1994) for a detailed analysis of the founda-tions of Utility Theory, with in particular a description of *decision trees*.

2.3 Utility and loss

Let us switch back to a purely statistical setting. From a decision-theoretic point of view, the statistical model now involves three spaces: $\mathcal{X}$, *observa-tion* space, Θ, *parameter* space, and $\mathcal{D}$, *decision* space (or *action* space). Statistical inference then consists of taking a decision $d \in \mathcal{D}$ related to the parameter $\theta \in \Theta$ based on the observation $x \in \mathcal{X}$, x and θ being related by the distribution $f(x|\theta)$. In most cases, the decision d will be to evalu-ate (or *estimate*) a function of θ, $h(\theta)$, as accurately as possible. Decision Theory assumes in addition that each action d can be evaluated (i.e., that its accuracy can be quantified) and leads to a reward r, with utility $U(r)$ (which exists under the assumption of rationality of the decision-makers). From now on, this utility is written as $U(\theta, d)$ to stress that it only depends on these two factors. When other random factors r are involved in U, we take $U(\theta, d) = \mathbb{E}_{\theta, d}[U(r)]$. Therefore, $U(\theta, d)$ can be seen as a measure of proximity between the proposed estimate d and the true value $h(\theta)$.

Once the utility function has been constructed (or approximated), we derive the corresponding *loss* function

$$\mathrm{L}(\theta, d) = -U(\theta, d).$$

In general, the loss function is supposed to be nonnegative, which implies that $U(\theta, d) \leq 0$, and therefore that there is no decision with infinite utility. The existence of a lower bound on L can be criticized as being too stringent, but it does avoid paradoxes such as those mentioned above. It can also be argued that, from a statistical point of view, the loss function L indeed represents the *loss* (or error) owing to a bad evaluation of the function of θ of interest, and therefore that even the best evaluation of this function, i.e., when θ is known, can induce at best a null loss. Otherwise, there would be a lack of continuity of the loss function in $d = \theta$ which could even prevent the choice of a decision procedure.

Obviously, except for the most trivial settings, it is generally impossible to uniformly minimize (in d) the loss function $L(\theta, d)$ when θ is unknown. In order to derive an effective comparison criterion from the loss function, the *frequentist* approach proposes to consider instead the average loss (or *frequentist risk*)

$$
\begin{aligned}
R(\theta, \delta) &= \mathbb{E}_\theta[L(\theta, \delta(x))] \\
&= \int_{\mathcal{X}} L(\theta, \delta(x)) f(x|\theta)\, dx,
\end{aligned}
$$

where $\delta(x)$ is the decision rule, i.e., the allocation of a decision to each outcome $x \sim f(x|\theta)$ from the random experiment. The function δ, from $\mathcal{X}$ in $\mathcal{D}$, is usually called *estimator* (while the *value* $\delta(x)$ is called *estimate* of θ). When there is no risk of confusion, we also denote the set of estimators by $\mathcal{D}$.

The frequentist paradigm relies on this criterion to compare estimators and, if possible, to select the best estimator, the reasoning being that estimators are evaluated on their long-run performance for all possible values of the parameter θ. Notice, however, that there are several difficulties associated with this approach.

(1) The error (loss) is averaged over the different values of x proportionally to the density $f(x|\theta)$. Therefore, it seems that the observation x is not taken into account any further. The risk criterion evaluates procedures on their long-run performance and not directly for the given observation, x. Such an evaluation may be satisfactory for the statistician, but it is not so appealing for a client, who wants optimal results for her data x, not that of another's!

(2) The frequentist analysis of the decision problem implicitly assumes that this problem will be met again and again, for the frequency evaluation to make sense. Indeed, $R(\theta, \delta)$ is approximately the average loss over i.i.d. repetitions of the same experiment, according to the Law of Large Numbers. However, on both philosophical and practical grounds, there is a lot of controversy over the very notion of *repeatability of experiments* (see Jeffreys (1961)). For one thing, if new observations come to the statistician, she should make use of them, and this could modify the way the experiment is conducted, as in, for instance, medical trials.

(3) For a procedure δ, the risk $R(\theta, \delta)$ is a *function* of the parameter θ. Therefore, the frequentist approach does not induce a total ordering on the set of procedures. It is generally impossible to compare decision procedures with this criterion, since two crossing risk functions prevent comparison between the corresponding estimators. At best, one may hope for a procedure δ_0 that uniformly minimizes $R(\theta, \delta)$, but such cases rarely occur unless the space of decision procedures is restricted.

Best procedures can only be obtained by restricting rather artificially the set of authorized procedures.

Example 2.3.1 Consider x_1 and x_2, two observations from

$$P_\theta(x = \theta - 1) = P_\theta(x = \theta + 1) = 0.5, \qquad \theta \in \mathbb{R}.$$

The parameter of interest is θ (i.e., $\mathcal{D} = \Theta$) and it is estimated by estimators δ under the loss

$$L(\theta, \delta) = 1 - \mathbb{I}_\theta(\delta),$$

often called $0 - 1$ *loss*, which penalizes errors of estimation, whatever their magnitude, by 1. Considering the particular estimator

$$\delta_0(x_1, x_2) = \frac{x_1 + x_2}{2},$$

its risk function is

$$\begin{aligned}
R(\theta, \delta_0) &= 1 - P_\theta(\delta_0(x_1, x_2) = \theta) \\
&= 1 - P_\theta(x_1 \neq x_2) = 0.5.
\end{aligned}$$

This computation shows that the estimator δ_0 is correct half of the time. Actually, this estimator is always correct when $x_1 \neq x_2$, and always wrong otherwise. Now, the estimator $\delta_1(x_1, x_2) = x_1 + 1$ also has a risk function equal to 0.5, as does $\delta_2(x_1, x_2) = x_2 - 1$. Therefore, δ_0, δ_1 and δ_2 cannot be ranked under the $0 - 1$ loss. ∥

On the contrary, the Bayesian approach to Decision Theory integrates on the space Θ since θ is unknown, instead of integrating on the space $\mathcal{X}$ as x is known. It relies on the *posterior expected loss*

$$\begin{aligned}
\varrho(\pi, d|x) &= \mathbb{E}^\pi[L(\theta, d)|x] \\
&= \int_\Theta L(\theta, d)\pi(\theta|x)\, d\theta,
\end{aligned}$$

which averages the error (i.e., the loss) according to the posterior distribution of the parameter θ, *conditionally on the observed value x*. Given x, the average error resulting from decision d is actually $\varrho(\pi, d|x)$. The posterior expected loss is thus a function of x but this dependence is not troublesome, as opposed to the frequentist dependence of the risk on the parameter because x, contrary to θ, is known.

Given a prior distribution π, it is also possible to define the *integrated risk*, which is the frequentist risk averaged over the values of θ according to their prior distribution

$$\begin{aligned}
r(\pi, \delta) &= \mathbb{E}^\pi[R(\theta, \delta)] \\
&= \int_\Theta \int_\mathcal{X} L(\theta, \delta(x))\, f(x|\theta)\, dx\, \pi(\theta)\, d\theta.
\end{aligned}$$

One particular interest of this second concept is that it associates a real number with every estimator, not a function of θ. It therefore induces a

total ordering on the set of estimators, i.e., allows for the direct comparison
of estimators. This implies that, while taking into account the prior information through the prior distribution, the Bayesian approach is sufficiently
reductive (in a positive sense) to reach an effective decision. Moreover, the
above two notions are equivalent in that they lead to the same decision.

Theorem 2.3.2   *An estimator minimizing the integrated risk $r(\pi, \delta)$ can
be obtained by selecting, for every $x \in \mathcal{X}$, the value $\delta(x)$ which minimizes
the posterior expected loss, $\varrho(\pi, \delta|x)$, since*

$$(2.3.1) \qquad\qquad r(\pi, \delta) = \int_{\mathcal{X}} \varrho(\pi, \delta(x)|x)m(x)\, dx.$$

Proof.   Equality (2.3.1) follows directly from Fubini's Theorem since, as
$L(\theta, \delta) \geq 0$,

$$
\begin{aligned}
r(\pi, \delta) &= \int_{\Theta} \int_{\mathcal{X}} L(\theta, \delta(x)) f(x|\theta)\, dx\ \pi(\theta)\, d\theta \\
&= \int_{\mathcal{X}} \int_{\Theta} L(\theta, \delta(x)) f(x|\theta) \pi(\theta)\, d\theta\, dx \\
&= \int_{\mathcal{X}} \int_{\Theta} L(\theta, \delta(x)) \pi(\theta|x)\, d\theta\ m(x)\, dx\ .
\end{aligned}
$$

□□

This result leads to the following definition of a Bayes estimator.

Definition 2.3.3   *A Bayes estimator associated with a prior distribution
π and a loss function L is any estimator δ^{π} which minimizes $r(\pi, \delta)$. For
every $x \in \mathcal{X}$, it is given by $\delta^{\pi}(x)$, argument of $\min_d \varrho(\pi, d|x)$. The value
$r(\pi) = r(\pi, \delta^{\pi})$ is then called the Bayes risk.*

Theorem 2.3.2 thus provides a constructive tool for the determination of
the Bayes estimators. Notice that, from a strictly Bayesian point of view,
only the posterior expected loss $\varrho(\pi, \delta|x)$ is important, as the Bayesian
paradigm is based on the conditional approach. To average over all possible values of x, when we know the observed value of x, seems to be a waste
of information. Nonetheless, the equivalence exhibited in Theorem 2.3.2 is
important because, on one hand, it shows that the conditional approach
is not necessarily as dangerous as frequentists may depict it. This is so
because, while the Bayesian approach works conditional upon the actual
observation x, it also incorporates the probabilistic properties of the dis-
tribution of the observation, $f(x|\theta)$. On the other hand, this equivalence
provides a connection between the classical results of Game Theory (see
Section 2.4) and the axiomatic Bayesian approach, based on the posterior
distribution. It also explains why Bayes estimators play an important role
in frequentist optimality criteria.

The result above is valid for proper and improper priors, *as long as the
Bayes risk $r(\pi)$ is finite.* Otherwise, the notion of a (decision-theoretic)

Bayes estimator is weakened: we then define a *generalized Bayes estimator* as the minimizer, for every x, of the posterior expected loss. In terms of frequentist optimality, we will see that the division between proper and improper priors is much less important than the division between regular and generalized Bayes estimators, since the formers are admissible. Notice that, for strictly convex losses, the Bayes estimators are unique.

We conclude this section with an example of construction of a loss function in an expert calibration framework. References on this topic are DeGroot and Fienberg (1983), Murphy and Winkler (1984), Bayarri and DeGroot (1988) and Schervish (1989). Smith (1988) also shows how forecaster evaluation can help improve the assessment of prior probabilities. See also Note 2.8.1 for an illustration in imaging.

Example 2.3.4 Meteorological forecasts are often given as probability statements such as *"the probability of rain for tomorrow is 0.4."* Such forecasts being quantified, it is of interest to evaluate weather forecasters through a loss function (for their employers as well as users).

For a given forecaster, let N be the number of different percentages predicted at least once in a year and let p_i $(1 \leq i \leq N)$ be the corresponding percentages. For instance, we may have $N = 5$ and

$$p_1 = 0, \quad p_2 = 0.45, \quad p_3 = 0.7, \quad p_4 = 0.9, \quad \text{and} \quad p_5 = 0.95.$$

In this case, the parameters θ_i are actually observed, i.e.,

$$\theta_i = \frac{\text{number of rainy days when } p_i \text{ is forecasted}}{\text{number of days when } p_i \text{ is forecasted}}$$

(more exactly, this ratio is a good approximation of θ_i).

If q_i denotes the proportion of days where p_i is forecasted, a possible loss function for the forecasters is

$$L(\theta, p) = \sum_{i=1}^{N} q_i (p_i - \theta_i)^2 + \sum_{i=1}^{N} q_i \log(q_i).$$

For a given set of θ_i's $(1 \leq i \leq N)$, the best forecaster is the perfectly calibrated forecaster, i.e., the one who satisfies $p_i = \theta_i$ $(1 \leq i \leq N)$. Moreover, among these perfect forecasters, the best one is the most well balanced, satisfying $q_i = 1/N$ $(1 \leq i \leq N)$, i.e., the more daring forecaster, as opposed to a forecaster which would always give the same forecast, p_{i_0}, because of the *entropy* term, $\sum_i q_i \log(q_i)$. However, the distance $(p_i - \theta_i)^2$ could be replaced by any other function taking its minimum at $p_i = \theta_i$ (see Exercises 2.12 and 2.14). The weight q_i in the first sum is also used to calibrate more properly forecasters, in order to prevent overpenalization of rare forecasts.

This loss has been constructed with a bias in favor of forecasters with large N, since the *entropy* $\log(N)$ increases with N. However, a better performance for a larger N requires that p_i is (almost) equal to θ_i and q_i is close to $1/N$.   ‖

2.4 Two optimalities: minimaxity and admissibility

This section deals with the two fundamental notions of frequentist Decision Theory, introduced by Wald (1950) and Neyman and Pearson (1933a,b). As mentioned above, and contrary to the Bayesian approach, the frequentist paradigm is not reductive enough to lead to a single optimal estimator. While we are mainly concerned in this book with the Bayesian aspects of Decision Theory, it is still necessary to study these frequentist notions in detail because they show that Bayes estimators are often optimal for the frequentist concepts of optimality, therefore should still be considered even when prior information is ignored. In other words, one can reject the Bayesian paradigm and ignore the meaning of the prior distribution and still obtain good estimators from a frequentist point of view when using this prior distribution. Therefore, in this technical sense, frequentists should also take into account the Bayesian approach, since it provides a *tool* for the derivation of optimal estimators (see Brown (1971, 2000), Strawderman (1974), Berger (1985a), or Berger and Robert (1990) for examples). Moreover, these properties can be helpful in the selection of a prior distribution, when prior information is not precise enough to lead to a single prior distribution (see Chapter 3).

2.4.1 Randomized estimators

Similar to the study of the utility function, where we extended the reward space from $\mathcal{R}$ to $\mathcal{P}$, we need to extend the decision space to the set of *randomized estimators*, taking values in $\mathcal{D}^*$, space of the probability distributions on $\mathcal{D}$. To use a randomized estimator δ^* means that the action is generated according to the distribution with probability density $\delta^*(x,.)$, once the observation x has been collected. The loss of a randomized estimator δ^* is then defined as the average loss

$$L(\theta, \delta^*(x)) = \int_{\mathcal{D}} L(\theta, a)\delta^*(x, a)\, da.$$

This extension is necessary to deal with minimaxity and admissibility. Obviously, such estimators are not to be used, if only because they contradict the Likelihood Principle, giving several possible answers for the same value of x (and thus of $\ell(\theta|x)$). Moreover, it seems quite paradoxical to add noise to a phenomenon in order to take a decision under uncertainty!

Example 2.4.1 (Example 2.3.1 continued) Consider the randomized estimator

$$\delta^*(x_1, x_2)(t) = \begin{cases} \mathbb{I}_{(x_1+x_2)/2}(t) & \text{if } x_1 \neq x_2, \\ [\mathbb{I}_{(x_1-1)}(t) + \mathbb{I}_{(x_1+1)}(t)]/2 & \text{otherwise,} \end{cases}$$

where $\mathbb{I}_v$ denotes the Dirac mass at v. Actually, if $x_1 = x_2$, the two values $\theta_1 = x_1 - 1$ and $\theta_2 = x_1 + 1$ have the same likelihood. Compared with δ_0 which never estimates θ correctly if $x_1 = x_2$, δ^* is exact with probability $1/2$. However, when δ^* misses θ, it is farther away from θ than δ_0. The choice of the estimator then depends on the loss function, i.e., the way the distance between the estimator and θ (or the error) is measured.   ‖

Randomized estimators are nonetheless necessary from a frequentist point of view, for instance, for the frequentist theory of tests, as they provide access to confidence levels otherwise unattainable (see Chapter 5). The set $\mathcal{D}^*$ thus appears as a completion of $\mathcal{D}$. However, this modification of the decision space does not modify the Bayesian answers, as shown by the following result (where $\mathcal{D}^*$ also denotes the set of functions taking values in $\mathcal{D}^*$).

Theorem 2.4.2 *For every prior distribution π on Θ, the Bayes risk on the set of randomized estimators is the same as the Bayes risk on the set of nonrandomized estimators, i.e.,*

$$\inf_{\delta \in \mathcal{D}} r(\pi, \delta) = \inf_{\delta^* \subset \mathcal{D}^*} r(\pi, \delta^*) = r(\pi).$$

Proof. For every $x \in \mathcal{X}$ and every $\delta^* \in \mathcal{D}^*$, we have

$$\int_\Theta \int_{\mathcal{D}} L(\theta, a) \delta^*(x, a) da\, \pi(\theta|x) d\theta$$
$$= \int_{\mathcal{D}} \int_\Theta L(\theta, a) \pi(\theta|x) d\theta\, \delta^*(x, a) da$$
$$\geq \int_{\mathcal{D}} \inf_a \left\{ \int_\Theta L(\theta, a) \pi(\theta|x) d\theta \right\} \delta^*(x, a) da$$
$$= \varrho(\pi, \delta^\pi|x).$$

□□

This result thus holds even when the Bayes risk $r(\pi)$ is infinite. The proof relies on the fact that a randomized procedure averages the risks of nonrandomized estimators and thus cannot improve on them. However, the fact that randomized procedures are not relevant does not hold for the frequentist risk unless some conditions, such as convexity, are imposed on the loss function.

2.4.2 Minimaxity

The minimax criterion we introduce now appears as an insurance against the worst case because it aims at minimizing the expected loss in the least favorable case. It also represents a frequentist effort to skip the Bayesian paradigm while producing a (weak) total ordering on $\mathcal{D}^*$.

Definition 2.4.3 *The minimax risk associated with a loss function* L *is the value*

$$\bar{R} = \inf_{\delta \in \mathcal{D}^*} \sup_\theta R(\theta, \delta) = \inf_{\delta \in \mathcal{D}^*} \sup_\theta \mathbb{E}_\theta[L(\theta, \delta(x))],$$

and a minimax estimator is any (possibly randomized) estimator δ_0 such that

$$\sup_\theta R(\theta, \delta_0) = \bar{R}.$$

This notion is validated by Game Theory, where two adversaries (here, "the statistician" and "Nature") are competing. Once the statistician has selected a procedure, Nature selects the state of nature (i.e., the parameter) that maximizes the loss of the statistician. (We will see below that this choice is usually equivalent to the choice of a prior distribution π. Therefore, the Bayesian approach does not really fit in that conflicting framework, since the prior distribution is also supposed to be known.) In general, it seems unfortunate to resort to such an antagonistic perspective in a statistical analysis. Indeed, to perceive Nature (or reality) as an enemy involves a bias toward the worst cases and prevents the statistician from using the available information (for an analysis and a defense of minimaxity, see Brown (1993) and Strawderman (2000)).

The notion of minimaxity provides a good illustration of the conservative aspects of the frequentist paradigm. Since this approach refuses to make any assumption on the parameter θ, it has to consider the worst cases as equally likely, and thus needs to focus on the maximal risk. In fact, from a Bayesian point of view, it is often equivalent to take a prior concentrated on these worst cases (see Section 2.4.3). In most settings, this point of view is thus too conservative because some values of the parameter are less likely than others.

Example 2.4.4 The first oil-drilling platforms in the North Sea were designed according to a minimax principle. In fact, they were supposed to resist the conjugate action of the worst gale and the worst storm ever observed, at the minimal record temperature. This strategy obviously gives a comfortable margin of safety, but is quite costly. For more recent platforms, engineers have taken into account the distribution of these weather phenomena in order to reduce the production cost.   ‖

Example 2.4.5 A waiting queue at a red light is usually correctly represented by a Poisson distribution. The number of cars arriving during the observation time, N, is thus distributed according to $\mathcal{P}(\lambda)$, with the mean parameter λ to be estimated. Obviously, the values of λ above a given limit are quite unlikely. For instance, if λ_0 is the number of cars in the whole city, the average number of cars waiting at a given traffic light will not exceed λ_0. However, it may happen that some estimators are not minimax because their risk are above $\bar{R}$ for the largest values of λ.   ‖

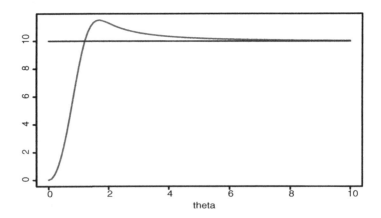

Figure 2.4.1. *Comparison of the risks of the estimators δ_1 and δ_2.*

The example above does not directly criticize the minimax principle but rather argues for the fact that some residual information is attached to most problems and that it should be used, even marginally. In a similar manner, Example 2.4.6 exhibits two estimators, δ_1 and δ_2, such that δ_1 has a constant minimax risk $\bar{R}$ and δ_2 has a risk which can be as low as $\bar{R}/10$ but goes slightly above $\bar{R}$ for the largest values of the parameter (see Figure 2.4.2). Therefore, according to the minimax principle, δ_1 should be preferred to δ_2, although the values of θ for which δ_1 dominates δ_2 are the most unlikely (see Exercise 2.28 for another striking example).

Example 2.4.6 For reasons explained in Note 2.8.2, we consider the following estimator

$$\delta_2(x) = \begin{cases} \left(1 - \dfrac{2p-1}{||x||^2}\right) x & \text{if } ||x||^2 \geq 2p-1, \\ 0 & \text{otherwise,} \end{cases}$$

to estimate θ when $x \sim \mathcal{N}_p(\theta, I_p)$. This estimator, called the *positive-part James–Stein estimator*, is evaluated under *quadratic loss*,

$$L(\theta, d) = ||\theta - d||^2.$$

Figure 2.4.2 gives a comparison of the respective risks of δ_2 and $\delta_1(x) = x$, maximum likelihood estimator, for $p = 10$. This figure shows that δ_2 cannot be minimax, since the maximum risk of δ_2 is above the (constant) risk of δ_1, that is, $R(\theta, \delta_2) = \mathbb{E}_\theta[||\theta - \delta_2(x)||^2] = p$. (We show in Section 2.4.3 that δ_1 is actually minimax in this case.) But the estimator δ_2 is definitely superior on the most interesting part of the parameter space, the additional loss being in perspective quite negligible.   ‖

Table 2.4.1. *Utility function* $U(\theta_i, a_j)$.

	a_1	a_2
θ_1	-4	-10
θ_2	8	30

The opposition between minimax and Bayesian analyses is illustrated by the following example, which borrows from Game Theory (since there is no observation or statistical model).

Example 2.4.7 Two persons, A and B, suspected of being accomplices in a robbery, have been apprehended and placed in separate cells. Both suspects are questioned and enticed to confess the burglary. Although they cannot be convicted unless one of them talks, the incentive is that the first person to cooperate will get a reduced sentence. Table 2.4.7 provides the rewards as perceived by A (in years of freedom), where a_1 (resp. θ_1) represents the fact that A (resp. B) talks. The two suspects have an optimal gain if they both remain silent. However, from A's point of view, the optimal strategy is to be the first one to talk, i.e., a_1, since $\max_\theta R(a_1, \theta) = 4$ and $\max_\theta R(a_2, \theta) = 10$. Therefore, both burglars will end up in jail!

On the contrary, if π is the (subjective) probability assigned by A to the event *"B talks,"* i.e., to θ_1, the Bayes risk of a_1 is

$$r(\pi, a_1) = \mathbb{E}^\pi [-U(\theta, a_1)] = 4\pi - 8(1 - \pi) = 12\pi - 8$$

and, for a_2,

$$r(\pi, a_2) = \mathbb{E}^\pi [-U(\theta, a_2)] = 10\pi - 30(1 - \pi) = 40\pi - 30.$$

It is straightforward to check that, for $\pi \leq 11/14$, $r(\pi, a_2)$ is smaller than $r(\pi, a_1)$. Therefore, unless A is convinced that B will talk, it is better for A to keep silent.   ‖

2.4.3 Existence of minimax rules and maximin strategy

An important difficulty related with minimaxity is that a minimax estimator does not necessarily exist. Ferguson (1967) and Berger (1985a, Chapter 5) give sufficient conditions. In particular, there exists a minimax strategy when Θ is finite and the loss function is continuous. More generally, Brown (1976) (see also Le Cam (1986) and Strasser (1985)) considers the decision space $\mathcal{D}$ as embedded in another space so that the set of risk functions on $\mathcal{D}$ is compact in this larger space. From this perspective and under additional assumptions, it is then possible to derive minimax estimators when the loss is continuous. However, these extensions involve topological techniques too advanced to be considered in this book. Therefore, we only give the following result (see Blackwell and Girshick (1954) for a proof).

Theorem 2.4.8 *If $\mathcal{D} \subset \mathbb{R}^k$ is a convex compact set and if $L(\theta, d)$ is continuous and convex as a function of d for every $\theta \in \Theta$, there exists a nonrandomized minimax estimator.*

The restriction to nonrandomized estimators when the loss is convex follows from *Jensen's inequality*, since

$$L(\theta, \delta^*) = \mathbb{E}^{\delta^*}[L(\theta, \delta)] \geq L(\theta, \mathbb{E}^{\delta^*}(\delta)).$$

This result is a special case of the *Rao–Blackwell Theorem* (see Lehmann and Casella (1998, p. 47)).

Example 2.4.9 (Example 2.4.1 continued) The randomized estimator δ^* is uniformly dominated for every convex loss by the nonrandomized estimator $\mathbb{E}^{\delta^*}[\delta^*(x_1, x_2)]$, i.e.,

$$\tilde{\delta}(x_1, x_2) = \begin{cases} \frac{1}{2}(x_1 + x_2) & \text{if } x_1 \neq x_2, \\ \frac{1}{2}(x_1 - 1) + \frac{1}{2}(x_1 + 1) = x_1 & \text{otherwise,} \end{cases}$$

which is actually identical to the estimator δ_0 considered originally. Notice that this is not true for the $0 - 1$ loss where δ^* dominates $\tilde{\delta}$.   ∥

The following result points out the connection between the Bayesian approach and the minimax principle. (The proof is straightforward and thus omitted.)

Lemma 2.4.10 *The Bayes risks are always smaller than the minimax risk, i.e.,*

$$\underline{R} = \sup_{\pi} r(\pi) = \sup_{\pi} \inf_{\delta \in \mathcal{D}} r(\pi, \delta) \leq \bar{R} = \inf_{\delta \in \mathcal{D}^*} \sup_{\theta} R(\theta, \delta).$$

The first value is called *maximin risk* and a distribution π^* such that $r(\pi^*) = \underline{R}$ is called a *least favorable distribution*, when such distributions exist. In general, the upper bound $r(\pi^*)$ is rather attained by an improper distribution, which can be expressed as a limit of proper prior distributions π_n, but this phenomenon does not necessarily deter from the derivation of minimax estimators (see Lemma 2.4.15). When they exist, least favorable distributions are those with the largest Bayes risk, thus the less interesting distributions in terms of loss performances if they are not suggested by the available prior information. The above result is quite logical, in the sense that prior information can only improve the estimation error, even in the worst case.

A particularly interesting case corresponds to the following definition.

Definition 2.4.11 *The estimation problem is said to have a value when $\underline{R} = \bar{R}$, i.e., when*

$$\sup_{\pi} \inf_{\delta \in \mathcal{D}} r(\pi, \delta) = \inf_{\delta \in \mathcal{D}^*} \sup_{\theta} R(\theta, \delta).$$

When the problem has a value, some minimax estimators are the Bayes estimators for the least favorable distributions. However, they may be randomized, as illustrated by the following example. Therefore, the minimax principle does not always lead to acceptable estimators.

Example 2.4.12 Consider[3] a Bernoulli observation, $x \sim \mathcal{B}e(\theta)$ with $\theta \in \{0.1, 0.5\}$. Four nonrandomized estimators are available,

$$\delta_1(x) = 0.1, \qquad \delta_2(x) = 0.5,$$
$$\delta_3(x) = 0.1 \, \mathbb{I}_{x=0} + 0.5 \, \mathbb{I}_{x=1}, \quad \delta_4(x) = 0.5 \, \mathbb{I}_{x=0} + 0.1 \, \mathbb{I}_{x=1}.$$

We assume in addition that the penalty for a wrong answer is 2 when $\theta = 0.1$ and 1 when $\theta = 0.5$. The *risk vectors* $(R(0.1, \delta), R(0.5, \delta))$ of the four estimators are then, respectively, $(0, 1)$, $(2, 0)$, $(0.2, 0.5)$, and $(1.8, 0.5)$. It is straightforward to see that the risk vector of any randomized estimator is a convex combination of these four vectors or, equivalently, that the *risk set*, $\mathcal{R}$, is the convex hull of the above four vectors, as represented by Figure 2.4.3.

In this case, the minimax estimator is obtained at the intersection of the diagonal of $\mathbb{R}^2$ with the lower boundary of $\mathcal{R}$. As shown by Figure 2.4.3, this estimator δ^* is randomized and takes the value $\delta_3(x)$ with probability $\alpha = 0.87$ and $\delta_2(x)$ with probability $1 - \alpha$. The weight α is actually derived from the equation

$$0.2\alpha + 2(1 - \alpha) = 0.5\alpha.$$

This estimator δ^* is also a (randomized) *Bayes estimator* with respect to the prior

$$\pi(\theta) = 0.22 \, \mathbb{I}_{0.1}(\theta) + 0.78 \, \mathbb{I}_{0.5}(\theta);$$

the prior probability $\pi_1 = 0.22$ corresponds to the slope between $(0.2, 0.5)$ and $(2, 0)$, i.e.,

$$\frac{\pi_1}{1 - \pi_1} = \frac{0.5}{2 - 0.2}.$$

Notice that every randomized estimator that is a combination of δ_2 and of δ_3 is a Bayes estimator for this distribution, but that δ^* only is also a minimax estimator.   ‖

Similar to minimax estimators, a least favorable distribution does not necessarily exist since its existence depends on a separating hyperplane theorem that does not always apply (see Pierce (1973), Brown (1976), Berger (1985a), and Chapter 8). In addition, Strawderman (1973) shows that, in the special case when $x \sim \mathcal{N}_p(\theta, I_p)$, there is no minimax proper Bayes estimator if $p \leq 4$. From a more practical point of view, Lemma 2.4.10 provides sufficient conditions of minimaxity.

Lemma 2.4.13 *If δ_0 is a Bayes estimator with respect to π_0 and if $R(\theta, \delta_0) \leq r(\pi_0)$ for every θ in the support of π_0, δ_0 is minimax and π_0 is the least favorable distribution.*

---

[3] The computations in this example are quite simple. See Chapter 8 for details.

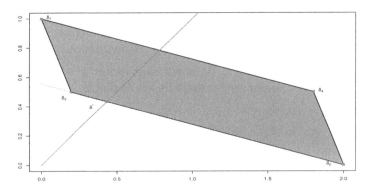

Figure 2.4.2. *Risk set for the estimation of the Bernoulli parameter.*

Example 2.4.14 (Berger (1985a)) Consider $x \sim \mathcal{B}(n, \theta)$ when θ is to be estimated under the quadratic loss,

$$\mathrm{L}(\theta, \delta) = (\delta - \theta)^2.$$

Bayes estimators are then given by posterior expectations (see Section 2.5) and, when $\theta \sim \mathcal{B}e\left(\frac{\sqrt{n}}{2}, \frac{\sqrt{n}}{2}\right)$, the posterior mean is

$$\delta^*(x) = \frac{x + \sqrt{n}/2}{n + \sqrt{n}}.$$

Moreover, this estimator has *constant risk*, $R(\theta, \delta^*) = 1/4(1+\sqrt{n})^2$. Therefore, integrating out θ, $r(\pi) = R(\theta, \delta^*)$ and δ^* is minimax according to Lemma 2.4.13. Notice the difference with the maximum likelihood estimator, $\delta_0(x) = x/n$, for the small values of n, and the unrealistic concentration of the prior around 0.5 for larger values of n.   ‖

Since minimax estimators usually correspond to *generalized Bayes estimators*, it is often necessary to use a limiting argument to establish minimaxity, rather than computing directly the Bayes risk as in Lemma 2.4.13.

Lemma 2.4.15 *If there exists a sequence (π_n) of proper prior distributions such that the generalized Bayes estimator δ_0 satisfies*

$$R(\theta, \delta_0) \leq \lim_{n \to \infty} r(\pi_n) < +\infty$$

for every $\theta \in \Theta$, then δ_0 is minimax.

Example 2.4.16 When $x \sim \mathcal{N}(\theta, 1)$, the maximum likelihood estimator $\delta_0(x) = x$ is a generalized Bayes estimator associated with the Lebesgue measure on $\mathbb{R}$ and the quadratic loss. Since $R(\delta_0, \theta) = \mathbb{E}_\theta(x - \theta)^2 = 1$, this risk is the limit of the Bayes risks $r(\pi_n)$ when π_n is equal to $\mathcal{N}(0, n)$, as $r(\pi_n) = \frac{n}{n+1}$. Therefore, the maximum likelihood estimator δ_0 is minimax. Note that this argument can be extended directly to the case $x \sim \mathcal{N}_p(\theta, I_p)$ to establish that δ_0 is minimax for every p.   ‖

When the space Θ is compact, minimax Bayes rules (or estimators) can be exactly described, owing to the *separated zeros principle* in complex calculus: if $R(\theta, \delta^\pi)$ is not constant and is analytic, the set of θ's where $R(\theta, \delta^\pi)$ is maximal is separated and, in the case of a compact set Θ, is necessarily finite.

Theorem 2.4.17 *Consider a statistical problem that simultaneously has a value, a least favorable distribution π_0, and a minimax estimator δ^{π_0}. Then, if $\Theta \subset \mathbb{R}$ is compact and if $R(\theta, \delta^{\pi_0})$ is an analytic function of θ, then either π_0 has a finite support or $R(\theta, \delta^{\pi_0})$ is constant.*

Example 2.4.18 Consider $x \sim \mathcal{N}(\theta, 1)$, with $|\theta| \leq m$, namely, $\theta \in [-m, m]$. Then, according to Theorem 2.4.17, least favorable distributions have necessarily a finite support, $\{\pm\theta_i, \ 1 \leq i \leq \omega\}$, with cardinal 2ω and supporting points θ_i depending on m. In fact, the only estimator with constant risk is $\delta_0(x) = x$, which is not minimax in this case. In general, the exact determination of n and of the points θ_i can only be done numerically. For instance, when $m \leq 1.06$, the prior distribution with weights $1/2$ at $\pm m$ is the *unique* least favorable distribution. Then, for $1.06 \leq m \leq 2$, the support of π contains $-m$, 0, and m. See Casella and Strawderman (1981) and Bickel (1981) for details, and Johnstone and MacGibbon (1992) for a similar treatment of the Poisson model.   ‖

The above examples show why, while being closely related to the Bayesian paradigm, the minimax principle is not necessarily appealing from a Bayesian point of view. Indeed, apart from the fact that minimax estimators are sometimes randomized, as in Example 2.4.12, Examples 2.4.14 and 2.4.18 show that the least favorable prior is often unrealistic because it induces a strong prior bias towards a few points of the sample space. For Example 2.4.18, Gatsonis et al. (1987) have shown that uniform priors are good substitutes to the point mass priors, although they are not minimax.

Extensions of Theorem 2.4.17 to the noncompact case are given in Kempthorne (1988). In multidimensional settings, when the problem is invariant under rotation, the least favorable distributions are uniform on a sequence of embedded spheres (see Robert et al. (1990)). The practical problem of determining the points of the support is considered in Kempthorne (1987) and Eichenauer and Lehn (1989).

In settings where the problem has a value, it is often difficult to derive the least favorable distribution and alternative methods are then necessary to produce a minimax estimator. Chapter 9 shows how the exhibition of some invariance structures of the model may lead to identify the best equivariant estimator and a minimax estimator (Hunt–Stein Theorem). Unfortunately, the conditions under which this theorem applies are difficult to check and often do not hold.

Lastly, when a minimax estimator has been derived, its optimality is still to be assessed: there may exist several minimax estimators and some may perform uniformly better than others. It is then necessary to introduce a second (and more local) criterion to compare minimax estimators, i.e., estimators that perform well globally.

2.4.4 Admissibility

This second frequentist criterion induces a partial ordering on $\mathcal{D}^*$ by comparing the frequentist risks of the estimators, $R(\theta, \delta)$.

Definition 2.4.19 *An estimator δ_0 is inadmissible if there exists an estimator δ_1 which dominates δ_0, that is, such that, for every θ,*
$$R(\theta, \delta_0) \geq R(\theta, \delta_1)$$
and, for at least one value θ_0 of the parameter,
$$R(\theta_0, \delta_0) > R(\theta_0, \delta_1).$$
Otherwise, δ_0 is said to be admissible.

This criterion is particularly interesting for its *reductive* action. Indeed, at least in theory, it seems logical to advocate that inadmissible estimators should not be considered at all since they can be uniformly improved. For instance, the Rao–Blackwell Theorem then implies that, for convex losses, randomized estimators are inadmissible. However, admissibility alone is not enough to validate the use of an estimator. For instance, constant estimators $\delta(x) = \theta_0$ are usually admissible because they produce the exact value at $\theta = \theta_0$. From a frequentist point of view, it is then important to look for estimators satisfying both optimalities, that is, minimaxity and admissibility. In this regard, two results can be mentioned.

Proposition 2.4.20 *If there exists a unique minimax estimator, this estimator is admissible.*

Proof. If δ^* is the only minimax estimator, for any estimator $\tilde{\delta} \neq \delta^*$,

$$\sup_\theta R(\theta, \tilde{\delta}) > \sup_\theta R(\theta, \delta^*).$$

Therefore, $\tilde{\delta}$ cannot dominate δ^*. □□

Notice that the converse to this result is false, since there can exist several minimax admissible estimators. For instance, in the $\mathcal{N}_p(\theta, I_p)$ case, there exist proper Bayes minimax estimators when $p \geq 5$ (Strawderman (1973) and Fourdrinier and Strawderman (1999)). When the loss function L is strictly convex (in d), it also allows for the following characterization.

Proposition 2.4.21 *If δ_0 is admissible with constant risk, δ_0 is the unique minimax estimator.*

Proof. For any $\theta_0 \in \Theta$, $\sup_\theta R(\theta, \delta_0) = R(\theta_0, \delta_0)$. Therefore, if there exists δ_1 such that $\bar{R} \leq \sup_\theta R(\theta, \delta_1) < R(\theta_0, \delta_0)$, δ_0 cannot be admissible. Similarly, if $\bar{R} = \sup_\theta R(\theta, \delta_1) = R(\theta_0, \delta_0)$ and if θ_1 is such that $R(\theta_1, \delta_1) < \bar{R}$,

δ_1 dominates δ_0. Therefore, when δ_0 is admissible, the only possible case is that there exists δ_1 such that $R(\theta, \delta_1) = R(\theta, \delta_0)$ for every $\theta \in \Theta$. And this is also impossible when δ_0 is admissible (see Exercise 2.36).   □□

Again, notice that the converse of this result is false. There may be minimax estimators with constant risk that are inadmissible: actually, they are certainly inadmissible if there are other minimax estimators. For instance, this is the case for $\delta_0(x) = x$ when $x \sim \mathcal{N}_p(\theta, I_p)$ and $p \geq 3$ (see Note 2.8.2). There also are cases when there is no minimax admissible estimator (this requires that there is no *minimal complete class*, see Chapter 8).

The previous section showed that minimaxity can sometimes be considered from a Bayesian perspective as the choice by Nature of a maximin strategy (least favorable distribution), π, therefore that *some* minimax estimators are Bayes. Admissibility is even more strongly related to the Bayes paradigm in the sense that, in most statistical problems, the Bayes estimators are "spanning" the class of admissible estimators, i.e., the latter can be expressed as Bayes estimators or generalized Bayes estimators or limits of Bayes estimators. Chapter 8 deals in more detail with the relations between Bayes estimators and admissibility. We only give here two major results.

Proposition 2.4.22 *If a prior distribution π is strictly positive on Θ, with finite Bayes risk and the risk function, $R(\theta, \delta)$, is a continuous function of θ for every δ, the Bayes estimator δ^π is admissible.*

Proof. Suppose δ^π is inadmissible and consider δ' which uniformly dominates δ^π. Then, for every θ, $R(\theta, \delta') \leq R(\theta, \delta^\pi)$ and, in an *open set* C of Θ, $R(\theta, \delta') < R(\theta, \delta^\pi)$. Integrating out this inequality, we derive that

$$r(\pi, \delta') < r(\pi, \delta^\pi) = \int_\Theta R(\theta, \delta^\pi)\pi(\theta)\, d\theta,$$

which is impossible.   □□

Proposition 2.4.23 *If the Bayes estimator associated with a prior π is unique, it is admissible.*

The proof of this result is similar to the proof of Proposition 2.4.20. Even if the Bayes estimator is not unique, it is still possible to exhibit at least *one* admissible Bayes estimator. When the loss function is strictly convex, the Bayes estimator is necessarily unique and thus admissible, according to the above proposition.

Example 2.4.24 (Example 2.4.14 continued) The estimator δ^* is a (proper) Bayes estimator, therefore admissible, and it has constant risk. Therefore, it is the *unique minimax estimator* under squared error loss. ‖

Notice that Proposition 2.4.22 contains the assumption that the Bayes risk is finite. Otherwise, every estimator is, in a way, a Bayes estimator (see

Exercise 2.43). On the other hand, some admissibility results can be established for improper priors. This is why we prefer to call *generalized Bayes* estimators the estimators associated with an infinite Bayes risk, rather those corresponding to an improper prior. This choice implies that the Bayes estimators of different quantities associated with the same prior distribution can be simultaneously regular Bayes estimators and generalized Bayes estimators, depending on what they estimate. This also guarantees that regular Bayes estimators will always be admissible, as shown by the following result.

Proposition 2.4.25 *If a Bayes estimator, δ^π, associated with a (proper or improper) prior π and a strictly convex loss function, is such that the Bayes risk,*

$$r(\pi) = \int_\Theta R(\theta, \delta^\pi)\pi(\theta)\,d\theta,$$

is finite, δ^π is admissible.

Example 2.4.26 Consider $x \sim \mathcal{N}(\theta, 1)$ and the null hypothesis $H_0 : \theta \leq 0$ is tested against the alternative hypothesis $H_1 : \theta > 0$. This testing problem is an *estimation* problem if we consider the estimation of the indicator function $\mathbb{I}_{H_0}(\theta)$. Under the quadratic loss

$$\left(\mathbb{I}_{H_0}(\theta) - \delta(x)\right)^2,$$

we can propose the following estimator

$$
\begin{aligned}
p(x) &= P_0(X > x) \qquad (X \sim \mathcal{N}(0,1)) \\
&= 1 - \Phi(x),
\end{aligned}
$$

called the *p-value*, which is considered as a good frequentist answer to the testing problem (see Kiefer (1977) and Casella and Berger (1987)). Using Example 1.5.1, it is easy to show that p is a generalized Bayes estimator under Lebesgue measure and quadratic loss, since $\pi(\theta|x)$ is the $\mathcal{N}(x, 1)$ distribution and

$$
\begin{aligned}
p(x) &= \mathbb{E}^\pi[\mathbb{I}_{H_0}(\theta)|x] = P^\pi(\theta < 0|x) \\
&= P^\pi(\theta - x < -x|x) = 1 - \Phi(x).
\end{aligned}
$$

Moreover, the Bayes risk of p is finite (Exercise 2.34). Therefore, the p-value, when taken as an estimator of $\mathbb{I}_{H_0}$, is admissible. (See Section 5.4 for an extended analysis of the properties of the p-value.)   ‖

Example 2.4.27 In the setting of the previous example, if θ is the parameter of interest, $\delta_0(x) = x$ is a generalized Bayes estimator under quadratic loss, but

$$
\begin{aligned}
r(\pi, \delta_0) &= \int_{-\infty}^{+\infty} R(\theta, \delta_0)\,d\theta \\
&= \int_{-\infty}^{+\infty} 1\,d\theta = +\infty.
\end{aligned}
$$

Therefore, Proposition 2.4.23 is useless in this case to assess the admissibility of δ_0. While δ_0 is actually admissible, its admissibility must be established through a sequence of proper priors, as shown in Chapter 8. ∥

Example 2.4.28 Consider $x \sim \mathcal{N}_p(\theta, I_p)$. If the parameter of interest is $||\theta||^2$ and the prior distribution is the Lebesgue measure on $\mathbb{R}^p$, since $\mathbb{E}^\pi[||\theta||^2|x] = \mathbb{E}[||y||^2]$, with $y \sim \mathcal{N}_p(x, I_p)$, the Bayes estimator under quadratic loss is

$$\delta^\pi(x) = ||x||^2 + p.$$

This generalized Bayes estimator is not admissible because it is dominated by $\delta_0(x) = ||x||^2 - p$ (Exercise 2.35). Since the classical risk is $R(\theta, \delta^\pi) = \text{var}(||x||^2) + 4p^2$, the Bayes risk is infinite. This phenomenon shows that the Lebesgue measure is not necessarily the best noninformative choice for a prior measure when the parameter of interest is a subvector of the parameter (see Chapter 3). ∥

2.5 Usual loss functions

When the setting of an experiment is such that the utility function cannot be determined (lack of time, limited information, etc.), a customary alternative is to resort to classical losses, which are mathematically tractable and well documented. Of course, this approach is an approximation of the underlying statistical model and should only be adopted when the utility function is missing. We conclude this section with a note on more intrinsic loss functions, although these are rarely used in practice. (See also Note 2.8.1 for a description of losses used in imaging.)

2.5.1 The quadratic loss

Proposed by Legendre (1805) and Gauss (1810), this loss is undoubtedly the most common evaluation criterion. Founding its validity on the ambiguity of the notion of *error* in statistical settings (i.e., measurement error versus random variation), it also gave rise to many criticisms, commonly dealing with the fact that the squared error loss

(2.5.1) $$L(\theta, d) = (\theta - d)^2$$

penalizes large deviations too heavily.

However, convex loss functions like (2.5.1) have the incomparable advantage of avoiding the paradox of *risk lovers* and to exclude randomized estimators. Another usual justification for the quadratic loss is that it provides a Taylor expansion approximation to more complex symmetric losses (see Exercise 4.14 for a counterexample). In his 1810 paper, Gauss already acknowledged the arbitrariness of the quadratic loss and was defending it on grounds of simplicity. Although the criticisms over a systematic use of the quadratic loss are quite valid, this loss is nonetheless extensively used because it gives intuitively sound Bayesian solutions, i.e., those one would naturally suggest as estimators for a non-decision-theoretic inference based on the posterior distribution. In fact, the Bayes estimators associated with the quadratic loss are the posterior means. However, note that the quadratic loss is not the only loss enjoying this property. Losses leading to posterior means as the Bayes estimators are called *proper losses* and characterized in Lindley (1985), Schervish (1989), van der Meulen (1992), and Hwang and Pemantle (1994). (See also Exercise 2.15.)

Proposition 2.5.1   *The Bayes estimator δ^π associated with the prior distribution π and with the quadratic loss (2.5.1), is the posterior expectation*

$$\delta^\pi(x) = \mathbb{E}^\pi[\theta|x] = \frac{\int_\Theta \theta f(x|\theta)\pi(\theta)\,d\theta}{\int_\Theta f(x|\theta)\pi(\theta)\,d\theta}.$$

Proof.   Since

$$\mathbb{E}^\pi[(\theta - \delta)^2|x] = \mathbb{E}^\pi[\theta^2|x] - 2\delta\mathbb{E}^\pi[\theta|x] + \delta^2,$$

the posterior loss actually attains its minimum at $\delta^\pi(x) = \mathbb{E}^\pi[\theta \mid x]$.   ☐☐

The following corollaries are straightforward to derive.

Corollary 2.5.2   *The Bayes estimator δ^π associated with π and with the weighted quadratic loss*

(2.5.2) $L(\theta, \delta) = \omega(\theta)(\theta - \delta)^2,$

where $\omega(\theta)$ is a nonnegative function, is

$$\delta^\pi(x) = \frac{\mathbb{E}^\pi[\omega(\theta)\theta|x]}{\mathbb{E}^\pi[\omega(\theta)|x]}.$$

Corollary 2.5.3   *When $\Theta \in \mathbb{R}^p$, the Bayes estimator δ^π associated with π and with the quadratic loss,*

$$L(\theta, \delta) = (\theta - \delta)^t Q(\theta - \delta),$$

is the posterior mean, $\delta^\pi(x) = \mathbb{E}^\pi[\theta|x]$, for every positive-definite symmetric $p \times p$ matrix Q.

Corollary 2.5.2 exhibits a (weak) *duality* between loss and prior distribution, in the sense that it is equivalent to estimate θ under (2.5.2) with the prior π, or under (2.5.1) with the prior $\pi_\omega(\theta) \propto \pi(\theta)\omega(\theta)$. Moreover, while admissibility is independent of the weight factor, the Bayes estimator strongly depends on the function ω. For instance, δ^π may not exist if ω increases too fast to $+\infty$. On the other hand, Corollary 2.5.3 shows that the Bayes estimators are robust with respect to the quadratic form Q. (Shinozaki (1975) has also proved that admissibility does not depend on Q.)

The quadratic loss is particularly interesting in the setting of bounded parameter spaces when the choice of a more subjective loss is impossible. In fact, this loss is quite tractable and the approximation error is usually negligible. Indeterminacy about the loss function (and thus its replacement by a quadratic approximation) often occurs in *accuracy evaluation*, including for instance *loss estimation* (see Rukhin (1988a,b), Lu and Berger (1989a,b), Hwang, Casella et al. (1992), Robert and Casella (1993, 1994), and Fourdrinier and Wells (1994)).

Example 2.5.4 (Example 2.4.9 continued)   We are looking for an evaluation of the performances of the estimator

$$\delta(x_1, x_2) = \begin{cases} \dfrac{x_1 + x_2}{2} & \text{if } x_1 \neq x_2, \\ x_1 + 1 & \text{otherwise,} \end{cases}$$

by $\alpha(x_1, x_2)$ under the quadratic criterion

$$\left[\mathbb{I}_\theta(\delta(x_1, x_2)) - \alpha(x_1, x_2) \right]^2,$$

where $\mathbb{I}_\theta(v)$ is 1 if $v = \theta$, 0 otherwise; the function α somehow evaluates the probability that δ takes the true value θ. (This is a special case of loss estimation, when the loss function is $1 - \mathbb{I}_\theta(\delta)$.) Two estimators can be proposed:

(i) $\alpha_0(x_1, x_2) = 0.75$, which is the expectation of $\mathbb{I}_\theta(\delta(x_1, x_2))$; and

(ii) $\alpha_1(x_1, x_2) = \begin{cases} 1 & \text{if } x_1 \neq x_2, \\ 0.50 & \text{if } x_1 = x_2. \end{cases}$

The risks of the two evaluators are then

$$\begin{aligned} R(\theta, \alpha_0) &= \mathbb{E}_\theta \left(\mathbb{I}_\theta(\delta(x_1, x_2)) - 0.75 \right)^2 \\ &= 0.75 - (0.75)^2 = 0.1875 \,; \end{aligned}$$

and

$$\begin{aligned} R(\theta, \alpha_1) &= \mathbb{E}_\theta \left(\mathbb{I}_\theta(\delta(x_1, x_2)) - \alpha_1(x_1, x_2) \right)^2 \\ &= (0.5)^2 \frac{1}{2} = 0.125 \,. \end{aligned}$$

Therefore, α_1 is a better estimator of the performances of δ than α_0. As mentioned in Berger and Wolpert (1988), this domination result is quite logical and it suggests that a conditional evaluation of estimators is more appropriate.   ‖

2.5.2 *The absolute error loss*

An alternative solution to the quadratic loss in dimension one is to use the absolute error loss,

(2.5.3) $L(\theta, d) = | \theta - d |,$

already considered by Laplace (1773) or, more generally, a multilinear function

$$(2.5.4) \qquad L_{k_1,k_2}(\theta, d) = \begin{cases} k_2(\theta - d) & \text{if } \theta > d , \\ k_1(d - \theta) & \text{otherwise.} \end{cases}$$

Such functions increase more slowly than the quadratic loss. Therefore, while remaining convex, they do not overpenalize large but unlikely errors. Huber (1964) also proposed a mixture of the absolute error loss and the quadratic loss, in order to keep a quadratic penalization around 0,

$$\tilde{L}(\theta, d) = \begin{cases} (d - \theta)^2 & \text{if } | d - \theta | < k, \\ 2k | d - \theta | - k^2 & \text{otherwise.} \end{cases}$$

Although a convex[4] loss, the mixed loss slows down the progression of the quadratic loss for large errors and has a robustifying effect. Unfortunately, there usually is no explicit derivation of Bayes estimators under this loss $\tilde{L}$.

Proposition 2.5.5 *A Bayes estimator associated with the prior distribution π and the multilinear loss (2.5.4) is a $(k_2/(k_1 + k_2))$ fractile of $\pi(\theta|x)$.*

Proof. The following classical equality

$$\begin{aligned} \mathbb{E}^\pi [L_{k_1,k_2}(\theta, d)|x] &= k_1 \int_{-\infty}^d (d - \theta)\pi(\theta|x)\, d\theta + k_2 \int_d^{+\infty} (\theta - d)\pi(\theta|x)\, d\theta \\ &= k_1 \int_{-\infty}^d P^\pi(\theta < y|x)\, dy + k_2 \int_d^{+\infty} P^\pi(\theta > y|x)\, dy, \end{aligned}$$

is obtained by an integration by parts. Taking the derivative in d, we get

$$k_1 P^\pi(\theta < d|x) - k_2 P^\pi(\theta > d|x) = 0,$$

i.e.,

$$P^\pi(\theta < d|x) = \frac{k_2}{k_1 + k_2}. \qquad \Box\Box$$

In particular, if $k_1 = k_2$, i.e., in the case of the absolute error loss, the Bayes estimator is the posterior median, which is the estimator obtained by Laplace (see Example 1.2.4). Notice that, when π has a nonconnected support, Proposition 2.5.5 provides examples of multiple Bayes estimators for some values of x (see Exercise 2.40).

2.5.3 The $0 - 1$ loss

This loss is mainly used in the classical approach to hypothesis testing, as formalized by Neyman and Pearson (see Section 5.3). More generally, this

---

[4] Again, if we insist so much on *convexity*, it is because it ensures that randomized estimators are suboptimal from a frequentist point of view. Therefore, a statistical decision-theoretic approach that would agree as much as possible with the Likelihood Principle necessarily calls for convex losses. This requirement obviously eliminates bounded losses.

is a typical example of a nonquantitative loss. In fact, for this loss, the penalty associated with an estimate δ is 0 if the answer is correct and 1 otherwise.

Example 2.5.6 Consider the test of $H_0 : \theta \in \Theta_0$ versus $H_1 : \theta \notin \Theta_0$. Then $\mathcal{D} = \{0, 1\}$, where 1 stands for acceptance of H_0 and 0 for rejection (in other words, the function of θ to be estimated is $\mathbb{I}_{\Theta_0}(\theta)$). For the $0 - 1$ loss, i.e.,

$$(2.5.5) \qquad \mathrm{L}(\theta, d) = \begin{cases} 1 - d & \text{if } \theta \in \Theta_0 \\ d & \text{otherwise,} \end{cases}$$

the associated risk is

$$\begin{aligned} R(\theta, \delta) &= \mathbb{E}_\theta[\mathrm{L}(\theta, \delta(x))] \\ &= \begin{cases} P_\theta(\delta(x) = 0) & \text{if } \theta \in \Theta_0 \text{ ,} \\ P_\theta(\delta(x) = 1) & \text{otherwise,} \end{cases} \end{aligned}$$

which are exactly the *type–one and type–two errors* underlying the Neyman–Pearson theory.   ‖

This loss is not very interesting because of its nonquantitative aspect, and we will consider in Chapter 5 some alternative theories for testing hypotheses. The associated Bayes estimators also reflect the primitive aspect of such a loss (see also Exercise 2.41).

Proposition 2.5.7 *The Bayes estimator associated with π and with the loss (2.5.5) is*

$$\delta^\pi(x) = \begin{cases} 1 & \text{if } P(\theta \in \Theta_0|x) > P(\theta \notin \Theta_0|x), \\ 0 & \text{otherwise,} \end{cases}$$

i.e., $\delta^\pi(x)$ is equal to 1 if and only if $P(\theta \in \Theta_0|x) > 1/2$.

2.5.4 Intrinsic losses

It may occur that some settings are so noninformative that not only the loss function is unknown, but there is not even a natural parameterization. Such cases happen when the distribution $f(x|\theta)$ itself is of interest, for instance, in prediction settings.

However, as we mentioned in the previous section, the choice of the parameterization is important because, contrary to the maximum likelihood estimation approach, if g is a one-to-one transformation of θ, the Bayes estimator of $g(\theta)$ is usually different from the transformation by g of the Bayes estimator of θ under the same loss (see Exercise 2.36). This lack of invariance, although often troubling to beginners, is not usually a concern for decision-makers because it shows how the Bayesian paradigm can adapt to the estimation problem at hand *and* the selected loss function, while maximum likelihood estimation is totally loss-blind. But the few cases where loss function and natural parameterization are completely unavailable may call for this kind of ultimate invariance. (See Wallace and Boulton (1975) for another approach.)

In such noninformative settings, it seems natural to use losses that compare directly the distributions $f(\cdot|\theta)$ and $f(\cdot|\delta)$ associated with the true parameter θ and the estimate δ. Such loss functions,

$$L(\theta, \delta) = d(f(\cdot|\theta), f(\cdot|\delta)),$$

are indeed parameterization-free. Two usual distribution distances are

(1) the *entropy distance*

(2.5.6) $$L_e(\theta, \delta) = \mathbb{E}_\theta \left[\log \left(\frac{f(x|\theta)}{f(x|\delta)} \right) \right],$$

which is also called the Kullback–Leibler divergence and which is not a distance in the mathematical sense because of its asymmetry; and

(2) the *Hellinger distance*

(2.5.7) $$L_H(\theta, \delta) = \frac{1}{2} \mathbb{E}_\theta \left[\left(\sqrt{\frac{f(x|\delta)}{f(x|\theta)}} - 1 \right)^2 \right].$$

Example 2.5.8 Consider $x \sim \mathcal{N}(\theta, 1)$. Then we have

$$\begin{aligned} L_e(\theta, \delta) &= \frac{1}{2} \mathbb{E}_\theta[-(x-\theta)^2 + (x-\delta)^2] = \frac{1}{2}(\delta - \theta)^2, \\ L_H(\theta, \delta) &= 1 - \exp\{-(\delta - \theta)^2/8\}. \end{aligned}$$

Considering the normal case when $\pi(\theta|x)$ is a $\mathcal{N}(\mu(x), \sigma^2)$ distribution, it is straightforward to show that the Bayes estimator is $\delta^\pi(x) = \mu(x)$ in both cases. ‖

The Hellinger loss is undoubtedly more intrinsic than the entropy loss, if only because it always exists (note that (2.5.7) is bounded above by 1). Unfortunately, while leading to explicit expressions of $L_H(\theta, \delta)$ for the usual distribution families, it does not allow for an explicit derivation of the Bayes estimators, except in the special case treated above. On the contrary, in *exponential families*, the entropy loss provides explicit estimators which are the posterior expectations for the estimation of the *natural parameter* (see Chapter 3). Moreover, although quite different from the Hellinger loss, the entropy loss provides similar answers for the usual distribution families (see Robert (1996b)). There are also various theoretical reasons to defend the use of the Kullback-Leibler distance, ranging from information theory (Exercise 2.48) to the relevance of logarithmic scoring rule and the location-scale invariance of the distance, as detailed in Bernardo and Smith (1994).

2.6 Criticisms and alternatives

Some criticisms about the frequentist notions of minimaxity and admissibility have been mentioned in the previous sections. These concepts are actually of secondary interest from a purely Bayesian point of view, since, on one hand, admissibility is automatically satisfied by most Bayes estimators. On the other hand, minimaxity is somehow incompatible with the Bayesian paradigm, since, under a prior distribution, each value of the parameter cannot be equally weighted. However, minimaxity may be relevant from a robustness point of view, that is, when the prior information is not precise enough to determine the prior distribution.

It may happen that the decision-maker cannot define a loss function exactly. For instance, when the decision-maker is a committee comprising several experts, it is often the case that they differ about the relevant loss function (and sometimes even about the prior distribution). Starting with Arrow (1951), the literature on these extensions of classical Decision Theory is quite extensive (see Genest and Zidek (1986), Rubin (1987), and Van Eeden and Zidek (1993) for details and references).

When the loss function has not been completely determined, it might be assumed to belong to a parametrized class of loss functions, the decision maker selecting the most accurate parameter. Apart from L_p losses, two other possible classes are

$$L_1(\theta, \delta) = \log(\alpha||\theta - \delta||^2 + 1), \qquad L_2(\theta, \delta) = 1 - \exp\{-c||\theta - \delta||^2\}.$$

An alternative approach more in tune with the Bayesian paradigm is to consider that, since the loss is partly unknown, this uncertainty can be represented by using a *random loss* $L(\theta, \delta)$. The evaluation of estimators is then done by integrating out with respect to this additional variable: If F is the distribution of the loss, the objective function to minimize (in δ) is

(2.6.1) $$\int_\Theta \int_\Omega L(\theta, \delta, \omega) dF(\omega) \, d\pi(\theta|x),$$

where F possibly depends on θ or even on x. This case is actually the only interesting extension because, otherwise, to minimize (2.6.1) is equivalent to using the average loss

$$\bar{L}(\theta, \delta) = \int_\Omega L(\theta, \delta, \omega) \, dF(\omega).$$

Another approach to the lack of precision on the loss function consists of considering simultaneously a set of losses and look for estimators performing well for all these losses. Obviously, this multidimensional criterion only induces a *partial* ordering on estimators.

Example 2.6.1 Consider $x \sim \mathcal{N}_p(\theta, I_p)$. The parameter θ is estimated under quadratic loss. If the loss matrix Q is not exactly determined, a robust alternative is to include the losses associated with the matrices Q such that $Q_1 \preceq Q \preceq Q_1$ (where $A \preceq B$ means that the matrix $B - A$ is nonnegative definite). Notice that, according to Corollary 2.5.3, the Bayes estimator is the same for all Q's.   ‖

Example 2.6.2 In the setting of the above example, Brown (1975) shows that a shrinkage estimator of the form $(1 - h(x))x$ dominates $\delta_0(x) = x$ for a class of quadratic losses, i.e., a class of matrices Q if and only if

$$(2.6.2) \qquad\qquad \text{tr}(Q) - 2\lambda_{\max}(Q) > 0$$

for every matrix in the class (where $\lambda_{\max}$ denotes the largest eigenvalue). Notice that this condition excludes the case $p \leq 2$, where δ_0 is actually admissible. The constant $\text{tr}(Q) - 2\lambda_{\max}(Q)$ also appears in the majorization constant of $||x||^2 h(||x||^2)$ (see Theorem 2.8.1). Therefore, (2.6.2) is both a necessary and sufficient condition for the Stein effect to occur.   ‖

The ultimate criterion in loss robustness is called *universal domination* and was introduced in Hwang (1985). It actually takes into account the set of all losses $\ell(||\delta - \theta||_Q)$, for a given norm $||x||_Q = x^t Q x$ and all nondecreasing functions ℓ. An estimator δ_1 will be said to *universally dominate* another estimator δ_2 if, for every ℓ,

$$\mathbb{E}_\theta[\ell(||\delta_1(x) - \theta||_Q)] \leq \mathbb{E}_\theta[\ell(||\delta_2(x) - \theta||_Q)].$$

A second criterion is called *stochastic domination*: δ_1 *stochastically dominates* δ_2 if, for every $c > 0$,

$$P_\theta(||\delta_1(x) - \theta||_Q \leq c) \geq P_\theta(||\delta_2(x) - \theta||_Q \leq c).$$

Although this criterion seems more intrinsic and less related to Decision Theory than universal domination, Hwang (1985) has shown that the two criteria are actually equivalent.

Theorem 2.6.3 *An estimator δ_1 universally dominates an estimator δ_2 if and only if δ_1 stochastically dominates δ_2 .*

Proof. The estimator δ_1 stochastically dominates δ_2 if, for every $c > 0$,

$$P_\theta(||\delta_1(x) - \theta||_Q \leq c) \geq P_\theta(||\delta_2(x) - \theta||_Q \leq c).$$

This can be rewritten as

$$\mathbb{E}_\theta \left[\mathbb{I}_{[c,+\infty[}(||\delta_1(x) - \theta||_Q) \right] \leq \mathbb{E}_\theta \left[\mathbb{I}_{[c,+\infty[}(||\delta_2(x) - \theta||_Q) \right].$$

Since $\ell(t) = \mathbb{I}_{[c,+\infty[}(t)$ is a nondecreasing function of t, universal domination implies stochastic domination. The converse follows from the fact that the first moments of two stochastically ordered random variables are also ordered.   □□

Moreover, these two criteria are not empty since Hwang (1985) has established the following domination result: If $x \sim \mathcal{T}_\alpha(\mu, \sigma^2)$, Student's t-distribution with α degrees of freedom, some shrinkage estimators universally dominate $\delta_0(x) = x$. If the dimension is not too small (usually,

$p = 4$ is sufficient), Brown and Hwang (1989) virtually showed that, if $x \sim \mathcal{N}_p(\theta, \Sigma)$, the estimator $\delta_0(x)$ is admissible for universal domination *if and only if* $Q = \Sigma$. For other choices of the matrix Q and p large enough, δ_0 is stochastically dominated. Therefore, even though this criterion is less discriminating than usual losses, it allows for comparison, and even for a Stein effect, since classical estimators are not necessarily optimal.

The study of multiple losses is not very developed from a Bayesian point of view, since Bayes estimators usually vary with a change in the loss function. However, in a very special case, Rukhin (1978) has shown that the Bayes estimators were *independent of the loss function*. Under some regularity assumptions, this case corresponds to the equation

$$\log f(x|\theta) + \log \pi(\theta) = A_1(x)e^{\alpha\theta} + A_2(x)e^{-\alpha\theta} + A_3(x),$$

where π is the prior distribution. Therefore, for this *exponential family* (see Section 3.3.3),

$$(2.6.3) \qquad f(x|\theta) = \frac{B(x)}{\pi(\theta)} \exp\{A_1(x)e^{\alpha\theta} + A_2(x)e^{-\alpha\theta}\},$$

the Bayes estimators are *universal*, because they do not depend on the loss. The next chapter covers in detail the case of exponential families, which are classes of distributions on $\mathbb{R}^k$ with densities

$$f(x|\theta) = c(\theta)h(x)\exp[R(\theta) \cdot T(x)],$$

where $R(\theta), T(x) \in \mathbb{R}^p$. However, notice that (2.6.3) is a rather special exponential family.

2.7 Exercises

Section 2.2

2.1 Show that, if the utility function U is convex, every $P \in \mathcal{P}_{\mathcal{E}}$ satisfies

$$\mathbb{E}^P[r] = \int_{\mathcal{R}} r \, dP(r) \preceq P.$$

Conclude that a concave loss is not realistic.

2.2 Consider four dice with respective numbers on their faces $(4,4,4,4,0,0)$, $(3,3,3,3,3,3)$, $(6,6,2,2,2,2)$, $(1,1,1,5,5,5)$. Two players roll one die each and compare their outcome. Show that the relation die $[i]$ beats die $[j]$ is intransitive, i.e., that for every choice of the first player the second player can choose a die so that the probability of winning is greater than 0.5. Relate this example to the Pitman closeness setting of Note 2.8.3.

2.3 Show that $\mathcal{P}_{\mathcal{B}} \subset \mathcal{P}_{\mathcal{E}}$, i.e., that bounded reward distributions have a finite expected utility.

2.4 Show Lemmas 2.2.2 and 2.2.3.

2.5 *(DeGroot (1970))* In order to show the extension of Theorem 2.2.4 from $\mathcal{P}_{\mathcal{B}}$ to $\mathcal{P}_{\mathcal{E}}$, consider a sequence s_m decreasing (for $\preceq$) in $\mathcal{R}$ such that, for every

$r \in \mathcal{R}$, there exists m with $s_m \preceq r$. If $P \in \mathcal{P}_\mathcal{E}$ and if $P(\{s_m \preceq r\}) > 0$, denote by P_m the conditional distribution

$$P_m(A) = \frac{P(A \cap \{s_m \preceq r\})}{P(\{s_m \preceq r\})}.$$

Similarly, if t_n is an increasing sequence in $\mathcal{R}$ such that, for every $r \in \mathcal{R}$, there exists n with $r \preceq t_n$, we define P^n as

$$P^n(A) = \frac{P(A \cap \{r \preceq t_n\})}{P(\{r \preceq t_n\})},$$

when $P(\{r \preceq t_n\}) > 0$. We assume that such sequences can be exhibited in $\mathcal{R}$.

a. Show that P^n and P_m are included in $\mathcal{P}_\mathcal{B}$.

We introduce the additional hypothesis:

(A_6) *For every P, $Q \in \mathcal{P}_\mathcal{E}$, such that there exists $r_0 \in \mathcal{R}$ satisfying $P(\{r \preceq r_0\}) = Q(\{r_0 \preceq r\}) = 1$, the ordering $P \preceq Q$ is necessarily satisfied.*

b. Show that (A_6) is actually satisfied in $\mathcal{P}_\mathcal{B}$.

c. Show that, for every $P \in \mathcal{P}_\mathcal{E}$,

$$\mathbb{E}^P[U(r)] = \lim_{m \to +\infty} \mathbb{E}^{P_m}[U(r)] = \lim_{n \to +\infty} \mathbb{E}^{P^n}[U(r)].$$

d. Consider $P \in \mathcal{P}_\mathcal{E}$ and $m < m_1$, $n < n_1$ such that $P(\{s_m \preceq r\}) > 0$ and $P(\{r \preceq t_n\}) > 0$. Show that

$$P^n \preceq P^{n_1} \preceq P \preceq P_{m_1} \preceq P_m.$$

The second additional hypothesis:

(A_7) *Consider P and Q in $\mathcal{P}_\mathcal{E}$. If there exists m_0 such that $P_m \succeq Q$ when $m \geq m_0$, then $P \succeq Q$. Moreover, if there exists n_0 such that $P^n \preceq Q$ when $n \geq n_0$, then $P \preceq Q$,*

is assumed to hold below.

e. Consider P and Q in $\mathcal{P}_\mathcal{E}$ with r_1, r_2 in $\mathcal{R}$ such that

$$P(\{r_1 \preceq r\}) = Q(\{r_2 \preceq r\}) = 1.$$

Show that $P \preceq Q$ if and only if $\mathbb{E}^P[U(r)] \leq \mathbb{E}^Q[U(r)]$. (*Hint:* Consider the sequences P^n, P_m, and $a_m = \mathbb{E}^{P_m}[U(r)]$, $b_n = \mathbb{E}^{P^n}[U(r)]$. Use hypothesis ($A_4$) and questions c. and d.)

f. Deduce from the above question that, if P, $Q \in \mathcal{P}_\mathcal{E}$, $P \preceq Q$ if and only if $\mathbb{E}^P[U(r)] \leq \mathbb{E}^Q[U(r)]$.

2.6 In the setup of Example 2.2.6 on the Saint Petersburg paradox, give the average utility of a player for Saint Petersburg paradox, give the average utility of a player for $\delta = 1$ and $\delta = 10$. Compute the average number of games a player is ready to play in the modified game.

2.7 *(Smith (1988)) An expert has a preference ordering such that the rewards $\alpha \mathbb{I}_{(x+h)} + (1 - \alpha)\mathbb{I}_{(x-h)}$ and x are equivalent, with α independent of x. Show that the utility function of this expert is either linear (when $\alpha = 1/2$), of the form e^{cx} ($c > 0$) ($\alpha < 1/2$), or of the form $1 - e^{-cx}$ ($\alpha > 1/2$).

2.8 (Raiffa (1968)) In a first setup, a person has to choose between a sure gain of \$10,000 (a_1) and a gain of \$50,000 with probability 0.89 (a_2). The second setup is such that a gain of \$50,000 with probability 0.1 (a_3) is opposed to a gain of \$10,000 with probability 0.11 (a_4). Show that, even if it seems natural to prefer a_1 to a_2 and a_3 to a_4, there is no utility function preserving the order $a_1 \preceq a_2$ and $a_3 \preceq a_4$.

2.9 In the setup of the Saint Petersburg paradox defined in Example 2.2.6, consider the following three classes of utility functions:

(i) $U(r) = \log(\delta + r)$;

(ii) $U(r) = (\delta + r)^\varrho$ $(0 < \varrho < 1)$; and

(iii) $U(r) = 1 - e^{\delta + r}$.

For each class, determine the maximum entrance fee and the optimal number of games.

Section 2.3

2.10 (Casella (1990)) Show that, if the function r, from $\mathbb{R}_+$ in $\mathbb{R}_+$, is concave, then $r(t)$ is nondecreasing and $r(t)/t$ is nonincreasing.

2.11 Considering the loss proposed in Example 2.3.4, show that a perfect expert for $N = 2$ dominates a perfect expert for $N = 1$. Does the same phenomenon occur for $N = 3$?

2.12 (Smith (1988)) Using the notations of Example 2.3.4, the *Brier score* is defined as the loss function

$$L(\theta, p) = \sum_{i=1}^{N} q_i (p_i - \theta_i)^2 + \bar{q}(1 - \bar{q}) - \sum_{i=1}^{N} q_i (p_i - \bar{q})^2,$$

with $\bar{q} = \sum_{i=1}^{N} q_i \theta_i$, the proportion of rainy days. Show that a perfect expert P_1 is better than a perfect expert P_2 if its (so-called) resolution

$$R = \sum_{i=1}^{N} q_i (\theta_i - \bar{q})^2$$

is larger. Comment on the form of the loss.

2.13 Show that, for a loss function $L(\theta, d)$ strictly increasing in $|d - \theta|$ such that $L(\theta, \theta) = 0$, there is no uniformly optimal statistical procedure. Give a counterexample when

$$L(\theta, \varphi) = \theta(\mathbb{I}_{\mathbb{R}^*}(\theta) - \varphi)^2.$$

2.14 In relation to Example 2.3.4, the *scoring rule* of a weather forecaster is the sum, over the year, of the errors $(\mathbb{I}_{A_{ij}} - p_i)^2$ for all the days for which the probability p_i was announced and for which A_{ij} is the event that it actually rained. If n_i is the number of days that p_i was forecasted, show that the scoring rule can be decomposed as

$$\sum_{i=1}^{N} \sum_{j=1}^{n_i} (\mathbb{I}_{A_{ij}} - \theta_i)^2 + \sum_{i=1}^{N} n_i (\theta_i - p_i)^2.$$

2.15 *(Schervish (1989)) Consider an inferential problem where the probability p of an event E is to be forecasted. The answer $\delta \in [0, 1]$ of a forecaster is evaluated through a *scoring rule* $L(E, \delta)$, which takes the value $g_i(\delta) \geq 0$ if $\mathbb{I}_E = i$ $(i = 0, 1)$. The scoring rule is said to be *proper* if the average error

$$m(\delta) = pg_1(\delta) + (1 - p)g_0(\delta)$$

is minimized for $\delta = p$.

a. Show that, for a proper scoring rule, g_0 is nondecreasing and g_1 is nonincreasing.

b. Show that, if the g_i are differentiable, the scoring rule is proper if and only if

$$-pg_1'(p) = (1 - p)g_0'(1 - p)$$

for every p in $[0, 1]$.

c. Deduce that, when the scoring rule is proper, there exists a nonnegative function h, integrable on $[0, 1]$ such that

$$g_0(r) = \int_{[0,r]} h(t)\, dt \qquad \text{and} \qquad g_1(r) = \int_{[1-r,1]} \frac{t}{1-t} h(t)\, dt.$$

2.16 Show through discrete and continuous examples that a Bayes estimator can correspond to several prior distributions for the same loss function.

2.17 Two experts must provide an estimate of $p \in [0, 1]$ under the loss $(\delta - p)^2$. They have the respective prior distributions π_1 and π_2, equal to $\mathcal{B}e(1, 2)$ and $\mathcal{B}e(2, 3)$.

a. Give both estimates δ_1 and δ_2 when the experts answer separately (with no observation).

b. Expert 1 knows δ_2. We assume that the quantity p is observed afterward and that the best expert is fined $(\delta_i - p)^2$ while the worst expert is fined a fixed amount A. Show that the loss function for expert 1 is

$$(\delta_1 - p)^2 \mathbb{I}_{|\delta_1 - p| \leq |\delta_2 - p|} + A\mathbb{I}_{|\delta_1 - p| > |\delta_2 - p|}.$$

Deduce that, if A is large enough, the optimal answer for expert 1 is $\delta_1 = \delta_2$.

c. Modify the above loss function in order to force expert 1 to give an honest answer, i.e., the original δ_1.

2.18 (Raiffa and Schlaifer (1961)) Given a loss function $L(\theta, d)$, define the optimal decision as the decision d_θ that minimizes $L(\theta, d)$ for a given θ. The *opportunity loss* is then defined as $L^*(\theta, d) = L(\theta, d) - L(\theta, d_\theta)$.

a. Show that this is equivalent to assume that $\inf_\theta L(\theta, d) = 0$ for every θ.

b. Show that the set of classical (frequentist) optimal procedures (admissible, minimax) is the same for L and L^*.

c. Show that the Bayes procedures are the same for L and L^*.

2.19 (Raiffa and Schlaifer (1961)) Given a loss function $L(\theta, d)$ and a prior distribution π, the *optimal prior decision* is d^π which minimizes $\mathbb{E}^\pi[L(\theta, d)]$.

a. Consider $\mathcal{D} = \{d_1, d_2\}$ and $L(\theta, d_1) = 0.5 + \theta$, $L(\theta, d_2) = 2 - \theta$. Give the optimal prior decisions when π is $\mathcal{B}e(1, 1)$ and $\mathcal{B}e(2, 2)$.

b. *The value of sample information* x is defined as

$$\nu(x) = \mathbb{E}^\pi[L(\theta, d^\pi)|x] - \mathbb{E}^\pi[L(\theta, \delta^\pi(x))|x],$$

where $\delta^\pi(x)$ is the regular Bayesian estimator of θ. Indicate why $\nu(x) \geq 0$ and give the value of sample information when $x \sim \mathcal{B}(n, \theta)$ for the above loss function and priors.

c. When $\Theta = \mathcal{D} = \mathbb{R}$, $x \sim \mathcal{N}(\theta, 1)$, and $\theta \sim \mathcal{N}(\theta_0, 10^2)$, show that the optimal prior decision under squared error loss is $d^\pi = \theta_0$ and that the value of sample information is $(\theta_0 - x)^2$. Conclude by discussing the coherence of this notion.

2.20 An investment strategy can be implemented according to two different strategies, d_1 and d_2. The benefit (or utility) of the investment depends on a rentability parameter $\theta \in \mathbb{R}$ and is $U(\theta, d_i) = k_i + K_i\theta$.

a. Given a prior distribution π on θ, what is the optimal prior decision?

b. Let $x \sim \mathcal{N}(\theta, 1)$ and let $\theta \sim \mathcal{N}(0, 10)$. Give the optimal prior and posterior strategies. Give the improvement brought about by the observation of x in terms of posterior utility and expected utility.

c. If there is a cost c_s for the observation of x, determine the maximum cost c_s when the advantage of observing x disappears.

2.21 (Raiffa and Schlaifer (1961)) In a setting similar to the above exercise, the decision space is $\mathcal{D} = \{d_1, d_2\}$ and the parameter $\theta \in [0, 1]$. The utility function is $L(\theta, d_i) = k_i + K_i\theta$.

a. Defining $\varphi = (k_1 - k_2)/(K_1 - K_2)$, show that $\varphi \notin (0, 1)$ implies that one of the two decisions is always optimal. In the following questions, we assume $\varphi \in (0, 1)$.

b. Let $x|\theta \sim \mathcal{B}(n, \theta)$ and let $\theta \sim \mathcal{B}e(r, n' - r)$. Compute the optimal prior and posterior decisions and the expected improvement (in utility) gained by using the observation x.

c. Given an observation cost of K for each Bernoulli random variable, determine the optimal sample size n for the expected utility.

Section 2.4.1

2.22 Prove Theorem 2.4.2 when $r(\pi)$ is finite.

2.23 Compare δ_0 and δ^* of Example 2.3.1 under $0 - 1$ loss. Does this result contradict the Rao–Blackwell Theorem (Theorem 2.4.8)?

Section 2.4.2

2.24 Produce an example similar to Example 2.4.7, but where A would be forced to confess from a Bayesian point of view.

2.25 Consider the case when $\Theta = \{\theta_1, \theta_2\}$ and $\mathcal{D} = \{d_1, d_2, d_3\}$, for the following loss structure

	d_1	d_2	d_3
θ_1	2	0	0.5
θ_2	0	2	1

a. Determine the minimax procedures.

b. Identify the least favorable prior distribution. (*Hint:* Represent the risk space associated with the three actions as in Example 2.4.12.)

2.26 Consider the following risk function for $\Theta = \{\theta_1, \theta_2\}$ and $\mathcal{D} = \{d_1, d_2, d_3\}$

	d_1	d_2	d_3
θ_1	1	2	1.75
θ_2	2	1	1.75

a. Draw the risk diagram as in Example 2.4.12 and deduce the minimax estimators.

b. Deduce from this example that minimaxity is not *coherent* in the following sense: d_1, d_2, d_3 may be such that $\max_\theta R(\theta, d_1) \geq \max_\theta R(\theta, d_3)$ and $\max_\theta R(\theta, d_2) \geq \max_\theta R(\theta, d_3)$, while the minimax estimator is of the form $\alpha d_1 + (1 - \alpha)d_2$.

Section 2.4.3

2.27 Prove Lemma 2.4.10.

2.28 Consider $x \sim \mathcal{B}(n, \theta)$, with n known.

a. If $\pi(\theta)$ is the beta distribution $\mathcal{B}e(\sqrt{n}/2, \sqrt{n}/2)$, give the associated posterior distribution $\pi(\theta|x)$ and the posterior expectation, $\delta^\pi(x)$.

b. Show that, when $L(\delta, \theta) = (\theta - \delta)^2$, the risk of δ^π is constant. Conclude that δ^π is minimax.

c. Compare the risk for δ^π with the risk function of $\delta_0(x) = x/n$ for $n = 10, 50,$ and 100. Conclude about the appeal of δ^π.

2.29 Prove Lemmas 2.4.13 and 2.4.15.

2.30 Consider $x \sim \mathcal{N}(\theta, 1)$ and $\theta \sim \mathcal{N}(0, n)$. Show that the Bayes risk is equal to $n/(n + 1)$. Conclude about the minimaxity of $\delta_0(x) = x$.

2.31 *Give the density of the uniform distribution on the sphere of radius c and derive the marginal distribution of $x \sim \mathcal{N}_p(0, I_p)$, when θ is uniformly distributed on this sphere. Compute the posterior expectation δ^π and study its properties.

2.32 Show the equivalent of Example 2.4.16 when $x \sim \mathcal{P}(\lambda)$, i.e., that $\delta_0(x) = x$ is minimax. (*Hint:* Notice that δ_0 is a generalized Bayes estimator for $\pi(\lambda) = 1/\lambda$ and use a sequence of $\mathcal{G}(\alpha, \beta)$ priors.)

2.33 Establish Propositions 2.4.20, 2.4.23, and 2.4.25.

Section 2.4.4

2.34 In the setting of Example 2.4.26, we want to show that the Bayes risk of $p(x)$ is finite.

a. Show that

$$\tau(\pi) = \int_{\mathbb{R}^2} \left\{ \Phi^2(x) - 2\Phi(x)\mathbb{I}_{\theta \leq 0} + \mathbb{I}_{\theta \leq 0} \right\} \frac{e^{-(x-\theta)^2/2}}{\sqrt{2\pi}} d\theta dx$$

when $\pi(\theta) = 1$.

b. Deduce that

$$\tau(\pi) \;=\; \int_{-\infty}^{+\infty} \Phi(x)\Phi(-x)dx$$

$$=\; 2\int_{0}^{+\infty} \Phi(x)\Phi(-x)dx$$

by integrating first in θ.

c. Show that

$$\int_{0}^{+\infty} \Phi(-x)dx = \int_{0}^{+\infty} y\frac{e^{-y^2}}{\sqrt{2\pi}}dy.$$

d. Conclude about the finiteness of $\tau(\pi)$.

2.35 Consider $x \sim \mathcal{N}_p(\theta, I_p)$. A class of estimators of $||\theta||^2$ is given by

$$\delta_c(x) = ||x||^2 + c, \quad c \in \mathbb{R}.$$

a. Show that, under quadratic loss, δ_{-p} minimizes the risk for every θ among the estimators δ_c. Does this estimation problem have a value?

b. How can we choose $w(\theta)$ so that the risk of δ_{-p} is uniformly bounded for the quadratic loss weighted by $w(\theta)$? Conclude about the minimaxity of δ_{-p}.

c. Show that δ_{-p} is not admissible, and propose an estimator that dominates δ_{-p} uniformly.

2.36 Show that, under squared error loss, if two real estimators δ_1 and δ_2 are distinct and satisfy

$$R(\theta, \delta_1) = (\theta - \delta_1(x))^2 = R(\theta, \delta_2) = (\theta - \delta_2(x))^2,$$

the estimator δ_1 is not admissible. (*Hint:* Consider $\delta_3 = (\delta_1 + \delta_2)/2$ or $\delta_4 = \delta_1^\alpha \delta_2^{1-\alpha}$.) Extend this result to all strictly convex losses and construct a counter-example when the loss function is not convex.

2.37 Let $\Theta = \{\theta_1, \theta_2\}$ and consider the case when the risk set is $\mathcal{R} = \{(r_1, r_2); (r_1 - 2)^2 + (r_2 - 2)^2 < 2, r_1 \leq 2, r_2 \leq 2\}$.

a. Draw $\mathcal{R}$ and deduce whether there exists a minimax point.

b. Exhibit the two admissible rules for this problem.

c. What can be said about the existence of Bayes procedures?

2.38 Two experts have different loss functions described in the following table for $\mathcal{D} = \{d_1, d_2, d_3\}$ and $\Theta = \{\theta_1, \theta_2\}$

L_1/L_2	d_1	d_2	d_3
θ_1	1/1	2.5/1.5	2/2.5
θ_2	1.5/4	2/3.5	3/3

a. Plot the risk sets for both experts and identify minimax and admissible procedures in each case.

b. There are several ways to combine the expert opinions, i.e., to derive a single loss function. For each of the following choices, derive the risk set and the optimal procedures:

(i) $L = (L_1 + L_2)/2$ (ii) $L = \sup(L_1, L_2)$ (iii) $L = \sqrt{L_1 L_2}$.

c. For what choice of L above are the admissible rules admissible for one of the two original losses? When is the risk set convex?

Section 2.5

2.39 *Establish Propositions 2.5.1, 2.5.2, and 2.5.3. Show Shinozaki's lemma (1975): *if δ is admissible for the usual quadratic loss, it is admissible for every quadratic loss.*

2.40 Consider $\pi(\theta) = (1/3)(\mathcal{U}_{[0,1]}(\theta) + \mathcal{U}_{[2,3]}(\theta) + \mathcal{U}_{[4,5]}(\theta))$ and $f(x|\theta) = \theta e^{-\theta x}$. Show that, under the loss (2.5.4), for every x, there exist values of k_1 and k_2 such that the Bayes estimator is not unique.

2.41 Establish Proposition 2.5.7 and show that the loss L considered in Example 2.5.8 is equivalent to estimate $\mathbb{I}_{H_0}(\theta)$ under the *absolute error loss*,

$$L(\theta, \delta) = |\theta - \delta|.$$

Derive the Bayes estimator associated with the quadratic loss.

2.42 *(Zellner (1986a)) Consider the LINEX loss in $\mathbb{R}$, defined by

$$L(\theta, d) = e^{c(\theta - d)} - c(\theta - d) - 1.$$

a. Show that $L(\theta, d) > 0$ and plot this loss as a function of $(\theta - d)$ when $c = 0.1, 0.5, 1, 2$.

b. Give the expression of a Bayes estimator under this loss.

c. If $x_1, \ldots, x_n \sim \mathcal{N}(\theta, 1)$ and $\pi(\theta) = 1$, give the associated Bayes estimator.

2.43 (Berger (1985a)) Consider $x \sim \mathcal{N}(\theta, 1)$, $\theta \sim \mathcal{N}(0, 1)$ and the loss

$$L(\theta, \delta) = e^{3\theta^2/2}(\theta - \delta)^2.$$

a. Show that $\delta^\pi(x) = 2x$.

b. Show that δ^π is uniformly dominated by $\delta_0(x) = x$ and that $r(\pi) = +\infty$.

2.44 Determine the Bayes estimator associated with the absolute error loss in $\mathbb{R}^k$,

$$L(\theta, \delta) = ||\theta - \delta||.$$

2.45 Consider the following questions for the entropic and the Hellinger intrinsic losses.

a. Show that L_e (resp. L_H) is nonnegative, is equal to 0 when $d = \theta$, and determine under which condition $d = \theta$ is the unique solution of $L_e(\theta, d) = 0$ (resp. of $L_H(\theta, d) = 0$).

b. Give the expressions of both losses when $x \sim \mathcal{N}(0, \theta)$ and $x \sim \mathcal{B}e(n, \theta)$.

c. Show that, if $x \sim \mathcal{G}(\alpha, \theta)$ and $\theta \sim \mathcal{G}(\nu, x_0)$, the Bayes estimator of θ under the Hellinger loss is of the form $k/(x_0 + x)$.

2.46 *(Wells (1992)) As mentioned in Section 2.5.4, the Bayes estimators are not invariant under arbitrary reparameterization. In the normal case, $x \sim \mathcal{N}(\theta, 1)$, examine whether the only transformations of θ for which the Bayes estimators are invariant under squared error loss are the affine transformations, $\eta = a\theta + b$. [*Note*: The answer is no.]

2.47 *(Efron (1992)) Derive the Bayes estimators of θ when $\theta|x \sim \mathcal{N}(\mu(x), 1)$ and when the loss function is the *asymmetric squared error loss*,

$$L(\theta, \delta) = \begin{cases} \omega(\theta - \delta)^2 & \text{if } \delta < \theta, \\ (1 - \omega)(\theta - \delta)^2 & \text{otherwise.} \end{cases}$$

2.48 (Robert (1996b)) Show that both the entropic and the Hellinger losses are locally equivalent to the quadratic loss associated with the Fisher information,

$$I(\theta) = \mathbb{E}_\theta \left[\frac{\partial \log f(x|\theta)}{\partial \log} \left(\frac{\partial \log f(x|\theta)}{\partial \log} \right)^t \right],$$

that is,

$$L_e(\theta, \delta) = L_e(\theta - \delta)^t I(\theta)^{-1}(\theta - \delta) + O(\|\theta - \delta\|^2)$$

and

$$L_H(\theta, \delta) = c_H(\theta - \delta)^t I(\theta)^{-1}(\theta - \delta) + O(\|\theta - \delta\|^2),$$

where c_e and c_H are constants.

2.49 Consider $y = x + \epsilon$ with ϵ and x independent random variables and $\mathbb{E}[\epsilon] = 0$.

a. Show that $\mathbb{E}[y|x] = x$.

b. Show that the reverse does not necessarily hold, i.e., that $\mathbb{E}[x|y]$ is not always equal to y. (*Hint:* Consider, for instance, the case when $x \sim p\mathcal{N}(\theta_1, 1) + (1 - p)\mathcal{N}(\theta_2, 1)$ and $\epsilon \sim \mathcal{N}(0, 1)$.)

Section 2.6

2.50 Show that, for the universal distributions of Rukhin (1978), the Bayes estimators are indeed independent of the loss function. In the particular case when $x \sim \mathcal{G}(\nu, 1/\nu)$ identify θ, $A_1(x)$, $A_2(x)$, and the universal prior $\pi(\theta)$.

Note 2.8.1

2.51 Show that that the Bayes estimator associated with the L^0 loss function is the MAP estimator.

2.52 Show that that the Bayes estimator associated with the L^1 loss function is the componentwise MAP estimator.

2.53 If $\mathcal{D}$ is a subset of $\{1, \ldots, N\}$, $\mathbf{e}$ denotes a misclassification vector and $m_{\mathcal{D}}$ the corresponding number of misclassifications.

a. Show that $p(m_{\mathcal{D}})$ can also be written as

$$p(m_{\mathcal{D}}) = 1 - \prod_{i \in \mathcal{D}} (1 - e_i).$$

b. Let $q(m_{\mathcal{D}})$ to be 1 if and only if $m_{\mathcal{D}} = |\mathcal{D}|$. Show that

$$q(m_{\mathcal{D}}) = \prod_{i \in \mathcal{D}} e_i.$$

c. Show that

$$p(m_{\mathcal{D}}) = \sum_{k=1}^{|\mathcal{D}|} (-1)^{k+1} \sum_{\omega \in \mathcal{P}_k(\mathcal{D})} q(m_\omega).$$

Note 2.8.2

2.54 Show that the Stein paradox cannot occur when δ_0 is a proper Bayes estimator, whatever the dimension p is. [*Note:* Brown (1971) shows that there are also generalized Bayes estimators that enjoy this property.]

2.55 Show that the majorizing constant in Theorem 2.8.1 can be replaced by

$$c = 2\frac{q - 2\alpha}{p - q + 4\beta}.$$

(*Hint:* Bound $h^2(t, u)$ by $c(u/t)h(t, u)$ first.) Compare the two bounds.

2.56 *(Stein (1973)) Establish Stein's lemma: *If $x \sim \mathcal{N}(\theta, 1)$ and f is continuous and a.e. differentiable, then*

$$\mathbb{E}_\theta[(x - \theta)f(x)] = \mathbb{E}_\theta[f'(x)].$$

Deduce that, if $x \sim \mathcal{N}_p(\theta, \Sigma)$, $\delta(x) = x + \Sigma\gamma(x)$, and $L(\theta, \delta) = (\delta - \theta)^t Q(\delta - \theta)$, with γ differentiable, then

$$R(\theta, \delta) = \mathbb{E}_\theta \left[\text{tr}(Q\Sigma) + 2\,\text{tr}(J_\gamma(x)Q^*) + \gamma(x)^t Q^* \gamma(x) \right],$$

where $\text{tr}(A)$ is the trace of A, $Q^* = \Sigma Q\Sigma$ and $J_\gamma(x)$ is the matrix with generic element $\frac{\partial}{\partial x_i}\gamma_j(x)$. [*Note:* This representation of the risk leads to the technique *of unbiased estimation of the risk*, which is quite influential in the derivation of sufficient conditions of domination of usual estimators. See Berger (1985a) and Johnstone (1988).]

Note 2.8.3

The following exercises (2.57–2.62) consider the Pitman closeness criterion. An estimator δ_1 of θ is said to Pitman-dominate an estimator δ_2, denoted $\delta_1 \overset{P}{\succ} \delta_2$, if, for every $\theta \in \Theta$,

$$P_\theta(|\delta_1(X) - \theta| < |\delta_2(X) - \theta|) > 0.5.$$

The notion of Pitman-admissibility follows directly.

2.57 *Consider a median unbiased estimator δ^M, i.e., δ^M such that

$$\forall \theta, \qquad P_\theta(\delta^M(x) \le \theta) = 0.5.$$

a. Show that δ^M is the best estimator (for the Pitman criterion) among the linear estimators $\delta^M(x) + K$, $K \in \mathbb{R}$.

b. If $\theta > 0$ and $\delta^M > 0$, show that δ^M is also the best estimator (for the Pitman criterion) among the estimators $K\delta^M$, $K > 0$.

2.58 *Consider $X = \theta U$, $\theta > 0$, $U \sim \mathcal{U}(-0.9, 1.1)$. Show that

$$X \overset{P}{\succ} 0.9|X| \overset{P}{\succ} 3.2|X| \overset{P}{\succ} X .$$

2.59 *(Robert et al. (1993)) Consider $X \sim f(x - \theta)$, with

$$\int_{-\infty}^0 f(u)\, du = 1/2$$

and $f(0) > 0$. If F is the c.d.f. of X for $\theta = 0$, the function $\epsilon(\theta)$ is defined by

$$F(-\theta) = \begin{cases} P_0(0 < X < \epsilon(\theta)) & \text{if } \theta > 0, \\ 1 - P_0(0 > X > -\epsilon(\theta)) & \text{if } \theta < 0, \end{cases}$$

and $\epsilon(0) = 0$. Consider

$$\theta_1 = \text{Arg}\{\min_{\theta>0} |\theta + \epsilon(\theta)|\}, \quad \theta_2 = \text{Arg}\{\min_{\theta<0} |\theta - \epsilon(\theta)|\}.$$

The truncated version of ϵ is defined by

$$\epsilon^*(\theta) = \begin{cases} \epsilon(\theta) & \text{if } \theta > \theta_1 \text{ or } \theta < \theta_2 \\ \theta_1 + \epsilon(\theta_1) - \theta & \text{if } 0 < \theta < \theta_1 \\ \theta + \epsilon(\theta_2) - \theta_2 & \text{if } 0 > \theta > \theta_2. \end{cases}$$

The set A satisfies

$$(x, \theta) \in A \quad \text{if and only if} \quad 0 < x \le \theta + \epsilon^*(\theta)$$

for $\theta > 0$, and

$$(x, \theta) \in A \quad \text{if and only if} \quad \theta - \epsilon^*(\theta) \le x < 0.$$

for $\theta < 0$.

a. Justify the truncation of ϵ in terms of A and represent A in a special case when the derivation of ϵ^* is manageable.

b. Show that, if $\delta(x)$ is a nondecreasing function such that $(x, \delta(x)) \in A$, then $\delta \overset{P}{\succ} \delta_0(x) = x$.

c. Show that, if $F(c) - F(-c) = 1/2$, every estimator δ such that

$$(2.7.1) \qquad\qquad \delta(x) = 0 \quad \text{when} \quad |x| < c$$

is Pitman-admissible.

d. When δ is monotone, satisfies (2.7.1), and belongs to A, show that δ is Pitman-admissible and Pitman-dominates δ_0. Show that

$$c < \theta_1 + \epsilon(\theta_1) \quad \text{and} \quad -c > \theta_2 - \epsilon(\theta_2)$$

and conclude about the existence of such estimators.

2.60 Consider a couple of random variables (x, y) with joint c.d.f.

$$F_\alpha(x, y) = \frac{xy}{1 + \alpha(1 - x)(1 - y)} \mathbb{I}_{[0,1]^2}(x, y).$$

a. Show that F_α is indeed a c.d.f. and deduce the density $f_\alpha(x, y)$.

b. Give the marginal distribution of x and y.

c. Suppose two estimators δ_1 and δ_2 are distributed according to $\theta^{-2} f_\alpha(\delta_1/\theta, \delta_2/\theta)$. What can be said about the Pitman closeness to θ? (*Hint:* Compute $P(|\delta_1 - \theta| < |\delta_2 - \theta|)$.)

2.61 *Show that, if $X_1, X_2 \sim f(x|\theta)$, $\overline{X} \overset{P}{\succ} X_1$. Apply this result to the case of the Cauchy distribution. Show that, for every real η, $\overline{X}$ is Pitman-closer to η than X_1, even if η is arbitrary. [*Note:* This property is not specific to Pitman closeness, since it is also satisfied by the quadratic loss.]

2.62 *(Robert et al. (1993)) Show (or use the result) that, if $\chi^2_\alpha(p, \lambda)$ is the α-quantile of a noncentral chi-squared distribution, $\chi^2_p(\lambda)$, it satisfies

$$p - 1 + \lambda \le \chi^2_{0.5}(p, \lambda) \le \chi^2_{0.5}(p, 0) + \lambda.$$

a. Deduce from this inequality that the James–Stein estimators

$$\delta_h(x) = \left(1 - \frac{h(x)}{||x||^2}\right) x$$

Pitman-dominate δ_0 when $x \sim \mathcal{N}(\theta, I_p)$ and

$$0 < h(x) \leq 2(p-1).$$

b. Show that this condition is also necessary when h is constant.

2.8 Notes

2.8.1 Loss functions for imaging

An image, as represented on a computer screen, is a two dimensional structure $\mathbf{x}$ made of pixels of different colors (or grey levels, for black and white images). Images are often observed with some noise, which can be resulting from imperfections in the imaging device, like a camera out of focus, to perturbations in the transmission process, like parasites on a telephone line, or to defects in the image itself, like clouds for a satellite image. Among other things, Bayesian imaging aims at reconstructing the true image.

The observed image, $\mathbf{x}$, can also be written as a vector $(x_1 \ldots, x_N)$, each x_i taking values in $\{0, 1, \ldots, C-1\}$, which is the set of colors. The true image is denoted θ and $\mathbf{x}$ is distributed as $\mathbf{x} \sim f(\mathbf{x}|\theta)$.

The most rudimentary loss function in this setting is the "0 1" dichotomous loss function, $L^0(\theta, \delta) = 0$ if $\theta = \delta$ and $L^0(\theta, \delta) = 1$ otherwise. Given a prior $\pi(\theta)$, the Bayes estimator δ^π associated with the 0–1 loss function is the image that maximizes the posterior distribution $\pi(\theta|\mathbf{x})$, the so-called *MAP estimator*. As pointed out by Rue (1995), since the loss function is extremely sensitive to misclassifications, this loss leads to oversmoothing, omitting small structures which are important in settings like pattern recognition.

The other standard loss function is the *misclassification rate*, that is, the number of misclassifications, based on the misclassification vector $\mathbf{e}$, which is defined, for an estimator δ and a true image θ, as $e_i = \mathbb{I}_{\delta_i \neq \theta_i}$ $(i = 1, \ldots, N)$. The number of misclassifications is thus

$$L^1(\theta, \delta) = \sum_{i=1}^{N} e_i.$$

Given the additive structure of this loss, the posterior loss is the sum of the site losses $\mathbb{E}[e_i|\mathbf{x}]$ and the Bayes estimator is therefore the vector of the marginal MAP estimators. The drawback of this loss function is thus the opposite of the 0–1 loss function: the estimation is too local and does not take into account interactions between neighboring sites.

Rue (1995) introduces a family of new loss functions for the construction of estimators in Bayesian imaging, in order to account for different features of the image. If $\mathcal{D}$ is a subset of $\{1, \ldots, N\}$, $m_{\mathcal{D}}$ denotes the number of misclassifications in $\mathcal{D}$,

$$m_{\mathcal{D}} = \sum_{i \in \mathcal{D}} e_i,$$

$p(m_{\mathcal{D}})$ is 0 if $m_{\mathcal{D}} = 0$ and 1 otherwise, $R_\phi \mathcal{D}$ is the set $\mathcal{D}$ rotated by an angle $\phi \in \{0, \pm\pi/2, \pi\}$ and $T_s \mathcal{D}$ is the set $\mathcal{D}$ translated by s (in its two-dimensional representation). If $\mathcal{P}_j(\mathcal{D})$ denotes the set of subsets of $\mathcal{D}$ of size j, the loss functions are constructed by selecting (i) a set of basic subsets of $\{1, \ldots, N\}$, $\mathcal{B}_1, \ldots, \mathcal{B}_n$, and (ii) penalty coefficients t_{ij}, such that the penalty associated with a region $\mathcal{B}_i$ is

$$P_i(m_{\mathcal{B}_i}) = \sum_{i=1}^{|\mathcal{B}_i|} t_{i,j} \sum_{\omega \in \mathcal{P}_j(\mathcal{B}_i)} p(m_\omega) \,.$$

The loss function is then

(2.8.1) $$\mathrm{L}(\theta, \delta) = \sum_{i=1}^{n} \sum_{s,\phi} P_i(m_{T_s R_\phi \mathcal{B}_i}) \,,$$

where the inner sum is restricted to the couples (s, ϕ) such that $T_s R_\phi \mathcal{B}_i$ is a subset of $\{1, \ldots, N\}$, that is, is inside the original image.

The motivation for using such a combination becomes clearer when, as in Rue (1995), $n = 1$, $\mathcal{B}_1$ is the 2×2 region made of the four neighbors of an arbitrary point. In this particular case, Rue (1995) propose to take $t_{1,1} = 1$ to penalize misclassifications at one site and to choose a $t_{1,2} > 0$ to penalize further cases when two neighbor sites simultaneously misclassify, while $t_{1,3} = t_{1,4} = 0$. The resulting loss is then the number of misclassified sites in the image, plus $t_{1,2}$ times the number of couples of simultaneously misclassified neighbors.

As detailed in Rue (1995), other examples include minimal resolution problems, pattern recognition and Ising models. For instance, the basic subsets $\mathcal{B}_i$ may include particular shapes of interest, like cars for traffic control vision or tumors in radiological image-processing. Obviously, the computation of the Bayes estimate associated with (2.8.1) is not so straightforward as with L^0 and L^1, and Rue (1995) proposes an iterative method based on a Markov chain (Chapter 6).

2.8.2 The Stein effect

If there is a unique minimax estimator, this estimator is admissible, according to Proposition 2.4.20. Conversely, if a minimax estimator δ_0 is inadmissible, there are other minimax estimators that improve upon δ_0 (under some minor regularity conditions, see Brown (1976)). In particular, if the constant risk minimax estimator is inadmissible, this is the worst minimax estimator in the sense that every other minimax estimator has a uniformly smaller risk. Until 1955, it was assumed that the least-squares estimator, $\delta_0(x) = x$, when $x \sim \mathcal{N}_p(\theta, I_p)$, was admissible and, since its risk is constant, that it was the unique minimax estimator. Stein (1955a) showed that this result only holds for $p = 1, 2$ and hence discovered "the Stein effect" phenomenon, that is, the exhibition of apparently paradoxical domination results for usual estimators.

Formally, the Stein paradox is as follows. If a standard estimator $\delta^*(x) = (\delta_0(x_1), \ldots, \delta_0(x_p))$ is evaluated under weighted quadratic loss

(2.8.2) $$\sum_{i=1}^{p} \omega_i (\delta_i - \theta_i)^2,$$

with $\omega_i > 0$ $(i = 1, \ldots, p)$, there exists p_0 such that δ^* is not admissible for $p \geq p_0$, although the components $\delta_0(x_i)$ are separately admissible to estimate the θ_i's. The Stein effect can be explained through the use of the joint loss (2.8.2), that allows the dominating estimator to borrow strength from the other components, even when they are independent and deal with totally different estimation problems. The literature on the Stein effect and related phenomena is now too extensive for us to give here a comprehensive covering of the results in this field. We refer the reader to Judge and Bock (1978), Lehmann (1983), and Berger (1985a) for a more detailed bibliography and we develop in Chapter 10 a Bayesian analysis of the Stein effect. This note briefly presents the main results about the Stein effect from a frequentist point of view.

First, while Stein's (1955a) proof of inadmissibility was nonconstructive, James and Stein (1961) exhibited an estimator that uniformly dominates $\delta_0(x) = x$ under quadratic loss for $p \geq 3$ in the normal case, i.e., such that, for every θ,

$$p = \mathbb{E}_\theta[||\delta_0(x) - \theta||^2] > \mathbb{E}_\theta[||\delta^{JS}(x) - \theta||^2].$$

This estimator,

(2.8.3) $$\delta^{JS}(x) = \left(1 - \frac{p-2}{||x||^2}\right)x,$$

is now called the *James–Stein* estimator. Note the strange behavior of δ^{JS} when x gets near 0. The factor $1 - \frac{p-2}{||x||^2}$ becomes negative and even goes to $-\infty$ as $||x||$ goes to 0. However, δ^{JS} still dominates δ_0 for all θ's. (This is a consequence of Theorem 2.8.1 below.) Baranchick (1970) corrected this paradoxical behavior by showing that the *truncated* estimators ($p - 2 \leq c \leq 2(p-2)$)

$$\delta_c^+(x) = \left(1 - \frac{c}{||x||^2}\right)^+ x$$

(2.8.4) $$= \begin{cases} (1 - \frac{c}{||x||^2})x & \text{if } ||x||^2 > c, \\ 0 & \text{otherwise,} \end{cases}$$

uniformly dominate their nontruncated counterparts and, in particular, that δ_{p-2}^+ was improving on δ^{JS}. They are, moreover, noncomparable (as c varies). This class of estimators is important because, although made of inadmissible estimators (see Chapter 8), estimators that dominate the truncated James–Stein estimator are quite difficult to derive and do not bring significant improvement in terms of risk (see Shao and Strawderman (1996)). On the contrary, the truncated (or *positive-part*) James–Stein estimators improve quite significantly on the least-squares estimator, as illustrated in Figure 2.4.2 for $p = 10$ and $c = 2p - 1$.

Following James and Stein (1961), more general classes of estimators dominating δ_0 have been proposed by Alam (1973), Berger and Bock (1976), Judge and Bock (1978), Stein (1981), George (1986a,b), and Brandwein et al. (1992). These estimators are called *shrinkage estimators* because, as (2.8.3) and (2.8.4), they shrink x toward 0. Stein effects have also been exhibited for distributions other than the normal distribution and losses other than the quadratic loss by Berger (1975), Brandwein and Strawderman (1980), Hwang (1982a), Ghosh et al. (1983), Bock (1985), Haff and Johnston (1987), Srivastava and Bilodeau (1988), Brandwein and Strawderman (1990). Some restrictions on the classes of shrinkage estimators have been proposed, in order to integrate the admissibility requirement (Brown (1971), Alam (1973),

Strawderman (1974), Brown (1975), Berger and Srinivasan (1978), Brown and Hwang (1982), Das Gupta and Sinha (1986), Brown (1988), and Fraisse et al. (1998)). Bondar (1987) has shown that the improvement (in terms of risk) brought by the shrinkage estimators is only significant on a limited part of the parameter space, but George (1986a,b) introduced the concept of *multiple shrinkage* estimators to extend the region where the improvement occurs (see Exercise 10.38).

The Stein effect is also *robust* in the sense that it depends mainly on the loss function, rather than on the exact distribution of the observations, as shown by Brown (1975), Shinozaki (1980, 1984), Berger (1980a,b), Das Gupta (1984), Bilodeau (1988), Cellier, Fourdrinier and Robert (1989), Brandwein and Strawderman (1990) or Kubokawa, Robert and Saleh (1991, 1992, 1993). It is not limited to point estimation, but also occurs for confidence regions (Stein (1962a), Hwang and Casella (1982, 1984), Casella and Hwang (1983, 1987), Robert and Casella (1990), Hwang and Ullah (1994)) and in accuracy (or loss) estimation (Johnstone (1988), Rukhin (1988a,b), Lu and Berger (1989a,b), Robert and Casella (1993), Fourdrinier and Wells (1994)), George and Casella (1994). However, Gutmann (1982) established that the Stein effect cannot occur in finite parameter spaces. Brown (1971) (see also Srinivasan (1981), Johnstone (1984), and Eaton (1992)) showed that admissibility is related to the recurrence of a stochastic process associated with the estimator and Brown (1980) shows the surprising result (called *Berger's phenomenon*, from Berger (1980b)) that there always exists a loss function such that the *boundary* between admissibility and inadmissibility for the usual estimator is an *arbitrary dimension* p_0.

This overview does not do justice to the richness of the work on the Stein effect. The advances realized in this field in the last thirty years have greatly benefitted Decision Theory in general, and Bayesian Decision Theory in particular. In fact, one of the major impacts of the Stein paradox has been to signify the end of a Golden Age for classical Statistics, since it shows that the quest for *the* best estimator, that is, the unique minimax admissible estimator, is hopeless, unless one restricts the class of estimators to be considered, or incorporates some prior information. The works on the Stein effect have thus led to the progressive abandonment of *unbiasedness*, to a deeper understanding of minimaxity and admissibility, and to the improvement of frequentist techniques of risk computation (with Stein's (1973) idea of *unbiased estimator of the risk*). However, its main consequence has been to reinforce the Bayesian-frequentist interface,[5] by inducing frequentists to call for Bayesian techniques (see, for instance, Bock's (1988) idea of *pseudo-Bayes estimators*) and Bayesians to robustify their estimators in terms of frequentist performances and prior uncertainty (Berger (1980b, 1982a, 1984a), George (1986a,b), Lu and Berger (1989a,b), Berger and Robert (1990)). We refer the reader to the books mentioned above as well as to Brandwein and Strawderman (1990) and Lehmann and Casella (1998) for additional references.

We conclude this note with the proof of the inadmissibility of $\delta_0(x) = x$ in the estimation of θ for *spherically symmetric* distributions, i.e., distributions with densities $f(||x - \theta||)$ in $\mathbb{R}^p$ ($p \geq 3$). References on these distributions which

---

[5] A typical example is provided by the development of the *empirical Bayes* techniques (see Chapter 10).

generalize the normal distribution in linear regression models are given in Kelker (1970), Eaton (1986) and Fang and Anderson (1990) (see also Exercise 1.1). This result was first established in Cellier et al. (1989).

Theorem 2.8.1 *Consider* $z = (x^t, y^t)^t \in \mathbb{R}^p$, *with distribution*

$$(2.8.5) \qquad\qquad z \sim f(||x - \theta||^2 + ||y||^2),$$

and $x \in \mathbb{R}^q$, $y \in \mathbb{R}^{p-q}$. *An estimator*

$$\delta_h(z) = (1 - h(||x||^2, ||y||^2))x$$

dominates δ_0 *under quadratic loss if there exist* $\alpha, \beta > 0$ *such that:*

(1) $t^\alpha h(t, u)$ *is a nondecreasing function of* t *for every* u;

(2) $u^{-\beta} h(t, u)$ *is a nonincreasing function of* u *for every* t; *and*

(3) $0 \leq (t/u)h(t, u) \leq \dfrac{2(q - 2)\alpha}{p - q - 2 + 4\beta}$.

The above conditions on h are thus independent of f in (2.8.5), which does not need to be known, and, moreover, they are identical to those obtained in the normal case (see Brown (1975)). The occurrence of the Stein effect is then robust in the class of spherically symmetric distributions with finite quadratic risk.

Proof. Conditions (1) and (2) imply

$$\begin{cases} t\frac{\partial}{\partial t}h(t, u) \geq -\alpha h(t, u), \\ u\frac{\partial}{\partial u}h(t, u) \leq \beta h(t, u). \end{cases}$$

The risk of δ_h can be developed as follows:

$$\begin{aligned} R(\theta, \delta_h) &= \mathbb{E}_\theta\left[\sum_{i=1}^q \left\{x_i - \theta_i - h(||x||^2, ||y||^2)x_i\right\}^2\right] \\ &= \mathbb{E}_\theta\left[\sum_{i=1}^q (x_i - \theta_i)^2\right] - 2\mathbb{E}_\theta\left[\sum_{i=1}^q h(||x||^2, ||y||^2)x_i(x_i - \theta_i)\right] \\ &\quad + \mathbb{E}_\theta\left[h^2(||x||^2, ||y||^2)||x||^2\right]. \end{aligned}$$

An integration by parts shows that

$$\begin{aligned} &\int_{-\infty}^{+\infty} h(||x||^2, ||y||^2)x_i(x_i - \theta_i)f(||x - \theta||^2 + ||y||^2)\, dx_i \\ &= \int_{-\infty}^{+\infty} \frac{\partial}{\partial x_i}\left[h(||x||^2, ||y||^2)x_i\right]\bar{F}(||x - \theta||^2 + ||y||^2)\, dx_i, \end{aligned}$$

with

$$\bar{F}(t) = \int_t^{+\infty} f(u)du.$$

Therefore,

$$\begin{aligned} &\mathbb{E}_\theta\left[\sum_{i=1}^q h(||x||^2, ||y||^2)x_i(x_i - \theta_i)\right] \\ &= \int_{\mathbb{R}^p} \left[qh(||x||^2, ||y||^2) + 2h_1'(||x||^2, ||y||^2)||x||^2\right]\bar{F}(||x - \theta||^2 + ||y||^2)\, dz, \end{aligned}$$

where $h'_1(t, u) = \frac{\partial}{\partial t} h(t, u)$. Similarly,

$$\mathbb{E}_\theta[h^2(||x||^2, ||y||^2)||x||^2] = \mathbb{E}_\theta\left[\frac{||x||^2}{||y||^2} h^2(||x||^2, ||y||^2)||y||^2\right]$$

$$= \int_{\mathbb{R}^p} ||x||^2 \sum_{j=1}^{p-q} \frac{\partial}{\partial y_j}\left(h^2(||x||^2, ||y||^2)\frac{y_j}{||y||^2}\right) \bar{F}(||x - \theta||^2 + ||y||^2)\, dz$$

$$= \int_{\mathbb{R}^p} ||x||^2 \left[4h(||x||^2, ||y||^2)h'_2(||x||^2, ||y||^2)||x||^2\right.$$

$$= + (p - q - 2)h^2(||x||^2, ||y||^2)\frac{1}{||y||^2}\left.\right] \bar{F}(||x - \theta||^2 + ||y||^2)\, dz,$$

where $h'_2(t, u) = \frac{\partial}{\partial u} h(t, u)$. The difference of the risks is then

$$R(\theta, \delta_0) - R(\theta, \delta_h) =$$

$$\int_{\mathbb{R}^p} \left\{2\left[qh(||x||^2, ||y||^2) + 2h'_1(||x||^2, ||y||^2)||x||^2\right]||x||^2 h(||x||^2, ||y||^2)\right.$$

$$\left[4h'_2(||x||^2, ||y||^2) - (p - q - 2)h(||x||^2, ||y||^2)\frac{1}{||y||^2}\right]\Big\}$$

$$\times \bar{F}(||x - \theta||^2 + ||y||^2)\, dz$$

$$\geq \int_{\mathbb{R}^p} h(||x||^2, ||y||^2)\left[-h(||x||^2, ||y||^2)\frac{||x||^2}{||y||^2}(p - q - 2 + 4\beta)\right.$$

$$\left. + 2(q - 2\alpha)\right] \bar{F}(||x - \theta||^2 + ||y||^2)\, dz > 0,$$

which concludes the proof.   □□

Notice that this domination result includes as a particular case the estimation of a normal mean vector when the variance is known up to a multiplicative factor, i.e., the problem originally considered in James and Stein (1961). When $h(t, u) = au/t$, the bound on a is $2(q - 2)/(p - q + 2)$, as obtained in James and Stein (1961).

2.8.3 Pitman closeness

An alternative to standard Decision Theory has been proposed by Pitman (1937). In order to compare two estimators of θ, δ_1 and δ_2, he indeed suggested the comparison of the distribution of their distance (or *closeness*) to θ, i.e.,

$$P_\theta\left(||\delta_1(x) - \theta|| \leq ||\delta_2(x) - \theta||\right).$$

If this probability is uniformly larger than 0.5, δ_1 is said *to dominate δ_2 in Pitman's sense*, with the implicit message that δ_1 should be preferred to δ_2 in this case. Even though formally close to stochastic domination, this criterion, called *Pitman closeness*, exhibits important flaws and we suggest it not be used as a comparison criterion. Nonetheless, the literature on the subject is quite important (see, e.g., Blyth (1972), Rao (1980, 1981), Blyth and Pathak (1985), Rao et al. (1986), Keating and Mason (1985), Peddada and Khattree (1986), Sen et al. (1989), Ghosh and Sen (1989)). These papers study the properties of Pitman closeness and stress its *intrinsic aspect*, since it involves the whole distribution of $||\delta_1(x) - \theta||$ (as opposed to the reductive evaluation through a loss, like the quadratic loss). On the other hand, Robert et al. (1993) expose the fundamental drawbacks of this criterion. We present here two characteristic points (see Exercises 2.57–2.62 for other illustrations).

A first major criticism of Pitman closeness deals with its *nontransitivity*. Indeed, it does not provide a mean of selecting an optimal estimator, or even to order estimators. Pitman (1937) already pointed out this defect, but some proponents of the criterion (see, e.g., Blyth (1993)) paradoxically assert that the property is an additional advantage of the criterion, since it better reflects the complexity of the world. As discussed above, it may indeed happen that realistic orderings of preferences are not always transitive. But, apart from the utmost necessity for reducing this complexity, note that the Pitman closeness criterion is advocated as a comparison criterion, an alternative to regular loss functions: when nontransitivity occurs, the ordering derived from the criterion is not absolute since, as the following example shows, there is always a chance for a preference cycle. In such cases, the criterion cannot provide a selection of a best estimator.

Example 2.8.2 Consider $U \sim \mathcal{U}_{[-0.9,1.1]}$ and $x = \theta U$. Then, it can be shown that, under the Pitman closeness criterion, $\delta_0(x) = x$ dominates $\delta_1(x) = 0.9|x|$, δ_1 dominates $\delta_2(x) = 3.2|x|$, and δ_2 dominates δ_0. If one of the three estimators δ_0, δ_1, and δ_2 has to be selected, the criterion is of no help.   ‖

Obviously, nontransitivity prevents the Pitman criterion from being equivalent to a loss; therefore, it does not pertain to Decision Theory. For the same reason, it cannot be equivalent to stochastic domination. Actually, Blyth and Pathak (1985) provide an example where the two criteria give opposed ordering. It is also impossible to define a (decision-theoretic) Bayes estimator for the Pitman criterion (although a posterior Pitman estimator may exist. See Bose (1991) and Ghosh et al. (1993).)

A second major defect of Pitman closeness is that it may exclude some classical estimators although these are admissible under most losses. For instance, Efron (1975) noticed that it is possible to dominate $\delta_0(x) = x$ for the Pitman closeness in the normal case, $x \sim \mathcal{N}(\theta, 1)$. Robert et al. (1993) show that there is a Stein effect for $\mathcal{N}_p(\theta, I_p)$ $(p \geq 2)$, and that the dominating condition only involves an upper bound on the shrinkage function h (see also Sen et al. (1989) and Exercise 2.62). The following result extends Efron (1975) to the general case when $x \sim f(x - \theta)$ and θ is the median of the distribution (see Exercise 2.59 for a proof).

Proposition 2.8.3 *Under the above conditions, the estimator $\delta_0(x) = x$ is inadmissible for the Pitman criterion.*

Moreover, the dominating estimators may exhibit strange behaviors, such as being 0 on large parts of the sample space (see Exercise 2.59).

These multiple drawbacks seem to indicate clearly that the Pitman closeness is not a viable alternative to Decision Theory. On the other hand, the failure of this substitution reinforces our belief that Decision Theory is the proper setting to take decisions under uncertainty.

As stressed in the Introduction, determination of the loss is an important step in the derivation of the model. This requirement is too often bypassed by resorting to classical losses, and it would be interesting to consider loss robustness analyses, similar to those conducted about the influence of the prior distribution (see Section 3.5). However, the difficulty of the task at hand is not enough to justify the abandonment of the coherence inherent to Decision Theory for exotic criteria like the Pitman closeness.

CHAPTER 3

From Prior Information
to Prior Distributions

In the meantime, there was so much information to gather, so many puzzles to solve. Their house was the perfect place for Moraine to find the information she needed. Except that it was not there.

Robert Jordan, *The Great Hunt, Book II of the Wheel of Time.*

3.1 The difficulty in selecting a prior distribution

Undoubtedly, the most critical and most criticized point of Bayesian analysis deals with the choice of the prior distribution, since, once this prior distribution is known, inference can be led in an almost mechanic way by minimizing posterior losses, computing higher posterior density regions, or integrating out parameters to find the predictive distribution. The prior distribution is the key to Bayesian inference and its determination is therefore the most important step in drawing this inference. To some extent, it is also the most difficult. Indeed, in practice, it seldom occurs that the available prior information is precise enough to lead to an exact determination of the prior distribution, in the sense that many probability distributions are compatible with this information. There are many reasons for this: the decision-maker, the client or the statistician do not necessarily have the time or resources (nor often the willingness) to hunt for an exact prior (which, anyway, may not exist given the information at hand) and they have to complete the partial information they gathered with a subjective input to build a prior distribution.

Most often, it is then necessary to make a (partly) arbitrary choice of the prior distribution, which can drastically alter the subsequent inference. In particular, the systematic use of parametrized distributions (like the normal, gamma, beta, etc., distributions) and the further reduction to *conjugate* distributions (defined below in Section 3.3) cannot be justified at all times, since they trade an improvement in the analytical treatment of the problem for the subjective determination of the prior distribution and may therefore ignore part of the prior information. Some settings nonetheless call for a partly automated determination of the prior distribution as, for instance, when prior information is totally lacking. We will consider two common techniques, the conjugate prior approach (Section 3.3), which requires a limited amount of information, and the noninformative approach (Section 3.5), which can be directly derived from the sampling distribution.

Historically, critics of the Bayesian paradigm have focused their criticisms on the choice of the prior distribution, starting with Laplace's as, while Bayes was able to justify his prior modeling on the billiard balls by a physical reasoning (see Section 1.2), the abstract modelings of Laplace on the distribution of white balls in an urn (Example 1.2.3), or on the proportion of boys (Example 1.2.4), *both based on a principle of insufficient reason*, were more apt to give an opening for criticisms which indeed appeared soon after (see Boole (1854), Venn (1866), Bertrand (1889), and Chrystal (1891)).

These attacks against the Bayesian approach had some validity in the sense that they pointed out that there is no unique way of choosing a prior distribution, and that the choice of the prior distribution has an influence on the resulting inference. This influence can be negligible, moderate or enormous: it is always possible to choose a prior distribution that gives the answer one wishes. The main point here is that, first, ungrounded prior distributions produce unjustified posterior inference and, second, that there is no such thing as *the* prior distribution, except for very special settings. After years of criticism, the work of Jeffreys (1946) on noninformative priors thus came as a blessing for the Bayesian community because it describes a method to derive the prior distribution directly from the sampling distribution, although some Bayesians disagree with the use of such automated methods (see, e.g., Lindley (1971, 1990)). More recently, theoretical developments on *robustness* and *sensitivity analysis* also provided a sounder basis for Bayesian analysis when it is faced with incomplete prior information, while the introduction of hierarchical modeling (Chapter 10) allowed to push the prior selection to higher levels, with an observed decrease in the influence on the resulting inference.

3.2 Subjective determination and approximations

3.2.1 Existence

Unless the decision-maker (or statistician) is informed about the (physical, economical, biological, etc.) mechanism underlying the generation of the parameter θ, it is generally quite difficult to propose an exact, or even a parametrized, form for the prior distribution on θ. Indeed, in most cases, θ does not have an (intrinsic) existence of its own, but rather corresponds to an indexing of the distributions describing the random phenomenon of

Table 3.2.1. *Capture and survival parameter prior information for different capture times and capture sites. (Source: Dupuis (1995).)*

Time	2	3	4	5	6
Mean	0.3	0.4	0.5	0.2	0.2
95% cred. int.	[0.1,0.5]	[0.2,0.6]	[0.3,0.7]	[0.05,0.4]	[0.05,0.4]

Site	A		B	
Time	$t = 1, 3, 5$	$t = 2, 4$	$t = 1, 3, 5$	$t = 2, 4$
Mean	0.7	0.65	0.7	0.7
95% cred. int.	[0.4,0.95]	[0.35,0.9]	[0.4,0.95]	[0.4,0.95]

interest. The prior π is then a tool summarizing the available information on this phenomenon, as well as the uncertainty related with this information. Such settings obviously imply approximations of the true prior distribution—if there is such a thing! In fact, as discussed earlier in Chapter 1, statistical models are most often reductive representations of these random phenomena and, since there is no true model—but only the model closest to the phenomenon for some appropriate distance— it is conceptually difficult to speak of the true value of θ and, a fortiori, of a true prior.

Example 3.2.1 (Dupuis (1995)) In a capture-recapture experiment (see Section 4.3.3) on lizards, biologists are interested in the migrations of lizards between zones of their habitat (located in a mountainous area in the south of France). After discussions with the biologists, the information available on capture and survival probabilities, p_t and q_{it} respectively, where t and i are indices for the various times and zones respectively, is represented in Table 3.2.1, as prior mean and prior 95% confidence intervals on these probabilities. Many prior distributions are compatible with this prior information, obviously, but, since the beta distribution $\mathcal{Be}(\alpha, \beta)$ can be characterized by its mean and a confidence interval (see Exercise 3.1), the statistician chose to use the beta prior distributions given in Table 3.2.1. ‖

Example 3.2.2 A decision-maker wants to model the distributions on both the observations and the parameter as normal distributions: $x_1, \ldots, x_n \sim \mathcal{N}(\theta, 1)$ and $\theta \sim \mathcal{N}(\mu, \tau)$. Since the posterior mean of θ is

$$\delta^{\pi}(x_1, \ldots, x_n) = \frac{\bar{x}\tau + \mu/n}{\tau + 1/n},$$

the hyperparameter τ^{-1} behaves like the sample size, n, while μ behaves like the sample average, $\bar{x}$. Therefore, these hyperparameters can be approximated by deriving a sample-equivalent of the amount of information brought through (μ, τ), for instance, by considering that the (known) mean μ is the average of a *virtual sample of size* $1/\tau$.   ‖

Table 3.2.2. *Capture and survival prior modeling corresponding to the information of Table 3.2.1 (Source: Dupuis (1995).)*

Time	2	3	4	5	6
Dist.	$\mathcal{B}e(6, 14)$	$\mathcal{B}e(8, 12)$	$\mathcal{B}e(12, 12)$	$\mathcal{B}e(3.5, 14)$	$\mathcal{B}e(3.5, 14)$

Site	A		B	
Time	t=1,3,5	t=2,4	t=1,3,5	t=2,4
Dist.	$\mathcal{B}e(6.0, 2.5)$	$\mathcal{B}e(6.5, 3.5)$	$\mathcal{B}e(6.0, 2.5)$	$\mathcal{B}e(6.0, 2.5)$

From a formal point of view, it is possible to build up a prior distribution the way utility functions were constructed in the previous chapter, by determining a scale of the respective likelihoods of the values of the parameter θ. When the scaling is *coherent*, i.e., respects axioms given below, the existence of a prior distribution can be deduced. The existence of subjective prior distributions as a consequence of an ordering of relative likelihoods is very important, since it allows us to escape the restrictive framework of frequentist justifications that are not always applicable. We describe in Note 3.8.1 the axioms underlying this derivation of the existence of a prior distribution from the ordering on the likelihoods, and refer the reader to DeGroot (1970, Chapter 6) for a more thorough treatment (see also Jeffreys (1961) and Bernardo and Smith (1994)).

It often occurs that the subjective determination of the prior distribution leads to incoherences in the likelihood ordering, for psychological reasons, but also because the ability of individuals to identify small probabilities is quite limited. On this topic, as well as on the practical construction of probability distributions and the assessment of forecasters, see DeGroot and Fienberg (1983), Dawid (1984), Lindley (1985) and Smith (1988).

Example 3.2.3 A study in the *New England Journal of Medicine* showed that 44% of the questioned individuals were ready to undertake a treatment against lung cancer when told that the *survival probability* was 68%. However, only 18% were still willing to undertake it when told that the probability of failure (death) was 32%. ||

3.2.2 Approximations to the prior distribution

When the parameter space, Θ, is *finite*, one can often obtain a subjective evaluation of the probabilities of the different values of θ. Sometimes, but not always, it is possible to use past experiments of the same type. Think, for instance, of constructing the probability of a nuclear war! (See Press (1989) for such a construction.) More fundamentally, this frequentist approach leads to the conceptual question of the *repeatability* of experiments

(Are experimental settings always the same? Can a previous experiment be without effect on the following one?); Jeffreys (1961) provides an extended criticism of this assumption.

When the parameter space Θ is noncountable, for instance, equal to an interval, the subjective determination of the prior π is obviously much more complicated. A first approximation of π is usually obtained through a partition of Θ in sets (e.g., intervals) and determination of the probability of each set; $\pi(\theta)$ is thus approached by an *histogram*. An alternative approach is to select significant elements of Θ, to evaluate their respective likelihoods and to deduce a likelihood curve proportional to π. In both cases, a major difficulty occurs when Θ is not *bounded*, due to the construction of the *tails* of the distribution, since it is quite complicated to evaluate subjectively the probabilities of the extreme regions of the parameter space, while the shape and properties of the resulting estimators deeply depend on them (see Example 3.2.6).

When no direct information is available on θ, an alternative approach is to use the *marginal distribution* of x,

$$m(x) = \int_{\Theta} f(x|\theta)\pi(\theta)\, d\theta$$

to derive information on π. Several techniques have been proposed in the literature (see Berger (1985a, §3.6)); apart from the moment method, we can mention the maximum entropy and the *ML-II* methods (Good (1983)). The basis for this derivation is that it may occur that the observed random phenomenon can be incorporated into a larger class (or a *meta-model*) about which information is available. For instance, if θ is the average daily milk production of a given dairy cow, information about θ can be gathered from the production of the herd it belongs to, although these observations originate from the marginal distribution. This perspective is at the core of *hierarchical models* (Chapter 10).

3.2.3 Maximum entropy priors

If some characteristics of the prior distribution (moments, quantiles, etc.) are known, assuming that they can be written as prior expectations,

$$(3.2.1) \qquad\qquad \mathbb{E}^{\pi}[g_k(\theta)] = \omega_k$$

$(k = 1, \ldots, K)$, a way to select a prior π satisfying these constraints is the *maximum entropy* method, developed in Jaynes (1980, 1983).

In a finite setting, the *entropy* is defined as

$$\mathcal{E}(\pi) = -\sum_i \pi(\theta_i) \log(\pi(\theta_i))\,.$$

This quantity has been introduced by Shannon (1948) as a measure of uncertainty in information theory and signal processing. The prior π maximizing the entropy is, in this information-theoretic sense, minimizing the prior information brought through π about θ. The *maximum entropy distribution*, under the moment constraints (3.2.1), is associated with the density

$$\pi^*(\theta_i) = \frac{\exp\left\{\sum_1^K \lambda_k g_k(\theta_i)\right\}}{\sum_j \exp\left\{\sum_1^K \lambda_k g_k(\theta_j)\right\}},$$

the numbers λ_k being derived from (3.2.1) as Lagrange multipliers. For instance, if there is no constraint on π, the maximum entropy distribution is the uniform distribution on Θ. (This is a problem because it means the maximum entropy priors are not invariant under reparameterization; see Section 3.5.1).

The extension to the continuous case is quite delicate, as it involves the choice of a reference measure, π_0, which can be seen as the completely noninformative distribution, i.e., the maximum entropy prior when no constraint is imposed. This reference measure can be selected in many ways, as explained in Section 3.5, while the maximum entropy distribution depends on this choice. When a group structure is available (and accepted as part of the prior information) for the problem of interest, it is usually agreed that the right-invariant Haar measure associated with this group is the natural choice for π_0. (Justifications for such a choice are given in Chapter 9.) The reference measure π_0 being selected, the entropy of π is defined as

$$\begin{aligned}
\mathcal{E}(\pi) &= \mathbb{E}^{\pi_0}\left[\log\left(\frac{\pi(\theta)}{\pi_0(\theta)}\right)\right] \\
&= \int \log\left(\frac{\pi(\theta)}{\pi_0(\theta)}\right)\pi_0(d\theta),
\end{aligned}$$

which is also the Kullback–Leibler distance between π and π_0. In this case, the maximum entropy distribution under (3.2.1) is given by the density

$$(3.2.2) \qquad \pi^*(\theta) = \frac{\exp\left\{\sum_1^K \lambda_k g_k(\theta)\right\}\pi_0(\theta)}{\int \exp\left\{\sum_1^K \lambda_k g_k(\eta)\right\}\pi_0(d\eta)},$$

thus showing the importance of π_0. Notice that the above distributions π^* are necessarily in an exponential family (see Section 3.3.3).

In addition to the dependence on π_0 exhibited in (3.2.2), another drawback of the maximum entropy method is that the constraints (3.2.1) are not always sufficient to derive a distribution on θ. Notice that this is often the case when characteristics (3.2.1) are related with the *quantiles*, since the functions $g_k(\theta)$ are of the form $\mathbb{I}_{(-\infty,a_k]}(\theta)$ or $\mathbb{I}_{(b_k,\infty]}(\theta)$.

Example 3.2.4 Consider θ, a real parameter such that $\mathbb{E}^\pi[\theta] = \mu$. If the reference measure π_0 is the Lebesgue measure on $\mathbb{R}$, the maximum entropy prior satisfies $\pi^*(\theta) \propto e^{\lambda\theta}$ and cannot be normalized into a probability distribution. On the contrary, if in addition it is known that $\mathrm{var}(\theta) = \sigma^2$, the corresponding maximum entropy prior is

$$\pi^*(\theta) \propto \exp\{\theta\lambda_1 + \theta^2\lambda_2\},$$

i.e., the normal distribution $\mathcal{N}(\mu, \sigma^2)$.   ‖

See Seidenfeld (1987) and Kass and Wasserman (1996) for criticisms of the maximum entropy approach (Exercise 3.2).

3.2.4 Parametric approximations

A frequently used alternative when building up a continuous prior consists of arbitrarily restricting the choice of π to a *parametrized* type of density and determining the corresponding parameters, either through the *moments*, or through the *quantiles*, since the latter are more robust. For instance, subjective evaluations of the median and the 75% quantile are enough to identify the two parameters of a normal distribution. (See also Example 3.2.1.)

Example 3.2.5   Let $X_i \sim \mathcal{B}(n_i, p_i)$ be the number of passing students in a freshman calculus course of n_i students. Over the previous years, the average of the p_i is 0.70, with variance 0.1. If we assume that the p_i's are all generated according to the same beta distribution, $\mathcal{B}e(\alpha, \beta)$, the parameters α and β can be estimated through

$$\frac{\alpha}{\alpha + \beta} = 0.7, \qquad \frac{\alpha\beta}{(\alpha + \beta)^2(\alpha + \beta + 1)} = 0.1,$$

i.e., $\alpha = 0.77$, $\beta = 0.33$, leading to the prior distribution

$$p \sim \mathcal{B}e(0.77, 0.33).$$

The choice of a beta distribution is motivated in this setting by conjugate prior arguments (see Section 3.3).   ‖

The moment method is often impractical and sometimes produces impossible values of the parameters; for instance, it can give negative variances. However, a deeper drawback of most parametric approaches is that the selection of the parametrized family is based on ease in the mathematical treatment, not on a subjective basis such as a preliminary histogram approximating π. These approaches may even lead to a partial rejection of the available information, on the grounds that it is not compatible with the parametric distribution. For instance, in Examples 3.2.1 and 3.2.5, the additional prior knowledge of the median may prevent the use of a beta distribution. Actually, the derivation of a distribution from a histogram may also be misleading since different families may fit the histogram and still lead to quite different inferences. (Nonetheless, we will study in the next section a particular parametrized prior determination in detail, since limited information settings call for parametrized prior distributions.)

Table 3.2.3. *Ranges of the posterior moments for fixed prior moments $\mu_1 = 0$ and μ_2. (Source: Goutis (1990).)*

μ_2	x	Minimum mean	Maximum mean	Maximum variance
3	0	-1.05	1.05	3.00
3	1	-0.70	1.69	3.63
3	2	-0.50	2.85	5.78
1.5	0	-0.59	0.59	1.50
1.5	1	-0.37	1.05	1.97
1.5	2	-0.27	2.08	3.80

Example 3.2.6 (Berger (1985a)) Let $x \sim \mathcal{N}(\theta, 1)$. Assume that the prior median of θ is 0, the first quartile is -1, and the third quartile is $+1$. Then, if the prior distribution on θ is of the form $\mathcal{N}(\mu, \tau)$, we must have $\theta \sim \mathcal{N}(0, 2.19)$. On the contrary, the choice of a Cauchy distribution implies $\theta \sim \mathcal{C}(0, 1)$. Under a quadratic loss, the Bayes estimator will be in the first case

$$\delta_1^\pi(x) = x - \frac{x}{3.19}$$

and, for $|x| \geq 4$,

$$\delta_2^\pi(x) \approx x - \frac{x}{1 + x^2}$$

in the second case (see Berger and Srinivasan (1978)). Therefore, for $x = 4$, which is an observation quite compatible with the prior information in both cases, the two estimations would be $\delta_1^\pi(4) = 2.75$ and $\delta_2^\pi(4) = 3.76$.   ‖

These posterior discrepancies call for some tests on the validity (or *robustness*) of the selected priors, depending on the observation, in order to evaluate how a slight change in the prior distribution is perceived in the inference about the parameter of interest. (Section 3.5 deals with this evaluation.) The example below illustrates again the fact that an item of information that is too vague can produce very different conclusions, depending on the way it is interpreted.

Example 3.2.7 (Goutis (1990, 1994)) Let $x \sim f(x|\theta)$, with $\theta \in \mathbb{R}$, and assume that the prior mean of θ, μ_1, is known. Too many prior distributions agree with this piece of information, since

$$\inf_\pi \mathbb{E}^\pi[\theta|x] = -\infty \quad \text{and} \quad \sup_\pi \mathbb{E}^\pi[\theta|x] = +\infty$$

and no useful inference can be derived from this single fact; notice that, in this setting, it is not possible to construct a maximum entropy distribution either (see Example 3.2.4). If, in addition, the prior variance μ_2, is fixed, the variability of the posterior answers is more restricted since

(3.2.3) $$-\infty < \inf_\pi \mathbb{E}^\pi[\theta|x] \leq \sup_\pi \mathbb{E}^\pi[\theta|x] < +\infty,$$

as long as $f(x|\theta)$ is positive in a neighborhood of μ_1 and bounded when $|\theta - \mu_1|$ is large. Under the same set of assumptions, we have, in addition,

(3.2.4) $$0 = \inf_\pi \mathrm{Var}^\pi[\theta|x] \leq \sup_\pi \mathrm{Var}^\pi[\theta|x] < +\infty.$$

Table 3.2.4 gives the exact range of the bounds (3.2.3) and (3.2.4) for a normal distribution $\mathcal{N}(\theta, 1)$ and $\mu_1 = 0$.   ∥

3.2.5 Other techniques

Empirical and *hierarchical* Bayes techniques are two rather antagonistic approaches that naturally incorporate uncertainty about the prior distribution and will be detailed in Chapter 9 (see also Carlin and Louis (2000a)). The empirical Bayes technique relies on the observations (and the marginal distribution) to estimate the parameters of the prior distribution; it is used by frequentists more often than by Bayesians because it does not belong to the Bayesian paradigm. Formally, it seems paradoxical to choose a *prior* distribution *a posteriori*! More fundamentally, the choice of π depending on x, the derived estimators do not enjoy the optimality properties of the true Bayes estimators. A last criticism is that too many choices are possible for the estimation techniques used in the construction of the prior distribution, thus leading to an important arbitrariness in the selection of the prior.

The hierarchical Bayes approach models the lack of information on the parameters of the prior distribution according to the Bayesian paradigm, i.e., through another prior distribution on these parameters (the parameters of this distribution are then called *hyperparameters* and this new prior a *hyperprior*). Although this choice may seem too abstract conceptually, Bayesians usually prefer this approach over the empirical Bayes alternative because it generally provides better estimators in practical and theoretical senses. (Chapter 10 presents and compares these two techniques.)

3.3 Conjugate priors

3.3.1 Introduction

When prior information about the model is too vague or unreliable, a full subjective derivation of the prior distribution is obviously impossible. Other reasons (time delays, cost prohibitions, lack of communication between the statistician and the decision maker, etc.) may explain the absence of a well-defined prior distribution. Moreover, objectivity requirements may force the statistician to provide an answer with as little subjective input as possible, in order to base the inference on the sampling model alone. These settings may seem to call for a non-Bayesian solution (maximum likelihood estimator, best unbiased estimator, etc.). However, keeping in mind the Bayesian foundations of the frequentist optimality criteria (see Chapters 2, 8 and 9), it appears preferable to follow a Bayesian approach, using an objective prior derived from the model as a technical tool. When no prior information at all is available, these priors are called noninformative and are considered in Section 3.5.

First, we study in this section a classical parametric approach which involves a subjective input as limited as possible, and which underlies both the hierarchical and empirical Bayes techniques of Chapter 10. Besides requiring a minimal subjective input, conjugate priors can indeed be considered a starting point to build up prior distributions based on limited prior input, whose imprecision can be assessed through additional prior distributions. The reader should be aware, however, from the beginning that the common impression that conjugate priors are noninformative priors is false: the choice of conjugate priors, while arguable as shown below, is still a choice and, therefore, influences the resulting inference to some extent. Moreover, it may force to ignore part of the prior information if the later is not entirely compatible with the conjugate prior format and also there are other priors with the same limited input, but with a more limited influence on the resulting inference (see Section 3.6).

Definition 3.3.1 *A family $\mathcal{F}$ of probability distributions on Θ is said to be conjugate (or closed under sampling) for a likelihood function $f(x|\theta)$ if, for every $\pi \in \mathcal{F}$, the posterior distribution $\pi(\theta|x)$ also belongs to $\mathcal{F}$.*

A trivial example of a conjugate family is the set $\mathcal{F}_0$ made of all distributions on Θ, which is, of course, useless for the choice of a prior distribution. The main interest of conjugacy becomes more apparent when $\mathcal{F}$ is as small as possible and *parametrized*. Indeed, when $\mathcal{F}$ is parametrized, switching from prior to posterior distribution is reduced to an updating of the corresponding parameters. This property alone can explain why conjugate priors are so popular, since the posterior distributions are always computable (at least to a certain extent). On the contrary, such a justification is rather weak from a subjective point of view and any other parametrized family would be as convenient. Notice that the goal to obtain the *minimal conjugate family* as the intersection of all conjugate families is unfortunately doomed to failure, since this intersection is empty (Exercise 3.13).

3.3.2 Justifications

The conjugate prior approach, which originated in Raiffa and Schlaifer (1961), can be partly justified through an *invariance* reasoning. Actually, when the observation of $x \sim f(x|\theta)$ modifies $\pi(\theta)$ into $\pi(\theta|x)$, the information conveyed by x about θ is obviously limited; therefore, it should not lead to a modification of the whole *structure* of $\pi(\theta)$, but simply of its *parameters*. In other words, the modification resulting from the observation of x should be of finite dimension. A more radical change of π is thus unacceptable and the choice of the prior distributions should always be made among conjugate distributions, whatever the prior information is. De Finetti (1974) somehow held similar views because he considered that the prior information could be translated into virtual past observations, as in Example 3.2.2, leading necessarily to conjugate priors for exponential families (see below). This requirement unfortunately becomes paradoxical in the extreme case when the whole prior distribution is already available! But conjugate priors are mainly used in limited information environments, since they only call for the determination of a few parameters. Another justification for using conjugate priors is that some Bayes estimators are then linear, as shown by Diaconis and Ylvisaker (1979) (see Proposition 3.3.13, below). Nonetheless, we must acknowledge that the main motivation for using conjugate priors remains their tractability.

This particular modeling through a parametrized family of priors is indeed attractive, since it allows for an explicit treatment of posterior distributions. These conjugate priors are sometimes called *objective* because the sampling model, $f(x|\theta)$, entirely determines the class of priors, but any method deriving priors automatically from sampling distributions would be similarly objective. A contrario, the use of conjugate priors is strongly suspicious for subjective Bayesians because it is mainly justified on technical grounds, rather than for its proper fitting of the available prior information. The role of conjugate priors is thus to provide a first approximation to the adequate prior distribution, which should be followed by a robustness analysis (see Section 3.5). We will see in Section 3.4 that they are more justified if treated as a basis (in the *functional* sense) for prior information modeling.

3.3.3 Exponential families

Conjugate prior distributions are usually associated with a particular type of sampling distribution that always allows for their derivation, and is even characteristic of conjugate priors, as we will see below. These distributions constitute what is called *exponential families*, studied in detail in Brown (1986).

Definition 3.3.2 *Let μ be a σ-finite measure on $\mathcal{X}$, and let Θ be the parameter space. Let C and h be functions, respectively, from $\mathcal{X}$ and Θ to $\mathbb{R}_+$, and let R and T be functions from Θ and $\mathcal{X}$ to $\mathbb{R}^k$. The family of distributions with densities (w.r.t. μ)*

(3.3.1) $$f(x|\theta) = C(\theta)h(x)\exp\{R(\theta) \cdot T(x)\}$$

is called an exponential family of dimension k. In the particular case when

$\Theta \subset \mathbb{R}^k$, $\mathcal{X} \subset \mathbb{R}^k$ and

(3.3.2) $$f(x|\theta) = C(\theta)h(x)\exp\{\theta \cdot x\},$$

the family is said to be natural.

Notice that a change of variables from x to $z = T(x)$ and a reparameterization from θ to $\eta = R(\theta)$ authorizes us to consider mainly the natural form (3.3.2), although the spaces $T(\mathcal{X})$ and $R(\Theta)$ may be difficult to describe and work with.

From an analytic point of view, exponential families have many interesting properties (see Brown (1986)). In particular, they are such that, for any sample from (3.3.1), there exists a sufficient statistic of constant dimension. Indeed, if $x_1, \ldots, x_n \sim f(x|\theta)$, with f satisfying (3.3.2),

$$\bar{x} = \frac{1}{n}\sum_{i=1}^{n} x_i \in \mathbb{R}^k$$

is sufficient for every n. The converse of this result has also been established by Koopman (1936) and Pitman (1936) (see also Jeffreys (1961, §3.7.1) for a proof).

Theorem 3.3.3 (Pitman–Koopman Lemma) *If a family of distributions $f(\cdot|\theta)$ is such that, for a sample size large enough, there exists a sufficient statistic of constant dimension, the family is exponential if the support of $f(\cdot|\theta)$ does not depend on θ.*

The restriction on the support of $f(x|\theta)$ is necessary for the lemma to hold because the uniform $\mathcal{U}([-\theta, \theta])$ and the Pareto $\mathcal{P}(\alpha, \theta)$ distributions also satisfy this property (see Example 3.3.10). These distributions could actually be called *quasi-exponential* families because they partake of many interesting properties of exponential families, including the existence of constant-dimension sufficient statistics and of conjugate priors (Exercise 3.15).

Many common continuous and discrete distributions belong to exponential families.

Example 3.3.4 If S is the simplex of $\mathbb{R}^k$,

$$S = \left\{\omega = (\omega_1, \ldots, \omega_k); \sum_{i=1}^{k} \omega_i = 1, \omega_i > 0\right\},$$

the *Dirichlet distribution* on S, $\mathcal{D}_k(\alpha_1, \ldots, \alpha_k)$, is an extension of the beta distribution, which is defined by

$$f(p|\alpha) = \frac{\Gamma(\alpha_1 + \cdots + \alpha_k)}{\Gamma(\alpha_1) \cdots \Gamma(\alpha_k)} \prod_{i=1}^{k} p_i^{\alpha_i - 1} \mathbb{I}_S(p),$$

where $p = (p_1, \ldots, p_k)$. Since

$$f(p|\alpha) = C(\alpha)h(p)\exp\left(\sum_{i=1}^{k}\alpha_i\log(p_i)\right),$$

the Dirichlet distributions constitute a natural exponential family for $T(p) = (\log(p_1), \ldots, \log(p_k))$.   ‖

Example 3.3.5 Let $x \sim \mathcal{N}_p(\theta, \sigma^2 I_p)$. Then

$$
\begin{aligned}
f(x|\theta) &= \frac{1}{\sigma^p}\frac{1}{(2\pi)^{p/2}}\exp\left(-\sum_{i=1}^{p}(x_i - \theta_i)^2/2\sigma^2\right) \\
&= C(\theta, \sigma)h(x)\exp\{x.(\theta/\sigma^2) + ||x||^2(-1/2\sigma^2)\}
\end{aligned}
$$

and the normal distribution belongs to an exponential family with natural parameters θ/σ^2 and $-1/2\sigma^2$. Similarly, if $x_1, \ldots, x_n \sim \mathcal{N}_p(\theta, \sigma^2 I_p)$, the joint distribution satisfies

$$
\begin{aligned}
f(x_1, \ldots, x_n) &= C'(\theta, \sigma)h'(x_1, \ldots, x_n) \\
&\times \exp\left\{n\bar{x}\cdot(\theta/\sigma^2) + \sum_{i=1}^{n}||x_i - \bar{x}||^2(-1/2\sigma^2)\right\}
\end{aligned}
$$

and the statistic $(\bar{x}, \sum_i ||x_i - \bar{x}||^2)$ is sufficient for all $n \geq 2$.   ‖

In the previous example, notice that the parameter space is of dimension $p + 1$, while the dimension of the observable, x, is p. While the dimension of an exponential family is not fixed, since it is always possible to add convex combinations of the original parameters as additional (and useless) parameters, there is an intrinsic minimal dimension associated with this family.

Definition 3.3.6 *Let $f(x|\theta) = C(\theta)h(x)\exp(\theta.x)$ be a natural exponential family. The natural parameter space is*

$$N = \left\{\theta; \int_{\mathcal{X}} e^{\theta\cdot x}h(x)\,d\mu(x) < +\infty\right\}.$$

The family is said to be regular if N is an open set and minimal if $\dim(N) = \dim(K) = k$, where K is the closure of the convex envelope of the support of μ.

It is always possible to reduce an exponential family to a standard and minimal form of dimension m, and this dimension m does not depend on the chosen parameterization (Brown (1986, pp. 13–16)). (See Exercise 3.23 for the example of a non-regular exponential family.)

Natural exponential families can also be rewritten under the form

(3.3.3) $f(x|\theta) = h(x)e^{\theta.x - \psi(\theta)}$

and $\psi(\theta)$ is called the *cumulant generating function* for the following reason, whose proof is left to the reader.

Lemma 3.3.7 *If $\theta \in \overset{\circ}{N}$, the interior set of N, the cumulant generating function ψ is $\mathcal{C}^\infty$ and*

$$\mathbb{E}_\theta[x] = \nabla \psi(\theta), \qquad \mathrm{cov}(x_i, x_j) = \frac{\partial^2 \psi}{\partial \theta_i \partial \theta_j}(\theta),$$

where ∇ denotes the gradient operator.

Example 3.3.8 Let $x \sim \mathcal{P}(\lambda)$. Then

$$f(x|\lambda) = e^{-\lambda}\frac{\lambda^x}{x!} = \frac{1}{x!}e^{\theta \cdot x - e^\theta}$$

and $\psi(\theta) = \exp(\theta)$ for the natural parameter $\theta = \log \lambda$. Therefore, $\mathbb{E}_\lambda[x] = e^\theta = \lambda$ and $\mathrm{var}(x) = \lambda$.   ‖

   The regular structures of exponential families allow for numerous statistical developments, as shown by the extensive literature on the subject. (See, for instance, the classification of exponential families according to the type of variance function: Morris (1982), Letac and Mora (1990), and Exercises 3.24 and 10.33.) We show in Section 3.3.4 that they also allow for a straightforward derivation of conjugate priors.

Example 3.3.9 If $x \sim \mathcal{N}(\theta, \theta^2)$ in a multiplicative model, the conjugate prior is not the normal distribution. The likelihood is proportional to

$$\frac{1}{|\theta|}\exp\left\{\frac{x}{\theta} - \frac{x^2}{2\theta^2}\right\}$$

and the distribution induces an exponential family of dimension 2. Therefore, the *generalized inverse normal* distributions $\mathcal{IN}(\alpha, \mu, \tau)$, with density

$$\pi(\theta) \propto |\theta|^{-\alpha}\exp\left\{-\left(\frac{1}{\theta} - \mu\right)^2 \Big/ 2\tau^2\right\}$$

constitute a conjugate family in this model. This family of distributions, which belongs to the exponential family, generalizes the distribution of the inverse of a normal observation (which corresponds to the case $\alpha = 2$). (See Exercise 3.33 for details.)   ‖

   Obviously, most distributions do not belong to an exponential family! For instance, Student's t-distribution, $\mathcal{T}_p(\nu, \theta, \sigma^2)$, cannot be written under the form (3.3.1). Definition 3.3.2 also excludes all distributions with non-constant support, while some do allow for conjugate priors with a finite number of parameters (or, more exactly, *hyperparameters*).

Example 3.3.10 Pareto distributions, $\mathcal{P}(\alpha, \theta)$, with density

$$f(x|\alpha, \theta) = \alpha\frac{\theta^\alpha}{x^{\alpha+1}}\mathbb{I}_{[\theta, +\infty[}(x) \qquad (\theta > 0),$$

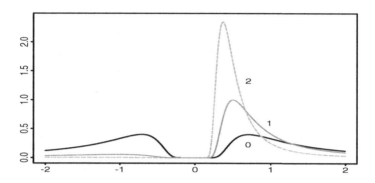

Figure 3.3.1. *Densities of* $\mathcal{IN}(\alpha, \mu, \tau)$ *for* $\alpha = 2$, $\tau = 1$, *and* $\mu = 0, 1, 2$.

are such distributions since, although outside the exponential family frame-work because their support depends on θ, they allow for simple conjugate distributions on θ, namely, Pareto distributions for $1/\theta$.   ‖

Other examples of families where conjugate priors are available are $\mathcal{U}_{[-\theta,\theta]}$ and $\mathcal{U}_{[0,\theta]}$ distributions; these distributions are also quasi-exponential, since they allow for sufficient statistics of constant dimension. For instance, if $x_1, \ldots, x_n \sim \mathcal{U}_{[-\theta,\theta]}$, a sufficient statistic is the *order statistic* $(x_{(1)}, x_{(n)})$, where $x_{(1)}$ denotes the smallest value in the sample and $x_{(n)}$ the largest.

Note that, in Example 3.3.9, the conjugate prior on θ depends on three hyperparameters, α, μ, and τ^2; therefore, its use induces a greater complexity than the sampling distribution. This phenomenon, that is, the fact that the structure of the model requires a larger number of hyperparameters, is often encountered for *curved exponential families*, for instance when a natural reparameterization by $\eta = R(\theta)$ is not useful because of the constraints on the natural parameters. It is obviously a drawback, since the values of these hyperparameters have to be determined to derive an inference on θ using conjugate priors.

When a distribution does not allow for conjugate families, except the trivial $\mathcal{F}_0$, it is sometimes possible to express the distribution as a *mixture* of distributions from exponential families; f is then said to be a *hidden mixture*, since this mixture representation is of no importance for the inferential problem, but is useful for the practical computation of posterior distributions and Bayes estimators, as shown in Chapter 6.

Example 3.3.11 (Dickey (1968)) For Student's t-distribution, there exists a hidden mixture representation through the normal distribution, since $f(x|\theta)$ is the mixture of a normal distribution and an inverse gamma distribution: If $x \sim \mathcal{T}_1(p, \theta, \sigma^2)$,

$$x|z \sim \mathcal{N}(\theta, z\sigma^2), \qquad z^{-1} \sim \mathcal{G}(p/2, p/2).$$

A technically interesting prior on θ is then $\mathcal{N}(\mu, \tau^2)$ and most of the computations can be done conditional upon z. This decomposition is more useful when x is multidimensional because some integrals can then be expressed as being one-dimensional. ‖

Example 3.3.12 Several noncentral distributions can be written as the (hidden) mixture of the corresponding central distribution by a Poisson distribution, due to an *infinite divisibility* property (see Feller (1971, Chapter 9)). For instance, this is the case for the noncentral chi-squared distribution, $\chi_p^2(\lambda)$: When $x \sim \chi_p^2(\lambda)$, the generation of x can also be decomposed as

$$x|z \sim \chi_{p+2z}^2, \qquad z \sim \mathcal{P}(\lambda/2).$$

This decomposition is used in James and Stein (1961) to express the risk of the James–Stein estimator and derive a sufficient condition of domination of the maximum likelihood estimator (see Note 2.8.2). ‖

3.3.4 Conjugate distributions for exponential families

Consider $f(x|\theta) = h(x)e^{\theta.x - \psi(\theta)}$, a generic distribution from an exponential family. It then allows for a conjugate family, as shown by the following result (whose proof is straightforward).

Proposition 3.3.13 *A conjugate family for $f(x|\theta)$ is given by*

(3.3.4) $$\pi(\theta|\mu, \lambda) = K(\mu, \lambda)\, e^{\theta.\mu - \lambda\psi(\theta)},$$

where $K(\mu, \lambda)$ is the normalizing constant of the density. The corresponding posterior distribution is $\pi(\theta|\mu + x, \lambda + 1)$.

The measure defined by (3.3.4) is σ-finite; it induces a probability distribution on Θ if and only if

(3.3.5) $$\lambda > 0 \qquad \text{and} \qquad \frac{\mu}{\lambda} \in \overset{\circ}{N}$$

(Exercise 3.35). It is only when (3.3.5) holds that $K(\mu, \lambda)$ is well defined. Therefore, there is an automated way to deduce a conjugate distribution from $f(x|\theta)$; this is why (3.3.4) is often called the *natural conjugate distribution* of f. Table 3.3.4 presents the conjugate distributions for the usual distributions belonging to an exponential family.[1] Obviously, Bayesian inference cannot be conducted unless the hyperparameters μ and λ are known.

---

[1] Since the conjugate distributions are also from an exponential family, Bar-Lev et al. (1994) have studied a reciprocal problem, namely the determination of the distributions $\pi(\theta)$ for which an exponential family has $\pi(\theta)$ as conjugate distribution.

Table 3.3.1. *Natural conjugate priors for some common exponential families*

$f(x|\theta)$	$\pi(\theta)$	$\pi(\theta|x)$
Normal $\mathcal{N}(\theta, \sigma^2)$	Normal $\mathcal{N}(\mu, \tau^2)$	$\mathcal{N}(\varrho(\sigma^2\mu + \tau^2 x), \varrho\sigma^2\tau^2)$ $\varrho^{-1} = \sigma^2 + \tau^2$
Poisson $\mathcal{P}(\theta)$	Gamma $\mathcal{G}(\alpha, \beta)$	$\mathcal{G}(\alpha + x, \beta + 1)$
Gamma $\mathcal{G}(\nu, \theta)$	Gamma $\mathcal{G}(\alpha, \beta)$	$\mathcal{G}(\alpha + \nu, \beta + x)$
Binomial $\mathcal{B}(n, \theta)$	Beta $\mathcal{B}e(\alpha, \beta)$	$\mathcal{B}e(\alpha + x, \beta + n - x)$
Negative Binomial $\mathcal{N}eg(m, \theta)$	Beta $\mathcal{B}e(\alpha, \beta)$	$\mathcal{B}e(\alpha + m, \beta + x)$
Multinomial $\mathcal{M}_k(\theta_1, \ldots, \theta_k)$	Dirichlet $\mathcal{D}(\alpha_1, \ldots, \alpha_k)$	$\mathcal{D}(\alpha_1 + x_1, \ldots, \alpha_k + x_k)$
Normal $\mathcal{N}(\mu, 1/\theta)$	Gamma $\mathcal{G}a(\alpha, \beta)$	$\mathcal{G}(\alpha + 0.5, \beta + (\mu - x)^2/2)$

The automatic aspect of conjugate priors is thus misleading because they still require a subjective input through the determination of these values. Notice also that (3.3.4) requires an additional parameter, compared with $f(x|\theta)$.

For natural exponential families, conjugate priors have an additional appeal, as shown by Diaconis and Ylvisaker (1979): If $\xi(\theta)$ is the expectation of $x \sim f(x|\theta)$, the posterior mean of $\xi(\theta)$ is linear in x for a conjugate prior distribution.

Proposition 3.3.14 *If Θ is an open set in $\mathbb{R}^k$ and θ has the prior distribution*

$$\pi_{\lambda,x_0}(\theta) \propto e^{\theta \cdot x_0 - \lambda\psi(\theta)}$$

with $x_0 \in \mathcal{X}$, then

$$\mathbb{E}^\pi[\xi(\theta)] = \mathbb{E}^\pi[\nabla\psi(\theta)] = \frac{x_0}{\lambda}.$$

Therefore, if $x_1, \ldots, x_n$ are i.i.d. $f(x|\theta)$,

(3.3.6) $$\mathbb{E}^\pi[\xi(\theta)|x_1, \ldots, x_n] = \frac{x_0 + n\bar{x}}{\lambda + n}.$$

This result is well known for the normal distributions and can be generalized for all exponential families. (The proof is straightforward.) Equation (3.3.6) shows that the parameter λ is similar to the sample size n. Its determination can therefore be achieved, if necessary, by considering that the prior information on x_0 originated from a virtual sample of size λ. Brown (1986) shows that Proposition 3.3.14 can be extended to the case where π_{λ,x_0} is improper, for instance when $\lambda = 0$ and $x_0 = 0$. In this case, the posterior expectation is $\bar{x}$, which is also the maximum likelihood estimator of $\xi(\theta)$.

Diaconis and Ylvisaker (1979) have shown, in addition, a reciprocal to this proposition, namely, that, under specific regularity conditions, if the dominating measure is continuous with respect to the Lebesgue measure, linearity of $\mathbb{E}^\pi[\xi(\theta)|x]$ as in (3.3.6) implies that the prior distribution is of the form (3.3.2). Discrete-case extensions are more delicate.

While exponential families usually allow for an easy processing and, in particular, for the convenient call to conjugate prior distributions and the analytical derivation of posterior means, as in Proposition 3.3.14, this is not always the case. For instance, when $x \sim \mathcal{Be}(\alpha, \theta)$ with known α, the distribution belongs to an exponential family because

$$f(x|\theta) \propto \frac{\Gamma(\alpha + \theta)(1 - x)^\theta}{\Gamma(\theta)},$$

but the conjugate distributions are not easily manageable, as

$$\pi(\theta|x_0, \lambda) \propto \left(\frac{\Gamma(\alpha + \theta)}{\Gamma(\theta)} \right)^\lambda (1 - x_0)^\theta$$

involves the Gamma function $\Gamma(\theta)$, which does not have a closed-form expression.

Example 3.3.15 *Logistic regression* is used to describe qualitative models as in Example 1.1.1. Given an indicator variable y, which only takes values in $\{0, 1\}$, and explanatory variables $x \in \mathbb{R}^k$, the distribution of y conditional on x is

$$(3.3.7) \qquad P_\alpha(y = 1) = 1 - P_\alpha(y = 0) = \frac{\exp(\alpha^t x)}{1 + \exp(\alpha^t x)}.$$

This model allows for the extension of the quite useful linear regression model to more qualitative settings. For a sample $(y_1, x_1), \ldots, (y_n, x_n)$ from (3.3.7), the model is indeed exponential *conditional upon the x_i's*, as

$$f(y_1, \ldots, y_n|x_1, \ldots, x_n, \alpha) = \exp\left(\alpha^t \sum_{i=1}^n y_i x_i \right) \prod_{i=1}^n (1 + e^{\alpha^t x_i})^{-1},$$

which only depends on the sufficient statistic $\sum_{i=1}^n y_i x_i$. In practice, the conjugate priors are rather difficult to handle because they are of the form

$$\pi(\alpha|y_0, \lambda) \propto e^{\alpha^t y_0} \prod_{i=1}^n (1 + e^{\alpha^t x_i})^{-\lambda}.$$

The normalizing constant for $\pi(\alpha|y_0, \lambda)$ is unknown and approximations of posterior quantities such as the posterior mean and posterior median can only be achieved through simulation techniques presented in Chapter 6. ∥

3.4 Criticisms and extensions

As seen above, the automated aspect of conjugate distributions is simultaneously an advantage and a nuisance. In addition to invariance and linearity
arguments, it has been argued that this is an objective approach, where the
subjective input is reduced to the choice of the hyperparameters. Barring
the fact that objectivity is a concept difficult to define, it can be countered
that any other prior distribution with the same number of hyperparameters
would seem equally objective. Moreover, the conjugate priors are not necessarily the most robust prior distributions (see Section 3.5) and, from this
point of view, alternative distributions could be preferred, if the imperative
is to minimize the influence of the prior input on the inferential output.
The following example shows how the choice of the prior can modify the
posterior distribution for small sample sizes.

Example 3.4.1 (Diaconis and Ylvisaker (1985)) When a coin is spun on
its edge, instead of being thrown in the air, the proportion of *heads* is rarely
close to $1/2$, but is rather $1/3$ or $2/3$ because of irregularities in the edge
that cause the game to favor one side or the other. When spinning, n times,
a given coin on its edge, we observe the number of heads, $x \sim \mathcal{B}(n, p)$. The
prior distribution on p is then likely to be bimodal; this cannot be modeled
through a conjugate prior π_1 like $\mathcal{B}e(1, 1)$. A mixture prior distribution π_2
such as

$$\frac{1}{2} \left[\mathcal{B}e(10, 20) + \mathcal{B}e(20, 10) \right]$$

is indeed more appropriate. It may also be the case that previous experiments with the same coin have already hinted at a bias toward *head* and
that they lead to the following alternative, π_3:

$$0.5 \, \mathcal{B}e(10, 20) + 0.2 \, \mathcal{B}e(15, 15) + 0.3 \, \mathcal{B}e(20, 10).$$

Figure 3.4.1 provides the graphs of the two prior densities above, along
with the neutral $\mathcal{B}e(1, 1)$ prior, the differences between the three prior
modelings being quite important. If, for $n = 10$, we observe $x = 3$, the
corresponding posterior distributions are

(i) $\mathcal{B}e(1 + x, 1 + n - x)$, i.e., $\mathcal{B}e(4, 8)$;

(ii) $0.84 \, \mathcal{B}e(13, 27) + 0.16 \, \mathcal{B}e(23, 17)$; and

(iii) $0.77 \, \mathcal{B}e(13, 27) + 0.16 \, \mathcal{B}e(18, 22) + 0.07 \, \mathcal{B}e(23, 17)$.

In (ii), the posterior probability weights are obtained as being proportional to

$$\frac{1}{2} \frac{B(13, 27)}{B(10, 20)} \qquad \text{and} \qquad \frac{1}{2} \frac{B(23, 17)}{B(20, 10)}$$

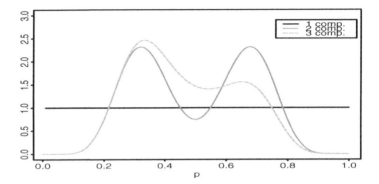

Figure 3.4.1. *Three prior distributions for a spinning-coin experiment.*

and, for (iii),

$$0.5\,\frac{B(13,27)}{B(10,20)}, \qquad 0.2\,\frac{B(18,22)}{B(15,15)}, \qquad \text{and} \qquad 0.3\,\frac{B(23,17)}{B(20,10)},$$

where

$$B(a,b) = \frac{\Gamma(a)\Gamma(b)}{\Gamma(a+b)}$$

is the inverse of the normalizing term in the beta density (defined in Appendix A.3), which must be approximated using numerical methods.

Therefore, for this sample, the three posterior means, $1/3$, 0.365, and 0.362 respectively, are fairly close but the shapes of the posterior distributions still differ (see Figure 3.4). Consider now a sample of size $n = 50$ with $x = 36$. The posterior distributions are

(i) $\mathcal{B}e(15,37)$;

(ii) $0.997\,\mathcal{B}e(24,56) + 0.003\,\mathcal{B}e(34,46)$; and

(iii) $0.95\,\mathcal{B}e(24,56) + 0.047\,\mathcal{B}e(29,51) + 0.003\,\mathcal{B}e(34,46)$.

They are then much closer than for $n = 10$, as shown by Figure 3.4.   ‖

Two general remarks logically stem from this example. First, it shows that prior modeling is indeed important for small sample sizes, but also that it becomes less important as the sample size increases. When the sample size goes to infinity, most priors will lead to the same inference and this inference will be equivalent to the one based only on the likelihood function, as pointed out in Note 1.8.4. Moreover, this example shows that *mixtures* of conjugate priors are as easy to manipulate as regular conjugate distributions, while leading to a greater freedom in the modeling of the prior information. In fact, mixtures of natural conjugate distributions also make conjugate families.

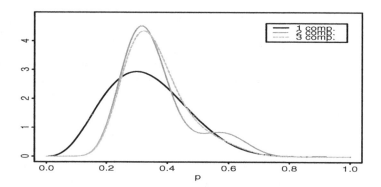

Figure 3.4.2. *Posterior distributions for the spinning model for 10 observations.*

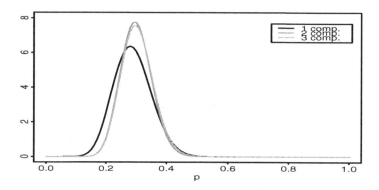

Figure 3.4.3. *Posterior distributions for 50 observations.*

Lemma 3.4.2 *Let $\mathcal{F}$ be the natural conjugate family of an exponential family (3.3.2). Then the set of mixtures of N conjugate distributions,*

$$\tilde{\mathcal{F}}_N = \left\{ \sum_{i=1}^{N} \omega_i \pi(\theta|\lambda_i, \mu_i); \sum_{i=1}^{N} \omega_i = 1, \, \omega_i > 0 \right\},$$

is also a conjugate family. Moreover, if

$$\pi(\theta) = \sum_{i=1}^{N} \omega_i \pi(\theta|\lambda_i, \mu_i),$$

the posterior distribution is a mixture

$$\pi(\theta|x) = \sum_{i=1}^{N} \omega_i'(x) \pi(\theta|\lambda_i + 1, \mu_i + x),$$

with

$$\omega_i'(x) = \frac{\omega_i K(\mu_i, \lambda_i)/K(\mu_i + x, \lambda_i + 1)}{\sum_{j=1}^{N} \omega_j K(\mu_j, \lambda_j)/K(\mu_j + x, \lambda_j + 1)}.$$

Mixtures can then be used as a basis to approximate any prior distribution, in the sense that the Prohorov distance between a distribution and a mixture representation can be made arbitrarily small. The *Prohorov distance* between two measures π and $\tilde{\pi}$, $d^P(\pi, \tilde{\pi})$ is defined as

$$d^P(\pi, \tilde{\pi}) = \inf_A \left\{ \epsilon \, ; \, \pi(A) \leq \tilde{\pi}(A^\epsilon) + \epsilon \right\},$$

where the infimum is taken on the Borel sets A, and the set A^ϵ denotes the set of points at distance at most ϵ of A (Le Cam (1986)).

Theorem 3.4.3 *If Θ is the natural parameter space for the exponential family $f(x|\theta)$ and π is a prior distribution on Θ, then, for any $\epsilon > 0$, there exist N and $\tilde{\pi} \in \tilde{\mathcal{F}}_N$ such that $d^P(\pi, \tilde{\pi}) < \epsilon$.*

The proof of this theorem can be related to the fact that finite mixtures of Dirac measures constitute a dense set for the Prohorov topology, and that Dirac masses can be approximated by mixtures of conjugate priors. (For more details, see Brown (1986, pp. 254–267).) This result justifies the use of conjugate priors much more strongly than the invariance, linearity, or simplicity arguments presented in the previous section. Whatever prior information is available, it could be modeled through a mixture from $\tilde{\mathcal{F}}_N$ with N as small as possible. However, this approximation result is also incomplete, since it does not show how the approximation is transferred to posterior quantities, while Bayesian inference only considers these posterior quantities. Berger (1985b) illustrates this difference through the following example.

Example 3.4.4 Consider $x \sim \mathcal{N}(\theta, 1)$ when the associated prior π_0 is a Cauchy distribution, $\mathcal{C}(0, 1)$. The natural conjugate priors being $\mathcal{N}(\mu, A)$, π_0 can be approximated by

$$\tilde{\pi} = \sum_{i=1}^{N} \lambda_i \pi_i,$$

where π_i is $\mathcal{N}(\mu_i, A_i)$, according to Theorem 3.4.3. Actually, when x goes to $+\infty$, $\pi_0(\theta|x)$ goes to $\mathcal{N}(x, 1)$ while $\tilde{\pi}(\theta|x)$ is roughly $\mathcal{N}(\mu(x), \varrho)$, with

$$\varrho = \frac{A^*}{1 + A^*}, \quad \mu(x) = \varrho x + (1 - \varrho)\mu^*, \quad A^* = \max_i \{A_i\}, \quad \mu^* = \max_{A_i = A^*} \mu_i.$$

Therefore, $\pi_0(\theta|x)$ and $\tilde{\pi}(\theta|x)$ will significantly differ for large values of x. One could argue that such values are not compatible with the prior information anyway and should lead to a modification of the prior modeling. But these differences still show that the prior approximation is not uniformly valid a posteriori. $\qquad \|$

Example 3.4.4 illustrates rather forcefully the following point: distributions with heavy tails will not be properly approximated by lighter tail distributions. This difficulty, and more generally, the problem of approximating the posterior distributions somehow disappears in the generalization of Dalal and Hall (1983), who consider continuous mixtures (in the continuous case). We briefly describe their approach in Note 3.8.3, but want to point out that their approximation by continuous mixtures misses the appeal of the previous approximation, since the latter most often calls for a numerical or Monte Carlo resolution.

3.5 Noninformative prior distributions

The previous section has shown that conjugate priors were useful as approximations of the true prior distributions. However, when no prior information is available, their sole justification is analytical, as they can lead to closed-form expressions for some posterior quantities. In such settings, it is impossible to justify the choice of prior distributions on a subjective basis and the hyperparameters of the conjugate prior cannot be determined. Instead of turning back to classical alternatives, like maximum likelihood estimation, or to use the data to approximate these hyperparameters, as in empirical Bayes analysis, it still is preferable to use Bayesian techniques, if only because they underlie classical optimality criteria (see Chapters 2, 8, and 9). Then, these particular prior distributions must be derived from the sample distribution, since this is the only available information. For obvious reasons, they are called *noninformative* priors. We describe below some of the most important techniques in the derivation of noninformative priors, referring the reader to Kass and Wasserman (1996) for an extensive treatment of these notions, and an annotated bibliography. Their main point is worth reproducing here, before we embark upon this description: noninformative priors cannot be expected to represent exactly total ignorance about the problem at hand, but should rather be taken as reference or default priors, upon which everyone could fall back when the prior information is missing. In this light, some noninformative priors may be more useful or more efficient than others, but they cannot be said to be less informative than others.

3.5.1 Laplace's prior

Historically, Laplace was the first to use noninformative techniques, since, although he had no information about the number of white balls in the urn or the proportion of male births (Examples 1.2.3 and 1.2.4), he put a prior distribution on these parameters that took into account his ignorance and gave the same likelihood to each value of the parameter, i.e., he used a *uniform prior*. His reasoning, later called the *Principle of Insufficient Reason*, was based on the *equiprobability* of elementary events and therefore appeared to be sound enough.

Three criticisms were later addressed to this choice. First, the resulting distributions are improper when the parameter space is not compact and some statisticians object to the use of improper priors, arguing that they lead to difficulties like the *marginalization paradoxes* (see Exercises 3.44–3.50). Such misgivings are not really justified, since it is actually possible to work with improper priors, as seen in Section 1.5, as long as we do not try to interpret them as probability distributions (see also Stone (1976)). As mentioned in Section 3.2, it may be argued that, on the contrary, a subjective determination of the prior distribution *should* lead to an improper prior.

Second, the Laplace principle of equiprobable events is not coherent under partitioning: if $\Theta = \{\theta_1, \theta_2\}$, Laplace's rule gives $\pi(\theta_1) = \pi(\theta_2) = 1/2$ but, if the definition of Θ is refined as $\Theta = \{\theta_1, \omega_1, \omega_2\}$, Laplace's rule leads to $\pi(\theta_1) = 1/3$, which is obviously not coherent with the first statement. As argued in Kass and Wasserman (1996), this incoherence is not such a serious problem: it can be evacuated by arguing that the level of partitioning must be fixed at some point in the analysis and that introducing a finer degree in the partitioning does not necessarily make sense.

The third criticism is more fundamental, since it deals with the problem of *invariance under reparameterization*. If we switch from $\theta \in \Theta$ to $\eta = g(\theta)$ by a one-to-one transformation g, prior information is still totally missing and should not be modified. However, if $\pi(\theta) = 1$, the corresponding prior distribution on η is $\pi^*(\eta) = \left| \frac{d}{d\eta} g^{-1}(\eta) \right|$ by the Jacobian formula. Therefore, $\pi(\eta)$ is usually not constant.

Example 3.5.1 If p, the proportion of male births, has a uniform prior distribution on [0,1], the odds parameter $\varrho = \frac{p}{1-p}$ has a prior distribution with density $1/(1 + \varrho)^2$ which is therefore not constant. ∥

Of course, it can sometimes be argued that there is a natural parameter of interest, and therefore that the choice of a uniform prior on the parameter of interest does not need to be invariant under reparameterization. But this argument does not hold if more than one inference about θ is necessary; for instance, we may need to derive the first two posterior moments of θ, but the latter is also the expectation of θ^2. Or, in Example 3.5.1, both the probability θ and the odds ratio ϱ may be of interest. Therefore, it seems that a more intrinsic and more acceptable notion of noninformative priors should satisfy *invariance under reparameterization*.

3.5.2 Invariant priors

A first solution is to take advantage of the invariance characteristics of the problem, that is, to use the groups $\mathcal{G}$ acting on $\mathcal{X}$ that induce groups $\mathcal{G}^*$ acting on Θ (that is, only the parameters the distribution of x change under transformations of x by elements of $\mathcal{G}$). Chapter 9 details the links between invariance structures and the Bayesian approach, since these structures allow for derivation of a noninformative prior that is compatible with the invariance requirement, namely, the right Haar measure on $\mathcal{G}^*$. See Kass and Wasserman (1996) for various arguments in favor of the right Haar measure.

Two introductory examples are presented below.

Example 3.5.2 The family $f(x - \theta)$ is *translation invariant*, i.e., $y = x - x_0$ has a distribution in the same family for every x_0, $f(y - (\theta - x_0))$; θ is then said to be a *location parameter* and an invariance requirement is that the prior distribution should be translation invariant, i.e., satisfy

$$\pi(\theta) = \pi(\theta - \theta_0)$$

for every θ_0. The solution is $\pi(\theta) = c$, the uniform distribution on Θ.   ‖

Example 3.5.3 If the distribution family is parametrized by a *scale parameter*, i.e., is of the form $1/\sigma f(x/\sigma)$ $(\sigma > 0)$, it is *scale-invariant*, i.e., $y = x/\sigma \sim f(y)$. A scale invariant prior π satisfies $\pi(A) = \pi(A/c)$ for every measurable set A in $(0, +\infty)$ and $c > 0$, i.e.

$$\pi(\sigma) = \frac{1}{c}\pi\left(\frac{\sigma}{c}\right).$$

This implies $\pi(\sigma) = \alpha/\sigma$, where α is a constant. Therefore, the invariant measure is not constant anymore.   ‖

The invariance approach is only partly satisfactory because it implies the reference to an invariance structure, which may be chosen in several ways, may not exist (see Chapter 9), or may be of no interest to the decision maker.

3.5.3 The Jeffreys prior

Jeffreys (1946, 1961) proposes an intrinsic approach that indeed obviates the need to take a potential invariance structure into account, while often being compatible with it when it exists. The *Jeffreys noninformative prior distributions* are based on *Fisher information*, given by

$$I(\theta) = \mathbb{E}_\theta\left[\left(\frac{\partial \log f(X \mid \theta)}{\partial \theta}\right)^2\right]$$

in the one-dimensional case. Under some regularity assumptions, this information can also be written as

(3.5.1) $$I(\theta) = -\mathbb{E}_\theta\left[\frac{\partial^2 \log f(X \mid \theta)}{\partial \theta^2}\right].$$

The Jeffreys prior distribution is

$$\pi^*(\theta) \propto I^{1/2}(\theta),$$

defined up to a normalization coefficient when π^* is proper. It actually satisfies the invariant reparameterization requirement, since, given a one-to-one transform h, we have the (Jacobian) transformation

$$I(\theta) = I(h(\theta))(h'(\theta))^2$$

(which explains the exponent $1/2$). Moreover, it also provides the invariant distributions obtained in Examples 3.5.2 and 3.5.3. More fundamentally, the choice of a prior depending on Fisher information is justified by the fact that $I(\theta)$ is widely accepted as an indicator of the amount of information brought by the model (or the observation) about θ (Fisher (1956)). Therefore, at least at a qualitative level, it seems intuitively justified that the values of θ for which $I(\theta)$ is larger should be more likely for the prior distribution. In other words, $I(\theta)$ can evaluate the ability of the model to discriminate between θ and $\theta + d\theta$ through the expected slope of $\log f(x|\theta)$. To favor the values of θ for which $I(\theta)$ is large is equivalent to minimizing the influence of the prior distribution and is therefore as noninformative as possible. In fact, the Jeffreys prior is usually improper but the developments of Section 1.5 show how to enhance a Bayesian analysis in this case.

Example 3.5.4 (Example 3.5.1 continued) If $x \sim B(n, p)$,

$$f(x|p) \;=\; \binom{n}{x} p^x (1-p)^{n-x},$$

$$\frac{\partial^2 \log f(x|p)}{\partial p^2} \;=\; \frac{x}{p^2} + \frac{n-x}{(1-p)^2},$$

and

$$I(p) \;=\; n\left[\frac{1}{p} + \frac{1}{1-p}\right] = \frac{n}{p(1-p)}.$$

Therefore, the Jeffreys prior for this model is

$$\pi^*(p) \propto [p(1-p)]^{-1/2}$$

and is thus proper, since it is a $\mathcal{Be}(1/2, 1/2)$ distribution.   ‖

When θ is a multidimensional parameter, the Fisher information matrix is defined as a generalization of (3.5.1). For $\theta \in \mathbb{R}^k$, $I(\theta)$ has the following elements,

$$I_{ij}(\theta) = -\mathbb{E}_\theta\left[\frac{\partial^2}{\partial\theta_i\partial\theta_j} \log f(x|\theta)\right] \qquad (i, j = 1, \ldots, k),$$

and the Jeffreys noninformative prior is then defined by

$$\pi^*(\theta) \propto [\det(I(\theta))]^{1/2}.$$

It is still reparameterization-invariant. Note that, if $f(x|\theta)$ belongs to an exponential family,

$$f(x|\theta) = h(x)\exp(\theta \cdot x - \psi(\theta)),$$

the Fisher information matrix is given by $I(\theta) = \nabla\nabla^t\psi(\theta)$ and

$$(3.5.2) \qquad\qquad \pi^*(\theta) \propto \left(\prod_{i=1}^{k} \psi''_{ii}(\theta)\right)^{1/2},$$

where $\psi''_{ii}(\theta) = \frac{\partial^2}{\partial\theta_i^2}\psi(\theta)$.

In the multidimensional case, the Jeffreys noninformative approach may lead to incoherences or even paradoxes (see Examples 3.5.6 and 3.5.9) and we must stress that Jeffreys (1961) was mainly emphasizing the use of these distributions in the one-dimensional case (see Berger and Bernardo (1992b)). However, his method provides one of the best automated techniques to derive noninformative prior distributions. Moreover, it is often related to some classical estimators.

Example 3.5.5 Consider $x \sim \mathcal{N}(\theta, I_p)$. Because this is a location family, the Jeffreys prior is constant. The corresponding generalized Bayes estimator is given by

$$\delta^{\pi^*}(x) = \frac{\int_{\mathbb{R}^p} \theta \exp(-||x-\theta||^2/2)\,d\theta}{\int_{\mathbb{R}^p} \exp(-||x-\theta||^2/2)\,d\theta} = x.$$

It is minimax for every p and admissible for $p \leq 2$. Notice that this estimator is, in addition, the best equivariant estimator for location parameters (see Chapter 9). ‖

Example 3.5.6 Consider $x \sim \mathcal{N}(\mu, \sigma^2)$ with $\theta = (\mu, \sigma)$ unknown. In this case,

$$\begin{aligned}
I(\theta) &= \mathbb{E}_\theta\left[\begin{pmatrix} 1/\sigma^2 & 2(x-\mu)/\sigma^3 \\ 2(x-\mu)/\sigma^3 & 3(\mu-x)^2/\sigma^4 - 1/\sigma^2 \end{pmatrix}\right] \\
&= \begin{pmatrix} 1/\sigma^2 & 0 \\ 0 & 2/\sigma^2 \end{pmatrix}
\end{aligned}$$

and the corresponding noninformative prior distribution is $\pi(\theta) \propto 1/\sigma^2$. If, however, μ and σ are assumed to be a priori independent, the corresponding noninformative prior would be $\pi(\mu, \sigma) = \sigma^{-1}$, which is also the invariant Haar measure for this location-scale model (see Example 3.5.3 and Chapter 9). ‖

This approach is criticized by some Bayesians as being merely a tool without subjective justifications in terms of prior information. However, the only alternative to an automated approach is to require that prior information be always available, a requirement that cannot hold in every setting. Another criticism of the Jeffreys method is that, although it meets the *reparameterization-invariance* requirement, it does not satisfy the Likelihood Principle. In fact, the Fisher information can differ for two experiments providing proportional likelihoods, as shown by the following example.

Example 3.5.7 We saw in Example 1.3.4 that binomial and negative binomial modelings could lead to the same likelihood. However, if $x \sim \mathcal{B}(n, \theta)$, the noninformative prior $\pi_1(\theta)$ is $\mathcal{B}e(1/2, 1/2)$ (Example 3.5.1) and, if $n \sim \mathcal{N}eg(x, \theta)$, the Jeffreys prior is

$$
\begin{aligned}
\pi_2(\theta) &= -\mathbb{E}_\theta\left[\frac{\partial^2}{\partial\theta^2}\log f(x|\theta)\right] \\
&= \mathbb{E}_\theta\left[\frac{x}{\theta^2} + \frac{n-x}{(1-\theta)^2}\right] = \frac{x}{\theta^2(1-\theta)},
\end{aligned}
$$

i.e., $\pi_2(\theta) \propto \theta^{-1}(1-\theta)^{-1/2}$, which is improper and, more importantly, differs from π_1.   ‖

As shown by the following example, it often occurs that Jeffreys' noninformative distribution is a limit of conjugate priors.

Example 3.5.8 If $x \sim \mathcal{U}([0, \theta])$, a conjugate prior is the Pareto distribution, $\mathcal{P}a(\theta_0, \alpha)$,

$$
\pi(\theta) = \alpha \theta_0^\alpha \theta^{-\alpha-1} \mathbb{I}_{[\theta_0, +\infty[}(\theta),
$$

leading to the posterior distribution $\mathcal{P}a(\max(\theta_0, x), \alpha+1)$. Under the invariant loss

$$
L(\theta, \delta) = \frac{(\theta - \delta)^2}{\theta^2},
$$

the Bayes estimator is, if $\theta_0 \vee x = \max(\theta_0, x)$,

$$
\delta^\pi(x) = \frac{\int_{\theta_0 \vee x}^{+\infty} \theta^{-1}(\alpha+1)\theta_0^{\alpha+1}\theta^{-\alpha-2}d\theta}{\int_{\theta_0 \vee x}^{+\infty} \theta^{-2}(\alpha+1)\theta_0^{\alpha+1}\theta^{-\alpha-2}d\theta} = \frac{\alpha+3}{\alpha+2}(\theta_0 \vee x),
$$

converging to the minimax estimator, $\delta_0(x) = (3/2)x$, when α and θ_0 go to 0. Because θ is a scale parameter, the noninformative distribution can be taken to be $\pi(\theta) = 1/\theta$, which is also the Jeffreys prior for this model. This distribution corresponds to $\theta_0 = 0$ and $\alpha = 0$ for an unnormalized Pareto distribution (that is, without the scaling factor $\alpha\theta_0^\alpha$). This representation can also be used to show that δ_0 is admissible, using *Stein's sufficient admissibility condition* (see Chapter 8).   ‖

A more important drawback of Jeffreys' noninformative prior distributions is that they do not necessarily perform satisfactorily for all inferential purposes, in particular when considering subvectors of interest. The following problem was pointed out by Stein (1959) (see also Tibshirani (1989)).

Example 3.5.9 If $x \sim \mathcal{N}_p(\theta, I_p)$, the noninformative prior is $\pi(\theta) = 1$. The resulting estimator of θ, x, is quite acceptable, as seen in Example

3.5.5. Now, as $\theta|x \sim \mathcal{N}_p(x, I_p)$, the posterior distribution of $\eta = ||\theta||^2$ is $\chi_p^2(||x||^2)$, the noncentral chi-squared distribution. When η is the parameter of interest, the posterior mean of η is $\delta^\pi(x) = \mathbb{E}^\pi[\eta|x] = ||x||^2 + p$. However, the best estimator among the estimators of the form $||x||^2 + c$ (for quadratic loss) is $||x||^2 - p$, which uniformly dominates the generalized Bayes estimator, δ^π (see Exercise 2.35). Therefore, the marginal distribution on η deduced from the Jeffreys noninformative prior on θ is definitely suboptimal. Moreover, the Jeffreys noninformative distribution derived from the reduced observation $z = ||x||^2$ is quite different from $\chi_p^2(||x||^2)$ and leads to an estimator of η with much more acceptable performance (see Exercise 3.52). ∥

Exercise 4.47 also illustrates the possible inconsistency of the Jeffreys prior in a linear calibration setting, with a resolution of the inconsistency by the reference prior method.

3.5.4 Reference priors

The type of problem mentioned at the end of the previous section was taken into account by Bernardo (1979), who proposed a modification of the Jeffreys approach called *the reference prior approach*. A major difference is that this method distinguishes between parameters of interest and nuisance parameters (for instance, $||\theta||^2$ and $\theta/||\theta||$ in Example 3.5.9). Therefore, the resulting prior distribution depends not only on the sample distribution, but also on the inferential problem at hand. The remainder of this section briefly presents the derivation of reference priors. For a detailed study, see Berger and Bernardo (1989, 1992a, 1992b) and Kass and Wasserman (1996).

When $x \sim f(x|\theta)$ and $\theta = (\theta_1, \theta_2)$, where θ_1 is the parameter of interest, the reference prior is obtained by first defining $\pi(\theta_2|\theta_1)$ as the Jeffreys prior associated with $f(x|\theta)$ when θ_1 is fixed, then deriving the marginal distribution

$$(3.5.3) \qquad \tilde{f}(x|\theta_1) = \int f(x|\theta_1, \theta_2)\pi(\theta_2|\theta_1)d\theta_2,$$

and computing the Jeffreys prior $\pi(\theta_1)$ associated with $\tilde{f}(x|\theta_1)$. The principle behind the reference prior is therefore to eliminate the nuisance parameter by using a Jeffreys prior where the parameter of interest remains fixed. (Notice that the integral in (3.5.3) is not necessarily defined, a problem that may require integrating first on a sequence of compact sets and then taking the limit.)

Example 3.5.10 The Neyman–Scott (1948) problem is related to the observation of x_{ij}'s distributed from $\mathcal{N}(\mu_i, \sigma^2)$, $i = 1, \ldots, n$, $j = 1, 2$. The usual Jeffreys prior for this model is $\pi(\mu_1, \ldots, \mu_n, \sigma) = \sigma^{-n-1}$ and an inconsistency arises because $\mathbb{E}[\sigma^2|x_{11}, \ldots, x_{n2}] = s^2/(2n-2)$, where

$$s^2 = \sum_{i=1}^{n} \frac{(x_{i1} - x_{i2})^2}{2},$$

and this posterior expectation converges to $\sigma^2/2$ with n. (Notice that this is a case where the number of parameters increases with the number of observations.) The reference prior associated with $\theta_1 = \sigma$ and $\theta_2 = (\mu_1, \ldots, \mu_n)$ gives a flat prior for $\pi(\theta_2|\theta_1)$, since θ_2 is a location parameter. Then

$$\tilde{f}(x|\theta_1) = \prod_{i=1}^{n} e^{-(x_{i1}-x_{i2})^2/4\sigma^2} \frac{1}{\sqrt{2\pi}2\sigma}$$

is a scale family and $\pi(\sigma) = 1/\sigma$. Therefore, $\mathbb{E}[\sigma^2|x_{11}, \ldots, x_{n2}] = s^2/(n-2)$, which is consistent.   ‖

The general derivation of the reference prior is as follows: Consider $x \sim f(x|\theta)$, with $\theta \in \Theta \subset \mathbb{R}^k$. Assume that the Fisher information matrix $I(\theta)$ exists and is of full rank. We denote $\mathbf{S} = I^{-1}(\theta)$. Parameters are now separated in m groups that correspond to their respective levels of importance,

(3.5.4) $\theta_{(1)} = (\theta_1, \ldots, \theta_{n_1}),$ $\ldots$ $\theta_{(m)} = (\theta_{N_{m-1}+1}, \ldots, \theta_k),$

with $N_i = \sum_{j=1}^{i} n_j$ (after a possible reindexing of the components of θ). The reference prior method derives a prior distribution on $(\theta_{(1)}, \ldots, \theta_{(m)})$ that takes into account this division, i.e., which truly separates between nuisance parameters and parameters of interest, even allowing a finer level of separation between the respective levels of importance of these parameters. We introduce the following notation: for $j = 1, \ldots, m$,

$$\theta_{[j]} = (\theta_{(1)}, \ldots, \theta_{(j)}) \quad \text{and} \quad \theta_{[\sim j]} = (\theta_{(j+1)}, \ldots, \theta_{(m)}).$$

The matrix $\mathbf{S}$ is decomposed according to the partition (3.5.4),

$$\mathbf{S} = \begin{pmatrix} \mathbf{A}_{11} & \mathbf{A}_{21}^t & \cdots & \mathbf{A}_{m1}^t \\ \mathbf{A}_{21} & \mathbf{A}_{22} & & \mathbf{A}_{m2}^t \\ & & \cdots & \\ \mathbf{A}_{m1} & & & \mathbf{A}_{mm} \end{pmatrix}$$

and S_j is the upper left (N_j, N_j) corner of $\mathbf{S}$. (For instance, $S_1 = \mathbf{A}_{11}$.) We denote $\mathbf{H}_j = S_j^{-1}$ and $\mathbf{h}_j$ is the lower right (n_j, n_j) corner of $\mathbf{H}_j$. (In particular, $\mathbf{h}_1 = \mathbf{A}_{11}^{-1}$.) The reference prior construction then proceeds as follows.

– *Initialization:*

$$\pi_m(\theta_{(m)}|\theta_{[m-1]}) = \frac{|\mathbf{h}_m(\theta)|^{1/2}}{\int |\mathbf{h}_m(\theta)|^{1/2} \, d\theta_{(m)}}.$$

– *Iteration:*   For $j = m - 1, \ldots, 1$,

$$\pi_j(\theta_{[\sim j-1]}|\theta_{[j-1]}) = \frac{\pi_{j+1}(\theta_{[\sim j]}|\theta_{[j]})\exp\{\frac{1}{2}\mathbb{E}_j[\log(|\mathbf{h}_j(\theta)|)|\theta_{[j]}]\}}{\int \exp\{\frac{1}{2}\mathbb{E}_j[\log(|\mathbf{h}_j(\theta)|)|\theta_{[j]}]\}\,d\theta_{(j)}},$$

where

$$\mathbb{E}_j[g(\theta)|\theta_{[j]}] = \int g(\theta)\pi_{j+1}(\theta_{[\sim j]}|\theta_{[j]})\,d\theta_{[\sim j]}.$$

– *Conclusion:*   The reference prior is $\pi(\theta) = \pi_1(\theta_{[\sim 0]}|\theta_{[0]})$.

Often, some of the integrals appearing in this algorithm are not defined. Berger and Bernardo (1989) then propose to derive the reference prior for compact subsets Θ_n of Θ and to consider the limit of the corresponding reference priors π_n as n goes to infinity and Θ_n goes to Θ. In general, the resulting limit does not depend on the choice of the sequence of compact sets.

Example 3.5.11 (Example 3.5.9 continued) Since $\eta = ||\theta||^2$ is the parameter of interest, θ can be written in polar coordinates $(\eta, \varphi_1, \ldots, \varphi_{p-1})$, with

$$
\begin{aligned}
\theta_1 &= \sqrt{\eta}\cos(\varphi_1),\\
\theta_2 &= \sqrt{\eta}\sin(\varphi_1)\cos(\varphi_2),\\
&\cdots\\
\theta_{p-1} &= \sqrt{\eta}\sin(\varphi_1)\cdots\cos(\varphi_{p-1}),\\
\theta_p &= \sqrt{\eta}\sin(\varphi_1)\cdots\sin(\varphi_{p-1}).
\end{aligned}
$$

The Fisher information matrix for $(\eta, \varphi_1, \ldots, \varphi_{p-1})$ is then $\mathbf{H} = \mathbf{J}\mathbf{J}^t$, where $\mathbf{J}$ is the Jacobian matrix $\frac{D(\theta_1,\ldots,\theta_p)}{D(\eta,\varphi_1,\ldots,\varphi_{p-1})}$. It can be shown that $\mathbf{J}$ is of the form

$$\mathbf{J} = \begin{bmatrix} A^t/\sqrt{\eta} \\ \sqrt{\eta}\mathbf{B} \end{bmatrix},$$

with $A \in \mathbb{R}^p$ and $\mathbf{B}$ $(p-1) \times p$ matrix. Thus, for the partition of θ in $\theta_{(1)} = \eta$, $\theta_{(2)} = (\varphi_1, \ldots, \varphi_{p-1})$, we get

$$\pi_2(\varphi_1, \ldots, \varphi_{p-1}|\eta) \propto |\mathbf{H}_{22}|^{1/2},$$

which does not depend on η. The marginal distribution of η is

$$\pi_1(\eta) \propto \exp\left\{\mathbb{E}\left[\log\left(\frac{1}{2}\frac{|\mathbf{H}|}{|\mathbf{H}_{22}|}\right)\Big|\eta\right]\right\}$$

and $\frac{|\mathbf{H}|}{|\mathbf{H}_{22}|} \propto (1/\eta)$. Therefore,

$$\pi_1(\eta) = 1/\sqrt{\eta},$$

which leads to an estimator of $||\theta||^2$ more interesting than $||x||^2 + p$ (see Exercise 3.52).

Actually, the same marginalization problem appears for maximum likelihood estimation. In fact, the maximum likelihood estimator of η based on the sample is $||x||^2$, which is also dominated by $||x||^2 - p$. On the contrary, the maximum likelihood estimator derived from

$$z = ||x||^2 \sim \chi_p^2(||\theta||^2)$$

behaves similarly to $(||x||^2 - p)^+$ (see Saxena and Alam (1982), Chow (1987), and Chow and Hwang (1990)). ||

This algorithm is justified as providing the prior distribution that maximizes the posterior information (see Bernardo (1979) and Berger and Bernardo (1992b)). More precisely, if $x_{1:n}$ denotes the sample $(x_1, \ldots, x_n)$ and if $K_n(\pi)$ is the Kullback–Leibler divergence between the prior π and the corresponding posterior,

$$K_n(\pi) = \int \pi(\theta|x_{1:n}) \log \left(\pi(\theta|x_{1:n})/\pi(\theta)\right) \, d\theta,$$

Bernardo's (1979) idea is to use $\mathbb{E}[K_n(\pi)]$, where the expectation is taken under the marginal distribution of $x_{1:n}$, as a measure of *missing information*, and he defines the reference prior as π maximizing

$$K^*(\pi) = \lim_{n \to \infty} \mathbb{E}[K_n(\pi)].$$

Barring the technical difficulties associated with possibly infinite integrals, the resulting prior is the Jeffreys prior for continuous parameter spaces and the uniform prior for finite spaces. See Ghosh and Mukerjee (1992a), Clarke and Wasserman (1993) and Kass and Wasserman (1996) for more motivations in terms of asymptotic optimality.

Reference prior distributions also depend on the way parameters are ordered (see Exercise 3.59), an advantage compared with the Jeffreys method because nuisance parameters are considered in a different way. Paradoxes like those of Example 3.5.9 are then avoided. It may seem unreasonable to modify the prior distributions according to the problem of interest, but one has to realize that, apart from the sample distribution $f(x|\theta)$, these inferential problems are the unique available information.[2] Notice that invariance under reparameterization is preserved only if the changes are bijective and within each of the groups in (3.5.4). However, the invariance requirement is of less importance in this setting because the ordering (3.5.4) somehow prohibits a reparameterization between the classes, as the different groups are not of the same type. When no such ordering can be proposed, Berger and Bernardo (1992a) suggest that one consider as a noninformative prior the reference prior for which each component of θ is considered separately. (On the contrary, the Jeffreys noninformative prior treats θ as a single group.)

---

[2] If a loss function L is available, it also contains some information about θ and one could use the duality between loss and prior distribution to derive a prior distribution adapted to this loss. (See Chapter 2 and Rubin (1987).) But very little work has been done about the derivation of a prior distribution from a loss function.

Example 3.5.12 (Berger and Bernardo (1992a)) Consider an analysis of variance model

$$x_{ij} = \mu + \alpha_i + \epsilon_{ij}, \quad i = 1, \ldots, p, \quad j = 1, \ldots, n,$$

with $\alpha_i \sim \mathcal{N}(0, \tau^2)$, $\epsilon_{ij} \sim \mathcal{N}(0, \sigma^2)$. For different orderings of the parameters, μ, τ^2, σ^2, we get the following reference priors:

$$\pi_1((\mu, \sigma^2, \tau^2)) \propto \sigma^{-2}(n\tau^2 + \sigma^2)^{-3/2}$$

$$\pi_2(\mu, \sigma^2, \tau^2) \propto \tau^{-C_n}\sigma^2 \left[(n-1) + (1 + n\tau^2/\sigma^2)^{-2}\right]^{1/2}$$

$$\pi_3(\mu, (\sigma^2, \tau^2)) \propto \sigma^{-2}(n\tau^2 + \sigma^2)^{-1}$$

$$\pi_4((\mu, \sigma^2), \tau^2) \propto \sigma^{-5/2}(n\tau^2 + \sigma^2)^{-1}$$

with $C_n = \{1 - \sqrt{n-1}(\sqrt{n} + \sqrt{n-1})^{-3}\}$.   ‖

3.5.5 Matching priors

A peculiar, not to say paradoxical, approach to noninformative priors is to look for good frequentist properties, that is, properties that hold on the average (in x), rather than conditional on x. First, as discussed in Chapters 2 and 8, there are priors that lead to optimal estimators in terms of frequentist optimality properties such as minimaxity or admissibility, and one may want to restrict the choice of a prior to these optimal distributions. It is, however, rarely the case that one specific prior emerges from this requirement. Either there is no such prior, as in small dimensions for estimation under the quadratic loss (Note 2.8.2), or there is an infinity of priors associated with, say, admissible minimax estimators (Fourdrinier, Strawderman and Wells (1998)). (The exception here is when invariance structures are present, in which case the right Haar measure is the appropriate choice, as discussed in Section 3.5.2.)

A more common approach is to require that some posterior probabilities coincide, to a certain degree of approximation, with the corresponding frequentist coverage. Hence the denomination of matching priors, which is often restricted in the literature to one-sided confidence intervals. Given a posterior confidence set on $g(\theta)$, C_x,

$$\pi(g(\theta) \in C_x | x) = 1 - \alpha,$$

which may be one- or two-sided, this set defines a confidence set in the frequentist sense, with a frequentist coverage

$$P_\theta(C_x \ni g(\theta)) = \int \mathbb{I}_{C_x}(g(\theta)) \, f(x|\theta) \, dx,$$

which is usually different from $1 - \alpha$. When there are pivotal quantities, as in the normal $\mathcal{N}(\theta, 1/n)$ case, the $1 - \alpha$ HPD region (Chapter 5) is given by

$$C_x = [\bar{x}_n - n^{-1/2}q_{\alpha/2}, \bar{x}_n + n^{-1/2}q_{\alpha/2}],$$

where $q_{\alpha/2}$ is the normal $1 - \alpha/2$ quantile and the frequentist coverage of C_x is also $1 - \alpha$. (Lindley (1958) generalizes this result to other location families and shows that it only holds for location families.) In a general (unidimensional) framework, Welch and Peers (1963, 1965) have shown that, when $C_x = (-\infty, k_\alpha(x)]$,

$$P_\theta(\theta \le k_\alpha(x)) = 1 - \alpha + O(n^{-1/2}),$$

and that, for the Jeffreys prior,

$$P_\theta(\theta \le k_\alpha(x)) = 1 - \alpha + O(n^{-1}),$$

which thus improves the agreement by a factor of $1/2$.

In the case there are nuisance parameters in the model, that is, when inference concentrates on a unidimensional component θ_1 of the parameter, things get more complicated. References to works in this area include Sweeting (1985), Severini (1991), Ghosh and Mukerjee (1992a,b, 1993), Mukerjee and Dey (1993), DiCiccio and Stern (1993, 1994), Liseo (1993), and Datta and Ghosh (1995a,b). We focus here on some results obtained in Rousseau (1997, 2000, 2002, 2005).

The *Edgeworth expansion* (see Bhattacharya and Rao (1986), Bickel and Ghosh (1990), and DiCiccio and Stern (1994)) of the frequentist coverage probability is given by

$$P_\theta(\theta_1 < k_n(\alpha)) = 1 - \alpha +$$
$$\frac{\varphi(\Phi^{-1}(1 - \alpha))}{\sqrt{n}} \left(\frac{I'(\theta)\nabla \log \pi(\theta)}{I''(\theta)^{1/2}} - \nabla^t \frac{I'(\theta)}{I''(\theta)^{1/2}} \right) + O(n^{-1}),$$

in the one-sided case, where φ and Φ are the density and c.d.f. of the normal distribution, respectively, and $I(\theta)$, $I'(\theta)$, and $I''(\theta)$ are Fisher information and its first and second derivatives, respectively. For the two-sided case of an HPD region at level $1 - \alpha$, $C_x^{HPD}(\alpha)$, on $\theta \in \mathbb{R}$, the corresponding expansion is

$$P_\theta(\theta \in C_x^{HPD}) = 1 - \alpha + n^{-1}q(\alpha)b(\pi, \theta) + O(n^{-3/2}),$$

where q is related to the χ^2 density and

$$b(\pi, \theta) = \frac{\mu_3' - \mu_2''}{I(\theta)^2} + 2\frac{\mu_2'(\mu_3 - \mu_2')}{I(\theta)^3} + \frac{\pi'(\theta)}{\pi(\theta)} \frac{\mu_3 - \mu_2'}{I(\theta)^2} - \frac{\pi''(\theta)}{\pi(\theta)I(\theta)} - \frac{\mu_2'\pi'(\theta)}{\pi(\theta)I(\theta)^2},$$

where the μ_j's are defined as $(j = 2, 3)$

$$\mu_j = \mathbb{E}_\theta \left[\frac{\partial^j \log f(x|\theta)}{\partial \theta^j} \right].$$

The choice of a matching prior is then dictated by the cancellation of the first order term, as in Welch and Peers' (1963) differential equation:

$$[I''(\theta)]^{-1/2}I'(\theta)\nabla \log \pi(\theta) + \nabla^t\{I'(\theta)[I''(\theta)]^{-1/2}\} = 0.$$

This differential equation may or may not have a solution. Moreover, as shown in the generalization of Rousseau (2002) to HPD regions, the solution, when it exists, will depend on the parameter of interest for those regions and most often does not produce Jeffreys' prior, although there always is a parameterization leading to Jeffreys' prior.

Example 3.5.13 (Rousseau (2000)) Consider the $\mathcal{G}(k,\theta)$ distribution. Then, if θ itself is the parameter of interest, the priors canceling the second order term for HPD regions are of the form

$$\pi(\theta) = \frac{c_1 + c_2\theta}{\theta}, \qquad c_1, c_2 > 0,$$

and thus include the Jeffreys prior as a particular case. If $\eta = c_1\theta^{5/3} + c_2\log(\theta)$ is the quantity of interest, corresponding to the χ^2 parameterization, the prior with the maximum matching property is

$$\pi(\eta) = I(\eta)^{-1},$$

instead of the Jeffreys prior, $I(\eta)^{1/2}$. Lastly, consider the mean parameterization, $\mu = k/\theta$. The matching priors are then of the form

$$\pi(\mu) = c_1\mu^2 + c_2/\mu, \qquad c_1, c_2 > 0,$$

and, again, do not include the Jeffreys prior. $\qquad\qquad\qquad\qquad\qquad\quad \|$

See also Rousseau (1997) for an extension to discrete settings where matching the coverage probabilities is not possible for orders higher than $n^{1/2}$ and randomization is necessary to achieve such orders.

Example 3.5.14 (Ghosh, Carlin and Srivastava (1995)) A simple version of the *linear calibration model* is to consider $(i = 1, \ldots, n, j = 1, \ldots, k)$,

$$(3.5.5) \qquad y_i = \alpha + \beta x_i + \varepsilon_i, \qquad y_{0j} = \alpha + \beta x_0 + \varepsilon_{0j},$$

where x_0 is unknown and is the quantity of interest (see Exercise 4.47 for more details on this model). For one-sided confidence intervals, the differential equation associated with (3.5.5) is then

$$|\beta|^{-1}s^{-1/2}\frac{\partial}{\partial x_0}\{e(x_0)\pi(\theta)\} - e^{-1/2}(x_0)\mathrm{sgn}(\beta)n^{-1}s^{1/2}\frac{\partial\pi(\theta)}{\partial x_0}$$

$$-e^{-1/2}(x_0)(x_0 - \bar{x})s^{-1/2}\frac{\partial}{\partial\beta}\{\mathrm{sgn}(\beta)\pi(\theta)\} = 0$$

with $\theta = (x_0, \alpha, \beta, \sigma^2)$ and

$$s = \Sigma(x_i - \bar{x})^2, \quad e(x_0) = [(n+k)s + nk(x_0 - \bar{x})^2]/nk.$$

The solutions to this differential equation are then of the form

$$(3.5.6) \qquad \pi(x_0, \alpha, \beta, \sigma^2) \propto e(x_0)^{(d-1)/2}|\beta|^d g(\sigma^2),$$

where g is arbitrary. For instance, if $g(\sigma^2) = (\sigma^2)^{-a/2}$, the corresponding posterior distribution is proper for $(n+k+a-2d-5) > 0$. In this case, the reference priors are also matching priors (3.5.6), as shown by Table 3.5.5 for four different orderings of the parameters. $\qquad\qquad\qquad\qquad\quad \|$

Table 3.5.1. *Matching reference priors associated with different orderings for the linear calibration model (3.5.5).*

Partition	Prior
$(x_0, \alpha, \beta, \sigma^2)$	$|\beta|(\sigma^2)^{-5/2}$
$x_0, \alpha, \beta, \sigma^2$	$e(x_0)^{-1/2}(\sigma^2)^{-1}$
$x_0, \alpha, (\sigma^2, \beta)$	$e(x_0)^{-1/2}(\sigma^2)^{-3/2}$
$x_0, (\alpha, \beta), \sigma^2$	$e(x_0)^{-1/2}(\sigma^2)^{-1}$
$x_0, (\alpha, \beta, \sigma^2)$	$e(x_0)^{-1/2}(\sigma^2)^{-2}$

In general, (reverse) reference priors are matching priors when the parameter of interest, λ, and the nuisance parameter, ω, are orthogonal in the sense of Fisher information

$$I(\lambda, \eta) = \begin{pmatrix} I_{11} & 0 \\ 0 & I_{22} \end{pmatrix},$$

as detailed in Tibshirani (1989), and also when the reverse order (ω, λ) is used for the construction of the reference prior as discussed in Berger (1992) and Berger et al. (1998).

Besides the technical difficulty one faces in handling matching priors, there is also a conceptual difficulty in asking for frequentist coverage when constructing a prior distribution, whose goal is to condition on the observation rather than to rely on frequentist long-term properties. Agreement between both approaches should not be systematically rejected, as shown in Chapter 5, but this shift of paradigm is fairly disturbing, as shown for instance in Rousseau (1997) by the need to call for randomization, in violation of the Likelihood Principle. Therefore, we do not recommend it.

3.5.6 Other approaches

Alternatives to noninformative Bayesian analysis are described in Berger (1985a, Chapter 3) and Kass and Wasserman (1996). For instance, we can mention Rissanen (1983, 1990), who uses transmission information theory as in Shannon (1948). Considering the transmission of a binary message by a physical device, the noninformative prior distribution for a model $f(x|\theta)$ is the minimum length of message necessary to describe this model. In the simplest case, these distributions are similar to the Jeffreys priors and this similarity should also hold in general because of the connections existing between statistical information and information theory. A recent survey of this theory of *stochastic complexity* is given in Dawid (1992). See also Hansen and Yu (2001).

Notice also that testing settings require special prior distributions, as pointed out by Jeffreys (1961) and Kass and Wasserman (1996). We will discuss this specific problem in Chapter 5.

3.6 Posterior validation and robustness

Even when prior information is available, it rarely occurs that this information leads to an exact determination of the prior distribution, $\pi(\theta)$, if only because the discriminating power of individuals is restricted and practical determination of the tails of a distribution is almost impossible. In most cases, there is therefore an uncertainty about the selected prior distribution used for Bayesian inference. Obviously, if the information is precise, the prior will be better defined than in a noninformative setting. However, it always matters that the influence of this indeterminacy in the prior distribution on posterior quantities is clearly assessed and that the arbitrary part of the prior distribution does not predominate. The assessment of the influence of the prior is called *sensitivity analysis* (or *robustness analysis*). The concern about robustness and the derivation of appropriate tools to deal with this problem appear in the works of Good (1983) and Berger (1982b, 1984a, 1985a, 1990a). Other references on this problem are Berger and Berliner (1986), Berger and Sellke (1987), Berger and Delampady (1987), O'Hagan and Berger (1988), Sivaganesan and Berger (1989), Walley (1991), and Wasserman (1992).

Following Berger's (1990a) classification, we consider that uncertainty about the prior distribution π can be represented by the assumption that the (unknown) prior belongs to a class of distributions, Γ. These classes can be determined from a subjective or practical perspective. The major types of robustness classes considered in the literature follow.

(i) *Conjugate prior classes.* Such classes are typically chosen for practical reasons, since they generally provide explicit bounds on the quantity of interest. For instance, Das Gupta and Studden (1988) consider the case when $x \sim \mathcal{N}_p(\theta, I_p)$ and $\theta \sim \mathcal{N}_p(0, \Sigma)$, where $\Sigma_1 \preceq \Sigma \preceq \Sigma_2$, the order relation $\preceq$ being that the difference of the two matrices is positive semidefinite. The above criticisms on conjugate prior distributions obviously apply in this case and even more strongly, since the resulting class is only made of convenient distributions, but does not lead to a wide set of prior distributions compatible with the prior information.

(ii) *Determined moment classes.* If we assume that the (limited) prior information is only providing bounds on some moments of π, the corresponding class is

$$\Gamma_M = \{\pi; \ a_i \leq \mathbb{E}^\pi[\theta^i] \leq b_i, i = 1, \ldots, k\}.$$

However, Γ_M is not really more satisfactory than the previous classes in (i), because it imposes strong conditions on the tails of the prior and contains unrealistic prior distributions, including finite support distributions.More specifically, in most cases, the bounds on the posterior quantities will be attained for finite support distributions, owing to convexity reasons.

(iii) *Neighborhood classes.* Following its introduction by Huber (1972) for outliers detection, a rather popular class in robustness studies is the *ϵ-contamination class* around a given prior distribution π_0,

$$\Gamma_{\epsilon,\mathcal{Q}} = \{\pi = (1 - \epsilon)\pi_0 + \epsilon q; \ q \in \mathcal{Q}\},$$

where $\mathcal{Q}$ is a class of distributions, to be chosen according to the precision of the prior information. Berger and Berliner (1986) and Sivaganesan and Berger (1989) provide examples where such classes can be used. The main drawback in using $\Gamma_{\epsilon,\mathcal{Q}}$ is that ϵ and $\mathcal{Q}$ both need to be determined and it is usually difficult to derive them from the degree of uncertainty about π_0. But *mixture estimation* techniques may be instrumental in this setting when the prior information is derived from a sample of (possibly virtual) previous observations (see Section 6.4). A related relation would be to consider a true neighborhood associated with a given functional distance like Hellinger or Kullback–Leibler distances (see Section 2.5.4 and Zucchini (1999)). The difficulty is then in the *scaling* of these neighborhoods.

(iv) *Underspecified classes.* Such classes result from a construction of the prior on a sub-σ-algebra, that is, on a coarser set of events than the one of interest. This approach is strongly related to the axiomatic developments of Note 3.8.1, since the ordering on the relative likelihoods does not necessarily lead to a prior distribution on the Borel σ-field. For instance, it may be the case that the prior distributions have some determined quantiles,

$$\Gamma_Q = \{\pi; \ \ell_i \leq \int_{I_i} \pi(\theta) \, d\theta \leq u_i, i = 1, \dots, m\}$$

where $I_1, \dots, I_m$ is a partition of Θ. These classes are preferable to (ii), but it may still be necessary to eliminate unrealistic prior distributions from Γ_Q as in O'Hagan and Berger (1988). However, it seems that this approach is the most realistic because, for instance, fractiles are usually easier to determine than moments, and the most apt to give rise to practical implementation among the classes we consider here.

(v) *Ratio of densities classes.* Considering as for (iv) a subjective derivation of the prior distribution, this may also be done by a histogram representation. In such a case, the uncertainty about the prior information can be represented by upper and lower bounds on the density π and leads to the class

$$\Gamma_R = \{\pi; \ L(\theta) \leq \pi(\theta) \leq U(\theta)\},$$

where L and U are specified. The choice of these functions is difficult but is quite influential because, if they are similar, all the distributions in Γ_R will have the same behavior in the tails. See DeRobertis and Hartigan (1981) for a related class.

Berger (1990a) and Wasserman (1992) also provide computational tools to derive bounds on posterior quantities for the above classes. In fact, the robustness point of view substitutes for the current estimator $\varrho(\pi)$ the range of possible values of the estimator when the prior π varies in the class Γ,

$$\varrho^L = \inf_{\pi \in \Gamma} \varrho(\pi), \qquad \varrho^U = \sup_{\pi \in \Gamma} \varrho(\pi).$$

Goutis (1990, 1994) (see Example 3.2.7) illustrates this approach for class (ii). Chapter 5 presents such a study in order to derive conservative bounds on posterior probabilities of a null hypothesis.

A more conservative approach to robustness requirements is to look for *robust prior distributions*, i.e., for parametrized distributions as insensitive as possible to small variations in the prior information. For instance, it can be shown that Student's t-distributions are preferable to normal priors in the normal case, although the latter distributions are conjugate and so are maximum entropy priors in some cases (see Zellner (1971), Angers (1987), and Angers and MacGibbon (1990)). Similarly, *poly-t* distributions, derived from the product of several Student's t densities, are used in the econometric analysis of *simultaneous equations* for the same reason (see Drèze and Morales (1976), Richard and Tompa (1980), and Bauwens (1984)). In general, and unlike conjugate distributions, these robust priors will have heavy tails.

Another approach robustifies the conjugate distributions by *hierarchical modeling*. The hierarchical Bayes approach is presented in Chapter 10, but it seems quite intuitive at this stage that an additional level in the prior modeling should increase the robustness of the prior distribution. Consider a conjugate prior for $f(x|\theta)$, $\pi_1(\theta|\lambda)$. As mentioned above, classes like (i) are not very efficient in terms of robustness and, moreover, require the specification of bounds on the hyperparameters λ. Because these hyperparameters are (partly or totally) unknown, a natural extension (in a Bayesian framework) is to introduce a noninformative prior on λ, π_2 (or a hyperprior compatible with the available information). This modeling induces the following hierarchical structure:

$$\begin{aligned} \lambda &\sim \pi_2(\lambda), \\ \theta|\lambda &\sim \pi_1(\theta|\lambda), \\ x|\theta &\sim f(x|\theta). \end{aligned}$$

The prior distribution on θ is then the marginal distribution derived from $\pi_1(\theta|\lambda)\pi_2(\lambda)$, i.e., by integrating λ out,

$$(3.6.1) \qquad\qquad \pi(\theta) = \int \pi_1(\theta|\lambda)\pi_2(\lambda)d\lambda.$$

This prior distribution is not conjugate in general, but the main purpose of the hierarchical extension was actually to avoid the restrictive framework of conjugate priors. By integrating out the hyperparameters λ, we derive a distribution (3.6.1) that usually enjoys heavier tails than conjugate priors.

For instance, the Student's t-distribution can be written as (3.6.1) with π_2 an inverse gamma distribution (see Example 3.3.11). Hierarchical settings are also quite interesting from a computational point of view, as shown in Chapter 6.

Other perspectives incorporate the loss function in the robustness study in order to select an estimator that is conservative with respect to all possible priors $\pi \in \Gamma$. For instance, δ^* can be the solution of

$$\inf_\delta \sup_{\pi \in \Gamma} r(\pi, \delta) \quad \text{or} \quad \inf_\delta \sup_{\pi \in \Gamma} [r(\pi, \delta) - r(\pi, \delta^\pi)],$$

the first quantity being the Γ-*minimax risk* and the second quantity the Γ-*minimax regret*, as developed in Robbins (1951) and Good (1952). See Berger and Berliner (1986), Berger (1985a), and Kempthorne (1988) for additional references.

The literature on Bayesian robustness has increased considerably in the last few years and we refer the reader to the papers mentioned above for additional references. To conclude this chapter, let us point out that the choice of the prior distribution determines the resulting Bayesian inference, that this choice may sometimes be trivial and sometimes quite delicate, but that it should be justified in term of the available prior information in all cases and, moreover, that a robustness analysis should be conducted in order to assess the amount of posterior modification a change in the prior distribution induces. Obviously, this assessment will also depend on the evaluation measures considered for the quantities of interest, as for instance on the *losses* used in the estimation process. This leads to the possibility of using the knowledge of the loss function to determine noninformative prior distributions, but little work has been done in this direction, even though many Bayesians point out that loss and prior are indistinguishable (see, e.g., Lindley (1985) and Exercise 3.57.) A final warning to the reader is that the prior influence is often underestimated by users, while it may have unexpected influence on the resulting inference. Therefore, whenever possible, other values of the hyperparameters, but also other types of distributions, should be used to assess the real effect of the prior choice on the resulting inference.

3.7 Exercises

Section 3.1

3.1 (Dupuis (1995)) Recall that the beta $\mathcal{Be}(\alpha, \beta)$ distribution has a density given by

$$\pi(\theta) = \frac{\Gamma(\alpha + \beta)}{\Gamma(\alpha)\Gamma(\beta)} \theta^{\alpha-1}(1 - \theta)^\beta, \qquad 0 \le \theta \le 1.$$

a. Give the mean of the $\mathcal{Be}(\alpha, \beta)$ distribution.

b. Show that there a one-to-one correspondence between (α, β) and the triplet $(\mu, \theta_0, \theta_1)$, where $\pi(\theta \in [\theta_0, \theta_1]) = p$ and μ is the mean of the distribution.

c. What are the conditions on $(\mu, \theta_0, \theta_1)$ for (α, β) to exist?

Section 3.2.3

3.2 (Seidenfeld (1987)) Consider a six-sided die, with θ the random variable corresponding to the upper face of the die.

    a. If π is the distribution of θ, give the maximum entropy prior associated with the information $\mathbb{E}[\theta] = 3.5$.

    b. Show that, if A is the event "θ is odd", the updated distribution $\pi(\cdot|A)$ is $(1/3, 0, 1/3, 0, 1/3, 0)$.

    c. Show that the maximum entropy prior associated with the constraints $\mathbb{E}[\theta] = 3.5$ and $\mathbb{E}[\mathbb{I}_A] = 1$ is $(.22, 0, .32, 0, .47)$.

[*Note:* Seidenfeld (1987) and Kass and Wasserman (1996) use this example to show that the maximum entropy approach is not always compatible with the Bayesian updating principle given in (1.4.1).]

3.3 Show that, if the constraints (3.2.1) are all associated with functions g_k of the form $g_k(\theta) = \mathbb{I}_{(-\infty, a_k]}(\theta)$, there is no maximum entropy prior distribution for $\Theta = \mathbb{R}$ and π_0 the Lebesgue measure on $\mathbb{R}$.

3.4 Consider $\theta \in \mathbb{R}$ and a prior π such that $\text{var}^\pi(\theta) = 1$, $\pi(\theta < -1) = 0.1$, and $\pi(\theta > 1) = 0.1$. Derive the maximum entropy prior associated with the Lebesgue measure on $\mathbb{R}$ if this is possible.

3.5 Let π_0 be a reference measure for the maximum entropy method and π'_0 a measure which is absolutely continuous with respect to π_0.

    a. Give examples where the maximum entropy priors associated with π_0 and π'_0 coincide.

    b. Apply this to the case when π_0 is the Lebesgue measure on $\mathbb{R}$, π'_0 is the $\mathcal{N}(0, 1)$ distribution, and the constraints (3.2.1) are $\mathbb{E}^\pi[\theta] = 0$, $\text{var}^\pi(\theta) = \sigma^2$, depending on the value of σ.

3.6 Consider $\theta \in \mathbb{R}_+$. Determine whether there exists a maximum entropy prior under the constraint $\mathbb{E}^\pi[\theta] = \mu$ for $\pi_0(\theta) = 1$ and $\pi_0(\theta) = 1/\theta$.

3.7 Let $x \sim \mathcal{P}(\lambda)$.

    a. Find the maximum entropy prior associated with $\pi_0(\theta) = 1/\sqrt{\theta}$ and $\mathbb{E}^\pi[\theta] = 2$.

    b. Determine the hyperparameters of the prior distribution π if π is

       (i) $\mathcal{E}xp(\mu)$;
       (ii) $\mathcal{G}(2, \varrho)$.

    c. Derive the three corresponding posterior distributions when $x = 3$ and compare the Bayes estimators of θ under the loss $L(\theta, \delta) = \theta(\theta - \delta)^2$.

Section 3.2.4

3.8 Determine the prior distributions in Example 3.2.6, when the first and third quartiles are 2 and -2 and the median is 0.

3.9 Let $x \sim \mathcal{B}(n, \theta)$ and $\theta \sim \mathcal{B}e(\alpha, \beta)$. Determine whether there exist values of α, β such that $\pi(\theta|x)$ is the uniform prior on $[0, 1]$, even for a single value of x.

3.10 Let $x \sim \mathcal{P}a(\alpha, \theta)$, a Pareto distribution, and $\theta \sim \mathcal{B}e(\mu, \nu)$. Show that, if $\alpha < 1$ and $x > 1$, a particular choice of μ and ν gives $\pi(\theta|x)$ as the uniform prior on $[0, 1]$.

Section 3.3.1

3.11 If π is a finite mixture of conjugate distributions, give the form of $\pi(\theta|x)$. In particular, derive the posterior weights. Deduce the results of Example 3.4.1.

3.12 Determine *symmetric distributions*, i.e., distributions such that conjugate distributions and sampling distributions belong to the same parametrized family.

3.13 This exercise shows why the notion of a minimal conjugate family is usually vacuous.

    a. Using the notations of Proposition 3.3.13, show that the set of λ's in $\pi(\theta|\mu, \lambda)$ can be restricted to vary in $\lambda_0 + \mathbb{N}$ for any $\lambda_0 > 0$.

    b. Deduce that, if $\lambda_0 - \lambda_0' \notin \mathbb{Z}$, the conjugate families associated with $\lambda_0 + \mathbb{N}$ and $\lambda_0' + \mathbb{N}$ are disjoint.

    c. Conclude that the intersection of all conjugate families is empty.

3.14 Consider a population divided into k categories (or *cells*) with probability p_i for an individual to belong to the ith cell ($1 \leq i \leq n$). A sequence (π_k) of prior distributions on $p^k = (p_1, \ldots, p_k)$, $k \in \mathbb{N}$, is called *coherent* if any grouping of cells into m categories leads to the prior π_m for the transformed probabilities.

    a. Determine coherence conditions on the sequence (π_k).

    b. In the particular case when π_k is a Dirichlet distribution $\mathcal{D}_k(\alpha_1, \ldots, \alpha_k)$, express these conditions in terms of the α_k's.

    c. Does the Jeffreys prior induce a coherent sequence?

    d. What about $\pi_k(p^k) \propto \prod_i p_i^{-1/k}$, proposed by Perk (1947)?

Section 3.3.3

3.15 Show that every distribution from an exponential family can be generalized into a pseudo-exponential family by adding parametrized constraints on the support of x. Elaborate on the modification in the sufficient statistics.

3.16 Show that, if the support of $f(x|\theta)$ does not depend on θ and if there exists a parametrized conjugate prior family $\mathcal{F} = \{\pi(\theta|\lambda), \lambda \in \Lambda\}$ with $\dim(\Lambda) < +\infty$, $f(x|\lambda)$ is necessarily from an exponential family. (*Hint:* This is a consequence of the Pitman–Koopman lemma.)

3.17 Give a sufficient statistic associated with a sample $x_1, \ldots, x_n$ from a Pareto $\mathcal{P}a(\alpha, \theta)$ distribution.

3.18 Give a sufficient statistic associated with a sample $x_1, \ldots, x_n$ from a truncated normal distribution

$$f(x|\theta) \propto e^{-(x-\theta)^2/2} \mathbb{I}_{[\theta-c, \theta+c]}(x),$$

when c is known.

3.19 *(Brown (1986)) Show that, for every exponential family, there exists a reparameterization which gives a natural exponential family. Show also that the dimension of a *natural* reparameterization does not depend on the choice of the reparameterization.

3.20 *(Dynkin (1951)) Show that the normal distributions and distributions of the form $c \log(y)$, when $y \sim \mathcal{G}(\alpha, \beta)$, are the only ones which can belong simultaneously to an exponential family and a location family. Deduce that the normal distribution is the only distribution from an exponential family that is also spherically symmetric (see Exercise 1.1).

3.21 *(Lauritzen (1996)) Consider $X = (x_{ij})$ and $\Sigma = (\sigma_{ij})$ symmetric positive-definite $m \times m$ matrices. The *Wishart distribution*, $\mathcal{W}_m(\alpha, \Sigma)$, is defined by the density

$$p_{\alpha, \Sigma}(X) = \frac{|X|^{\frac{\alpha - (m+1)}{2}} \exp(-\mathrm{tr}(\Sigma^{-1} X)/2)}{\Gamma_m(\alpha)|\Sigma|^{\alpha/2}},$$

with $\mathrm{tr}(A)$ the trace of A and

$$\Gamma_m(\alpha) = 2^{\alpha m/2} \pi^{m(m-1)/4} \prod_{i=1}^{m} \Gamma\left(\frac{\alpha - i + 1}{2}\right).$$

a. Show that this distribution belongs to an exponential family. Give its natural representation and derive the mean of $\mathcal{W}_m(\alpha, \Sigma)$.

b. Show that, if $z_1, \ldots, z_n \sim \mathcal{N}_m(0, \Sigma)$,

$$\sum_{i=1}^{n} z_i z_i' \sim \mathcal{W}_m(n, \Sigma).$$

c. Show that the moments are given by

$$\mathbb{E}[X|\alpha, \Sigma] = \alpha\Sigma, \qquad \mathrm{Cov}(X) = 2\alpha\Sigma \otimes \Sigma.$$

d. Show that the mean of the inverse X^{-1} is given by

$$\mathbb{E}[X^{-1}|\alpha, \Sigma] = \frac{1}{\alpha - p - 1}\Sigma, \qquad \alpha > p + 1.$$

3.22 *(Pitman (1936)) Show the *Pitman–Koopman lemma*: If, for $n \geq n_0$, there exists T_n from $\mathbb{R}^n$ in $\mathbb{R}^k$ such that $T_n(x_1, \ldots, x_n)$ is sufficient when $x_1, \ldots, x_n$ are i.i.d. $f(x|\theta)$, the distribution f necessarily belongs to an exponential family if the support of f does not depend on θ. Study the case when the support of f depends on θ.

3.23 *(Brown (1986)) A natural exponential family $f(x|\theta) = \exp(\theta \cdot x - \psi(\theta))$ is said to be *steep* if, for every $\theta_0 \in \overset{\circ}{N}$, the interior set of N, $\theta_1 \in N/\overset{\circ}{N}$, and $\theta_\varrho = \varrho\theta_1 + (1 - \varrho)\theta_0$, it satisfies

$$\lim_{\varrho \to 1} \frac{\partial \psi}{\partial \varrho}(\theta_\varrho) = +\infty.$$

a. Show that the family is steep if and only if

$$\mathbb{E}_\theta[\|x\|] = +\infty$$

for every $\theta \in N/\overset{\circ}{N}$.

b. Show that the *Inverse Gaussian* distribution, with density

$$(\pi)^{-1/2} z^{-3/2} \exp\{\theta_1 z + \theta_2(1/z) - (2\theta_1\theta_2)^{1/2} + (1/2)\log(-2\theta_2)\}$$

where $z \in \mathbb{R}_+$ and $\theta_1, \theta_2 \in \mathbb{R}_-$, is exponential and steep but not regular.

c. Show that a minimal and steep exponential family can be re parametrized by $\xi(\theta) = \mathbb{E}_\theta[x] = \nabla\psi(\theta)$ and that this function defines a one-to-one transformation from $\overset{\circ}{N}$ to $\overset{\circ}{K}$.

d. Show that, for minimal and steep exponential families, the maximum likelihood estimator of θ, $\hat{\theta}(x)$, satisfies

$$\xi(\hat{\theta}(x)) = x.$$

3.24 *(Morris (1982)) A *restricted natural exponential* family on $\mathbb{R}$ is defined by

(3.7.1) $\qquad P_\theta(x \in A) = \int_A \exp\{\theta x - \psi(\theta)\}\, dF(x), \qquad \theta \in \Theta.$

a. Show that, if $0 \in \Theta$, F is necessarily a cumulative distribution function. Otherwise, show that the transformation of F into

$$dF_0(x) = \exp\{\theta_0 x - \psi(\theta)\}\, dF(x),$$

for an arbitrary $\theta_0 \in \Theta$ and the replacement of θ by $\theta - \theta_0$, provides this case.

b. Show that, in this restricted sense, $\mathcal{B}e(m\mu, m(1 - \mu))$ and the lognormal distribution $\mathcal{L}og\mathcal{N}(\alpha, \sigma^2)$ do not belong to an exponential family.

c. If $\mu = \psi'(\theta)$ is the mean of the distribution (3.7.1), the *variance function* of the distribution is defined by $V(\mu) = \psi''(\theta) = \mathrm{var}_\theta(x)$. Show that V is indeed a function of μ and, moreover, that if the variation space of μ, Ω, is known, the couple (V, Ω) completely characterizes the family (3.7.1) by

$$\psi\left(\int_{\mu_0}^\mu \frac{dm}{V(m)}\right) = \int_{\mu_0}^\mu \frac{m\, dm}{V(m)}.$$

(Notice that $\theta = \int_{\mu_0}^\mu dm/V(m)$.) Show that $V(\mu) = \mu^2$ defines *two* families, depending on whether $\Omega = \mathbb{R}^-$ or $\Omega = \mathbb{R}^+$.

d. Show that $V(\mu) = \mu(1 - \mu)/(m + 1)$ corresponds simultaneously to the binomial distribution $\mathcal{B}(m, \mu)$ and to $\mathcal{B}e(m\mu, m(1 - \mu))$. Deduce that the characterization by V is only valid for natural exponential families.

e. Show that exponential families with *quadratic variance functions*, i.e.,

(3.7.2) $\qquad\qquad\qquad V(\mu) = v_0 + v_1\mu + v_2\mu^2,$

include the following distributions: normal, $\mathcal{N}(\mu, \sigma^2)$, Poisson, $\mathcal{P}(\mu)$, gamma, $\mathcal{G}(r, \mu/r)$, binomial, $\mathcal{B}(m, m\mu)$, and negative binomial, $\mathcal{N}eg(r, p)$, defined in terms of the number of successes before the rth failure, with $\mu = rp/(1-p)$.

f. Show that the normal distribution (respectively, the Poisson distribution) is the unique natural exponential distribution with a constant (respectively, of degree one) variance function.

g. Assume $v_2 \neq 0$ in (3.7.2) and define $d = v_1^2 - 4v_0v_2$, discriminant of (3.7.2), $a = 1$ if $d = 0$ and $a = \sqrt{dv_2}$ otherwise. Show that $x^* = aV'(x)$ is a linear transformation of x with the variance function

(3.7.3) $\qquad\qquad\qquad V^*(\mu^*) = s + v_2(\mu^*)^2,$

where $\mu^* = aV'(\mu)$ and $s = -\mathrm{sign}(dv_2)$. Show that it is sufficient to consider V^* to characterize natural exponential families with a quadratic variance function, in the sense that other families are obtained by inverting the linear transform.

h. Show that (3.7.3) corresponds to six possible cases depending on the sign of v_2 and the value of s $(-1, 0, 1)$. Eliminate the two impossible cases and identify the families given in e, above. Show that the remaining case is $v_2 > 0$, $s = 1$. For $v_2 = 1$, show that this case corresponds to the distribution of $x = \log\{y/(1-y)\}/\pi$, where

$$y \sim \mathcal{B}o\left(\frac{1}{2} \,\Big|\, \frac{\theta}{\pi}, \frac{1}{2} - \frac{\theta}{\pi}\right), \qquad |\theta| < \frac{\pi}{2},$$

and

(3.7.4) $$f(x|\theta) = \frac{\exp[\theta x + \log(\cos(\theta))]}{2\cosh(\pi x/2)}.$$

(The reflection formula $B(0.5 + t, 0.5 - t) = \pi/\cos(\pi t)$ can be of use.) The distributions spanned by the linear transformations of (3.7.4) are called GHS(r, λ) (meaning *generalized hyperbolic secant*), with $\lambda = \tan(\theta)$, $r = 1/v_2$, and $\mu = r\lambda$. Show that the density of GHS(r, λ) can be written

$$f_{r,\lambda}(x) = \left(1 + \lambda^2\right)^{-r/2} \exp\{x \arctan(\lambda)\} f_{r,0}(x)$$

(do not try to derive an explicit expression for $f_{r,0}$).

[*Note:* Exercise 10.33 exhibits additional properties of the quadratic variance exponential families in terms of conjugate families and Bayes estimators. Exercise 6.2.6 shows how orthogonal polynomials can be related to each distribution in the quadratic variance exponential families.]

3.25 Compare usual exponential families with the distributions (2.6.1) obtained in Chapter 2 and check whether they give *universal estimators*.

3.26 Show that, for every exponential family, the natural space N is convex.

3.27 Show the decomposition of Example 3.3.11

(i) directly; and

(ii) through a usual representation of Student's t-distribution.

3.28 An alternative to the logistic regression introduced in Example 3.3.15 is the *probit* model, where

$$P_\alpha(y_i = 1) = 1 - P_\alpha(y_i = 0) = \Phi(\alpha^t x_i), \qquad i = 1, \ldots, n,$$

and Φ is the c.d.f. of the standard normal distribution.

a. Show that this alternative does not belong to an exponential family, even conditional upon the x_i's.

b. The observation y_i can be considered as the indicator function $\mathbb{I}_{z_i \leq \alpha^t x_i}$ where z_i is an unobserved $\mathcal{N}(0, 1)$ random variable. Show that, if the z_i's are known, the Lebesgue measure provides an explicit posterior distribution. [*Note:* The interesting aspect of this remark will be made clearer in Chapter 6, since the *missing data* $z_1, \ldots, z_n$ can be simulated.]

Section 3.3.4

3.29 For an arbitrary exponential family distribution, determine the constraints such that a maximum entropy prior is also a conjugate prior.

3.30 A classical linear regression can be written as $y \sim \mathcal{N}_p(X\beta, \sigma^2 I_p)$ with X a $p \times q$ matrix and $\beta \in \mathbb{R}^q$. When X is known, give the natural parameterization of this exponential family and derive the conjugate priors on (β, σ^2). Generalize to $\mathcal{N}_p(X\beta, \Sigma)$ with Σ known.

3.31 Consider $x \sim \mathcal{N}(\theta, \theta)$ with $\theta > 0$.

  a. Determine the Jeffreys prior $\pi^J(\theta)$.

  b. Say whether the distribution of x belongs to an exponential family and derive the conjugate priors on θ.

  c. Use Proposition 3.3.14 to relate the hyperparameters of the conjugate priors with the mean of θ.

3.32 Show that, if $x \sim \mathcal{B}e(\theta_1, \theta_2)$, there exist conjugate priors on $\theta = (\theta_1, \theta_2)$ but that they do not lead to tractable posterior quantities, except for the computation of $\mathbb{E}^\pi[\theta_1/(\theta_1 + \theta_2)|x]$, according to Proposition 3.3.14.

3.33 *(Robert (1991)) The generalized inverse normal distribution $\mathcal{IN}(\alpha, \mu, \tau)$ has the density

$$K(\alpha, \mu, \tau)|\theta|^{-\alpha} \exp\left\{-(\frac{1}{\theta} - \mu)^2/2\tau^2\right\},$$

with $\alpha > 0$, $\mu \in \mathbb{R}$, and $\tau > 0$.

  a. Show that this density is well defined and that the normalizing factor is

$$K(\alpha, \mu, \tau)^{-1} = \tau^{\alpha-1} e^{-\mu^2/2\tau^2} 2^{(\alpha-1)/2} \, \Gamma(\frac{\alpha-1}{2}) \, {}_1F_1\left(\frac{\alpha-1}{2}; 1/2; \frac{\mu^2}{2\tau^2}\right),$$

where ${}_1F_1$ is the *confluent hypergeometric function* (see Abramowitz and Stegun (1964)).

  b. Show that this distribution generalizes the distribution of $y = 1/x$ when $x \sim \mathcal{N}(\mu, \tau^2)$. Check that the above normalizing constant is correct in this particular case.

  c. Deduce that the mean of $\mathcal{IN}(\alpha, \mu, \tau)$ is defined for $\alpha > 2$ and is

$$\mathbb{E}_{\alpha,\mu,\tau}[\theta] = \frac{\mu}{\tau^2} \frac{{}_1F_1(\frac{\alpha-1}{2}; 3/2; \mu^2/2\tau^2)}{{}_1F_1(\frac{\alpha-1}{2}; 1/2; \mu^2/2\tau^2)}.$$

  d. Show that these distributions $\mathcal{IN}(\alpha, \mu, \tau)$ constitute a conjugate family for the multiplicative model $\mathcal{N}(\theta, \theta^2)$.

3.34 Show that a Student's t-distribution $\mathcal{T}_p(\nu, \theta, \tau^2)$ does not allow for a conjugate family, apart from the trivial family $\mathcal{F}_0$.

3.35 Proposition 3.3.13 exhibits a conjugate family for every exponential family, of the form (3.3.4),

$$\pi(\theta|\lambda, \mu) = \exp\{\theta \cdot \mu - \lambda\psi(\theta)\}K(\mu, \lambda).$$

  a. Show that the distribution (3.3.4) is actually well defined when $\lambda > 0$ and $(\mu/\lambda) \in \overset{\circ}{N}$.

  b. Give the constant K for normal, gamma, and negative binomial distributions.

  c. Deduce that the likelihood function $\ell(\theta|x)$ is a particular prior distribution for exponential families (by mean of a reparameterization) and give the corresponding prior for the above families.

  d. Is this property characterizing exponential families? Give a counter-example.

3.36 *Show Proposition 3.3.14 and its reciprocal in the continuous case. Apply to the distributions in Table 3.3.4.

3.37 Show that the distributions in Table 3.3.4 are actually conjugate

(i) directly; and

(ii) through Proposition 3.3.14.

3.38 Consider $x \sim \mathcal{G}(\theta, \beta)$, i.e., $f_\beta(x|\theta) = \frac{\beta^\theta}{\Gamma(\theta)} x^{\theta-1} e^{-\beta x}$.

a. Can you derive a conjugate family for this distribution?

b. Consider the case where $\theta \in \mathbb{N}$.

c. Same question for $x \sim \mathcal{B}e(1, \theta)$.

3.39 Show that, for exponential families, a multiplication of the number of hierarchical levels does not modify the conjugate nature of the resulting prior if conjugate distributions with fixed scale parameters are used at every level of the hierarchy. (Consider, for instance, the normal case.)

3.40 *(Robert (1993a)) Consider $f(x|\theta)$ from an exponential family,

$$f(x|\theta) = e^{\theta \cdot x - \psi(\theta)} h(x), \qquad x \in \mathbb{R}^k,$$

and $\pi_0(\theta|x_0, \lambda)$ a conjugate prior distribution,

$$\pi_0(\theta|x_0, \lambda) = e^{\theta \cdot x_0 - \lambda \psi(\theta)}.$$

We are looking for a so-called objective estimation of $\nabla \psi(\theta)$, based on an arbitrary prior distribution $\pi_0(\theta|x_0, \lambda)$. To this effect, we replace the distribution π_0 by $\pi_1(\theta|x_1, \lambda)$ defined by the relation

(3.7.5) $$\mathbb{E}^{\pi_1}[\nabla \psi(\theta)] = \mathbb{E}^{\pi_0}[\nabla \psi(\theta)|x],$$

in order to reduce the influence of x_0.

a. Deduce the relation between x_1 and x_0.

b. We iterate the updating process (3.7.5) in order to eliminate, as much as possible, the influence of x_0 and we construct in this way a sequence $\pi_n(\theta|x_n, \lambda)$ of conjugate priors. Give the relation between x_n and x_{n-1} and deduce the limit of the sequence (x_n).

c. Give the corresponding limit of the Bayes estimators of $\nabla \psi(\theta)$. How do you characterize the resulting estimator? Is it still a Bayes estimator?

d. In the particular case when $x \sim \mathcal{N}(\theta, 1)$, the parameter of interest is $h(\theta) = e^{-\theta}$. Give the estimator $h(\theta)$ obtained this way using the iterative updating

$$\mathbb{E}^{\pi_n}[h(\theta)] = \mathbb{E}^{\pi_{n-1}}[h(\theta)|x].$$

e. Consider the case $x \sim \mathcal{G}(\alpha, \theta)$ and $h(\theta) = \theta^k$ to show that this iterative method, called *Prior Feedback*, does not always converge to the maximum likelihood estimator.

f. Show that the limit of the Prior Feedback estimate when λ goes to $+\infty$ is the maximum likelihood estimator of $h(\theta)$, for an arbitrary function h and every exponential family.

Section 3.4

3.41 In the setting of Example 3.4.1, build up a prior distribution by observing a few coins and imposing a mixture of beta distributions as in Diaconis and Ylvisaker (1985). Select one of these coins and derive a posterior distribution on θ, probability of heads, after 10 trials and 50 trials.

3.42 Consider $x \sim \mathcal{N}(0,1)$ and $\theta \sim \mathcal{T}_1(5,0,1)$.

a. Devise a method to approximate this prior distribution by a mixture of: (i) two normal distributions; and (ii) five normal distributions.

b. In each case, give the approximating posterior expectation of θ when $x = 1$ and compare it with the exact value.

Section 3.5.1

3.43 Consider $x_1, \ldots, x_n \sim \mathcal{N}(\mu + \nu, \sigma^2)$, with $\pi(\mu, \nu, \sigma) \propto 1/\sigma$.

a. Show that the posterior distribution is not defined for every n.

b. Extend this result to overparametrized models with improper priors.

The following exercises (3.44–3.50) present the marginalization paradox through several examples and show that it can only occur for improper priors. Dawid et al. (1973), Stone (1976), and Jaynes (1980) give some partial resolutions of these paradoxes. Notice that a fundamental explanation is that the improper distribution $\pi(d\eta, d\theta) = \pi(\eta)\, d\eta\, d\theta$ does not induce the pseudo-marginal distribution $\pi(d\eta) = \pi(\eta)\, d\eta$.

3.44 *(Dawid et al. (1973)) Consider n random variables $x_1, \ldots, x_n$, such that the first ξ of these variables has an $\mathcal{E}xp(\eta)$ distribution and the $n - \xi$ other have a $\mathcal{E}xp(c\eta)$ distribution, where c is known and ξ takes its values in $\{1, 2, \ldots, n - 1\}$.

a. Give the shape of the posterior distribution of ξ when $\pi(\xi, \eta) = \pi(\xi)$ and show that it only depends on $z = (z_2, \ldots, z_n)$, with $z_i = x_i/x_1$.

b. Show that the distribution of z, $f(z|\xi)$, only depends on ξ.

c. Show that the posterior distribution $\pi(\xi|x)$ cannot be written as a posterior distribution for $z \sim f(z|\xi)$, whatever $\pi(\xi)$, although it only depends on z. How do you explain this?

d. Show that the paradox does not occur when $\pi(\xi, \eta) = \pi(\xi)\eta^{-1}$.

3.45 *(Dawid et al. (1973)) Consider u_1, u_2, s^2 such that

$$u_1 \sim \mathcal{N}(\mu_1, \sigma^2), \qquad u_2 \sim \mathcal{N}(\mu_2, \sigma^2), \qquad s^2 \sim \sigma^2 \chi_\nu^2/\nu,$$

and $\zeta = (\mu_1 - \mu_2)/(\sigma\sqrt{2})$ is the parameter of interest. The prior distribution is

$$\pi(\mu_1, \mu_2, \sigma) = \frac{1}{\sigma}.$$

a. Show that the posterior distribution $\pi(\zeta|x)$ only depends on

$$z = \frac{u_1 - u_2}{s\sqrt{2}}.$$

b. Show that the distribution of z only depends on ζ, but that a paradox occurs; it is still impossible to derive $\pi(\zeta|x)$ from $f(z|\zeta)$, even though $\pi(\zeta|x)$ only depends on z.

c. Show that the paradox does not occur when

$$\pi(\mu_1, \mu_2, \sigma) = \frac{1}{\sigma^2}.$$

3.46 *(Dawid et al. (1973)) Consider

$$x_{11}, \ldots, x_{1n} \sim \mathcal{N}(\mu_1, \sigma^2),$$
$$x_{21}, \ldots, x_{2n} \sim \mathcal{N}(\mu_2, \sigma^2),$$

$2n$ independent random variables.

a. The parameter of interest is $\xi = (\xi_1, \xi_2) = (\mu_1/\sigma, \mu_2/\sigma)$ and the prior distribution is

$$\pi(\mu_1, \mu_2, \sigma) = \sigma^{-p}.$$

Show that $\pi(\xi|x)$ only depends on $z = (z_1, z_2) = (\bar{x}_1/s, \bar{x}_2/s)$ and that the distribution of z only depends on ξ. Derive the value of p that avoids the paradox.

b. The parameter of interest is now $\zeta = \xi_1$. Show that $\pi(\zeta|x)$ only depends on z_1 and that $f(z_1|\xi)$ only depends on ζ. Give the value of p that avoids the paradox.

c. Consider the previous questions when $\sigma \sim \mathcal{P}a(\alpha, \sigma_0)$.

3.47 *(Dawid et al. (1973)) Consider (x_1, x_2) with the following distribution:

$$f(x_1, x_2|\theta) \propto \int_0^{+\infty} t^{2n-1} \exp\left[-\frac{1}{2}\left\{ t^2 + n(x_1 t - \zeta)^2 + n(x_2 t - \xi)^2 \right\} \right] dt,$$

with $\theta = (\zeta, \xi)$. Justify this distribution by considering the setting of Exercise 3.46. The prior distribution on θ is $\pi(\theta) = 1$.

a. Show that $\pi(\zeta|x)$ only depends on x_1 and that $f(x_1|\theta)$ only depends on ζ, but that $\pi(\zeta|x)$ cannot be obtained from $x_1 \sim f(x_1|\zeta)$.

b. Show that, for any distribution $\pi(\theta)$ such that $\pi(\zeta|x)$ only depends on x_1, $\pi(\zeta|x)$ cannot be proportional to $\pi(\zeta)f(x_1|\zeta)$.

3.48 *(Jaynes (1980)) In the setting of Exercise 3.44, consider $\pi(\xi, \eta) = \pi(\xi)\pi(\eta)$.

a. Show that

$$\pi(\xi|x) \propto \pi(\xi)c^{-\xi} \int_0^{+\infty} \eta^{-n} \exp(-\eta x_1 Q)\pi(\eta)\, d\eta,$$

with

$$Q = \sum_{i=1}^{\xi} z_i + c \sum_{\xi+1}^{n} z_i.$$

b. Examine whether the paradox occurs for $\pi(\eta) = \eta^{-k}$ $(k > -n - 1)$.

c. Same question for $\eta \sim \mathcal{P}a(\alpha, \eta_0)$.

3.49 *(Jaynes (1980)) Consider

$$f(y, z|\eta, \zeta) \propto \frac{\zeta^z \eta^y (1 - \eta)^{z-y}}{y!(z - y)!} \qquad (0 \le y \le z),$$

with $0 < \eta < 1$.

a. Show that $f(z|\eta, \zeta)$ only depends on ζ and derive the distribution $f(y, z|\eta, \zeta)$ from $f(y|z, \eta, \zeta)$.

b. Show that, for every $\pi(\eta)$, the paradox does not occur.

3.50 *(Dawid et al. (1973)) Consider $x = (y, z)$ with distribution $f(x|\theta)$ and $\theta = (\eta, \xi)$. Assume that $\pi(\xi|x)$ only depends on z and that $f(z|\theta)$ only depends on ξ.

a. Show that the paradox does not occur if $\pi(\theta)$ is proper.
b. Generalize to the case where $\int \pi(\eta, \xi)\, d\eta = \pi(\xi)$ and examine whether the paradox is eliminated.

Section 3.5.3

3.51 In relation to Example 3.5.7, if $x \sim \mathcal{B}(n, p)$, find a prior distribution on n such that $\pi(n|x)$ is $\mathcal{N}eg(x, p)$.

3.52 *In relation to Example 3.5.9,

a. Show that the Bayes estimator of $\eta = ||\theta||^2$ under quadratic loss for $\pi(\eta) = 1/\sqrt{\eta}$ and $x \sim \mathcal{N}(\theta, I_p)$ can be written as

$$\delta^\pi(x) = \frac{{}_1F_1(3/2; p/2; ||x||^2/2)}{{}_1F_1(1/2; p/2; ||x||^2/2)},$$

where ${}_1F_1$ is the confluent hypergeometric function.
b. Deduce from the series development of ${}_1F_1$ the asymptotic development of δ^π (for $||x||^2 \to +\infty$).
c. Compare δ^π with $\delta_0(x) = ||x||^2 - p$.
d. Study the behavior of these estimators under the weighted quadratic loss

$$L(\delta, \theta) = \frac{(||\theta||^2 - \delta)^2}{2||\theta||^2 + p}$$

and conclude.

3.53 Find a transform of θ, $\eta = g(\theta)$, such that the Fisher information $I(\eta)$ is constant for:

(i) the Poisson distribution, $\mathcal{P}(\theta)$;
(ii) the gamma distribution, $\mathcal{G}(\alpha, \theta)$, with $\alpha = 1, 2, 3$; and
(iii) the binomial distribution, $\mathcal{B}(n, \theta)$.

3.54 Assuming that $\pi(\theta) = 1$ is an acceptable prior for real parameters, show that this generalized prior leads to $\pi(\sigma) = 1/\sigma$ if $\sigma \in \mathbb{R}^+$ and to $\pi(\varrho) = 1/\varrho(1 - \varrho)$ if $\varrho \in [0, 1]$ by considering the natural transformations $\theta = \log(\sigma)$ and $\theta = \log(\varrho/(1 - \varrho))$.

3.55 *(Saxena and Alam (1982)) In a setting identical to that in Exercise 3.52:

a. Give the maximum likelihood estimator of $||\theta||^2$ when $x \sim \mathcal{N}(\theta, I_p)$.
b. Show that the maximum likelihood estimator derived from $z = ||x||^2$ satisfies the implicit equation

$$1 = \frac{z}{\sqrt{\lambda z}} \frac{I_{p/2}(\sqrt{\lambda z})}{I_{(p-1)/2}(\sqrt{\lambda z})} \qquad (z > p),$$

where I_ν is the *modified Bessel function* (see Abramowitz and Stegun (1964) or Exercise 4.35).
c. Use the series expansion of I_ν to show that the maximum likelihood estimator $\hat{\lambda}$ satisfies

$$\hat{\lambda}(z) = z - p + 0.5 + O(1/z).$$

d. Show that $\hat{\lambda}$ is dominated by $(z - p)^+$ under quadratic loss.

3.56 The Fisher information is not defined when the support of $f(x|\theta)$ depends on θ. Consider the following cases:

(i) $x \sim \mathcal{U}_{[-\theta,\theta]}$;   (ii) $x \sim \mathcal{P}a(\alpha, \theta)$;   (iii) $f(x|\theta) \propto e^{-(x-\theta)^2/2} \mathbb{I}_{[0,\theta]}(x)$.

3.57 Show that a second order approximation of both the entropy and Hellinger losses introduced in Section 2.5.4 is $(\theta - \delta)^2 I(\theta)$. Does this result give additional support to the use of the Jeffreys prior?

3.58 Consider $x \sim \mathcal{P}(\theta)$.

a. Determine the Jeffreys prior π^J and discuss whether the scale-invariant prior $\pi_0(\theta) = 1/\theta$ is preferable.

b. Find the maximum entropy prior for the reference measure π_0^J and the constraints $\mathbb{E}^\pi[\theta] = 1$, $\text{var}^\pi(\theta) = 1$. What about using π_0 instead?

c. Actually, x is the number of cars crossing a railroad in a period T. Show that x is distributed according to a Poisson distribution $\mathcal{P}(\theta)$ if the interval between two arrivals is distributed according to $\mathcal{E}xp(\lambda)$. Note that $\theta = \lambda T$.

d. Use the above derivation of the Poisson distribution to justify the use of π_0.

Section 3.5.4

3.59 If $x \sim \mathcal{N}(\theta, \sigma^2)$, give the reference prior distributions for $\{\theta, \sigma\}$ and $\{\sigma, \theta\}$.

3.60 Consider $\theta \in [a, b]$ and $\pi(\theta) \propto 1/\theta$.

a. Determine the normalizing factor of π.

b. Compute $p_i = \pi(i \leq \theta < i + 1)$ for $a \leq i \leq b - 1$.

c. Deduce the limit of p_i as a goes to 0 or b goes to ∞. [*Note:* This exercise is related to the *table entry problem*, namely, that in many numerical tables the frequency of the first significant digit is $\log_{10}(1 + i^{-1})$ $(1 \leq i \leq 9)$. See Berger (1985a, p. 86) for a detailed account.]

3.61 *(Kass and Wasserman (1996)) Show that the reference prior derived from the Jeffreys prior for θ_1 fixed, $\pi(\theta_2|\theta_1)$, and the Jeffreys prior on the marginal (3.5.3) can also be written as

$$\pi(\theta_1, \theta_2) \propto \pi(\theta_2|\theta_1) \exp\left\{\int \pi(\theta_2|\theta_1) \log \sqrt{|\mathbf{I}|/|\mathbf{I}_{22}|} d\theta_2\right\},$$

where $\mathbf{I}$ is the Fisher information and $\mathbf{I}_{22}$ is the part of $\mathbf{I}$ associated with θ_2.

Section 3.6

3.62 *(Berger (1990a)) Consider the class $\Gamma_{\epsilon,\mathcal{Q}}$ defined in §3.6 (iii), with

$$\mathcal{Q} = \{ \text{unimodal distributions symmetric around } \theta_0\}.$$

When π varies in $\mathcal{Q}$, the marginal

$$m(\pi) = \int f(x|\theta)\pi(\theta) \, d\theta$$

varies between upper and lower bounds, m^U and m^L.

a. Show that every unimodal distribution symmetric around θ_0 can be written as a mixture of uniform distributions symmetric around θ_0, $\mathcal{U}_{[\theta_0-a,\theta_0+a]}$.

b. Deduce that

$$m^U = \sup_{\pi\in\Gamma_{\epsilon,\mathcal{Q}}} m(\pi) = (1-\epsilon)m(\pi_0) + \epsilon\sup_{z>0}\int_{\theta_0-z}^{\theta_0+z}\frac{f(x|\theta)}{2z}\,d\theta.$$

c. If the quantity of interest is the *Bayes factor*,

$$B(\pi) = \frac{f(x|\theta_0)}{\int_{\theta\neq\theta_0}f(x|\theta)\pi_1(\theta)\,d\theta},$$

where π_1 is π conditioned by $\theta\neq\theta_0$ and π_0 is the Dirac mass at θ_0, show that

$$B^L = \inf_{\pi\in\Gamma_{\epsilon,\mathcal{Q}}} B(\pi) = \frac{f(x|\theta)}{\epsilon\sup_z\int_{\theta_0-z}^{\theta_0+z}(f(x|\theta)/2z)\,d\theta}.$$

3.63 Consider the class of prior distributions

$$\Gamma = \{\mathcal{N}(\mu,\tau^2),\, 0\le\mu\le 2,\, 2\le\tau^2\le 4\}$$

when $x\sim\mathcal{N}(\theta,1)$.

a. Study the variations of $\mathbb{E}^\pi[\theta|x]$ and $\operatorname{var}^\pi(\theta|x)$ when $\pi\in\Gamma$.

b. Study $\varrho(\pi,\delta^{\pi'})$ when $\pi,\pi'\in\Gamma$ and $\delta^\pi(x) = \mathbb{E}^\pi[\theta|x]$, $L(\theta,\delta) = (\theta-\delta)^2$ in order to determine the Γ-minimax estimator.

3.64 *(Walley (1991))* Suppose that, instead of defining a prior distribution π on the σ-algebra of Θ, one defines some upper and lower bound measures on π, $\overline{\pi}$ and $\underline{\pi}$. For any event A, $\underline{\pi}(A)$ represents the largest amount one is willing to bet to receive one unit if A occurs. Similarly, $1-\overline{\pi}(A)$ is the smallest amount one is ready to bet against the event A.

a. Show that, if a prior distribution π is available, $\underline{\pi} = \pi = \overline{\pi}$.

b. Show that $\underline{\pi}(A) + \underline{\pi}(A^c) \le 1 \le \overline{\pi}(A) + \overline{\pi}(A^c)$ for every A if *sure loss* is to be avoided.

c. If $\underline{\pi}(A\cup B)$ is the maximum bet one is ready to place on $A\cup B$, show that $\underline{\pi}(A\cup B) \ge \underline{\pi}(A) + \underline{\pi}(B)$ and, similarly, that $\overline{\pi}(A\cup B) \le \overline{\pi}(A) + \overline{\pi}(B)$.

3.65 *(**Exercise 3.64 cont.**)* If, instead, one considers *gambles*, i.e., bounded real-valued functions X on a measured space Ω corresponding to variable rewards depending on the uncertainty state $\omega\in\Omega$, it is also possible to define *upper and lower previsions*, $\overline{P}$ and $\underline{P}$, where $\underline{P}(X)$ denotes the maximum price acceptable to get the reward X and $\overline{P}(X)$ the maximum selling price.

a. A gamble is *desirable* if there is a chance one gambles for it. Justify the following axioms:

　　(A) If $\sup_\omega X(\omega) < 0$, then X is not desirable;
　　(B) If $\inf_\omega X(\omega) > 0$, then X is desirable;
　　(C) If X is desirable and $\lambda > 0$, then λX is desirable; and
　　(D) If X and Y are both desirable, then $X+Y$ is desirable.

b. Justify the following *coherence* axioms on $\underline{P}$ and show that they correspond to (B), (C) and (D) above:

$\quad(P_1)$ $\underline{P}(X) \geq \inf_\omega X(\omega)$;

$\quad(P_2)$ $\underline{P}(\lambda X) = \lambda \underline{P}(X)$; and

$\quad(P_3)$ $\underline{P}(X + Y) \geq \underline{P}(X) + \underline{P}(Y)$.

c. Given a lower prevision $\underline{P}$, the *conjugate upper prevision* is defined by $\overline{P}(X) = -\underline{P}(-X)$. Show that, if $\underline{P}$ is coherent and $\overline{P}$ is the conjugate of $\underline{P}$, they satisfy

$$\inf_\omega X(\omega) \leq \underline{P}(X) \leq \overline{P}(X) \leq \sup_\omega X(\omega),$$

and deduce that $\overline{P}$ is a convex function.

d. Show that, when $\underline{P}$ is *self-conjugate*, then $\underline{P}(X) = \overline{P}(X)$ and it satisfies the linearity requirements

$$\underline{P}(X + Y) = \underline{P}(X) + \underline{P})Y) \quad \text{and} \quad \underline{P}(\lambda X) = \lambda \underline{P}(X), \ \lambda \in \mathbb{R}.$$

3.66 *(Exercise 3.65 cont.) A lower prevision $\underline{P}$ is said to *avoid sure loss* if, for every $n \geq 1$ and every set of gambles $X_1, \ldots, X_n$,

$$\sup_\omega \sum_{i=1}^n X_i - \underline{P}(X_i) \geq 0.$$

a. Show that $\underline{P}$ avoids sure loss if and only if

$$\sup_\omega \sum_{i=1}^n \lambda_i (X_i - \underline{P}(X_i)) \geq 0$$

for every $n \geq 1$, every set of gambles $X_1, \ldots, X_n$ and every $\lambda_i \geq 0$.

b. Assuming that $\underline{P}$ avoids sure loss, show that, for every $\lambda \geq 0$,

$$\underline{P}(\lambda X) \leq \lambda \overline{P}(X), \qquad \overline{P}(\lambda X) \geq \lambda \underline{P}(X),$$
$$\underline{P}(\lambda X + (1 - \lambda)Y) \leq \lambda \overline{P}(X) + (1 - \lambda)\overline{P}(Y).$$

when $\overline{P}$ is the conjugate upper prevision.

c. A lower prevision is *coherent* if

$$\sup_\omega \left[\sum_{i=1}^n (X_i - \underline{P}(X_i)) - m(X_0 - \underline{P}(X_0)) \right] \geq 0$$

for every m, n and every set of gambles $X_0, \ldots, X_n$. Show that $\underline{P}$ is coherent if and only if it satisfies axioms (P_1), (P_2), and (P_3).

d. Show that linearity is equivalent to coherence plus self-conjugacy, if *linearity* is defined by

$$\sup_\omega \left\{ \sum_{i=1}^n X_i(\omega) - \sum_{j=1}^m Y_i(\omega) \right\} \geq \sum_{i=1}^n \underline{P}(X_i) - \sum_{j=1}^m \underline{P}(Y_i)$$

for every n, m and every set of gambles $X_1, \ldots, X_n, Y_1, \ldots, Y_m$.

e. Show that $\underline{P}$ is a linear prevision if and only if $\underline{P}(X + Y) = \underline{P}(X) + \underline{P}(Y)$ and $\underline{P}(X) \geq \inf_\omega X(\omega)$. Deduce that $\underline{P}$ is a linear prevision if and only if it satisfies linearity, (P_2), and

$\quad(P_4)$ if $X \geq 0$, then $\underline{P}(X) \geq 0$; and

$\quad(P_5)$ $\underline{P}(1) = 1$.

Note 3.8.3

3.67 Apply Dalal and Hall's (1983) decomposition to the following cases:

(i) $x \sim \mathcal{N}(\theta, I_p)$, $\theta \sim \mathcal{T}_p(m, 0, \tau^2)$; and

(ii) $x \sim \mathcal{N}eg(N, p)$, $p/(1 - p) \sim \mathcal{G}(1/2, 1/2)$.

3.68 *Find the natural measures ν_m of Dalal and Hall (1983) for the distributions of Table 3.8.3.

3.8 Notes

3.8.1 Axiomatic derivation of prior distributions

To derive the existence of a prior distribution, we need, as in the utility case (see Section 2.2), to start from an ordering on events rather than rewards. Consider, thus, that the decision-maker, the client or the statistician is able to determine an order relation on a σ-algebra $\mathcal{B}(\Theta)$. This order relation, denoted by $\preceq$, is such that $B \prec A$ means that A is more likely than B, $B \preceq A$, that A is at least as likely as B, and $B \sim A$, that A and B are equally likely. Obviously, if there exists a probability distribution on $(\Theta, \mathcal{B}(\Theta))$, P, P automatically induces an order relation on $\mathcal{B}(\Theta)$. We consider below under what hypotheses the reciprocal can be established. A first assumption is that the order relation is *total*:

(A₁) For all measurable sets A and B, one and only one of the following relations is satisfied:
$$A \prec B, \quad B \prec A \quad \text{or} \quad A \sim B.$$

Another assumption is:

(A₂) If A_1, A_2, B_1, B_2 are four measurable sets satisfying $A_1 \cap A_2 = B_1 \cap B_2 = \emptyset$ and $A_i \preceq B_i$ $(i = 1, 2)$, then $A_1 \cup A_2 \preceq B_1 \cup B_2$. Moreover, if $A_1 \prec B_1$, $A_1 \cup A_2 \prec B_1 \cup B_2$.

This natural hypothesis implies *transitivity* for the order relation. The following assumption ensures that there is no measurable set with a negative likelihood (i.e., less likely than the empty set):

(A₃) For every event A, $\emptyset \preceq A$ and $\emptyset \prec \Theta$.

The additional condition $\emptyset \prec \Theta$ avoids the trivial case where all events are equivalent. It is also necessary to allow for the comparison of an infinite sequence of events.

(A₄) If $A_1 \supset A_2 \supset \cdots$ is a decreasing sequence of measurable sets and B is a given event such that $B \preceq A_i$ for every i, then

$$B \preceq \bigcap_{i=1}^{+\infty} A_i.$$

This assumption somehow ensures the continuity of the preference ordering and is related to the σ-additivity property of probability measures. However, these axioms (A₁)–(A₄) are still not sufficient to derive the existence of a probability distribution from the likelihood ordering. In fact, a last assumption is also necessary to move from a qualitative comparison scaling to a quantitative comparison.

(A$_5$) There exists a random variable X on $(\Theta, \mathcal{B}(\Theta))$ with *uniform distribution* on $[0, 1]$, i.e., such that, for all I_1, I_2, intervals on $[0, 1]$, it satisfies

$$\{X \in I_1\} \preceq \{X \in I_2\}$$

if and only if

$$\lambda(I_1) \leq \lambda(I_2),$$

where λ is the Lebesgue measure.

This additional hypothesis is then sufficient to establish the following existence result (see DeGroot (1970) for a proof).

Theorem 3.8.1 *Under the axioms* (A$_1$)–(A$_5$), *there exists a distribution* P *such that* $P(A) \leq P(B)$ *if and only if* $A \preceq B$.

Compared with the utility function derivation in Chapter 2, the previous developments on the axiomatic foundations of the prior distribution are more limited. A first reason for this brevity is that the above hypotheses and the surrounding setting are more difficult to justify. Indeed, when a statistician is able to talk about the *likelihood* of an event, it implies that she has, consciously or not, built up an underlying probabilistic model and, therefore, the previous construction is rather tautological. Assumption (A$_5$) is particularly demanding and can seldom be verified in practice. Notice, however, that to some extent a similar criticism could be addressed to the derivation of the utility function.

A second reason for this limitation takes place at a more pragmatic level; in fact, according to Theorem 3.8.1, the decision-maker can recover a prior distribution from her likelihood ordering. However, it is most likely, especially when Θ is not finite, that this ordering will be *coarse*, i.e., that the derived σ-algebra $\mathcal{B}(\Theta)$ will not correspond to the usual Borelian σ-algebra on Θ, thus preventing the use of classical distributions on θ. Nevertheless, it is comforting to be able to justify the use of a prior distribution on an alternative basis than on the frequentist model implying repeatability of experiments, even though it is of limited use in practice.

3.8.2 Exchangeability and conjugate priors

Bernardo and Smith (1994, Section 4.3) justify to some extent the existence of prior distributions by the notion of *exchangeability*:

Definition 3.8.2 *A sequence* $(x_1, \ldots, x_n)$ *of random variables is finitely exchangeable if the joint distribution* $p(x_1, \ldots, x_n)$ *is invariant under any permutation of the indices of the random variables, that is,*

$$p(x_1, \ldots, x_n) = p(x_{(1)}, \ldots, x_{(n)}),$$

An infinite sequence $(x_n)_n$ *is infinitely exchangeable if any finite subsequence is finitely exchangeable.*

While the assumption of exchangeability is not always reasonable (see Bernardo and Smith (1994, Section 4.2.2) for examples), there are also many settings where the order in which the data is obtained is indeed unimportant. The consequences of this assumption of infinite exchangeability are, moreover, quite interesting.

For instance, if $(x_n)_n$ is an infinite sequence of random variables taking values on $\{0, 1\}$, de Finetti (1930) has shown that there exists a probability measure $\pi(\theta)$ such that, for every n, the joint distribution of $(x_1, \ldots, x_n)$ writes down as

$$p(x_1, \ldots, x_n) = \int_0^1 \prod_{i=1}^n \theta^{x_i}(1-\theta)^{1-x_i} d\pi(\theta),$$

that is, conditional on θ, the x_i's are i.i.d. Bernoulli $\mathcal{B}(\theta)$ random variables. As shown by Bernardo and Smith (1994, Section 4.3.2), this property extends to random variables taking value in a finite set, $\{1, 2, \ldots, k\}$ say, as being multinomial, conditional on a vector $\theta = (\theta_1, \ldots, \theta_k)$.

In the general case where the x_i's are real valued, and infinitely exchangeable, there also exists an interesting representation, under the form

$$p(x_1, \ldots, x_n) = \int \prod_{i=1}^n F(x_i) d\pi(F),$$

where F is a c.d.f. and π a probability measure on the space of distribution functions (see Chow and Teicher (1988) for a more precise formulation of this result, whose measure-theoretic subtleties are beyond the level of this book). This representation is intrinsically non-parametric (see Note 1.8.2), but Bernardo and Smith (1994, Section 4.6) elaborate on other notions of exchangeability to come back to the parametric setting.

3.8.3 Approximation by continuous mixtures of conjugate priors

Consider a density from an exponential family written in the form

$$f(x|\theta) = \exp\{x \cdot \tau(\theta) - \gamma(\theta)\},$$

with $\mathbb{E}[x] = \theta$. (This parameterization is called *mean parameterization*; see Brown (1986, Chapter 3).) A sequence of natural conjugate distributions is given by $(m \in \mathbb{N})$

(3.8.1) $h_m(\theta|s) = \exp\{s \cdot \tau(\theta) - m\gamma(\theta)\}c_m(s),$

where $c_m(s)$ is the normalizing factor. Recall that the prior distribution (3.8.1) corresponds to an update of a flat prior on θ for m *fictious* (or *virtual*) observations $\tilde{x}_1, \ldots, \tilde{x}_m$ from $f(x|\theta)$, such that $s = \sum_{i=1}^m \tilde{x}_i$.

Now, (3.8.1) can also be considered as the density of s for a measure, $d\nu_m$, called the *natural measure*. If $\mathcal{S}_m$ is the space in which s varies, and dQ_m is a probability measure on $\mathcal{S}_m$,

(3.8.2) $$\nu_m(\theta) = \int_{\mathcal{S}_m} h_m(\theta|s) \, dQ_m(s)$$

constitutes a (continuous) mixture of conjugate priors. For a prior distribution π on Θ, defining

$$dQ_m(s) = \frac{\pi(s/m) \, d\nu_m(s)}{\int_{\mathcal{S}_m} \pi(t/m) \, d\nu_m(t)},$$

we get an approximation of π, as shown by the following lemma.

Theorem 3.8.3 *If ν_m is absolutely continuous with respect to the Lebesgue measure or if ν_m is absolutely continuous with respect to the counting measure*

Table 3.8.1. *Approximation of prior distributions by mixtures of conjugate priors.*
(Source: Dalal and Hall (1983).)

Distrib. $f(x, \theta)$	$\tau(\theta), \gamma(\theta)$	$c_m(s)$	$h_m(\theta\vert s)$ prior distribution
Normal $\mathcal{N}(\theta, 1)$	$\theta, \theta^2/2$	$\sqrt{m}\varphi(s/\sqrt{m})$	$\theta \sim \mathcal{N}\left(\frac{s}{m}, \frac{1}{m}\right)$
Gamma $\mathcal{G}\left(\frac{\beta}{\theta}, \theta\right)$	$-\frac{\beta}{\theta}, -\beta\log\left(\frac{\beta}{\theta}\right)$	$\frac{s^{m\beta-1}}{\beta\,\Gamma(m\beta-1)}$	$\frac{1}{\theta} \sim \mathcal{G}(s\beta, m\beta-1)$
Poisson $\mathcal{P}(\theta)$	$\log\theta, \theta$	$m^{s+1}/\Gamma(s+1)$	$\theta \sim \mathcal{G}(m, s+1)$
Bernoulli $\mathcal{B}(1, \theta)$	$\log\frac{\theta}{1-\theta}, \log\frac{1}{1-\theta}$	$\frac{(m+1)!}{s!(m-s)!}$	$\theta \sim \mathcal{Be}(s+1, m-s+1)$
Neg. bin. $\mathcal{N}eg\left(r, \frac{r}{r+\theta}\right)$	$\log\frac{\theta}{r+\theta}, r\log(r+\theta)$	$\frac{r^{mr}(mr+s-1)!}{rs!(mr-2)!}$	$\frac{r}{r+\theta} \sim \mathcal{Be}(mr-1, s+1)$

on $\mathcal{S}_m$ with density $f_m(s)$, and if $f_m(s)$ uniformly converges to 1 on $\mathcal{S}_m$ when m goes to $+\infty$, then

$$v_m(\theta) \longrightarrow \pi(\theta)$$

pointwise and for the L_1 convergence.

Moreover, the approximation remains valid a posteriori, using the total variation distance as a measure of discrepancy. The total variation norm is defined as

$$||\pi - \tilde{\pi}||_{TV} = \sup_A |\pi(A) - \tilde{\pi}(A)|,$$

and is therefore always bounded by 1. It is thus somehow a weaker approximation result, compared with the L_1 norm of Theorem 3.8.3.

Theorem 3.8.4 *If p_m, marginal distribution of x under h_m, is finite and if $\pi(\theta)$ and $\pi(\theta\vert x)$ are proper, $v_m(\theta\vert x)$ converges to $\pi(\theta\vert x)$, pointwise and for the total variation norm.*

The approximative posterior distribution is, for n observations and $t = \sum_{i=1}^{n} x_i$,

$$v_m(\theta\vert n, t) = \frac{\int_{\mathcal{S}_m} h_{m+n}(\theta\vert s+t)\frac{c_m(s)}{c_{m+n}(s+t)}\pi(s/m)\,d\nu_m(s)}{\int_{\mathcal{S}_m}\frac{c_m(s')}{c_{m+n}(s'+t)}\pi(s'/m)\,d\nu_m(s')}$$

and Table 3.8.3 provides the values of τ, γ and c_m for some usual distributions. Compared with the results of Diaconis and Ylvisaker (1985), Theorems 3.8.3 and 3.8.4 are indeed more general and ensure, in addition, convergence of the posterior distributions. The drawback, however, is that this approach does not preserve the advantage of conjugate priors, namely their simplicity. Simulation methods such as those presented in Chapter 6 are thus necessary to derive these Bayes estimators.

3.8.4 Bartlett corrections

In standard asymptotic theory, the likelihood ratio statistic

$$\varpi_n = 2\left\{\sum_{i=1}^{n} f(x_i|\hat{\theta}) - \sum_{i=1}^{n} f(x_i|\hat{\theta}_0)\right\},$$

is approximately distributed from a χ_k^2 distribution, where $\hat{\theta}$ and $\hat{\theta}_0$ denote the maximum likelihood estimator and the constrained maximum likelihood estimator, and k is the number of constraints (Gouriéroux and Monfort (1996)). Bartlett (1937) noticed that the fit to a χ_k^2 distribution was improved when ϖ_n was replaced with $k\varpi_n/\mathbb{E}_\theta[\varpi_n]$, in the sense that (Lawley (1956))

$$P_\theta\left(\frac{k\varpi_n}{\hat{E}} \leq t\right) = \chi_k^2(t) + O\left(n^{-2}\right),$$

where $\hat{E}$ is an appropriate estimate of $\mathbb{E}_\theta[\varpi_n]$ and $\chi_k^2(t)$ denotes the χ_k^2 c.d.f. The *Bartlett correction* thus leads to an improvement from $O\left(n^{-1}\right)$ to $O\left(n^{-2}\right)$ in the χ_k^2 approximation.

As noted in DiCiccio and Stern (1994), if $\theta = (\psi, \varphi)$ and the constraint on θ is that ψ is fixed, the likelihood ratio depends on ψ, $\varpi_n = \varpi_n(\psi)$. Bickel and Ghosh (1990) have established that the Bartlett correction extends to the posterior distribution of $\varpi_n(\psi)$, that is, that there exists a correction to $\varpi_n(\psi)$ such that

$$P\left(\varpi_n(\psi) \times \left(1 - \frac{A_B}{k}\right) \leq t|x, \ldots, x_n\right) = \chi_k^2(t) + O\left(n^{-2}\right),$$

where A_B is deduced from an expansion of the posterior expectation

$$\mathbb{E}[\varpi_n(\psi)|x_1, \ldots, x_n] = k + A_B + O\left(n^{-3/2}\right),$$

and is of order $O\left(n^{-1}\right)$. DiCiccio and Stern (1994) also established that this second order χ_k^2 approximation holds for adjusted likelihood-ratio statistics, $\varpi_n(\psi) + \omega_n(\psi)$, where $\omega_n(\psi)$ is of order $O(1)$. For instance, Kass et al. (1989) use

$$\omega_n(\psi) = \frac{-1}{2}\log\left(\frac{\det \ell_{\varphi\varphi}(\hat{\theta}(\psi))}{\det \ell_{\varphi\varphi}(\hat{\theta})}\right) + \log\left(\frac{\pi(\hat{\theta}(\psi))}{\pi(\hat{\theta})}\right),$$

where $\hat{\theta}$ and $\hat{\theta}(\psi)$ are the maximum likelihood estimator and the constrained maximum likelihood estimator, and $\ell_{\varphi\varphi}$ is the matrix of the second derivatives of the log-likelihood for the nuisance parameter φ. DiCiccio and Stern (1994) provide the corresponding correction A_B, while DiCiccio and Stern (1993) give the correction factor for the posterior ratio statistic

$$\kappa^\pi = 2\left\{\log \pi(\hat{\psi}|x) - \log \pi(\psi|x)\right\}.$$

where $\hat{\psi}$ is the marginal MAP estimator of ψ.

Example 3.8.5 (DiCiccio and Stern (1993)) Consider the normal regression

$$y_i \sim \mathcal{N}\left(\sum_{j=1}^{k} u_{ij}\beta_j, \sigma^2\right), \qquad i = 1, \ldots, n,$$

associated with a flat improper prior on (β, η), for $\eta = \log \sigma$. If the parameter of interest is η (or σ^2), then

$$\kappa^{\pi} = (n - k + 2) \left[\frac{n\varrho}{n - k + 2} - \log \frac{n\varrho}{n - k + 2} - 1 \right],$$

where $\varrho = \hat{\sigma}^2 / \sigma^2$ and the correcting term $\aleph^{\pi}$, such that $(1 + \aleph^{\pi})^{-1}\kappa^{\pi}$ is χ_p^2 up to the order $O(n^{-2})$, is $\aleph^{\pi}(\eta) = n^{-1}/3$. When $\xi = (\beta_1, \ldots, \beta_p)$ is the parameter of interest, $\aleph^{\pi}(\xi) = (1 + p/2)n^{-1}$.   ‖

CHAPTER 4

Bayesian Point Estimation

"There is always something new from you," Perrin growled. "Can't you tell us what to expect once in a while, instead of explaining after it happens?" Uno looked as though he was trying to think of a reason to leave.

Moraine gave Perrin a flat look. "You want me to share a lifetime of knowledge with you in a single afternoon? Or even a single year?"

Robert Jordan, *The Dragon Reborn, Book III of the Wheel of Time.*

4.1 Bayesian inference

4.1.1 Introduction

When the prior distribution $\pi(\theta)$ is available, the posterior distribution $\pi(\theta|x)$ can be formally derived from the observation x with distribution $f(x|\theta)$. This updated distribution is then the extensive summary of the information available on the parameter θ, integrating simultaneously prior information and information brought by the observation x. (The same obviously holds when a sample $x_1, \ldots, x_n$ is available, but it can usually be reduced to the above situation through a sufficient statistic.) The Bayesian version of the Likelihood Principle thus implies that the inference on θ should rely entirely on the posterior distribution $\pi(\theta|x)$. Even though θ is not necessarily a *random variable*, the distribution $\pi(\theta|x)$ can be used as a regular probability distribution to describe the properties of θ. Summarizing indices for $\pi(\theta|x)$ such as the posterior mean, the posterior mode, the posterior variance, and the posterior median, can be used. For instance, when the quantity of interest is $h(\theta)$, a possible estimator of $h(\theta)$ is the posterior mean $\mathbb{E}^\pi[h(\theta)|x]$. (As mentioned in Section 3.5, when the distribution π is a noninformative prior, some *marginalization* difficulties may occur, and it is sometimes necessary to derive a new reference prior for the parameter of interest, $h(\theta)$.)

4.1.2 MAP estimator

If a choice must be made among the above summaries, there is no way of selecting a best estimator, short of using a loss criterion. Nonetheless, a possible estimator of θ based on $\pi(\theta|x)$ is the *maximum a posteriori (MAP) estimator*, defined as the posterior mode. Notice that the MAP estimator also maximizes $\ell(\theta|x)\pi(\theta)$, thus bypassing the computation of the marginal distribution.

It is associated with the $0-1$ loss, as seen in Section 2.5.3 for the special case $\theta \in \{0,1\}$. In continuous settings, since, for every $\delta \in \Theta$,

$$\int_\Theta \mathbb{I}_{\delta \neq \theta} \pi(\theta|x) d\theta = 1 \ ,$$

the $0-1$ loss must be replaced by a sequence of losses, $L_\varepsilon(d,\theta) = \mathbb{I}_{||\theta-d||>\varepsilon}$, and the MAP estimate is then the limit of the Bayes estimates associated with L_ε, when ε goes to 0.

This natural estimator can be expressed as a *penalized maximum likelihood estimator* in the classical sense (see Akaike (1978, 1983)). Notice that the asymptotic optimality properties of the regular maximum likelihood estimator (consistency, efficiency) are preserved for these Bayesian extensions, under a few regularity conditions on f and π (see Note 1.8.4 and Ibragimov and Has'minskii (1981)). This extension of the asymptotic properties of the maximum likelihood estimator is intuitively sound since, as the sample size grows to infinity, the information contained in this sample becomes predominant compared with the *fixed* information brought by the prior distribution π. Therefore, MAP estimators are asymptotically equivalent to the classical maximum likelihood estimators but, furthermore, have the advantage of being available for finite sample sizes; the latter are mainly justified on asymptotic grounds.

Example 4.1.1 Consider $x \sim \mathcal{B}(n,p)$. We saw in the previous chapter that the Jeffreys prior in this setting is the beta distribution $\mathcal{Be}(1/2,1/2)$, i.e.,

$$\pi^*(p) = \frac{1}{B(1/2,1/2)} p^{-1/2}(1-p)^{-1/2} \ ,$$

omitting the indicator function $\mathbb{I}_{[0,1]}(p)$ to simplify the notations. Two other noninformative distributions have been proposed in the literature by Laplace and Haldane (1931) (see also Exercise 4.4),

$$\pi_1(p) = 1 \qquad \text{and} \qquad \pi_2(p) = p^{-1}(1-p)^{-1} \ .$$

The corresponding MAP estimators are then, for $n > 2$,

$$\delta^*(x) = \max\left(\frac{x-1/2}{n-1}, 0\right),$$

$$\delta_1(x) = \frac{x}{n},$$

$$\delta_2(x) \quad = \quad \max\left(\frac{x-1}{n-2}, 0\right).$$

When $n = 1$, δ^* and δ_2 are equal to δ_1. For $n = 2$ and $x = 1$, the estimator δ_2 is also equal to δ_1, which is the regular maximum likelihood estimator. Notice that, when n is large, the three estimators are indeed equivalent. ‖

Example 4.1.2 Consider $x \sim \mathcal{C}(\theta, 1)$, i.e.,

$$f(x|\theta) = \frac{1}{\pi}\left[1 + (x - \theta)^2\right]^{-1},$$

and $\pi(\theta) = \frac{1}{2}e^{-|\theta|}$. The MAP estimator of θ is then $\delta^*(x) = 0$, as the maximum of $\exp(-|\theta|)[1 + (x - \theta)^2]^{-1}$ is attained for $\theta = 0$, whatever the value of x. This behavior may be explained by the flatness of the likelihood function, which is not informative enough, compared with the sharp prior distribution. However, from a practical point of view, this estimator is useless (see also Exercise 4.6). ‖

4.1.3 Likelihood Principle

Bayesian inference appears as a way to efficiently implement the Likelihood Principle, since it provides an estimator, by selecting for instance, as in Example 4.1.3 below, one of several maxima of the likelihood function. As stressed by Savage (1954) and Berger and Wolpert (1988), there are many philosophical and practical considerations linking the Likelihood Principle to a robust Bayesian implementation. In particular, it allows for the elimination of some classical paradoxes, such as those of Stein (1962b), Stone (1976), Fraser, Monette and Ng (1984), and Le Cam (1990). The following example illustrates the resolution of the paradox of Fraser et al. (1984). (See also Joshi (1990) for an extended analysis of the phenomenon.)

Example 4.1.3 (Berger and Wolpert (1988)) Consider $\mathcal{X} = \Theta = \mathbb{N}^*$ and

$$(4.1.1) \quad f(x|\theta) = \frac{1}{3} \text{ for } \begin{cases} \theta/2, 2\theta, 2\theta + 1 & \text{if } \theta \text{ is even,} \\ (\theta - 1)/2, 2\theta, 2\theta + 1 & \text{if } \theta \text{ is odd and } \theta \neq 1, \\ 1, 2, 3 & \text{if } \theta = 1. \end{cases}$$

The likelihood function is then

$$\ell(\theta|x) = \frac{1}{3} \text{ for } \begin{cases} x/2, 2x, 2x + 1 & \text{if } x \text{ is even,} \\ (x - 1)/2, 2x, 2x + 1 & \text{if } x \text{ is odd and } x \neq 1, \\ 1, 2, 3 & \text{if } x = 1, \end{cases}$$

and the three values of θ where $\ell(\theta|x) \neq 0$ are equally weighted by the likelihood function. Consider the three following estimators:

$$\delta_1(x) = \begin{cases} x/2 & \text{if } x \text{ is even,} \\ (x - 1)/2 & \text{if } x \text{ is odd and } x \neq 1, \\ 1 & \text{if } x = 1, \end{cases}$$

and
$$\delta_2(x) = 2x, \qquad \delta_3(x) = 2x + 1.$$
They are equivalent from the Likelihood Principle perspective, since the likelihood function is flat on its support, but δ_2 and δ_3 are quite suboptimal estimators because
$$P(\delta_2(x) = \theta) = P(x = \theta/2) = \begin{cases} 1/3 & \text{if } \theta \text{ is even,} \\ 0 & \text{otherwise,} \end{cases}$$

$$P(\delta_3(x) = \theta) = P(x = (\theta - 1)/2) = \begin{cases} 1/3 & \text{if } \theta \neq 1 \text{ is odd,} \\ 0 & \text{otherwise,} \end{cases}$$
while
$$P(\delta_1(x) = \theta) = \begin{cases} 1 & \text{if } \theta = 1, \\ 2/3 & \text{otherwise.} \end{cases}$$
The estimator δ_1 is therefore preferable under losses such as the $0 - 1$ loss. When the information available on the model is reduced to the likelihood function (4.1.1), a possible noninformative distribution on θ is $\pi(\theta) = 1/\theta$, since θ can approximately be considered as a scale parameter. In this case,
$$\pi(\theta|x) \propto \frac{1}{3\theta} \left[\mathbb{I}_{\delta_1(x)}(\theta) + \mathbb{I}_{\delta_2(x)}(\theta) + \mathbb{I}_{\delta_3(x)}(\theta) \right]$$
and this posterior distribution gives $\delta_1(x)$ as being four times more likely than δ_2 or δ_3. It can also be shown that $P^\pi(\theta = \delta_1(x)|x) \simeq 2/3$ for large x's. This gives a good justification to the choice of δ_1. A more informative prior modeling would lead to a similar conclusion (since a proper distribution $\pi(\theta)$ must be decreasing for θ large enough).   ‖

Berger and Wolpert (1988) provide similar resolutions of the paradoxes exhibited by Stein (1962b) and Stone (1976). An immediate advantage of the Bayes approach over other implementations of the Likelihood Principle is that it takes care of the nuisance parameters in the likelihood function by integrating them out. In fact, if $\ell(\theta, \tau|x)$ also depends on a nuisance parameter τ, a natural derivation of an estimate $\hat{\theta}$ of θ is to consider the maximum of the integrated likelihood
$$\int \ell(\theta, \tau|x)\pi(\theta, \tau)\, d\tau$$
instead of the more classical "profile" likelihood,
$$\max_\tau \ell(\theta, \tau|x)\pi(\theta, \tau).$$
See also Basu (1988) for an extensive analysis of the treatment of the nuisance parameters.

4.1.4 Restricted parameter space

Berger (1985a) points out the interest of a noninformative Bayesian approach for *restricted parameter spaces*, since the prior distribution is just the truncation of the nonrestricted noninformative distribution. From a classical point of view, the derivation of restricted maximum likelihood estimators is often complicated, especially when the constraints are nonlinear (see Robertson et al. (1988)). On the contrary, the implementation of the Bayesian approach through a Monte Carlo simulation scheme (see Chapter 6) allows for an easy derivation of the Bayes estimators. (This advantage can even be used to compute restricted maximum likelihood estimators via Bayesian techniques. See Geyer and Thompson (1992) and Robert and Casella (2004, Chapter 5)).

Example 4.1.4 Consider the estimation of the *linear regression model*

$$(4.1.2) \qquad\qquad y = b_1 X_1 + b_2 X_2 + \epsilon,$$

which relates *direct incomes* (X_1), *saving incomes* (X_2), and *savings* (y). A careful estimation of the *saving rates* b_1 and b_2 helps enable the government to determine its fiscal and interest rates policies. The saving rates are obviously constrained by $0 \le b_1, b_2 \le 1$. Consider a sample $(y_1, X_{11}, X_{21}), \ldots,$ (y_n, X_{1n}, X_{2n}) from (4.1.2) and assume that the errors ϵ_i are independent and distributed according to $\mathcal{N}(0, 1)$, i.e., that $y_i \sim \mathcal{N}(b_1 X_{1i} + b_2 X_{2i}, 1)$. The corresponding noninformative distribution is then the proper distribution

$$\pi(b_1, b_2) = \mathbb{I}_{[0,1]}(b_1) \mathbb{I}_{[0,1]}(b_2)$$

and the posterior mean is given by $(i = 1, 2)$

$$\mathbb{E}^{\pi}[b_i | y_1, \ldots, y_n] = \frac{\int_0^1 \int_0^1 b_i \prod_{j=1}^n \varphi(y_j - b_1 X_{1j} - b_2 X_{2j}) \, db_1 \, db_2}{\int_0^1 \int_0^1 \prod_{j=1}^n \varphi(y_j - b_1 X_{1j} - b_2 X_{2j}) \, db_1 \, db_2},$$

where φ is the density of the standard normal distribution. If we denote by $(\hat{b}_1, \hat{b}_2)$ the unconstrained least-squares estimator of (b_1, b_2), which is also the regular maximum likelihood estimator of (b_1, b_2), the unconstrained posterior distribution on (b_1, b_2) is

$$(4.1.3) \qquad\qquad \begin{pmatrix} b_1 \\ b_2 \end{pmatrix} \sim \mathcal{N}_2 \left(\begin{pmatrix} \hat{b}_1 \\ \hat{b}_2 \end{pmatrix}, (X^t X)^{-1} \right),$$

with

$$X = \begin{pmatrix} X_{11} & X_{21} \\ \ldots & \ldots \\ X_{1n} & X_{2n} \end{pmatrix}.$$

Therefore, the restricted Bayes estimator is given by $(i = 1, 2)$

$$\delta_i^{\pi}(y_1, \ldots, y_n) = \frac{\mathbb{E}^{\pi}\left[b_i \mathbb{I}_{[0,1]^2}(b_1, b_2) | y_1, \ldots, y_n\right]}{P^{\pi}\left((b_1, b_2) \in [0, 1]^2 | y_1, \ldots, y_n\right)},$$

where the right-hand term is computed under the distribution (4.1.3). If we denote

$$\Sigma = (X^t X)^{-1} = \begin{pmatrix} \sigma_{11}^2 & \sigma_{12} \\ \sigma_{12} & \sigma_{22}^2 \end{pmatrix},$$

the conditional distribution of b_1 is

$$b_1|b_2 \sim \mathcal{N}\left(\hat{b}_1 + \sigma_{12}(b_2 - \hat{b}_2)/\sigma_{22}^2, \sigma_{11}^2 - \sigma_{12}^2\sigma_{22}^{-2}\right).$$

Then

$$P^\pi\left((b_1, b_2) \in [0,1]^2|y_1, \ldots, y_n\right)$$

$$= \int_0^1 \left\{ \Phi\left(\frac{1 - \hat{b}_1 - \sigma_{12}(b_2 - \hat{b}_2)/\sigma_{22}^2}{\sqrt{\sigma_{11}^2 - \sigma_{12}^2\sigma_{22}^{-2}}}\right)\right.$$

$$\left. - \Phi\left(\frac{-\hat{b}_1 - \sigma_{12}(b_2 - \hat{b}_2)/\sigma_{22}^2}{\sqrt{\sigma_{11}^2 - \sigma_{12}^2\sigma_{22}^{-2}}}\right)\right\} \sigma_{22}^{-1}\varphi\left(\frac{b_2 - \hat{b}_2}{\sigma_{22}}\right) db_2$$

and

$$\mathbb{E}^\pi[b_i \mathbb{I}_{[0,1]^2}(b_1, b_2)|y_1, \ldots, y_n] = \int_0^1 \left[\hat{b}_1 + \frac{\sigma_{12}}{\sigma_{22}^2}(b_2 - \hat{b}_2)\right.$$

$$+ (\sigma_{11}^2 - \sigma_{12}^2\sigma_{22}^{-2})^{1/2}\left\{\varphi\left(\frac{1 - \hat{b}_1 - \sigma_{12}(b_2 - \hat{b}_2)/\sigma_{22}^2}{\sqrt{\sigma_{11}^2 - \sigma_{12}^2\sigma_{22}^{-2}}}\right)\right.$$

$$\left.\left. - \varphi\left(\frac{-\hat{b}_1 - \sigma_{12}(b_2 - \hat{b}_2)/\sigma_{22}^2}{\sqrt{\sigma_{11}^2 - \sigma_{12}^2\sigma_{22}^{-2}}}\right)\right\}\right] \sigma_{22}^{-1}\varphi\left(\frac{b_2 - \hat{b}_2}{\sigma_{22}}\right) db_2.$$

Notice that this second integral can be obtained in closed form using the standard normal c.d.f. Φ, but that the denominator cannot be computed analytically. It is therefore more efficient to compute both integrals by a Monte Carlo simulation (see Chapter 6).

If b_1 and b_2 are independent a posteriori, i.e., when $\sigma_{12} = 0$, the Bayes estimators are explicit and given by ($i = 1, 2$)

$$\mathbb{E}^\pi[b_i|y_1, \ldots, y_n] = \hat{b}_i - \sigma_{ii}\frac{\exp\{-(1 - \hat{b}_i)^2/2\sigma_{ii}^2\} - \exp\{-\hat{b}_i^2/2\sigma_{ii}^2\}}{\sqrt{2\pi}\{\Phi((1 - \hat{b}_i)/\sigma_{ii}) - \Phi(-\hat{b}_1/\sigma_{ii})\}}.$$

$\|$

Notice that a Bayesian modeling is also quite appropriate to incorporate *vague* information, i.e., cases where a restriction on the parameter space is likely but not certain. Chapter 10 demonstrates that a typical way to do this is to use a hierarchical or empirical Bayes modeling.

4.1.5 Precision of the Bayes estimators

Since the whole posterior distribution $\pi(\theta|x)$ is available, it is possible to associate to an estimator $\delta^\pi(x)$ of $h(\theta)$ an evaluation of the *precision* of the estimation, through, for instance, the *posterior squared error*,

$$\mathbb{E}^\pi[(\delta^\pi(x) - h(\theta))^2|x],$$

equal to $\mathrm{var}^{\pi}(h(\theta)|x)$ when $\delta^{\pi}(x) = \mathbb{E}^{\pi}[h(\theta)|x]$. Similarly, in a multidimensional setting, the covariance matrix characterizes the performances of estimators. These additional indications provided by the posterior distribution illustrate the operational advantage of the Bayesian approach, since the classical approach usually has difficulties motivating the choice of these evaluations. Moreover, Bayesian evaluation measures are always conditional[1], while the frequentist approach usually relies on upper bounds through the minimax principle, since the parameter θ is unknown (see Berger and Robert (1990) for a comparison of both approaches).

Example 4.1.5 (Example 4.1.1 continued) Consider the maximum likelihood estimator of p, $\delta_1(x) = x/n$. Then

$$
\begin{aligned}
\mathbb{E}^{\pi}[(\delta_1(x) - p)^2|x] &= \mathbb{E}^{\pi}[(p - x/n)^2|x] \\
&= \left(\frac{x + 1/2}{n + 1} - \frac{x}{n}\right)^2 + \frac{(x + 1/2)(n - x + 1/2)}{(n + 1)^2(n + 2)} \\
&= \frac{(x - n/2)^2}{(n + 1)^2 n^2} + \frac{(x + 1/2)(n - x + 1/2)}{(n + 1)^2(n + 2)},
\end{aligned}
$$

(4.1.4)

since $\pi(p|x)$ is the beta distribution $\mathcal{B}e(x + 1/2, n - x + 1/2)$. From a frequentist viewpoint, the risk of the maximum likelihood estimator is

$$
\mathbb{E}_p[(\delta_1(x) - p)^2] = \mathrm{var}(x/n) = \frac{p(1 - p)}{n}
$$

and

$$
\sup_p p(1 - p)/n = 1/4n.
$$

Developing (4.1.4), it is easy to verify that the maximum of (4.1.4) is

$$
1/[4(n + 2)],
$$

always smaller than $1/4n$. The major advantage of the quantity (4.1.4), though, is still to provide an *adjustable* answer for the evaluation of δ_1, since (4.1.4) varies between $1/[4(n+2)]$ and $3/[4(n+1)(n+2)]$. Obviously, a frequentist approximation of $p(1 - p)/n$ can also be proposed, namely, $(x/n)(1 - x/n)/n$. This evaluation has then the opposite drawback of $1/4n$ since it varies too widely, as shown by Figure 4.1.5. It can even take the value 0 when $x = 0, n$. A similar behavior is discussed in Berger (1990b) in a general framework.   ‖

4.1.6 Prediction

Furthermore, Bayesian inference can operate as well in *prediction* problems. If $x \sim f(x|\theta)$ and $z \sim g(z|x, \theta)$, where z does not necessarily depend on x,

---

[1] In fact, there are Bayesian counterparts to the Cramér-Rao inequalities used in the evaluation of unbiased estimators. These are the *Van Trees* bounds (Gill and Levit (1995)), whose use can be found in signal processing and other domains, as shown in Bergman et al. (2001).

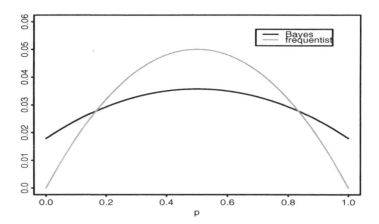

Figure 4.1.1. *Comparison of Bayesian and frequentist evaluations of the estimation error in the binomial case* $(n = 3)$.

the *predictive distribution* of z after the observation of x is given by

$$(4.1.5) \qquad g^\pi(z|x) = \int_\Theta g(z|x, \theta)\pi(\theta|x)\, d\theta.$$

The distribution of z is thus quite logically averaged over the values of θ according to the posterior distribution, which is also the actualized distribution of θ. It is possible to use (4.1.5) to derive the predictive mean and variance of the random variable z. In Section 4.3.1, we consider a particular example of determination of a discrete predictive distribution (see also Exercise 4.40).

Example 4.1.6 A particular case of AR(1) model, where AR stands for *autoregressive*, defines the distribution of a stochastic process $(x_t)_{1 \le t \le T}$ by a linear dynamic representation conditional on the previous variable x_{t-1}, as

$$x_t = \varrho x_{t-1} + \epsilon_t,$$

where the ϵ_t's are i.i.d. $\mathcal{N}(0, \sigma^2)$. (This model will be considered in detail in Section 4.5.) Given a sequence of observations till time $T-1$, $x_{1:(T-1)} = (x_1, \dots, x_{(T-1)})$, the predictive distribution of x_T is then given by

$$x_T|x_{1:(T-1)} \sim \int \frac{1}{\sqrt{2\pi}}\sigma^{-1}\exp\{-(x_T - \varrho x_{T-1})^2/2\sigma^2\}\pi(\varrho, \sigma|x_{1:(T-1)})d\varrho d\sigma,$$

where $\pi(\varrho, \sigma|x_{1:(T-1)})$ can be expressed in closed form (Exercise 4.13). ||

Notice that the decision-theoretic approach developed in the following sections is also applicable in a prediction setting, even though we do not mention it later. Indeed, if a prediction loss $L(z, \delta)$ is available, a predictor $\delta(x)$ can be chosen in order to minimize the expected prediction error (the expectation being with respect to the predictive distribution (4.1.5)). (See Exercise 4.45.)

4.1.7 Back to Decision Theory

Given the scope of possible uses of the posterior distribution, some Bayesians consider that clients should be provided with the posterior distribution so that they might use it at their convenience. Although the communication of $\pi(\theta|x)$ is indeed of interest for small dimensions, the information provided by $\pi(\theta|x)$ gets blurred by its own complexity for large dimensions. The posterior distribution is obviously essential in the decision-making process, but it is part of the statistician's role to assist the decision-maker further, in order to extract the features of interest from $\pi(\theta|x)$. We thus confront again the major problem of selecting among estimators and we saw in Chapter 2 that this selection is efficient and coherent only when based on a *loss* criterion. The following sections set forth Bayesian Decision Theory, with particular attention to the normal and sampling cases. Testing perspectives and confidence-set estimation are treated separately in Chapter 5.

4.2 Bayesian Decision Theory

4.2.1 Bayes estimators

Let us recall here that, given a loss function $L(\theta, \delta)$ and a prior distribution (or a measure) π, the *Bayes rule* $\delta^\pi(x)$ is solution of

$$\min_\delta \mathbb{E}^\pi[L(\theta, \delta)|x].$$

Depending on the complexity of the loss L and the posterior distribution $\pi(\theta|x)$, the estimator δ^π will be determined analytically or numerically, similar to maximum likelihood estimation.

   As shown in Chapter 2, the solutions associated with classical losses are formally known and correspond to the natural indicators associated with a distribution (mean, median, mode, quantiles, etc.). For instance, the Bayes estimator associated with the quadratic loss is the posterior mean (Proposition 2.5.1 and Corollary 2.5.2). Of course, this formal derivation of classical Bayes estimators does not always avoid a numerical approximation of the estimators, especially in multidimensional settings.

Example 4.2.1   Consider $x \sim \mathcal{N}_p(\theta, I_p)$. As mentioned in Section 3.6, Student's t-distribution provides a robust alternative to the conjugate normal prior for estimating θ. Consider thus $\theta \sim \mathcal{T}_p(\alpha, 0, \tau^2 I_p)$, i.e.,

$$\pi(\theta|\alpha, \tau) = \frac{\Gamma((\alpha+p)/2)}{(\alpha\tau\pi)^{p/2}\Gamma(\alpha/2)}\left(1 + \frac{||\theta||^2}{\alpha\tau^2}\right)^{-(\alpha+p)/2}.$$

Therefore,

$$\pi(\theta|x) \propto \left(1 + \frac{||\theta||^2}{\alpha\tau^2}\right)^{-(\alpha+p)/2} e^{-||x-\theta||^2/2},$$

which does not lead to a closed-form expression for the posterior distribution. However, it is still possible to reduce the computational problem to a *single integration*, for every value of p, as shown by Dickey (1968). In fact, if $\theta \sim \mathcal{T}_p(\alpha, 0, \tau^2 I_p)$, the posterior distribution of θ can be expressed as a hidden mixture (see Example 3.3.11),

$$\begin{aligned}
\theta|z &\sim \mathcal{N}_p(0, \tau^2 z I_p), \\
z^{-1} &\sim \mathcal{G}(\alpha/2, \alpha/2),
\end{aligned}$$

where z is an auxiliary random variable. Conditional upon z, the posterior distribution of θ is

$$\theta|x, z \sim \mathcal{N}_p\left(\frac{x}{1+\tau^2 z}, \frac{\tau^2 z}{1+\tau^2 z}I_p\right)$$

and, since

$$\pi(z|x) \propto (1+\tau^2 z)^{-p/2} e^{-||x||^2/2(1+\tau^2 z)} \pi(z),$$

we derive the Bayes estimator as

$$\begin{aligned}
\delta^\pi(x) &= \int_0^{+\infty} \mathbb{E}^\pi[\theta|x, z]\pi(z|x)\, dz \\
&= x\, \frac{\int_0^{+\infty}(1+\tau^2 z)^{-(p+2)/2} e^{-||x||^2/2(1+\tau^2 z)} z^{-(\alpha+2)/2} e^{-\alpha/2z}\, dz}{\int_0^{+\infty}(1+\tau^2 z)^{-p/2} e^{-||x||^2/2(1+\tau^2 z)} z^{-(\alpha+2)/2} e^{-\alpha/2z}\, dz}.
\end{aligned}$$

This estimator can therefore be expressed through a single integral for every value of p. ‖

However, decomposition subtleties as in the above example are not always available, and the computation of a Bayes estimator then calls for general approximation methods such as those described in Chapter 6. An interesting result is that, when the marginal distribution $m(x)$ is available, the posterior expectation of the natural parameter can easily be derived for exponential families.

Lemma 4.2.2 *Consider $f(x|\theta) = h(x)e^{\theta \cdot x - \psi(\theta)}$, a distribution from an exponential family. For every prior distribution π, the posterior mean of θ is given by*

(4.2.1) $$\delta^\pi(x) = \nabla \log m_\pi(x) - \nabla \log h(x),$$

where ∇ denotes the gradient operator and m_π is the marginal distribution associated with π.

Proof. The posterior expectation is given by

$$\mathbb{E}^\pi[\theta_i|x] = \frac{\int_\Theta \theta_i h(x)e^{\theta \cdot x - \psi(\theta)}\pi(\theta)\, d\theta}{m_\pi(x)}$$

$$= \left(\frac{\partial}{\partial x_i} \int_\Theta h(x) e^{\theta \cdot x - \psi(\theta)} \pi(\theta)\, d\theta \right) \frac{1}{m_\pi(x)} - \left(\frac{\partial}{\partial x_i} h(x) \right) \frac{1}{h(x)}$$

$$= \frac{\partial}{\partial x_i} \left[\log m_\pi(x) - \log h(x) \right].$$

$\square\square$

Notice that this lemma is satisfied for every π; it appears as the dual result of the derivation of the moments of $f(x|\theta)$ from the derivative of ψ in exponential families (see Lemma 3.3.7). Its practical interest is rather limited since the derivation of the marginal distribution is usually quite delicate and to know $m_\pi(x)$ explicitly is equivalent to knowing $\pi(\theta|x)$ explicitly. From a theoretical viewpoint, a consequence of this lemma is to show that Bayes estimators are *analytic* (or holomorphic) functions for exponential families such that the function h spanning the exponential family is holomorphic (since m_π/h is the Laplace transform of $e^{-\psi(\theta)}\pi(\theta)$). Chapter 8 derives an inadmissibility criterion from this property.

Example 4.2.3 We introduced in Note 2.8.2 the truncated James–Stein estimator,

$$\delta^{\mathrm{JS}}(x) = \left(1 - \frac{p-2}{||x||^2} \right)^+ x$$

when $x \sim \mathcal{N}_p(\theta, I_p)$. In the normal case, (4.2.1) reduces to

$$\delta^\pi(x) = x + \nabla \log m_\pi(x).$$

Although there exists a function m such that δ^{JS} can be written as above (see Bock (1988)), m is not a marginal distribution and this estimator cannot be a Bayes estimator: it is equal to 0 on the open set $\{||x||^2 < p-2\}$ and should be null everywhere if it were analytic. ‖

The representation (4.2.1) of the Bayes estimators is also useful in techniques underlying the *Stein effect*, either to establish domination conditions as in Stein (1981), George (1986a), Berger and Robert (1990), and Brandwein and Strawderman (1990), or to point out the inadmissibility of some estimators as in Bock (1988) and Brown (1988). (See Exercise 4.43.)

4.2.2 Conjugate priors

In the particular case of conjugate distributions, the posterior expectations of the natural parameters can obviously be expressed analytically and this is practically the only case where closed-form expressions are available in such generality. Table 4.2.1 presents the Bayes estimators associated with the usual distributions and their conjugate priors. Notice that when several observations from $f(x|\theta)$ are available the conjugate distributions are the same, and that only the parameters in the estimator are modified by virtue of the sufficiency properties of exponential families (see Section 3.3.3).

Table 4.2.1. *The Bayes estimators of the parameter θ under quadratic loss for conjugate distributions in the usual exponential families.*

Distribution	Conjugate prior	Posterior mean
Normal $\mathcal{N}(\theta, \sigma^2)$	Normal $\mathcal{N}(\mu, \tau^2)$	$\dfrac{\mu\sigma^2 + \tau^2 x}{\sigma^2 + \tau^2}$
Poisson $\mathcal{P}(\theta)$	Gamma $\mathcal{G}(\alpha, \beta)$	$\dfrac{\alpha + x}{\beta + 1}$
Gamma $\mathcal{G}(\nu, \theta)$	Gamma $\mathcal{G}(\alpha, \beta)$	$\dfrac{\alpha + \nu}{\beta + x}$
Binomial $\mathcal{B}(n, \theta)$	Beta $\mathcal{Be}(\alpha, \beta)$	$\dfrac{\alpha + x}{\alpha + \beta + n}$
Negative binomial $\mathcal{N}eg(n, \theta)$	Beta $\mathcal{Be}(\alpha, \beta)$	$\dfrac{\alpha + n}{\alpha + \beta + x + n}$
Multinomial $\mathcal{M}_k(n; \theta_1, \ldots, \theta_k)$	Dirichlet $\mathcal{D}(\alpha_1, \ldots, \alpha_k)$	$\dfrac{\alpha_i + x_i}{\left(\sum_j \alpha_j\right) + n}$
Normal $\mathcal{N}(\mu, 1/\theta)$	Gamma $\mathcal{G}(\alpha/2, \beta/2)$	$\dfrac{\alpha + 1}{\beta + (\mu - x)^2}$

Example 4.2.4 If $x_1, ..., x_n$ are independent observations from $\mathcal{N}eg(m, \theta)$ and if $\theta \sim \mathcal{Be}(\alpha, \beta)$, the posterior distribution of θ is the beta distribution $\mathcal{Be}\left(\alpha + mn, \sum_{i=1}^{n} x_i + \beta\right)$ and

$$\delta^\pi(x_1, ..., x_n) = \frac{\alpha + mn}{\alpha + \beta + mn + \sum_{i=1}^{n} x_i}.$$

This result is a straightforward consequence of $\sum_{i=1}^{n} x_i \sim \mathcal{N}eg(mn, \theta)$. ∥

Example 4.2.5 Consider n observations $x_1, \ldots, x_n$ from $\mathcal{U}([0, \theta])$ and $\theta \sim \mathcal{P}a(\theta_0, \alpha)$. Then

$$\theta | x_1, ..., x_n \sim \mathcal{P}a(\max(\theta_0, x_1, ..., x_n), \alpha + n)$$

and

$$\delta^\pi(x_1, ..., x_n) = \frac{\alpha + n}{\alpha + n - 1} \max(\theta_0, x_1, ..., x_n).$$

Therefore, compared with the maximum likelihood estimator,

$$\delta_0(x_1, ..., x_n) = \max(x_1, ..., x_n),$$

the Bayes estimator gives a more optimistic estimation of θ, since

$$\frac{\alpha + n}{\alpha + n - 1} > 1.$$

On the contrary, the best equivariant estimator of θ under squared error loss is $\frac{n}{n-1}\delta_0(x_1, \ldots, x_n)$ (see Chapter 9), larger than δ^π when θ_0 is small. This shrinking behavior of δ^π is explained by the choice of π, which decreases with θ, thus favoring values of θ close to θ_0. ‖

Similarly, recall that the estimation of a function of θ, $g(\theta)$, under quadratic loss gives $\delta^\pi(x) = \mathbb{E}^\pi[g(\theta)|x]$ as the Bayes estimator.

Example 4.2.6 Consider $x \sim \mathcal{G}(\nu, \theta)$, where the shape parameter ν is known, and $\theta \sim \mathcal{G}(\alpha, \beta)$. The parameter of interest is $1/\theta$, the expectation of x. Under the quadratic loss

$$L(\theta, \delta) = \left(\delta - \frac{1}{\theta}\right)^2,$$

the Bayes estimator is then

$$\begin{aligned}
\delta_1^\pi(X) &= \frac{(\beta + x)^{\alpha + \nu}}{\Gamma(\alpha + \nu)} \int_0^{+\infty} \frac{1}{\theta} \theta^{\alpha + \nu - 1} e^{-(\beta + x)\theta} d\theta \\
&= \frac{\beta + x}{\alpha + \nu - 1}.
\end{aligned}$$ ‖

Under a renormalized (or weighted) quadratic loss,

$$L(\theta, \delta) = w(\theta) \| \delta - \theta \|_Q^2,$$

where Q is a $p \times p$ nonnegative symmetric matrix, the corresponding Bayes estimator is

$$\delta^\pi(x) = \frac{\mathbb{E}^\pi[\theta w(\theta)|x]}{\mathbb{E}^\pi[w(\theta)|x]}.$$

Example 4.2.7 (Example 4.2.6 continued) A scale-invariant loss does not depend on the unit of measurement and may be more relevant for the estimation of $1/\theta$. For instance, the loss

$$L(\theta, \delta) = \theta^2 \left(\delta - \frac{1}{\theta}\right)^2$$

gives the Bayes estimator

$$\begin{aligned}
\delta_2^\pi(x) &= \frac{\mathbb{E}^\pi\left[\theta^2/\theta \mid x\right]}{\mathbb{E}^\pi[\theta^2 \mid x]} \\
&= \frac{\int_0^{+\infty} \theta\theta^{\alpha+\nu-1} e^{-(\beta+x)\theta} d\theta}{\int_0^{+\infty} \theta^{\alpha+\nu+1} e^{-(\beta+x)\theta} d\theta} \\
&= \frac{\beta + x}{\alpha + \nu + 1} = \frac{\alpha + \nu - 1}{\alpha + \nu + 1} \delta_1^\pi(x).
\end{aligned}$$ ‖

Let us stress again that, even for conjugate distributions, the fact that the Bayes estimator of any function of θ involves posterior expectations does not necessarily avoid numerical computations because analytical integration may be impossible, especially in multidimensional problems.

Example 4.2.8 Consider $x \sim \mathcal{N}_p(\theta, I_p)$ and $h(\theta) = ||\theta||^2$. The loss considered in Saxena and Alam (1982) is

$$L(\theta, \delta) = \frac{(\delta - ||\theta||^2)^2}{2||\theta||^2 + p}$$

since, if $\delta_0(x) = ||x||^2 - p$,

$$R(\delta_0, \theta) = \frac{1}{2||\theta||^2 + p} \, \mathbb{E}(||x||^2 - ||\theta||^2 - p)^2 = 2$$

and δ_0 has a constant risk. Without this renormalization, all estimators have a maximum risk equal to $+\infty$, while under L, the estimator δ_0 is minimax. Then, even for the conjugate distributions, $\mathcal{N}_p(0, \tau^2 I_p)$, the computation of

$$\delta^\pi(x) = \frac{\mathbb{E}^\pi[||\theta||^2/(2||\theta||^2 + p)|x]}{\mathbb{E}^\pi[1/(2||\theta||^2 + p)|x]}$$

cannot be conducted analytically.   ‖

In the previous examples, we resorted rather heavily to the quadratic loss since it constitutes a standard loss and allows, as much as possible, for explicit computations. We refer the reader to Chapter 2 for criticisms about the arbitrariness of standard losses and the opposition between bounded concave and unbounded convex losses, the former leading to the risk-lover paradox and the latter to more instability in the resulting procedures (see Kadane and Chuang (1978), Smith (1988), and Exercises 4.1 and 4.14). Even so, it bears mentioning that, when the loss function is truly determined by the decision-maker, it is usually complex and most often calls for a numerical minimization to determine the Bayes estimator.

4.2.3 Loss estimation

For a given loss function, $L(\theta, \delta)$, it may also be of interest to assess the performance of the Bayes estimator $\delta^\pi(x)$. This evaluation can be perceived from a decision-theoretic point of view as the estimation of the loss $L(\theta, \delta^\pi(x))$ by $\gamma(x)$ under a loss function, like

(4.2.2) $\tilde{L}(\theta, \delta^\pi, \gamma) = [\gamma(x) - L(\theta, \delta^\pi(x))]^2.$

Again, the quadratic loss (4.2.2) is no more justified as an automatic loss in this context than in other estimation settings. But, apart from tractability reasons, the choice of the quadratic loss can be grounded on a lack of utility foundations, and thus a closer perception of the error as a variance. Under

(4.2.2), the Bayesian evaluation of the performances of δ^π is given by the following result.

Proposition 4.2.9 *The Bayes estimator of the loss $L(\theta, \delta^\pi(x))$ under (4.2.2) for the prior distribution π is*

$$\gamma^\pi(x) = \mathbb{E}^\pi[L(\theta, \delta^\pi(x))|x].$$

This result directly follows from Proposition 2.5.1, since, conditional upon x, the purpose is to estimate a particular function of θ under quadratic loss. Notice that the dependence of this function on x does not matter from a Bayesian perspective since, once x is *observed*, x is *fixed*. Similarly, for the absolute error loss, the Bayes estimator of the loss is the median of the posterior distribution of $L(\theta, \delta^\pi(x))$, which is usually more difficult to obtain. When L is the quadratic loss, the posterior variance, $\mathrm{var}^\pi(x)$, is thus the Bayes estimator of the loss associated with δ^π.

From a frequentist perspective, loss estimation has been studied by Johnstone (1988) and Rukhin (1988a,b), the former establishing that, for a minimax estimator with constant risk p, the evaluation $\gamma(x) = p$ is not necessarily admissible under (4.2.2). Berger (1984b, 1985b) (see also Lu and Berger (1989a,b)) develops an additional concept for loss estimation, called *frequentist validity*: an estimator γ of the loss $L(\theta, \delta(x))$ is said to be *frequency valid* if

$$\mathbb{E}_\theta[\gamma(x)] \geq R(\theta, \delta(x)), \qquad \theta \in \Theta,$$

i.e., if, in the long run, this estimator never underestimates the error resulting from the use of δ. Such a restriction may seem intuitively appealing, but it is based on an intuition at the basis of the notion of *unbiased estimation*, and this restriction comes to contradict the Likelihood Principle. Robert and Casella (1994) propose a purely decision-theoretic approach of loss estimation in the setting of confidence regions (see Chapter 5). If $C(x)$ is a confidence region for θ, the usual loss underlying the estimation process is the $0-1$ loss,

$$L(C(x), \theta) = 1 - \mathbb{I}_{C(x)}(\theta),$$

and a loss estimator $\gamma(x)$ thus evaluates the coverage rate of $C(x)$, i.e., somehow approximates the coverage probability of the confidence region. Hwang and Brown (1991) have shown that, for the usual confidence regions C_0, in the normal setting, the constant estimator

$$\alpha = P(\theta \notin C_0(x))$$

is admissible among frequency valid estimators, while inadmissible for $p > 5$ without the restriction (see Section 5.5).

Example 4.2.10 Consider $x \sim \mathcal{N}_p(\theta, \sigma^2 I_p)$ and $\theta \sim \mathcal{N}_p(0, \tau^2 I_p)$. Under quadratic loss,

$$\delta^\pi(x) = \frac{\sigma^2}{\sigma^2 + \tau^2}\, x \quad \text{and} \quad V^\pi(x) = \frac{\sigma^2 \tau^2}{\sigma^2 + \tau^2}\, p.$$

On the contrary, the frequentist approach gives $+\infty$ as the maximal risk for δ^{π} and is thus poorly adapted to this problem.   ‖

4.3 Sampling models

In this section, we consider three sampling problems in which a Bayesian approach is easy to implement. Notice that, in general, *discrete models* require less prior information to build up a prior distribution. The first problem we consider is related to the *Laplace succession rule*, introduced in 1774 by Laplace. The second problem was studied under the name of the *tramcar problem* by Neyman in the 1930s. The last section studies *capture-recapture* models, which are of interest in animal biology and other population estimation modelings. The three problems have in common that they involve an inference on a finite population or subpopulation. These are cases where some prior information is usually available, or where the choice of a noninformative prior can be made rather unambiguously.

4.3.1 Laplace succession rule

Consider a population of known size N divided into two subpopulations, of respective sizes N_1 and $N_2 = N - N_1$, which are unknown. When sampling without replacement x individuals from this population, x_1 individuals belong to the first subpopulation and $x_2 = x - x_1$ to the second. When no information is available on N_1, a noninformative distribution is

$$\pi(N_1) = \frac{1}{N+1} \mathbb{I}_{\{0,1,\ldots,N\}}(N_1)$$

and the corresponding posterior distribution of N_1 is ($x_1 \le N_1 \le N - (x - x_1)$)

$$\pi(N_1|x_1) = \frac{\binom{N_1}{x_1}\binom{N-N_1}{x-x_1}}{\sum_{i=0}^{N}\binom{i}{x_1}\binom{N-i}{x-x_1}} = \frac{\binom{N_1}{x_1}\binom{N-N_1}{x-x_1}}{\binom{N+1}{x+1}}.$$

Let E be the event that the next draw will give an individual from the first subpopulation, p being the probability of E. Then

$$P(E|N_1,x_1) = \frac{N_1 - x_1}{N - x}.$$

Therefore,

$$
\begin{aligned}
P(E,N_1|x_1) &= \frac{N_1 - x_1}{N - x} \frac{\binom{N_1}{x_1}\binom{N-N_1}{x-x_1}}{\binom{N+1}{x+1}} \\
&= \frac{x_1 + 1}{N - x} \frac{\binom{N_1}{x_1+1}\binom{N-N_1}{x-x_1}}{\binom{N+1}{x+1}},
\end{aligned}
$$

and

$$p = P(E|x_1) = \frac{x_1 + 1}{N - x} \frac{\binom{N+1}{x+2}}{\binom{N+1}{x+1}} = \frac{x_1 + 1}{x + 2},$$

which does not depend on N. Therefore, the predictive distribution of the outcome of the $(x+1)$th draw is a Bernoulli distribution, $\mathcal{B}(1, (x_1+1)/(x+2))$. Laplace considered the special case $x = x_1$ and derived his succession rule: *If n first draws all give an outcome from the same subpopulation, the probability that the next draw will also give an outcome from this population is $\frac{n+1}{n+2}$.* A consequence of the Laplace succession rule is that the probability that the whole population is of the same kind as the n first observations is $\frac{n+1}{N+1}$. Some criticized this succession rule as being biased in favor of the most common subpopulation, since rare populations will not be detected (see also Popper (1983)). On the contrary, Jeffreys (1961, §3.3.3) maintains that, in Physics at least, this rule leads quite often to rejection of the proposed distributions.

4.3.2 The tramcar problem

Jeffreys (1961, p. 238) mentions the following problem, attributed to Neyman: *"A man traveling in a foreign country has to change trains at a junction, and goes into the town, the existence of which he has only just heard. He has no idea of its size. The first thing that he sees is a tramcar numbered 100. What can he infer about the number of tramcars in the town? It may be assumed that they are numbered consecutively from 1 upwards."* Clearly, this problem has less anecdotal applications. For instance, it is related to many *coincidence* problems as described in Diaconis and Mosteller (1989).

Example 4.3.1 Consider a *cyclic* phenomenon with unknown period T and K possible states (stock market crises, comet occurrences, genetic mutations, traffic lights, etc.) and the observation that, at times t_1 and t_2, the phenomenon is in the same state. The inferential problem is to derive T from the observation of the difference $t_2 - t_1$. ‖

In the case of the tramcar problem, the total number N can take the values $1, 2, \ldots$. Instead of a uniform distribution on $\mathbb{N}^*$, it is preferable to consider the noninformative distribution

$$\pi(N) = \frac{1}{N},$$

since N can be interpreted as a scale parameter. Moreover, the uniform prior does not lead to a defined posterior distribution. If T is the observed tramcar number, it is distributed as

$$f(t|N) = P(T = t|N) = \frac{1}{N} \qquad (t = 1, 2, \ldots, N).$$

Therefore,

$$\pi(N|T) \propto \frac{1}{N^2} \mathbb{I}_{(N \geq T)}$$

Table 4.3.1. *Probability parameters for a capture-recapture experiment.*

| | | Sample 2 | |
		captured	missed
Sample 1	captured	p^i_{11}	p^i_{12}
	missed	p^i_{21}	p^i_{22}

and

$$P^\pi(N \geq n_0|T) = \frac{\sum_{n=n_0}^{+\infty} 1/n^2}{\sum_{n=T}^{+\infty} 1/n^2} \approx \frac{\int_{n_0}^{+\infty} (1/x^2)dx}{\int_T^{+\infty} (1/x^2)dx} = \frac{T}{n_0}.$$

In this case, the posterior median is approximately $N^\pi(T) \approx 2T$, which is the estimator usually considered in the tramcar problem. In fact, notice that the mean of T conditional upon N is $\frac{N-1}{2} \approx \frac{N}{2}$.

4.3.3 Capture-recapture models

When working with an *hypergeometric distribution* $\mathcal{H}(N, n, p)$, the parameter p is most often the parameter of interest but it may also occur that the population size, N, is unknown and has to be estimated. More generally, in cases where the census of a population is impossible (or too costly), it is necessary to provide estimation methods for its size.

Example 4.3.2 An herd of deer is living on an island of Newfoundland that is isolated from any predator. To prevent the deer from disturbing the ecological equilibrium of the island, it is necessary to keep the number of deer under 40 by culling. An annual census of all deer is, however, too time-consuming.   ‖

We could mention many examples in biology, sociology, psychology, meteorology, ecology, etc., where an evaluation of a population size is necessary. For instance, the capture-recapture methods exposed here are used in the U.S. census to account for *undercount* on some populations badly estimated by usual census techniques, such as nomads, homeless people or illegal immigrants. The usual approach is called *capture-recapture* because it relies on taking at least two successive samples from the population of interest, and because it was first implemented in animal biology, where individuals were actually captured. See Seber (1983, 1986) and Pollock (1991) for surveys.

We use in this section the general framework of Wolter (1986), who shows that most capture-recapture models can be described by a multinomial distribution for each individual i in the population ($1 \leq i \leq N$). Table 4.3.3 describes the probabilities of capture, with $p^i_{11} + p^i_{12} + p^i_{21} + p^i_{22} = 1$. For instance, p^i_{12} represents the probability of being captured only in the first

Table 4.3.2. *Population division corresponding to the model of Table 4.3.3.*

| | | Sample 2 | |
		captured	missed
Sample 1	captured	n_{11}	n_{12}
	missed	n_{21}	n_{22}

sample. After the two capture experiments, the population is divided as in Table 4.3.3, with $n_{11}+n_{12}+n_{21}+n_{22} = N$ (the fourth sample size n_{22} being unknown). For the simplest model, called *uniform*, each individual has the same probability p of being captured in both experiments. Therefore, $p_{11} = p^2$, $p_{12} = p_{21} = p(1-p)$, and $p_{22} = (1-p)^2$. The likelihood can be written

$$L(N, p|n_{11}, n_{12}, n_{21}) = \binom{N}{n_{11}\ n_{21}\ n_{21}} p^{n.}(1-p)^{2N-n.},$$

where $n. = 2n_{11} + n_{12} + n_{21}$ is the total number of captured individuals and

$$\binom{N}{n_{11}\ n_{12}\ n_{21}} = \frac{N!}{n_{11}!\,n_{21}!\,n_{12}!\,n_{22}!}$$

is the *multinomial coefficient*. For $\pi(N, p) = \pi(N)\pi(p)$ with $\pi(p)$ a beta distribution $\mathcal{B}e(\alpha, \beta)$, the conditional posterior distribution on p is

$$\pi(p|N, n_{11}, n_{12}, n_{21}) \propto p^{\alpha+n.-1}(1-p)^{\beta+2N-n.-1},$$

i.e.,

$$p|N, n. \sim \mathcal{B}e(\alpha + n., \beta + 2N - n.).$$

Unfortunately, the marginal posterior distribution of N is quite complicated. For instance, if $\pi(N) = 1$, it satisfies

(4.3.1) $$\pi(N|n.) \propto \binom{N}{n_+}\frac{B(\alpha + n., \beta + 2N - n.)}{B(\alpha, \beta)},$$

where $n_+ = n_{11}+n_{12}+n_{21}$ is the number of different captured individuals. This distribution is sometimes called a *beta-Pascal* distribution (see Raiffa and Schlaifer (1961)), but it is intractable. The same complexity occurs if $\pi(N) = 1/N$ as in Castledine (1981) or if $\pi(N)$ is a Poisson distribution $\mathcal{P}(\lambda)$ as in Raftery (1988), George and Robert (1992), and Dupuis (1995). Obviously, N being an integer, it is always possible to approximate the normalizing factor in (4.3.1) by summing out on N. But, apart from the required computing time, the approximation errors can become important when N and n_+ are large. Notice that, for the Poisson prior,

$$N - n_+|n_+, p \sim \mathcal{P}((1-p)^2\lambda),$$

therefore the conditional posterior distributions are available (Chapter 6 makes use of this property). Extensions of the uniform model are described in Wolter (1986), George and Robert (1992) and Dupuis (1995).

A simpler model used in capture-recapture settings is the hypergeometric model, also called the *Darroch model* (Darroch (1958)), in which the two sample sizes $n_1 = n_{11} + n_{12}$ and $n_2 = n_{11} + n_{21}$ are fixed. In this case, the above description does not apply and the only remaining random variable is n_{11}, with distribution $\mathcal{H}(N, n_2, \frac{n_1}{N})$. In fact, the values n_1 and n_2 are not fixed in advance, but rather determined by a stopping rule that is generally unknown. However, if the prior distribution on N is noninformative and with a countable support, the computation of the Bayes estimators is of the same order of complexity. But, since the Darroch model can be written as a particular case of the Wolter model (see Exercise 4.34), it is then possible to extend the approximation techniques developed for Wolter's modeling to this setting (see Dupuis (1995) and Chapter 6).

For the Darroch model, the classical estimator of N is the *maximum likelihood estimator*

$$\hat{N} = \frac{n_1}{(n_{11}/n_2)},$$

which identifies the proportion in the population (n_1/N) and the proportion in the sample (n_{11}/n_2). This estimator presents an important drawback: it cannot be used when $n_{11} = 0$. This situation would call for a third draw of n_3 individuals, and the observation of n_{111} individuals common with the first or the second sample. Since the number of tagged individuals increases with the number of samples, the probability of observing only new individuals at each draw decreases. It is nonetheless unreasonable to call for additional sampling when the initial objective of the statistical modeling was to reduce sampling costs.

A Bayesian analysis does not suffer from this defect because it reaches a conclusion even when $n_{11} = 0$. Given a prior distribution π on N, it is formally easy to derive the posterior $\pi(N = n|n_{11})$ and draw an inference on N.

Example 4.3.3 (Example 4.3.2 continued) Birth and death patterns for the deer imply that the number of deer varies between 36 and 50. A more thorough biological study of the deer life expectancy could certainly help to refine the prior modeling on N, but we will use here a uniform distribution on $\{36, \ldots, 50\}$. Considering $n_1 = n_2 = 5$, the Bayes formula,

$$\pi(N = n|n_{11}) = \frac{\binom{n_1}{n_{11}}\binom{n_2}{n_2 - n_{11}}/\binom{n}{n_2}\pi(N = n)}{\sum\limits_{k=36}^{50}\binom{n_1}{n_{11}}\binom{n_2}{n_2 - n_{11}}/\binom{k}{n_2}\pi(N = k)},$$

leads to Table 4.3.3, which provides the posterior distribution of N.

Since the complete posterior distribution of N is available, we can derive the posterior mean, median, or mode of N (or any other Bayes estimator). Table 4.3.3 gives the posterior means for different values of n_{11} (to be compared with the classical estimator $25/n_{11}$ for $n_{11} \neq 0$, which varies

Table 4.3.3. *Posterior distribution of the deer population size, $\pi(N|n_{11})$.*

N	0	1	n_{11} 2	3	4	5
36	0.058	0.072	0.089	0.106	0.125	0.144
37	0.059	0.072	0.085	0.098	0.111	0.124
38	0.061	0.071	0.081	0.090	0.100	0.108
39	0.062	0.070	0.077	0.084	0.089	0.094
40	0.063	0.069	0.074	0.078	0.081	0.082
41	0.065	0.068	0.071	0.072	0.073	0.072
42	0.066	0.068	0.067	0.067	0.066	0.064
43	0.067	0.067	0.065	0.063	0.060	0.056
44	0.068	0.066	0.062	0.059	0.054	0.050
45	0.069	0.065	0.060	0.055	0.050	0.044
46	0.070	0.064	0.058	0.051	0.045	0.040
47	0.071	0.063	0.056	0.048	0.041	0.035
48	0.072	0.063	0.054	0.045	0.038	0.032
49	0.073	0.062	0.052	0.043	0.035	0.028
50	0.074	0.061	0.050	0.040	0.032	0.026

Table 4.3.4. *Posterior mean of the deer population size, N.*

n_{11}	0	1	2	3	4	5
$\mathbb{E}(N|n_{11})$	43.32	42.77	42.23	41.71	41.23	40.78

much more widely with n_{11}).

If, instead of the squared error, we use the loss

$$(4.3.2) \qquad L(N,\delta) = \begin{cases} 10(\delta - N) & \text{if } \delta > N, \\ N - \delta & \text{otherwise}, \end{cases}$$

in order to avoid an overestimation of the number of deer (this would be more dramatic for the future of the herd than an underestimation), the Bayes estimator is the $(1/11)$-quantile of $\pi(N|n_{11})$, given in Table 4.3.3 for different values of n_{11}. Notice that, in this case, the estimators are necessarily integers. ||

A very interesting Bayesian application of capture-recapture inference is given by Mosteller and Wallace (1984). It deals with *author identification by statistical linguistics* when the origin of some literary work is uncertain. For instance, Mosteller and Wallace (1984) study the *Federalist Papers*, a collection of articles written in 1787 in order to support the new American constitution. Twelve of these articles can be attributed to either Hamilton or Madison. Using authenticated writings of the two authors, Mosteller and

Table 4.3.5. *Estimator of the deer population size under asymmetric loss (4.3.2).*

n_{11}	0	1	2	3	4	5
$\delta^\pi(n_{11})$	37	37	37	36	36	36

Wallace derive the frequencies of thirty current words and deduce from a capture-recapture approach that the twelve articles should be attributed to Madison. Efron and Thisted (1976) have also used capture-recapture to study Shakespeare's, vocabulary and later identify in Thisted and Efron (1987) a recently discovered poem as quite likely to have been written by Shakespeare.

4.4 The particular case of the normal model

4.4.1 Introduction

When Gauss introduced the normal distribution around 1810, Laplace thought this was actually the *ideal* error distribution (see Example 1.2.5). Later, relying on the Central Limit Theorem, statisticians in the first half of the nineteenth century were almost always referring to the normal distribution (Stigler (1986)). There are obviously many phenomena for which a normal model is not applicable, but it is still extensively used, in particular, in econometrics and in fields where the Central Limit Theorem approximation can be justified (particle physics, etc.). In fact, the normal approximation is often justified for asymptotic reasons (see also Cox and Reid (1987)). Therefore, it is of interest to study in detail this particular distribution from a Bayesian viewpoint.

Given one observation from a multivariate normal distribution, $\mathcal{N}_p(\theta, \Sigma)$, with known covariance matrix Σ, the conjugate distribution is also normal, $\mathcal{N}_p(\mu, A)$, and the posterior distribution $\pi(\theta|x)$ is

$$\mathcal{N}_p\left(x - \Sigma(\Sigma + A)^{-1}(x - \mu), (A^{-1} + \Sigma^{-1})^{-1}\right).$$

Under quadratic loss, the Bayes estimator is then the posterior mean

$$\begin{aligned} \delta^\pi(x) &= x - \Sigma(\Sigma + A)^{-1}(x - \mu) \\ &= \left(\Sigma^{-1} + A^{-1}\right)^{-1}\left(\Sigma^{-1}x + A^{-1}\mu\right); \end{aligned}$$

notice that $\delta^\pi(x)$ can be written as a convex combination of the observation, x, and of the prior mean, μ, the weights being proportional to the inverses of the covariance matrix.

The more accurate the prior information on θ is, the closer to μ the Bayes estimator is. Notice also that the prior information (resp., the observation of x) brings a reduction of the variance from Σ (respectively, from A) to $\left(\Sigma^{-1} + A^{-1}\right)^{-1}$. For repeated observations of the above normal model,

$x_1, ..., x_n$, the sufficient statistic

$$\bar{x} = \frac{1}{n} \sum_{i=1}^{n} x_i \sim \mathcal{N}_p \left(\theta, \frac{1}{n} \Sigma \right)$$

directly extends the previous analysis.

A criticism already mentioned in Chapter 3 is that the conjugate normal prior distributions are not sufficiently robust, and that it would be preferable to use Student's t-distributions for $\pi(\theta)$. The Cauchy distribution, the limiting case of a Student's t-distribution, can then be used because of its heavier tails, but it still prohibits exact computation (see Example 4.2.1), although Angers (1992) proposes an analytical resolution using confluent hypergeometric functions.

4.4.2 Estimation of variance

In most cases, the variance of the model is partly or totally unknown. It is then necessary to consider prior distributions on the parameter (θ, Σ). If the variance is known up to a multiplicative constant, σ^2, it is usually possible to get back to the unidimensional case, that is, when $x_1, \ldots, x_n$ are i.i.d. $\mathcal{N}(\theta, \sigma^2)$, by sufficiency reasons. (The particular case where σ^2 only is unknown is treated in Tables 3.3.4 and 4.3.3). If we define the statistics $\bar{x} = \frac{1}{n} \sum_{i=1}^{n} x_i$ and $s^2 = \sum_{i=1}^{n} (x_i - \bar{x})^2$, the likelihood can be written

$$\ell(\theta, \sigma \mid \bar{x}, s^2) \propto \sigma^{-n} \exp \left[-\frac{1}{2\sigma^2} \left\{ s^2 + n (\bar{x} - \theta)^2 \right\} \right]$$

and the Bayes estimators only depend on $\bar{x}$ and s^2. We showed in Example 3.5.5 that the *Jeffreys distribution* for this model is $\pi^*(\theta, \sigma) = \frac{1}{\sigma^2}$ and mentioned that it is better to consider the alternative $\tilde{\pi}(\theta, \sigma) = \frac{1}{\sigma}$ for invariance reasons. In this case,

$$(4.4.1) \quad \ell(\theta, \sigma \mid \bar{x}, s^2) \tilde{\pi}(\theta, \sigma) \propto \sigma^{-n-1} \exp \left[-\frac{1}{2\sigma^2} \left\{ s^2 + n (\bar{x} - \theta)^2 \right\} \right].$$

Therefore,

Proposition 4.4.1 *If $x_1, \ldots, x_n$ are i.i.d. $\mathcal{N}(\theta, \sigma^2)$, the posterior distribution of (θ, σ) associated with $\tilde{\pi}$ is*

$$\theta \mid \sigma, \bar{x}, s^2 \quad \sim \quad \mathcal{N} \left(\bar{x}, \frac{\sigma^2}{n} \right),$$

$$(4.4.2) \qquad\qquad \sigma^2 \mid \bar{x}, s^2 \quad \sim \quad \mathcal{IG} \left(\frac{n-1}{2}, \frac{s^2}{2} \right).$$

Equation (4.4.2) really defines the posterior distribution of (θ, σ^2), since it provides the marginal distribution of σ^2 and the distribution of θ conditional on σ^2. The proof of this proposition is a direct consequence of (4.4.1), as

$$\tilde{\pi}(\theta, \sigma^2 | \bar{x}, s^2) \propto \sigma^{-1} e^{-n(\bar{x}-\theta)^2/2\sigma^2} \sigma^{-n} e^{-s^2/2\sigma^2} \sigma^{-1},$$

and the *inverse gamma* distribution $\mathcal{IG}(\alpha, \beta)$ has the density

(4.4.3) $$\pi(x|\alpha, \beta) = \frac{\beta^\alpha}{\Gamma(\alpha)x^{\alpha+1}} e^{-\beta/x} \mathbb{I}_{(0,+\infty)}(x).$$

Thus, the marginal posterior distribution of σ^2 is of the same type than when θ is known. On the contrary, the marginal posterior distribution of θ is modified, since it follows from (4.4.2) that

$$\tilde{\pi}(\theta | \bar{x}, s^2) \propto \left\{ s^2 + n(\bar{x} - \theta)^2 \right\}^{-n/2},$$

i.e.,

(4.4.4) $$\theta | \bar{x}, s^2 \sim \mathcal{T}_1 \left(n - 1, \bar{x}, \frac{s^2}{n(n-1)} \right).$$

For the Jeffreys distribution, π^*, the equivalent of (4.4.4) is a Student's t-distribution with n degrees of freedom, which is always defined, while (4.4.4) is only defined for $n \geq 2$. (Notice that the exclusion of $n = 1$ could be seen as an additional argument in favor of $\tilde{\pi}$ since, in a noninformative setting, it seems difficult to propose an inference about the whole parameter (θ, σ) with a *single* observation.)

Conjugate posterior distributions have the same form as (4.4.2). They exhibit a puzzling peculiarity, namely, that θ and σ^2 are not a priori independent. Therefore, the prior distribution on the mean θ depends on the precision associated with the measure of the mean. Some settings can justify this dependence,[2] but it cannot hold for every estimation problem, and can even less be argued to be a representative standard prior distribution (see Berger (1985a)). However, these subjective criticisms are not seconded by particularly negative properties of the resulting estimators.

Consider then

$$\pi(\theta, \sigma^2) = \pi_1(\theta | \sigma^2) \pi_2(\sigma^2),$$

where π_1 is a normal distribution $\mathcal{N}(\mu, \sigma^2/n_0)$ and π_2 is a inverse gamma distribution $\mathcal{IG}(\nu/2, s_0^2/2)$. The posterior distribution satisfies

$$\pi(\theta, \sigma^2 | x) \propto \sigma^{-n-\nu-3} \exp\left\{ -\frac{1}{2} \left[s^2 + s_0^2 + n_0(\theta - \mu)^2 + n(\bar{x} - \theta)^2 \right] / \sigma^2 \right\}$$

$$= \sigma^{-n-\nu-3} \exp\left\{ -\frac{1}{2} \left[s_1^2 + n_1(\theta - \theta_1)^2 \right] / \sigma^2 \right\},$$

where

$$n_1 = n + n_0, \quad \theta_1 = \frac{1}{n_1} \left(n_0 \theta_0 + n\bar{x} \right),$$

$$s_1^2 = s^2 + s_0^2 + \left(n_0^{-1} + n^{-1} \right)^{-1} \left(\theta_0 - \bar{x} \right)^2.$$

---

[2] When the prior distribution is built from previous observations, it makes sense that the prior variance of θ involves σ^2 (conditionally).

These distributions are actually conjugate since

$$\pi\left(\theta|\bar{x}, s^2, \sigma\right) \propto \frac{1}{\sigma}\exp\left\{-\frac{n_1(\theta - \theta_1)^2}{2\sigma^2}\right\},$$

$$\pi\left(\sigma^2|\bar{x}, s^2\right) \propto \sigma^{-n-\nu-2}\exp\left\{-s_1^2/2\sigma^2\right\}.$$

As in the noninformative case, the marginal posterior distribution of θ is a Student's t-distribution. Notice that, except when π is built from previous (or virtual) observations, n_0 is not a sample size. Rather, n_0/n characterizes the relative precision of the determination of the prior distribution as compared with the precision of the observations. In general, n_0 is smaller than the sample size n. Notice also that, if n_0/n goes to 0, we get the limiting case $\theta|\bar{x}, \sigma^2 \sim \mathcal{N}(\bar{x}, \sigma^2/n)$, corresponding to the posterior distribution associated with the Jeffreys prior. This fact is yet another illustration of the phenomenon that noninformative distributions often occur as limits of conjugate distributions.

Statistical inference based on the above conjugate distribution requires a careful determination of the hyperparameters θ_0, s_0^2, n_0, ν, in order to express the Bayes estimators. If the determination of θ_0 and n_0 is rather classical, it is usually more difficult to get a prior information on σ^2. Recall that, if $\sigma^2 \sim \mathcal{IG}(\nu/2, s_0^2/2)$, the first two moments are given by

$$\mathbb{E}^\pi\left[\sigma^2\right] = \frac{s_0^2}{\nu - 2}, \quad \mathrm{var}^\pi(\sigma^2) = \frac{2s_0^4}{(\nu - 2)^2(\nu - 4)}.$$

These formulas can then be used to model a prior information into the conjugate form, i.e., to determine s_0^2 and ν.

When the parameter (θ, Σ) is totally unknown, it is still possible to derive conjugate prior distributions. Given n observations $x_1, \ldots, x_n$ of $\mathcal{N}_p(\theta, \Sigma)$, a sufficient statistic is

$$\bar{x} = \frac{1}{n}\sum_{i=1}^n x_i, \quad S = \sum_{i=1}^n (x_i - \bar{x})(x_i - \bar{x})^t,$$

and

$$\ell(\theta, \Sigma|\bar{x}, S) \propto |\Sigma|^{-n/2}\exp - \frac{1}{2}\left\{n(\bar{x} - \theta)^t\Sigma^{-1}(\bar{x} - \theta) + \mathrm{tr}(\Sigma^{-1}S)\right\}.$$

The form of the likelihood function then suggests the following conjugate distributions:

$$\theta|\Sigma \sim \mathcal{N}_p\left(\mu, \frac{\Sigma}{n_0}\right),$$

(4.4.5)
$$\Sigma^{-1} \sim \mathcal{W}_p(\alpha, W),$$

where $\mathcal{W}_p$ denotes the Wishart distribution, defined in Exercise 3.21. The posterior distributions are then

$$\theta|\Sigma, \bar{x}, S \sim \mathcal{N}_p\left(\frac{n_0\mu + n\bar{x}}{n_0 + n}, \frac{\Sigma}{n_0 + n}\right),$$

$$\Sigma^{-1}|\bar{x}, S \sim W_p(\alpha + n, W_1(\bar{x}, S)),$$

with

$$W_1(\bar{x}, S)^{-1} = W^{-1} + S + \frac{nn_0}{n + n_0}(\bar{x} - \mu)(\bar{x} - \mu)^t.$$

Notice that this multidimensional case is the generalization of the unidimensional case considered above, since the Wishart distribution W_p is the generalization in dimension p of chi-squared distributions. Let us recall here that the first two moments of $\Xi = (\xi_{ij}) \sim W_p(\alpha, W)$ are

$$\mathbb{E}[\Xi] = \alpha W, \qquad \text{var}(\xi_{ij}) = 2\alpha w_{ij}^2,$$

and that the hyperparameters of the prior distribution of Σ can be derived from

$$\mathbb{E}[\Sigma] = \frac{W^{-1}}{\alpha - p - 1}, \qquad \text{var}(\sigma_{ij}) = \frac{2(w^{ij})^2}{(\alpha - p - 3)(\alpha - p - 1)^2},$$

if $\Sigma^{-1} \sim W_p(\alpha, W)$ and $W^{-1} = (w^{ij})$ (see Eaton (1982) and Anderson (1984)).

In this setting, the Jeffreys distribution is also a limiting case of conjugate distributions since Geisser and Cornfield (1963) have shown that it is

$$\pi^J(\theta, \Sigma) = \frac{1}{|\Sigma|^{(p+1)/2}},$$

therefore corresponding to the limit of the Wishart distribution $W_p(\alpha, W)$ on Σ^{-1} when W^{-1} goes to $\mathbf{O}$ and α to 0. In fact, the density of Σ when $\Sigma^{-1} \sim W_p(\alpha, W)$ is

$$f(\Sigma|\alpha, W) \propto |\Sigma|^{-(\alpha+p+1)/2} \exp\left\{-\frac{1}{2}\text{tr}(W^{-1}\Sigma^{-1})\right\}$$

(see Anderson (1984)).

4.4.3 Linear models and G–priors

The usual regression model,

(4.4.6) $$y = X\beta + \epsilon,$$

with $\epsilon \sim \mathcal{N}_k(0, \Sigma)$, $\beta \in \mathbb{R}^p$, can be analyzed as above if the covariance matrix Σ is known, when working conditional upon X. In fact, a sufficient statistic is then $\hat{\beta} = (X^t\Sigma^{-1}X)^{-1}X^t\Sigma^{-1}y$, the maximum likelihood estimator and the least-squares estimator of β. It is distributed according to a $\mathcal{N}_p(\beta, (X^t\Sigma^{-1}X)^{-1})$ distribution. Lindley and Smith (1972) have studied conjugate distributions of the type

$$\beta \sim \mathcal{N}_p(A\theta, C),$$

where $\theta \in \mathbb{R}^q$ ($q \leq p$). In this model, the regressor matrix X is considered to be constant. In other words, the inference is made conditional upon X. (Usually, X is also partly random, but this conditioning is justified by the

Likelihood Principle as long as the distribution of X does not depend on the parameters of the regression model.) Therefore, A, C, or θ can depend on X (see below for the example of the *G-priors* of Zellner (1971)). When the stochastic nature of X *has* to be considered, the usual approach is to study a random-effect model,

$$y = X_1\beta_1 + X_1X_2\beta_2 + \epsilon,$$

which can be decomposed as

$$
\begin{aligned}
y|\theta_1 &\sim \mathcal{N}_k(X_1\theta_1, \Sigma_1), \\
\theta_1|\theta_2 &\sim \mathcal{N}_p(X_2\theta_2, \Sigma_2),
\end{aligned}
$$

with a prior distribution

$$\theta_2|\theta_3 \sim \mathcal{N}_q(X_3\theta_3, \Sigma_3).$$

Smith (1973) analyzes this model and shows that

$$\theta_1|y, \theta_3 \sim \mathcal{N}_p(\theta_1^*, D_1),$$

with

$$
\begin{aligned}
\theta_1^* &= D_1\left[\hat{D}_1^{-1}\hat{\theta}_1 + (\Sigma_2 + X_2\Sigma_3X_2^t)^{-1}X_2X_3\theta_3\right], \\
D_1^{-1} &= \hat{D}_1^{-1} + (\Sigma_2 + X_2\Sigma_3X_2^t)^{-1},
\end{aligned}
$$

involving the classical least-squares estimators

$$
\begin{aligned}
\hat{D}_1^{-1} &= X_2^t\Sigma_1^{-1}X_2, \\
\hat{\theta}_1 &= \hat{D}_1X_2^t\Sigma_1^{-1}y.
\end{aligned}
$$

Therefore, the Bayes estimator θ_1^* is a convex combination of the least-squares estimator, $\hat{\theta}_1$, and of the prior mean, $X_2X_3\theta_3$.

We introduce below an example where an unknown variance structure still allows for closed-form derivation of the Bayes estimators. However, if the variance Σ is totally unknown, it is not possible to derive conjugate prior distributions, as noticed by Lindley and Smith (1972). Press (1989) proposes a resolution in a particular case when *independent* observations are available for every component of β. In the general case, the Jeffreys prior is again (see Geisser and Cornfield (1963))

$$\pi^J(\beta, \Sigma) = \frac{1}{|\Sigma|^{(k+1)/2}}.$$

The likelihood

$$\ell(\beta, \Sigma|y) \propto |\Sigma|^{-n/2}\exp\left\{-\frac{1}{2}\mathrm{tr}\left[\Sigma^{-1}\sum_{i=1}^n(y_i - X_i\beta)(y_i - X_i\beta)^t\right]\right\},$$

then suggests the use of Wishart distributions, but the posterior marginal distributions on β are only defined for sufficiently large sample sizes and, moreover, are not explicit, whatever the sample size (see Exercise 4.44).

In the particular case when the variance of the model (4.4.6) is known up to a multiplicative factor σ^2, it is possible to rewrite the model as $\epsilon \sim \mathcal{N}_k(0, \sigma^2 I_k)$ and the least-squares estimator $\hat{\beta}$ has a normal distribution $\mathcal{N}_p(\beta, \sigma^2(X^t X)^{-1})$. A family of conjugate distributions on (β, σ^2) is then

$$\beta | \sigma^2 \;\sim\; \mathcal{N}_p\left(\mu, \frac{\sigma^2}{n_0}(X^t X)^{-1}\right),$$

(4.4.7)
$$\sigma^2 \;\sim\; \mathcal{IG}(\nu/2, s_0^2/2),$$

since, if $s^2 = ||y - X\hat{\beta}||^2$, the posterior distributions are

$$\beta | \hat{\beta}, s^2, \sigma^2 \;\sim\; \mathcal{N}_p\left(\frac{n_0 \mu + \hat{\beta}}{n_0 + 1}, \frac{\sigma^2}{n_0 + 1}(X^t X)^{-1}\right),$$

$$\sigma^2 | \hat{\beta}, s^2 \;\sim\; \mathcal{IG}\left(\frac{k - p + \nu}{2}, \frac{s^2 + s_0^2 + \frac{n_0}{n_0+1}(\mu - \hat{\beta})^t X^t X(\mu - \hat{\beta})}{2}\right).$$

Indeed,

$$\pi(\beta, \sigma^2 | \hat{\beta}, s^2) \propto (\sigma^2)^{-k/2} \exp\left[-\frac{1}{2\sigma^2}\left\{(\beta - \hat{\beta})^t X^t X(\beta - \hat{\beta}) + s^2\right\}\right]$$

$$\times \exp\left(-\frac{n_0}{2\sigma^2}(\beta - \mu)^t X^t X(\beta - \mu)\right)(\sigma^2)^{-\nu/2-1}\exp\left(-\frac{s_0^2}{2\sigma^2}\right)$$

$$\propto (\sigma^2)^{-p/2}\exp\left\{-\frac{n_0 + 1}{2\sigma^2}\left(\beta - \frac{n_0 \mu + \hat{\beta}}{n_0 + 1}\right)^t X^t X\left(\beta - \frac{n_0 \mu + \hat{\beta}}{n_0 + 1}\right)\right\}$$

$$\times \sigma^{-(k-p+\nu+2)}\exp\left[-\frac{1}{2\sigma^2}\left\{s_0^2 + s^2 + \frac{n_0}{n_0 + 1}(\mu - \hat{\beta})^t X^t X(\mu - \hat{\beta})\right\}\right].$$

Although (4.4.7) is only a particular case of conjugate distribution, many criticisms have been raised about this choice, developed by Zellner (1971, 1986b) under the name of *G-priors*. These criticisms do not usually address the problem of the reductive aspect of conjugate modeling, a quite legitimate argument already mentioned in Chapter 3, but rather the dependency of the prior distribution on X. It can be argued that X is also a random variable and therefore that an a priori modeling should not depend on X. In fact, the alternative prior distributions

$$\beta | \sigma \sim \mathcal{N}_p(\beta_0, \sigma^2 A)$$

also constitute a conjugate family which is less arguable if A is fixed. However, we consider the debate to be rather vacuous because:

(1) The whole regression model is conditional on the explanatory variables. The prior distribution (4.4.7) can then be seen as a posterior distribution with respect to these variables (or to extend the usual hypothesis of independence between the explanatory variables and the errors to the hypothesis of Bayesian independence with the parameters). This approach is thus justified from both the conditional and Bayesian points

of view, the conditioning being operated in two steps.

(2) The *G-prior* distribution suggests a constant distribution on the mean of y, $\theta = \mathbb{E}_\theta[y|X]$, rather than on β. The prior distribution is then determined with respect to the subspace spanned by the columns of X, not with respect to a special basis of this subspace.

(3) This modeling is appropriate to take into account the problems of *multicollinearity*, since it allows for a large prior variance on the components affected by multicollinearity (hence, the most difficult to estimate). (See Zellner (1971), Casella (1985b), or Steward (1987), for references on multicollinearity.)

(4) From both practical and subjective viewpoints, the determination of a matrix A instead of a scalar n_0 requires a larger amount of prior information. Because the use of conjugate distributions indicates a setting where the prior information is sparse and where the derivation of hyperparameters is quite difficult, the resort to the covariance matrix $\sigma^2(X^tX)^{-1}/n_0$ prevents a possibly unrealistic determination of A.

Again, notice that these attacks on *G-priors* hardly ever consider its major drawback, namely, that its choice is not based on prior information. For applications of *G-priors* in regression settings, see Ghosh et al. (1989), or Blattberg and George (1991). See Bauwens, Lubrano and Richard (1999, Chapter 4) for alternatives to the conjugate priors in the linear model such as *poly-t* priors (see Note 4.7.5 below).

4.5 Dynamic models

4.5.1 Introduction

Dynamic (or *time series*) models appear as a particular case of a parametric model where the distribution of the observed variables $x_1, \ldots, x_T$ varies over time, that is,

$$(4.5.1) \qquad f(x_1, \ldots, x_T|\theta) = \prod_{t=1}^{T} f_t(x_t|x_{1:(t-1)}, \theta),$$

where $x_{1:(t-1)}$ denotes the vector of the previous variables $x_1, \ldots, x_{t-1}$, with the convention that $x_{1:0}$ is either empty or represents the initial value x_0 of the series of observations (it is then implicit on the lhs of (4.5.1)). While the representation (4.5.1) seems to be unnecessarily restrictive, the inclusion of unobserved components in x_t provides a fairly large scope for this model, as will be made clearer in the paragraph on state-space representations.

These models are obviously special cases of parametric models and, as such, they can be processed like other parametric models by Bayesian tools, once a prior distribution has been chosen, following the guidelines provided in the previous sections. They are singled out in this section for several reasons: first, they are some of the most commonly used models in applications, ranging from finance and economics to reliability to medical experiments to ecology. Most of the models encountered in practice enjoy a temporal dimension that can sometimes be concealed, but which must most often be taken into account. This is clearly the case, for instance, for series of pollution data, such as ozone concentration levels, or stock market prices, whose value at time t depends on the previous value at time $t-1$ and also on the previous values, for instance through their *trend*.

Example 4.5.1 (Example 4.1.6 continued) A finite time version of the autoregressive AR(1) model is generally defined through the distribution of x_t conditional on $x_{1:(t-1)}$ $(1 \leq t \leq T)$ as

$$(4.5.2) \qquad x_t = \mu + \varrho(x_{t-1} - \mu) + \epsilon_t \,,$$

where ϵ_t is independent from $x_{1:(t-1)}$ and can be chosen as, say, a normal $\mathcal{N}(0, \sigma^2)$ perturbation. This conditional distribution only depends on x_{t-1}, which shows that $(x_t)_{t \geq 1}$ is a *Markov chain* (Meyn and Tweedie (1993)).

The likelihood function of the AR(1) model is then

$$\sigma^{-T} \exp \left\{ \frac{-1}{2\sigma^2} \sum_{i=1}^{T} (x_t - \mu - \varrho(x_{t-1} - \mu))^2 \right\} \,,$$

and thus depends on the initial condition x_0. Either x_0 is known and the model is then conditional on x_0, or x_0 has to be integrated out by taking a prior distribution $\pi(x_0|\theta)$, since x_0 is then an additional parameter of the model. For instance, if $x_0 = 0$, it is straightforward to see that $\mathbb{E}_\theta[x_t] = 0$ and that $\text{var}(x_t) = \varrho^2 \text{var}(x_{t-1}) + \sigma^2$, thus, if $\varrho^2 \neq 1$,

$$(4.5.3) \qquad \text{var}(x_t) = \frac{1 - \varrho^{2t}}{1 - \varrho^2} \sigma^2 \,,$$

which implies that $\text{var}(x_t)$ converges to $\sigma^2/(1 - \varrho^2)$ if $\varrho^2 < 1$ and goes to $+\infty$ otherwise. $\qquad \|$

The second reason for studying dynamic models is that they are more challenging than the static models studied above, owing to *stationarity* constraints. While we cannot present a rigorous introduction to the notion of stationarity for stochastic processes chains (and refer the reader to Meyn and Tweedie (1993) for a general coverage of Markov processes and to Box and Jenkins (1976) or Brockwell and Davis (1998) for the special case of time series), let us recall here that a process (x_t) is *stationary* (or, more exactly, strictly stationary) if the distribution of $(x_{t+1}, \ldots, x_{t+d})$ is the same as the distribution of $(x_1, \ldots, x_d)$ for every (t, d). The stationarity problem can be illustrated in the setting of Example 4.5.1: when $\varrho^2 \geq 1$, not only the variance $\text{var}(x_t)$ goes to infinity with t, but the limiting behavior of the chain (x_t) cannot be characterized unless a dependence between x_t and ϵ_t is introduced (Brockwell and Davis (1998)). There is no limiting distribution for the process (x_t), because the Markov chain has no *stationary* distribution, that is, there is no density f such that, if $x_t \sim f$,

$x_{t+1} \sim f$ (Exercise 4.50). For instance, if $\varrho = 1$, (x_t) is the *random walk* on $\mathbb{R}$ and, on average, it takes the chain an infinite time to come back to the set it started from (Meyn and Tweedie (1993)).

To impose the *stationarity* condition on a model is objectionable on the ground that the data itself should indicate whether the underlying model is stationary. However, for reasons ranging from asymptotics to causality to identifiability (see below) to common practice, it is customary to impose stationarity constraints, even though a Bayesian inference on a non stationary process can be conducted in principle (see Note 4.7.2). Such constraints appear in the prior distribution as a restriction on the values of θ. For instance, for the AR(1) model of Example 4.5.1, the constraint is $|\varrho| < 1$. The practical difficulty is that, for more complex models, the stationarity constraints may become much more involved and even be unknown in some cases, as in general threshold models (Tong (1991)).

Example 4.5.2 The AR(p) model generalizes the AR(1) model by increasing the lag dependence on the past values, that is $(1 \leq t \leq T)$,

$$(4.5.4) \qquad x_t - \mu = \sum_{i=1}^{p} \varrho_i(x_{t-i} - \mu) + \epsilon_t, \qquad \epsilon_t \sim \mathcal{N}(0, \sigma^2).$$

The stochastic process defined by (4.5.4) can then be stationary[3] only if the roots of the polynomial

$$\mathcal{P}(x) = 1 - \sum_{i=1}^{p} \varrho_i x^i$$

are all outside the unit circle in the complex plane. While this condition is clearly defined, it is also *implicit* with respect to the vector $(\varrho_1, \ldots, \varrho_p)$: to verify that a given vector satisfies this condition, it is necessary either to find the roots of the pth degree polynomial $\mathcal{P}$ and check that they all are of modulus greater than 1, or to compute the partial autocorrelations (see below in Section 4.5.2) and apply *Schur's lemma* to check they are all between -1 and 1. ‖

Example 4.5.3 A switching AR(p) model is defined as an AR(p) model whose parameters change with time according to a hidden (unobserved) finite state Markov process, that is

$$(4.5.5) \qquad x_t = \sum_{i=1}^{p} \varrho_i(z_t)x_{t-i} + \sigma(z_t)\epsilon_t, \qquad \epsilon_t \sim \mathcal{N}(0, 1),$$

with (z_t) being the unobserved Markov chain,

$$P(z_t = i | z_{t-1} = j, z_{t-2}, \ldots) = \pi_{j,i}, \qquad i, j = 1, \ldots, K.$$

---

[3] That (4.5.4) is stationary obviously depends on the distribution of the initial values $x_{(-p+1):-1}$. We refer the reader to Brockwell and Davis (1998, §3.1) for a more detailed coverage of this issue, where the case of a stochastic process $(x_t)_{t \in \mathbb{Z}}$ is also studied, resulting in different stationarity conditions.

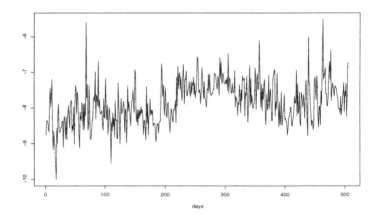

Figure 4.5.1. *Plot of the log transforms of averaged IBM stock prices over the period 1992-1997.*

This model was introduced by Hamilton (1989) as a way to represent series with different dynamics over time, such as the series in Figure 4.5.1 that plots a transformation of IBM stock prices from 1992 to 1997. The difficulty with the model (4.5.5) is that there was no known necessary and sufficient condition for stationarity when the number of states K of the hidden Markov chain (z_t) is larger than 2, till the recent derivation by Francq and Zakoïan (2001) and Yao and Attali (2000) of such conditions.   ∥

We develop in Sections 4.5.2–4.5.4 some features of the standard dynamic models, namely the AR, MA and ARMA models, focusing on representation problems and prior modeling under stationarity. Notes 4.7.3 and 4.7.4 present two other dynamic models often encountered in practice. See West and Harrison (1997) for a general approach to the Bayesian processing of time series and Bauwens et al. (1999) for an econometric monograph on this topic. Marin and Robert (2007, Chapter 7) also cover the practical aspects of prior determination and Bayesian inference for those models.

4.5.2 The AR model

As introduced in Example 4.5.2, the AR(p) model expresses the conditional distribution of x_t given the past $x_{1:(t-1)}$ as a normal linear regression on the p more recent variables, that is ($t = 1, 2, \ldots$),

$$(4.5.6) \qquad x_t \sim \mathcal{N}\left(\mu + \sum_{i=1}^{p} \varrho_i(x_{t-i} - \mu), \sigma^2\right),$$

where the location parameter μ is introduced for more generality. This model is thus Markovian, since the distribution of x_t only depends on a fixed number of past values, $x_{(t-p):(t-1)}$. It can be expressed as a regular Markov chain when considering the vector $\mathbf{z}_t = x_{t:(t-p+1)}$, since

$$(4.5.7) \qquad \mathbf{z}_t = \mu \mathbf{1} + B(\mathbf{z}_{t-1} - \mu \mathbf{1}) + \varepsilon_t \,,$$

where

$$\mathbf{1} = (1, \ldots, 1)^t \,, \quad B = \begin{pmatrix} \varrho_1 & \varrho_2 & \cdots & \varrho_p \\ 1 & 0 & \cdots & 0 \\ & & \ddots & \\ 0 & 0 & & 0 \end{pmatrix} \quad \text{and} \quad \varepsilon_t - (\epsilon_t, 0, \ldots, 0)^t \,.$$

Because the likelihood conditional on the negative time values $x_0, \ldots, x_{-p+1}$, can be written as

$$(4.5.8) \quad L(\mu, \varrho_1, \ldots, \varrho_p, \sigma | x_{1:T}, x_{0:(-p+1)}) =$$

$$\sigma^{-T} \prod_{t=1}^{T} \exp\left\{ -\left(x_t - \mu + \sum_{i=1}^{p} \varrho_i(x_{t-i} - \mu) \right)^2 \Big/ 2\sigma^2 \right\} \,,$$

a natural conjugate prior can be found for the parameter $\theta = (\mu, \varrho_1, \ldots, \varrho_p, \sigma^2)$, that is, a normal distribution on $(\mu, \varrho_1, \ldots, \varrho_p)$ and an inverse gamma distribution on σ^2. Instead of the Jeffreys prior, which is controversial in this setting (see Note 4.7.2), we can also propose a more conventional non-informative prior like $\pi(\mu, \sigma, \varrho) = 1/\sigma$.

If we impose the stationarity constraint that all the roots of $\mathcal{P}$ are outside the unit circle, the parameter space is too complex for values of p larger than 3 to propose as a prior distribution the normal conjugate prior restricted to this space. For instance, to simulate from such a prior is too costly. A solution, called the *Durbin–Levinson recursion* (see Monahan (1984)), is to propose a *reparameterization* from the parameters ϱ_i to the *partial autocorrelations* ψ_i (Exercise 4.53) which satisfy, under the stationarity constraint,

$$\psi_i \in (-1, 1) \,, \quad i = 1, \cdots, p \,,$$

and thus allow for a uniform prior.[4] The following result provides a constructive connection between $(\varrho_1, \ldots, \varrho_p)$ and $(\psi_1, \ldots, \psi_p)$.

Lemma 4.5.4 *Under stationarity of the model (4.5.6), the coefficients ϱ_i are deduced from the coefficients ψ_i by the algorithm*

---

0. Define $\varphi^{ii} = \psi_i$ and $\varphi^{ij} = \varphi^{(i-1)j} - \psi_i \varphi^{(i-1)(i-j)}$, for $i > 1$ and $j = 1, \cdots, i - 1$.

1. Take $\varrho_i = \varphi^{pi}$ for $i = 1, \cdots, p$.

---

[4] The partial autocorrelations, also called *reflection coefficients* in the signal processing literature, can be used to test for stationarity since, according to Schur's lemma they must all be between -1 and 1 for the Markov kernel (4.5.6) to have a stationary distribution.

While the resulting prior (and posterior) distribution on $(\varrho_1, \ldots, \varrho_p)$ is not explicit, in the sense that the computation of the prior (or of the posterior) at a given value of the parameter is quite time-consuming, this representation can be exploited for simulation purposes, as in Chapter 6 (see Barnett et al. (1996)), because of the linearity of the relation between the ϱ_j's and a given ψ_i, conditional upon the other ψ_ℓ's. Huerta and West (2000) propose a different approach via the real and complex roots of the polynomial $\mathcal{P}$, whose inverses are also within the unit circle. (See also Marin and Robert (2007, Chapter 7) for an implementation of this approach.)

4.5.3 The MA model

A fundamental result in the theory of stochastic processes is the *Wold decomposition*, which states that most stationary processes $(x_t)_{t \geq 1}$ can be represented as

$$(4.5.9) \qquad x_t = \mu + \sum_{i=0}^{\infty} \psi_i \epsilon_{t-i} \, ,$$

where $\psi_0 = 1$ and $(\epsilon_t)_{t \in \mathbb{Z}}$ is a *white noise*, that is, a sequence of random variables with zero mean, fixed variance and zero covariance. See Box and Jenkins (1976) for theoretical details.

Example 4.5.5 (Example 4.1.6 continued) If $x_t = \varrho x_{t-1} + \epsilon_t$, x_t can also be written as

$$x_t = \epsilon_t + \varrho \epsilon_{t-1} + \varrho^2 \epsilon_{t-2} + \ldots$$

if $|\varrho| < 1$. $\qquad \qquad \qquad \qquad \qquad \qquad \qquad \qquad \qquad \qquad \qquad \quad \|$

The *MA(q) model*, where MA stands for *moving average*, is a special case of (4.5.9) when the ψ_i's are equal to 0 for $i > q$, that is,

$$(4.5.10) \qquad x_t = \mu + \epsilon_t - \sum_{j=1}^{q} \vartheta_j \epsilon_{t-j} \, , \quad \epsilon_t \sim \mathcal{N}(0, \sigma^2)$$

In contrast with the AR(1) model, where the covariance between the terms of the series is exponentially decreasing to 0 but always remains different from 0, the MA(q) process is then such that the autocovariances

$$\gamma_s = \mathrm{cov}(x_t, x_{t+s})$$

are equal to 0 for $|s| > q$. According to the Wold decomposition, the MA(q) process is stationary, whatever the vector $(\vartheta_1, \ldots, \vartheta_q)$. However, invertibility and identifiability considerations (see Exercise 4.58) lead to the condition that the polynomial

$$\mathcal{Q}(x) = 1 - \sum_{j=1}^{q} \vartheta_j x^j$$

must have all its roots outside the unit circle.

Example 4.5.6 In the particular case of the MA(1) model, $x_t = \mu + \epsilon_t - \vartheta_1 \epsilon_{t-1}$ and $\mathrm{var}(x_t) = (1 + \vartheta_1^2)\sigma^2$, while $\gamma_1 = \vartheta_1 \sigma^2$. Then x_t can also be written as

$$x_t = \mu + \tilde{\epsilon}_{t-1} - \frac{1}{\vartheta_1}\tilde{\epsilon}_t, \quad \tilde{\epsilon} \sim \mathcal{N}(0, \vartheta_1^2\sigma^2),$$

which shows that the couples (ϑ_1, σ) and $(1/\vartheta_1, \vartheta_1\sigma)$ lead to two alternative representations of the same model. This somehow justifies the restriction to $|\vartheta_1| < 1$.   ∥

Contrary to the AR(p) model, this model is not Markovian per se (even though it can be represented as a Markov process using the state-space representation introduced below). While the whole vector $x_{1:T}$ is a normal random variable with constant mean μ and covariance matrix

$$\Sigma = \begin{pmatrix} \sigma^2 & \gamma_1 & \gamma_2 & \cdots & \gamma_q & 0 & \cdots & 0 & 0 \\ \gamma_1 & \sigma^2 & \gamma_1 & \cdots & \gamma_{q-1} & \gamma_q & \cdots & 0 & 0 \\ & & & \ddots & & & & & \\ 0 & 0 & 0 & \cdots & 0 & 0 & \cdots & \gamma_1 & \sigma^2 \end{pmatrix},$$

with $(|s| \leq q)$

$$(4.5.11) \qquad\qquad \gamma_s = \sigma^2 \sum_{i=0}^{q-|s|} \vartheta_i \vartheta_{i+|s|},$$

and thus provides an explicit likelihood function, the computation and obviously the integration (or maximization) of this likelihood for a given value of the parameter is quite costly, since it involves inverting the (n, n) matrix Σ. A more manageable representation is to use the likelihood of $x_{1:T}$ conditional on $(\epsilon_0, \ldots, \epsilon_{-q+1})$,

$$(4.5.12) \quad L(\mu, \vartheta_1, \ldots, \vartheta_q, \sigma | x_{1:T}, \epsilon_0, \ldots, \epsilon_{-q+1}) =$$

$$\sigma^{-T} \prod_{t=1}^{T} \exp\left\{ -\left(x_t - \mu + \sum_{j=1}^{q} \vartheta_j \hat{\epsilon}_{t-j} \right)^2 \Big/ 2\sigma^2 \right\},$$

where $(t > 0)$

$$(4.5.13) \qquad\qquad \hat{\epsilon}_t = x_t - \mu + \sum_{j=1}^{q} \vartheta_j \hat{\epsilon}_{t-j}$$

and $\hat{\epsilon}_0 = \epsilon_0$, ..., $\hat{\epsilon}_{1-q} = \epsilon_{1-q}$. This recursive definition of the likelihood is still costly, since it involves T sums of q terms. Nonetheless, even though the problem of the conditioning values $(\epsilon_0, \ldots, \epsilon_{-q+1})$ must be treated separately, for instance via an MCMC step (Chapter 6), the complexity of this representation is much more manageable than the normal representation given above.

Another approach of interest to the representation of the MA(q) model is to use the so-called *state-space representation*, inspired by the *Kalman filter*, which provides recursive linear formulas for *prediction*, smoothing and *filtering.* Brockwell and Davis (1996, Chapter 8) provide a general coverage of this technique, while West and Harrison (1997) present its Bayesian version, but the general idea is to represent a time series $(x_t)_{t \geq 1}$ as a system of two equations,

$$(4.5.14) \qquad\qquad x_t \;=\; G_y \mathbf{y}_t + \varepsilon_t \,,$$

$$(4.5.15) \qquad\qquad \mathbf{y}_{t+1} \;=\; F_t \mathbf{y}_t + \xi_t \,,$$

where the vectors ε_t and ξ_t are multivariate normal vectors with general covariance matrices that may depend on t and $\mathbb{E}[\varepsilon_u \xi_v'] = 0$ for all (u,v)'s. Equation (4.5.14) is called the *observation equation* and (4.5.15) is called the *state equation*. This representation projects the process of interest $(x_t)_{t \geq 1}$ into a larger space, the *state space*, where the process $(\mathbf{y}_t)_{t \geq 1}$ is Markovian and linear. For instance, (4.5.7) is a *state-space representation* of the AR(p) model.

The MA(q) model can be written that way by defining $\mathbf{y}_t = (\epsilon_{t-q}, \ldots, \epsilon_{t-1}, \epsilon_t)'$. Then the state equation is

$$(4.5.16) \qquad \mathbf{y}_{t+1} = \begin{pmatrix} 0 & 1 & 0 & \ldots & 0 \\ 0 & 0 & 1 & \ldots & 0 \\ & & & \ldots & \\ 0 & 0 & 0 & \ldots & 1 \\ 0 & 0 & 0 & \ldots & 0 \end{pmatrix} \mathbf{y}_t + \epsilon_{t+1} \begin{pmatrix} 0 \\ 0 \\ \vdots \\ 0 \\ 1 \end{pmatrix}$$

while the observation equation is

$$x_t = \mu - (\, \vartheta_q \quad \vartheta_{q-1} \quad \ldots \quad \vartheta_1 \quad -1 \,)\, \mathbf{y}_t \,.$$

This decomposition thus involves no vector ε_t in the observation equation, while ξ_t is degenerate in the state equation. This degeneracy phenomenon is quite common in state-space representations, but this is not a hindrance in conditional uses of the model, as in the MCMC algorithms of Chapter 6. Notice also that the state-space representation of a model is not unique.

Example 4.5.7 (Example 4.5.6 continued) For the MA(1) model, the observation equation can also be $x_t = (\, 1 \quad 0 \,)\mathbf{y}_t$ with $\mathbf{y}_t = (\, y_{1t} \quad y_{2t} \,)'$ directed by the state equation

$$\mathbf{y}_{t+1} = \begin{pmatrix} 0 & 1 \\ 0 & 0 \end{pmatrix} \mathbf{y}_t + \epsilon_{t+1} \begin{pmatrix} 1 \\ \vartheta_1 \end{pmatrix}. \qquad\qquad \|$$

Whatever the representation chosen for the MA(q) model, the identifiability condition on $\mathcal{Q}(x)$ imposes to the ϑ_j's to vary in a complex space, which cannot be directly described for values of q larger than 3. The reparameterization described in Lemma 4.5.4 also formally applies in this case, with a different interpretation of the ψ_i's which are then the *inverse partial auto-correlations* (Jones (1987)). A uniform prior on the ψ_i's can be used for the estimation of the ϑ_i's, which necessarily goes through an MCMC step (see Chapter 6, Chib and Greenberg (1994), Barnett et al. (1996), Billio et al. (1999), Marin and Robert (2007, Chapter 7)).

4.5.4 The ARMA model

A straightforward extension of the previous model is the $\text{ARMA}(p, q)$ model, where $(t = 1, 2, \ldots)$

$$(4.5.17) \qquad x_t = \mu + \sum_{i=1}^{p} \varrho_i(x_{t-i} - \mu) + \epsilon_t - \sum_{j=1}^{q} \vartheta_j \epsilon_{t-j},$$

where the ϵ_t's are i.i.d. $\mathcal{N}(0, \upsilon^2)$. The role of such models, as compared with both AR and MA models, is to aim towards parsimony, that is, to use much smaller values of p and q than in a pure AR or a pure MA modeling.

As detailed in Box and Jenkins (1976), the stationarity and identifiability conditions still correspond to the roots of the polynomials $\mathcal{P}$ and $\mathcal{Q}$ being outside the unit circle, with a further condition that both polynomials have no common root. (But this almost surely never happens under a continuous prior on the parameters.) The reparameterization of Lemma 4.5.4 (1984) can therefore be implemented for both the ϑ_i's and the ϱ_j's, still calling for MCMC techniques owing to the complexity of the posterior distribution.

State-space representations also exist for $\text{ARMA}(p, q)$ models, one possibility being (Brockwell and Davis (1997, Example 8.3.2))

$$x_t = \mu - (\,\vartheta_{r-1} \quad \vartheta_{r-2} \quad \ldots \quad \vartheta_1 \quad -1\,)\,\mathbf{y}_t$$

for the observation equation and

$$(4.5.18) \quad \mathbf{y}_{t+1} = \begin{pmatrix} 0 & 1 & 0 & \ldots & 0 \\ 0 & 0 & 1 & \ldots & 0 \\ & & & \ldots & \\ 0 & 0 & 0 & \ldots & 1 \\ \varrho_r & \varrho_{r-1} & \varrho_{r-2} & \ldots & \varrho_1 \end{pmatrix} \mathbf{y}_t + \epsilon_{t+1} \begin{pmatrix} 0 \\ 0 \\ \vdots \\ 0 \\ 1 \end{pmatrix},$$

for the state equation, with $r = \max(p, q + 1)$ and the convention that $\varrho_t = 0$ if $t > p$ and $\vartheta_t = 0$ if $t > q$. As for the $\text{MA}(q)$ models, this representation is handy to devise MCMC algorithms (see Chapter 6) to handle the simulation of the posterior distribution of the parameters of the $\text{ARMA}(p, q)$ model.

4.6 Exercises

Section 4.1

4.1 (Smith (1988)) Consider x, a random variable with mean μ, c.d.f. F, and density f. The two functions f and f' are supposed to be bounded. Define a sequence of random variables y_n with c.d.f.'s

$$G_n(y) = \left(1 - \frac{1}{n}\right) F(y) + \frac{1}{n} H_n(y),$$

satisfying

(i) $\mathbb{E}^{H_n}[y] = n^2$; and

(ii) $H'_n = h_n$ and h'_n are bounded.

Show that $G_n \to F$, $G'_n = g_n \to f$, and $g'_n \to f'$, but that $|\mu - \mathbb{E}[y_n]| \to \infty$.

4.2 If $\psi(\theta|x)$ is a posterior distribution associated with $f(x|\theta)$ and a (possibly improper) prior distribution π, show that

$$\frac{\psi(\theta|x)}{f(x|\theta)} = k(x)\pi(\theta).$$

a. Deduce that, if f belongs to an exponential family, the posterior distribution also belongs to an exponential family, whatever π is.

b. Show that if ψ belongs to an exponential family, the same holds for f.

4.3 *(Berger and Wolpert (1988)) In the following setting, Stein (1962b) points out some limitations of the Likelihood Principle. Assume that a value $\theta > 0$ can be assessed either by $x \sim \mathcal{N}(\theta, \sigma^2)$ (with known σ^2) or by

$$y \sim f(y|\theta) = cy^{-1}\exp\left\{-\frac{d^2}{2}\left(1 - \frac{\theta}{y}\right)^2\right\}\mathbb{I}_{[0,b\theta]}(y),$$

where b is huge and d large (50, say).

a. Show that the two maximum likelihood estimators of θ are $\delta_1(x) = x$ and $\delta_2(y) = y$.

b. Consider the special case $x = y = \theta d$. Explain why the inference on θ should be the same in both cases.

c. Explain why

$$[x - 1.96\,\sigma, x + 1.96\,\sigma]$$

could be proposed as a confidence interval on θ at the level 95%.

d. Deduce that

$$[y - (1.96)(y/d), y + (1.96)(y/d)]$$

can be used as a confidence interval if y is observed.

e. Show that

$$P(y - (1.96)(y/d) < \theta < y + (1.96)(y/d))$$

can be made as small as wished for an adequate choice of b.

f. Conclude that the above confidence interval is not appropriate for large values of $x = y$ and σ and discuss the relevance of a confidence interval for the Likelihood Principle.

g. Study the problem with the prior distribution $\pi(\theta) = 1/\theta$.

4.4 Show that, if $p \in [0, 1]$, $\theta = p/(1-p)$ and if $\pi(\theta) = 1/\theta$, the prior distribution $\pi(p)$ is Haldane distribution.

4.5 Show that a setting opposite to Example 4.1.2 may happen, namely, a case when the prior information is negligible. (*Hint:* Consider $\pi(\theta)$ to be $\mathcal{C}(\mu, 1)$ and $f(x|\theta) \propto \exp -|x - \theta|$, and show that the MAP estimator does not depend on μ.)

4.6 In the setting of Example 4.2, consider $\pi(\theta) \propto \exp -a|\theta|$ and show that, for a small enough, the MAP estimator is not always equal to 0.

4.7 A *contingency table* is a $k \times \ell$ matrix such that the (i, j)-th element is n_{ij}, the number of simultaneous occurrences of the ith modality of a first characteristic, and of the jth modality of a second characteristic in a population of n individuals $(1 \leq i \leq k, 1 \leq j \leq \ell)$. The probability of this occurrence is denoted by p_{ij}.

    a. Show that these distributions belong to an exponential family.

    b. Determine the distributions of the margins of the table, i.e., of $n_{i\cdot} = n_{i1} + \ldots + n_{i\ell}$ and $n_{\cdot j} = n_{1j} + \ldots + n_{kj}$. Deduce the distributions of $(n_{1\cdot}, \ldots, n_{k\cdot})$ and of $(n_{\cdot 1}, \ldots, n_{\cdot \ell})$.

    c. Derive conjugate priors on $p = (p_{ij})$ and the Jeffreys prior.

    d. In the particular case of *independence* between the two variables, the parameters are supposed to satisfy the relations $p_{ij} = p_{i\cdot}p_{\cdot j}$ where $(p_{1\cdot}, \ldots, p_{k\cdot})$ $(p_{\cdot 1}, \ldots, p_{\cdot \ell})$ are two vectors of probabilities. Relate these vectors to the distributions derived in b. and construct the corresponding conjugate priors.

    e. Compare the posterior expectations of p_{ij} for the conjugate priors of c. and d. [*Note:* See Santner and Duffy (1989) for a detailed presentation of the Bayesian processing of these models.]

4.8 Determine whether the following distributions are possible posterior distributions:

    (i) $\mathcal{T}_1(k, \mu(x), \tau^2(x))$ when $x \sim \mathcal{N}(0, \sigma^2)$ and σ^2 is known;

    (ii) a truncated normal distribution $\mathcal{N}(\mu(x), \tau^2(x))$ when $x \sim \mathcal{P}(\theta)$; and

    (iii) $\mathcal{P}a(\alpha(x), \mu(x))$ when $x \sim \mathcal{B}(n, 1/\theta)$.

4.9 *(**Exercise 4.8 cont.**) Given a sample distribution $f(x|\theta)$ and a conditional distribution $g(\theta|x)$, give a necessary and sufficient condition for $g(\theta|x)$ to be a posterior distribution associated with $f(x|\theta)$ and an arbitrary prior distribution $\pi(\theta)$.

4.10 Let $(x_n)_n$ be a Markov chain with finite state space $\{1, \ldots, p\}$ and transition matrix P.

    a. If the sample is $x_1, \ldots, x_n$, express the likelihood function and derive conjugate priors for the components of P.

    b. The Markov chain is now observed at random times $t_1 < \cdots < t_n$. Give the likelihood function $\ell(P|x_{t_1}, \ldots, x_{t_n})$ when the distribution of the t_i's does not depend on P and examine whether the above prior distributions are still manageable for posterior computations.

    c. A random variable y_t is observed for $t = 1, \ldots, n$ with conditional distribution $f(y|\theta_{x_t})$. We assume the y_t's to be independent conditional on the x_t's. Show that the marginal distribution of the y_t's is a mixture of the distributions $f(y|\theta_k)$.

    d. If only the y_t's are observed, the model is called a *hidden Markov chain*. When $f(y|\theta)$ belongs to an exponential family, give the likelihood function and the conjugate priors on $(P, \theta_1, \ldots, \theta_p)$.

    e. Consider the special case $p = 2$ and $f(y|\theta) = \theta \exp(-\theta y)\mathbb{I}_{\mathbb{R}_+}(y)$ to examine whether the above priors are manageable.

4.11 Consider $x \sim \mathcal{B}(m, p)$ and $p \sim \mathcal{B}e(1/2, 1/2)$.

a. Show that this prior is equivalent to an uniform prior on $\theta = \arcsin(\sqrt{p})$. How can you justify this transformation? [*Note:* See Feller (1970) for details on the arcsine distribution.]

b. Let $y \sim \mathcal{B}(n, q)$ an independent observation with $q \sim \mathcal{B}e(1/2, 1/2)$. Use the approximation $\arcsin x \sim \mathcal{N}(\theta, 1/4m)$ to give an approximate posterior distribution of $\arcsin(\sqrt{p}) - \arcsin(\sqrt{q})$.

c. Deduce an approximation of

$$\pi(|\arcsin(\sqrt{p}) - \arcsin(\sqrt{q})| < 0.1 | x, y).$$

4.12 The *logistic distribution* is defined by the density

$$e^{-(x-\theta)}/(1 + e^{-(x-\theta)})^2$$

on $\mathbb{R}$.

a. Show that the above function is truly a density and derive the maximum likelihood estimator of θ.

b. Show that this distribution does not belong to an exponential family (i) directly; and (ii) using Exercise 3.20. Deduce that there is no associated conjugate prior and propose a noninformative prior.

c. What is the maximum likelihood estimator of θ for a sample $x_1, \ldots, x_n$? Show through an example that the likelihood can be multimodal.

d. Relate logistic regression and logistic distribution by exhibiting latent logistic random variables in the logistic regression model. Is there a contradiction between b. and the fact that the logistic regression model belongs to an exponential family, as shown in Example 3.3.15?

4.13 For the AR model of Example 4.1.6, show that the joint posterior distribution $\pi(\varrho, \sigma^2 | x_{1:(T-1)})$ is explicit under the conjugate prior

$$\varrho \sim \mathcal{N}(0, \kappa \sigma^2), \qquad \sigma^2 \sim \mathcal{IG}(\alpha, \beta).$$

Deduce the predictive density $\pi(x_T | x_{1:(T-1)})$.

Section 4.2

4.14 (Smith (1988)) A usual justification of quadratic losses is that they provide a second-order approximation for symmetric losses. Consider the loss

$$L(\theta, \delta) = 1 - e^{-(\delta - \theta)^2/2},$$

and $\pi(\theta | x) = (1/2)\{\varphi(\theta; 8, 1) + \varphi(\theta; -8, 1)\}$, a mixture of two normal distributions with means 8 and -8, and variance 1.

a. Show that $\pi(\theta | x)$ can actually be obtained as a posterior distribution.

b. Show that $\mathbb{E}^\pi[\theta | x]$ is a local maximum of the posterior loss.

c. Relate the loss $L(\theta, \delta)$ with the intrinsic losses of §2.5.4.

4.15 Consider $x \sim \mathcal{P}(\lambda)$ and $\pi(\lambda) = e^{-\lambda}$. The purpose of the exercise is to compare the estimators $\delta_c(x) = cx$ under the quadratic loss $L(\lambda, \delta) = (\delta - \lambda)^2$.

a. Compute $R(\delta_c, \lambda)$ and show that δ_c is not admissible for $c > 1$.

b. Compute $r(\pi, \delta_c)$ and deduce the optimal c^π.

c. Derive the best estimator δ_c for the minimax criterion.

d. Solve the above questions for the loss

$$L'(\lambda, \theta) = \left(\frac{\delta}{\lambda} - 1\right)^2.$$

4.16 Show that a Bayes estimator associated with a quadratic loss and a proper prior distribution cannot be unbiased. Does this result hold for generalized Bayes estimators? For other losses?

4.17 Consider $x \sim \mathcal{B}(n, p)$ and $p \sim \mathcal{B}e(\alpha, \beta)$.

a. Derive the posterior and marginal distributions. Deduce the Bayes estimator under quadratic loss.

b. If the prior distribution is $\pi(p) = [p(1-p)]^{-1} \mathbb{I}_{(0,1)}(p)$, give the generalized Bayes estimator of p (when it is defined).

c. Under what condition on (α, β) is δ^π unbiased? Is there a contradiction with Exercise 4.16?

d. Give the Bayes estimator of p under the loss

$$L(p, \delta) = \frac{(\delta - p)^2}{p(1-p)}.$$

4.18 Using the estimators in Table 4.2.1, show that the estimators corresponding to noninformative prior distributions can be written as limits of conjugate estimators. Does this convergence extend to other posterior quantities for the same sequence of conjugate hyperparameters? Try to derive a general result.

4.19 Consider $x \sim \mathcal{N}(\theta, 1)$, $\theta \sim \mathcal{N}(0, 1)$, and $L(\theta, \delta) = \mathbb{I}_{\{\delta < \theta\}}$. Show that, in this case, there is no Bayes estimator.

4.20 Consider $x \sim \mathcal{P}(\theta)$, with $\Theta = \{\theta_1, \theta_2\}$ and $\mathcal{D} = \{d_1, d_2, d_3\}$. The loss function is defined by the matrix

$$L = \begin{pmatrix} 0 & 20 & 10 \\ 50 & 0 & 20 \end{pmatrix}$$

(where $L_{ij} = L(\theta_i, d_j)$, $i = 1, 2$, $j = 1, 2, 3$). Show that the Bayes estimators are of the form

$$\delta^\pi(x) = \begin{cases} d_1 & \text{if } x < k - \log_2 3, \\ d_2 & \text{if } x > k - 1, \\ d_3 & \text{otherwise,} \end{cases}$$

and define k in terms of the prior distribution π.

4.21 (Ferguson (1967)) Consider x from the renormalized negative binomial distribution,

$$f(x|\theta) = \binom{r + x - 1}{x} \theta^x (1 + \theta)^{-(r+x)}, \qquad x = 0, 1, \ldots, \qquad \theta \in \mathbb{R}_+^*,$$

so that $\mathbb{E}_\theta[x] = r\theta$ (i.e., $\theta = p/(1-p)$). The loss function is the weighted squared error

$$L(\theta, \delta) = \frac{(\theta - \delta)^2}{\theta(1 + \theta)}.$$

a. Give the maximum likelihood estimator of θ.

b. Show that $\delta_0(x) = x/r$ has a constant risk and is a generalized Bayes estimator for $\pi(\theta) = 1$ if $r > 1$. What happens for $r = 1$?

c. Show that
$$\delta_{\alpha,\beta}(x) = \frac{\alpha + x - 1}{\beta + r + 1}$$

is a Bayes estimator for
$$\pi(\theta|\alpha,\beta) \propto \theta^{\alpha-1}(1+\theta)^{-(\alpha+\beta)}$$

and that this distribution is conjugate for $f(x|\theta)$.

d. Deduce that $\delta_1(x) = x/(r+1)$ is a minimax estimator.

4.22 (Ferguson (1967)) Consider $\Theta = [0,1]$ and $L(\theta,\delta) = \frac{(\theta-\delta)^2}{1-\theta}$, for the geometric distribution
$$f(x|\theta) = \theta^x(1-\theta) \qquad (x \in \mathbb{N}).$$

a. Give a power series representation of $R(\theta,\delta)$.

b. Show that the unique *nonrandomized* estimator with constant risk is δ_0 such that
$$\delta_0(0) = 1/2, \qquad \delta_0(x) = 1 \text{ if } x \geq 1.$$

c. Show that, if δ^π is the Bayes estimator associated with π, $\delta^\pi(n) = \mu_{n-1}/\mu_n$, where μ_i is the ith moment of π.

d. Show that δ_0 is minimax.

4.23 *(Casella and Strawderman (1981)) Consider $x \sim \mathcal{N}(\theta,1)$ with $|\theta| \leq m$ $(m < 1)$.

a. Show that $\delta^m(x) = m\tanh(mx)$ is a Bayes estimator associated with
$$\pi^m(\theta) = \frac{1}{2}\mathbb{I}_{\{-m,m\}}(\theta).$$

b. Show that, for the quadratic loss, $r(\pi^m, \delta^m) = R(\delta^m, \pm m)$ and deduce that δ^m is minimax. [*Note:* This is actually the unique minimax estimator in this case.]

c. Compare with the estimator δ^U associated with the uniform prior
$$\pi(\theta) = \frac{1}{2m}\mathbb{I}_{[-m,m]}(\theta),$$

in terms of m. [*Note:* Gatsonis et al. (1987) give a detailed study of the performance of δ^U in term of minimaxity.]

4.24 (Casella and Berger (1990)) Consider $x \sim \mathcal{U}_{\{1,2,...,\theta\}}$ and $\theta \in \Theta = \mathbb{N}^*$.

a. If $\mathcal{D} = \Theta$, show that, under quadratic loss, $\mathbb{E}^\pi[\theta|x]$ is not necessarily the Bayes estimator.

b. If $\mathcal{D} = [1,+\infty)$, show that $\mathbb{E}^\pi[\theta|x]$ is the Bayes estimator (when it exists).

c. Show that $\delta_0(x) = x$ is admissible, for every choice of $\mathcal{D}$. (*Hint:* Start with $R(1,\delta_0)$.)

d. Show that δ_0 is a Bayes estimator and that there exist other Bayes estimators for this prior distribution with different risk functions.

4.25 Consider x_1, x_2 i.i.d. with distribution $f(x|\theta) = (1/2)\exp(-|x-\theta|)$ and $\pi(\theta) = 1$. Determine the Bayes estimators associated with the quadratic and absolute error losses. Same question for an additional observation. [*Note:* See Example 1.2.5 for an historical motivation.]

Section 4.3.1

4.26 Chrystal (1891) writes: *"No one would say that, if you simply put two white balls into a bag containing one of unknown color, equally likely to be black or white, this action raised the odds that the unknown ball is white from even to 3 to 1,"* as an argument against the Laplace succession rule. Do you consider this criticism as acceptable? (See Zabell (1989).)

4.27 (Jeffreys (1961))

a. Show that

$$\sum_{i=1}^{N} \binom{i}{x_1} \binom{N-i}{x-x_1} = \binom{N+1}{x+1}$$

   (i) algebraically, and
   (ii) using combinatorics.

b. If the sample contains $x = x_1 + x_2$ individuals, show that the probability that the following $y = y_1 + y_2$ draws will contain y_1 individuals from the first and y_2 from the second population is

$$P(y_1, y_2 | x_1, x_2) = \frac{y!}{y_1! y_2!} \frac{(x_1+1)\ldots(x_1+y_1)(x_2+1)\ldots(x_2+y_2)}{(x+2)\ldots(x+y+1)}.$$

c. If $x = x_1$, deduce that the probability that the y next draws are of the same type is

$$\frac{x+1}{x+y+1}.$$

4.28 Generalize the Laplace succession rule for a multinomial model.

Some problems similar to the Laplace succession rule were considered by Lewis Carroll in his Pillow Problems. *Seneta (1993) gives a detailed commentary on these problems, two of which are given below.*

4.29 Consider two bags, H and K, with two balls each. Each ball is either black or white. A white ball is added to bag H and a hidden ball is transferred at random from bag H to bag K.

a. What is the chance of drawing a white ball from bag K?

b. A white ball is then added to bag K and a hidden ball is transferred from bag K to bag H. What is the chance now of drawing a white ball from bag H?

4.30 *"If an infinity of rods is broken, find the chance that one at least is broken in the middle."* While this question is poorly formulated, a discrete solution is proposed below.

a. Assume that each rod has $2m+1$ breaking points and that there are exactly $2m + 1$ rods. Give the probability that no rod breaks in the middle and derive the limiting value when m goes to infinity.

b. Study the dependence of this limit upon the assumption that the number of breaking points is equal to the number of rods.

Section 4.3.2

4.31 In the setting of Example 4.3.1, develop a Bayesian model for the distribution of $(t_2 - t_1)$. Extend to the following problem: Given that a traffic light has been red for one minute, what is the probability that it will turn green in the next minute?

4.32 Show that, for the tramcar problem, the maximum likelihood estimator $\hat{N} = T$ is admissible under any loss function of the form $L(|\hat{N} - N|)$ where L is strictly increasing. (*Hint:* Consider the case where $N = 1$ first.)

Section 4.3.3

4.33 During the launch of a new campus journal, $n_1 = 220$ and $n_2 = 570$ persons bought the two test issues -1 and 0. The number of persons who bought both issues is $n_{11} = 180$. Give a Bayes estimator of N, total number of readers, assuming that a capture-recapture modeling applies here and that $\pi(N)$ is $\mathcal{P}(1000)$.

4.34 (Castledine (1981)) For the Wolter modeling introduced in Section 4.3.3, i.e., when n_1 and n_2 are random variables, the *temporal model* considers the case when all individuals have the same probability of capture for a given experiment, but where the probability varies between the first and the second captures. These two probabilities are denoted p_1 and p_2 .

a. Give the likelihood and the maximum likelihood estimator associated with this model when p_1 and p_2 are *known*.

b. Show that the posterior distribution of N given p_1 and p_2 only depends on $n_+ = n_1 + n_2 - n_{11}$ and $\mu = 1 - (1 - p_1)(1 - p_2)$. If the prior distribution of N is $\pi(N) = 1/N$, show that $\pi(N|n_+, \mu)$ is $\mathcal{N}eg(n_+, \mu)$.

c. Give the posterior marginal distribution of N if $p_1 \sim \mathcal{B}(\alpha, \beta)$ and $p_2 \sim \mathcal{B}(\alpha, \beta)$.

d. Show that, if $\alpha = 0$, $\beta = 1$, we recover the Darroch model as the marginal distribution of N. Does this decomposition facilitate the derivation of a Bayes estimator?

Section 4.4.1

4.35 *(Robert (1990a)) The modified Bessel function I_ν ($\nu \geq 0$) is a solution to the differential equation $z^2 f'' + z f' - (z^2 + \nu^2) f(z) = 0$ and can be represented by the power series

$$I_\nu(z) = \left(\frac{z}{2}\right)^\nu \sum_{k=0}^{\infty} \frac{(z/2)^{2k}}{k!\,\Gamma(\nu + k + 1)}.$$

a. Show that the above series converges in $\mathbb{R}$.

b. Developing

$$\int_0^\pi e^{z\cos(\theta)} \sin^{2\nu}(\theta)\, d\theta$$

in a power series, show that I_ν can be written as

(4.6.1) $\qquad I_\nu(z) = \frac{(z/2)^\nu}{\pi^{1/2}\Gamma(\nu + \frac{1}{2})} \int_0^\pi e^{z\cos(\theta)} \sin^{2\nu}(\theta)\, d\theta.$

c. Establish the following recurrence formulas:

$$\begin{cases} I_{\nu+1}(z) = I_{\nu-1}(z) - (2\nu/z)I_\nu(z), \\ I'_\nu(z) = I_{\nu-1}(z) - (\nu/z)I_\nu(z). \end{cases}$$

d. Derive from (4.6.1) by an integration by parts that, for $z > 0$,

$$I_{\nu+1}(z) \leq I_\nu(z).$$

e. Derive from the power series representation of I_ν that $t^{-\nu}I_\nu(t)$ is increasing in t. If we define r_ν as

$$r_\nu(t) = \frac{I_{\nu+1}(t)}{I_\nu(t)},$$

show that r_ν is increasing, concave, and that $r_\nu(t)/t$ is decreasing.

f. Show that

$$\lim_{t \to 0} r_\nu(t) = 1, \qquad \lim_{t \to \infty} \frac{r_\nu(t)}{t} = \frac{1}{2(\nu+1)},$$

and that

$$r'_\nu(t) = 1 - \frac{2\nu+1}{t}r_\nu(t) - r_\nu^2(t).$$

g. Show that the density of the noncentral chi-squared distribution with non-centrality parameter λ and ν degrees of freedom can be expressed using a modified Bessel function, namely,

$$p_{\lambda,\nu}(x) = \frac{1}{2}\left(\frac{x}{\lambda}\right)^{\frac{\nu-2}{4}} I_{\frac{\nu-2}{2}}\left(\sqrt{\lambda x}\right)e^{-\frac{x+\lambda}{2}}.$$

4.36 *(Bock and Robert (1991)) On $\mathbb{R}^p$, the sphere of radius c is defined by

$$S_c = \left\{z \in \mathbb{R}^p; \ ||z||^2 = c\right\}.$$

a. If $x \sim \mathcal{N}_p(\theta, I_p)$, with $p \geq 3$, and θ has the prior distribution π_c, uniform distribution on S_c, show that the marginal density of x is proportional to

$$m_c(x) = e^{-||x||^2/2}e^{-c^2/2}\frac{I_{\frac{p-2}{2}}(||x||c)}{(c||x||)^{\frac{p-2}{2}}}.$$

b. Show that the proportionality coefficient is independent of c and recall why it does not appear in the posterior distribution.

c. Derive from a. the posterior mean δ_c by a differential computation. (*Hint:* See Lemma 4.2.2.)

d. Show that, if $c \geq \sqrt{p}$, δ_c is a shrinkage estimator outside the ball $\{x; \ ||x|| \leq \varrho\}$ and an expander within this ball. Determine the boundary value ϱ.

e. Show that δ_c cannot be minimax. Is this estimator admissible?

f. Explain why δ_c never belongs to the sphere S_c while π_c is concentrated on S_c. Is δ_c the "true" Bayes estimator then?

g. Using the recurrence relations of Exercise 4.35, show that

$$\delta_c(x) = \left(1 - \frac{p-2}{||x||^2}\right)x + h_c(||x||^2)x,$$

where $h_c(t) > 0$ when $t \leq \max(c^2, p-2)$. Try to propose a more interesting estimator.

4.37 Consider $x_1, \ldots, x_{10}$ i.i.d. $\mathcal{N}(\theta, \theta^2)$, with $\theta > 0$, which represents ten observations of the speed of a star. Justify the choice $\pi(\theta) = 1/\theta$ and determine the generalized Bayes estimator associated with the invariant loss

$$L(\theta, \delta) = \left(\frac{\delta}{\theta} - 1\right)^2.$$

(*Hint:* Use Exercise 3.33.)

4.38 *(Lindley (1965)) Consider $x_1, \ldots, x_n$ a sample from $\mathcal{N}(\theta, \sigma^2)$ with σ^2 known. The prior density $\pi(\theta)$ is such that there exist ϵ, M, and c such that $c(1 - \epsilon) \leq \pi(\theta) \leq c(1 + \epsilon)$ for $\theta \in I = [\bar{x} - 1.96\,\sigma/\sqrt{n}, \bar{x} + 1.96\,\sigma/\sqrt{n}]$ and $\pi(\theta) \leq Mc$ otherwise.

  a. Show that these constraints are compatible, i.e., that there exists such a prior distribution.

  b. Show that

$$(1 - \epsilon)[0.95(1 + \epsilon) + 0.05M]^{-1}\frac{e^{-(x-\theta)^2 n/2\sigma^2}}{\sqrt{2\pi\sigma^2/n}} \leq \pi(\theta|x)$$

$$\leq (1 + \epsilon)[(1 - \epsilon)0.95]^{-1}\frac{e^{-(x-\theta)^2 n/2\sigma^2}}{\sqrt{2\pi\sigma^2/n}}$$

if $\theta \in I$ and

$$\pi(\theta|x) \leq \frac{M}{0.95(1 - \epsilon)}\frac{e^{-1.96^2/2}}{\sqrt{2\pi\sigma^2/n}}$$

otherwise.

  c. Discuss the interest of the approximations when $\theta \in I$ and when $\theta \notin I$. Can you derive a conservative confidence region?

4.39 Consider a normal random variable, $x \sim \mathcal{N}(\theta, 1)$ and the one-to-one transform $\eta = \sinh(\theta)$.

  a. When $\pi(\eta) = 1$, show that the resulting posterior distribution on θ is

$$\pi(\theta|x) \propto e^x \mathcal{N}(x + 1, 1) + e^{-x}\mathcal{N}(x - 1, 1).$$

  b. Compare the behavior of this posterior distribution with the usual Jeffreys posterior $\mathcal{N}(x, 1)$ in term of posterior variance, posterior quantiles and modes. In particular, determine the values of x for which the posterior distribution is bimodal and those for which there are two global maxima.

  c. Consider the behavior of $\pi(\theta|x)$ for large values of x and conclude that the prior $\pi(\eta) = 1$ is unreasonable.

Section 4.4.2

4.40 (Jeffreys (1961)) Consider $x_1, \ldots, x_{n_1}$ i.i.d. $\mathcal{N}(\theta, \sigma^2)$. Let $\bar{x}_1, s_1^2$ be the associated statistics. For a second sample of observations, give the predictive distribution of $(\bar{x}_2, s_2^2)$ under the noninformative distribution $\pi(\theta, \sigma) = \frac{1}{\sigma}$. If $s_2^2 = s_1^2/y$ and $y = e^z$, deduce that z follows a Fisher's F distribution.

4.41 Show that, if $x \sim \mathcal{G}(\alpha, \beta)$, $1/x \sim \mathcal{IG}(\alpha, \beta)$ as defined in (4.4.3).

4.42 *(Ghosh and Yang (1996)) As in Exercise 3.46, consider $x_{11}, \ldots, x_{1n_1}$ and $x_{21}, \ldots, x_{2n_2}$, two independent samples with $x_{ij} \sim \mathcal{N}(\mu_i, \sigma^2)$.

a. Show that the Fisher information matrix is

$$\mathbf{I}(\mu_1, \mu_2, \sigma) = \sigma^{-2} \begin{pmatrix} n_1 & 0 & 0 \\ 0 & n_2 & 0 \\ 0 & 0 & 2(n_1 + n_2) \end{pmatrix}.$$

b. Welch and Peers's (1963) matching prior (see Section 3.5.5) for the quantity of interest $\theta = (\mu_1 - \mu_2)/\sigma$ is

(4.0.2)
$$\frac{\partial}{\partial \mu_1}(\eta_1 \pi) + \frac{\partial}{\partial \mu_2}(\eta_2 \pi) + \frac{\partial}{\partial \sigma}(\eta_3 \pi) = 0,$$

where

$$(\eta_1, \eta_2, \eta_3) = \mathbf{I}^{-1} \nabla\theta / (\nabla\theta^t \mathbf{I}^{-1} \nabla\theta)^{1/2}.$$

Show that a class of solutions to (4.6.2) are of the form

(4.6.3)
$$\left[n_1^{-1} + n_2^{-1} + \frac{1}{2}(\mu_1 - \mu_2)^2 / \{(n_1 + n_2)\sigma^2\} \right]^{1/2} g(\mu_1, \mu_2, \sigma)$$

where

$$g(\mu_1, \mu_2, \sigma) \propto \left[d_1(\mu_1 - \mu_2)^2 + d_2(n_1 \mu_1^2 + n_2 \mu_2^2) + d_3 \sigma^2 \right]^c,$$

c is an arbitrary constant, and (d_1, d_2, d_3) satisfy

$$d_1(n_1^{-1} + n_2^{-1}) + d_2 = \frac{1}{2} d_3 (n_1 + n_2)^{-1}.$$

c. Deduce that the matching prior for (θ, μ_2, σ) is

$$\pi(\theta, \mu_2, \sigma) \propto \sigma^{2c+1} \left[n_1^{-1} + n_2^{-1} + \frac{1}{2}\theta^2 (n_1 + n_2)^{-1} \right]^{c+1/2}.$$

d. Show that

$$\pi(\theta | \bar{x}_1, \bar{x}_2, s) \propto \left[n_1^{-1} + n_2^{-1} + \frac{1}{2}\theta^2 (n_1 + n_2)^{-1} \right]^{c+1/2}$$
$$\int_0^\infty v^{n_1 + n_2 - 2c - 4} \exp\left\{ \frac{-1}{2}\left(v^2 + \frac{n_1 n_2}{n_1 + n_2}(vz - \theta)^2 \right) \right\} dv$$

where $z = (\bar{x}_1 - \bar{x}_2)/s$.

e. Show that that the distribution of z only depends on θ.

f. Show that the only choice of c which avoids the marginalization paradox of Exercises 3.44–3.50 is $c = -1$.

Section 4.4.3

4.43 a. If $x \sim \mathcal{N}_p(\theta, \Sigma)$, show that, for every prior distribution π,

$$\delta^\pi(x) = x + \Sigma \nabla \log m_\pi(x).$$

b. (Bock (1988)) *Pseudo-Bayes estimators* are defined as the estimators of the form

$$\delta(x) = x + \nabla \log m(x)$$

when $x \sim \mathcal{N}_p(\theta, I_p)$. Show that the truncated James–Stein estimator given in Example 4.2.3 is a pseudo-Bayes estimator (i.e., define the corresponding m). Can this estimator be a Bayes estimator?

4.44 *For a normal model $\mathcal{N}_k(X\beta, \Sigma)$ where the covariance matrix Σ is totally unknown, give the noninformative Jeffreys prior.

a. Show that the posterior distribution of Σ conditional upon β is a Wishart distribution and deduce that there is no proper marginal posterior distribution on β when the number of observations is smaller than k.

b. Explain why it is not possible to derive a conjugate distribution in this setting. Consider the particular case when Σ has a Wishart distribution.

c. What is the fundamental difference in this model which prevents what was possible in Section 4.4.2?

4.45 *Consider a linear regression prediction setting, where $y = X\beta + \epsilon$ has been observed with $\beta \in \mathbb{R}^k$ and $\epsilon \sim \mathcal{N}_p(0, \Sigma)$, and $z = T\beta + \epsilon'$ is to be predicted (with T known and $\epsilon' \sim \mathcal{N}_p(0, \Sigma)$ independent of ϵ).

a. If δ is the predictor and the prediction error is evaluated through the loss $L(z, \delta) = ||z - \delta||^2$, show that the frequentist expected error is

$$\mathbb{E}^{z,x}[L(z, \delta(x))] = \mathrm{tr}(\Sigma) + \mathbb{E}^x[||\delta(x) - T\beta||^2].$$

b. Show that the problem can be expressed as the estimation of β under the quadratic loss associated with $Q = T^t T$. (*Hint:* Show first that $\delta(x)$ is necessarily of the form $T\gamma(x)$, with $\gamma(x) \in \mathbb{R}^k$, or dominated by an estimator of this form.)

c. Deduce from the fact that Q is degenerate with a single eigenvalue different from 0 that a Stein effect cannot occur in this setting.

d. Consider now that T is a random matrix with mean 0 and $\mathbb{E}[T^t T] = M$. Show that, when $\delta(x) = T\gamma(x)$, the frequentist risk is

$$\mathbb{E}^{z,x,T}[L(z, \delta(x))] = \mathrm{tr}(\Sigma) + \mathbb{E}^x[(\gamma(x) - \beta)^t M(\gamma(x) - \beta)],$$

and therefore that a Stein effect is possible when M has at least three positive eigenvalues. [*Note:* This phenomenon is related to the ancillarity paradoxes developed in Brown (1990). See also Foster and George (1996).]

e. Let $\beta \sim \mathcal{N}_k(0, \sigma^2 I_k)$. Derive the Bayes predictor of z when T is fixed and when T is random. Conclude.

4.46 *Tobit models* are used in econometrics (see Gouriéroux and Monfort (1996)) to represent truncated settings. Consider $y|x \sim \mathcal{N}(\beta^t x, \sigma^2)$, which is only reported when y is positive, x being an explanatory variable in $\mathbb{R}^p$.

a. Show that tobit models are a mixture of probit models (when $y < 0$) and of regular regression models (when $y \geq 0$).

b. Give the likelihood function $\ell(\beta, \sigma^2 | y_1, \ldots, y_n)$ associated with a sample $y_1, \ldots, y_n, x_1, \ldots, x_n$ and derive sufficient statistics for this model.

c. Conditional upon $(x_1, \ldots, x_n)$, show that this model belongs to an exponential family and propose a conjugate prior on (β, σ). Are the corresponding computations tractable?

4.47 *The *inverse regression* (or *calibration*) model is given by

$$y \sim \mathcal{N}_p(\beta, \sigma^2 I_p), \qquad z \sim \mathcal{N}_p(\lambda_0 \beta, \sigma^2 I_p), \qquad s^2 \sim \sigma^2 \chi_q^2,$$

with $\beta \in \mathbb{R}^p$, $\lambda_0 \in \mathbb{R}$.

a. Give the maximum likelihood estimator of λ and show that its quadratic risk can be infinite.

b. Compute the Jeffreys prior on $(\beta, \sigma^2, \lambda_0)$ and show that the corresponding posterior expectation of λ_0 is the *inverse regression estimator*, $\delta^I(y, z, s) = y^t z/(s + \|y\|^2)$.

c. Using the technique of the *reference prior* introduced in Section 3.5, propose an alternative prior distribution $\pi(\lambda_0, (\beta, \sigma^2))$ when (β, σ^2) is considered as a nuisance parameter. Derive the corresponding posterior expectation of λ_0, $\delta^R(y, z, s)$.

d. Show that, when q goes to infinity, δ^I a.s. converges to 0, but that δ^R is free of this inconsistency. [*Note:* See Osborne (1991) for a survey of calibration models and Kubokawa and Robert (1994) for decision-theoretic perspectives on these estimators.]

Section 4.5.1

4.48 For the AR(1) model (4.5.2), give the covariance matrix of $(x_1, \ldots, x_T)$.

4.49 (**Exercise 4.48 cont.**)

a. Show that the variance of x_t is given by (4.5.3).

b. What happens in the case $\varrho = 1$, since (4.5.3) has no meaning?

c. Extend to the case when x_0 is an arbitrary value.

4.50 (**Exercise 4.49 cont.**) We want to establish that there exists no stationary distribution for our AR(1) model when $|\varrho| \geq 1$, that is, no density f such that, if $x_t \sim f$, then $x_{t+1} \sim f$.

a. Show that, when $|\varrho| < 1$, the stationary distribution is the normal distribution $\mathcal{N}(0, \sigma^2/(1 - \varrho^2))$.

b. For the case $|\varrho| = 1$, show that the Lebesgue measure is the stationary measure of the chain (x_t), that is, for every measurable set A,

$$\int_A dx = \int_A \int f(y|x) dx dy,$$

where $f(y|x)$ denotes the conditional distribution of x_t given x_{t-1}, that is, $\mathcal{N}(x_{t-1}, \sigma^2)$ in this case. Deduce from the unicity of the stationary measure that there cannot exist a stationary distribution.

c. Extend to the case when $|\varrho| \geq 1$, by writing x_t as

$$x_t = \sum_{i=0}^{t-1} \varrho^i \epsilon_{t-i} + \varrho^t x_0$$

and deducing that x_t is almost surely infinite when t goes to infinity. (*Hint:* When $x_0 = 0$, replace the above decomposition with the corresponding decomposition conditional on x_1.)

Section 4.5.2

4.51 (Bernardo and Smith (1994)) Show that, for a two dimensional vector,

$$(x_1 \quad x_2)^t \sim \mathcal{N}_2 \left((\mu_1 \quad \mu_2)^t, \begin{bmatrix} \sigma_1^2 & \varrho\sigma_1\sigma_2 \\ \varrho\sigma_1\sigma_2 & \sigma_2^2 \end{bmatrix} \right),$$

the Jeffreys prior is $\pi(\theta) \propto (1 - \varrho^2)^{-1}/\sigma_1\sigma_2$.

4.52 (Bauwens et al. (1999)) For the AR(1) model (4.5.2),

a. Show that μ is a location parameter and, therefore, that it does not appear in the Jeffreys prior.

b. Show that

$$\mathbb{E}\left[\frac{\partial^2 \log L(\theta|x_{1:T})}{\partial \sigma^2}\right] = \frac{-T}{2\sigma^4}, \quad \mathbb{E}\left[\frac{\partial^2 \log L(\theta|x_{1:T})}{\partial \varrho^2}\right] = \frac{-1}{\sigma^2}\mathbb{E}\left[\sum_{t=0}^{T-1} x_t^2\right].$$

c. Using the stationary distribution of the y_t's, deduce from $\mathbb{E}[y_t^2] = \sigma^2/(1-\varrho^2)$ that the Jeffreys prior $\pi_1^J(\sigma^2,\varrho) = 1/\sigma^2\sqrt{1-\varrho^2}$.

4.53 *(Brockwell and Davis (1996)) The Durbin–Levinson algorithm derives the partial autocorrelations as follows: define $\phi_{n1},\ldots,\phi_{nn}$ recursively from the autocovariances $\gamma(s)$ as

$$\phi_{nn} = \left(\gamma(n) - \sum_{j=1}^{n-1}\phi_{(n-1)j}\gamma(n_j)\right)v_{n-1}^{-1}$$

and

$$\begin{pmatrix}\phi_{n1} \\ \vdots \\ \phi_{n(n-1)}\end{pmatrix} = \begin{pmatrix}\phi_{(n-1)1} \\ \vdots \\ \phi_{(n-1)(n-1)}\end{pmatrix} - \phi_{nn}\begin{pmatrix}\phi_{(n-1)(n-1)} \\ \vdots \\ \phi_{(n-1)1}\end{pmatrix},$$

where $v_n = v_{n-1}(1-\phi_{nn})^2$, $\phi_{11} = \gamma(1)/\gamma(0)$ and $v_0 = \gamma(0)$.

a. Show that, if $\psi_n = \phi_{nn}$, the reverse of the Durbin–Levinson algorithm is given by Lemma 4.5.4.

b. Show that the partial autocorrelations ψ_n of an MA(q) process are 0 for $n > q$.

c. Show that the partial autocorrelations ψ_n of a AR(1) process are given by $\psi_n = (-1)^{n+1}\vartheta_1^n/(1+\vartheta_1^2+\ldots+\vartheta_1^{2n})$.

4.54 (Bauwens et al. (1999)) For the AR(1) model (4.5.2),

a. Using the Wold decomposition (4.5.9) obtained in Example 4.5.5, show that

$$\mathbb{E}\left[x_t^2\right] = \mathbb{E}\left[\left(\varrho^t x_0 + \sum_{i=0}^{t-1}\varrho^i\epsilon_{t-i}\right)^2\right] = \frac{1-\varrho^{2t}}{1-\varrho^2}\sigma^2$$

when $x_0 = 0$.

b. Deduce the Jeffreys prior π_2^J.

Section 4.5.3

4.55 *Give the Wold decomposition for the stationary AR(p) model. (*Hint:* Use the *lag polynomial representation* of the AR(p) model, that is, $\mathcal{P}(B)x_t = \epsilon_t$, where $B^d x_t = x_{t-d}$.)

4.56 Show that the autocorrelations γ_s for the MA(q) model are given by (4.5.11).

4.57 Show that that the representation (4.5.16) holds. Give an extension of the representation in Example 4.5.7 to the general MA(q) model.

Section 4.5.4

4.58 *An ARMA(p,q) model $(t \geq 1)$

$$x_t - \mu = \sum_{i=1}^{p} \varrho_i(x_{t-i} - \mu) + \sum_{j=1}^{q} \vartheta_j \epsilon_{t-j} + \epsilon_t,$$

is *invertible* (Brockwell and Davis (1996, §3.1)) if there exists a sequence $(\varpi_j)_j$ such that

$$\sum_j |\varpi_j| < \infty \quad \text{and} \quad \epsilon_t = \sum_{j=0}^{\infty} \varpi_j x_{t-j}.$$

Show that invertibility is equivalent to the condition that $\mathcal{Q}(x)$ must have its roots outside the unit circle. (*Hint:* Use the *lag polynomial representation* of the ARMA(p,q) model, that is, $\mathcal{P}(B)x_t = \mathcal{Q}(B)\epsilon_t$, where $B^d x_t = x_{t-d}$.)

4.59 *An ARMA(p,q) model is *causal* (Brockwell and Davis (1996, §3.1)) if there exists a sequence $(\varphi_j)_j$ such that

$$\sum_j |\varphi_j| < \infty \quad \text{and} \quad x_t = \sum_{j=0}^{\infty} \varphi_j \epsilon_{t-j}.$$

Show that causality is equivalent to the condition that $\mathcal{P}(x)$ must have its roots outside the unit circle. (*Hint:* Use the lag polynomial representation of Exercise 4.58.)

4.60 Show that the representation (4.5.18) holds. Propose an alternative representation.

4.61 Propose a state-space representation similar to (4.5.18) for the *ARIMA model*

(4.6.4) $$z_t - \mu = \sum_{i=1}^{p} \varrho_i(z_{t-i} - \mu) + \sum_{j=1}^{q} \vartheta_j \epsilon_{t-j} + \epsilon_t,$$

where z_t is the difference of the observations, $z_t = x_t - x_{t-d}$, $d \in \mathbb{N}^*$. [*Note:* As detailed in Brockwell and Davis (1996, §6.5), the general ARIMA(p,d,q) is given as an ARMA(p,q) model on the difference $x_t - \Psi_1 x_{t-d} - \ldots - \Psi_P x_{t-Pd}$.]

Note 4.7.1

4.62 (Deely and Gupta (1968)) Consider $x_1 \sim \mathcal{N}(\theta_1, \sigma_1^2), \ldots, x_k \sim \mathcal{N}(\theta_k, \sigma_k^2)$ when the quantity of interest is $\theta_{[k]}$, the largest mean among $\theta_1, \ldots, \theta_k$. The loss function is $L(\theta, \varphi) = \theta_{[k]} - \varphi$.

a. Show that, if $\sigma_1 = \ldots = \sigma_k$ are known and $\pi(\theta_1) = \ldots = \pi(\theta_k) = 1$, the Bayes estimator selects the population with the largest observation.

b. Generalize to the case where the θ_i's have an exchangeable prior distribution $\mathcal{N}(0, \tau^2)$.

4.63 *(Goel and Rubin (1977)) Show that the s_j^*'s truly constitute a complete class when the prior distribution on $\theta = (\theta_1, \ldots, \theta_k)$ is symmetric. (*Hint:* Show that s_j^* is optimal among the subsets of size $|s_j^*|$.)

4.64 (**Exercise 4.63 cont.**) Extend this result to the distributions $f(x|\theta)$ with monotone likelihood ratio property in θ.

4.65 (Chernoff and Yahav (1977)) Extend the complete class result of Exercise 4.63 to the loss function

$$L(\theta, s) = c(\theta_{[k]} - \theta_s) - \frac{1}{s} \sum_{j \in s} \theta_j.$$

(*Hint:* Show that, if $\theta_{i_1} \leq \ldots \leq \theta_{i_j}$, $s = \{i_1, \ldots, i_j\}$ is dominated by the set $\{i_j\}$.)

Note 4.7.2

4.66 *For the AR(1) model (4.5.2), assume the quantity of interest is x_0, the starting value of the chain. Compute a reference prior for $(x_0, (\varrho, \sigma^2))$ and derive an estimator of x_0 under quadratic loss.

Note 4.7.3

4.67 Consider the factor model $(t = 1, \ldots, T)$,

(4.6.5)
$$\begin{cases} y_t^* = [\alpha + \beta(y_{t-1}^*)^2]^{1/2} \epsilon_t^* \\ y_t = y_t^* \mu + \sigma \epsilon_t, \end{cases}$$

with $\epsilon_t^* \sim \mathcal{N}(0, 1)$, and only the $y_t \in \mathbb{R}^p$ are observed.

a. Write the (complete) likelihood associated with the couples (y_t, y_t^*).

b. Show that the y_t^*'s cannot be analytically integrated out.

c. Deduce that the factor model cannot be expressed as a special case of an ARCH model (4.7.1).

4.68 (Bauwens et al. (1999)) Show that the ARCH(p) model is useless when $\alpha = 0$. (*Hint:* Show that var(y_t) = 0.)

4.7 Notes

4.7.1 Ranking and selection

A great deal of effort has been dedicated to the problem of estimating and comparing several normal means. We mention briefly a few approaches here to illustrate the interest of the Bayes treatment, referring the reader to the literature for a more developed discussion. See, for instance, Gibbons et al. (1977), Gupta and Panchapakesan (1979), and Dudewicz and Koo (1982), following the introductory papers of Bechofer (1954) and Gupta (1965). As described by Berger and Deely (1988), ranking and selection techniques also appear as substitutes for the *analysis of variance* approach (Chapter 10).

Given $x_1 \sim \mathcal{N}(\theta_1, \sigma_1^2)$, ..., $x_k \sim \mathcal{N}(\theta_k, \sigma_k^2)$, the problem is to select the population with the highest mean, $\theta_{[k]}$. The variances $\sigma_1^2, \ldots, \sigma_k^2$ are supposed to be known but the more general setting where they are estimated by $\hat{\sigma}_1^2, \ldots, \hat{\sigma}_k^2$ can be treated similarly by the Bayesian paradigm. Berger and Deely (1988) rewrite the problem to answer the following questions: (a) Can we accept the hypothesis $H_0 : \theta_1 = \cdots = \theta_k$? (b) In the event of a negative answer to (a), what is the largest mean? They solve the first question by computing the Bayes factor against H_0 and then the posterior probabilities p_j that θ_j is the largest mean $(1 \leq j \leq k)$. (Chapter 5 covers definition and computation of these quantities.) The prior distributions they use are hierarchical

$$\theta_i | \beta, \sigma_\pi^2 \sim \mathcal{N}(\beta, \sigma_\pi^2), \quad \beta \sim \mathcal{N}(\beta^0, A) \quad \text{and} \quad \sigma_\pi^2 \sim \gamma \mathbb{I}_0(\sigma_\pi^2) + (1 - \gamma)\pi_{22}^*(\sigma_\pi^2).$$

The special form of the prior distribution on σ_π^2 is necessary to test whether the θ_i's are actually identical. For π_{22}^*, Berger and Deely (1988) suggest the *informative* choice

$$\pi_{22}^*(\sigma_\pi^2) = (m-1)C(1+C\sigma_\pi^2)^{-m},$$

where C and m can be derived from prior quantiles. For a *noninformative* choice, possibilities are $\pi_{22}^*(\sigma_\pi^2) = 1$ and

$$\pi_{22}^*(\sigma_\pi^2) = \prod_{i=1}^{k}(\sigma_i^2 + \sigma_\pi^2)^{-1/k},$$

although these priors may lead to difficulties in the derivation of the posterior probability of H_0 (see Chapter 5).

Goel and Rubin (1977) adopt a more decision-theoretic perspective, considering as decision space $\mathcal{D}$ the set of all nonempty parts of $\{1,\ldots,k\}$, denoted $\{s_1, s_2, \ldots, s_K\}$ where $K = 2^k - 1$. They introduce the loss function

$$L(\theta, s) = c|s| + \theta_{[k]} - \theta_s,$$

where $|s|$ denotes the cardinality of s and $\theta_s = \max_{j \in s} \theta_j$. This loss includes a penalization cost c for each population involved in the decision set s. This is quite logical because a parsimony requirement forces the decision set s to be as small as possible, the ideal setting being $|s| = 1$. Goel and Rubin (1977) first show that a Bayes rule associated with this loss and a symmetric prior distribution has to be chosen among the sets $s_j^* = \{\omega_k, \ldots, \omega_{k-j+1}\}$ $(1 \le j \le k)$, where ω_j denotes the population of $x_{(j)}$. The Bayes rule s^π is then a solution of

$$\varrho(\pi, s^\pi | x) = \min_{j=1,\ldots,k} \varrho(\pi, s_j^* | x),$$

where $\varrho(\pi, s | x) = c|s| + \mathbb{E}^\pi[\theta_{[k]} - \theta_s | x]$. Introducing

$$\Delta_m = \varrho(\pi, s_{m+1}^* | x) - \varrho(\pi, s_m^* | x) \qquad (1 \le m \le k-1),$$

the Bayes rule is s_k^* if $A = \{j; \Delta_j \ge 0\}$ is empty and s_m^* otherwise, where $m = \min(A)$. A difficult point in the derivation of s^π is obviously the computation of the posterior expectations $\mathbb{E}^\pi[\theta_{[k]} - \theta_s | x]$. The authors detail the particular case of a normal exchangeable prior on the θ_j's, which still depends on the function

$$t_m(z) = \int_{-\infty}^{+\infty} \Phi^m(z+x)\Phi(-x)\,dx.$$

Nonetheless, they show that, in the noninformative case, when $\sigma_1 = \cdots = \sigma_k$, the Bayes rule is s_1^* when $c/\sigma_1 \ge 1/\pi^2$.

4.7.2 Jeffreys prior for the AR(1) model

There is a controversy surrounding the Jeffreys prior in this case, owing to the debate of whether to include the stationarity condition and the discrepancy between both cases. If we assume $x_t = \mu + \varrho(x_{t-1} - \mu) + \epsilon_t$ with $x_0 = 0$, the Jeffreys prior associated with the stationary representation is (Exercise 4.52)

$$\pi_1^J(\mu, \sigma^2, \varrho) \propto \frac{1}{\sigma^2}\frac{1}{\sqrt{1-\varrho^2}}.$$

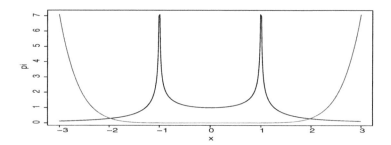

Figure 4.7.1. *Graphs of the priors $\pi_1^J(\varrho)$ and $\pi^B(\varrho)$ for $T = 10$.*

When moving to the non-stationary region $|\varrho| > 1$, Phillips (1991) has shown that the Jeffreys prior is then (Exercise 4.54)

$$\pi_2^J(\mu, \sigma^2, \varrho) \propto \frac{1}{\sigma^2} \frac{1}{\sqrt{|1 - \varrho^2|}} \sqrt{\left| 1 - \frac{1 - \varrho^{2T}}{T(1 - \varrho^2)} \right|}.$$

While $\pi_2^J(\mu, \sigma^2, \varrho)$ is equivalent to $\pi_1^J(\mu, \sigma^2, \varrho)$ for large values of T and $|\varrho| < 1$, the dominant part of the prior is the non-stationary region, since it is equivalent to ϱ^{2T} (Bauwens et al. (1999)). Berger and Yang (1994a) have also shown that the reference prior is π_1^J, and is only defined when the stationary constraint holds. They then suggest to symmetrize this prior to the region $|\varrho| > 1$, taking

$$\pi^B(\mu, \sigma^2, \varrho) \propto \frac{1}{\sigma^2} \begin{cases} 1/\sqrt{1 - \varrho^2} & \text{if } |\varrho| < 1, \\ 1/|\varrho|\sqrt{\varrho^2 - 1} & \text{if } |\varrho| > 1, \end{cases}$$

which has a more reasonable shape that π_2^J, as shown by Figure 4.7.2.

As detailed in Bauwens et al. (1999, §6.8), it is also possible to come up with the Jeffreys priors in the stationary and non stationary cases, when including the distribution of the initial value x_0, with priors similar to π_1^J and π_2^J.

4.7.3 ARCH models

ARCH models, as introduced by Engle (1982), are used to represent processes, particularly in finance, with independent errors with time-dependent variances, as in the ARCH(p) process ($t \geq 1$),

$$(4.7.1) \qquad x_t = \sigma_t \epsilon_t, \qquad \sigma_t^2 = \alpha + \sum_{i=1}^{p} \beta_i x_{t-i}^2,$$

where the ϵ_t's are i.i.d. $\mathcal{N}(0, 1)$. The acronym ARCH stands for *autoregressive conditional heterocedasticity*, heterocedasticity being a term used by econometricians for heterogeneous variance. Gouriéroux (1997) provides a general reference on these models, as well as classic inferential methods of estimation. See Bauwens et al. (1999, §7.4) for Bayesian extensions to GARCH processes. As shown in Nelson (1991) and Kleibergen and van Dijk (1993), a stationarity condition for the ARCH(1) model is that $\mathbb{E}[\log(\beta_1 \epsilon_t^2)] < 0$, which is equivalent to $\beta_1 < 3.4$.

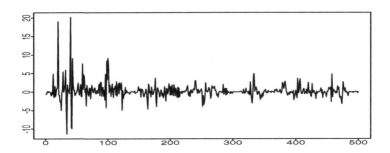

Figure 4.7.2. *Simulated sample from the stochastic volatility model (4.7.2) with* $\sigma = 1$ *and* $\varrho = .9$. *(Source: Robert and Casella (2004).)*

Contrary to the stochastic volatility models of Note 4.7.4, ARCH(p) models enjoy closed-form likelihoods, when conditioning on the initial values $x_1, \ldots, x_p$. However, because of the non-linearities in the variance terms, approximations methods as those of Chapter 6 must be used.

4.7.4 Stochastic volatility models

Stochastic volatility models are modeling the *volatility*, $\log(\sigma_t^2)$, of a series $(x_t)_{t \geq 1}$ as a random variable. While such models are more complex to study than their ARCH counterparts, they are often used in finance to represent series with sudden variations in scale (see, for instance, Jacquier et al. (1994)). A simple illustration of such models is the SV(1) model, where $(t = 1, \ldots, T)$

$$(4.7.2) \qquad \begin{cases} y_t^* = \alpha + \varrho y_{t-1}^* + \sigma \epsilon_{t-1}^* \,, \\ y_t = e^{y_t^*/2} \epsilon_t \,, \end{cases}$$

and the ϵ_t's and ϵ_t^*'s are i.i.d. $\mathcal{N}(0,1)$. The unobserved quantity (y_t^*) thus represents the *volatility*. (An usual assumption on the initial condition is that $y_0^* \sim \mathcal{N}(\alpha, \sigma^2)$.) Figure 4.7.4 plots a simulated series of stochastic volatilities for $\sigma = 1$ and $\varrho = .9$.

The difficulty with this model is that the unobserved volatilities contain the information about the parameters $(\alpha, \varrho, \sigma)$ because, conditional upon y_t^*, they are independent from y_t. (Obviously, the parameters do depend on the data, albeit marginally.) Moreover, the observed likelihood $L(\alpha, \varrho, \sigma | y_0, \ldots, y_T)$ cannot be expressed in closed form because the y_t^*'s cannot be integrated out analytically. On the opposite, the completed likelihood is explicit, that is,

$$(4.7.3) \qquad L^c(\alpha, \varrho, \sigma | y_0, y_0^* \ldots, y_T, y_T^*) \propto$$

$$\sigma^{-T+1} \exp - \left\{ (y_0^* - \alpha)^2 + \sum_{t=1}^{T} (y_t^* - \alpha - \varrho y_{t-1}^*)^2 \right\} \bigg/ 2\sigma^2$$

$$\exp - \sum_{t=0}^{T} \left\{ y_t^2 e^{-y_t^*} + y_t^* \right\} / 2 \,.$$

It can then be used in simulation based methods (Chapter 6), in alternance with a simulation of the unobserved volatilities y_t^*. Figure 4.7.4 illustrates this

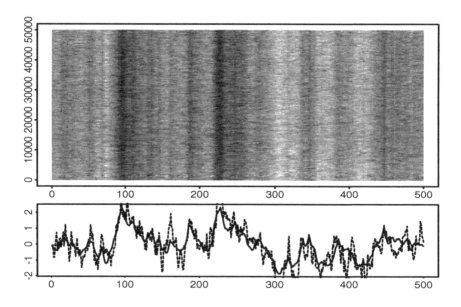

Figure 4.7.3. *Allocation map (top) and average allocations versus true volatilities (bottom) for the model (4.7.2). The true volatilities are represented by broken lines. (Source: Mengersen et al. (1998).)*

simulation for the simulated dataset of Figure 4.7.4, where the true values of the y_t^*'s are known. (The blurred image on top of the graph is called an *allocation map* and represents successive values of the y_t^*'s as grey levels along iterations of the sampling method.)

4.7.5 Poly-t priors

Poly-t priors were proposed by Drèze (1978) and Richard and Tompa (1980) as robust alternatives to the conjugate priors for linear regressions. Their motivation is given by the following example, developed in Bauwens et al. (1999, §4.5). Consider two independent regressions,

$$y_1 = X_1\beta + \sigma_1\varepsilon_1, \ y_2 = X_2\beta + \sigma_2\varepsilon_2, \ \varepsilon_1 \sim \mathcal{N}_{T_1}(0, I_{T_1}), \ \varepsilon_2 \sim \mathcal{N}_{T_2}(0, I_{T_2}).$$

If $\pi(\beta, \sigma_1, \sigma_2) = 1/\sigma_1\sigma_2$, integration of the σ_i's leads to the marginal posterior

$$\begin{aligned}\pi(\beta|y_1, y_2) \ \propto \ & [S_1 + (\beta - \hat{\beta}_1)^t M_1(\beta - \hat{\beta}_1)]^{-T_1/2} \\ & \times [S_2 + (\beta - \hat{\beta}_2)^t M_2(\beta - \hat{\beta}_2)]^{-T_2/2},\end{aligned}$$

where $\hat{\beta}_i$ is the OLS estimator $(X_i^t X_i)^{-1} X_i y_i$, $M_i = (X_i^t X)$ and $S_i = ||y_i - X_i - \hat{\beta}_i||^2$ $(i = 1, 2)$. This distribution is a $2 - 0$ *poly-t distribution*.

In general, an $m - n$ *poly-t distribution* is defined as a product of m Student's t densities divided by n Student's t densities,

$$\varphi_{m,n}(x) \ \propto \ \prod_{j=1}^{n}\left[1 + (x - \mu_j^0)^t P_j^0(x - \mu_j^0)\right]^{\nu_j^0/2}$$

$$\bigg/ \prod_{j=1}^{m} \left[1 + (x - \mu_j^1)^t P_j^1 (x - \mu_j^1)\right]^{\nu_j^1/2} .$$

As shown in Bauwens et al. (1999, Theorem A.21), the $\varphi_{m,0}$ density can be expressed as a (continuous) mixture of regular Student's t densities against $(m-1)$ auxiliary variables, a property that can be exploited either for direct simulation as in Bauwens (1984), or for MCMC implementation (see Chapter 6), since the direct computation of the normalization constant of $\varphi_{m,n}$, or of the corresponding posterior mean, is impossible. An additional difficulty with poly-t priors is that, when compared with conjugate priors, they require the determination of a much larger number of hyperparameters.

CHAPTER 5

Tests and Confidence Regions

Twenty-six more tests were going to take the rest of daylight, maybe more. Heat or no heat, the days still grew shorter as if winter really was coming on, and a failed test would take a few minutes longer that one passed, just to make certain.

Robert Jordan, *Lord of Chaos, Book VI of the Wheel of Time.*

5.1 Introduction

Although testing theory can be perceived as a special case of Decision Theory for a restricted decision space (and even as an estimation problem), we consider testing inference in a separate chapter because there is much more ambiguity about the real inferential purpose of testing than when estimating a regular function of the parameter. In fact, this part of statistical inference is still incomplete, in the sense that many other answers have been proposed, none being entirely satisfactory. In particular, there are pronounced differences between frequentist and Bayesian testing theories. This is nonetheless a setting where a Bayesian approach is quite appealing because the notion of the probability of a hypothesis, like $\pi(\theta \in \Theta_0|x)$, can only be defined through this approach.

Some Bayesians actually think that there should be no testing, or, at least, that there should be no point-null hypothesis testing (see, for instance, Gelfand et al. (1992)), and we will see in this chapter several philosophical reasons that somehow motivate this radical perspective. These reasons range from the reductive aspect of modeling *(no model is right, but some models are less wrong than others)*, to the unnatural modification of the prior imposed by point-null hypotheses, to the lack of problem-motivated decision-theoretic structure, to the subsequent call to rudimentary conventional $0 - 1$ losses and acceptance levels, to the impossibility of using improper priors in both point-null hypotheses and model choice settings (Chapter 7). But pragmatic considerations are such that the Bayesian toolbox must also include testing devices, if only because users of Statistics have been accustomed to testing as a formulation of their problems, given their strong tendency to take problems at face value.

We first consider in Section 5.2 the usual Bayesian approach to testing, i.e., through an evaluation of decisions by $0 - 1$ losses, and compare the Bayesian procedures with their frequentist counterparts in Section 5.3. We then propose in Section 5.4 an alternative decision-theoretic approach through more adaptive losses which emphasize the "postdata" evaluation of testing procedures (as opposed to Neyman–Pearson procedures for which the evaluation is operated in a "predata" spirit).

This chapter exhibits a strong contrast between Bayesian and frequentist approaches under different evaluation tools; it is revealing because this shows the incompleteness of the classical modeling, which relies on artificial concepts to derive its optimal procedures. Contrary to point estimation settings, these optimal frequentist procedures are no longer limits of Bayes procedures and differ numerically from their Bayesian counterparts. However, we moderate this rejection in Section 5.3 by showing that classical answers may sometimes lead to similar conclusions than noninformative Bayes procedures. Chapter 7 deals with *model choice*, which can be seen as a particular case of testing point-null hypotheses, but also enjoys enough peculiarities to deserve a chapter on its own (besides calling almost uniformly for the computational methods presented in Chapter 6).

5.2 A first approach to testing theory

5.2.1 Decision-theoretic testing

Consider a statistical model $f(x|\theta)$ with $\theta \in \Theta$. Given a subset of interest of Θ, Θ_0, which sometimes consists of a single point $\{\theta_0\}$, the question to be answered is whether the true value of the parameter θ belongs to Θ_0, i.e., to *test* the hypothesis[1]

$$H_0 : \ \theta \in \Theta_0,$$

usually called the *null hypothesis*. In linear models, Θ_0 may be a *subspace* of the vector space Θ and the testing problem is then a particular case of a *model choice* problem (see also Chapter 7).

Example 5.2.1 Consider a *logistic regression* model,

$$P_\alpha(y = 1) = 1 - P_\alpha(y = 0) = \exp(\alpha^t x)/(1 + \exp(\alpha^t x)), \qquad \alpha, x \in \mathbb{R}^p,$$

which models the probability of developing a prostate cancer in a lifetime in terms of explanatory variables $x = (x_1, \ldots, x_p)$. Of particular interest is the dependence on work environment variables such as asbestos concentration x_{i_0} and a worker union may want to test whether the coefficient α_{i_0} corresponding to x_{i_0} is null.   ‖

---

[1] There is a certain amount of ambiguity involved in the terminology: *Test* simultaneously denotes the question, and the procedure used to answer the question.

In the Neyman–Pearson perspective (Section 5.3), the testing problem is formalized through a decision space $\mathcal{D}$ restricted to {yes, no} or, equivalently, to $\{1, 0\}$. In fact, it makes sense to perceive testing problems as an inference about the indicator function $\mathbb{I}_{\Theta_0}(\theta)$ and therefore to propose answers in $\mathbb{I}_{\Theta_0}(\Theta) = \{0, 1\}$. Of course, the relevance of such a restriction is less obvious when considering that testing settings often occur as components (or preliminary steps) or more complex inferential structures and, in particular, that the answer to a test question has also consequences in terms of (regular) estimation errors. It may then be more interesting to propose procedures taking values in $[0, 1]$. (We discuss this approach in Section 5.4.)

In some cases, additional information is available about the support of θ, namely, that $\theta \in \Theta_0 \cup \Theta_1 \neq \Theta$. In such settings, we define the *alternative* hypothesis against which we test H_0 as

$$H_1 : \ \theta \in \Theta_1.$$

Under this formalization, every test procedure φ appears as an estimator of $\mathbb{I}_{\Theta_0}(\theta)$ and we only need a loss function $L(\theta, \varphi)$ to derive the Bayes estimators. For instance, the loss function proposed by Neyman and Pearson is the $0 - 1$ loss

$$L(\theta, \varphi) = \begin{cases} 1 & \text{if } \varphi \neq \mathbb{I}_{\Theta_0}(\theta), \\ 0 & \text{otherwise,} \end{cases}$$

introduced in Chapter 2. For this loss, the Bayesian solution is

$$\varphi^\pi(x) = \begin{cases} 1 & \text{if } P^\pi(\theta \in \Theta_0|x) > P^\pi(\theta \in \Theta_0^c|x), \\ 0 & \text{otherwise.} \end{cases}$$

This estimator is easily justified on an intuitive basis since it chooses the hypothesis with the largest posterior probability. A generalization of the above loss is to penalize differently errors when the null hypothesis is true or false. The weighted $0 - 1$ losses

$$(5.2.1) \qquad L(\theta, \varphi) = \begin{cases} 0 & \text{if } \varphi = \mathbb{I}_{\Theta_0}(\theta), \\ a_0 & \text{if } \theta \in \Theta_0 \text{ and } \varphi = 0, \\ a_1 & \text{if } \theta \notin \Theta_0 \text{ and } \varphi = 1, \end{cases}$$

are called "$a_0 - a_1$" for obvious reasons. The associated Bayes estimator is then given by the following result.

Proposition 5.2.2 *Under the loss (5.2.1), the Bayes estimator associated with a prior distribution π is*

$$\varphi^\pi(x) = \begin{cases} 1 & \text{if } P^\pi(\theta \in \Theta_0|x) > \dfrac{a_1}{a_0 + a_1}, \\ 0 & \text{otherwise.} \end{cases}$$

Proof. Since the posterior loss is

$$\begin{aligned} L(\pi, \varphi|x) &= \int_\Theta L(\theta, \varphi)\pi(\theta|x)d\theta \\ &= a_0 P^\pi(\theta \in \Theta_0|x)\mathbb{I}_{\{0\}}(\varphi) + a_1 P^\pi(\theta \notin \Theta_0|x)\mathbb{I}_{\{1\}}(\varphi), \end{aligned}$$

the Bayes estimator can be derived directly.   □□

For this class of losses, the null hypothesis H_0 is rejected when the posterior probability of H_0 is too small, the *acceptance level* $a_1/(a_0 + a_1)$ being
determined by the choice of the loss function. Notice that φ^π only depends
on a_0/a_1 and that the larger a_0/a_1 is, i.e., the more important a wrong
answer under H_0 is relative to H_1, the smaller the posterior probability of
H_0 needs to be for H_0 to be accepted.

Example 5.2.3 Consider $x \sim \mathcal{B}(n, p)$ and $\Theta_0 = [0, 1/2]$. Under the uniform prior distribution $\pi(p) = 1$, the posterior probability of H_0 is

$$P^\pi(p \le 1/2 | x) = \frac{\int_0^{1/2} p^x (1 - p)^{n-x} dp}{B(x + 1, n - x + 1)}$$

$$= \frac{(1/2)^{n+1}}{B(x + 1, n - x + 1)} \left\{ \frac{1}{x + 1} + \frac{n - x}{(x + 1)(x + 2)} + \ldots + \frac{(n - x)! x!}{(n + 1)!} \right\}$$

which can be easily computed and compared to the acceptance level.   ‖

Example 5.2.4 Consider $x \sim \mathcal{N}(\theta, \sigma^2)$ and $\theta \sim \mathcal{N}(\mu, \tau^2)$. Then $\pi(\theta | x)$ is
the normal distribution $\mathcal{N}(\mu(x), \omega^2)$ with

$$\mu(x) = \frac{\sigma^2 \mu + \tau^2 x}{\sigma^2 + \tau^2} \quad \text{and} \quad \omega^2 = \frac{\sigma^2 \tau^2}{\sigma^2 + \tau^2}.$$

To test $H_0 : \theta < 0$, we compute

$$P^\pi(\theta < 0 | x) = P^\pi \left(\frac{\theta - \mu(x)}{\omega} < \frac{-\mu(x)}{\omega} \right)$$

$$= \Phi(-\mu(x)/\omega).$$

If z_{a_0, a_1} is the $a_1/(a_0 + a_1)$ quantile, i.e., if it satisfies $\Phi(z_{a_0, a_1}) = a_1/(a_0 + a_1)$, H_0 is accepted when

$$-\mu(x) > z_{a_0, a_1} \omega,$$

the upper acceptance bound then being

$$-\frac{\sigma^2}{\tau^2} \mu - (1 + \frac{\sigma^2}{\tau^2}) \omega z_{a_0, a_1}.$$

‖

Again, notice that, from a Bayesian point of view, it seems natural to
base the decision upon the posterior probability that the hypothesis is true.
In Section 5.4, we show that an alternative decision-theoretic approach
leads to this posterior probability as the Bayesian estimator itself and thus
avoids the comparison to a predetermined acceptance level. In fact, a difficulty with the losses (5.2.1) is the choice of the weights a_0 and a_1, since
they are usually selected automatically rather than determined from utility
considerations.

5.2.2 The Bayes factor

While from a decision-theoretic point of view the *Bayes factor* is only a one-to-one transform of the posterior probability, this notion came out to be considered on its own ground in Bayesian testing.

Definition 5.2.5 *The Bayes factor is the ratio of the posterior probabilities of the null and the alternative hypotheses over the ratio of the prior probabilities of the null and the alternative hypotheses, i.e.,*

$$B_{01}^\pi(x) = \frac{P(\theta \in \Theta_0 \mid x)}{P(\theta \in \Theta_1 \mid x)} \Big/ \frac{\pi(\theta \in \Theta_0)}{\pi(\theta \in \Theta_1)}.$$

This ratio evaluates the modification of the odds of Θ_0 against Θ_1 due to the observation and can naturally be compared to 1, although an exact comparison scale can only be based upon a loss function. In the particular case where $\Theta_0 = \{\theta_0\}$ and $\Theta_1 = \{\theta_1\}$, the Bayes factor simplifies to the usual *likelihood ratio*

$$B_{01}^\pi(x) = \frac{f(x|\theta_0)}{f(x|\theta_1)}.$$

In general, the Bayes factor depends on prior information, but is still proposed as an "objective" Bayesian answer, since it partly eliminates the influence of the prior modeling and emphasizes the role of the observations. Actually, it can be perceived as a Bayesian likelihood ratio since, if π_0 is the prior distribution under H_0 and π_1 the prior distribution under H_1, $B_{01}^\pi(x)$ can be written as

$$(5.2.2) \qquad B_{01}^\pi(x) = \frac{\int_{\Theta_0} f(x|\theta_0)\pi_0(\theta)\,d\theta}{\int_{\Theta_1} f(x|\theta_1)\pi_1(\theta)\,d\theta} = \frac{m_0(x)}{m_1(x)},$$

thus replacing the likelihoods with the marginals under both hypotheses. Alternatively, if $\hat\theta_0$ is the maximum likelihood estimator on Θ_0 and $\hat\theta_1$ the maximum likelihood estimator on Θ_1, the likelihood ratio

$$R(x) = \frac{f(x|\hat\theta_0)}{f(x|\hat\theta_1)} = \frac{\sup_{\Theta_0} f(x|\theta)}{\sup_{\Theta_1} f(x|\theta)}$$

appears as a particular case of $B_{01}^\pi(x)$ when π_0 and π_1 are Dirac masses at $\hat\theta_0$ and $\hat\theta_1$. This does not legitimate the use of $R(x)$ in the least, though, because π_0 and π_1 both depend on x.

As indicated above, the Bayes factor is, from a decision-theoretic point of view, completely equivalent to the posterior probability of the null hypothesis as, under (5.2.1), H_0 is accepted when

$$(5.2.3) \qquad B_{01}^\pi(x) > \frac{a_1}{a_0} \Big/ \frac{\varrho_0}{\varrho_1} = \frac{a_1 \varrho_1}{a_0 \varrho_0},$$

where

$$(5.2.4) \qquad \begin{aligned} \varrho_0 &= \pi(\theta \in \Theta_0) \quad \text{and} \\ \varrho_1 &= \pi(\theta \in \Theta_1) = 1 - \varrho_0. \end{aligned}$$

This alternative version of Proposition 5.2.2 thus provides an illustration of the duality existing between loss and prior distribution, already mentioned in Chapter 2. Indeed, (5.2.3) shows that it is equivalent to weight both hypotheses equally, $\varrho_0 = \varrho_1 = 1/2$, and to modify the error penalties into $a_i' = a_i \varrho_i$ $(i = 0, 1)$ or to penalize similarly both types of errors ($a_1 = a_0 = 1$), when the prior distribution incorporates the actual weights in the weighted prior probabilities,

$$\varrho_0' = \frac{a_0 \varrho_0}{a_0 \varrho_0 + a_1 \varrho_1}, \quad \varrho_1' = \frac{a_1 \varrho_1}{a_0 \varrho_0 + a_1 \varrho_1}.$$

Following Good (1958) and Jeffreys (1961), many Bayesians now consider the Bayes factor on its own ground (see, e.g., Kass and Raftery (1995) for a detailed review). In particular, Jeffreys (1961) developed a scale to judge the evidence in favor of or against H_0 brought by the data, *outside a true decision-theoretic setting*. The scale goes as follows:

– if $\log_{10}(B_{10}^\pi)$ varies between 0 and 0.5, the evidence against H_0 is *poor*,
– if it is between 0.5 and 1, it is is *substantial*,
– if it is between 1 and 2, it is *strong*, and
– if it is above 2 it is *decisive*.

Obviously, this scaling of the Bayes factor gives some indication of the strength of the evidence, but the precise bounds separating one strength from another are a matter of convention and they can be arbitrarily changed, as shown in Kass and Raftery (1995). This is a consequence of the lack of decision-theoretic backup via a loss function. (But the same criticism applies to the conventional α levels of 0.05 or 0.01 used for $a_0/(a_0 + a_1)$ in (5.2.1).)

Chapter 6 will give precisions on methods used to approximate Bayes factors when the integral in (5.2.2) cannot be computed analytically, which is often the case.

Example 5.2.6 (Kass and Raftery (1995)) The "hot hand" in basketball is the belief that players have good and bad days, rather than a fixed probability to win a shoot. For a given player, the model under the null hypothesis *(no hot hand)* is then $H_0 : y_i \sim \mathcal{B}(n_i, p)$ $(i = 1, \dots, G)$, where G denotes the number of games and n_i (y_i) the number of shoots (of good shoots) during the ith game. The model under the general alternative is $H_1 : y_i \sim \mathcal{B}(n_i, p_i)$, the probability p_i varying from game to game. Under a conjugate prior $p_i \sim \mathcal{B}(\xi/\omega, (1 - \xi)/\omega)$, the average $\mathbb{E}[p_i | \xi, \omega] = \xi$ is distributed from a uniform prior $\mathcal{U}([0, 1])$, as is p under H_0, and ω is fixed. The Bayes factor is then

$$B_{10} = \frac{\int_0^1 \prod_{i=1}^G \int_0^1 p_i^{y_i}(1 - p_i)^{n_i - y_i} p_i^{\alpha - 1}(1 - p_i)^{\beta - 1} d\, p_i}{}$$

$$\frac{(\Gamma(1/\omega)/[\Gamma(\xi/\omega)\Gamma((1 - \xi)/\omega)])^G\, d\xi}{\int_0^1 p^{\sum_i y_i}(1 - p)^{\sum_i (n_i - y_i)} d\, p}$$

$$= \frac{\int_0^1 \prod_{i=1}^{G} [\Gamma(y_i + \xi/\omega)\Gamma(n_i - y_i + (1-\xi)/\omega)/\Gamma(n_i + 1/\omega)]}{}$$

$$\frac{(\Gamma(1/\omega)/[\Gamma(\xi/\omega)\Gamma((1-\xi)/\omega)])^G \, d\xi}{\Gamma(\sum_i y_i + 1)\Gamma(\sum_i (n_i - y_i) + 1)/\Gamma(\sum_i n_i + 2)},$$

where $\alpha = \xi/\omega$ and $\beta = (1-\xi)/\omega$. Formally, the numerator can be computed exactly, despite the gamma functions, because of the simplification

$$\Gamma(y_i + \xi/\omega)/\Gamma(\xi/\omega) = \prod_{j=1}^{y_i} (j - 1 + \xi/\omega),$$

$$\Gamma(n_i - y_i + (1-\xi)/\omega)/\Gamma((1-\xi)/\omega)] = \prod_{j=1}^{n-i-y_i} (j - 1 + (1-\xi)/\omega),$$

but the function of ξ to be integrated is then a high-degree polynomial. The resolution of the integral thus calls for a software like `Maple` or `Mathematica`. For a given player, the value of B_{10} is 0.16 for $\omega = 0.005$ and $G = 138$, which does not indicate any conclusive evidence in favor of the hot hand hypothesis.   ‖

5.2.3 Modification of the prior

The notion of Bayes factor is also instrumental in pointing out an important aspect of Bayesian testing. In fact, this factor is only defined when $\varrho_0 \neq 0$ and $\varrho_1 \neq 0$. This implies that, if H_0 or H_1 are a priori impossible, the observations will not modify this absolute information: Null probabilities are absorbing states! Therefore, a point-null hypothesis $H_0 : \theta = \theta_0$ cannot be tested under a *continuous* prior distribution. More generally, model choice is incompatible with prior distributions that are absolutely continuous with respect to the Lebesgue measure on the largest space.

Testing of point-null hypotheses and the like thus impose a drastic modification of the prior distribution, since it requires to derive a prior distribution on both subsets Θ_0 and Θ_1, for instance, the distributions π_0 and π_1 with densities

$$g_0(\theta) \propto \pi(\theta)\mathbb{I}_{\Theta_0}(\theta), \quad g_1(\theta) \propto \pi(\theta)\mathbb{I}_{\Theta_1}(\theta),$$

(with respect to the natural measures on Θ_0 and Θ_1) although this definition is not always free of ambiguity (see Exercise 5.5). Joined with the prior probabilities ϱ_0 and ϱ_1 of Θ_0 and Θ_1 given by (5.2.4), π_0 and π_1 define the prior π. In other words,

$$\pi(\theta) = \varrho_0\pi_0(\theta) + \varrho_1\pi_1(\theta).$$

(When $\Theta_0 = \{\theta_0\}$, the prior distribution on Θ_0 is just the Dirac mass at θ_0.)

This modification of the prior is puzzling from a measure-theoretic point of view, since it puts some weight on a set previously of measure 0. It also highlights the dichotomy imposed by the usual approach to testing for which the null hypothesis is either right or wrong. However, unless the decision maker is adamant about the prior distribution π, in which case H_0 should be rejected if π does not give any weight to Θ_0, the testing problem can be considered as providing some additional (although vague) information about θ. Indeed, to test for $\theta \in \Theta_0$ implies that there is a chance that θ truly belongs to Θ_0 and therefore that some possibly ill-defined indication has been provided about this fact.

To consider testing settings as sources of information is even more convincing if the final decision is not the answer to the test but the estimation of a function of θ, that is, if the test appears as the choice of a submodel. A preliminary test about the vague information may then improve the estimation step. Moreover, keeping this model choice perspective as the real purpose of the analysis, it also makes sense to build up a separate prior distribution for each subspace because only one of the two Θ_i will be considered after the testing step. For instance, given a *point-null hypothesis*, $H_0 : \theta = \theta_0$, the noninformative distribution $\pi(\theta) = 1$ cannot be considered as an acceptable prior on Θ because the particular value θ_0 has been singled out as a possible value for θ. (In Chapter 7, we shall further defend the perspective that similar parameters appearing in two different models must be considered as separate entities.) In general, to consider that the testing problem occurs because of (unavailable) additional observations may help in the derivation of a noninformative prior, even though there is no consensus on noninformative Bayes modeling for tests (see Section 5.3.5).

5.2.4 Point-null hypotheses

A usual criticism of point-null hypothesis settings is that they are not *realistic* (see, e.g., Casella and Berger[2] (1987)). For instance, as pointed out by Good (1980), it does not actually make sense to test whether the probability of rain for tomorrow is $0.7163891256\ldots$[3] However, some statistical problems definitely call for point-null hypothesis testing. For instance, in mixture estimation (see Section 1.1 and Section 6.4), it may be important to know whether a mixture distribution has two or three components, so one may test whether one of the component weights is 0. Similarly, in linear regression, tests on the nullity of the regression coefficients are useful for the elimination of useless exogenous variates, as in Example 5.2.1. Even more to the point, testing whether the Universe is expanding, contracting or stable is akin to testing whether a certain constant is larger, smaller or equal to a specific value.

More generally, two-sided hypotheses such as $H_0 : \theta \in \Theta_0 = [\theta_0 - \epsilon, \theta_0 +$

---

[2] Roger Berger, not James Berger!

[3] But it would still make sense to test whether the prediction of 75% given by the local weather forecaster is exact, that is, whether the probability of rain for the given day is 0.75, or another of the probabilities announced by the forecaster (see Example 2.3.4).

ϵ] can be approximated by $H_0 : \theta = \theta_0$, with hardly any modification of the posterior probabilities when ϵ is small enough. For instance, this happens when the likelihood is constant in a neighborhood of θ_0 (see Berger (1985a) and Berger and Delampady (1987)). Point-null hypotheses are also quite important in practice; for instance, while it makes sense to determine whether a medical treatment has a positive or negative effect, the first issue may be to decide whether it has any effect at all.

Considering the point-null hypothesis $H_0 : \theta = \theta_0$, we denote by ϱ_0 the prior probability that $\theta = \theta_0$ and by g_1 the prior density under the alternative. The prior distribution is then $\pi_0(\theta) = \varrho_0 \mathbb{I}_{\Theta_0}(\theta) + (1 - \varrho_0)g_1(\theta)$ and the posterior probability of H_0 is given by

$$\pi(\Theta_0|x) = \frac{f(x|\theta_0)\varrho_0}{\int f(x|\theta)\pi(\theta)\,d\theta} = \frac{f(x|\theta_0)\varrho_0}{f(x|\theta_0)\varrho_0 + (1 - \varrho_0)m_1(x)},$$

the marginal distribution on H_1 being

$$m_1(x) = \int_{\Theta_1} f(x|\theta)g_1(\theta)\,d\theta.$$

This posterior probability can also be written as

$$\pi(\Theta_0|x) = \left[1 + \frac{1 - \varrho_0}{\varrho_0}\frac{m_1(x)}{f(x|\theta_0)}\right]^{-1}.$$

Similarly, the Bayes factor is

$$B_{01}^{\pi}(x) = \frac{f(x|\theta_0)\varrho_0}{m_1(x)(1 - \varrho_0)} \Big/ \frac{\varrho_0}{1 - \varrho_0} = \frac{f(x|\theta_0)}{m_1(x)}$$

and we derive the following general relation between the two quantities:

$$\pi(\Theta_0|x) = \left[1 + \frac{1 - \varrho_0}{\varrho_0}\frac{1}{B_{01}^{\pi}(x)}\right]^{-1}.$$

Example 5.2.7 (Example 5.2.3 continued) Consider the test of $H_0 : p = 1/2$ against $p \neq 1/2$. For $g_1(p) = 1$, the posterior probability is then given by

$$\begin{aligned}\pi(\Theta_0|x) &= \left[1 + \frac{1 - \varrho_0}{\varrho_0}2^n B(x + 1, n - x + 1)\right]^{-1}\\ &= \left[1 + \frac{1 - \varrho_0}{\varrho_0}\frac{x!(n - x)!}{(n + 1)!}2^n\right]^{-1},\end{aligned}$$

since $m(x) = \binom{n}{x}B(x + 1, n - x + 1)$. For instance, if $n = 5$, $x = 3$, and $\varrho_0 = 1/2$, the posterior probability is

$$\left(1 + \frac{1}{120}2^5\right)^{-1} = \frac{15}{19}$$

and the corresponding Bayes factor is $15/8$, close to 2. So, in the most supporting cases, the posterior probabilities tend to favor H_0. When the

Table 5.2.1. *Posterior probabilities of $p = 1/2$ when $n = 10$.*

x	0	1	2	3	4	5
$P(p = 1/2|x)$	0.0106	0.0970	0.3259	0.5631	0.6928	0.7302

Table 5.2.2. *Posterior probabilities of $\theta = 0$ for different values of $z = x/\sigma$ and for $\tau = \sigma$.*

z	0	0.68	1.28	1.96
$\pi(\theta = 0|z)$	0.586	0.557	0.484	0.351

sample size increases, the range of the possible answers also widens. For instance, if $\pi(p)$ is $\mathcal{Be}(1/2, 1/2)$ and $n = 10$, the posterior probabilities are given in Table 5.2.4 and support H_0 for x close to 5, even though the prior distribution is rather biased against the null hypothesis (since it heavily weights the extreme values, 0 and 1).   ‖

Example 5.2.8 (Example 5.2.4 continued) Consider the test of $H_0 :$ $\theta = 0$. It seems reasonable to choose π_1 as $\mathcal{N}(\mu, \tau^2)$ and $\mu = 0$, if no additional information is available. Then

$$\frac{m_1(x)}{f(x|0)} = \frac{\sigma}{\sqrt{\sigma^2 + \tau^2}} \frac{e^{-x^2/2(\sigma^2 + \tau^2)}}{e^{-x^2/2\sigma^2}}$$

$$= \sqrt{\frac{\sigma^2}{\sigma^2 + \tau^2}} \exp\left\{\frac{\tau^2 x^2}{2\sigma^2(\sigma^2 + \tau^2)}\right\},$$

and the posterior probability can be derived as

$$\pi(\theta = 0|x) = \left[1 + \frac{1 - \varrho_0}{\varrho_0} \sqrt{\frac{\sigma^2}{\sigma^2 + \tau^2}} \exp\left(\frac{\tau^2 x^2}{2\sigma^2(\sigma^2 + \tau^2)}\right)\right]^{-1}.$$

In the special case when $\varrho_0 = 1/2$ and $\tau = \sigma$, Table 5.2.4 gives the posterior probabilities in terms of $z = x/\sigma$.   ‖

Consider now the alternative case $\tau^2 = 10\sigma^2$; it is supposed to indicate a more diffuse prior information on θ. The posterior probabilities of H_0 are then modified as shown in Table 5.2.4.

5.2.5 Improper priors

The recourse to noninformative prior distributions for testing hypotheses is rather delicate, and DeGroot (1973) states that improper priors should not be used *at all* in tests. In fact, as noticed previously, the testing setting

Table 5.2.3. *Posterior probabilities of $\theta = 0$ for $\tau^2 = 10\sigma^2$ and $z = x/\sigma$.*

z	0	0.68	1.28	1.96
$\pi(\theta = 0\|x)$	0.768	0.729	0.612	0.366

Table 5.2.4. *Posterior probabilities of $\|\theta\| < 1$.*

x	0.0	0.5	1.0	1.5	2.0
$\pi(\|\theta\| \leq 1\|x)$	0.683	0.625	0.477	0.302	0.157

is not coherent with an absolute lack of information, since it implies at least a division of the parameter space into two subsets, out of which at least one is of measure zero under the conventional Jeffreys prior. However, the inconvenience with the use of improper prior distributions goes deeper, since they are incompatible with most tests of point-null hypotheses.

We illustrate this difficulty in the normal setting, $x \sim \mathcal{N}(\theta, 1)$, for the point-null hypothesis $H_0 : \theta = 0$ to test against $H_1 : \theta \neq 0$. If we use the improper prior $\pi(\theta) = 1$ on $\{\theta \neq 0\}$, i.e., if π is

$$\pi(\theta) = \frac{1}{2}\mathbb{I}_0(\theta) + \frac{1}{2} \cdot 1,$$

the posterior probability of H_0 is

$$\pi(\theta = 0|x) = \frac{e^{-x^2/2}}{e^{-x^2/2} + \int_{-\infty}^{+\infty} e^{-(x-\theta)^2/2} \, d\theta} = \frac{1}{1 + \sqrt{2\pi}e^{x^2/2}}.$$

(The particular choice of the constant 1 in the prior is crucial for the following discussion, while being arbitrary, as seen below.) Therefore, this posterior probability of H_0 is bounded from above by $1/(1+\sqrt{2\pi}) = 0.285$. This implies that the posterior distribution is rather biased against H_0, even in the most favorable case. Unless the scale of comparison, that is, the loss, is modified to account for these low values, the null hypothesis will quite often be rejected . A similar phenomenon occurs when Θ_0 is compact. For instance, the test of $H_0 : |\theta| \leq 1$ versus $H_1 : |\theta| > 1$ leads to the following posterior probability:

$$\begin{aligned}
\pi(|\theta| \leq 1|x) &= \frac{\int_{-1}^{1} e^{-(x-\theta)^2/2} \, d\theta}{\int_{-\infty}^{+\infty} e^{-(x-\theta)^2/2} \, d\theta} \\
&= \Phi(1 - x) - \Phi(-1 - x) \\
&= \Phi(x + 1) - \Phi(x - 1),
\end{aligned}$$

whose numerical values are given in Table 5.2.5. Therefore, the maximal support of H_0, 0.683, is still moderate.

Table 5.2.5. *Posterior probabilities of $\theta = 0$ for the Jeffreys prior $\pi(\theta) = 1$.*

x	0.0	1.0	1.65	1.96	2.58
$\pi(\theta = 0\|x)$	0.285	0.195	0.089	0.055	0.014

An interesting feature of the Lebesgue prior distribution can be exhibited for the point-null hypothesis $H_0 : \theta = 0$. The resulting procedure agrees with the corresponding classical answer, as shown by Table 5.2.5. The posterior probability $\pi(\theta = 0|x)$ is indeed quite close to the classical significance levels 0.10, 0.05, and 0.01 when x is 1.65, 1.96, or 2.58 (it will be demonstrated in Section 5.3.4 that this comparison is meaningful). This coincidence does not hold for all values of x but shows that, for usual significance levels (and testing purposes), the classical answer could be considered as a noninformative Bayes answer, even though it corresponds to a hardly defendable prior.

Another illustration of the delicate issue of improper priors in testing settings is provided by the *Jeffreys–Lindley paradox*. In fact, limiting arguments are not valid in testing settings and prevent an alternative derivation of noninformative answers. For instance, considering the conjugate prior distributions introduced in Example 5.2.8, the posterior probabilities are

$$\pi(\theta = 0|x) = \left\{ 1 + \frac{1 - \varrho_0}{\varrho_0} \sqrt{\frac{\sigma^2}{\sigma^2 + \tau^2}} \exp\left[\frac{\tau^2 x^2}{2\sigma^2(\sigma^2 + \tau^2)} \right] \right\}^{-1},$$

which converge to 1 when the prior variance τ goes to $+\infty$, for every x. This limit differs from the "noninformative" answer derived previously $[1 + \sqrt{2\pi} \exp(x^2/2)]^{-1}$ and, more importantly, is totally useless. This phenomenon can also be observed by comparing Tables 5.2.4 and 5.2.4, since the probability is larger when $\tau^2 = 10\sigma^2$ than when $\tau = \sigma$ for all the values of z considered in the tables. See Aitkin (1991) and Robert (1993b) for recent discussions on this paradox.

Paradoxes associated with improper priors like the Jeffreys–Lindley example are actually because of a weighting indeterminacy that does not occur for point estimation, or even for one-sided tests.

Example 5.2.9 Consider $x \sim \mathcal{N}(\theta, 1)$ and $H_0 : \theta \leq 0$ to test versus $H_1 : \theta > 0$. For the diffuse distribution $\pi(\theta) = 1$,

$$
\begin{aligned}
\pi(\theta \leq 0|x) &= \frac{1}{\sqrt{2\pi}} \int_{-\infty}^{0} e^{-(x-\theta)^2/2} \, d\theta \\
&= \Phi(-x).
\end{aligned}
$$

In this case, the generalized Bayes answer is also the classical procedure, called the *p-value* (see Section 5.3.4). ‖

For two-sided problems, if g_0 and g_1 are σ-finite measures corresponding

Table 5.2.6. *Posterior probabilities of $\theta = 0$ for the Jeffreys prior $\pi(\theta) = 10$.*

x	0.0	1.0	1.65	1.96	2.58
$\pi(\theta = 0\|x)$	0.0384	0.0236	0.0101	0.00581	0.00143

to truncated noninformative priors on the subspaces Θ_0 and Θ_1, the choice of the normalizing constants will influence the Bayesian estimator. In fact, if g_i is replaced by $c_i g_i$ $(i = 0, 1)$, the Bayes factor is multiplied by c_0/c_1. For instance, if the Jeffreys prior is uniform and $g_0 = c_0$, $g_1 = c_1$, the posterior probability is

$$
\pi(\theta \in \Theta_0|x) = \frac{\varrho_0 c_0 \int_{\Theta_0} f(x|\theta)\, d\theta}{\varrho_0 c_0 \int_{\Theta_0} f(x|\theta)\, d\theta + (1 - \varrho_0) c_1 \int_{\Theta_1} f(x|\theta)\, d\theta}
$$

$$
= \frac{\varrho_0 \int_{\Theta_0} f(x|\theta)\, d\theta}{\varrho_0 \int_{\Theta_0} f(x|\theta)\, d\theta + (1 - \varrho_0)[c_1/c_0] \int_{\Theta_1} f(x|\theta)\, d\theta},
$$

which does depend on the ratio c_1/c_0.

It is therefore necessary to extend the noninformative perspective to these testing settings by developing a technique able to derive the weights c_i in a noninformative and acceptable way. Bernardo (1980), Spiegelhalter and Smith (1980), Smith and Spiegelhalter (1982), Aitkin (1991), Pettit (1992), Robert (1993b) and Berger and Pericchi (1996a,b) have made proposals in this direction, as detailed in Section 5.2.6. Notice that Jeffreys (1961) proposed instead to use proper priors in such settings, like $\mathcal{C}(0, \sigma^2)$ or $\mathcal{N}(0, 10\sigma^2)$ in the case of $x \sim \mathcal{N}(\theta, \sigma^2)$ and $H_0 : \theta = 0$. The problem is then that the choice of the proper prior distribution will influence the answer to the test. For instance, the equivalent of Table 5.2.5 for $\pi(\theta) = 10$ are given in Table 5.2.5, with important differences for most values of x, since they vary by an order of magnitude.

Before introducing in Section 5.2.6 some of the recent developments linked to the use of improper priors, let us make the following point: it still does not feel right to use improper priors, like the Jeffreys priors, for two-sided tests because they seem to lead to too much arbitrariness, in the sense that many competing solutions abound, with about the same theoretical background and with different numerical values, in contradiction with the Likelihood Principle. In other words, although the solutions proposed in the next section are interesting and overcome, as working principles, the difficulties related to the use of improper priors, they are not, strictly speaking, pertaining to the Bayesian paradigm. We consider in Section 5.3 an alternative approach which defines a least favorable Bayesian answer as a lower bound on the (proper) Bayes estimators.

The difficulties encountered with noninformative priors in testing settings also point out that a testing problem cannot be treated in a coherent way

if no prior information is available, that is, that the information brought by
the observations alone is usually not enough to infer about the truth of a
hypothesis in a categorical fashion (*yes/no*). This obviously reinforces the
motivation for a Bayesian treatment of such testing problems, since it is
the only coherent approach taking advantage of the residual information.

5.2.6 Pseudo-Bayes factors

Most[4] of the solutions proposed to overcome the difficulties related to the
use of improper priors (see Section 5.2.5) are based on the idea to use part
of the data/information to transform the priors into proper distributions,
or to call to imaginary observations to obtain the same effect.

Definition 5.2.10 *Given an improper prior π, a sample $(x_1, \ldots, x_n)$ is a
training sample if the corresponding posterior $\pi(\cdot|x_1, \ldots, x_n)$ is proper and
is a minimal training sample if no subsample is a training sample.*

Example 5.2.11 For the $\mathcal{N}(\mu, \sigma^2)$ model, the minimal training sample
size associated with the improper prior $\pi_0(\mu, \sigma^2) = 1/\sigma^2$ is 2 since

$$\int e^{-\{(x_1-\mu)^2+(x_2-\mu)^2\}/2\sigma^2} \sigma^{-4} d\mu \, d\sigma^2$$

$$= \int_0^\infty \sigma^{-3} e^{-s^2/2\sigma^2} d\sigma^2$$

$$= \int_0^\infty \omega^{3/2-2} e^{-s^2\omega/2} d\omega \,,$$

while

$$\int e^{-(x_1-\mu)^2/2\sigma^2} \sigma^{-3} d\mu \, d\sigma^2 = \infty \,.$$

If we now consider the prior $\pi_1(\mu, \sigma^2) = 1/\sigma$, the minimal training sample
size is 3 since

$$\int e^{-\{(x_1-\mu)^2+(x_2-\mu)^2\}/2\sigma^2} \sigma^{-3} d\mu \, d\sigma^2$$

$$= \int_0^\infty \sigma^{-2} e^{-s^2/2\sigma^2} d\sigma^2$$

$$= \int_0^\infty \omega^{-1} e^{-s^2\omega/2} d\omega = \infty \,,$$

which is a good argument in favor of π_0 against π_1.   ‖

The idea is then to use a minimal training sample, $x_{(\ell)}$ say, to "properize"
the improper prior π into $\pi(\cdot|x_{(\ell)})$, and then to use this posterior distribu-
tion *as if* it were a regular proper prior for the remainder of the sample,

---

[4] This section, which may be skipped on a first reading, contains more advanced material on the so-called intrinsic priors developed by Berger and Pericchi (1996a,b). It
will not be used in the rest of the book, except in Chapter 7. See Berger and Pericchi
(2001) for a much more detailed review, upon which this section is based.

$x_{(-\ell)}$ say, in order to avoid the data twice, as in the proposal of Aitkin (1991). When facing a hypothesis H_0 with a prior distribution π_0, with a broader alternative H_1 with a prior distribution π_1, if the minimal training sample under H_1 is such that $\pi_0(\cdot|x_{(\ell)})$ is also proper, the *pseudo-Bayes factor*

$$(5.2.5) \qquad B_{10}^{(\ell)} = \frac{\int_{\Theta_1} f_1(x_{(-\ell)}|\theta_1)\pi_1(\theta_1|x_{(\ell)})d\theta_1}{\int_{\Theta_0} f_0(x_{(-\ell)}|\theta_0)\pi_0(\theta_0|x_{(\ell)})d\theta_0}$$

is then independent from the normalizing constants used in both π_0 and π_1. To see this more clearly, consider the following representation proposed in Berger and Pericchi (2001).

Lemma 5.2.12 *In the case of independent distributions, the pseudo-Bayes factor can be written as*

$$(5.2.6) \qquad B_{10}^{(\ell)} = B_{10}(x) \times B_{01}(x_{(\ell)}),$$

with

$$B_{10}(x) = \frac{\int_{\Theta_1} f_1(x|\theta_1)\pi_1(\theta_1)d\theta_1}{\int_{\Theta_0} f_0(x|\theta_0)\pi_0(\theta_0)d\theta_0}$$

and

$$B_{01}(x_{(\ell)}) = \frac{\int_{\Theta_0} f_0(x_{(\ell)}|\theta_0)\pi_0(\theta_0)d\theta_0}{\int_{\Theta_1} f_1(x_{(\ell)}|\theta_1)\pi_1(\theta_1)d\theta_1}.$$

In this decomposition, $B_{10}(x)$ and $B_{01}(x_{(\ell)})$ are the Bayes factors computed for the unnormalized priors π_1 and π_0, for the whole sample x and the training sample $x_{(\ell)}$, respectively, as if these were regular priors. It is then straightforward to see that multiplying π_0 by c_0 and π_1 by c_1 has no influence on $B_{10}^{(\ell)}$ because these constants cancel. Notice the interesting inversion from $B_{10}(x)$ to $B_{01}(x_{(\ell)})$: the training sample effect is removed from the Bayes factor $B_{10}(x)$.

While the normalizing constant problem disappears, a remaining difficulty is that the solution $B_{10}^{(\ell)}$ is not strictly Bayesian. Moreover, besides sequential sampling, there is no obvious choice for $x_{(\ell)}$, while this choice of the training sample influences the resulting value of $B_{10}^{(\ell)}$ (and thus violates the Likelihood Principle).

Example 5.2.13 (Example 5.2.11 continued) If $H_0 : \mu = 0$, with $\pi_0(\sigma^2) = 1/\sigma^2$, and $H_1 : \mu \neq 0$, with $\pi_1(\mu,\sigma^2) = 1/\sigma^2$, the minimal training sample of size 2 under H_1. Then

$$\pi_1(\mu,\sigma^2|x_1,x_2) = \frac{1}{\sigma}\exp\{-2(\mu-\bar{x}_1)^2/2\sigma^2\}s_1^5\sigma^{-3}e^{-s_1^2/2\sigma^2}$$

and

$$\pi_0(\sigma^2|x_1,x_2) = \frac{s_0^6}{\sigma^4}e^{-s_0^2/2\sigma^2},$$

with the following notations:

$$\bar{x}_1 = \frac{x_1+x_2}{2}, \quad s_1^2 = \frac{(x_1-x_2)^2}{2}, \quad s_0^2 = x_1^2 + x_2^2.$$

Then

$$B_{10}^{(2)} = \frac{s_1^5 \int e^{-\{(n-2)(\bar{x}_2-\mu)^2-2(\mu-\bar{x}_1)^2-s_2^2-s_1^2\}/2\sigma^2}\sigma^{-n-2}d\mu d\sigma^2}{s_0^6 \int_0^\infty e^{-\{-s_3^2-s_0^2\}/2\sigma^2}\sigma^{-n-2}d\sigma^2}$$

depends on the choice of (x_1, x_2) through $(\bar{x}_1 - \bar{x}_2)^2$, s_1^2, s_0^2 (see Exercise 5.14). ‖

A way to remove this dependence on the training sample is to average the different pseudo-Bayes factors (5.2.6) over all the possible training samples $x_{(\ell)}$. The next difficulty is to decide what kind of averaging should be used here. For instance, Berger and Pericchi (1996a, 1998, 2001) list

- the *arithmetic intrinsic Bayes factor,*

$$(5.2.7) \qquad B_{10}^A = \frac{1}{L}\sum_{x_{(\ell)}} B_{10}^{(\ell)} = B_{10}(x)\frac{1}{L}\sum_{x_{(\ell)}} B_{01}(x_{(\ell)}),$$

  where L is the number of different training samples;
- the *geometric intrinsic Bayes factor,*

$$(5.2.8)\, B_{10}^G = \exp\frac{1}{L}\sum_{x_{(\ell)}} \log B_{10}^{(\ell)} = B_{10}(x)\exp\frac{1}{L}\sum_{x_{(\ell)}} \log B_{01}(x_{(\ell)});$$

  and
- the *median intrinsic Bayes factor,*

$$(5.2.9) \qquad B_{10}^M = \text{med}\, B_{10}^{(\ell)} = B_{10}(x)\text{med}\, B_{01}(x_{(\ell)}),$$

  where med $B_{10}^{(\ell)}$ denotes the median of the $B_{10}^{(\ell)}$ over the different training samples.

Although these solutions all come close to a Bayesian answer, in particular by using the data only once (Exercise 5.15) by separating the part used to "properize" the improper prior and the part used to run the test, none of the above is a true Bayesian answer. We will discuss below more serious drawbacks to these different intrinsic Bayes factors, but it appears that they sometimes correspond to genuine Bayes factors under different prior distributions, called *intrinsic priors* in Berger and Pericchi (1996a, 1998).[5] (This phenomenon will also occur in Section 5.3.5 with Berger and Sellke's (1987) lower bounds.)

Example 5.2.14 (Berger and Pericchi (1998)) In the $\mathcal{N}(\theta, 1)$ case, when $H_0 : \theta = 0$ and $\pi_1(\theta) = 1$, for a sample $(x_1, \ldots, x_n)$, the arithmetic intrinsic Bayes factor,

$$B_{10}^A = B_{10}(x)\frac{1}{\sqrt{2\pi}}\frac{1}{n}\sum_{i=1}^n e^{-x_i^2/2},$$

---

[5] The name *intrinsic* associated with the Bayes factor and the corresponding prior tries to convey an idea of quantities solely derived from the distribution, but the diversity of possible answers shows this is somehow a misnomer!

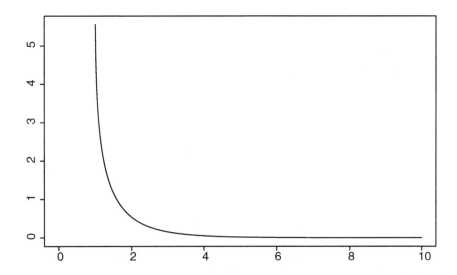

Figure 5.2.1. *Graph of the intrinsic prior associated with the exponential testing,* when $\theta_0 = 1$.

is almost identical to the regular Bayes factor associated with a normal $\mathcal{N}(0,2)$ prior under H_1.   ‖

Example 5.2.15 (Berger and Pericchi (1998)) For $x_1, \ldots, x_n$ i.i.d. observations from a translated exponential distribution, with density $\exp(\theta - x)\mathbb{I}_{x \geq \theta}$, if $H_0 : \theta = \theta_0$ and $H_1 : \theta > \theta_0$, with $\pi_1(\theta) = 1$,

$$B_{10}^A = B_{10}(x)\frac{1}{n}\sum_{i=1}^{n}\left[e^{x_i - \theta_0} - 1\right]^{-1},$$

which corresponds to the regular Bayes factor associated with the proper prior

$$\pi_2(\theta) = e^{\theta_0 - \theta}\left\{1 - \log\left(1 - e^{\theta_0 - \theta}\right)\right\},$$

which behaves as indicated on Figure 5.2.6.   ‖

O'Hagan (1995) introduces an alternative to intrinsic Bayes factors that avoids the selection of—and the subsequent averaging over—training samples. His idea is to use a *fraction b* of the likelihood to properize the prior, that is, to take $0 < b < 1$ such that

$$\int_{\Theta_0} f_0(x|\theta_0)^b \pi_0(\theta_0)d\theta_0 < \infty$$

and
$$\int_{\Theta_1} f_1(x|\theta_1)^b \pi_1(\theta_1)d\theta_1 < \infty .$$

The remaining fraction $(1-b)$ of the likelihood is then used to run the test, as in the intrinsic Bayes factor case. The *fractional Bayes factor* is thus defined as

$$B_{10}^F = \frac{\int_{\Theta_1} f_1(x|\theta_1)^{1-b}\pi_1^b(\theta_1|x)d\theta_1}{\int_{\Theta_0} f_0(x|\theta_0)^{1-b}\pi_0^b(\theta_0|x)d\theta_0}$$

(5.2.10)
$$= B_{10}(x)\frac{\int_{\Theta_0} f_0(x|\theta_0)^b\pi_0(\theta_0)d\theta_0}{\int_{\Theta_1} f_1(x|\theta_1)^b\pi_1(\theta_1)d\theta_1} ,$$

where $\pi_0^b(\theta_0|x)$ and $\pi_1^b(\theta_1|x)$ denote the pseudo-posteriors associated with $f_0(x|\theta_0)^b$ and $f_1(x|\theta_1)^b$, respectively. For exponential families, the fraction b clearly corresponds to a training sample size, since for n observations from an exponential family with sufficient statistic T ,

$$\exp\{\theta \cdot n\, T(x) - n\Psi(\theta)\}^b = \exp\{\theta \cdot [bn]\, T(x) - [bn]\Psi(\theta)\} .$$

For other distributions, the fraction b must be determined through a more ad-hoc approach (see O'Hagan (1995, 1997)).

As in the intrinsic Bayes factor case, there exist occurrences where this solution is equal to a regular Bayes factor with a corresponding "intrinsic" prior.

Example 5.2.16 (Example 5.2.14 continued) For any $0 < b < 1$,

$$B_{10}^F = \frac{\int e^{-n(1-b)(\bar{x}-\theta)^2/2}\sqrt{b}e^{-nb((\bar{x}-\theta)^2/2}d\theta}{\sqrt{2\pi}e^{-n(1-b)\bar{x}^2/2}}$$

(5.2.11)
$$= \sqrt{b}e^{n(1-b)\bar{x}^2/2} ,$$

which is equal to the Bayes factor associated with the proper $\theta \sim \mathcal{N}(0, (1-b)/nb)$ under H_1. ‖

There are, however, enough difficulties with these pseudo-Bayes factors to make us question their use in testing and model choice problems:

(i) Bayes factors, when associated with proper priors, do satisfy some *coherence* properties such as

$$B_{12} = B_{10}B_{02} \qquad \text{and} \qquad B_{01} = 1/B_{10} .$$

Most pseudo-Bayes factors do not, even though the fractional Bayes factor satisfies $B_{01}^F = 1/B_{10}^F$.

(ii) When the pseudo-Bayes factors can be expressed as true Bayes factors, the corresponding intrinsic priors are not necessarily appealing, as shown in Example 5.2.14 for the arithmetic Bayes factor and in Example 5.2.16 for the fractional Bayes factor, and these priors do depend on the choice of the improper reference priors π_0 and π_1, hence are hardly intrinsic.

(iii) Following the above point, the pseudo-Bayes factors may also exhibit a bias in favor of one of the hypotheses, in the sense that they can be expressed as a true Bayes factor multiplied by a certain factor.

Example 5.2.17 (Example 5.2.15 continued) For the median intrinsic Bayes factor,

$$B_{10}^M = B_{10}(x) \left[e^{\text{med}(x_i)} - \theta_0 \right]^{-1}$$

(5.2.12)
$$= 0.69 \tilde{B}_{10}(x)$$

where $\tilde{B}_{10}(x)$ is the Bayes factor associated with the prior $\pi_3(\theta) \propto (2\exp\{\theta - \theta_0\} - 1)^{-1}$, which, while being similar to π_2, does not provide exactly the same coverage of regions near 1.   ‖

In such cases, the pseudo-Bayes factors can be seen as modifying the probabilities of both hypotheses from the reference value $1/2$, a feature we will also encounter for least favorables bounds in Section 2.5.2.

(iv) Most often, however, the pseudo-Bayes factors do not correspond to any true Bayes factor, and they may give strongly biased solutions. For instance, Berger and Pericchi (2001) state that arithmetic intrinsic Bayes factors do not have intrinsic priors for most one-sided testing problems.

Example 5.2.18 (Example 5.2.15 continued) The fractional Bayes factor,

(5.2.13)
$$B_{10}^F = B_{10}(x) b n \left\{ e^{-bn(x_{(1)} - \theta_0)} - 1 \right\}^{-1},$$

is always larger than 1, thus, always favors the alternative hypothesis, according to Jeffreys' scale. This paradoxical behavior can be related to the fact that the fraction b does not modify the indicator function.   ‖

(v) Pseudo-Bayes factors may simply not exist for a whole class of models.

Example 5.2.19 Mixtures of normal distributions

$$p\mathcal{N}(\mu_1, \sigma_1^2) + (1 - p)\mathcal{N}(\mu_2, \sigma_2^2)$$

were presented in Example 1.1.6. As shown in Exercise 1.57, improper priors of the form $\pi_1(\mu_1, \sigma_1)\pi_2(\mu_2, \sigma_2)$ cannot be used in this setting, whatever the sample size n. (The fundamental reason for this is that there is a probability $(1 - p)^n$ that no observation is associated with the first component $\mathcal{N}(\mu_1, \sigma_1^2)$.) Therefore, standard noninformative priors do not allow for training samples and intrinsic Bayes factors cannot be constructed. The same applies to fractional Bayes factors (see Exercise 5.21).   ‖

(vi) As shown in this section, there are many ways of defining pseudo-Bayes factors and, while most are arguably logical, there is no coherent way of ordering them. Pseudo-Bayes factors, as defined here, do agree with the Likelihood Principle, but the multiplication of possible answers, even if those are close, is not a good signal to users.[6] Similarly, there is no clear-cut procedure for the choice of the fraction b in the fractional Bayes factors, since the minimal training sample size is not always clearly defined.

(vii) The issue of computing pseudo-Bayes factors has not been mentioned so far, for lack of appropriate tools, which will be introduced in Chapters 6 and 7. But each Bayes factor $B_{10}^{(\ell)}$ may be a complex integral and the derivation of the averaged intrinsic Bayes factor may involve $\binom{m}{n}$ integrals of this kind, if m is the minimal training sample size. Fractional Bayes factors are easier to compute in exponential settings, but other distributions are much more difficult to handle (Exercise 5.22).

5.3 Comparisons with the classical approach

5.3.1 UMP and UMPU tests

The classical approach to testing theory is the theory of Neyman–Pearson, as presented, for instance, in Lehmann (1986). For the $0 - 1$ loss, denoted L below, the frequentist notion of optimality is based on the power of a test, defined as follows:

Definition 5.3.1 *The power of a testing procedure φ is the probability of rejecting H_0 under the alternative hypothesis, that is, $\beta(\theta) = 1 - \mathbb{E}_\theta[\varphi(x)]$ when $\theta \in \Theta_1$. The quantity $1 - \beta(\theta)$ is called type–two error, while the type–one error is $\mathbb{E}_\theta[\varphi(x)]$ when $\theta \in \Theta_0$.*

Optimal frequentist tests are then those that minimize the risk $\mathbb{E}_\theta[\mathrm{L}(\theta, \varphi(x))]$ *only under H_1:*

Definition 5.3.2 *If $\alpha \in]0,1[$ and $\mathcal{C}_\alpha$ is the class of the procedures φ satisfying the following constraint on the type I error:*

$$\sup_{\theta \in \Theta_0} \mathbb{E}_\theta[\mathrm{L}(\theta, \varphi(x))] = \sup_{\theta \in \Theta_0} P_\theta(\varphi(x) = 0) \le \alpha,$$

a test procedure φ is said to be uniformly most powerful at level α (UMP) if it minimizes the risk $\mathbb{E}_\theta[\mathrm{L}(\theta, \varphi(x))]$ uniformly on Θ_1 in $\mathcal{C}_\alpha$.

This optimality is much weaker than the notion of admissibility developed in Section 2.4. In fact, the loss is *bidimensional* in this setting because of the restriction on the *type I error*, namely, $\sup_{\Theta_0} \mathbb{E}_\theta[\mathrm{L}(\theta, \varphi)] \le \alpha$. This

---

[6] Berger and Pericchi (2001) argue that the multiplicity of intrinsic Bayes factors is no more worrying than the multiplicity of possible default priors. The parallel is slightly flawed, though, because *every* default prior induces a multiplicity of intrinsic Bayes factors!

restriction is usually necessary to obtain an optimal test procedure, since the risk functions of admissible procedures do cross, but:

(i) It produces a situation of asymmetry between the null and the alternative hypotheses, which induces an unnatural behavior for the test procedures. In fact, since the type I error is fixed, a balance between the two types of error (acceptance under H_1 and rejection under H_0) is impossible, hence a much bigger type II error. This asymmetry is also responsible for the theory bypassing minimaxity considerations. For instance, this is the case when the two hypotheses H_0 and H_1 are *contiguous*, that is, when it is possible to go from Θ_0 to Θ_1 by a continuous transformation.

(ii) It implies the selection of a *significance level* α by the decision maker, in addition to the choice of the loss function L, and this generally leads to the call to "standard" levels, 0.05 or 0.01, and the drawbacks of these "universal" levels (see below).

(iii) It does not necessarily imply a sufficient reduction of the class of test procedures and does not always allow for the selection of a unique optimal procedure. It is sometimes necessary to impose further constraints on these classes.

In the simplest case, in which null and alternative hypotheses are point hypotheses, $H_0 : \theta = \theta_0$ versus $H_1 : \theta = \theta_1$, the *Neyman–Pearson lemma* establishes that there exist UMP test procedures and that they are of the form[7]

$$\varphi(x) = \begin{cases} 1 & \text{if } f(x|\theta_1) < kf(x|\theta_0), \\ 0 & \text{otherwise,} \end{cases}$$

k being related to the selected significance level α. Obviously, the fact that Θ_1 is reduced to $\{\theta_1\}$ is quite helpful, since it allows for a total ordering of the procedures of $\mathcal{C}_\alpha$. For *monotone likelihood ratio* families, that is, parametrized families for which there exists a statistic $T(x)$ such that

$$\frac{f(x|\theta')}{f(x|\theta)}$$

is increasing in $T(x)$ for $\theta' > \theta$, Karlin and Rubin (1956) have established the following extension of the Neyman–Pearson lemma (see Lehmann (1986, p. 79) for a proof).

Proposition 5.3.3 *Consider $f(x|\theta)$ with a monotone likelihood ratio in $T(x)$. For $H_0 : \theta \leq \theta_0$ and $H_1 : \theta > \theta_0$ there exists a UMP test such that*

$$\varphi(x) = \begin{cases} 1 & \text{if } T(x) < c, \\ \gamma & \text{if } T(x) = c, \\ 0 & \text{otherwise,} \end{cases}$$

---

[7] Conserving the interpretation that a test procedure is an estimator of $\mathbb{I}_{\Theta_0}(\theta)$, the test procedures in this book are complements to 1 of the classical Neyman–Pearson procedures, for which a value of 1 corresponds to the *rejection* of H_0.

γ *and* c *being determined by the constraint*

$$\mathbb{E}_{\theta_0}[\varphi(x)] = \alpha.$$

Karlin and Rubin (1956) have also shown that, for the loss functions of the class (5.2.1), the test procedures provided in Theorem 5.3.3 form an *essentially complete class*, that is, a class of procedures large enough to be at least as good as any other procedure (see Chapter 8). Moreover, if the support of the distribution $f(x|\theta)$ is independent of θ, the class obtained in Proposition 5.3.3 is *minimal essentially complete*: it cannot be reduced any further (see Lehmann (1986, pp. 82-83)), and therefore only contains optimal procedures.

Notice that an important class of monotone likelihood ratio families consists of the exponential families, since

$$\frac{f(x|\theta')}{f(x|\theta)} = \frac{e^{\theta'x - \psi(\theta')}}{e^{\theta x - \psi(\theta)}} = \frac{e^{(\theta'-\theta)x}}{e^{\psi(\theta') - \psi(\theta)}}$$

is increasing in x. Pfanzagl (1968) has also established a reciprocal to Proposition 5.3.3 in the spirit of the Pitman–Koopman lemma (see Section 3.3.3), namely, that the existence of a UMP test for every sample size and a given level α implies that the distribution belongs to an exponential family.

Example 5.3.4 Consider $x \sim \mathcal{P}(\lambda)$ and $H_0 : \lambda \leq \lambda_0$, $H_1 : \lambda > \lambda_0$. For m independent observations from this distribution, a sufficient statistic is $s = \sum_i x_i \sim \mathcal{P}(m\lambda)$ and, according to Proposition 5.3.3, a UMP test is given by

$$\varphi(x) = \begin{cases} 1 & \text{if } s < k, \\ \gamma & \text{if } s = k, \\ 0 & \text{otherwise,} \end{cases}$$

for $\mathbb{E}_{\lambda_0}[\varphi(x)] = P_{m\lambda_0}(s > k) + \gamma P_{m\lambda_0}(s = k) = \alpha$.   ‖

Proposition 5.3.3 and the above example stresses a major difficulty with the Neyman–Pearson approach, namely, that arbitrary significance levels are not necessarily attainable unless one calls for *randomization*. Indeed, as the decision space is $\mathcal{D} = \{0, 1\}$, $\varphi(x) = \gamma$ means that $\varphi(x) = 1$ with probability γ (and 0 otherwise). Such procedures are obviously incompatible with the Likelihood Principle, although they only appear for discrete cases. Lehmann (1986) indicates that the significance level α should be modified until randomization is avoided, but this modification induces another drawback: the choice of the significance level depends on the observation, not on a utility function.

Moreover, Proposition 5.3.3 only applies to one-sided hypotheses. In a particular case of two-sided hypotheses, we can exhibit an optimality result (see Lehmann (1986, pp. 101–103)).

Proposition 5.3.5 *Consider an exponential family*

$$f(x|\theta) = e^{\theta T(x) - \psi(\theta)} h(x)$$

and $H_0 : \theta \leq \theta_1$ *or* $\theta \geq \theta_2$, $H_1 : \theta_1 < \theta < \theta_2$. *There exists a UMP test of the form*

$$\varphi(x) = \begin{cases} 0 & \text{if } c_1 < T(x) < c_2, \\ \gamma_i & \text{if } T(x) = c_i \quad (i = 1, 2), \\ 1 & \text{otherwise,} \end{cases}$$

with $(i = 1, 2)$

$$\mathbb{E}_{\theta_i}[\varphi(x)] = \alpha.$$

However, there is no corresponding UMP test for the opposite case, i.e., $H_0 : \theta_1 \leq \theta \leq \theta_2$. This paradox illustrates quite forcibly the lack of symmetry—and therefore of coherence—of the UMP criterion, but is also quite puzzling, and casts a doubt on the validity of the Neyman–Pearson analysis or the relevance of a symmetric loss like the $0 - 1$ loss. In these cases, the Neyman–Pearson solution is to propose an additional reduction of the class of test procedures by considering *unbiased* tests, i.e., those also satisfying

$$\sup_{\Theta_0} P_\theta(\varphi(x) = 0) \leq \inf_{\Theta_1} P_\theta(\varphi(x) = 0).$$

In other words, φ must also satisfy

$$\inf_{\Theta_0} \mathbb{E}_\theta[\varphi(x)] \geq \sup_{\Theta_1} \mathbb{E}_\theta[\varphi(x)].$$

The notion of *uniformly most powerful unbiased* tests (UMPU) then follows. However, this restriction induces a further asymmetry between H_0 and H_1. Although intuitively acceptable, this notion of unbiasedness is yet another example of the restrictions imposed by the frequentist approach to optimality which denaturate the true purpose of Decision Theory.

Example 5.3.6 If, for $x \sim \mathcal{N}(\theta, 1)$, we test $H_0 : \theta = 0$ versus $H_1 : \theta \neq 0$, there is no UMP test. A UMPU test at level $\alpha = 0.05$ is

$$\varphi(x) = \begin{cases} 1 & \text{if } |x| \leq 1.96, \\ 0 & \text{otherwise.} \end{cases} \qquad \parallel$$

5.3.2 Least favorable prior distributions

When no UMPU test is available, it gets quite difficult for the classical approach to defend, or even construct, a specific testing procedure. Apart from restricting any further the class of acceptable procedures, a customary approach is to consider the likelihood ratio

(5.3.1)
$$\frac{\sup_{\theta \in \Theta_0} f(x|\theta)}{\sup_{\theta \in \Theta_1} f(x|\theta)}$$

and its distribution, or to base the test upon the asymptotic distribution of (5.3.1). This ratio exhibits a link with the Bayesian approach, since, as mentioned above, it appears as a Bayes factor for a prior distribution π that is supported by $\hat{\theta}_0$ and $\hat{\theta}_1$, maximum likelihood estimators of θ on Θ_0 and Θ_1. This analysis is rather formal, since Dirac masses are artificial priors and, moreover, the $\hat{\theta}_i$'s depend on the observation. However, it also indicates that the likelihood ratio has a Bayesian motivation.

Relations between Bayesian testing procedures and Neyman–Pearson optimal tests are given in Lehmann (1986) through the notion of a *least favorable distribution*, described below.[8] Consider $H_0 : \theta \in \Theta_0$, $H_1 : \theta = \theta_1$ with π a prior distribution on Θ_0. From a Bayesian point of view, the test problem can be represented as the test of $H_\pi : x \sim m_\pi$ versus $H_1 : x \sim f(x|\theta_1)$, where m is the marginal distribution under H_0

$$m_\pi(x) = \int_{\Theta_0} f(x|\theta)\pi(\theta)\, d\theta.$$

Since both hypotheses (H_π and H_1) are point hypotheses, the Neyman–Pearson lemma ensures the existence of a UMP test φ_π, at significance level α, with power $\beta_\pi = P_{\theta_1}(\varphi_\pi(x) = 0)$. This test is of the form

$$\varphi_\pi(x) = \begin{cases} 1 & \text{if } m_\pi(x) > kf(x|\theta_1), \\ 0 & \text{otherwise.} \end{cases}$$

Definition 5.3.7 *A least favorable distribution is any prior distribution π which maximizes the power β_π.*

This definition is used in the following result (Lehmann (1986, p. 105)).

Theorem 5.3.8 *Consider $H_0 : \theta \in \Theta_0$ to be tested against a point alternative $H_1 : \theta = \theta_1$. If the UMP test φ_π at level α for H_π versus H_1 satisfies*

$$\sup_{\theta \in \Theta_0} \mathbb{E}_\theta[L(\theta, \varphi_\pi)] \le \alpha,$$

then

(i) *φ_π is UMP at level α;*

(ii) *if φ_π is the unique α-level test of H_π versus H_1, φ_π is the unique UMP test at level α to test H_0 versus H_1; and*

(iii) *π is a least favorable distribution.*

The constraint in the above theorem may seem unnecessary, but notice that φ_π is defined by

$$\int_{\{m_\pi(x) > kf(x|\theta_1)\}} m_\pi(x)\, dx = \alpha.$$

---

[8] The remainder of Section 5.3.2 is not used later. The connection stressed here is of lesser importance than the corresponding relation obtained in minimaxity theory (see Section 2.4.3). Moreover, it only applies to specific cases and does not validate any further the classical answers, which still cannot be obtained as limits of Bayesian procedures (see Section 5.4).

This relation does not guarantee that $\mathbb{E}_\theta[\mathrm{L}(\theta, \varphi_\pi)] \leq \alpha$ for every $\theta \in \Theta_0$.

5.3.3 Criticisms

Theorem 5.3.8 exhibits a connection between Bayesian and UMP tests in the same vein that least favorable distributions lead to minimax estimators in point estimation problems with a value (see Section 2.4), although the Bayes procedure corresponds to a modified testing problem involving π. We do not pursue this connection any further because, like other authors, we oppose the Neyman–Pearson approach as a whole. Indeed, in addition to the randomization problems mentioned above, a major drawback of this perspective is to restrict the decision space to the couple $\{0, 1\}$, thus to force a categorical decision. It seems to us that a more adaptive answer is preferable. Moreover, UMP (and UMPU) tests, when they exist, depend on an evaluation measure (the significance level α) not reconsidered after the observation. For instance, in Example 5.3.6, the level being fixed at 0.05, the classical answer is identical for $x = 1.96$ and $x = 100$. From a purely decision-theoretic point of view, it also seems paradoxical to force the inferential procedures into a restricted framework, since it can (and does) lead to suboptimal procedures. In particular, the notion of *unbiasedness*, which has been successfully removed from point estimation tools by the Stein effect (Note 2.8.2), should also disappear from the testing machinery.

A more fundamental criticism of the Neyman–Pearson approach (and, basically, of every frequentist approach) is that it bases rejection of H_0 on *unlikely events that did not occur*, to quote Jeffreys (1961). In fact, a UMP rejection region is of the form

$$\{T(X) \geq T(x)\}$$

if the distribution has monotone likelihood ratio in T, since, under the null hypothesis,

(5.3.2) $$P(T(X) \geq T(x)) < \alpha.$$

However, the event which actually occurs is $\{T(X) = T(x)\}$. There is therefore a loss of information in the (classical) decision process which is usually biased against the null hypothesis. Indeed, the region $\{T(X) \geq T(x)\}$ is relatively more unlikely than a neighborhood of $T(x)$, thus explaining the more optimistic values of the Bayesian answers (see Section 5.3.5). Of course, the only coherent approach which allows for conditioning on $\{T(X) = T(x)\}$, that is, on the observation, is the Bayesian approach. On the contrary, to choose a procedure on the basis of (5.3.2) involves the whole distribution of x, and thus potentially contradicts the Likelihood Principle, as shown by Examples 1.3.4 and 1.3.6. In fact, the Stopping Rule Principle cannot allow for a frequency-based theory of tests because the distribution of the sample size should not be relevant for the selection of a testing procedure. There is indeed this apparent paradox with the Likelihood Principle that a procedure based on a likelihood ratio is acceptable provided it does not involve the distribution of this ratio.

Example 5.3.9 The *chi-squared test* is a simple (if not always justified) procedure to test the goodness-of-fit of a sample to a distribution (or to a family of distributions). If the sample of size n is divided into k classes, with theoretical sizes $N_i = np_i$ and observed sizes n_i, it follows from the Central Limit Theorem that

$$D^2 = \sum_{i=1}^{k} \frac{(n_i - N_i)^2}{N_i}$$

is approximately distributed as a χ_ℓ^2 random variable, the degrees of freedom ℓ depending on the problem. (It is usually $k-1$ minus the number of estimated parameters.) As pointed out by Jeffreys (1961), the classical approach rejects the null hypothesis (of goodness-of-fit to the proposed family of distributions) if D^2 is too large, for instance, if

$$P(z > D^2) < 0.05$$

for $z \sim \chi_\ell^2$. However, there is no reason to accept the null hypothesis (that is, that D^2 is approximately distributed as χ_ℓ^2) if

$$P(z < D^2) \le 0.05,$$

since such values of D^2 are not more compatible with the distribution than when $P(z > D^2) \le 0.05$. From this point of view, it would also be justified to reject the validity of the null hypothesis, but the classical approach fails to do so. ‖

Example 5.3.10 A well known Bayesian criticism of Neyman–Pearson theory is the following opposition exhibited by Lindley (1957, 1961). Consider $\bar{x}_n \sim \mathcal{N}(0, 1/n)$ the average of a normal sample and $\theta \sim \mathcal{N}(0,1)$. To test $H_0 : \theta = 0$ versus $H_1 : \theta \ne 0$, the corresponding UMPU tests only depend on $z_n = |x_n|\sqrt{n}$. Consider $z_n = 1.97$. At the significance level 5%, the test procedure rejects H_0 for every n. On the contrary, the Bayesian posterior probability of H_0 is (see Example 5.2.4)

$$\pi(\theta = 0 | z_n) = \left(1 + \frac{1 - \varrho_0}{\varrho_0} \frac{1}{\sqrt{n+1}} \exp\{z_n^2 n / 2(n+1)\} \right)^{-1},$$

and thus goes to 1 as n goes to infinity. Actually, this result holds for most prior distributions, owing to the asymptotic normality of the posterior distributions (see Hartigan (1983)). This paradox can be related to *Kepler's problem* (see Jeffreys (1961) or Berger (1985a)), which is that, in astronomy, a null hypothesis—for instance, the elliptical nature of planet trajectories—is always rejected from a frequentist point of view for a sample size large enough, i.e., when enough observations have accumulated. ‖

Another major difficulty with the Neyman–Pearson approach is that the selection of the level α should be equivalent to the selection of the weights

a_0 and a_1 in the loss function, and thus should be based on utility considerations. Instead, the practice is to completely bypass this selection and, following a suggestion by Fisher (1956), it became a formal rule to adapt "classical" α-levels of 5% or 1%, no matter what the problem, the sample size, or the type II error were. Since the Neyman–Pearson approach is quite predominant nowadays, this dogmatic attitude created a publication bias, since results of experiments which are not *"significant at level 5%"* are most often rejected by editors or even censored by the authors themselves in many fields, including biology, medicine, and the social sciences.

5.3.4 The p-values

Frequentists (and practitioners) tried to compensate for the drawbacks of the Neyman–Pearson approach by removing the significance level α and proposing an answer taking values in $[0, 1]$, and more importantly depending on the observations on a more adaptive way than comparing $T(x)$ with a given threshold separating acceptance and rejection. The following notion was first introduced by Fisher (1956).

Definition 5.3.11   *The p-value associated with a test is the smallest significance level α for which the null hypothesis is rejected.*

A general definition for point-null hypotheses (see Thompson (1989)) considers that a p-value is any statistic with a uniform distribution under the null hypothesis, but it leads to the difficult problem of selecting one of these statistics, even though the same can be said about the *test* introduced in the above definition. Actually, if a test with rejection region R_α is available for every significance level α and if these regions are nested (that is, $R_\alpha \subset R_\beta$ if $\beta > \alpha$), the procedure

$$p(x) = \inf\{\alpha;\ x \in R_\alpha\}$$

is uniformly distributed if $\mathbb{E}_{\theta_0}[\mathbb{I}_{R_\alpha}(x)] = \alpha$ (see Goutis et al. (1996)). In the event of several contending tests, we suggest using the distribution of the likelihood ratio under the null hypothesis, if it is a point-null hypothesis.

Example 5.3.12 (Example 5.3.6 continued)   Since the critical region (i.e., the rejection region for H_0) of the UMPU test is $\{|x| > k\}$, an usual p-value is

$$
\begin{aligned}
p(x) &= \inf\{\alpha;\ |x| > k_\alpha\} \\
&= P^X(|X| > |x|), \qquad X \sim \mathcal{N}(0,1) \\
&= 1 - \Phi(|x|) + \Phi(|x|) = 2[1 - \Phi(|x|)].
\end{aligned}
$$

Therefore, if $x = 1.68$, $p(x) = 0.10$ and, if $x = 1.96$, $p(x) = 0.05$.   ‖

Example 5.3.13 Consider $x \sim \mathcal{B}(n, p)$, when the hypothesis to be tested is $H_0 : p = 1/2$ and $H_1 : p \neq 1/2$. The p-value associated with the likelihood ratio

$$\frac{f(x|1/2)}{\sup_p f(x|p)} = \frac{(1/2)^n}{\left(\frac{x}{n}\right)^x \left(1 - \frac{x}{n}\right)^{n-x}} \propto x^{-x}(n-x)^{-(n-x)}$$

is the function

$$\tilde{p}(x) = P_{1/2}\left(X^X(n-X)^{(n-X)} \leq x^x(n-x)^{(n-x)}\right),$$

where $X \sim \mathcal{B}(n, 1/2)$.   ‖

P-values are thus adaptive procedures that can be acceptable from a frequentist point of view and that, furthermore, meet the requirements of Kiefer (1977) and Robinson (1979) for a *conditional frequentist approach*. However, they are still exposed to criticisms because

(i) p-values also evaluate the wrong quantity, namely, the probability of *overpassing* the observed value of the test statistic. They therefore contradict the Likelihood Principle by involving the whole distribution of the observation.

(ii) Even if derived from optimal test procedures, p-values have no intrinsic optimality because they are not evaluated under a loss function. In fact, as shown in Section 5.4, they may even be suboptimal.

(iii) The new decision space, $\mathcal{D} = [0, 1]$, lacks a decision-theoretic foundation and thus the use of p-values is not made explicit. In particular, the p-values are often perceived as providing a frequentist approximation to $P(\theta \in \Theta_0|x)$, even though this expression is meaningless in a non-Bayesian setting.

(iv) From a classical perspective, p-values do not summarize all the information about the testing problem, since they should be compared with *type–two errors*, which are usually omitted from the analysis. Berger and Wolpert (1988) illustrate the danger of using only p-values by the following example. If $x \sim \mathcal{N}(\theta, 1/2)$, to test $\theta = -1$ versus $\theta = 1$ when $x = 0$ leads to an (UMP) p-value of 0.072, seemingly indicating a strong rejection of the null hypothesis, although the corresponding p-value for the test reversing H_0 and H_1 takes exactly the same value. In fact, while a rejection of H_0 should not always imply acceptance of H_1, practitioners often consider p-values as *the* testing procedure and they assume that it encompasses all the information about the testing problem, thus ending with this implication.

5.3.5 Least favorable Bayesian answers

The problem of evaluating p-values under an adapted loss is considered in Section 5.4. We conclude this section with a comparison of p-values with their Bayesian counterparts, the posterior probabilities. To consider the lower posterior probability on a class of prior distributions provides a Bayesian *least favorable answer* with respect to the null hypothesis. This lower bound cannot be considered as a noninformative procedure, since it enhances the prior most opposed to the null hypothesis, and is both biased against H_0 and dependent on the observation. It should be interpreted as an indicator of the *range* of the posterior probabilities, the most favorable answer being 1. An extensive literature is now available on this approach and we refer to Berger and Sellke (1987), Berger and Delampady (1987) and Berger and Mortera (1991) for additional references. Note 5.7.4 presents a different perspective of Berger, Boukai and Wang (1997) that reconciles frequentist and Bayesian testing by modifying the decision-theoretic framework.

Berger and Sellke (1987) and Berger and Delampady (1987) consider the case of a point-null hypothesis, $H_0 : \theta = \theta_0$, against the alternative hypothesis $H_1 : \theta \neq \theta_0$. For a family G of prior distributions on the alternative hypothesis, the evaluation measures of the veracity of H_0 are given by the lower bounds

$$\underline{B}(x, G) = \inf_{g \in G} \frac{f(x|\theta_0)}{\int_\Theta f(x|\theta)g(\theta)\,d\theta},$$

$$\underline{P}(x, G) = \inf_{g \in G} \frac{f(x|\theta_0)}{f(x|\theta_0) + \int_\Theta f(x|\theta)g(\theta)\,d\theta}$$

on the Bayes factors and posterior probabilities (for $\varrho_0 = 1/2$, considered to give equal weights to both hypotheses). These bounds can also be written as

$$\underline{B}(x, G) = \frac{f(x|\theta_0)}{\sup_{g \in G} \int_\Theta f(x|\theta)g(\theta)d\theta}, \quad \underline{P}(x, G) = \left[1 + \frac{1}{\underline{B}(x, G)}\right]^{-1}.$$

They obviously vary, depending on the class G considered. In the more general case, i.e., when G is G_A, the set of all prior distributions, the following result is straightforward.

Lemma 5.3.14 *If there exists a maximum likelihood estimator of θ, $\hat{\theta}(x)$, the lower bounds on the Bayes factors and posterior probabilities of H_0 are, respectively,*

$$\underline{B}(x, G_A) = \frac{f(x|\theta_0)}{f(x|\hat{\theta}(x))}, \quad \underline{P}(x, G_A) = \left[1 + \frac{f(x|\hat{\theta}(x))}{f(x|\theta_0)}\right]^{-1}.$$

A consequence of Lemma 5.3.14 is that the Bayesian answer will never *strongly* favor the null hypothesis, since

$$\underline{B}(x, G_A) \leq 1, \qquad \underline{P}(x, G_A) \leq \frac{1}{2}.$$

This behavior is not particularly surprising because the lower bounds correspond to the worst possible choice of g with respect to H_0. A more surprising

Table 5.3.1. *Comparison between p-values and Bayesian answers in the normal case. (Source: Berger and Sellke (1987).)*

p-value	0.10	0.05	0.01	0.001
$\underline{P}$	0.205	0.128	0.035	0.004
$\underline{B}$	0.256	0.146	0.036	0.004

phenomenon is that the decrease of these bounds when $|x|$ increases is much slower than for the p-values, as shown by the following example.

Example 5.3.15 (Example 5.3.6 continued) In the normal case, the lower bounds associated with $H_0 : \theta_0 = 0$ are

$$\underline{B}(x, G_A) = e^{-x^2/2} \quad \text{and} \quad \underline{P}(x, G_A) = \left(1 + e^{x^2/2}\right)^{-1},$$

leading to Table 5.3.5, which compares the p-values with the Bayesian least favorable answers. ‖

Therefore, the difference with the frequentist answers is quite important. P-values are smaller for the significance levels of interest, thus reject the null hypothesis H_0 "too often." Of course, for smaller values of x, the p-values are larger than the lower bounds, but what matters is that, in the range of values of x where the decision is the most difficult to take, i.e., for significance levels between 0.01 and 0.1, there is such a discrepancy between the Bayesian and frequentist answers.

Results such as those above are quite surprising because classical procedures usually belong to the range of Bayesian answers. Moreover, the class G_A is rather unreasonable, including point masses leading to the lower bound. The only justification for this class of priors relates to the minimax principle and the corresponding least favorable distribution. The above example shows that p-values are not minimax in this sense. Obviously, the discrepancy is more important for smaller classes of distributions. For instance, if G is G_S, the set of distributions which are symmetric around θ_0, the equivalent of Lemma 5.3.14 is:

Lemma 5.3.16 *The smallest Bayes factor when $g \in G_S$ is*

$$\underline{B}(x, G_S) = \frac{f(x|\theta_0)}{\sup_\xi \frac{1}{2}[f(x|\theta_0 - \xi) + f(x|\theta_0 + \xi)]},$$

which leads to the corresponding lower bound on posterior probabilities.

This result is derived from the fact that every symmetric distribution is a mixture of distributions with a two-point support of the form $\{\theta_0 - \xi, \theta_0 + \xi\}$. For multidimensional extensions, the supremum is to be taken on uniform distributions on spheres centered at θ_0 (see Berger and Delampady (1987)). Discrete settings call for some refinements, if only to define the

Table 5.3.2. *Comparison between p-values and Bayesian answers in the binomial case. (Source: Berger and Delampady (1987).)*

p-value	0.0093	0.0507	0.1011
$\underline{P}$	0.0794	0.2210	0.2969

notion of a symmetric distribution. For instance, in the binomial case, the corresponding class is G_S, made of the distributions which are symmetric in

$$\frac{p - p_0}{\sqrt{p(1-p)}}.$$

Example 5.3.17 (Example 5.3.13 continued) For H_0 : $p = 1/2$, Table 5.3.5 provides p-values and Bayesian lower bounds associated with G_S ($p_0 = 1/2$). ∥

Notice that in this case the p-values are not the standard levels because of the discrete nature of the binomial distribution.

Another interesting class of priors is that of unimodal distributions symmetric around θ_0, G_{SU}. These distributions can be written as mixtures of uniform symmetric distributions in dimension 1 (Berger and Sellke (1987)). Therefore, the computation of the lower bounds is still tractable. It is necessary to use such classes in multidimensional settings as the lower bounds associated with more general classes like G_A are close to 0 for most values of the observation.

Example 5.3.18 (Example 5.3.6 continued) In the normal case, if $|x| \leq 1$, $\underline{B}(x, G_{SU}) = 1$ and $\underline{P}(x, G_{SU}) = 1/2$. However, if $|x| > 1$, defining $g(\theta) = (1/2K)\mathbb{I}\{|\theta| < K\}$, we get

$$\int f(x|\theta)g(\theta)\,d\theta = \frac{1}{2K}[\Phi(K-x) - \Phi(-K-x)]$$

and the lower bound is associated with K maximizing this expression. Table 5.3.5 gives the values of $\underline{B}$ and $\underline{P}$ corresponding to p-values of 0.1 and 0.01, exhibiting a significant discrepancy with the frequentist answer. ∥

A first consequence of these comparisons is that, from a Bayesian viewpoint, p-values are not a valid tool for conducting testing experiments on null hypotheses. Contrary to regular point estimation settings such as those developed in Chapter 4, frequentist answers do not seem to be expressible as limits of Bayesian answers, and we give in Section 5.4 a formal proof of this fact. Since p-values are strictly smaller than Bayesian answers (for levels that really matter in a testing decision-theoretic process), the null hypothesis H_0 is rejected more often under the frequentist approach, while the Bayesian approach shows that the ratio of the posterior likelihoods of

Table 5.3.3. *Bayesian answers for p-values of* 0.01 *(above) and* 0.1 *(below) in the normal case. (Source: Berger and Delampady (1987).)*

dim.	1	3	5
$\underline{P}$	0.109	0.083	0.076
	0.392	0.350	0.339
$\underline{B}$	0.123	0.090	0.082
	0.644	0.540	0.531

H_0 and H_1 is quite moderate for the usual significance levels (0.05 or 0.01). This important discrepancy between the two approaches definitely calls for Bayesian modeling, since this approach includes more naturally the notion of the probability of a hypothesis. It also shows that the argument of *frequentist validity*, i.e., the long-run justification provided by a significance level of 5% or of 1%, is rather illusory and that the division introduced by the Neyman–Pearson theory in the treatment of H_0 and H_1 (between type I and type II errors) leads to a bias in favor of the alternative hypothesis for larger values of x or $T(x)$.

5.3.6 The one-sided case

One-sided hypotheses (e.g., $H_0 : \theta \leq \theta_0$ versus $H_1 : \theta > \theta_0$) do not exhibit such contrasts between frequentist and Bayesian solutions. Indeed, as shown in Example 5.2.9, the p-value can be written as a generalized Bayes estimator and, therefore, as a limit of Bayesian answers (since renormalizing does not matter). Thus, it is impossible to exhibit a dichotomy between both approaches as in the two-sided case. Casella and Berger (1987) consider this setting and generalize the above "reconciliation" phenomenon.

Theorem 5.3.19 *Let $x \sim f(x - \theta)$, with f symmetrical around 0. The null hypothesis to be tested is $H_0 : \theta \leq 0$. If f is a monotone likelihood ratio distribution, the p-value $p(x)$ is equal to the lower bound on the posterior probabilities, $\underline{P}(x, G_{SU})$, when this bound is computed for the class G_{SU} of unimodal symmetric prior distributions and when $x > 0$.*

Proof. In this case, the p-value is

$$p(x) = P_{\theta=0}(X > x) = \int_x^{+\infty} f(t)\, dt$$

and

$$\underline{B}(x, G_{SU}) = \inf_{\pi \in G_{SU}} P^{\pi}(\theta \leq 0 | x)$$

$$= \inf_{\pi \in G_{SU}} \frac{\int_{-\infty}^0 f(x - \theta)\pi(\theta)\, d\theta}{\int_{-\infty}^{+\infty} f(x - \theta)\pi(\theta)\, d\theta}$$

Table 5.3.4. *Comparison between p-values and Bayesian posterior probabilities in the case of a Cauchy distribution. (Source: Casella and Berger (1987).)*

p-value	0.437	0.102	0.063	0.013	0.004
$\underline{P}$	0.429	0.077	0.044	0.007	0.002

$$(5.3.3) \qquad -\inf_K \frac{\int_{-K}^{0} f(x-\theta)\,d\theta}{\int_{-K}^{K} f(x-\theta)\,d\theta},$$

owing to the representation of symmetric unimodal prior distributions as mixtures of uniform distributions on $[-K, K]$. The monotone likelihood ratio property then implies that (5.3.3) is attained for $K = +\infty$. □□

A consequence of Theorem 5.4.1 is that the lower bound of the Bayesian answers over all prior distributions is smaller than the p-value.

Example 5.3.20 Consider $X \sim \mathcal{C}(\theta, 1)$, the Cauchy distribution, when the hypothesis to be tested is $H_0 : \theta \leq 0$ versus $H_1 : \theta > 0$. If the prior distribution of θ is assumed to be in the class of distributions symmetric around 0, the lower bounds on the Bayesian answers and the corresponding p-values are given in Table 5.3.6. The differences in the numerical values are not as striking as in the previous examples. ‖

The distinction between one-sided and two-sided cases calls for the following comments:

(i) As mentioned several times above, Bayesian modeling in a two-sided setting is usually quite delicate, especially for point-null hypotheses, since it implies a *modification of the prior distribution imposed by the inferential problem*. This does not contravene the Bayesian principles if we consider that this modification results from additional (if vague) information; but how to use this information remains unclear. An illustration of this difficulty is found in the case of noninformative distributions, where several (and not entirely compatible) Bayesian approaches are competing, as detailed in Section 5.2.6.

(ii) That the p-value is close to the lower bound in the one-sided case illustrates the conservative (or minimax) behavior of the procedure. As it may be written as a generalized Bayes answer, this induces us to think that the p-value could also be expressed as a noninformative answer in the two-sided cases. Obviously, this does not necessarily imply that this answer should be used, since an effective use of the information contained in the testing problem itself is generally possible.

(iii) *P*-values are derived from UMP or UMPU tests by an ad-hoc empirical construction. The comparisons in Berger and Sellke (1987) and Casella and Berger (1987) show that they differ (or do not differ) from their

Bayesian counterparts. While these studies point out the existence of a theoretical problem, they are not, from a frequentist viewpoint, sufficient to reject the use of p-values. It is thus necessary to use a decision-theoretic perspective adapted to the evaluation of p-values. The next section deals with this comparison. It also provides theoretically grounded explanations for the two-sided/one-sided dichotomy exhibited above.

(iv) A different perspective, which allows for a larger decision space by including a no-decision possibility, brings both frequentist and Bayesian answers much closer, at both the conceptual as well and the numerical levels. It is detailed in Note 5.7.4.

5.4 A second decision-theoretic approach

As just stressed[9], p-values have no intrinsic justification, since they derive their claimed "optimality" from the optimality of the test procedures they are built on. In a sense, the same comment holds for the posterior probabilities since, although they are intuitively justifiable, they are not validated by a decision process. In this section, we construct an alternative to the Neyman–Pearson approach in order to justify the posterior probabilities and evaluate the p-values.

As shown in Section 5.2, the testing problem formalized by Neyman and Pearson can be expressed as estimating the indicator function $\mathbb{I}_{\Theta_0}(\theta)$ under the $0 - 1$ loss or, equivalently, the *absolute error loss*

$$(5.4.1) \qquad \mathrm{L}_1(\theta, \varphi) = |\varphi - \mathbb{I}_{\Theta_0}(\theta)| .$$

Indeed, if the estimators φ are only taking the values 0 and 1, there are many ways to write the $0 - 1$ loss, (5.4.1) being one of them. But, as indicated above, the Neyman–Pearson theory is predominantly a predata theory that does not provide a postdata (or more adaptive) solution. We then turn to a less restrictive theory, according to which estimators take values in $\mathcal{D} = [0, 1]$ and can be considered as indicators of the degree of evidence in favor of H_0.

Parallel to Schaafsma et al. (1989), Hwang et al. (1992) examine this approach to testing, in which the estimators of $\mathbb{I}_{\Theta_0}(\theta)$ belong to $[0, 1]$. When the restriction to $\{0, 1\}$ is dropped, the choice of the loss gets more important. For instance, (5.4.1) is too similar to the $0 - 1$ loss function as it provides the same Bayes procedures

$$\varphi^{\pi}(x) = \begin{cases} 1 & \text{if } P^{\pi}(\theta \in \Theta_0 | x) > P^{\pi}(\theta \notin \Theta_0 | x), \\ 0 & \text{otherwise.} \end{cases}$$

In the opposite, strictly convex losses, such as the quadratic loss

$$(5.4.2) \qquad \mathrm{L}_2(\theta, \varphi) = (\varphi - \mathbb{I}_{\Theta_0}(\theta))^2 ,$$

lead to more adaptive estimators.

[9] This section is at a more advanced level and can be skipped on a first reading.

Proposition 5.4.1 *Under the loss* (5.4.2), *the Bayes estimator associated with* π *is the posterior probability*

$$\varphi^\pi(x) = P^\pi(\theta \in \Theta_0 | x).$$

Indeed, the posterior expectation of $\mathbb{I}_{\Theta_0}(\theta)$ is nothing but the posterior probability of Θ_0. The quadratic loss (5.4.2) thus provides a decision-theoretic foundation to the use of posterior probabilities as Bayesian answers. Such losses are said to be *proper* (see Lindley (1985) and Schervish (1080); Exercise 2.15 characterizes proper losses). There exist other proper losses in addition of the quadratic loss, but Hwang and Pemantle (1995) have shown that it is sufficient to consider the quadratic loss in terms of *admissibility* and *complete classes* (see Chapter 8).

We consider in this section the special case of natural exponential families,

$$f(x|\theta) = e^{\theta x - \psi(\theta)}, \qquad \theta \in \Theta \subset \mathbb{R},$$

and we introduce the following definition, due to Farrell (1968a), which allows us to to evaluate procedures on an interval when they are constant outside this interval.

Definition 5.4.2 *For a one-sided test, i.e., for hypotheses of the form* $H_0 : \theta \leq \theta_0$ *versus* $H_1 : \theta > \theta_0$, *an interval* $[t_1, t_2]$ *is said to be a truncation set for the estimator* φ *if* $\varphi(t) = 1$ *when* $t < t_1$ *and* $\varphi(t) = 0$ *when* $t > t_2$. *For a two-sided test of* $H_0 : \theta \in [\theta_1, \theta_2]$, *the interval* $[t_1, t_2]$ *is said to be a truncation set for the estimator* φ *if* $\varphi(t) = 0$ *when* $t \notin [t_1, t_2]$.

The following results have been obtained in Hwang et al. (1992), based on a result of Brown (1986), which shows that every admissible estimator is a pointwise limit of Bayes estimators for a sequence of measures with finite supports (see Section 8.3.4).

Theorem 5.4.3 *For the two-sided problem*

$$\begin{aligned} H_0 : & \quad \theta \in [\theta_1, \theta_2] \quad versus \\ H_1 : & \quad \theta \notin [\theta_1, \theta_2], \end{aligned}$$

an estimator φ *with truncation set* $[t_1, t_2]$ *is admissible if there exist a probability measure* π_0 *on* $[\theta_1, \theta_2]$ *and a* σ-*finite measure* π_1 *on* $[\theta_1, \theta_2]^c$ *such that*

(5.4.3) $$\varphi(x) = \frac{\int f(x|\theta)\pi_0(\theta)\, d\theta}{\int f(x|\theta)\pi_0(\theta)d\theta + \int f(x|\theta)\pi_1(\theta)\, d\theta},$$

for $x \in [t_1, t_2]$. *Conversely, if* φ *is admissible, there exist* $[t_1, t_2]$, π_0, *and* π_1 *such that* (5.4.3) *is satisfied.*

In the one-sided case, we can only propose an admissibility necessary condition, but it implies that the generalized Bayes estimators form a complete class.

Theorem 5.4.4 *For the one-sided problem*

(5.4.4) $H_0 : \ \theta \leq \theta_0 \qquad versus \qquad H_1 : \ \theta > \theta_0,$

if φ is admissible, there exists an increasing procedure φ' such that φ' is (risk) equivalent to φ. If φ is an increasing admissible procedure and $[t_1, t_2]$ is a truncation set such that $0 < \varphi(x) < 1$ on $[t_1, t_2]$, there exist two σ-finite measures on $(-\infty, \theta_0]$ and $[\theta_0, +\infty)$, π_0, and π_1, such that

$$1 = \int e^{t_0 \theta - \psi(\theta)} (\pi_0(\theta) + \pi_1(\theta)) \, d\theta$$

for $t_1 < t_0 < t_2$ and φ is given by (5.4.3) on $[t_1, t_2]$.

These two *complete class* theorems show that it is sufficient to consider the generalized Bayes estimators to obtain admissible estimators under quadratic loss. Theorem 5.4.4 shows in addition that the monotone estimators form an *essentially complete class*. These results can be used to evaluate p-values. Notice again that the Bayes estimators are underlying (classical) optimal estimators. (Chapter 8 exposes more thoroughly the Bayesian foundations of admissibility.)

Recall also that Casella and Berger (1987) showed that p-values were within the variation range of Bayesian posterior probabilities in one-sided settings. It is therefore natural to examine the admissibility of p-values. The examples below show that they are admissible for most one-sided tests.

Example 5.4.5 Consider again $x \sim \mathcal{N}(\theta, 1)$ and H_0 of the form (5.4.4). We showed in Example 5.2.9 that

$$p(x) = P_{\theta_0}(X > x) = 1 - \Phi(x - \theta_0)$$

is a generalized Bayes estimator with respect to the Lebesgue measure. Moreover, the risk of the p-value is

$$
\begin{aligned}
r(\pi, p) \ &= \ \int_{-\infty}^{+\infty} R(p, \theta) \, d\theta \\
&= \ \int_{-\infty}^{+\infty} \int_{-\infty}^{+\infty} (p(x) - \mathbb{I}_{\Theta_0}(\theta))^2 f(x|\theta) \, dx \, d\theta \\
&= \ \int_{-\infty}^{\theta_0} \int_{-\infty}^{+\infty} (1 - \Phi(x - \theta_0))^2 f(x|\theta) \, dx \, d\theta \\
&\quad + \int_{\theta_0}^{+\infty} \int_{-\infty}^{+\infty} \Phi(x - \theta_0)^2 f(x|\theta) \, dx \, d\theta \\
&= \ 2 \int_{-\infty}^{+\infty} (1 - \Phi(x - \theta_0))^2 \Phi(x - \theta_0) \, dx
\end{aligned}
$$

by the Fubini Theorem. This integral is finite. Therefore, $r(\pi) < +\infty$ and p is admissible under (5.4.2) (see Section 2.4).   ‖

Example 5.4.6 Consider $x \sim \mathcal{B}(n, \theta)$. The p-value for the test of (5.4.3) is then

$$p(x) = P_{\theta_0}(X \geq x) = \sum_{k=x}^{n} \binom{n}{k} \theta_0^k (1 - \theta_0)^{n-x},$$

which is also a generalized Bayes estimator under the prior distribution $\pi(\theta) = 1/\theta$. It is again possible to show that p has a finite Bayes risk and is thus admissible. A similar result can be established for a Poisson distribution, $\mathcal{P}(\theta)$ (see Hwang et al. (1992)). $\qquad\qquad\qquad\|$

In two-sided settings, on the contrary, p-values are not admissible, as suggested by the comparisons of Section 5.3.5.

Theorem 5.4.7 *For the test of* (5.4.3), *when the sampling distribution is continuous with respect to the Lebesgue measure, the p-value is inadmissible for the loss* (5.4.2).

Proof. The result relies on the fact that the p-value p is equal to the value 1 with positive probability (see Hwang et al. (1992, §4.1.2)). In fact, if p is admissible, it can be written under the form (5.4.3). Since it is positive,

$$\int f(x|\theta)\pi_1(\theta)\,d\theta < +\infty.$$

Therefore, by continuity, the equality (5.4.3) holds everywhere and $p(x_0) = 1$ implies $\pi = \pi_0$, i.e., $p(x) = 1$ for every x, which cannot be true. $\qquad\square\square$

This result agrees with the observations of Berger and Sellke (1987), who showed that p-values do not belong to the range of Bayesian answers. It thus justifies the rejection of p-values for two-sided hypotheses. Furthermore, Hwang and Pemantle (1995) show that the inadmissibility of p-values can be extended to most bounded proper losses. As a concluding remark, let us point out that it is now necessary to construct estimators dominating the p-values. In the normal case, Hwang et al. (1992) show that it cannot be done using a proper Bayes estimator, but Hwang and Pemantle (1995) give numerical arguments in favor of an explicit dominating estimator.

5.5 Confidence regions

Apart from providing a decision-maker with approximations of the "true" value of a parameter θ, namely, point estimators, and with answers to questions about the inclusion of θ in a specific domain, that is, testing procedures, it is sometimes necessary to give in addition *confidence regions* on θ, i.e., subsets C_x of the parameter space Θ where θ should be with high probability (in the frequentist or in the Bayesian sense). This notion also extends to non-bijective transforms of θ. It is also of considerable interest in *forecasting* and prediction settings.

Example 5.5.1 Consider the IBM stock prices of Example 4.5.3, represented in Figure 4.5.1. If the series (x_t) has been observed till time T, the value at time $T + 1$, x_{T+1}, is obviously of considerable interest and it is important to present the investor not only with the most likely value of x_{T+1} given the past series, but also with a range of possible values of x_{T+1} so that she can take a decision against the corresponding possible gains. ‖

Once again, the Bayesian formulation that θ has a given probability to belong to a fixed region C_x is more appealing than the frequentist interpretation that a random region C_x has a given probability to contain the unknown parameter θ.

5.5.1 Credible intervals

As in the testing setting, the Bayesian paradigm proposes a notion of confidence regions that is more natural that its frequentist counterpart since, again, the notation $P(\theta \in C_x)$ is meaningful even conditional upon x.

Definition 5.5.2 *For a prior distribution π, a set C_x is said to be an α-credible set if*

$$P^\pi (\theta \in C_x | x) \geq 1 - \alpha.$$

This region is called an HPD α-credible region (for highest posterior density) if it can be written under the form[10]

$$\{\theta;\ \pi(\theta|x) > k_\alpha\} \subset C_x^\pi \subset \{\theta;\ \pi(\theta|x) \geq k_\alpha\},$$

where k_α is the largest bound such that

$$P^\pi (\theta \in C_x^\alpha | x) \geq 1 - \alpha.$$

To consider only HPD regions is motivated by the fact that they *minimize the volume among α-credible regions* and, therefore, can be envisioned as optimal solutions in a decision setting.

Example 5.5.3 If $\theta \sim \mathcal{N}(0, \tau^2)$, the posterior distribution of θ is $\mathcal{N}(\mu(x), \omega^{-2})$ with $\omega^2 = \tau^{-2} + \sigma^{-2}$ and $\mu(x) = \tau^2 x / (\tau^2 + \sigma^2)$. Then

$$C_\alpha^\pi = \left[\mu(x) - k_\alpha \omega^{-1}, \mu(x) + k_\alpha \omega^{-1} \right],$$

where k_α is the $\alpha/2$-quantile of $\mathcal{N}(0, 1)$. In particular, if τ goes to $+\infty$, $\pi(\theta)$ converges to the Lebesgue measure on $\mathbb{R}$ and gives

$$C_\alpha = \left[x - k_\alpha \sigma, x + k_\alpha \sigma \right],$$

i.e., the usual confidence interval, as a generalized Bayes estimator.   ‖

---

[10] This formulation allows for coverage of the special case when $\{\theta; \pi(\theta|x) = k_\alpha\}$ is not empty.

Table 5.5.1. *Confidence intervals for the binomial distribution.*

x	0	1	2
$\alpha = 5\%$	$[0.000, 0.38]$	$[0.022, 0.621]$	$[0.094, 0.791]$
$\alpha = 10\%$	$[0.000, 0.308]$	$[0.036, 0.523]$	$[0.128, 0.74]$

Example 5.5.4 Consider $x \sim B(n, p)$ and the noninformative distribution $p \sim Be(1/2, 1/2)$. Then $p|x \sim Be(x + 1/2, n - x + 1/2)$ and confidence intervals on p can be derived from the c.d.f. of the beta distribution. Table 5.5.1 gives these intervals for $n = 5$ and $\alpha = 5\%$, 10%. ‖

Notice the significant advantage of using a Bayesian approach in this setting of *discrete distributions*, as compared with a classical approach. In fact, the usual confidence intervals involve a *randomization* step to attain nominal confidence levels (see Blyth and Hutchinson (1961) for an illustration in the binomial case). Prior modeling avoids this addition of random noise and, on the contrary, takes advantage of the available prior information. Notice also that *improper priors* can be used in this setting, and do not encounter the same difficulties as when testing point-null hypotheses. In fact, posterior credible regions can be derived provided the posterior distribution is defined. Some classical confidence regions can be expressed as credible regions associated with generalized distributions.

Example 5.5.5 Consider $x_1, \ldots, x_n$ i.i.d. $\mathcal{N}(\theta, \sigma^2)$. The prior distribution is the noninformative prior

$$\pi(\theta, \sigma^2) = \frac{1}{\sigma^2}.$$

We showed in Section 4.4.2 that the marginal posterior distribution for $1/\sigma^2$ is a gamma distribution $\mathcal{G}\left((n - 1)/2, s^2/2\right)$ with $s^2 = \sum(x_i - \bar{x})^2$. Therefore

$$\frac{s^2}{\sigma^2}|\bar{x}, s^2 \sim \chi^2_{n-1},$$

and we get the same confidence interval as in the classical approach, but it is now justified conditional upon s^2. ‖

Example 5.5.6 Consider $x \sim B(n, p)$ and $p \sim Be(\alpha, \beta)$. In this case, $\pi(p|x)$ is the beta distribution $Be(\alpha + x, \beta + n - x)$. Depending on the values of α, β, n, and x, the confidence regions are of four types:

(i) $0 \leq p \leq K(x)$;

(ii) $K(x) \leq p \leq 1$;
(iii) $K_1(x) \leq p \leq K_2(x)$; and
(iv) $0 \leq p \leq K_1(x)$ or $K_2(x) \leq p \leq 1$.

The last region is quite artificial and rather useless. Notice that it corresponds to the case

$$\alpha + x < 1 \quad \text{and} \quad \beta + n - x < 1,$$

thus implying that α and β must be sufficiently negative since $\alpha + \beta < 2 - n$. Hence, this feature disappears for n large enough, unless α and β depend on n, which is not desirable from a Bayesian point of view. Notice that the limiting case $\alpha = \beta = 0$, which corresponds to Haldane's (1931) distribution

$$\pi(p) = [p(1-p)]^{-1},$$

already leads to regions of types (i)–(iii), although the posterior distribution is not defined for all x's (Example 1.5.3). ‖

When phenomena such as case (iv) of Example 5.5.6 occur, that is, when the confidence region is not connected (see also Exercise 5.5), the usual solution is to replace the HPD α-credible region by an interval with equal tails, i.e., $[C_1(x), C_2(x)]$ such that

$$P^\pi(\theta < C_1(x)|x) = P^\pi(\theta > C_2(x)|x) = \alpha/2.$$

Berger (1985a) remarks that the occurrence of nonconnected HPD regions also points out a discrepancy between the prior distribution and the observations and that this phenomenon should question the choice of the prior or of the sampling distribution. It may also exhibit a non-identifiable structure which is responsible for the multimodality of the posterior distribution.

If, conceptually, the determination of credible sets is rather straightforward, the practical derivation of these regions can be quite involved, especially when the dimension of Θ is large or when the posterior distribution is not available in a closed form. A first solution is to use numerical methods similar to those developed in Chapter 6, the problem being to assess the impending error (which can be much larger than the point estimation approximation errors). (Notice that equal tail credible regions are usually easier to approximate than HPD regions; see Eberly and Casella (1999).) A second solution, used in Berger (1980a, 1985a), is to build up a normal approximation, i.e., to consider that the posterior distribution of θ is roughly $\mathcal{N}_p(\mathbb{E}^\pi(\theta|x), \text{Var}^\pi(\theta|x))$, and to derive from this approximation the confidence region

$$C_\alpha = \left\{ \theta; (\theta - \mathbb{E}^\pi(\theta \mid x))^t \, \text{Var}^\pi(\theta|x)^{-1}(\theta - \mathbb{E}^\pi(\theta|x)) \le k_\alpha^2 \right\},$$

where k_α^2 is the α-quantile of χ_p^2. This approximation is only justified for a large sample size (see Hartigan (1983)), but it still provides fast and rather efficient confidence regions.

5.5.2 Classical confidence intervals

In the Neyman–Pearson theory, confidence regions can be deduced from UMPU tests by a *duality* argument: If

$$C_\theta = \{x; \varphi_\theta(x) = 1\}$$

is the acceptance region for the null hypothesis $H_0 : \theta = \theta_0$, φ_{θ_0} being a UMPU test at level α, the corresponding confidence region is

$$
\begin{aligned}
C_x &= \{\theta; x \in C_\theta\} \\
    &= \{\theta; \varphi_\theta(x) = 1\}
\end{aligned}
$$

and $P(\theta \in C_x) = 1 - \alpha$. More generally, a region C_x is said to be a *confidence region at level* α (in the frequentist sense) if, for every $\theta \in \Theta$, $P(\theta \in C_x) \geq 1 - \alpha$.

Example 5.5.7 (Example 5.5.3 continued) If $x \sim \mathcal{N}(\theta, \sigma^2)$, the 95% UMPU test is $\varphi_\theta(x) = \mathbb{I}_{[0,1.96]} (|x - \theta|/\sigma)$ and the corresponding confidence region, when σ is known, is

$$C_x = [x - 1.96\sigma, x + 1.96\sigma]. \qquad \|$$

Example 5.5.8 Consider $x \sim \mathcal{T}_p(N, \theta, I_p)$, a t-distribution with N degrees of freedom, and density

$$f(x \mid \theta) \propto \left(1 + \frac{1}{N} \| x - \theta \|^2\right)^{-(N+p)/2}.$$

Since $\frac{1}{p} \| x - \theta \|^2 \sim \mathcal{F}(p, N)$, we can derive a $1 - \alpha\%$ *confidence ball*

$$C_x = \left\{\theta; \| x - \theta \|^2 \leq p f_\alpha(p, N)\right\},$$

where $f_\alpha(p, N)$ is the α-quantile of $\mathcal{F}(p, N)$. $\qquad \|$

These confidence regions, although used quite extensively in practice (for instance, in the case of the linear regression), have been criticized on frequentist, conditional, and Bayesian grounds. First, as seen in the previous sections, the Neyman–Pearson approach itself is not free of drawbacks, and the optimality of UMPU tests can be contested. Therefore, confidence regions derived from these tests (called *uniformly most accurate regions* by Lehmann (1986)) do not necessarily have a proper behavior. Moreover, even from a frequentist perspective, the inversion of optimal test procedures into confidence regions does not automatically grant these regions with a derived optimality, despite the above denomination.

Besides conditional criticisms of confidence regions (see Note 5.7.3), there also exist frequentist criticisms. Following Stein (1962a) and Lindley (1962), Brown (1966) and Joshi (1967) have indeed established that these regions C_x^0 are not always optimal because there may exist another set C_x' such that

$$P_\theta(\theta \in C_x') \geq P_\theta(\theta \in C_x^0) \quad \text{and} \quad \text{vol}(C_x') \leq \text{vol}(C_x^0).$$

Therefore, the set C'_x is to be preferred to C^0_x since, for a smaller volume, it has a larger probability of containing the true value of the parameter. For instance, in the normal case, Joshi (1969) has established that, if $x \sim \mathcal{N}_p(\theta, I_p)$, the confidence region

$$C^0_x = \left\{ \theta;\ ||\theta - x||^2 \leq c_\alpha \right\}$$

is admissible (in the above sense) if and only if $p \leq 2$ (see also Cohen and Strawderman, 1973). For larger dimensions, it is possible to exhibit more efficient confidence regions.

This phenomenon pertains to the Stein effect, establishing the inadmissibility of the maximum likelihood estimator for $p \geq 3$ (see Note 2.8.2). Hwang and Casella (1982) have taken advantage of this analogy to show that, if

$$\delta^{\mathrm{JS}}(x) = \left(1 - \frac{a}{||x||^2} \right)^+ x$$

is a truncated James–Stein estimator, the *recentered confidence region*

$$C^{\mathrm{JS}}_x = \left\{ \theta;\ ||\theta - \delta^{\mathrm{JS}}(x)||^2 \leq c_\alpha \right\},$$

has the same volume as the usual ball C^0_x and satisfies

(5.5.1) $$P_\theta(\theta \in C^{\mathrm{JS}}_x) > P_\theta(\theta \in C^0_x) = 1 - \alpha$$

for a small enough. Therefore, C^{JS}_x dominates C^0_x in the above sense.

An extensive amount of literature on this subject of recentered confidence regions has been initiated by Hwang and Casella (1982, 1984), similar to the growth in point-estimation literature associated with the Stein effect (see Section 2.8.2). New recentered regions have been proposed in Hwang and Casella (1984) and Casella and Hwang (1983, 1987). Hwang and Chen (1986) and Robert and Casella (1990) have extended domination results to spherically symmetric distributions, although the case of the unknown variance normal problem is still unsolved (see Hwang and Ullah (1994)). Shinozaki (1990) has also devised a confidence region with exactly the same coverage probability, but with a smaller volume, taking advantage of the inadmissibility of the usual region the opposite way to (5.5.1). Lu and Berger (1989a), George and Casella (1994), and Robert and Casella (1993) have also taken advantage of (5.5.1) to propose improved confidence reports for the usual and usual and recentered sets. For the problem of estimating a normal variance, similar improvements can be found in Cohen (1972), Shorrock (1990), and Goutis and Casella (1991).

5.5.3 Decision-theoretic evaluation of confidence sets

As the reader may have noticed, the above construction of confidence regions has been done in a rather off-handed manner, with limited decision-theoretic justifications. The choice of HPD regions is usually related to a volume minimization requirement, under the coverage constraint

$$P(\theta \in C_\alpha | x) \geq 1 - \alpha.$$

Several authors have proposed differing derivations of confidence regions according to a purely decision-theoretic criterion. They consider loss functions that incorporate at once volume and coverage requirements. (In a way, the above approach corresponds to a bidimensional loss with components $\mathrm{vol}(C)$ and $1 - \mathbb{I}_C(\theta)$.) For instance, a simple version of this decision-theoretic perspective is to consider the linear combination

(5.5.2) $$L(C, \theta) = \mathrm{vol}(C) + c\mathbb{I}_{\theta \notin C},$$

leading to the risk

$$R(C, \theta) = \mathbb{E}[\mathrm{vol}(C_x)] + cP(\theta \notin C_x).$$

(The constant c can be related to a particular confidence level.) In addition, Cohen and Sackrowitz (1984) have shown that the above bidimensional loss can be related to the linear loss (5.5.2) when c is treated as an additional parameter of the model. Meeden and Vardeman (1985) also propose different evaluations of Bayesian confidence regions. They show that admissible and Bayesian confidence sets are equivalent for some criteria.

An important defect of the loss (5.5.2) has been pointed out by James Berger (see Casella et al. (1993a,b)). The problem results as a consequence of the unequal penalization between volume and coverage. In fact, the indicator function varies between 0 and 1 while the volume can increase to infinity and this asymmetry leads to a bias in favor of small confidence sets.

Example 5.5.9 Consider $x_1, \ldots, x_n$ i.i.d. $\mathcal{N}(\theta, \sigma^2)$. The classical t-interval on θ,

$$C_k(\bar{x}, s) = \left(\bar{x} - k\frac{s}{\sqrt{n}}, \bar{x} + k\frac{s}{\sqrt{n}} \right),$$

is an HPD region when

$$\bar{x} = \sum_{i=1}^{n} x_i/n, \quad s^2 = \sum_{i=1}^{n} (x_i - \bar{x})^2/(n-1), \quad \text{and} \quad \pi(\theta, \sigma^2) = \frac{1}{\sigma^2},$$

Jeffreys noninformative distribution. Indeed, in this case,

$$\sqrt{n} \, \frac{\theta - \bar{x}}{s} \mid \bar{x}, s \sim \mathcal{T}_{n-1},$$

Student's t-distribution with $n - 1$ degrees of freedom. Under (5.5.2), the posterior loss is

$$\begin{aligned}
\varrho(\pi, C_k(\bar{x}, s) | \bar{x}, s) &= 2k\frac{s}{\sqrt{n}} - cP^\pi(\theta \in C_k(\bar{x}, s) | \bar{x}, s) \\
&= 2k\frac{s}{\sqrt{n}} - cP(|T_{n-1}| \leq k).
\end{aligned}$$

Then, it is easy to see that the HPD region is dominated by the truncated

region
$$C_t'(\bar{x}, s) = \begin{cases} C_t(\bar{x}, s) & \text{if } s < \sqrt{n}c/(2k), \\ \{\bar{x}\} & \text{otherwise.} \end{cases}$$

This domination is counterintuitive: C_t' proposes the single point $\{\bar{x}\}$ (or equivalently $\emptyset$), seemingly indicating certainty, when the empirical variance increases, indicating growing uncertainty. A similar phenomenon occurs when k depends on s, i.e., the size of the credible region decreases to 0 as s increases (see Casella et al. (1993a,b)).   ‖

The above paradox exposes the limitations of the linear loss (5.5.2). Casella et al. (1993a) propose an alternative class of loss functions which avoid the paradox. The simplest of these losses are the so-called *rational losses*
$$\mathrm{L}(C, \theta) = \frac{\mathrm{vol}(C)}{\mathrm{vol}(C) + k} + \mathbb{I}_{\theta \notin C} \qquad (k > 0),$$

where both terms are then bounded by one. The Bayes estimators associated with these losses are still HPD regions but remain nonempty for all conjugate priors in the normal case. The parameter k can be obtained through techniques similar to those developed for regular losses, namely, by comparing the volume penalizations associated with different regions and approximating the utility function.

We do not pursue any further the decision-theoretic study of Bayesian confidence regions. Indeed, an important aspect usually overlooked in the derivation of confidence regions deals with how they will be used, although this very use is essential in the construction of the loss function. In fact, the decision-maker's purpose can be

(1) to consider set estimation as a preliminary step to *point estimation* (and, for instance, to derive a prior distribution with support equal to the estimated confidence region);

(2) to rely on the obtained confidence region to solve a *testing* problem (and reject a null hypothesis if the confidence region does not contain a specific value);

(3) to derive from the size (volume) of the confidence region an indicator of the *performances* of an associated estimator, for instance, the center of the region. A *performance curve* for this estimator can then be derived by relating size and confidence levels.

These three perspectives of confidence region estimation definitely lead to different loss functions, and it may be illusory to try to build up a global loss function unifying such contrasted purposes. In fact, separate losses are preferable because, in accordance with the foundations of Decision Theory, the decision-maker should select a loss function according to her needs. Notice also that the three purposes considered above correspond to inferential problems already studied previously, and thus that a specific approach to confidence regions may be partly useless. Therefore, it seems to us that,

at least, a more *conditional* approach should be used in the construction
of confidence regions (see Note 5.7.3). Following Kiefer (1977), we suggest
associating to a given set C_x a *confidence index* $\gamma(x)$, evaluated under the
loss

(5.5.3) $$L(C, \gamma, \theta) = (\mathbb{I}_C(\theta) - \gamma)^2.$$

The confidence region is thus replaced by a confidence procedure, related
to the conditional perspective of Robinson (1979). From this point of view,
the procedure $[\Theta, 1]$ is unfortunately perfect, a drawback suggesting that
an additional evaluation of C_x should be included in the loss function,
as in Rukhin (1988a,b). Similarly, the Bayesian procedure associated with
an HPD region C_α is $[C_\alpha, 1 - \alpha]$, as can be verified by minimizing the
posterior loss. For an arbitrary region, C_x, the corresponding procedure is
$[C_x, \gamma^\pi(x)]$, where

$$\gamma^\pi(x) = P^\pi(\theta \in C_x | x).$$

Introducing a global loss function combining volume, coverage, and confidence report as in (5.5.3), the optimal procedures would then be those
that minimize the maximal posterior (or frequentist) error. However, this
direction has not yet been treated in the literature.

5.6 Exercises

Section 5.2.1

5.1 In the setting of Example 5.2.4, study the modification of the posterior probability of H_0 when $x = 0$ and τ/σ goes to $+\infty$. Compare with the noninformative answer associated with $\pi(\theta) = 1$.

Section 5.2.2

5.2 Consider $x \sim \mathcal{N}(\theta, 1)$. The hypothesis to test is $H_0 : |\theta| \leq c$ versus $H_1 :$
$|\theta| > c$ when $\pi(\theta) = 1$.

a. Give the graph of the maximal probability of H_0 as a function of c.

b. Determine the values of c for which this maximum is 0.95 and the Bayes
   factor is 1. Are these values actually appealing?

5.3 A professor at Cornell University has to give an examination on two different days. Since students are sitting next to each other, she gives two different
examinations alternatively to students in order to reduce cheating. She then
repeats the same technique with a different class and the same two examinations the next day. The results are as follows: $n_{1A} = 17$ students took
examination A the first day and $n_{2A} = 19$ the second day, $n_{1B} = 15$ took
examination B the first day and $n_{2B} = 19$ the second day. The average grades
(out of 20) are $\hat{\mu}_{1A} = 10.3$, $\hat{\mu}_{2A} = 10.2$, $\hat{\mu}_{1B} = 7.9$, and $\hat{\mu}_{2B} = 8.7$, with
standard deviations $\hat{\sigma}_{1A} = 2.67$, $\hat{\sigma}_{2A} = 2.89$, $\hat{\sigma}_{1B} = 2.98$, and $\hat{\sigma}_{2B} = 2.91$.

a. Test whether there is a class effect, an examination effect, or a class-and-
   examination effect by modeling the results in an *analysis of variance* setting, namely, by assuming that each student grade x is normally distributed
   with mean $\mu_0 + \mu_e + \mu_c$ and variance σ_{ec}^2 ($e = A, B, c = 1, 2$) where
   $\mu_A + \mu_B = 0$, $\mu_A + \mu_B = 0$.

b. A student forgot to give back his copy of examination A the first day. Can you detect a cheating effect on the second day?

Section 5.2.3

5.4 (Pearl (1988)) After communicating a rumor to a neighbor, you hear it again from the same neighbor a few days after. Build up a model to test whether your neighbor has also heard this rumor from another person.

5.5 *Consider two independent standard normal observations x and y. The polar coordinates of (x, y) are (r, θ), with $x = r \cos \theta$ and $y = r \sin \theta$.

a. Given that $2r^2 = (x-y)^2 + (x+y)^2$ and that $x-y$ and $x+y$ are independent, show that the distribution of r^2 given $x = y$ is $\mathcal{G}(1/2, 1)$.

b. Show that r and θ are independent and deduce that the distribution of r^2 given $\theta = \pi/4, 5\pi/4$ is $\mathcal{G}(1/2, 1/2)$.

c. Since $\{x = y\} = \{\theta = \pi/4, 5\pi/4\}$, explain the apparent paradox, called *Borel paradox*, of the two different conditional distributions for a same event. (*Hint:* Replace the conditioning in a proper perspective of σ-algebras and compare the σ-algebras spanned by $x - y$ and by θ.)

Section 5.2.4

5.6 When $x \sim \mathcal{N}(\theta, 1)$ and $\theta \sim \mathcal{N}(0, \sigma^2)$, compare the Bayesian answers for the two testing problems

$$
\begin{aligned}
H_0^1 : & \quad \theta = 0 \text{ versus } H_1^1 : \theta \neq 0, \\
H_0^2 : & \quad |\theta| \leq \epsilon \text{ versus } H_1^2 : |\theta| > \epsilon,
\end{aligned}
$$

when ϵ and σ vary.

5.7 In the setting of Example 5.2.3, if $x \sim \mathcal{B}(n, p)$ and $H_0 : p = 1/2$ is to be tested, study the variation of the Bayesian answers as a function of n for $x = 0$ and $x = n/2$ if the prior distribution is the Jeffreys distribution.

5.8 * (Berger and Delampady (1987)) Consider $x \sim \mathcal{N}(\theta, 1)$. The purpose of the exercise is to compare $H_0 : |\theta - \theta_0| \leq \epsilon$ with the approximation $H_0^* : \theta = \theta_0$. Denote by g_0 and g_1 the prior densities on $\{|\theta - \theta_0| \leq \epsilon\}$ and $\{|\theta - \theta_0| > \epsilon\}$. Let g be a density on $\mathbb{R}$ such that

$$
g(\theta) \propto g_1(\theta) \quad \text{if } |\theta - \theta_0| > \epsilon,
$$

and

$$
\lambda = \int_{|\theta - \theta_0| \leq \epsilon} g(\theta)\, d\theta,
$$

is small enough. We denote

$$
B = \frac{\int_{|\theta - \theta_0| \leq \epsilon} f(x|\theta) g_0(\theta)\, d\theta}{\int_{|\theta - \theta_0| > \epsilon} f(x|\theta) g_1(\theta)\, d\theta} \quad \text{and} \quad \hat{B} = \frac{f(x|\theta_0)}{m_g(x)} = \frac{f(x|\theta)}{\int f(x|\theta) g(\theta)\, d\theta},
$$

$t = (x - \theta_0)$ and

$$
\gamma = \frac{1}{2\epsilon\varphi(t)} [\Phi(t + \epsilon) - \Phi(t - \epsilon)] - 1.
$$

Show that, if $|t| \geq 1$, $\epsilon < |t| - 1$, and $\hat{B} \leq (1 + \gamma)^{-1}$, then

$$
B = \hat{B}(1 + \varrho)
$$

with

$$-\lambda \le \frac{\lambda(\hat{B} - 1)}{1 - \lambda\hat{B}} \le \varrho \le \frac{\gamma + \lambda(1 + \gamma)(\hat{B} - 1)}{1 - \lambda\hat{B}(1 + \gamma)} \le \gamma.$$

Section 5.2.5

5.9 Consider $x \sim \mathcal{P}(\lambda)$. The hypothesis to test is $H_0 : \lambda \le 1$ versus $H_1 : \lambda > 1$. Give the posterior probability of H_0 for $x = 1$ and $\lambda \sim \mathcal{G}(\alpha, \beta)$.

    a. How does this probability get modified when α and β go to 0? Does this answer depend on the rates of convergence of α and β to 0?

    b. Compare with the probability associated with the noninformative distribution $\pi(\lambda) = 1/\lambda$. Is it always possible to use this improper prior?

5.10 Consider $x \sim \mathcal{B}(n,p)$, $H_0 : p = 1/2$, and $H_1 : p \neq 1/2$. The prior distribution $\pi(p)$ is a beta distribution $\mathcal{B}e(\alpha, \alpha)$. Determine the limiting posterior probability of H_0 when $n = 10$, $x = 5$ and $n = 15$, $x = 7$ as α goes to $+\infty$. Are these values intuitively logical? Give the posterior probabilities for Laplace, Jeffreys, and Haldane noninformative priors.

5.11 Solve Exercises 5.9 and 5.10 for the Bayes factors instead of the posterior probabilities.

5.12 In a normal setting, determine whether there exists a normalization problem associated with noninformative prior distributions for tests of one-sided hypotheses such as

$$H_0 : \theta \in [0, 1] \qquad \text{versus} \qquad H_1 : \theta > 1.$$

Replace 1 by ϵ and consider the evolution of the optimal answer as ϵ goes to 0.

Section 5.2.6

5.13 Establish the decomposition (5.2.6) from the original definition (5.2.5) of $B_{10}^{(\ell)}$. (*Hint:* Use Bayes formula to develop $\pi_1(\theta_1|x_{(\ell)})$ and $\pi_0(\theta_1|x_{(\ell)})$.)

5.14 In the setting of Example 5.2.13, show how $B_{10}(2)$ depends on the choice of (x_1, x_2) by computing the normalizing constants of both $\pi_0(\sigma^2|x_1, x_2)$ and $\pi_1(\mu, \sigma^2|x_1, x_2)$ and by completing the integration in $B_{10}(2)$.

5.15 Aitkin (1991) suggests removing the difficulty with improper priors by using the data *twice*: given $x \sim f(x|\theta)$ and an improper prior π, if $H_0 : \theta = \theta_0$ is the hypothesis to be tested, take $\tilde{\pi}(\theta) = \pi(\theta|x)$ and use $\tilde{\pi}$ as the prior in the Bayes factor.

    a. If $f(\cdot|\theta)$ is the $\mathcal{N}(\theta, 1)$ density and $\pi(\theta) = 1$, compute the corresponding pseudo-Bayes factor.

    b. Same question as a. when $f(\cdot|\theta)$ is the $\mathcal{P}(\lambda)$ probability mass function and $\pi(\lambda) = 1/\lambda$.

    c. Analyze the limiting behavior of the pseudo-Bayes factor when the procedure is iterated, that is, when π is iteratively updated in $\tilde{\pi}$. [*Note:* From a computational point of view, this technique can be of interest to compute maximum likelihood and MAP estimates. See Robert and Casella (2004, §5.2.4).]

5.16 In the setting of Example 5.2.14, compute the Bayes factor when $\pi_1(\theta)$ is a $\mathcal{N}(0, 2)$ distribution and compare with the arithmetic intrinsic Bayes factor. (*Hint:* Compute $\mathbb{E}[\exp(-x^2/2)]$.)

5.17 (Exercise 5.16 cont.) For the fractional Bayes factor (5.2.11),

a. Show that the minimum value of b is $1/n$.

b. Show that (5.2.11) corresponds to the intrinsic prior $\mathcal{N}(0, (1 - b)/nb)$.

c. Show that that a fixed value of b leads to a variance shrinking to 0 in the intrinsic prior.

d. Compare the numerical values of the arithmetic intrinsic and of the fractional Bayes factors.

e. Determine whether there exists a value of b such that these two pseudo-Bayes factors agree.

5.18 In the setting of Example 5.2.15,

a. Show that π_2 does integrate to 1.

b. Show that B_{10}^A corresponds to a genuine Bayes factor under π_2.

5.19 Coherence conditions for Bayes factors are given by

$$B_{12} = B_{10} B_{02} \quad \text{and} \quad B_{01} = 1/B_{10},$$

when considering three hypotheses, H_0, H_1 and H_2, under corresponding priors π_0, π_1 and π_2.

a. Show that these conditions are satisfied when the π_i's are proper priors.

b. Show that the fractional Bayes factor satisfies $B_{01} = 1/B_{10}$ but not $B_{12} = B_{10} B_{02}$

c. Show that neither the arithmetic not the geometric intrinsic Bayes factors satisfy these conditions.

5.20 For the intrinsic prior considered in Example 5.2.17,

a. Show that

$$\int_{\theta_0}^{\infty} \left(2e^{\theta - \theta_0} - 1\right)^{-1} d\theta = \log(2).$$

(*Hint:* Use a change of variables from θ to $\omega = \exp(\theta - \theta_0)$, a fractional decomposition of $1/\omega(2\omega - 1)$.)

b. Deduce the expression (5.2.12).

5.21 In the setting of Example 5.2.19,

a. Show that

$$\int \left(\prod_{t=1}^{n} \left\{\frac{p}{\sigma_1} e^{-(x_t - \mu_1)^2/2\sigma_1^2} + \frac{1 - p}{\sigma_2} e^{-(x_t - \mu_2)^2/2\sigma_2^2}\right\}\right)^b d\pi(\mu, \sigma)$$

$$\geq \int \left(\prod_{t=1}^{n} \frac{p}{\sigma_1} e^{-(x_t - \mu_1)^2/2\sigma_1^2}\right)^b d\pi(\mu, \sigma).$$

b. Deduce that the fractional Bayes factor does not exist for this model.

5.22 Consider n observations $x_1, \ldots, x_n$ from Student's t-distribution $\mathcal{T}(\nu, \mu, \sigma)$ and the null hypothesis $H_0 : \mu = 0$.

a. Determine the minimal training sample size for the priors $\pi_0(\sigma) = 1/\sigma$ and $\pi_1(\mu, \sigma) = 1/\sigma$.

b. Show that that the fractional Bayes factor cannot be computed analytically in this case.

Section 5.3.1

5.23 Let f and g be two nondecreasing real functions.

a. Show that
$$\mathbb{E}_\theta[f(x)g(x)] \geq \mathbb{E}_\theta[f(x)]\mathbb{E}_\theta[g(x)]$$
for every distribution P_θ on x.

b. Use a. to show that, if $f(x|\theta)$ is a monotone likelihood ratio density in $T(x)$, the expectation $\mathbb{E}_\theta[g(T(x))]$ is a nondecreasing function of θ. (*Hint:* Use $f(x) = 1 - f(x|\theta')/f(x|\theta)$ and show that $\mathbb{E}_\theta[f(x)] = 0$.)

5.24 Show that Student's t- and noncentral χ^2 distributions have monotone likelihood ratio.

Section 5.3.4

5.25 For the p-value $\tilde{p}$ defined in Example 5.3.13, determine the values of $\tilde{p}(x)$ for $n = 15$ and compare with
$$p(x) = P_{1/2}[f(X|1/2) > f(x|1/2)].$$

5.26 (Johnstone and Lindley (1995)) Consider a point-null hypothesis $H_0 : \theta = \theta_0$ where the p-value φ is well defined. The only available information is that the data is significant at level α, i.e., that $\varphi(x) < \alpha$.

a. Give the Bayes factor R_α of H_0 versus $H_1 : \theta \neq \theta_0$ when the data is significant at level α, for an arbitrary prior distribution π.

b. Given a second significance level β with $\beta < \alpha$, we assume that $R_\alpha < R_\beta$. Establish a sufficient condition on π for this condition to be satisfied.

c. If $R_{\alpha|\beta}$ is the Bayes factor based on the information $\beta < \varphi(x) < \alpha$, show that $R_\alpha = \omega R_\beta + (1 - \omega)R_{\alpha|\beta}$ and deduce that $R_\beta > R_\alpha > R_{\alpha|\beta}$.

d. In the particular case when $\pi(\theta)$ is $\varrho_0 \mathbb{I}_{\theta_0}(\theta) + (1 - \varrho_0)\mathcal{N}(\theta_0, \tau^2)$ and $x_1, \ldots, x_n \sim \mathcal{N}(\theta, \sigma^2)$, show that R_α converges to $(1 - \varrho_0)/\varrho_0\alpha$ when n goes to infinity but that $R_{\alpha|\beta}$ goes to 0.

Section 5.3.5

5.27 For $x \sim \mathcal{N}(\theta, 1)$ and $H_0 : \theta = 0$, determine when the p-value crosses the lower bounds $\underline{P}(x, G_A)$ and $\underline{P}(x, G_S)$.

5.28 (Berger and Delampady (1987)) Consider the case $x \sim \mathcal{B}(n, p)$ when $H_0 : p = 1/2$. For the following class of prior distributions
$$G_C = \{\text{conjugate distributions with mean } 1/2\},$$
show that
$$\underline{P}(x, G_C) = \inf_{g \in G_C} P(H_0|x)$$
$$= \left[1 + \frac{1 - \pi_0}{\pi_0} \sup_{c > 0} \frac{\Gamma(c)\Gamma(x + c/2)\Gamma(n - x + c/2)}{\Gamma(c/2)^2\Gamma(n + c)}\right]^{-1}$$
and derive a table of these lower bounds and the corresponding p-values for $n = 10, 20, 30$ and x going from 0 to $n/2$.

5.29 *(Casella and Berger (1987)) Establish the following lemma, used in Lemma 5.3.16 and Theorem 5.3.19: *if the family G is constituted of the mixtures*

$$g(\theta) = \int_{\Xi} g_\xi(\theta)h(\xi)\,d\xi,$$

for every density h on Ξ, with $g_\xi \in G_0$ and

$$G_0 = \{g_\xi;\ \xi \in \Xi\},$$

then, for any f,

$$\sup_{g \in G} \int f(x|\theta)g(\theta)\,d\theta = \sup_{\xi \in \Xi} \int f(x|\theta)g_\xi(\theta)\,d\theta.$$

5.30 In the case when $x \sim \mathcal{N}(\theta, 1)$ and $H_0 : \theta \le 0$, determine the lower bound

$$\underline{P}(x, G_{SU}) \quad = \quad \inf_{g \in G_{SU}}\ P^g(\theta \le 0|x)$$

$$= \quad \inf_{g \in G_{SU}} \frac{\int_{-\infty}^0 f(x-\theta)g(\theta)\,d\theta}{\int_{-\infty}^{+\infty} f(x-\theta)g(\theta)\,d\theta}$$

for $x < 0$. Does the conclusion of Casella and Berger (1987) still hold? Can you explain why?

5.31 *(Casella and Berger (1987)) Consider a bounded symmetric unimodal function, g. The family of the *scale mixtures of g* is defined by

$$G_g = \{\pi_\sigma;\ \pi_\sigma(\theta) = (1/\sigma)g(\theta/\sigma),\ \sigma > 0\}.$$

If the sample density is $f(x - \theta)$, with f symmetric in 0, if it satisfies the monotone likelihood ratio property, and if $x > 0$, show that

$$\underline{P}(x, G_g) = p(x)$$

for the test of $H_0 : \theta \le 0$.

5.32 *(Casella and Berger (1987)) Consider the test of $H_0 : \theta \le 0$ versus $H_1 : \theta > 0$ when $x \sim f(x - \theta)$. Let h and g be densities on $(-\infty, 0]$ and $(0, +\infty)$.

a. Show that, if $\pi(\theta) = \varrho_0 h(\theta) + (1 - \varrho_0)g(\theta)$,

$$\sup_h P^\pi(\theta \le 0|x) = \frac{\varrho_0 f(x)}{\varrho_0 f(x) + (1 - \varrho_0)\int_0^{+\infty} f(x-\theta)g(\theta)\,d\theta}$$

and deduce that the supremum actually favors H_0 by putting all the mass at the boundary $\theta = 0$.

b. If

$$\pi(\theta) = \varrho_0 h(\theta/\sigma_1)\frac{1}{\sigma_1} + (1 - \varrho_0)g(\theta/\sigma_2)\frac{1}{\sigma_2},$$

show that, when σ_1 is fixed

$$\lim_{\sigma_2 \to \infty}\ P^\pi(\theta \le 0|x) = 1$$

and that, when σ_2 is fixed,

$$\lim_{\sigma_1 \to \infty}\ P^\pi(\theta \le 0|x) = 0.$$

5.33 *(Caron (1994)) In order to alleviate criticisms directed toward point-null hypotheses, $H_0 : \theta = \theta_0$, the formulation of the null hypothesis can be modified according to the prior distribution. For instance, given a prior distribution π on Θ with mode in θ_0 which does not give any prior weight to θ_0, we can propose the transformed hypothesis $H_0^\pi : \pi(\theta) > k^\pi$, where the size of the HPD region is determined by the "objective" requirement $\pi(\pi(\theta) > k^\pi) = 0.5$. Consider the case $x \sim \mathcal{N}(\theta, 1)$ and $\theta_0 = 0$.

a. When π belongs to the family of the $\mathcal{N}(0, \sigma^2)$ distributions, determine k^π and derive the lower bound on the Bayesian answers within this family Compare with the posterior probabilities of Berger and Sellke (1987) for the values of interest.

b. Determine whether the Jeffreys–Lindley paradox occurs in this approach.

c. For the alternative families $\mathcal{U}_{[-c,c]}$ $(c > 0)$ and $\pi(\theta|\lambda) \propto \exp(-\lambda|\theta|)$ $(\lambda > 0)$, derive the corresponding lower bounds.

5.34 *(**Exercise 5.33 cont.**) Consider the case $x \sim \mathcal{C}(\theta, 1)$ when $H_0 : \theta = 0$.

a. Under Berger and Sellke (1987) approach, show that the posterior probability of H_0 when π_c is $\mathcal{U}_{[-c,c]}$ is

$$\pi_c(H_0|x) = \left[1 + (1 + x^2)(\arctan(c - x) + \arctan(c + x))/2c\right]^{-1}.$$

b. Under the approach developed in the previous exercise, show that the corresponding probability is

$$\pi_c(H_0^\pi|x) = \frac{\arctan(c/2 - x) + \arctan(c/2 + x))}{\arctan(c - x) + \arctan(c + x))}.$$

c. Compute and compare the lower bounds for both approaches.

d. Show that

$$\lim_{x \to \infty} \frac{\inf_c \pi_c(H_0^\pi|x)}{\inf_c \pi_c(H_0|x)} = \frac{2}{3}.$$

Section 5.4

5.35 (Hwang et al. (1992)) Show that, under the loss (5.4.2), the p-values defined in Example 5.4.6 are indeed admissible. (*Hint:* Show that the Bayes risks are finite.)

5.36 (Hwang et al. (1992)) The goal of this exercise is to show that, for the two-sided test (5.4.3), the p-value $p(x)$ can take the value 1. (*Hint:* Remember that the UMPU test in this setting is of the form

$$\varphi(x) = \begin{cases} 0 & \text{if } T(x) < c_0 \text{ or } T(x) > c_1, \\ 1 & \text{otherwise}, \end{cases}$$

with $c_0 = c_0(\alpha)$ and $c_1 = c_1(\alpha)$.)

a. Consider $\theta_1 \neq \theta_2$ and

$$c^* = \inf\{T(x); \ f(x|\theta_2) > f(x|\theta_1)\}.$$

Show that $c^* \in [c_0(\alpha), c_1(\alpha)]$ for every $0 < \alpha < 1$.

b. Consider $\theta_1 = \theta_2$. Apply the previous result to

$$f(x|\theta^*) = \mathbb{E}_{\theta_1}[T(x)]f(x|\theta_1), \qquad f(x|\theta^{**}) = T(x)f(x|\theta_1),$$

and conclude.

5.37 (Hwang et al. (1992)) In the normal setting, consider the point-null hypothesis $H_0 : \theta = 0$. Show that, under the loss (5.4.2), the p-value cannot be dominated by a proper posterior probability. (*Hint:* Show first that, for every a and ϵ,

$$\frac{P_\theta(a < |x| < a + \epsilon)}{P_\theta(|x| < a)} \longrightarrow +\infty$$

when θ goes to infinity.)

5.38 (Hwang et al. (1992)) Under the loss (5.4.2), show that $\varphi(x) = 1/2$ *is the unique minimax estimator*. Extend to all strictly convex losses. In this setting, does there exist least favorable distributions?

5.39 (Robert and Casella (1994)) A modification of the loss function (5.4.1) introduces a distance weight in order to penalize in a different manner errors made in the vicinity of the boundary between H_0 and H_1, and those made far away from this boundary.

a. If the null hypothesis is $H_0 : \theta \leq \theta_0$ for $x \sim \mathcal{N}(\theta, 1)$ and the loss function is

$$L(\theta, \varphi) = (\theta - \theta_0)^2 (\mathbb{I}_{H_0}(\theta) - \varphi)^2,$$

give the general form of the Bayes estimators.

b. If $\pi(\theta) = 1$, show that the Bayes estimator is smaller than the p-value if $x > \theta_0$ and larger if $x < \theta_0$.

5.40 (Robert and Casella (1994)) From a model-choice perspective, the loss function incorporates the consequences of an acceptance or of a rejection of the null hypothesis $H_0 : \theta = \theta_0$ in terms of estimation.

a. For the loss function

$$L_1(\theta, (\varphi, \delta)) = d(\theta - \delta)|1 - \varphi| + d(\theta_0 - \theta)|\varphi|,$$

show that the Bayes estimators are $(0, \delta^\pi(x))$ where $\delta^\pi(x)$ is the regular Bayes estimator of θ under $d(\theta - \delta)$ for every d and π.

b. For the loss function

$$L_2(\theta, (\varphi, \delta)) = d(\theta - \delta)|1 - \varphi| + d(\theta_0 - \delta)|\varphi|,$$

show that the Bayes rule is $(1, \theta_0)$ for every π and d.

c. For the loss function

$$L_3(\theta, (\varphi, \delta)) = (\delta - \theta)^2 (\mathbb{I}_{H_0}(\theta) - \varphi)^2,$$

show that the associated Bayes rule is $(0, \theta_0)$, i.e., that the Bayes procedure always rejects the null hypothesis $H_0 : \theta = \theta_0$, but always uses θ_0 as an estimator of θ.

d. Study the Bayes procedures under the modified loss

$$L_4(\theta, (\varphi, \delta)) = \left[1 + (\delta - \theta)^2\right] \left[1 + (\mathbb{I}_{H_0}(\theta) - \varphi)^2\right],$$

to examine whether they are less paradoxical.

e. Show that the loss function

$$L_5(\theta, (\varphi, \delta)) = \xi(\delta - \theta)^2 |1 - \varphi| + \{(\delta - \theta_0)^2 + (\theta - \theta_0)^2\}|\varphi|,$$

provides a reasonable pre-test Bayes procedure that avoids the paradoxes of L_1, L_2, and L_3 if and only if $\xi > 1$.

Section 5.5.1

5.41 Consider two independent observations x_1, x_2 from a Cauchy distribution $\mathcal{C}(\theta, 1)$. For $\pi(\theta) = 1$, give the shape of the α-credible HPD region. What alternative (and more convincing) α-credible region could you propose?

5.42 Give the α-credible region when $x \sim \mathcal{P}(\lambda)$ and $\lambda \sim \mathcal{G}(\delta, \beta)$. Study the evolution of this region as a function of δ and β. Examine the particular case of the noninformative distribution.

5.43 *An alternative notion of α-credible regions is studied in this exercise. The best Bayes center at level α is the estimator $\delta_\alpha^\pi(x)$, center of the ball of smallest radius with coverage $1 - \alpha$, i.e.,

$$P^\pi(||\theta - \delta_\alpha^\pi(x)|| < k|x) = \sup_\delta P^\pi(||\theta - \delta(x)|| < k|x) = 1 - \alpha.$$

a. Show that, if the posterior distribution is spherically symmetric and unimodal, the corresponding region is the HPD region.

b. Consider $x \sim \mathcal{N}(\theta, 1)$, $\theta \sim \mathcal{N}(0, \tau^2)$, and $\pi(\tau^2) = 1/\tau^{3/2}$. Determine the posterior distribution. Show that this distribution is unimodal when $0 < x^2 < 2$ and bimodal otherwise, with second mode

$$\delta(x) = \left(1 - \frac{1 - \sqrt{1 - (2/x^2)}}{2}\right) x.$$

Derive the best Bayes center and show that, if α is large enough, δ_α^π is discontinuous and close to

$$\phi(x) = (1 - \frac{1}{2x^2})^+ x,$$

i.e., that this Bayes estimator replicates the James–Stein estimator.

c. Generalize b. when $\pi(\tau^2) = \tau^{-\upsilon}$.

d. Show that the best Bayes center associated with a proper prior distribution π is admissible under the loss

$$L(\theta, \delta) = \mathbb{I}_{(k,+\infty)}(||\theta - \delta||^2).$$

5.44 *(Thatcher (1964)) Consider $x \sim \mathcal{B}(n, \theta)$ and, for $0 < \alpha < 1$ and a prior π on θ, define θ_x^π by $P^\pi(\theta \leq \theta_x^\pi|x) = \alpha$.

a. If $\pi(\theta) = (1 - \theta)^{-1}$, show that $P_\theta(\theta \leq \theta_x^\pi) \leq \alpha$ for $\theta > 0$.

b. If $\pi(\theta) = \theta^{-1}$, show that $P_\theta(\theta \leq \theta_x^\pi) \geq \alpha$ for $\theta < 1$.

c. Define θ_x^λ associated with $\pi(\theta) = \theta^{\lambda-1}(1 - \theta)^{-\lambda}$, $0 \leq \lambda \leq 1$. Show that θ_x^λ is increasing in λ and deduce that

$$\lim_{\theta \uparrow \theta_x^\lambda} P_\theta(\theta \leq \theta_x^\lambda) \geq \alpha \geq \lim_{\theta \downarrow \theta_x^\lambda} P_\theta(\theta \leq \theta_x^\lambda).$$

5.45 *(Hartigan (1983)) Consider $x \sim \mathcal{P}(\lambda)$ and for $0 < \alpha < 1$, and a prior π on λ, define λ_x^π by

$$P^\pi(0 \leq \lambda \leq \lambda_x^\pi|x) = \alpha.$$

a. Show that, if $\pi(\lambda) = 1/\lambda$, $P_\lambda(\lambda \leq \lambda_x^\pi) \leq \alpha$ for every λ.

b. Show that, if $\pi(\lambda) = 1$, $P_\lambda(\lambda \leq \lambda_x^\pi) \geq \alpha$ for every λ. (*Hint:* Use the following relation:

$$\sum_{x=x_0}^\infty e^{-\lambda}\frac{\lambda^x}{x!} = \int_0^\infty \frac{u^{x_0-1}}{(x_0-1)!}e^{-u}du. \Big)$$

5.46 A famous problem in classical Statistics is the *Behrens–Fisher* problem. It stems from a simple setting of two normal populations with unknown means and variances because there is no UMP or UMPU test to compare the means. Consider $x_1,\ldots,x_n$ a sample from $\mathcal{N}(0,\sigma^2)$ and $y_1,\ldots,y_m$ a sample from $\mathcal{N}(\mu,\tau^2)$ where θ,μ,τ,σ are unknown.

a. *Show that there is no UMPU test procedure for the hypothesis $H_0 : \theta = \mu$. (*Hint:* Condition on s_x^2 and s_y^2, given below, to show that the UMPU procedures vary with s_x^2 and s_y^2.)

b. Explain why a reasonable test should be based on the pivotal quantity

$$T = \frac{(\theta-\mu)-(\bar{x}-\bar{y})}{\sqrt{s_x^2/n + s_y^2/m}}$$

with $\bar{x} = \sum_i x_i/n$, $\bar{y} = \sum_j y_j/m$, $s_x^2 = \sum_i(x_i-\bar{x})^2/n-1$, and $s_y^2 = \sum_j(y_j-\bar{y})^2/m-1$.

c. Show that the distribution of T depends on σ/τ even when $\theta = \mu$, and that this distribution *is not* a Student's t-distribution.

d. Give the posterior of T when $\pi(\theta,\mu,\sigma,\tau) = 1/\sigma^2\tau^2$ and show that it depends only on $(s_x/\sqrt{n})(s_y/\sqrt{m})$. [*Note:* See Robinson (1982) for a detailed survey of the different issues related to this problem.]

Section 5.5.2

5.47 (Casella and Berger (1990)) Consider $x \sim \mathcal{N}(\mu,1)$ and

$$C_a(x) = \{\mu;\ \min(0,x-a) \leq \mu \leq \max(0,x+a)\}.$$

a. Consider $a = 1.645$. Show that C_a is a confidence interval at level 95% with

$$P_0(0 \in C_a(x)) = 1.$$

b. If $\pi(\mu) = 1$ and $a = 1.645$, show that C_a is also a 0.1-credible region and that

$$P^\pi(\mu \in C_a(x)|x) = 0.90$$

if $|x| \leq 1.645$ and

$$\lim_{|x|\to+\infty} P^\pi(\mu \in C_a(x)|x) = 1.$$

5.48 Consider $x \sim f(x|\theta)$ with $\theta \in \mathbb{R}$ and π a prior distribution on θ. If we define the α-credible set $(-\infty,\theta_x)$ by $P^\pi(\theta \geq \theta_x|x) = \alpha$, show that this one-sided interval cannot be at level α in the frequentist sense. (*Hint:* Show that $P(\theta \geq \theta_x|\theta \leq \theta_0) > \alpha$ for some θ_0.)

5.49 *(Fieller (1954)) In the setting of *calibration* (see Exercise 4.47), confidence sets need to have infinite length to maintain a fixed confidence level, as shown by Gleser and Hwang (1987). Consider $(x_1,y_1),\ldots,(x_n,y_n)$ a sample from $\mathcal{N}_2(\mu,\Sigma)$. The parameter of interest is θ, the ratio of the two means μ_x and μ_y.

a. Define $\bar{z}_\theta = \bar{y} - \theta\bar{x}$. Show that

$$\bar{z}_\theta \sim \mathcal{N}\left(0, \frac{1}{n}(\sigma_y^2 - 2\theta\sigma_{xy} + \theta^2\sigma_x^2)\right)$$

and that

$$\hat{v}_\theta = \frac{1}{n-1}(s_y^2 - 2\theta s_{xy} + \theta^2 s_x)$$

is an unbiased estimator of v_θ, the variance of $\bar{z}_\theta$, when $\bar{x}$, $\bar{y}$, s_x^2, s_{xy}, and s_y^2 denote the usual empirical moments and

$$\Sigma = \begin{pmatrix} \sigma_r^2 & \sigma_{ry} \\ \sigma_{xy} & \sigma_y^2 \end{pmatrix}.$$

b. Show that $\bar{z}_\theta$ and $\hat{v}_\theta$ are independent and that $(n-1)\hat{v}_\theta/v_\theta \sim \chi_{n-1}^2$. Deduce that $\{\theta; \bar{z}_\theta/\hat{v}_\theta \leq t_{n-1,\alpha/2}^2\}$ defines a $(1-\alpha)$ confidence set.

c. Show that this confidence set defines a parabola in θ and can be an interval, the complement of an interval or the whole real line.

Section 5.5.3

5.50 *Domination of the usual estimator as center of a confidence region does not necessarily follow from the corresponding domination for the quadratic loss. Show that, in the normal case, if

$$\delta_a^{JS}(x) = \left(1 - \frac{a}{||x||^2}\right)x,$$

the recentered confidence region

$$C_a^{JS}(x) = \{\theta; ||\theta - \delta_a^{JS}(x)||^2 \leq c_\alpha\},$$

does not dominate the usual confidence region, even though δ_a^{JS} dominates δ_0 when $a \leq 2(p-2)$. (*Hint:* Consider $\theta = 0$.)

5.51 (Casella et al. (1993b)) Show that the rational loss given in Section 5.5,

$$L(\theta, C) = \frac{\text{vol}(C)}{k + \text{vol}(C)} - \mathbb{I}_C(\theta),$$

does not lead to Berger's paradox in the normal case.

5.52 *(Casella et al. (1993b)) Consider a general loss function of the form

$$L(\theta, C) = S(\text{vol}(C)) - \mathbb{I}_C(\theta),$$

with S increasing, $0 \leq S(t) \leq 1$.

a. Show that the Bayes estimators are the HPD regions.

b. Show that, if $x \sim \mathcal{N}_p(\theta, I_p)$ and $\theta \sim \mathcal{N}_p(\mu, \tau^2 I_p)$, the Bayesian credible sets C^π are not empty if $S(t) = t/(a+t)$.

c. Determine the smallest radius of C^π as τ varies.

d. Consider $\bar{x} \sim \mathcal{N}(\theta, \sigma^2/n)$ and $s^2 \sim \sigma^2\chi_q^2$. Under the rational loss, show that

$$C^\pi(\bar{x}, s^2) = \left\{\theta; |\theta - \bar{x}| \leq \frac{t^* s}{\sqrt{n}}\right\},$$

where t^* is the solution of

$$\min_t \left(\frac{2ts/\sqrt{n}}{a + 2ts/\sqrt{n}} - P(|T_{n-1}| < t)\right).$$

Deduce that $P(|T_{n-1}| < t^*(s)|s) \geq 1/2$.

5.53 (Walley (1991)) Consider the double-exponential distribution, $f(x|\theta) = (1/2)\exp(-|x - \theta|)$.

  a. Show that $C_x = (-\infty, x]$ is a 50% confidence interval.

  b. Show that $P_\theta(\theta \in C_x | x < 0) < 0.5$ for every θ.

  c. Let $\varphi(x) = (e^{2x}/2)\mathbb{I}_{x<0}$. Show that

$$\mathbb{E}_\theta[\mathbb{I}_{x<0}(\mathbb{I}_{C_x}(\theta) - 1/2) + \varphi(x)] \geq 0$$

     and deduce that $\gamma(x) = 1/2$ is not an admissible confidence report under squared error loss for C_x.

Note 5.7.3

5.54 *(Brown (1967)) In the setting of Example 5.5.9, show that

$$P\left(\sqrt{n}|\bar{x} - \theta| \leq ks|s \leq 1\right) \leq \alpha > P\left(\sqrt{n}|\bar{x} - \theta| \leq ks|s > 1\right)$$

and derive a positively relevant subset. (*Hint:* Show that

$$P\left(\sqrt{n}|\bar{x} - \theta| \leq ks|s\right)$$

is increasing in s.)

5.55 (Walley (1991)) Consider a sample $x_1, \ldots, x_n$ from $\mathcal{U}_{[\theta, \theta+1]}$.

  a. Show that uniformly more accurate one-sided confidence intervals are of the form $C_x = [(x_{(1)} + 1 - K) \wedge (x_{(n)} - 1), x_{(1)} + 1]$ and check that the confidence level is $\gamma = 1 - (1 - K/2)^n$.

  b. For $n = 1$ and $\gamma = 1/2$, show that $C_x = [x, x + 1]$. Consider a strictly decreasing bounded function f and $\varphi(x) = (f(x) - f(x + 1)) \wedge (f(x - 1) - f(x))$. Verify that

$$\mathbb{E}_\theta[f(\mathbb{I}_{C_x}(\theta) - 0.5)] = 0.25 \int_\theta^{\theta+1} (f(x - 1) - f(x))\, dx$$

     and

$$\mathbb{E}_\theta[\varphi(x)] \leq \frac{1}{8} \int_\theta^{\theta+1} (f(x - 1) - f(x))\, dx.$$

  c. Deduce that

$$\mathbb{E}_\theta[f(\mathbb{I}_{C_x}(\theta) - 0.5) - \varphi(x)] \geq 0$$

     for every θ and that $\gamma = 1/2$ is not an admissible report.

  d. For $n \geq 2$, we define

$$B = \{(x_1, \ldots, x_n); x_{(n)} - x_{(1)} \geq 2 - K\}.$$

     Show that

$$P_\theta(\theta \in C(x_1, \ldots, x_n)|(x_1, \ldots, x_n) \in B) = 1$$

     and conclude that B is a relevant subset.

Note 5.7.4

5.56 (Berger et al. (1997)) For the report $\gamma(x)$ given in (5.7.1), show that

$$\gamma(x) = \frac{s}{1 + s} \quad \text{if} \quad s < r, \qquad \gamma(x) = \frac{1}{1 + s} \quad \text{if} \quad s > a.$$

5.57 Show that, in the setting of Example 5.7.4, $\Psi(1) > 1$ and give the Bayes factor in favor of H_0.

5.58 (Lindley (1997)) When introducing the third decision -1 in the testing problem, consider the following extension to the $0 - 1$ loss function:

$$L(\theta, \varphi) = \begin{cases} \ell_i & \text{if } \varphi = 1 - i \text{ and } H_i \text{ is true,} \\ m_i & \text{if } \varphi = -1 \text{ and } H_i \text{ is true.} \end{cases}$$

Compute the posterior losses and show that $\varphi = -1$ if

$$\frac{m_1 \varrho}{\ell_0 - m_0} < D_{10}(x) < \frac{(\ell_1 - m_1)\varrho}{m_0},$$

where ϱ denotes the prior odds, π_1/π_0.

5.59 (Lindley (1997)) Show that $S(x)$ given in (5.7.2) is not an ancillary statistic, except when

$$\tau(t) + \varrho = 1 + \frac{\varrho}{t}, \qquad t > c,$$

where c is defined by $F_0(c) = 1 - \varrho F_1(c)$ and $\tau(t)$ is given by $F_0(t) = 1 - \varrho F_1(\tau(t))$. Show that this property holds when $B_{10}(x)$ has the same distribution under m_1 than $B_{01}(x)$ under m_0. [*Note:* See Berger et al. (1994, p. 1798).]

5.7 Notes

5.7.1 P-values and Bayesian decisions

A common criticism of the comparison of Section 5.3.5 is that it is meaningless: the two types of answers are conceptually different and p-values are not probabilities. The answer to this criticism is that, besides being used *as probabilities* by practionners, p-values are addressing the same inferential problem as Bayesian posterior probabilities, from a decision-theoretic point of view. It thus does make sense to compare them directly.

Consider an $a_0 - a_1$ loss function as in (5.2.1). The UMPU *minimax test* is then

$$\varphi(x) = \begin{cases} 1 & \text{if } p(x) > \frac{a_1}{a_0 + a_1}, \\ 0 & \text{otherwise.} \end{cases}$$

In fact, when power functions are continuous and hypotheses are contiguous (see Lehmann (1986, Chapter 4)), a UMPU test satisfies

$$\sup_{\Theta_0} P_\theta(\varphi(x) = 0) = \alpha = \inf_{\Theta_1} P_\theta(\varphi(x) = 0) = 1 - \sup_{\Theta_1} P_\theta(\varphi(x) = 1).$$

Moreover, when φ is minimax under this loss, it satisfies

$$\sup_{\Theta_0} R(\theta, \varphi) = a_0 \sup_{\Theta_0} P_\theta(\varphi(x) = 0)$$

$$= \sup_{\Theta_1} R(\theta, \varphi) = a_1 \sup_{\Theta_1} P_\theta(\varphi(x) = 1).$$

Therefore, under regularity assumptions satisfied, for instance by exponential families, φ is such that

$$\sup_{\Theta_0} P_\theta(\varphi = 0) = \frac{a_1}{a_1 + a_0}.$$

It then follows from Proposition 5.2.2 that it is legitimate to compare the p-value $p(x)$ with posterior probabilities, since the Bayesian decision procedure is given by

$$\gamma^\pi(x) = \begin{cases} 1 & \text{if } P^\pi(\theta \in \Theta_0|x) > \frac{a_0}{a_0+a_1}, \\ 0 & \text{otherwise} \end{cases}$$

and both approaches compare a continuous evaluation (p-values or posterior probability) to the same bound.

5.7.2 Unequal prior probabilities

Another criticism of the lower bound evaluation of Section 5.3.5, found, for instance, in Casella and Berger (1987), is that the lower bound is not computed on the set of *all* prior distributions, since it only considers the prior probability $\varrho_0 = 1/2$. Obviously, if ϱ_0 can be modified too, it is always possible to find a Bayesian answer smaller than the p-value, since the lower bound over all Bayesian answers is then 0 for every x (corresponding to $\varrho_0 = 0$). However, for a fixed value of $\varrho_0 > 0$, there always are values of x for which the lower bound on posterior probabilities is higher than the p-value.

A refined version of this criticism is to consider that the weight $\varrho_0 = 1/2$ is not necessarily the most objective probability, and that it should be determined in terms of the prior π itself. In fact, as mentioned above, priors of the form $\pi(\theta) = \varrho_0 \mathbb{I}_{\theta_0}(\theta) + (1 - \varrho_0)\pi_1(\theta)$ are quite artificial. While such priors are necessary to solve the testing problem, it is more natural to think of π as a modification of the original prior π_1 in the light of this problem. The inferential problem, that is, the fact that θ_0 is of interest, contains some residual information strong enough to justify a modification of the prior distribution (otherwise, the test question should be modified to become compatible with the prior information). It thus makes sense to require that the weight ϱ_0 should appear as a function of π_1. (This point will reappear in Chapter 7 with the case of nested models in model choice: the upper model, which contains all others, should be more likely than the others.)

Example 5.7.1 (Example 5.3.6 continued) Since $H_0 : \theta = 0$ is to be tested, the prior probability of H_0 is null under any continuous prior π_1. Nonetheless, it is reasonable to require that H_0 should have a larger prior probability if π_1 is $\mathcal{N}(0, 1)$ than if π_1 is $\mathcal{N}(0, 10)$, since every neighborhood of 0 is less probable under the latter distribution. This is why the Jeffreys–Lindley paradox is deemed a "paradox": increase in the probabilities from Table 5.2.4 to Table 5.2.4 seems counterintuitive.   ‖

Unfortunately, a determination of the weight ϱ_0 as a function of π_1 is quite controversial and we only briefly mention a solution proposed in Robert and Caron (1996) (see Spiegelhalter and Smith (1980) for another approach based on the most favorable virtual observations). The basic idea is that the weight ϱ_0 should satisfy

$$(1 - \varrho_0)\pi_1(\theta_0) = \varrho_0,$$

in order for θ_0 to be equally weighted under both hypotheses. Of course, we are comparing a weight under the Dirac mass at 0, ϱ_0, with an instantaneous weight with respect to the Lebesgue measure, $(1 - \varrho_0)\pi_1(\theta_0)$, and the comparison is not justified from a mathematical point of view (since the value of the density π_1 at a given point like θ_0 is arbitrary). Moreover, the above equation does not always allow for a solution.

Example 5.7.2 (Example 5.3.6 continued) When $\pi_1(\theta)$ is a normal prior $\mathcal{N}(0, n)$, the above equality leads to the weight

$$\varrho_0 = \frac{\pi_1(0)}{1 + \pi_1(0)} = \frac{1}{1 + \sqrt{2\pi n}},$$

and the corresponding posterior probability of H_0 is

$$\left(1 + \frac{1}{\varrho_0} \frac{\varrho_U}{1} \frac{m_1(x)}{\varphi(x)}\right)^{-1} = \left(1 + \sqrt{2\pi \frac{n}{n+1}} e^{x^2/2 - x^2/2(n+1)}\right)^{-1}$$

$$= \left(1 + \sqrt{\frac{2\pi n}{n+1}} e^{\frac{n}{2(n+1)} x^2}\right)^{-1}.$$

Notice that this approach avoids the Jeffreys–Lindley paradox, since the limiting probability (when n goes to $+\infty$) is

$$\left(1 + \sqrt{2\pi} e^{x^2/2}\right)^{-1}.$$

This value also happens to be the posterior probability associated with the Lebesgue prior, $\pi(\theta) = 1$.   ‖

5.7.3 Conditional evaluation of confidence regions

A critical assessment of the Neyman–Pearson confidence regions (and more generally of frequentist procedures) follows from the *conditional* analysis of Kiefer (1977) and Robinson (1979). Lehmann (1986, Chapter 10) gives an overview of this approach (see also Buehler (1959), Pierce (1973), Casella (1987, 1992), Maatta and Casella (1990), and Goutis and Casella (1991, 1992)). These works have shown that the classical confidence procedures are often suboptimal when considered from a conditional viewpoint.

Definition 5.7.3 *Consider C_x, a confidence region at significance level α. A set $A \subset \mathcal{X}$ is said to be a negatively biased relevant subset for the confidence region C_x if there exists $\epsilon > 0$ such that*

$$P_\theta(\theta \in C_x | x \in A) \leq 1 - \alpha - \epsilon$$

for every $\theta \in \Theta$.

We can similarly define *positively biased relevant subsets*. This notion is generalized in Robinson (1979) into the notion of *relevant betting procedures*. The existence of such sets questions the notion of a confidence level α itself because, depending on the conditioning set, the coverage probability may vary and even fall below the nominal minimal confidence statement. Obviously, this criticism can be transfered to testing procedures by a duality argument.

In the setting of Example 5.5.8, while working on t-tests, Brown (1967) establishes that there exist positively biased relevant sets of the form $\{|x| < k\}$ and this implies

$$P_\theta(\theta \in C_x | |x| > k) \leq 1 - \alpha$$

(see also Exercise 5.54). Such phenomena led Kiefer (1977) to suggest a partition of the sample space $\mathcal{X}$ into subsets, and to allocate each of these subsets with a different confidence level (see also Brown (1978)). Following Fisher's analysis, he suggested that these subsets should be indexed by ancillary statistics. For instance, the adequate ancillary statistic for Example 2.3.1 is $x_1 - x_2$.

Unfortunately, in most settings, the choice of the ancillary statistic modifies the confidence report, and Berger and Wolpert (1988) give an example in which different ancillary statistics lead to different confidence statements, a setting incompatible with the Likelihood Principle. We consider that, fundamentally, the problem exhibited by the existence of relevant biased sets is not related to the confidence region C_x itself but rather with the associated confidence level α, which should be replaced by a more adaptive (or more conditional) confidence statement $\alpha(x)$ (see Section 4.2). In fact, the existence of relevant betting procedures is equivalent to the domination of the constant confidence report under quadratic loss (Robinson (1979)).

5.7.4 Reconciliation perspective

While Section 5.3 has shown that the frequentist answers, that is, the p-values, were intrinsically *and* numerically different from their Bayesian counterparts (see also Note 5.7.1), a modification of the decision setting, brought up by Berger, Brown and Wolpert (1994), allows for a partial reconciliation of both perspectives. While, from a Bayesian point of view, reconciliation is not an important feature—the basic purpose of a procedure is to act in an optimal way when faced with a decision problem, rather than to enjoy a long run stability—it offers several advantages from a practical point of view: firstly, it makes statisticians more comfortable with using a Bayesian procedure if this procedure simultaneously enjoys frequentist properties. Secondly, it eliminates the difficulty with the interpretation of a p-value as a posterior probability.

The modification is to introduce a "no-decision" alternative to the "accept" and "reject" answers used in classical tests. Even though this additional possibility may seem ludicrous from a decision-making point of view, it certainly makes sense from a statistical point of view: there are indeed cases where the data is too inconclusive to answer satisfactorily about a given hypothesis H_0 and when we want to ask the client for more observations, or for more precise prior information. In fact, this perspective has been around for sequential testing since Wald's *sequential probability ratio test* (see Lehmann (1986)). (Notice, however, that the procedure of Berger et al. (1994) does not account for repeated testing, which affects the acceptance levels. See also Example 1.3.6.)

In the case of two simple hypotheses,

$$H_0 : x \sim m_0(x) \quad \text{versus} \quad H_1 : x \sim m_1(x),$$

where m_0 and m_1 are known densities, the Bayes factor B_{10} is equal to the likelihood ratio $m_1(x)/m_0(x)$. If -1 denotes the no-decision answer, the modified Bayesian test of Berger et al. (1994) is of the form

$$(5.7.1) \qquad \varphi(x) = \begin{cases} 1 & \text{if } B_{10}(x) \geq a, \\ 0 & \text{if } B_{10}(x) \leq r, \\ -1 & \text{if } r < B_{10}(x) < a, \end{cases}$$

with an associated probability error report given by

$$\gamma(x) = \begin{cases} 1/(1 + B_{10}(x)) & \text{if } B_{10}(x) \geq a, \\ B_{10}(x)/(1 + B_{10}(x)) & \text{if } B_{10}(x) \leq r. \end{cases}$$

Notice that $\gamma(x)$ is the posterior probability of the rejected hypothesis and is thus optimal under squared error loss. (But the procedure φ is arguably not a decision procedure. See Exercise 5.58.)

Now, if we denote F_0 and F_1 the c.d.f.'s of $B_{10}(x)$ associated with m_0 and m_1 respectively, and if we define $\Psi(b) = F_0^{-1}(1 - F_1(b))$, then $\Psi^{-1}(b) = F_1^{-1}(1 - F_0(b))$, Berger et al (1994) take

$$r = 1 \quad \text{and} \quad a = \Psi(1) \quad \text{if } \Psi(1) > 1$$
$$r = \Psi^{-1}(1) \quad \text{and} \quad a = 1 \quad \text{if } \Psi(1) < 1.$$

They show that the report $\gamma(x)$ is then correct from a conditional frequentist perspective: conditional on

(5.7.2) $$S(x) = \min\{B_{10}(x), \Psi^{-1}(B_{10}(x))\},$$

the procedure (φ, γ) satisfies

$$P_0(B_{10}(x) \geq a | S(x) = s) = \gamma(s), \qquad P_1(B_{10}(x) \leq r | S(x) = s) = \gamma(s),$$

where $\gamma(x)$ only depends on s (Exercise 5.56). Notice however that $S(x)$ is not an ancillary statistic, except in special cases (Exercise 5.59).

The extension of this result to composite hypotheses,

$$H_0 : \theta = \theta_0 \quad \text{versus} \quad H_1 : \theta \in \Theta_1$$

follows by representing H_1 as in Section 5.3.5, that is, with

$$H_1 : x \sim m_1(x) = \int_{\Theta_1} f(x|\theta)\pi_1(\theta)d\theta.$$

Then Berger, Boukai and Wang (1997) show that the conditional frequentist evaluation under H_0 still coincides with the Bayesian report, in a weaker sense that, while being satisfactory from a Bayesian perspective, is less compelling from a frequentist perspective (Hinkley (1997), Louis (1997)).

Example 5.7.4 (Berger et al. (1997)) For $x_1, \ldots, x_n$ i.i.d. $\mathcal{N}(\theta, \sigma^2)$, with σ known, consider the test of $H_0 : \theta = \theta_0$ under the conjugate prior $\theta \sim \mathcal{N}(\mu, k\sigma^2)$. Then, if $z = \sqrt{n}(\bar{x}_n - \theta_0)/\sigma$,

$$m_0(z) = \frac{1}{\sqrt{2\pi}} \exp\{-z^2/2\}$$

and

$$m_1(z) = \frac{1}{\sqrt{2\pi}\sqrt{1+kn}} \exp\left\{\frac{-(z + \sqrt{kn}\Delta)^2}{2(1+kn)}\right\},$$

where $\Delta = (\theta_0 - \mu)/\sqrt{k}\sigma$. Then the Bayes factor is given by

$$B_{10}(x) = \sqrt{1+kn} \exp\left\{-\frac{kn}{2(1+kn)}\left[z - \frac{\Delta}{\sqrt{kn}}\right]^2 + \frac{\Delta^2}{2}\right\},$$

$\Psi(1) > 1$, $r = 1$ and $a = F_0^{-1}(1 - F_1(1))$. ‖

Bayesian Calculations

The contraption began to quiver, steam hissing out from two or three places. The hiss grew to a shriek, and the thing began trembling. It groaned ominously. The shriek became ear-piercing. It shook so hard the table moved. The balding man threw himself at the table, fumbling a plug loose on the largest cylinder. Steam rushed out in a cloud, and the thing went still.

Robert Jordan, *Lord of Chaos, Book VI of the Wheel of Time.*

6.1 Implementation difficulties

At this stage of the book, we need to discuss a practical aspect of the Bayesian paradigm, namely, the computation of Bayes estimators. The ultimate simplicity of the Bayesian approach is that, given a loss function L and a prior distribution π, the Bayes estimate associated with an observation x is the (usually unique) decision d minimizing the posterior loss

$$(6.1.1) \qquad L(\pi, d|x) = \int_\Theta L(\theta, d)\pi(\theta|x)\, d\theta.$$

However, minimizing (6.1.1) can be hindered by two difficulties in practice:

(i) the explicit computation of the posterior distribution, $\pi(\theta|x)$, may be impossible; and

(ii) even if $\pi(\theta|x)$ is known, this does not necessarily imply that minimizing (6.1.1) is an easy task; indeed, when analytic integration is impossible, numerical minimization sometimes calls for a formidable amount of computing time, especially when Θ and $\mathcal{D}$ have large dimensions.

Point (i) may seem to be a minor and formal difficulty since minimizing (6.1.1) is actually equivalent to minimizing

$$\int_\Theta L(\theta, d)\pi(\theta)f(x|\theta)\, d\theta,$$

which does not require an evaluation of $\pi(\theta|x)$. However, we saw in Chapters 2 and 4 that classical losses, like the quadratic losses, lead directly to estimators expressed through the posterior distribution, like the posterior mean

$$\delta^\pi(x) = \int_\Theta \theta\,\pi(\theta|x)\,d\theta$$
$$= \frac{\int_\Theta \theta\,\pi(\theta)f(x|\theta)\,d\theta}{\int_\Theta \pi(\theta)f(x|\theta)\,d\theta},$$

for the quadratic loss; they thus necessitate direct computation of posterior moments or other posterior quantities. A similar comment applies for the derivation of other posterior quantities of interest, such as posterior quantiles, Bayes factors, or confidence regions.

A simplifying answer to these computational difficulties is to only use sampling models, prior distributions, and losses which lead to explicit solutions for the minimization of (6.1.1). This restrictive approach was technically justified when the computational tools described below were not available, but is unacceptable on subjective grounds, since loss functions and prior distributions should be constructed according to the decision problem, not because they provide closed-form answers, as already stressed in Chapter 3.[1]

This chapter is thus intended to avoid a systematic recourse to simple prior distributions and losses by providing the reader with an array of recent and sophisticated approximation methods that can be used when no analytical expression of the posterior distribution or estimators is available. This chapter is only an introduction to these methods; the reader is referred to Robert and Casella (1999, 2004) for a more thorough treatment.

Although optimization problems like loss minimization or computation of a MAP estimator can also be addressed by simulation techniques (see Geyer and Thompson (1992), Geyer (1996), Robert and Casella (2004, Chapter 5), or Doucet and Robert (2000)), we will focus in this chapter on approximations to $\pi(\theta|x)$ and integrals involving $\pi(\theta|x)$ because this is the cornerstone of computational difficulties with Bayesian inference. In addition, if $\pi(\theta|x)$ can be correctly approximated, it is usually possible to derive an approximation of $L(\pi, d|x)$ for an arbitrary d, and then to use a classical minimization method.

We now introduce a series of examples used throughout this chapter to illustrate different computational methods.

Example 6.1.1 Consider $x_1, \ldots, x_n$ a sample from $\mathcal{C}(\theta, 1)$, a Cauchy distribution with location parameter θ, and $\theta \sim \mathcal{N}(\mu, \sigma^2)$, with known hyper-

---

[1] Classical illustrations resort to such simple settings because they allow for a clearer and more concise presentation of points of interest, and this book has made intensive use of exponential families, conjugate priors, and quadratic losses for this reason. Nevertheless, a more adaptive approach, relying for instance on mixtures of conjugate priors, should be adopted in practical settings.

parameters μ and σ^2. The posterior distribution of θ is then

$$\pi(\theta|x_1, \ldots, x_n) \propto e^{-(\theta-\mu)^2/2\sigma^2} \prod_{i=1}^{n}[1 + (x_i - \theta)^2]^{-1},$$

which cannot be integrated analytically. When δ^π is the posterior mean,

$$\delta^\pi(x_1, \ldots, x_n) = \frac{\int_{-\infty}^{+\infty} \theta e^{-(\theta-\mu)^2/2\sigma^2} \prod_{i=1}^{n}[1 + (x_i - \theta)^2]^{-1}d\theta}{\int_{-\infty}^{+\infty} e^{-(\theta-\mu)^2/2\sigma^2} \prod_{i=1}^{n}[1 + (x_i - \theta)^2]^{-1}d\theta},$$

its calculation requires two numerical integrations (one for the numerator and one for the denominator). The computation of the variance calls for an additional integration. Moreover, the usually *multimodal* structure of this distribution (see Exercise 1.28) may require special tuning for standard integration packages.   ‖

As we have seen before, the computational problem may result from the choice of the loss, even when the prior distribution is conjugate.

Example 6.1.2 Let $x|\theta \sim \mathcal{N}_p(\theta, \sigma^2 I_p)$ and $\theta|\mu, \tau \sim \mathcal{N}_p(\mu, \tau^2 I_p)$, with known hyperparameters μ and τ. The posterior distribution on θ is then quite manageable since

$$\theta|x \sim \mathcal{N}_p\left(\frac{\sigma^2\mu + \tau^2 x}{\sigma^2 + \tau^2}, \frac{\sigma^2\tau^2}{\sigma^2 + \tau^2}I_p\right).$$

When $||\theta||^2$ is the parameter of interest, the usual rescaled quadratic loss is

$$L(\theta, \delta) = \frac{(\delta - ||\theta||^2)^2}{2||\theta||^2 + p},$$

as in Saxena and Alam (1982). It leads to the following Bayes estimator:

$$\delta^\pi(x) = \frac{\mathbb{E}^\pi[||\theta||^2/(2||\theta||^2 + p)|x]}{\mathbb{E}^\pi[1/(2||\theta||^2 + p)|x]}.$$

Although $(\sigma^{-2} + \tau^{-2})||\theta||^2$ is distributed a posteriori as a $\chi_p^2(\lambda)$ random variable, with

$$\lambda = \frac{||\sigma^2\mu + \tau^2 x||^2}{\sigma^2\tau^2(\sigma^2 + \tau^2)},$$

an analytic version of δ^π does not exist and numerical approximation is again necessary. Notice that, in this case, numerical integration is more complicated than for Example 6.1.1 because the density of $\chi_p^2(\lambda)$ (see Appendix A) involves a modified Bessel function, $I_{(p-2)/2}(t)$, which must be approximated in most settings by a series of weighted central chi-squared densities or by a continued fraction approximation (see Exercise 4.35). An alternative approach is to integrate instead over θ, but this is only feasible for small p's.   ‖

Chapter 10 will also provide settings where approximations of Bayes estimators are necessary. Indeed, most *hierarchical Bayes estimators* cannot be computed analytically; for instance, this is the case for normal observations (see Lemma 10.2.16) and graphical models (see Note 10.7.1). Moreover, a numerical approximation of these estimators can get quite involved, especially for higher dimensions.

Example 6.1.3 The call to an *auxiliary variable* in a multivariate Student's t model reduces the number of integrations to one, as pointed out by Dickey (1968). Let us recall that, if

$$x \sim \mathcal{N}_p(\theta, \sigma^2 I_p), \qquad \theta \sim \mathcal{T}_p(\nu, \mu, \tau^2 I_p),$$

we can write

$$\theta | \xi, x \ \sim \ \mathcal{N}_p \left(\xi(x), \frac{\tau^2 \sigma^2}{\sigma^2 \xi + \tau^2} I_p \right),$$

$$\pi(\xi|x) \ \propto \ \frac{\xi^{(p+\nu)/2 - 1}}{(\xi \sigma^2 + \tau^2)^{p/2}} \exp \left\{ \frac{-1}{2} \left(\frac{||x - \mu||^2 \xi}{\tau^2 + \xi \sigma^2} + \xi^2 \nu \right) \right\},$$

with

$$\xi(x) = \frac{\xi \sigma^2 \mu + \tau^2 x}{\xi \sigma^2 + \tau^2}$$

(see Example 10.2.2). Consider the following generalization:

$$x | \theta, \Lambda \sim \mathcal{N}_p(\theta, \Lambda),$$

when θ and $\Lambda = \text{diag}(\sigma_1^2, \dots, \sigma_p^2)$ are unknown, with prior distributions $(1 \leq i \leq p)$

$$\theta_i | \sigma_i \sim \mathcal{N} \left(\mu_i, \frac{\sigma_i^2}{n_i} \right), \qquad \sigma_i^2 \sim \mathcal{IG}(\nu_i/2, s_i^2/2),$$

where the n_i's, s_i's and ν_i's are known hyperparameters. In this case $(1 \leq i \leq p)$,

$$\theta_i | x_i \sim \mathcal{T} \left(\nu_i + 1, \frac{x_i + n_i \mu_i}{n_i + 1}, \right.$$

$$\left. (\nu_i + 1)^{-1} (n_i + 1)^{-1} \left[s_i^2 + \frac{n_i}{n_1 + 1} (x_i - \mu_i)^2 \right] \right),$$

and the call to an auxiliary variable ξ_i for each component θ_i does not modify the complexity of the estimation problem, since it does not change the number of integrals.   ‖

The two examples below are paradoxical, in the sense that a formal explicit expression of the Bayes estimator is available, but it cannot readily be used in practice, either because it induces numerical instability and thus unreliability of the result (Example 6.1.4), or because the actual computation of the resulting Bayes estimator is impossible, meaning it cannot be done in a reasonable amount of time for realistic sizes (Example 6.1.5).

Example 6.1.4 In the setting of *capture-recapture* models, we consider the following temporal model (see Section 4.3.3) with conjugate priors

$$x_i|N, p_i \sim \mathcal{B}(N, p_i),$$
$$\pi(N) = 1/N, \quad p_i \sim \mathcal{B}e(\alpha, \beta) \quad (1 \le i \le n).$$

If x_+ denotes the number of *different* individuals captured at least once during the n captures, the posterior distribution on N and $p = (p_1, \dots, p_n)$ is, for $x = (x_1, \dots, x_n, x_+)$,

$$\pi(N, p|x) \propto \frac{(N-1)!}{(N-x_+)!} \prod_{i=1}^{n} p_i^{\alpha+x_i-1} (1-p_i)^{\beta+N-x_i-1}$$

and the marginal distribution of N can be derived as

$$\pi(N|x) \propto \frac{(N-1)!}{(N-x_+)!} \prod_{i=1}^{n} B(\alpha+x_i, \beta+N-x_i)$$

$$\propto \frac{(N-1)!}{(N-x_+)!} \prod_{i=1}^{n} \frac{\Gamma(\beta+N-x_i)}{\Gamma(\alpha+\beta+N)}.$$

Therefore, the posterior distribution $\pi(N|x)$ can be written in the "explicit" form

(6.1.2)
$$\frac{\frac{(N-1)!}{(N-x_+)!} \prod_{i=1}^{n} \Gamma(\beta+N-x_i)/\Gamma(\alpha+\beta+N)}{\sum_{M=x_+}^{+\infty} \frac{(M-1)!}{(M-x_+)!} \prod_{i=1}^{n} \Gamma(\beta+M-x_i)/\Gamma(\alpha+\beta+M)}.$$

Actually, because of the ratios in the numerator and denominator, formula (6.1.2) does not require any computation of the gamma function, but only the use of the recursive formula $\Gamma(x+1) = x\Gamma(x)$. Nonetheless, if n is large, that is, if many captures have been undertaken, and if, moreover, the resulting capture sizes x_i are very different, the computation of the posterior distribution (6.1.2) will be quite difficult. The quantities (6.1.2) can fluctuate widely and the stopping rule for the computation of the infinite sum in (6.1.2) must be devised accordingly, lest it ignore the significant terms corresponding to larger values of M. Moreover, the computation of the sequence (6.1.2) through the following recurrence formula:

$$\frac{\pi(N+1|x)}{\pi(N|x)} = \frac{N}{N+1-x_+} \prod_{i=1}^{n} \frac{\beta+N-x_i}{\alpha+\beta+N},$$

although possible, can be quite damaging because the approximation error increases at each step, especially when the x_i's are very different.

The same criticism applies for the computation of the posterior mean

(6.1.3) $$\delta^{\pi}(x) = \frac{\sum_{N=x_+}^{+\infty} \frac{N!}{(N-x_+)!} \prod_{i=1}^{n} \Gamma(\beta+N-x_i)/\Gamma(\alpha+\beta+N)}{\sum_{M=x_+}^{+\infty} \frac{(M-1)!}{(M-x_+)!} \prod_{i=1}^{n} \Gamma(\beta+M-x_i)/\Gamma(\alpha+\beta+M)}.$$

Therefore, even though this discrete model seems analytically tractable, the explicit formulas above can only be used for the simplest examples.

Table 6.1.1. *Parameters statistics for a lung radiograph model. (Source: Plessis (1989).)*

	μ_1	μ_2	σ_1	σ_2	p
Average	105.33	188.9	32.3	18.2	0.5
Standard deviation	11.18	7.38	5.62	4.5	0.08

When the numbers of observations and of captures get large, numerical alternatives become necessary. Furthermore, the appeal of these formulas disappears for a hierarchical extension, since they cannot be used when a hyperprior on (α, β) is considered (see George and Robert (1992)).   ‖

Example 6.1.5 Consider a sample $x_1, \ldots, x_n$ from

$$(6.1.4) \qquad f(x|\theta) = p\varphi(x; \mu_1, \sigma_1) + (1 - p)\varphi(x; \mu_2, \sigma_2),$$

that is, from a mixture of two normal distributions with means μ_i, variances σ_i^2 ($i = 1, 2$) and weight p ($0 < p < 1$). We introduced a radiological application of this model in Example 1.1.6. A study on a first set of lung radiographs showed that images were distributed with parameters varying according to Table 6.1.5.

As a first approximation, given the information provided by Table 6.1.5, a prior modeling is to use "conjugate" priors on $\theta = (\mu_1, \sigma_1^2, p, \mu_2, \sigma_2^2)$,

$$\mu_i|\sigma_i \sim \mathcal{N}(\xi_i, \sigma_i^2/n_i), \qquad \sigma_i^2 \sim \mathcal{IG}(\nu_i/2, s_i^2/2), \qquad p \sim \mathcal{Be}(\alpha, \beta),$$

and to derive the hyperparameters ξ_i, n_i, ν_i, s_i and (α, β) from Table 6.1.5 by the moments method[2]. In fact, these distributions are not conjugate in the sense of Definition 3.3.1, but the corresponding posterior distribution is

$$(6.1.5) \quad \pi(\theta|x_1, \ldots, x_n) \propto \prod_{j=1}^n \{p\varphi(x_j; \mu_1, \sigma_1) + (1 - p)\varphi(x_j; \mu_2, \sigma_2)\} \pi(\theta).$$

A straightforward rewriting of (6.1.5) is to represent it as a weighted sum (that is, a mixture) of conjugate distributions,

$$(6.1.6) \qquad \pi(\theta|x_1, \ldots, x_n) = \sum_{\ell=0}^n \sum_{(k_t)} \omega(k_t)\pi(\theta|(k_t)),$$

where ℓ denotes the number of observations attributed to the first component and where the second sum takes into account every permutation

---

[2] Notice that this modeling differs from an *empirical Bayes* modeling (Chapter 10). Indeed, although the resulting prior is only an approximation and the hyperparameters are estimated by classical means, this distribution is based on *previous* observations, which can be considered *prior information*, not on the observed sample for which the parameter θ is unknown.

(k_t) of $\{1, 2, \ldots, n\}$ which gives a different partition of $\{x_1, \ldots, x_n\}$ into $\{x_{k_1}, \ldots, x_{k_\ell}\}$ and $\{x_{k_{\ell+1}}, \ldots, x_{k_n}\}$, thus characterizing the ℓ observations attributed to the first component. The posterior weight of a partition (k_t) is (see below for notation)

$$
\omega(k_t) \propto \frac{\Gamma(\alpha + \ell)\,\Gamma(\beta + n - \ell)\,\Gamma([\nu_1 + \ell]/2)}{\left(s_1^2 + \hat{s}_1(k_t) + \frac{n_1\ell}{n_1+\ell}(\xi_1 - \bar{x}_1(k_t))^2\right)^{(\nu_1+\ell)/2}}
$$

$$
\times \frac{\Gamma([\nu_2 + n - \ell]/2)/\sqrt{(n_1 + \ell)(n_2 + n - \ell)}}{\left(s_2^2 + \hat{s}_2(k_t) + \frac{n_2(n-\ell)}{n_2+n-\ell}(\xi_2 - \bar{x}_2(k_t))^2\right)^{(\nu_2+n-\ell)/2}},
$$

normalized so that

$$
\sum_{\ell=0}^{n} \sum_{(k_t)} \omega(k_t) = 1.
$$

For a given permutation (k_t), the conditional posterior distribution is

$$
\pi(\theta | (k_t)) = \mathcal{N}\left(\xi_1(k_t), \frac{\sigma_1^2}{n_1 + \ell}\right) \times \mathcal{IG}((\nu_1 + \ell)/2, s_1(k_t)/2)
$$

$$
\times \mathcal{N}\left(\xi_2(k_t), \frac{\sigma_2^2}{n_2 + n - \ell}\right) \times \mathcal{IG}((\nu_2 + n - \ell)/2, s_2(k_t)/2)
$$

$$
\times \mathcal{Be}(\alpha + \ell, \beta + n - \ell),
$$

where

$$
\begin{aligned}
\bar{x}_1(k_t) &= \tfrac{1}{\ell}\sum_{t=1}^{\ell} x_{k_t}, & \hat{s}_1(k_t) &= \sum_{t=1}^{\ell}(x_{k_t} - \bar{x}_1(k_t))^2, \\
\bar{x}_2(k_t) &= \tfrac{1}{n-\ell}\sum_{t=\ell+1}^{n} x_{k_t}, & \hat{s}_2(k_t) &= \sum_{t=\ell+1}^{n}(x_{k_t} - \bar{x}_2(k_t))^2
\end{aligned}
$$

are the usual statistics for the two subsamples induced by the permutation and

$$
\xi_1(k_t) = \frac{n_1\xi_1 + \ell\bar{x}_1(k_t)}{n_1 + \ell}, \qquad \xi_2(k_t) = \frac{n_2\xi_2 + (n - \ell)\bar{x}_2(k_t)}{n_2 + n - \ell},
$$

$$
s_1(k_t) = s_1^2 + \hat{s}_1^2(k_t) + \frac{n_1\ell}{n_1 + \ell}(\xi_1 - \bar{x}_1(k_t))^2,
$$

$$
s_2(k_t) = s_2^2 + \hat{s}_2^2(k_t) + \frac{n_2(n - \ell)}{n_2 + n - \ell}(\xi_2 - \bar{x}_2(k_t))^2,
$$

are the posterior updates of the hyperparameters, conditional upon the partition (k_t).

This decomposition is quite interesting because it shows that, behind a seemingly inextricable formula, the Bayesian analysis of the mixture distribution (6.1.4) is quite logical. Indeed, the posterior distribution takes into account *every* possible partition of the sample, specifying from which component each observation originated through the corresponding permutation (k_t). It then attributes a weight $\omega(k_t)$ to the partition, which can be interpreted as the posterior probability of the selected partition, and operates as if each observation was actually coming from its selected component, since the (conditional) posterior distributions $\pi(\theta|(k_t))$ are identical to the usual posterior distributions on (μ_1, σ_1) and (μ_2, σ_2) resulting from the *separate* observation of $x_{k_1}, \ldots, x_{k_\ell}$ and $x_{k_{\ell+1}}, \ldots, x_{k_n}$. Similar comments apply to the posterior distribution of p since, conditional upon the partition (k_t), this distribution corresponds to the posterior distribution associated with the observation of a binomial random variable $\mathcal{B}(n, p)$, which is the number of observations attributed to the first component.

The decomposition (6.1.6) provides the following Bayes estimator of θ:

$$\delta^\pi(x_1, \ldots, x_n) = \sum_{\ell=0}^{n} \sum_{(k_t)} \omega(k_t) \mathbb{E}^\pi[\theta|\mathbf{x}, (k_t)],$$

the weighted sum of the Bayes estimators for each partition. For instance, the Bayes estimator of μ_1 is

(6.1.7) $$\mu_1^\pi(x_1, \ldots, x_n) = \sum_{\ell=0}^{n} \sum_{(k_t)} \omega(k_t) \xi_1(k_t).$$

These developments are very satisfactory from a theoretical point of view because the resulting estimators are easy to interpret and intuitively convincing. Quite naturally, since the origin of each observation in the sample is unknown, the posterior distribution takes into account the possibility that this observation was generated by the first or second component. However, the practical calculation of (6.1.7) involves two sums with 2^n terms each, which exactly correspond to the different partitions of the sample. It is therefore impossible to compute a Bayes estimator this way for most sample sizes.[3]   ‖

Example 6.1.5 is representative of a class of statistical models where similar problems occur, including most *missing data* (or *latent variable*) models such as mixtures, censored models, classification and clustering (see Robert and Casella (1999, Chapter 9)). They are *paradoxical* in the sense that explicit derivations of the Bayes estimators may be formally available, but are practically useless because of the computing time involved. Moreover, the computational difficulty increases with the sample size, leading to what could be called an information paradox, since the more information one gets the more difficult it becomes to draw an inference[4] about θ. In such settings, numerical approximation methods are seldom appropriate and tailored solutions are necessary, as developed in Sections 6.3 and 6.4.

---

[3] For instance, if it takes one second of CPU time to evaluate (6.1.7) for a sample of size 20, the computation of the corresponding estimator for a sample of size 40 would require twelve days.

[4] Strictly speaking, the computational difficulty is always increasing with the sample size, even in settings where there exists a sufficient statistic. However, in the case of Example 6.1.5, the difficulty is increasing so fast (at an exponential rate) that it completely prevents the actual computation. (Such problems are called *NP-hard* in Operation Research.)

6.2 Classical approximation methods

This section briefly covers some classical techniques that can facilitate Bayesian calculations, while the next section deals with a recent simulation method that seems particularly adapted to some requirements of Bayesian computation. A more detailed survey of these techniques is provided in Robert and Casella (2004, Chapters 2–5), while Berger (2000) and Carlin and Louis (2000a) survey available Bayesian software.

6.2.1 Numerical integration

Starting with the simple Simpson's method,[5] many approaches have been devised by applied mathematicians to approximate integrals numerically. For instance, *polynomial quadrature* is intended to approximate integrals involving distributions close to the normal distribution (see Naylor and Smith (1982), Smith et al. (1985), or Verdinelli and Wasserman (1998) for a detailed introduction). The basic approximation is given by

$$\int_{-\infty}^{+\infty} e^{-t^2/2} f(t)\, dt \approx \sum_{i=1}^{n} \omega_i f(t_i),$$

where

$$\omega_i = \frac{2^{n-1} n! \sqrt{n}}{n^2 [H_{n-1}(t_i)]^2}$$

and t_i is the ith zero of the nth *Hermite polynomial*, $H_n(t)$.

Other related integral approximations are also available, based upon different classical orthogonal bases (see Abramowitz and Stegun (1964)), or the *wavelets* (see Note 1.8.2 and Müller and Vidakovic (1999, Chapter 1)), but these methods usually require regularity assumptions on the function f, as well as preliminary studies in order to determine what basis is the most appropriate basis, and how accurate the approximation is. For instance, transformations of the model may be necessary to apply the above Hermite approximation (see Naylor and Smith (1982) and Smith and Hills (1992)); Morris (1982) (see also Diaconis and Zabell (1991)) shows how distributions from the quadratic variance exponential families (Exercises 3.24 and 10.33) can be associated with particular orthogonal bases (Exercise 6.15).

However, it seems that, no matter the numerical integration method used, its accuracy dramatically decreases as the dimension of Θ increases. More specifically, the error associated with numerical methods evolves as a power of the dimension of Θ. In fact, an empirical rule of thumb is that most standard methods should not be used for integration in dimensions larger than 4, although they keep improving over the years. But the size of the part of the space irrelevant for the computation of a given integral

---

[5] See Stigler (1986) for a closer connection between Simpson (1710–1761) and Bayesian Statistics.

grows considerably with the dimension of the space. This problem is often called *the curse of dimensionality*; see Robert and Casella (2004, Chapter 3) for details.

6.2.2 Monte Carlo methods

In a statistical problem, the approximation of the integral

$$(6.2.1) \qquad \int_\Theta g(\theta) f(x|\theta) \pi(\theta)\, d\theta,$$

should take advantage of the special nature of (6.2.1), namely, the fact that π is a probability density (assuming it is a proper prior) or, instead, that $f(x|\theta)\pi(\theta)$ is proportional to a density. A natural way to do so is the *Monte Carlo method*, introduced by Metropolis and Ulam (1949) and Von Neuman (1951). For instance, if it is possible to generate random variables $\theta_1, \ldots, \theta_m$ from $\pi(\theta)$, the average

$$(6.2.2) \qquad \frac{1}{m} \sum_{i=1}^{m} g(\theta_i) f(x|\theta_i)$$

converges (almost surely) to (6.2.1) when m goes to $+\infty$, according to the Law of Large Numbers. Similarly, if an i.i.d. sample of θ_i's from $\pi(\theta|x)$ can be produced, the average

$$(6.2.3) \qquad \frac{1}{m} \sum_{i=1}^{m} g(\theta_i)$$

converges to

$$\frac{\int_\Theta g(\theta) f(x|\theta) \pi(\theta)\, d\theta}{\int_\Theta f(x|\theta) \pi(\theta)\, d\theta}.$$

In addition, if the posterior variance $\mathrm{var}(g(\theta)|x)$ is finite, the Central Limit Theorem applies to the average (6.2.3), which is then asymptotically normal with variance $\mathrm{var}(g(\theta)|x)/m$. Confidence regions can then be built from this normal approximation and, most importantly, the magnitude of the error remains of order $1/\sqrt{m}$, whatever the dimension of the problem, in opposition with numerical methods.

The implementation of the method requires the production of the θ_i's by computer, using a deterministic pseudo-random generator to mimic generation from $\pi(\theta)$ or $\pi(\theta|x)$ by first replicating i.i.d. sampling from a uniform $\mathcal{U}([0,1])$ distribution (see Note 6.6.1) and then transforming the uniforms into the variables of interest (see Robert and Casella (2004, Chapter 2)).[6] Standard statistical techniques can also be used to ascertain the error in the approximation of (6.2.1) by the average (6.2.2).

---

[6] It is not surprising that Monte Carlo methods emerged exactly at the time of the first computer. They are simply not operational without a computer, and correspond to some of the first computer programs ever written.

The Monte Carlo method actually applies in a much wider generality than the above simulation from π. For instance, because (6.2.1) can be represented in many ways, there is no need to simulate from the distributions $\pi(\cdot|x)$ or π to get a good approximation of (6.2.1). Indeed, if h is a probability density with supp(h) including the support of $g(\theta)f(x|\theta)\pi(\theta)$, the integral (6.2.1) can also be represented as an expectation against h, namely

$$\int \frac{g(\theta)f(x|\theta)\pi(\theta)}{h(\theta)}h(\theta)\,d\theta.$$

This representation leads to the *Monte Carlo method with importance function* h: generate $\theta_1,\ldots,\theta_m$ according to h and approximate (6.2.1) through

$$\frac{1}{m}\sum_{i=1}^{m}g(\theta_i)\omega_i(\theta_i),$$

with the weights $\omega(\theta_i) = f(x|\theta_i)\pi(\theta_i)/h(\theta_i)$. Again, by the Law of Large Numbers, this approximation almost surely converges to (6.2.1). And an approximation to $\mathbb{E}^\pi[g(\theta)|x]$ is given by

(6.2.4)
$$\frac{\sum_{i=1}^{m}g(\theta_i)\omega(\theta_i)}{\sum_{i=1}^{m}\omega(\theta_i)}.$$

since the numerator and denominator converge to

$$\int_\Theta g(\theta)f(x|\theta)\pi(\theta)\,d\theta \qquad \text{and} \qquad \int_\Theta f(x|\theta)\pi(\theta)\,d\theta,$$

respectively, if supp(h) includes supp($f(x|\cdot)\pi$). Notice that the ratio (6.2.4) does not depend on the normalizing constants in either $h(\theta)$, $f(x|\theta)$ or $\pi(\theta)$. The approximation (6.2.4) can therefore be used in settings when some of these normalizing constants are unknown.

Although (6.2.4) theoretically converges to $\mathbb{E}^\pi[g(\theta)|x]$ for all functions h satisfying the condition on its support (Exercise 6.8), the choice of the importance function is crucial. First, simulation according to h must be easily implemented, requiring a fast and reliable pseudo-random generator. (See Exercises 6.9–6.10, Devroye (1985), Fishman (1996), Gentle (1998), or Robert and Casella (2004, Chapter 2). Moreover, the function $h(\theta)$ must be close enough to $g(\theta)\pi(\theta|x)$, in order to reduce the variability of (6.2.4) as much as possible (Exercise 6.12); otherwise, most of the weights $\omega(\theta_i)$ will be quite small and a few will be overly influential. In fact, if $\mathbb{E}^h[g^2(\theta)\omega^2(\theta)]$ is not finite, the variance of the estimator (6.2.4) is infinite. Obviously, the dependence on g of the importance function h can be avoided by proposing generic choices such as the posterior distribution $\pi(\theta|x)$ (which is not necessarily the best choice, as shown by Exercises 6.11 and 6.12).

Example 6.2.1 (Example 6.1.2 continued) The posterior distribution of $\eta = ||\theta||^2$ is well known, since $\pi(\eta|x)$ is a noncentral chi-squared distribution $\chi_p^2(\lambda)$ rescaled by $\sigma^2\tau^2/(\sigma^2+\tau^2)$. Simulating a sample $\eta_1,\ldots,\eta_m$

from $\pi(\eta|x)$ is straightforward: simulate

$$\xi_1, \ldots, \xi_n \sim \mathcal{N}(\sqrt{\lambda}, 1), \qquad \zeta_1, \ldots, \zeta_n \sim \mathcal{G}\left(\frac{p-1}{2}, \frac{1}{2}\right)$$

and take $\eta_i = \sigma^2\tau^2(\xi_i^2 + \zeta_i)/(\sigma^2 + \tau^2)$ $(i = 1, \ldots, n)$. We can then approximate (6.1.3) by

(6.2.5) $$\hat{\delta}^\pi(x) = \frac{\sum_{i=1}^m \eta_i/(2\eta_i + p)}{\sum_{i=1}^m 1/(2\eta_i + p)}.$$

Moreover, the variance of (6.2.5) controls the precision of the approximation (and the choice of m). ‖

When the posterior distribution is not available, another simple choice of importance function is the prior distribution π. It is obviously interesting when π is not necessarily explicit, but easy to simulate, for instance, in hierarchical models where both levels are proper. The same call for caution still applies, though, as π must be close enough to $\pi(\theta|x)$ and the variance of the estimator (6.2.4) finite. (Notice that this finiteness condition is usually satisfied since $\pi(\theta)$ often has fatter tails than $\pi(\theta|x)$.)

Example 6.2.2 (Example 6.1.1 continued) Because $\pi(\theta)$ is the normal distribution $\mathcal{N}(\mu, \sigma^2)$, it is possible to simulate a normal sample $\theta_1, \ldots, \theta_M$ and to approximate the Bayes estimator by

(6.2.6) $$\hat{\delta}^\pi(x_1, \ldots, x_n) = \frac{\sum_{t=1}^M \theta_t \prod_{i=1}^n [1 + (x_i - \theta_t)^2]^{-1}}{\sum_{t=1}^M \prod_{i=1}^n [1 + (x_i - \theta_t)^2]^{-1}}.$$

In the case where the x_i's are all far from μ, this choice may be detrimental since both the denominator and the weights of the θ_t's in the numerator are small for most θ_t's, and the approximation $\hat{\delta}^\pi$ is therefore quite unstable. Consider Figure 6.2.2, which represents the result of 500 parallel estimations (6.2.6) based on $M = 1000$ simulations each as the 90% central range of the $\hat{\delta}^\pi$'s minus the overall mean. The variation of δ^π increases rapidly between $\mu = 3$ and $\mu = 4$. This shows that, when $\mu > 3$, small changes in the simulated θ_t's can induce drastic changes in $\hat{\delta}^\pi$. ‖

Example 6.2.3 Consider the model

$$x \sim \mathcal{N}_p(\theta, I_p), \quad \theta|c \sim \mathcal{U}_{\{||\theta||^2=c\}}, \quad \text{and} \quad c \sim \mathcal{G}(\alpha, \beta).$$

(The justification for this setting will be made clear in Example 10.3.6.) Although

$$\pi(\theta|x) = \int_0^{+\infty} \pi_1(\theta|x, c)\pi_2(c|x)\, dc$$

leads to an explicit posterior distribution and an explicit Bayes estimator (see Example 10.3.6), it might be more interesting to generate $c_1, \ldots, c_m$ according to $\mathcal{G}(\alpha, \beta)$, then the θ_i's according to $\mathcal{U}_{\{||\theta||^2=c_i\}}$ $(1 \le i \le m)$ and

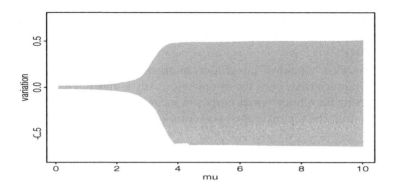

Figure 6.2.1. *90% range of variation of the approximation (6.2.6) as μ varies, for $n = 10$ observations from a Cauchy $\mathcal{C}(0,1)$ distribution and $M = 1000$ Monte Carlo simulations of θ from a $\mathcal{N}(\mu, 1)$ distribution.*

to approximate the posterior mean by

$$\hat{\delta}^\pi(x) = \frac{\sum_{i=1}^m \theta_i \exp\{-||x - \theta_i||^2/2\}}{\sum_{i=1}^m \exp\{-||x - \theta_i||^2/2\}},$$

since this alternative avoids the computation of confluent hypergeometric functions.   ∥

When the likelihood $\ell(\theta|x)$ can be normalized into a density, a possible choice of importance function is $h(\theta) \propto \ell(\theta|x)$. This choice makes sense when $\pi(\theta|x)$ is almost proportional to the likelihood—as it is for large sample sizes, or for almost constant prior distributions. For instance, this may occur in exponential settings since, if

$$f(x|\theta) \propto e^{\theta.x - \psi(\theta)},$$

a sample $\theta_1, \ldots, \theta_m$ from

$$h(\theta) \propto e^{\theta.x - \psi(\theta)}$$

can easily be obtained in general (see Exercise 6.20 for a limitation to this approach).

A final remark about the choice of the importance function is that there is generally a trade-off between preliminary studies leading to a "good" h and fast algorithms. For instance, when h is chosen because it makes the simulation of the θ_i's easier, attention should be paid to the tails of h so that they are heavier than the tails of $\pi(\theta|x)$, in order to avoid slow convergence and infinite variances. On the other hand, if h is specially tuned for the computation of a specific integral (Exercise 6.12), it may not work so well for another integral, despite the fact that, in principle, the same sample of θ_i's can be used for the computation of arbitrary integrals. However, barring these potential difficulties, importance sampling methods constitute a very general tool, which often ends up being competitive with Monte Carlo Markov chain techniques (Section 6.3), as shown for instance by the *particle filter* method (see Doucet et al. (2001) and Robert and Casella (2004)).

Compared with numerical integration methods, Monte Carlo methods have the advantage that, once the sample $\theta_1, \ldots, \theta_n$ is generated, it can be used repeatedly for all inferential purposes, including the derivation of the Bayes rules from the approximated posterior loss

$$\hat{L}(\pi, d|x) = \frac{1}{m} \sum_{i=1}^{m} L(\theta_i, d|x).$$

However, if the dimension of the problem is small and if the functions to be integrated are fairly regular, numerical integration methods tend to yield smaller errors with better convergence controls. Additional references and more detailed discussions about Monte Carlo methods, including the improved techniques of antithetic and control variates, and of their application to Bayesian Statistics, can be found in Robert and Casella (1999, 2004), Chen, Shao and Ibrahim (2000), and Marin and Robert (2007).

6.2.3 Laplace analytic approximation

When the function to integrate in (6.2.1) is regular enough, there exists an analytic—although asymptotic—alternative to Monte Carlo simulations. This method was introduced by Laplace and is thus called *Laplace approximation*. Consider the posterior expectation of interest

$$\mathbb{E}^{\pi}[g(\theta)|x] = \frac{\int_{\Theta} g(\theta) f(x|\theta) \pi(\theta)\, d\theta}{\int_{\Theta} f(x|\theta) \pi(\theta)\, d\theta}.$$

This ratio of integrals can be written as

$$(6.2.7) \qquad \mathbb{E}^{\pi}[g(\theta)|x] = \frac{\int_{\Theta} b_N(\theta) \exp\{-n h_N(\theta)\}\, d\theta}{\int_{\Theta} b_D(\theta) \exp\{-n h_D(\theta)\}\, d\theta},$$

where the dependence on x is suppressed for simplicity's sake and where n is usually the sample size (although it may sometimes correspond to the inverse prior variance, as in Robert (1993a), or in Example 6.2.6). When $h_N(\theta) = h_D(\theta)$, $\mathbb{E}^{\pi}[g(\theta)|x]$ is said to be written in *standard form*; when $b_N(\theta) = b_D(\theta)$, the posterior expectation (6.2.7) is written in *fully exponential form*, in the words of Tierney and Kadane (1986). Given a function h with a single minimum $\hat{\theta}$, the *Laplace expansion* of a general integral is given by

$$\int b(\theta) e^{-n h(\theta)}\, d\theta = \sqrt{2\pi} \sigma e^{-n\hat{h}} \left\{ \hat{b} + \frac{1}{2n} \left[\sigma^2 \hat{b}'' - \sigma^4 \hat{b}' \hat{h}''' \right. \right.$$
$$\left. \left. + \frac{5}{12} \hat{b}(\hat{h}''')^2 \sigma^6 - \frac{1}{4} \hat{b} \hat{h}^{(4)} \sigma^4 \right] \right\} + \mathrm{O}(n^{-2}),$$

where $\hat{b}$, $\hat{h}$, etc., denote the values of b, h, and of their derivatives for $\theta = \hat{\theta}$, and $\sigma^2 = [h''(\hat{\theta})]^{-1}$ (see Olver (1974) and Schervish (1995)). This second-order approximation only requires computation of the first two derivatives of g, as opposed to a similar approach proposed by Lindley (1980). Assuming, in addition, that h_N and h_D satisfy $\hat{h}_N - \hat{h}_D = O(n^{-1}), \ldots, \hat{h}_N^{(4)} - \hat{h}_D^{(4)} = O(n^{-1})$ (as is obviously the case for the standard form), Laplace expansion leads to the following approximation of $\mathbb{E}^\pi[g(\theta)|x]$ (with $\hat{b}_D = b_D(\hat{\theta}_D)$, $\hat{b}_N = b_N(\hat{\theta}_N)$, and so on):

Lemma 6.2.4 *If $b_D \neq 0$,*

$$\frac{\int_\Theta b_N(\theta) \exp\{-nh_N(\theta)\}\, d\theta}{\int_\Theta b_D(\theta) \exp\{-nh_D(\theta)\}\, d\theta} = \frac{\sigma_N}{\sigma_D} e^{-n(\hat{h}_N - \hat{h}_D)} \left[\frac{\hat{b}_N}{\hat{b}_D} + \frac{\sigma_D^2}{2n\hat{b}_D^2} \left\{ \hat{b}_D \hat{b}_N'' \right.\right.$$
$$\left.\left. - \hat{b}_N \hat{b}_D'' - \sigma_D^2 \hat{h}_D'''(\hat{b}_D \hat{b}_N' - \hat{b}_N \hat{b}_D') \right\} \right] + O(n^{-2}).$$

A proof of this result is given in Tierney et al. (1989) (see also Exercise 6.14). Lemma 6.2.4 thus implies the following development for the two forms of the ratio (6.2.7):

Corollary 6.2.5 *When $\mathbb{E}^\pi[g(\theta)|x]$ is written in standard form,*

$$(6.2.8) \quad \mathbb{E}^\pi[g(\theta)|x] = \hat{g} + \frac{\sigma_D^2 \hat{b}_D' \hat{g}'}{n\hat{b}_D} + \frac{\sigma_D^2 \hat{g}''}{2n} - \frac{\sigma_D^4 \hat{h}''' \hat{g}'}{2n} + O(n^{-2}).$$

For the fully exponential form, if g is positive and $g(\hat{\theta}_D)$ is uniformly (in n) bounded away from 0,

$$(6.2.9) \quad \mathbb{E}^\pi[g(\theta)|x] = \frac{\hat{b}_N}{\hat{b}_D} \frac{\sigma_N^2}{\sigma_D^2} e^{-n(\hat{h}_n - \hat{h}_D)} + O(n^{-2}).$$

Proof. For the standard form, $h_N = h_D$; therefore, $b_N = gb_D$, $\hat{\theta}_D = \hat{\theta}_N$. Thus,

$$\frac{\hat{b}_D \hat{b}_N' - \hat{b}_N \hat{b}_D'}{\hat{b}_D^2} = \left(\frac{b_N}{b_D} \right)' \bigg|_{\theta = \hat{\theta}_D} = \hat{g}'$$

and

$$\frac{\hat{b}_D \hat{b}_N'' - \hat{b}_N \hat{b}_D''}{\hat{b}_D^2} = \hat{g}'' + 2\frac{\hat{b}_D'}{\hat{b}_D} \hat{g}'.$$

The result then follows from Lemma 6.2.4.

In the fully exponential case, $h_N = h_D - (1/n)\log(g)$. Because we assume that $g(\hat{\theta}_D) \geq c > 0$ for every n, $\hat{\theta}_N - \hat{\theta}_D = O(n^{-1})$. Because $b_D = b_N$, this implies $\hat{b}_N^{(i)} - \hat{b}_D^{(i)} = O(n^{-1})$ $(i = 0, 1, 2)$. Additional terms in Lemma 6.2.4 can therefore be ignored. □□

Corollary 6.2.5 clearly points out the advantage of the fully exponential interpretation of (6.2.7), since it avoids computation of the first and second derivatives, $\hat{g}'$ and $\hat{g}''$, appearing in (6.2.8). Notice that (6.2.9) can also be written

$$\mathbb{E}^{\pi}[g(\theta)|x] = \frac{\sigma_N^2}{\sigma_D^2} \frac{g(\hat{\theta}_N) f(x|\hat{\theta}_N) \pi(\hat{\theta}_N)}{f(x|\hat{\theta}_D) \pi(\hat{\theta}_D)} + \mathrm{O}(n^{-2}).$$

The assumption on g, namely, that g is positive and bounded away from 0 in $\hat{\theta}_D$, is however quite restrictive. Moreover, the usual decomposition $g = g^+ - g^-$ does not work in this setting. Tierney et al. (1989) overcome this drawback by first evaluating the moment generating function of $g(\theta)$,

$$M(s) = \mathbb{E}^{\pi}[\exp\{sg(\theta)\}|x],$$

obviously positive, by $\hat{M}(s)$ through (6.2.9). They derived $\mathbb{E}^{\pi}[g(\theta)|x]$ as

$$\mathbb{E}^{\pi}[g(\theta)|x] = \frac{d}{ds}(\log \hat{M}(s))\big|_{s=0} + \mathrm{O}(n^{-2}).$$

They also establish the rather surprising result that this approach provides the standard development (6.2.8) without requiring an evaluation of the first and second derivatives of g (see Exercise 6.15).

Example 6.2.6 (Tierney et al. (1989)) Let $\pi(\theta|x)$ be a $\mathcal{Be}(\alpha, \beta)$ distribution, the posterior expectation of θ is then

$$\delta^{\pi}(x) = \frac{\alpha}{\alpha + \beta}.$$

This exact computation can be compared with the approximations (6.2.8),

$$\delta^{\pi}(x) = \frac{\alpha^2 + \alpha\beta + 2 - 4\alpha}{(\alpha + \beta - 2)^2} + \mathrm{O}((\alpha + \beta)^{-2}),$$

and (6.2.9),

$$\delta^{\pi}(x) = \frac{\alpha}{\alpha + \beta - 1} \left(\frac{\alpha}{\alpha - 1}\right)^{\alpha - 0.5} \left(\frac{\alpha + \beta - 2}{\alpha + \beta - 1}\right)^{\alpha + \beta - 0.5} + \mathrm{O}((\alpha + \beta)^{-2}).$$

Denoting $p = \alpha/(\alpha + \beta)$ and $n = \alpha + \beta$, the approximation error is

$$\Delta^S = 2\frac{1 - 2p}{n^2} + \mathrm{O}(n^{-3})$$

in the standard case, and

$$\Delta^E = 2\frac{1 - 13p^2}{12pn^2} + \mathrm{O}(n^{-3})$$

in the fully exponential case. The second development is then better for the median values of p. $\quad\|$

The reader is referred to Leonard (1982), Tierney and Kadane (1986), Tierney et al. (1989), and Kass and Steffey (1989) for additional results and comments. A reservation made in Smith et al. (1985) about Laplace approximation is that it is only justified *asymptotically*; the specific verifications conducted in the different papers cannot provide a global justification of the method, even though it seems to perform quite well in most cases. Other criticisms of this approach are that

(1) analytical methods always imply delicate preliminary studies about the regularity of the integrated function that are not necessarily feasible:;

(2) the posterior distribution should be similar enough to the normal distribution (for which Laplace approximation is exact); and

(3) such methods cannot be used in settings like those of Example 6.1.5, where the computation of the maximum likelihood estimator is quite difficult.

Extensions of Laplace methods to *saddle point approximations* are reviewed in Kass (1989) (see also Rousseau (1997, 2000, 2002)).

6.3 Markov chain Monte Carlo methods

In this section we consider a more general Monte Carlo method that approximates the generation of random variables from a posterior distribution $\pi(\theta|x)$ when this distribution cannot be directly simulated. Its advantage over the classical Monte Carlo methods described in Section 6.2.2 is that it does not require the precise construction of an importance function, while taking into account the characteristics of $\pi(\theta|x)$. This extension, called *Markov chain Monte Carlo* (and abbreviated as MCMC), has almost unlimited applicability, even though its performance varies widely, depending on the complexity of the problem. It derives its name from the idea that, to produce acceptable approximations to integrals and to other functionals depending on a distribution of interest, it is enough to generate a *Markov chain* $(\theta^{(m)})_m$ with limiting distribution the distribution of interest.[7] This idea of using the limiting behavior of a Markov chain came almost as early as the original Monte Carlo technique, at least in the particle Physics literature (Metropolis et al. (1953)), but it requires a computational power that was not available in those early days.

After a brief discussion on the appeal of using a Markov chain in simulation (Section 6.3.1), we introduce the two major techniques devised to create Markov chains with a given distribution, namely, Metropolis–Hastings algorithms (Section 6.3.2) and Gibbs sampling (Sections 6.3.3–6.3.6). We refer the reader to Gilks et al. (1996) and Robert and Casella (2004) for broader perspectives on this topic.

---

[7] This section minimizes the recourse to Markov chain theory, although some notions like ergodicity cannot be skipped. We refer the reader to Meyn and Tweedie (1993) for a deep and pedagogical introduction to this topic. See also Robert and Casella (2004, Chapter 6) for a more cursory treatment of the notions necessary for the understanding of MCMC methods.

6.3.1 MCMC in practice

The apparent paradox with using Markov chains for simulation purposes is that we seem to be calling *twice* for an asymptotic argument: first, the chain must converge to its stationary distribution; second, empirical averages such as (6.2.2) must converge to the corresponding expectation $\mathbb{E}^\pi[g(\theta)|x]$. We now describe why this is not so, thanks to the Ergodic Theorem.

By their very nature, if the Markov chains $(\theta^{(m)})_m$ produced by MCMC algorithms are irreducible, that is, if they are guaranteed to visit any set A such that $\pi(A|x) > 0$, then these chains are positive recurrent with stationary distribution $\pi(\theta|x)$. That is, the average number of visits to an arbitrary set A with positive measure is infinite. These Markov chains are also *ergodic*, which means that the distribution of $\theta^{(m)}$ converges to $\pi(\cdot|x)$ for almost every starting value $\theta^{(0)}$, that is, the influence of the starting value vanishes. (Under fairly general conditions, the chains are even *Harris-recurrent*, which implies that the "almost" in the above condition disappears.)

Therefore, for k large enough, the resulting $\theta^{(k)}$ is approximately distributed from $\pi(\theta|x)$, no matter what the starting value $\theta^{(0)}$ is. The problem in practice is then to determine what a "large" k means, since it governs the number of simulations to run: is it 200 or 10^{10}? The speed of convergence, that is, the type of decrease in the difference (distance) between the distribution of $\theta^{(k)}$ and its limit, brings an answer to this problem, but so far it has been mainly studied from a theoretical point of view. Moreover, this rate of convergence often depends on the starting point (otherwise, the chain is *uniformly ergodic*) and a given k does not provide the same quality of approximation for different values of $\theta^{(0)}$. There are thus practical hindrances in the use of Markov chains for simulation since we often ignore whether the chain has been run long enough. But, as detailed in Robert and Casella (2004, Chapter 12), there now are diagnostic tests and a corresponding software, `CODA` (see Note 6.6.2), that provide different indicators on the stationarity of the chain, and thus reduce this difficulty.

Once $\theta_1 = \theta^{(k)}$ is obtained, a naïve way to build an i.i.d. sample $\theta_1, \ldots, \theta_m$ from $\pi(\theta|x)$ is to use the same algorithm with another initial value $\theta_2^{(0)}$ and another sequence of k Markov transition moves to get θ_2, and so on until θ_m. As shown above, the speed of convergence often depends on the starting value and it is thus preferable (in terms of convergence) to take the current $\theta^{(k)}$ as the new starting value, even though this introduces dependence between the θ_i's. However, independence is not crucial if we are mainly interested in functionals of $\pi(\theta|x)$, since the *Ergodic Theorem* implies that the average

$$\frac{1}{K} \sum_{k=1}^{K} g(\theta^{(k)})$$

converges to $\mathbb{E}^\pi[g(\theta)|x]$ (as long as $\mathbb{E}^\pi[|g(\theta)||x]$ is finite) when K goes to infinity (see Meyn and Tweedie (1993)). Therefore, the influence of the starting value also vanishes in the average (hence the ergodicity). Moreover, this property is similarly satisfied by any subsequence of $(\theta^{(k)})$.

The Ergodic Theorem thus solves the paradox of the two asymptotics mentioned at the beginning of this paragraph, since it extends the Law of Large Numbers to dependent sequences of random variables and removes the need to first produce an i.i.d. sample, which would, moreover, be only approximate if we used the method proposed above. Indeed, as already noted in Geyer (1992), the available Markov chain theory does not indicate when stationarity is attained, since, from a mathematical point of view, this is only an asymptotic property of the chain. Therefore, it is better to consider a *single sequence* $(\theta^{(k)})$, as each simulation step brings us closer (in probability) to a realization from the stationary distribution, $\pi(\theta|x)$. In addition, multiple-starts simulation produces a considerable waste by rejecting most of the simulated values. However, the call to multiple chains is quite useful in studying convergence of a Markov chain (and thereby determining the proper k) and thus frequently appears in monitoring techniques, as in the within-between method of Gelman and Rubin (1992) (see Robert and Casella (2004, Section 12.3.4)).

When required, quasi-independence can be achieved by batch sampling, that is, by keeping only one member of the chain out of t iterations for the effective simulated sample, with $t = 5$ or $t = 10$ say. Raftery and Lewis (1992) propose a more advanced determination of the batch size t, which is chain-induced and based on a binarization of the chain. (See Robert and Casella (2004, Section 12.4.1) for an appraisal of this method.)

6.3.2 Metropolis–Hastings algorithms

Once the principle of using a Markov chain with stationary distribution π—instead of i.i.d. variables exactly distributed from π—to approximate quantities like (6.2.1) is accepted, the implementation of this principle requires the construction of a generation mechanism to produce such Markov chains. Surprisingly, an almost universal algorithm satisfying this constraint does exist: it has been constructed by Metropolis et al. (1953), originally for mechanical physics, and generalized by Hastings (1970) in a more statistical setting. It actually applies to a wide variety of problems, since its main restriction is that the distribution of interest be known up to a constant, but we will see later that this constraint can be lifted in many ways.

In its modern version, the *Metropolis–Hastings algorithm* can be described as follows. Given a density $\pi(\theta)$, known up to a normalizing factor, and a conditional density $q(\theta'|\theta)$, the algorithm generates the chain $(\theta^{(m)})_m$ by:

(i) Start with an arbitrary initial value $\theta^{(0)}$

(ii) Update from $\theta^{(m)}$ to $\theta^{(m+1)}$ $(m = 1, 2, \ldots)$ by

(a) Generate $\xi \sim q(\xi|\theta^{(m)})$

(b) Define

$$\varrho = \frac{\pi(\xi)\,q(\theta^{(m)}|\xi)}{\pi(\theta^{(m)})\,q(\xi|\theta^{(m)})} \wedge 1$$

(c) Take

$$\theta^{(m+1)} = \begin{cases} \xi & \text{with probability } \varrho, \\ \theta^{(m)} & \text{otherwise.} \end{cases}$$

The distribution with density $\pi(\theta)$ is often called the *target* or *objective distribution*, whereas the distribution with density $q(\cdot|\theta)$ is the *proposal distribution*. An astounding thing about this algorithm is the infinite number of proposal distributions that yield a Markov chain that converges to the distribution of interest.

Theorem 6.3.1 *If the chain* $(\theta^{(m)})_m$ *is irreducible, that is, such that, for any subset* A *with* $\pi(A) > 0$, *there exists* M *such that* $P_{\theta^{(0)}}(\theta^{(M)} \in A) > 0$, *then* π *is the stationary distribution of the chain. If, in addition, the chain is aperiodic, it is also ergodic with limiting distribution* π, *for almost every initial value* $\theta^{(0)}$, *that is,*

$$\lim_{m \to \infty} \sup_A \left| P_{\theta^{(0)}}(\theta^{(m)} \in A) - \pi(A) \right| = 0 \qquad (a.s.)$$

The property at the core of this result is the *detailed balance condition*, that is, the fact that the transition kernel of the Markov chain associated with the above algorithm, $K(\theta'|\theta)$ say, satisfies

(6.3.1) $\pi(\theta)K(\theta'|\theta) = \pi(\theta')K(\theta|\theta')$.

When integrating both sides of (6.3.1) against θ, the rhs provides $\pi(\theta')$ because $K(\theta|\theta')$ is a (conditional) density in θ, while the lhs gives the density of the Markov chain after one step, when $\theta^{(0)} \sim \pi$. Therefore, the distribution π is indeed stationary for the transition kernel $K(\theta'|\theta)$. (See Exercise 6.17 and Robert and Casella (2004, Section 7.3) for more details.)

If we write down the kernel of the Metropolis–Hastings algorithm as

$$K(\theta'|\theta) = \varrho(\theta, \theta')q(\theta'|\theta) + \int [1 - \varrho(\theta, \xi)]q(\xi|\theta)d\xi\, \delta_\theta(\theta'),$$

where δ denotes the Dirac mass, it is then straightforward to verify that (6.3.1) holds.

The irreducibility condition in Theorem 6.3.1 is obviously necessary for the chain to explore the support of π. Sufficient conditions for irreducibility to hold are, for instance, that the support of $q(\cdot|\theta)$ contain the support of π for every θ or, more generally, that $q(\cdot|\theta)$ be positive in a neighborhood of θ of fixed radius (see Robert and Casella (2004, Lemma 7.6)).

While Theorem 6.3.1 gives a formal condition for the chain to converge, which covers an immense class of proposal distributions, practical selection of this distribution is much more delicate because a poor overlap between the support of π and $q(\cdot|\theta)$ may considerably slow convergence.

Example 6.3.2 *Weibull distributions* are used extensively in reliability and other engineering applications, partly for their ability to describe different hazard rate behaviors, and partly for historic reasons. Because they do not belong to any exponential family, being of the form

(6.3.2) $$f(x) \propto \alpha \eta x^{\alpha-1} e^{-x^\alpha \eta},$$

they cannot lead to explicit posterior distributions on the parameters α and η. For $\theta = (\alpha, \eta)$, consider the prior

$$\pi(\theta) \propto e^{-\alpha} \eta^{\beta-1} e^{-\xi\eta}$$

and observations $x_1, \ldots, x_n$ from (6.3.2). A Metropolis–Hastings algorithm for the simulation of $\pi(\theta|x_1, \ldots, x_n)$ can be based upon the conditional distribution

$$q(\theta'|\theta) = \frac{1}{\alpha\eta} \exp\left(-\frac{\alpha'}{\alpha} - \frac{\eta'}{\eta}\right),$$

that is, on two independent exponential distributions with means α and η. The resulting acceptance probability is then

$$\varrho = 1 \wedge \left(\frac{\eta'}{\eta}\right)^{n+\beta} \left(\frac{\alpha'}{\alpha}\right)^{n-1} \left(\prod_{i=1}^{n} x_i\right)^{\alpha'-\alpha} \prod_{i=1}^{n} e^{\eta x_i^\alpha - \eta' x_i^{\alpha'}} e^{\alpha-\alpha'+\eta-\eta'+\alpha'/\alpha+\eta'/\eta},$$

if (α', η') is the simulated value and (α, η) is the current value of the parameters.   ‖

The most common choice for q, starting with Hastings (1970), is the *random-walk proposal*, where $q(\theta'|\theta)$ is of the form $f(\|\theta'-\theta\|)$. The proposed value ξ in the Metropolis–Hastings algorithm is thus of the form

$$\xi = \theta^{(m)} + \varepsilon,$$

with ε distributed as a symmetric random variable. The natural idea behind this proposal is to perturb the current value of the chain at random, while staying in a neighborhood of this value, and then see if the new value ξ is likely for the distribution of interest. For this random-walk proposal, the Metropolis–Hastings acceptance ratio is

$$\varrho = \frac{\pi(\xi)}{\pi(\theta^{(m)})} \wedge 1.$$

The chain $(\theta^{(m)})_m$ will thus stay longer in a given point ξ if the corresponding posterior value $\pi(\xi)$ is higher and, conversely, will never visit points ξ such that $\pi(\xi) = 0$. Standard choices for the proposal q are uniform, normal or Cauchy distributions. (Notice that Example 6.3.2 is a particular case of the random-walk Metropolis–Hastings algorithm, since the proposal is a random walk on $(\log\alpha, \log\eta)$.)

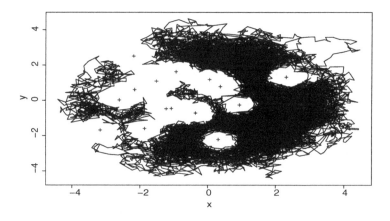

Figure 6.3.1. *Path of the Markov chain* $(\theta^{(m)})_m$ *for the posterior distribution* $\pi(\theta|x)$ *of Example 6.3.3 and repulsive points* μ_j *indicated by crosses for* $x = 0$ *and* $p = 15$ *(5000 iterations).*

Example 6.3.3 For $\theta, x \in \mathbb{R}^2$, consider the modified normal distribution

$$\pi(\theta|x) \propto \exp\{-||\theta - x||^2/2\} \prod_{i=1}^{p} \exp\left\{\frac{-1}{||\theta - \mu_i||^2}\right\},$$

where the μ_i's are repulsive points, that is, unlikely (or prohibited) values of θ. A random-walk Metropolis–Hastings algorithm based on a $\mathcal{N}_2(0, 0.2\,I_2)$ perturbation leads to the result described in Figure 6.3.2 for $x = 0$ and $p = 15$. The μ_j's, which are represented by crosses, are correctly avoided by the Markov chain, which also recover the shape of the normal p.d.f., as shown by the picture on the cover of this book.   ‖

   This algorithm is clearly widely applicable and, moreover, has limited tune-up requirements, since the distribution of the perturbation can be chosen almost independently of the true density π. (In fact, this distribution depends on a scale factor that should be calibrated by the average acceptance rate of the algorithm.[8] See Robert and Casella (2004, Section 7.6.1 and Note 7.8.4).) While it cannot enjoy convergence properties stronger than *geometric convergence* because of the heavy tails of the proposal (see Mengersen and Tweedie (1996)), the random-walk Metropolis–Hastings algorithm still appears to be the "gold standard" of MCMC techniques.

   Another class of proposals, more akin to standard Monte Carlo techniques, are the *independent proposals*, that is, densities $q(\cdot|\theta)$ that do not depend on θ,

$$q(\theta'|\theta) = h(\theta').$$

---

[8] The scale factor in Example 6.3.3 was deliberately chosen for being too small, towards a better representation of the way the Markov chain avoids the repulsive points μ_i.

(Because there is a positive probability of rejecting the proposal, the output of the algorithm still is a Markov chain.) Although their theoretical properties are often better than those of random-walk Metropolis–Hastings algorithms (see Mengersen and Tweedie (1996)), these methods have a more limited applicability because the proposal has to fit the target distribution π in some sense. This proposal may sometimes be the prior distribution, or it may be based on an asymptotic expansion of the distribution π, for instance a *saddlepoint approximation* (Robert and Casella (2004, Example 7.12)), or on an approximative accept-reject algorithm as in the ARMS algorithm of Gilks et al. (1995) (see also Exercise 6.10). (Notice the similarity with the importance sampling method of Section 6.2.2: the choice of the proposal distribution h is crucial for the practical implementation of the method, although being almost irrelevant theoretically.)

6.3.3 The Gibbs sampler

The Metropolis–Hastings technique presented in the previous section is appealing for its universality, but, in contrast, the lack of connection between the proposal q and the target distribution π may be detrimental to the convergence properties of the method and, in practice, may easily prevent convergence if the probability of reaching far-away parts of the distribution π is too small. Using a different perspective, the *Gibbs sampling* approach is actually based on the distribution π. This method takes its name from Gibbs random fields, where it was used for the first time by Geman and Geman (1984). (See Robert and Casella (2004, Note 10.6.1) for a brief account of the early history of the Gibbs sampler.)

From a general point of view, the Gibbs sampler takes advantage of *hierarchical structures*, i.e. when a Bayesian model can be written as

$$(6.3.3) \qquad \pi(\theta|x) = \int \pi_1(\theta|x,\lambda)\pi_2(\lambda|x)\,d\lambda.$$

The idea is then to simulate from the joint distribution $\pi_1(\theta|x,\lambda)\pi_2(\lambda|x)$ to recover $\pi(\theta|x)$ as the marginal: when both distributions $\pi_1(\theta|x,\lambda)$ and $\pi_2(\lambda|x)$ are known and can be simulated, the generation of θ from $\pi(\theta|x)$ is equivalent to the generation of λ from $\pi_2(\lambda|x)$, and of θ from $\pi_1(\theta|x,\lambda)$.

Example 6.3.4 (Casella and George (1992)) Consider $(\theta,\lambda) \in \mathbb{N} \times [0,1]$ and

$$\pi(\theta,\lambda|x) \propto \binom{n}{\theta}\lambda^{\theta+\alpha-1}(1-\lambda)^{n-\theta+\beta-1},$$

where the parameters α and β actually depend on x. This model can be written in the hierarchical form (6.3.3), with $\pi_1(\theta|x,\lambda)$ a binomial distribution, $\mathcal{B}(n,\lambda)$, and $\pi_2(\lambda|x)$ a beta distribution, $\mathcal{B}e(\alpha,\beta)$. The marginal distribution of θ is then

$$\pi(\theta|x) = \binom{n}{\theta}\frac{B(\alpha+\theta,\beta+n-\theta)}{B(\alpha,\beta)},$$

that is, a *beta-binomial distribution*. This distribution is not particularly easy to work with. For instance, the computation of $\mathbb{E}[\theta/(\theta+1)|x]$, or of the posterior distribution of $\eta = \exp(-\theta^2)$, cannot be done explicitly and may involve intricate computations, even from a numerical point of view, when α, β, and n are large. Therefore, depending on the inferential problem, it may be more advantageous to simulate $(\lambda^{(1)}, \theta^{(1)}), \ldots, (\lambda^{(m)}, \theta^{(m)})$ with $\lambda^{(i)} \sim \mathcal{B}e(\alpha, \beta)$ and $\theta^{(i)} \sim \mathcal{B}(n, \lambda^{(i)})$; for instance, $\mathbb{E}[\theta/(\theta+1)|x]$ can be then approximated with

$$\frac{1}{m} \sum_{i=1}^{m} \theta^{(i)}/(\theta^{(i)}+1).$$

$\parallel$

When, in contrast with Example 6.3.4, the marginal distribution $\pi_2(\lambda|x)$ is not always available (in analytical or algorithmic forms), the classical Monte Carlo method cannot be implemented. It is more often the case that both *conditional posterior distributions*, $\pi_1(\theta|x, \lambda)$ and $\pi_2(\lambda|x, \theta)$, can be simulated. Since they are sufficiently informative about the joint distribution, $\pi(\theta, \lambda|x)$, and since $\pi(\theta, \lambda|x)$ can be recovered from the conditional densities (see Exercises 6.23 and 6.24), it seems conceptually possible to base the simulation of $\pi(\theta|x)$ on these conditional distributions only.

Example 6.3.5 (Example 6.1.4 continued) For the temporal capture-recapture model, the two conditional posterior distributions are ($1 \leq i \leq n$)

$$\begin{aligned}
p_i|x, N &\sim \mathcal{B}e(\alpha + x_i, \beta + N - x_i) \\
N - x_+|x, p &\sim \mathcal{N}eg(x_+, \varrho),
\end{aligned}$$

with

$$\varrho = 1 - \prod_{i=1}^{n}(1 - p_i).$$

On the contrary, the posterior marginal distribution $\pi_2(p|x)$ cannot be obtained in a closed form or directly simulated. $\parallel$

A first *Gibbs sampling* technique called *data augmentation* was introduced by Tanner and Wong (1987) to take advantage of the conditional distributions according to the following iterative algorithm

---

Initialization: Start with an arbitrary value $\lambda^{(0)}$.

Iteration t: Given $\lambda^{(t-1)}$, generate

a. $\theta^{(t)}$ according to $\pi_1(\theta|x, \lambda^{(t-1)})$
b. $\lambda^{(t)}$ according to $\pi_2(\lambda|x, \theta^{(t)})$.

It is then straightforward to show that $\pi(\theta, \lambda|x)$ is a stationary distribution for the above transition: if $(\theta^{(i-1)}, \lambda^{(i-1)})$ is distributed from the joint distribution, $\lambda^{(i-1)}$ is distributed from the marginal distribution $\pi_2(\lambda|x)$ and, therefore, $(\theta^{(i)}, \lambda^{(i-1)})$ is still distributed from the joint distribution. (Actually, this requires the support of the joint distribution to be equal to the Cartesian product of the supports of π_1 and π_2. See Robert and Casella (2004, Example 10.7 and Figure 10.1) for a counter-example.) The same reasoning applies to the second step in the algorithm and the chain $(\theta^{(t)}, \lambda^{(t)})$ is ergodic with limiting distribution π. In addition, the dual structure of the above algorithm leads to good convergence properties, as pinpointed in Diebolt and Robert (1994):

Lemma 6.3.6 *If $\pi_1(\theta|x, \lambda) > 0$ on Θ ($\pi_2(\lambda|x, \theta) > 0$ on Λ, resp.), both sequences $(\theta^{(m)})$ and $(\lambda^{(m)})$ are ergodic Markov chains with invariant distributions $\pi(\theta|x)$ and $\pi(\lambda|x)$.*

Moreover, it can be shown that, if the convergence is uniformly geometric for one of the two chains, e.g., if it takes values in a finite space, the convergence to the stationary distribution is also *uniformly geometric* for the other chain. This property is now known as the *Duality Principle* (see Exercise 6.25).

Example 6.3.7 (Example 6.3.4 continued) The conditional distributions are

$$\theta|x, \lambda \sim \mathcal{B}(n, \lambda), \qquad \lambda|x, \theta \sim \mathcal{B}e(\alpha + \theta, \beta + n - \theta)$$

and allow for Gibbs sampling. Figure 6.3.3 gives a comparison of the histogram of a sample of 5000 observations obtained by batch sampling (with $t = 10$) with the histogram of a sample of 5000 observations θ simulated directly from the beta-binomial distribution. The strong similarity shows that the Gibbs sampling approximation is quite acceptable. ‖

6.3.4 Rao–Blackwellization

As discussed in Section 6.3.1, the sample $\theta^{(1)}, \ldots, \theta^{(m)}$ produced by the Gibbs sampler can be used similarly to those obtained by the classical Monte Carlo method, but Gelfand and Smith (1990) remark that the conditional structure of the sampling algorithm and the dual sample, $\lambda^{(1)}, \ldots, \lambda^{(m)}$, should be exploited. Indeed, if the quantity of interest is $\mathbb{E}^\pi[g(\theta)|x]$, one can use the average of the conditional expectations

$$\delta_2 = \frac{1}{m} \sum_{i=1}^{m} \mathbb{E}^\pi[g(\theta)|x, \lambda^{(m)}],$$

when they can be computed easily, instead of using the direct average

$$\delta_1 = \frac{1}{m} \sum_{i=1}^{m} g(\theta^{(i)}).$$

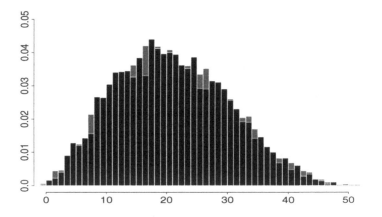

Figure 6.3.2. *Histograms for samples of size 5000 from the beta-binomial distribu-*
tion with parameters $n = 54$, $\alpha = 3.4$, *and* $\beta = 5.2$: *(dark grey) directly simulated;*
(light grey) obtained by Gibbs sampling.

This modification is based on the Rao–Blackwell theorem (see Theorem
2.4.8). Were the $\lambda^{(i)}$'s and $\theta^{(i)}$'s independent,

$$
\begin{aligned}
\mathbb{E}^\pi \left[(\delta_1 - \mathbb{E}^\pi[g(\theta)|x])^2 |x \right] &= \frac{1}{m}\mathrm{var}^\pi(g(\theta)|x) \\
&\geq \frac{1}{m}\mathrm{var}^\pi\left(\mathbb{E}^\pi[g(\theta)|x,\lambda]|x\right) \\
&= \mathbb{E}^\pi \left[(\delta_2 - \mathbb{E}^\pi[g(\theta)|x,\lambda])^2 |x \right].
\end{aligned}
$$

Liu et al. (1994) show that this inequality also holds for Data Augmenta-
tion because $\mathrm{cov}(\theta^{(0)}, \theta^{(m)})$ is then positive and decreasing with m (Exercise
6.27). The estimator δ_2, christened *Rao–Blackwellization*, therefore dominates δ_1. (But this domination does not necessarily extend to other MCMC
schemes, see Liu et al. (1994) and Geyer (1995).)

Example 6.3.8 (Casella and George (1992)) Consider the following conditional distributions (with x omitted):

$$
\begin{aligned}
\pi(\theta|\lambda) &\propto \lambda e^{-\theta\lambda}, && 0 < \theta < B, \\
\pi(\lambda|\theta) &\propto \theta e^{-\lambda\theta}, && 0 < \lambda < B.
\end{aligned}
$$

The marginal distribution of θ (or of λ) cannot be computed, but the conditional distributions are easy to simulate, being truncated exponential dis-
tributions. Since $\mathbb{E}^\pi[\theta|\lambda] \simeq 1/\lambda$ for B large, $\mathbb{E}^\pi[\theta|x]$ can be approximated
by

$$
\frac{1}{m}\sum_{i=1}^m \theta_i \quad \text{or} \quad \frac{1}{m}\sum_{i=1}^m \frac{1}{\lambda_i}.
$$

For this particular example, the complete symmetry existing between the two conditional distributions implies that the two estimators have exactly the same probabilistic properties, besides converging to the same value. ‖

The same argument leads us to propose the approximation of the posterior density $\pi(\theta|x)$ by the average of the conditional densities

$$\frac{1}{m} \sum_{i=1}^{m} \pi(\theta|x, \lambda_i),$$

instead of using regular *kernel estimation* methods (see Tanner and Wong (1987) and Gelfand and Smith (1990)).

6.3.5 The general Gibbs sampler

A generalization of the data augmentation algorithm is to consider several groups of parameters, $\theta, \lambda_1, \ldots, \lambda_p$, such that

$$(6.3.4) \qquad \pi(\theta|x) = \int \ldots \int \pi(\theta, \lambda_1, \ldots, \lambda_p|x) \, d\lambda_1 \cdots d\lambda_p.$$

This generalization corresponds to the introduction of additional levels in the hierarchical model (6.3.3), either for modeling or for simulation reasons, or it may occur because of the decomposition of the hyperparameter λ or of the parameter θ into components of smaller dimensions.

As mentioned in Section 6.3.3 with the data augmentation scheme, the Gibbs sampler provides simulations from the joint distribution $\pi(\theta, \lambda_1, \ldots, \lambda_p|x)$, when some conditional distributions associated with π are available. Obviously, when $\pi(\theta|x)$ itself can be decomposed into conditionals, there is no need to introduce the additional parameters λ_i ($1 \leq i \leq p$).

Example 6.3.9 (Example 6.3.4 continued) If the population size n has a Poisson prior, $\mathcal{P}(\xi)$, the overall joint posterior distribution is

$$\pi(\theta, \lambda, n|x) \propto \binom{n}{\theta} \lambda^{\theta+\alpha-1}(1-\lambda)^{n-\theta+\beta-1} e^{-\xi} \frac{\xi^n}{n!}$$

and the marginal distribution of θ cannot be derived. On the contrary, the full conditional distributions have explicit expressions, since

$$\begin{aligned}
\theta|x, \lambda, \xi &\sim \mathcal{B}(n, \lambda), \\
\lambda|x, \theta, \xi &\sim \mathcal{B}e(\theta+\alpha, n-\theta+\beta), \\
n-\theta|x, \theta, \lambda &\sim \mathcal{P}(\xi(1-\lambda)).
\end{aligned}$$

Simulation from these three conditionals is thus possible.   ‖

Example 6.3.10 (Tanner and Wong (1987)) Consider a multinomial model,

$$y \sim \mathcal{M}_5 \left(n; a_1\mu + b_1, a_2\mu + b_2, a_3\eta + b_3, a_4\eta + b_4, c(1-\mu-\eta)\right),$$

parametrized by μ and η, where

$$0 \leq a_1 + a_2 = a_3 + a_4 = 1 - \sum_{i=1}^{4} b_i = c \leq 1$$

and $c, a_i, b_i \geq 0$ are known. This model stems from sampling according to

$$x \sim \mathcal{M}_9(n; a_1\mu, b_1, a_2\mu, b_2, a_3\eta, b_3, a_4\eta, b_4, c(1 - \mu - \eta)),$$

and aggregating some coordinates:

$$y_1 = x_1 + x_2, \quad y_2 = x_3 + x_4, \quad y_3 = x_5 + x_6, \quad y_4 = x_7 + x_8, \quad y_5 = x_9.$$

A conjugate prior distribution on (μ, η) for the model on x is the Dirichlet distribution $\mathcal{D}(\alpha_1, \alpha_2, \alpha_3)$,

$$\pi(\mu, \eta) \propto \mu^{\alpha_1 - 1} \eta^{\alpha_2 - 1} (1 - \eta - \mu)^{\alpha_3 - 1},$$

where $\alpha_1 = \alpha_2 = \alpha_3 = 1/2$ corresponds to a noninformative modeling. In this setting, the posterior distribution of (μ, η) cannot be derived explicitly. However, if we introduce the *missing data* $z = (x_1, x_3, x_5, x_7)$, which is not observed (hence, is missing), x is in one-to-one correspondence with (y, z) and

$$\begin{aligned} \pi(\eta, \mu | y, z) &= \pi(\eta, \mu | x) \\ &\propto \mu^{z_1} \mu^{z_2} \eta^{z_3} \eta^{z_4} (1 - \eta - \mu)^{y_5 + \alpha_3 - 1} \mu^{\alpha_1 - 1} \eta^{\alpha_2 - 1}, \end{aligned}$$

where we denote the coordinates of z as (z_1, z_2, z_3, z_4). Therefore,

$$\mu, \eta | y, z \sim \mathcal{D}(z_1 + z_2 + \alpha_1, z_3 + z_4 + \alpha_2, y_5 + \alpha_3).$$

Moreover,

$$z_i | y, \mu, \eta \sim \mathcal{B}\left(y_i, \frac{a_i\mu}{a_i\mu + b_i}\right) \qquad (i = 1, 2),$$

$$z_i | y, \mu, \eta \sim \mathcal{B}\left(y_i, \frac{a_i\eta}{a_i\eta + b_i}\right) \qquad (i = 3, 4).$$

Defining $\theta = (\mu, \eta)$ and $\lambda = z$, it thus appears that conditional distributions can be simulated in this case. Notice that the missing data z does not appear in the original formulation of the problem and may be artificial, in the sense that the model at hand does not necessarily correspond to an aggregated multinomial model. However, it greatly facilitates the simulation of the θ's. Similar behavior will occur in other missing-data models. ‖

In this general hierarchical setting, implementation of the Gibbs sampler can be done in many ways. If the decomposition of (θ, λ) in $(\theta, \lambda_1, \ldots, \lambda_p)$ corresponds to a division of the model into its hierarchical levels, that is,

$$(6.3.5) \quad \pi(\theta | x) = \int \cdots \int \pi_1(\theta | \lambda_1, x) \pi_2(\lambda_1 | \lambda_2) .. \pi_{p+1}(\lambda_p) \, d\lambda_1 \cdots d\lambda_p,$$

it seems logical to simulate according to the conditional distributions

$$
\begin{aligned}
\pi(\theta|x, \lambda_1, \ldots, \lambda_p) &= \pi_1(\theta|\lambda_1, x), \\
\pi(\lambda_i|x, \theta, (\lambda_j)_{j \neq i}) &= \pi(\lambda_i|\lambda_{i-1}, \lambda_{i+1}) \qquad (1 < i < p), \\
(6.3.6) \quad \pi(\lambda_1|x, \theta, (\lambda_j)_{j \neq 1}) &= \pi(\lambda_1|\theta, \lambda_2), \\
\pi(\lambda_p|x, \theta, (\lambda_j)_{j \neq p}) &= \pi(\lambda_p|\lambda_{p-1}),
\end{aligned}
$$

whatever the dimensions of θ and λ_j are (Exercise 6.29). For instance, in Example 6.3.10, (μ, η) could thus be generated conditional upon (y, z) according to a Dirichlet distribution and z conditional upon (μ, η).

An alternative Gibbs sampler also proposed in Gelfand and Smith (1990) is the *one-at-a-time* Gibbs sampler, which neglects hierarchical divisions and consider only one-dimensional parameters, to generate them conditional upon the other parameters.

Example 6.3.11 (Example 6.3.10 continued) Since

$$
\begin{aligned}
\frac{\mu}{1 - \eta}|y, z, \eta &\sim \mathcal{B}e(z_1 + z_2 + \alpha_1, y_5 + \alpha_3), \\
\frac{\eta}{1 - \mu}|y, z, \mu &\sim \mathcal{B}e(z_3 + z_4 + \alpha_2, y_5 + \alpha_3),
\end{aligned}
$$

this version of Gibbs sampling leads to the iterative simulation of

$$
\begin{aligned}
\mu^{(t)} &\sim (1 - \eta^{(t-1)})\mathcal{B}e\left(z_1^{(t-1)} + z_2^{(t-1)} + \alpha_1, y_5 + \alpha_3\right), \\
\eta^{(t)} &\sim (1 - \mu^{(t)})\mathcal{B}e\left(z_3^{(t-1)} + z_4^{(t-1)} + \alpha_2, y_5 + \alpha_3\right), \\
(6.3.7) \quad z_j^{(t)} &\sim \mathcal{B}\left(y_j, \frac{a_j \mu^{(t)}}{a_j \mu^{(t)} + b_j}\right) \qquad (j = 1, 2), \\
z_j^{(t)} &\sim \mathcal{B}\left(y_j, \frac{a_j \eta^{(t)}}{a_j \eta^{(t)} + b_j}\right) \qquad (j = 3, 4).
\end{aligned}
$$

The difference with the simulation in Example 6.3.10 is thus minor.   ‖

The general formulation of the *Gibbs sampling algorithm* for a joint distribution $\pi(\theta_1, \ldots, \theta_p)$ with full conditionals $\pi_1, \ldots, \pi_p$ is set forth below.

Given $(\theta_1^{(t)}, \ldots, \theta_p^{(t)})$, simulate

1. $\theta_1^{(t+1)} \sim \pi_1(\theta_1|\theta_2^{(t)}, \ldots, \theta_p^{(t)})$,
2. $\theta_2^{(t+1)} \sim \pi_2(\theta_2|\theta_1^{(t+1)}, \theta_3^{(t)}, \ldots, \theta_p^{(t)})$,

   $\vdots$

p. $\theta_p^{(t+1)} \sim \pi_p(\theta_p|\theta_1^{(t+1)}, \ldots, \theta_{p-1}^{(t+1)})$.

The above validation of the data-augmentation algorithm extends to this case: the joint distribution π is stationary at each step of this algorithm, since the π_j's are the full conditionals of π. Under the *positivity constraint* that the support of π is the Cartesian product of the supports of the π_i's, the resulting chain is ergodic.

Compared with the Metropolis–Hastings algorithm, the number of possible implementations of the Gibbs sampler is small and, moreover, the differences in the convergence properties are often minor. Still, the approach of (6.3.6) (also called *substitution sampling* in Gelfand and Smith (1990)) should be preferred over the one-at-a-time alternative, since the former respects the initial hierarchical modeling and often converges more rapidly to the stationary distribution (see Liu et al. (1994) and Roberts and Sahu (1997)). Data augmentation is the only case of Gibbs sampling to produce a Markov chain for both $(\theta^{(t)})$ and $(\lambda^{(t)})$; in every other scheme, the subchains are not Markov chains (Exercise 6.30).

However, in order to be able to use data augmentation or even substitution sampling, one needs the conditional distributions for each hierarchical level (as $\pi(\eta, \mu | y, z)$ in Example 6.3.10), and they may be more difficult to derive than full conditional distributions (see Exercise 6.46). Moreover, the Gibbs sampler does not actually require the θ_i's to be one-dimensional, and the choice of the decomposition can then be entirely based on simulation reasons. Notice also that, when reduced conditional distributions, like $\pi(\theta | x, \lambda_{i_0})$, are available for simulation, it is obviously preferable to use these distributions, since they increase convergence speed by reducing the dependency on the other parameters. (This is called *blocking*; see for instance Roberts and Sahu (1997).) A last important remark is that, whenever simulation from a given conditional distribution $\pi_i(\theta_i | \theta_j, \ j \neq i)$ is difficult, this simulation step can be replaced with a *single* Metropolis–Hastings step with target distribution $\pi_i(\theta_i | \theta_j, \ j \neq i)$. This may appear to be a crude approximation device, but this is not the case: the replacement of a simulation from $\pi_i(\theta_i | \theta_j, \ j \neq i)$ with a Metropolis–Hastings step does not modify the stationary distribution of the chain, and is thus entirely valid from an MCMC point of view.

Example 6.3.12 (Example 6.1.4 continued) When N, the size of the population, is the parameter of interest, Gibbs sampling provides a sample, $N_1, \ldots, N_m$, starting with an initial value of $p = (p_1^{(0)}, \ldots, p_n^{(0)})$ and iteratively generating according to

$$N^{(j)} - x_+ | x, p^{(j-1)} \ \sim \ \mathcal{N}eg(x_+, \varrho^{(j-1)}),$$
$$p_i^{(j)} | x, N^{(j)} \ \sim \ \mathcal{B}e(\alpha + x_i, \beta + N^{(j)} - x_i) \qquad (1 \leq i \leq n).$$

(This is actually a data augmentation scheme.) The sample $N_1, \ldots, N_m$ is then obtained by taking $N_1 = N^{(k_0+T)}$, $N_2 = N^{(k_0+2T)}$, ..., $N_m = N^{(k_0+mT)}$, where k_0 represents the "burn-in" time, that is, the number of iterations to come reasonably close to stationarity, and T is the batch size, that is, the number of iterations to achieve approximate independence between the points of the sample. Gibbs sampling simultaneously gives a sample $p^1, \ldots, p^m$. The expectation $\mathbb{E}^\pi[N|x]$ can then be approximated by

$$
\begin{aligned}
\hat{\delta}^\pi(x) &= \frac{1}{m} \sum_{t=1}^{m} \mathbb{E}^\pi[N|x, p^t] \\
&= \frac{1}{m} \sum_{t=1}^{m} \left(1 - \prod_{i=1}^{n} (1 - p_i^t) \right)^{-1} x_+,
\end{aligned}
$$

according to the "Rao–Blackwellization" argument mentioned above. George and Robert (1992) provide hierarchical extensions in this setting by considering different families of hyperpriors on (α, β).   ∥

A more general comparison between Metropolis–Hastings and Gibbs sampling algorithms is meaningless: depending on the problem at hand and on the choice of the proposals/hierarchical decompositions, one algorithm may converge faster than the other. The only warning we can provide here is that, contrary to common belief, the Gibbs sampler is not necessarily the optimal solution for an MCMC implementation. Indeed, whereas on the one hand the Gibbs sampler is constructed directly from the target distribution π, and thus does not involve subjective input from the experimenter, on the other hand the fact that the Gibbs sampler updates the chain one component (or one block) at a time may induce very poor convergence properties if the distribution has a very narrow or multimodal support. In contrast, a Metropolis–Hastings algorithm using a random-walk proposal may be inefficient if the shape or scale of the proposal is at odds with the support of π, but it also allows for big jumps that may reach far-away modes of π. We could then classify Gibbs samplers as *local*, and random-walk Metropolis–Hastings algorithms as *global*, in the crude sense that the former often provide a better picture of the neighborhood of the starting point, while the latter explore the support of π on a larger scale (see Besag (2000) for a more detailed discussion). The best solution to this dilemma is then to take advantage of the positive features of these different samplers by combining them in an *hybrid* sampler calling for several MCMC algorithms in a deterministic or random manner.

6.3.6 The slice sampler

Gibbs sampling may appear at this stage as a particular MCMC method that only applies in a quite restricted setting: it involves hierarchical structures, as in (6.3.3), thus does not apply to unidimensional problems; it also requires full conditionals, thus cannot cover complex models.

This perception of the Gibbs sampler is mistaken: as we will now see, the Gibbs sampler also applies in unidimensional problems, it does not require simulation from the full conditionals, and it covers the same models as other MCMC methods. In fact, the hierarchical decomposition (6.3.3) is not particularly restrictive; and indeed, numerous distributions (on the observation or the parameter) can be written as *hidden mixtures*, with totally artificial parameter λ (see Note 6.6.3). Therefore, even when the hierarchical structure is missing in the original problem, it can often be reintroduced in order to improve the computation of the Bayes estimators, or even the choice of the prior distribution.

The generality of the Gibbs sampler is actually exposed in the particular version called the *slice sampler* (Wakefield et al. (1991), Besag and Green (1993), and Damien et al. (1999)). Consider a distribution $\pi(\theta)$ on a general set Θ, which can be one-dimensional or multidimensional, and write π as the product

$$(6.3.8) \qquad \pi(\theta) = \prod_{i=1}^{k} \varpi_i(\theta),$$

where the ϖ_i's are positive functions, but are not necessarily densities. Then $\pi(\theta)$ can be written as the marginal

$$\pi(\theta) = \int \prod_{i=1}^{k} \mathbb{I}_{0 \leq \omega_i \leq \varpi_i(\theta)} \, d\omega_1 \cdots d\omega_k .$$

The corresponding *slice sampler* is then straightforward:

At iteration t, simulate

1. $\omega_1^{(t+1)} \sim \mathcal{U}_{[0,\varpi_1(\theta^{(t)})]}$

 $\vdots$

k. $\omega_k^{(t+1)} \sim \mathcal{U}_{[0,\varpi_k(\theta^{(t)})]}$

k+1. $\theta^{(t+1)} \sim \mathcal{U}_{A^{(t+1)}}$, with

$$A^{(t+1)} = \{\xi; \ \varpi_i(\xi) \geq \omega_i^{(t+1)}, \ i = 1,\ldots,k\}.$$

The ω_j's are a particular type of *auxiliary variables*, which are meaningless for the statistical problem. Notice that there are many possible representations (6.3.8) for the same distribution π, including the simple

$$\pi(\theta) = \int_0^1 \mathbb{I}_{0 \leq \omega \leq \pi(\theta)} d\omega ,$$

and that the choice is purely dictated by convenience. In fact, the last step (k+1) in the above algorithm may be far from simple to implement, since the set $A^{(t)}$ is often difficult to construct, but this decomposition shows that the Gibbs sampler can provide, at least formally, a representation of all distributions. (See Roberts and Rosenthal (1998) and Mira and Tierney (2002) for theoretical properties of the slice sampler.)

Example 6.3.13 (Example 6.3.2 continued) The joint distribution of (α, η) being

$$\pi(\alpha, \eta | x_1, \ldots, x_n) \propto \alpha^n \eta^{n+\beta-1} \left(\prod_{i=1}^{n} x_i \right)^\alpha \exp \left\{ -\eta \sum_{i=1}^{n} x_i^\alpha - \alpha - \xi\eta \right\},$$

the conditional distribution $\pi_1(\eta|\alpha, x_1, \ldots, x_n)$ is simply a

$$\mathcal{G}(\beta + n, \xi + \sum_i x_i^\alpha)$$

distribution, which can be simulated easily. The conditional distribution $\pi_2(\alpha|\eta, x_1, \ldots, x_n)$ is much more complex because of the exponential part involving the x_i^α's. If we write this distribution as $\alpha^n \chi^\alpha \exp(-\eta \sum_{i=1}^n x_i^\alpha)$, it can be expressed as the marginal (in α) of

$$\alpha^n \mathbb{I}_{0 \leq \omega_0 \leq \chi^\alpha} \prod_{i=1}^n \mathbb{I}_{0 \leq \omega_i \leq \exp(-\eta x_i^\alpha)} \cdot$$

The conditional distribution of α given η and the ω_i's is then proportional to

$$\alpha^n \mathbb{I}_{\alpha \log(\chi) \leq \log(\omega_0)} \prod_{i=1}^n \mathbb{I}_{\alpha \log(x_i) \leq \log\{-\log(\omega_i)/\eta\}} \,,$$

that is, a simple power distribution α^n on an interval $(\underline{\alpha}, \overline{\alpha})$. The overall Gibbs sampler for the Weibull posterior distribution then goes through the iterative simulation of η, of the ω_i's and of α. ∥

Example 6.3.14 (Example 6.1.5 continued) Because the posterior distribution of $\theta = (\mu_1, \sigma_1^2, p, \mu_2, \sigma_2^2)$ is available in closed form,

$$\pi(\theta|x) \propto \tilde{\pi}(\theta|x) = \pi(\theta) \prod_{i=1}^n \{p\varphi(x_i; \mu_1, \sigma_1) + (1-p)\varphi(x_i; \mu_2, \sigma_2)\} \,,$$

a formal slice sampler with a single auxiliary variable ω can be proposed, with $\theta \sim \mathcal{U}_{\tilde{\pi}(\theta|x) \geq \omega}$. But it is impossible to simulate from this uniform distribution, since the constraint $\tilde{\pi}(\theta|x) \geq \omega$ cannot be inverted into a constraint on θ. A manageable version of the slice sampler in this setting can be constructed by introducing instead n auxiliary variables ω_i so that $\tilde{\pi}(\theta|x)$ reads like the marginal distribution of

$$\pi(\theta) \prod_{i=1}^n \mathbb{I}_{p\varphi(x_i; \mu_1, \sigma_1) + (1-p)\varphi(x_i; \mu_2, \sigma_2) \geq \omega_i \geq 0} \cdot$$

Although the joint distribution of θ conditional on the ω_i's is still intractable, the full conditionals of the parameters μ_1, σ_1^2, p, μ_2 and σ_2^2 can be easily simulated. (As we will see in Section 6.4, which concentrates on mixtures, the original Gibbs sampling solution also goes through the simulation of n auxiliary variables.) ∥

6.3.7 The impact of MCMC methods on Bayesian Statistics

This section has very briefly presented the bases of MCMC methods, with a few illustrations taken from Bayesian computational problems. It is important to stress at this point that the intrusion of these MCMC tools into the

Bayesian field has had a "devastating" effect! Indeed, it has radically modified the way people work with models and prior assumptions, by allowing much more complex structures to be proposed, as for instance in the case of *graphical models* where relations between variables are only defined at a local level, the joint distribution being impossible to envision (see Spiegelhalter et al. (1993) and Note 10.7.1). Similarly, *latent variable models* such as ARMA, hidden Markov, or stochastic volatility models can now be correctly analyzed (see Note 6.6.5 and Robert and Casella (2004, Note 9.7.2)) where only crude approximations could be used in the past, and this has had a tremendous effect in Bayesian signal processing, econometrics and mathematical finance.

The "devastation" mentioned above has also occurred at the level of the rigid structures previously imposed by the need for an analytical processing; for instance, conjugate priors are not compulsory anymore, even if they are still quite useful as basic priors in the different stages of a hierarchical modeling (see Chapter 10). Similarly, much more fluid representations can be proposed in model choice, as we will see in Chapter 7, where the possibility of envisioning many models at once leads the statistician away from strict testing towards *model averaging*, where the most likely models get the heavier weights but none are a priori excluded. See also Berger (2000), Cappé and Robert (2000), and Gelfand (2000) for short reviews on the impact of MCMC methods on the field.

As usual, a significant increase in the facility of using a given technique is accompanied by the corresponding increase in potential misuses of this technique! In the case of Bayesian analysis, this means that the effect of a prior modeling is more difficult to assess through the conditional distributions used in a Gibbs sampler. More crucially, the propriety of the posterior distribution discussed in Section 1.5 may fail to hold without the user being aware of it (see Note 6.6.4). But these drawbacks cannot compete with the explosion of the range and number of Bayesian applications, or the resolution of inferential problems never considered before.

6.4 An application to mixture estimation

We conclude this chapter by showing how MCMC methods are appropriate for the derivation of the Bayes estimators of the parameters of a mixture of normal distributions, considered in Example 6.1.5. Extensions to other mixtures of distributions from an exponential family or to hidden Markov models are straightforward (see Gruet et al. (1999) and Robert and Casella (1999, Section 9.5.1)). As detailed in Section 6.1, a Bayesian analysis of a mixture model leads to the information paradox that an explicit estimator is available and intuitively justified, but cannot be computed when the number of observations gets too large. Moreover, the maximum likelihood estimators of the parameters of (6.1.4) are not clearly defined, solving the likelihood equations is difficult, and there are even problems with analytical

approximations of the Bayes estimator. (See Crawford et al. (1992) for
an approach using Laplace approximation.) Similarly, a standard Monte
Carlo processing of mixture models is arduous, even though Casella et
al. (2000) have proposed an importance sampling solution in a conjugate
setting (Exercise 6.39). See Note 6.6.6 for further references and details on
the early history of mixture estimation.

Gibbs sampling for mixtures relies on a *missing data* representation, as
in Dempster et al. (1977), to build up a hierarchical structure similar to
(6.3.3). Consider

$$(6.4.1) \qquad x \sim f(x|\theta) = \sum_{i=1}^{k} p_i \varphi(x; \mu_i, \sigma_i),$$

a mixture of k normal distributions with means μ_i and variances σ_i^2 ($1 \leq i \leq k$) and $\sum_i p_i = 1$ ($p_i > 0$). Given a sample $x_1, \ldots, x_n$, from (6.4.1), we
define the missing values z_j ($1 \leq j \leq n$) as the *component indicator vectors*
for the x_j's, that is,

$$z_{ij} = \begin{cases} 1 & \text{if } x_j \sim \varphi(x; \mu_i, \sigma_i), \\ 0 & \text{otherwise,} \end{cases}$$

and $\sum_i z_{ij} = 1$. This vector can also be considered as an additional parameter and it provides the following joint distributions ($1 \leq j \leq n$):

$$z_j|\theta \quad \sim \quad \mathcal{M}_p(1; p_1, \ldots, p_k),$$

$$x_j|z_j, \theta \quad \sim \quad \mathcal{N}\left(\prod_{i=1}^{k} \mu_i^{z_{ij}}, \prod_{i=1}^{k} \sigma_i^{2z_{ij}}\right).$$

A convenient prior distribution on $\theta = (\mu_1, \sigma_1, p_1, \ldots, \mu_k, \sigma_k, p_k)$ is the
product of conjugate distributions $\pi_i(\mu_i, \sigma_i)$, with $\pi_i(\mu_i|\sigma_i)$ a normal distri-
bution $\mathcal{N}(\xi_i, \sigma_i^2/n_i)$, $\pi_i(\sigma_i^2)$ an inverse gamma distribution $\mathcal{IG}(\nu_i/2, s_i^2/2)$,
and $\pi(p)$ a Dirichlet distribution, $\mathcal{D}(\alpha_1, \ldots, \alpha_k)$, as in Example 6.1.5.

Notice that, once the allocation vectors z_j ($1 \leq j \leq n$) are known,
the mixture structure disappears, since this additional information breaks
down the sample into subsamples according to the values of z_{ij}. Although
the posterior distribution of θ cannot be used per se, as shown in Example
6.1.5, conditioning on $\mathbf{z} = (z_1, \ldots, z_n)$ removes the difficulty. Indeed, we
get the following posterior distributions ($1 \leq j \leq n$):

$$(6.4.2) \qquad z_j|x_j, \theta \sim \mathcal{M}_k(1; p_1(x_j, \theta), \ldots, p_k(x_j, \theta)),$$

with ($1 \leq i \leq k$)

$$p_i(x_j, \theta) = \frac{p_i \varphi(x_j; \mu_i, \sigma_i)}{\sum_{t=1}^{k} p_t \varphi(x_j; \mu_t, \sigma_t)},$$

and

$$(6.4.3) \qquad \mu_i|\mathbf{x}, \mathbf{z}, \sigma_i \sim \mathcal{N}(\xi_i(\mathbf{x}, \mathbf{z}), \sigma_i^2/(n + \sigma_i^2)),$$

$$\sigma_i^2|\mathbf{x}, \mathbf{z} \sim \mathcal{IG}\left(\frac{\nu_i + n_i}{2}, \frac{1}{2}\left[s_i^2 + \hat{s}_i^2(\mathbf{x}, \mathbf{z}) + \frac{n_i m_i(\mathbf{z})}{n_i + m_i(\mathbf{z})}(\bar{x}_i(\mathbf{z}) - \xi_i)^2\right]\right),$$

$$p|\mathbf{x}, \mathbf{z} \sim \mathcal{D}_k(\alpha_1 + m_1(\mathbf{z}), \ldots, \alpha_k + m_k(\mathbf{z})),$$

where

$$m_i(\mathbf{z}) = \sum_{j=1}^{n} z_{ij}, \qquad \bar{x}_i(j) = \frac{1}{m_i(\mathbf{z})} \sum_{j=1}^{n} z_{ij} x_j,$$

and

$$\xi_i(\mathbf{x}, \mathbf{z}) = \frac{n_i \xi_i + m_i(\mathbf{z}) \bar{x}_i(\mathbf{z})}{n_i + m_i(\mathbf{z})}, \qquad \hat{s}_i^2(\mathbf{x}, \mathbf{z}) = \sum_{j=1}^{n} z_{ij}(x_j - \bar{x}_i(\mathbf{z}))^2.$$

Conditional upon $\mathbf{z}$, the posterior distributions only take into account the subsamples related with each component, in a manner similar to the decomposition (6.1.6) of the true posterior distribution. In addition, simulations from (6.4.2) and (6.4.3) are particularly straightforward. Therefore, it is much easier to produce a sample $\theta_1, \ldots, \theta_m$ from $\pi(\theta|\mathbf{x})$ by Gibbs sampling than to use the true posterior distribution directly.

The remark following Lemma 6.3.6 implies that Gibbs sampling leads to uniform geometric convergence for the chain $(\theta^{(m)})$, since $\mathbf{z}$ has a finite support. The one-dimensional Gibbs sampling version of this algorithm only modifies the simulation for σ_i, which is then conditional on μ_i ($1 \leq i \leq k$),

$$\sigma_i^2|\mathbf{x}, \mathbf{z}, \mu_i \sim \mathcal{IG}\left(\frac{\nu_i + n_i + 1}{2}, \frac{1}{2}\left[s_i^2 + \sum_{j=1}^{n} z_{ij}(x_j - \xi_i)^2\right]\right).$$

Although the change is minor compared with (6.4.3), convergence results are much more difficult to establish in this setting. In particular, geometric convergence cannot be established without imposing restrictions on the σ_i's (see Diebolt and Robert (1994)). Moreover, since data augmentation can be implemented in this setting and is usually preferable to the one-dimensional Gibbs sampler, there is no reason to use the latter. In the setting of *hidden Markov chains*, which generalize mixture models like (6.4.1) by introducing Markovian dependence between the z_j's, this is not so straightforward (even though there exist in some settings closed-form representations of the likelihood that integrate out the latent variables; see Exercises 6.46 and 6.47, and Robert et al. (1999a)).

We stress as a final remark that Gibbs sampling is not the unique solution for the simulation of the posterior distribution $\pi(\theta|\mathbf{x})$. Indeed, as shown by Example 6.3.14, this distribution is available in closed form and can thus be used with any Metropolis–Hastings proposal (besides the slice sampler produced in Example 6.3.14). For instance, Celeux et al. (2000) demonstrate that the random-walk proposal can be used efficiently in this setting, with better mixing properties than the Gibbs sampler.

6.5 Exercises

Section 6.1

6.1 For a mixture of two normal distributions, as studied in Example 6.1.5 and the data of Table 6.1.5, identify the hyperparameters of the conjugate distributions by the moments method.

6.2 In the setting of Example 6.1.5, show that the posterior distribution can actually be written as (6.1.6) and develop $w(k_t)$ and $\pi(\theta|(k_t))$. Give the expressions of the Bayes estimators of μ_1, σ_1, and p for the hyperparameters obtained in Table 6.1.5.

6.3 Establish the equivalent of Exercise 6.2 for

(i) a mixture of two exponential distributions; and

(ii) a mixture of three uniform distributions.

6.4 In the setting of Exercise 6.2, how does the computing time evolve with the sample size when

(i) only the weight p is unknown; and

(ii) all the parameters are unknown.

6.5 *(Smith and Makov (1978)) Consider

$$x \sim f(x|p) = \sum_{i=1}^{k} p_i f_i(x),$$

where $p_i > 0$, $\sum_i p_i = 1$, and the densities f_i are known. The prior $\pi(p)$ is a Dirichlet distribution $\mathcal{D}(\alpha_1, \ldots, \alpha_k)$.

a. Show that the computing time becomes prohibitive as the sample size increases.

A sequential alternative which approximates the Bayes estimator is to replace $\pi(p|x_1, \ldots, x_n)$ by $\mathcal{D}(\alpha_1^{(n)}, \ldots, \alpha_k^{(n)})$, with

$$\alpha_1^{(n)} = \alpha_1^{(n-1)} + P(z_{n1} = 1|x_n), \ldots, \alpha_k^{(n)} = \alpha_k^{(n-1)} + P(z_{nk} = 1|x_n),$$

and z_{ni} $(1 \leq i \leq k)$ is the component indicator vector of x_n as defined in Section 6.4.

b. Justify this approximation and compare it with the updating of $\pi(\theta|x_1, \ldots, x_{n-1})$ when x_n is observed.

c. Examine the performances of this approximation for a mixture of two normal distributions $\mathcal{N}(0, 1)$ and $\mathcal{N}(2, 1)$ when $p = 0.1, 0.25, 0.5$.

d. If $\pi_i^n = P(z_{ni} = 1|x_n)$, show that

$$\hat{p}_i^{(n)}(x_n) = \hat{p}_i^{(n-1)}(x_{n-1}) - a_{n-1}\{\hat{p}_i^{(n-1)} - \pi_i^n\},$$

where $\hat{p}_i^{(n)}$ is the quasi-Bayesian approximation of $\mathbb{E}^\pi(p_i|x_1, \ldots, x_n)$.

6.6 In the setting of Example 6.1.4, determine the posterior distribution of $\pi(N|x)$: (a) for $n = 10$ and similar x_i's; and (b) for $n = 30$ and very different x_i's. Consider the same problem when $\pi(N)$ is a Poisson distribution $\mathcal{P}(\lambda)$ and λ varies. Pay particular attention to the potential problems linked with the direct evaluation.

Section 6.2.1

6.7 *(Morris (1982)) Given the quadratic variance natural exponential families studied in Exercises 3.24 and 10.33, consider

$$P_m(x,\mu) = V^m(\mu) \left\{ \frac{d^m}{d\mu^m} f(x|\mu) \right\} \Big/ f(x|\mu).$$

a. Show that P_m is a polynomial of degree m in x and in μ.
b. Show that $(m > 1)$

$$P_{m+1}(x,\mu) =$$
$$[P_1(x,\mu) - mV'(\mu)]P_m(x,\mu) - m[1 + (m-1)v_2]V(\mu)P_{m-1}(x,\mu),$$

where $V(\mu) = v_0 + v_1\mu + v_2\mu^2$.
c. Show that the polynomials P_m are orthogonal, and that $\mathbb{E}_\mu[P_m^2(x,\mu)] = a_m V^m(\mu)$.
d. Give the polynomials associated with the normal, Poisson, gamma, binomial, and negative binomial distributions. [*Note:* They are called *Hermite, Poisson–Charlier, generalized Laguerre, Krawtchouk*, and *Meixner* polynomials, respectively.]

Section 6.2.2

6.8 Show that, if the support of h does not contain the support of $f(x|\theta)\pi(\theta)$, the importance sampling approximation (6.2.4) does not converge.

6.9 The regular *accept–reject* simulation algorithm is defined as follows: If f and g are densities such that there exists M with $f(x) \le Mg(x)$,

---

1. Generate $y \sim g(y)$ and $u \sim \mathcal{U}_{[0,1]}$;
2. If $u > f(y)/Mg(y)$, go back to 1.
3. Take $x = y$.

Show that this algorithm actually provides an observation x from $f(x)$.

6.10 *(Gilks and Wild (1992)) A general *accept–reject* simulation method is proposed for *log-concave* densities on $\mathbb{R}$. The method is based on adaptive upper and lower bounds on the density, which are updated after each simulation step.

a. Given $f(x)$ proportional to the density to be simulated, we assume there exist $u(x)$ and $\ell(x)$, upper and lower bounds on $f(x)$ such that u is a density. The *envelope simulation algorithm* is as follows:

Iterate

(a) Generate $x \sim u(x)$ and $U \sim \mathcal{U}_{[0,1]}$
(b) Accept x if $U \le \ell(x)/u(x)$

(c) Otherwise, accept x if $U \le f(x)/u(x)$

till x is accepted

Show that this method actually produces a random variable with distribution f.

b. The two bounding functions can be constructed automatically for f log-concave as follows. For the first simulation, take three arbitrary values x_1, $x_2 > x_1$, and $x_3 > x_2$ such that at least one is on each side of the mode of f. (Explain why this can be done without requiring an explicit derivation of the mode.) Show that the lower bound $\log \ell(x)$ on $\log f(x)$ can be derived by joining the three points $(x_i, \log f(x_i))$ and $\ell(x)$ to be 0 outside the interval $[x_1, x_3]$. The upper bound $\log u(x)$ is constructed by taking the complements of the segments used for $\log \ell(x)$ until they meet: the tails are thus made of the extensions of the chords (x_1, x_2) and (x_2, x_3); $\log u(x)$ is completed by adding the vertical lines going through x_1 and x_3 until they meet the two chords.

c. Derive a way to update the upper and lower bounds after each simulation requiring the computation of $f(x)$.

d. Show that the two functions $u(x)$ and $\ell(x)$ are piecewise exponential and indicate how one can simulate from distributions proportional to these functions.

e. Illustrate the above algorithm for the simulation from a $\mathcal{N}(0,1)$ distribution. When does it become more time-consuming to evaluate and simulate from a better upper bound than to keep the current upper bound?

6.11 *(Rubinstein (1981)) Consider the integral

$$I = \int_a^b f(x)\, dx,$$

approximated by a Monte Carlo method with importance function h:

$$\hat{I} = \frac{1}{m} \sum_{i-1}^{m} f(x_i)/h(x_i).$$

a. Show that the variance of $\hat{I}$ is

$$\mathrm{var}(\hat{I}) = \frac{1}{n} \int_a^b \left(\frac{f(x)}{h(x)} - I \right)^2 h(x)\, dx$$

and deduce that it is minimized by $h \propto |f|$.

b. Consider $0 \le f(x) \le c$, $v_1, \ldots, v_m \sim \mathcal{U}_{[0,c]}$, and $u_1, \ldots, u_m \sim \mathcal{U}_{[a,b]}$. We define

$$\hat{I} = (b-a)\frac{1}{m}\sum_{i=1}^{m} f(u_i) \quad \text{and} \quad \tilde{I} = c(b-a)\frac{1}{m}\sum_{i=1}^{m} \mathbb{I}_{v_i \le f(u_i)}.$$

Show that

$$I = c(b-a)P(V \le f(U))$$

for $U \sim \mathcal{U}_{[a,b]}$ and $V \sim \mathcal{U}_{[0,c]}$.

c. Deduce that $\mathbb{E}[\tilde{I}] = I$ and $\mathrm{var}(\tilde{I}) \le \mathrm{var}(\hat{I})$.

d. Discuss the relevance of the notion of a "best" importance function h. (*Hint:* Consider a sequence of normal distributions centered at the value of interest, that is, at x^* such that $f(x^*) = I$ and with variance decreasing to 0.)

6.12 Show that, for a given function $g(\theta)$ and a distribution of interest $\pi(\theta)$, the optimal choice of the importance density h (in terms of variance of the estimator (6.2.4)) is

$$h(\theta) \propto |g(\theta)|\pi(\theta).$$

Express the corresponding estimator and deduce that, if g is of a constant sign, the resulting variance is 0. (*Hint:* See Robert and Casella (2004, Theorem 3.3.2) for a proof.)

Section 6.2.3

6.13 Justify the Laplace approximation when $h(\theta) = (\theta - \mu)^2$ and $b(\theta)$ is a polynomial of degree 2. What happens if b is of higher degree? Derive the general Laplace expansion by using Taylor series for b and h.

6.14 *(Tierney et al. (1989)) Deduce from the Laplace approximation that

$$\frac{\int b_N(\theta)e^{-nh_N(\theta)}\,d\theta}{\int b_D(\theta)e^{-nh_D(\theta)}\,d\theta} = \frac{A(N)}{A(D)} + O(\sigma^{-2}),$$

where

$$A(K) = \sigma_K \exp\{-n\hat{h}_k\} \quad \Big[\hat{b}_K + \frac{1}{2n}\{\sigma_K^2\hat{b}_K''$$
$$-\hat{h}_K'''\hat{b}_K'\sigma_K^2 + \frac{5}{12}\hat{b}_K(\hat{h}_K''')^2\sigma_K^6 - \frac{1}{4}\hat{b}_K\sigma_K^2 h_K^{(4)}\}\Big]$$

and $K = N, D$, if $\hat{h}_K = h(\hat{\theta}_K)$, etc., and $\hat{\theta}_K$ is minimizing h_K. Deduce Lemma 6.2.4 under the assumption that $\hat{h}_N^{(i)} - \hat{h}_D^{(i)} = O(n^{-1})$ for $i = 0, \ldots, 4$ and $\hat{b}_D \neq 0$. What happens if $\hat{b}_D = 0$?

6.15 *(Tierney et al. (1989)) If $M(s)$ is the moment generating function of $g(\theta)$ and $\hat{M}$ is the Laplace approximation of M for (6.2.9), with $b_N = b_D = b > 0$ and

$$h_D(\theta) = \{\log[f_x|\theta)] + \log[\pi(\theta)] - \log[b(\theta)]\}/n,$$
$h_N(\theta) = h_D(\theta) - sg(\theta)/n$, we define

$$\hat{\mathbb{E}}(g) = \hat{M}'(0).$$

a. Show that $\mathbb{E}^\pi[g(\theta)|x] = \hat{\mathbb{E}}(g) + O(n^{-2})$.

b. Let $\hat{\theta}$ be the minimum of h_D, let $\hat{\theta}_s$ be the minimum of h_N, and let $\sigma_s^2 = h_N^{(2)}(\hat{\theta}_s)$. Show that

$$\hat{\mathbb{E}}(g) = g(\hat{\theta}) + \frac{d}{ds}\log\sigma_s\Big|_{s=0} + \frac{d}{ds}\log b(\hat{\theta}_s)\Big|_{s=0}.$$

c. Deduce that

$$\hat{\mathbb{E}}(g) = \hat{g} + \frac{\sigma_D^2\hat{g}''}{2n} - \frac{\sigma_D^4\hat{h}_D'''\hat{g}'}{2n} + \frac{\sigma_D^2\hat{b}_D'\hat{g}'}{n\hat{b}_D},$$

and therefore that this method actually gives the approximation (6.2.8) for the standard form.

6.16 In the setting of Example 6.2.6, choose the standard and fully exponential representations leading to the proposed approximations.

Section 6.3.2

6.17 *Consider the *Metropolis–Hastings algorithm* of Section 6.3.2 that simulates a density $\pi(\theta)$ from a proposal density $q(\theta'|\theta)$.

a. Show that this algorithm reduces to regular simulation from π when $q(\theta'|\theta) = \pi(\theta')$.

b. Give the simplified version of the Metropolis–Hastings algorithm when $q(\theta|\theta')$ is symmetric in its arguments, that is, when $q(\theta|\theta') = q(\theta'|\theta)$.

c. Show directly, that is, without the balance condition (6.3.1), that $\pi(\theta)$ is a stationary distribution for this algorithm when the support of q contains the support of π. (*Hint:* Compute the probability distribution function of $\theta^{(m+1)}$ when $\theta^{(m)} \sim \pi(\theta)$ by breaking up the integral into four parts, and exchange the dummy variables θ and ξ in two of the four integrals.)

d. In the particular case when π is a $\mathcal{N}(0,1)$ distribution and $q(\theta|\theta')$ is a $\mathcal{N}(\theta', \sigma^2)$ distribution, study the probability of acceptance of ξ in the mth step as a function of σ. What is the exact distribution of $\theta^{(m)}$? Deduce the best choice of σ.

6.18 Show that the detailed balance condition (6.3.1) holds for the Metropolis–Hastings algorithm.

6.19 Examine whether the Metropolis–Hastings algorithm produces a reversible Markov chain, that is, such that the distribution of $(x^{(t)}, x^{(t+1)})$ is the same as the distribution of $(x^{(t+1)}, x^{(t)})$ under stationarity.

6.20 (Robert (1993a)) Consider n observations $y_1, \ldots, y_n$ from a general *logistic regression* model with

$$P(y_i = 1) = 1 - P(y_i = 0) = \frac{\exp(\theta^t x_i)}{1 + \exp(\theta^t x_i)},$$

and $x_i, \theta \in \mathbb{R}^p$.

a. Show that, conditional upon the x_i's, this distribution belongs to an exponential family and that $\sum_i y_i x_i$ is a sufficient statistic.

b. Give the general form of the conjugate prior distributions for this model and show that the normalization factor cannot be computed explicitly. Give an interpretation of the hyperparameters (ξ, λ) of the conjugate prior in terms of previous observations.

c. Show that the maximum likelihood estimator of θ, $\hat{\theta}$, cannot be obtained explicitly, and that it satisfies the following implicit equations ($j = 1, \ldots, p$):

(6.5.1)
$$\sum_{i=1}^{n} \frac{\exp(\hat{\theta}^t x_i)}{1 + \exp(\hat{\theta}^t x_i)} x_{ij} = \sum_{i=1}^{n} y_i x_{ij}.$$

d. Approximate a conjugate distribution by the Metropolis–Hastings algorithm. [*Note:* If a normal conditional distribution is used, attention should be paid to the variance factor.]

e. Explain why (6.5.1) can be used to control the convergence of the algorithm for some special values of the hyperparameter vector, (ξ, λ), namely, those for which

$$\mathbb{E}_{\xi,\lambda}^{\pi}\left[\sum_{i=1}^{n}\frac{\exp(\theta^t x_i)}{1+\exp(\theta^t x_i)}x_i\right] = \mathbb{E}_{\xi,\lambda}^{\pi}\left[\sum_{i=1}^{n}\frac{\exp(\theta^t x_i)}{1+\exp(\theta^t x_i)}x_i \,\Big|\, y_1,\ldots,y_n\right]$$

$$= \sum_{i=1}^{n} y_i x_i.$$

6.21 *Given a density of interest, π, and an available density f such that $\pi/f \leq M$, samples from π can be produced either by accept/reject, $\theta_1^{(1)},\ldots,\theta_p^{(1)}$, or by Metropolis–Hastings with proposal f, $\theta_1^{(2)},\ldots,\theta_n^{(2)}$, or else an importance sampling sample, $\theta_1^{(3)},\ldots,\theta_n^{(3)}$, can be generated from f. Compare the variances of

$$\frac{1}{p}\sum_{i=1}^{p}\theta_i^{(1)}, \qquad \frac{1}{n}\sum_{i=1}^{n}\theta_i^{(2)}, \qquad \frac{1}{n}\sum_{i=1}^{n}\frac{\pi(\theta_i^{(3)})}{f(\theta_i^{(3)})}\theta_i^{(3)}.$$

[*Note:* p denotes the random number of observations produced after n proposals in the accept/reject algorithm.]

6.22 Consider a probability distribution P and a function ϱ such that $0 \leq \varrho(x) \leq 1$ and $\mathbb{E}^P[1/\varrho(x)] < \infty$. A Markov chain $x^{(n)}$ is derived as follows. Update $x^{(n)}$ into $x^{(n+1)}$ by generating $y \sim P$ and take

$$x^{(n+1)} = \begin{cases} y & \text{with probability } \varrho(x^{(n)}), \\ x^{(n)} & \text{with probability } 1 - \varrho(x^{(n)}). \end{cases}$$

a. Show that this variation of the Metropolis–Hastings algorithm is converging to the stationary distribution with density

$$\varrho(x)^{-1}/\mathbb{E}^P[\varrho(x)^{-1}]$$

with respect to P.

b. Apply to the case when P is a $\mathcal{B}e(\alpha + 1, 1)$ distribution and $\varrho(x) = x$.

c. Study the performances of the method when $\alpha = 0.2$. [*Note:* See Robert and Casella (1999, Example 8.2.8, or 2004, Problems 7.5 and 7.6) for an illustration of the poor performances of this generator.]

Section 6.3.3

6.23 The *data augmentation* algorithm is based on the conditional distributions, $\pi(\theta|\lambda)$ and $\pi(\lambda|\theta)$. As described in Section 6.3, it successively simulates from $\pi(\theta|\lambda)$ and from $\pi(\lambda|\theta)$. This exercise shows why such a simulation of $\pi(\theta, \lambda)$ is justified from a probabilistic point of view.

a. Derive the joint distribution $\pi(\theta, \lambda)$ in terms of these conditional distributions.

b. Given two functions $q(\theta|\lambda)$ and $s(\lambda|\theta)$, what is a necessary and sufficient condition for q and s to be proportional to conditional distributions?

c. Consider the above questions in the case of n levels for the completed models, that is, when conditional distributions are available for $\theta, \lambda_1, \ldots, \lambda_{n-1}$.

6.24 (Exercise 6.23 cont.) The Hammersley–Clifford theorem establishes that the joint distribution $\pi(\vartheta)$ of a vector $\vartheta = (\theta_1, \ldots, \theta_p)$ can be derived from the full conditional distributions, $\pi_j(\theta_j | \ldots, \theta_{j-1}, \theta_{j+1}, \ldots, \theta_p)$. Show that

$$\pi(\vartheta) \propto \prod_{j=1}^{p} \frac{\pi_{\ell_j}(\theta_{\ell_j} | \theta_{\ell_1}, \ldots, \theta_{\ell_{j-1}}, \theta'_{\ell_{j+1}}, \ldots, \theta'_{\ell_p})}{\pi_{\ell_j}(\theta'_{\ell_j} | \theta_{\ell_1}, \ldots, \theta_{\ell_{j-1}}, \theta'_{\ell_{j+1}}, \ldots, \theta'_{\ell_p})}$$

for every permutation ℓ on $\{1, 2, \ldots, p\}$ and every $\theta' \in \Theta$. [*Note:* Clifford and Hammersley never published their result. See Hammersley (1974) and Robert and Casella (2004, Section 9.1.4) for details.]

6.25 *(Diebolt and Robert (1994)) Consider the two Markov chains $(\theta^{(m)})$ and $(\lambda^{(m)})$ used in data augmentation with conditional distributions $\pi_1(\theta|x, \lambda)$ and $\pi_2(\lambda|x, \theta)$.

a. Show that the respective transition kernels of these chains are

$$K(\theta'|\theta) = \int_\Lambda \pi_1(\theta'|x, \lambda)\pi_2(\lambda|x, \theta) \, d\lambda,$$

$$H(\lambda'|\lambda) = \int_\Theta \pi_2(\lambda'|x, \theta)\pi_1(\theta|x, \lambda) \, d\theta.$$

b. Show that $\pi_1(\theta|x)$ and $\pi_2(\lambda|x)$ are indeed stationary for these kernels.

c. Establish that, if $\theta^{(m)} \sim \pi_1^m(\theta|x, \lambda^{(0)})$ and $\lambda^{(m)} \sim \pi_2^m(\lambda|x, \lambda^{(0)})$,

$$||\pi_1^m(\cdot|x, \lambda^{(0)}) - \pi_1(\cdot|x)||_1 \leq ||\pi_2^m(\cdot|x, \lambda^{(0)}) - \pi_2(\cdot|x)||_1.$$

d. Derive Lemma 6.3.6 from c., and from the fact that irreducible Markov chains with stationary distributions are ergodic. Show that, if $(\lambda^{(m)})$ is geometrically ergodic with rate ϱ, $(\theta^{(m)})$ is also converging with rate ϱ, that is,

$$||\pi_1^m(\cdot|x, \lambda^{(0)}) - \pi_1(\cdot|x)||_1 \leq C\varrho^m.$$

e. The chain $(\lambda^{(m)})$ is φ-*mixing* if there exists φ, geometrically decreasing, and a finite measure μ such that

$$\left|\pi_2^m(\lambda|x, \lambda^{(0)}) - \pi_2(\lambda|x)\right| \leq \varphi(m)\mu(\lambda).$$

Show that, when $(\lambda^{(m)})$ is φ-mixing,

$$|\pi_1^m(\theta|x, \lambda^{(0)}) - \pi_1(\theta|x)| \leq \varphi(m) \int_\Lambda \pi_1(\theta|x, \lambda)\mu(d\lambda)$$

and deduce that, if Λ is compact, $(\theta^{(m)})$ is also φ-mixing.

f. Similarly, show that geometric convergence of $(\lambda^{(m)})$ and compactness of Λ are sufficient to ensure that, for every function h satisfying

$$\mathbb{E}^\pi[||h(\theta)||^2|x, \lambda] < \infty,$$

there exists C_h such that

$$|| \mathbb{E}^{\pi_1^m}[h(\theta)|x, \lambda^{(0)}] - \mathbb{E}^{\pi_1}[h(\theta)|x] ||^2 \leq C_h \varrho^m.$$

g. Take advantage of the fact that, *when Λ is finite, the chain $(\lambda^{(m)})$ is necessarily geometrically converging and φ-mixing* (Billingsley (1986)). Assess the importance of the above results in the setting of mixture estimation.

h. Extend the duality principle to the case of a hierarchical model with multiple levels, using the fact that conditional distributions only depend on the neighboring levels.

6.26 Two machines are run in parallel with breakdown times $x \sim f(x|\theta)$ and $y \sim g(y|\eta)$. The defective machine is supposed to be known when a breakdown occurs.

a. Give the distribution of z, breakdown time of the system, and derive a Gibbs sampling algorithm to get posterior estimators of θ and η when a sample $z_1, \ldots, z_n$ is available, and when conjugate priors are used on both θ and η.

b. Implement this algorithm in the special cases when (a) f and g are normal densities with means θ and η, and variance 1; (b) f and g are exponential distributions with parameters θ and η.

Section 6.3.4

6.27 For a chain $(\theta^{(t)}, \lambda^{(t)})$ produced by data augmentation,

a. Show that, for every function h,

$$\mathrm{cov}(h(\theta^{(1)}), h(\theta^{(2)})) = \mathrm{var}\left\{\mathbb{E}[h(\theta)|\lambda]\right\} \ .$$

b. Give a corresponding representation for $\mathrm{cov}(h(\theta^{(1)}), h(\theta^{(t)}))$.

c. Deduce that $\mathrm{cov}(h(\theta^{(1)}), h(\theta^{(t)}))$ is always positive and decreasing with t.

d. Conclude about the domination of the usual average by its Rao–Blackwellized version.

6.28 Show that, in the setting of Example 6.3.4, the marginal distributions on θ and λ cannot be derived explicitly and that, moreover, the restriction $B < +\infty$ is necessary for the marginal distributions to be defined.

Section 6.3.5

6.29 For a hierarchical model such as (6.3.5), show that the distribution of a given λ_i conditional on all the other parameters of the model $\pi(\lambda_i|x, \theta, (\lambda_j)_{j \neq i})$ $(1 \leq i \leq p)$ only depends on its two nearest neighbors in the vector $(x, \theta, \lambda_1, \ldots, \lambda_p)$. (*Hint:* Draw a graphical representation of the model.)

6.30 Show that, if the Gibbs sampler is implemented with more than two full conditional levels, as in, for instance, (6.3.7), the resulting subchains corresponding to the various levels are not Markov chains.

6.31 Considering the multinomial model of Example 6.3.10, explain why simulating from $\pi((\mu, \eta)|x)$ rather than from $\pi(\mu|x)$ and $\pi(\eta|x)$ should speed up convergence. (*Hint:* Study the correlation between $\mu^{(t)}$ and $\mu^{(t+1)}$ in both cases.)

6.32 Show that, in a Gibbs sampling algorithm, if an arbitrary simulation step, like the simulation from $\pi(\theta_1|\theta_2, \ldots, \theta_k)$ say, is replaced by a *single* step of a Metropolis–Hastings algorithm, the validity of the algorithm is preserved. Discuss the crucial interest of this property for practical issues.

6.33 Consider a distribution $\pi(\theta_1, \theta_2)$, which is not available under closed form, but such that the two conditional distributions $\pi(\theta_1|\theta_2)$ and $\pi(\theta_2|\theta_1)$ are known and can be simulated from.

a. Show that the Metropolis–Hastings algorithm can be implemented. (*Hint:* Show that the only difficulty is to simulate from $\pi(\theta_1)$ or from $\pi(\theta_2)$ and use Exercise 6.23.)

b. Deduce that in every setting Gibbs sampling can be used, the same applies to the general form of the Metropolis–Hastings algorithm.

6.34 Show that a Gibbs sampling step is a special case of the Metropolis algorithm where the acceptance probability is always equal to 1.

Section 6.3.6

6.35 A truncated normal distribution $\mathcal{N}_p(0, I_p)$ restricted to the polygon $\theta_i^t x_i \leq z_i$ $(1 \leq i \leq n)$ is to be simulated.

a. Give the distribution of θ_j conditional on θ_k $(k \neq j)$ and derive a Gibbs implementation for the simulation of this truncated normal distribution. (*Hint:* See Geweke (1991) or Robert (1995a) for accept-reject algorithms to simulate from a one-dimensional truncated normal distribution).

b. Propose a Metropolis–Hastings alternative based upon a $\mathcal{N}_p(\mu, \Sigma)$ simulation, when μ and Σ are derived from the polygon boundaries.

c. Propose a slice sampler based on a single auxiliary variable and a slice sampler based on p auxiliary variables.

d. Compare these different algorithms.

Section 6.3.7

6.36 (Rubin et al. (1992)) A study was conducted on the campus of Cornell University to model the sexual behavior of undergraduates. Out of a population of R_m (R_f) male (female) undergraduates, r_m (r_f) answered the survey and t_m (t_f) were found to be sexually active (in the previous two months).

a. The first quantities of interest are T_f and T_m, numbers of female and male undergraduates who were sexually active. Using a hypergeometric model on t_m and t_f and taking r_m and r_f as fixed, derive a Bayes estimator of T_f and T_m when

$$T_i \sim \mathcal{B}(R_i, p_i), \qquad p_i \sim \mathcal{B}e(\alpha, \beta), \qquad \pi(\alpha, \beta) = 1/\alpha\beta \qquad (i = f, m).$$

(*Numerical application:* $R_f = 5211$, $r_f = 253$, $t_f = 111$, $R_m = 6539$, $r_m = 249$ and $t_m = 22$.)

b. During the study, sexually active respondents were asked about the number of partners they had had during the two last months, y_f and y_m, as well as the number of Cornell undergraduate partners, x_m and x_f.

Assuming a Poisson distribution $\mathcal{P}(\lambda_i)$ for the number of additional partners $y_i - 1$ and a binomial distribution $\mathcal{B}(y_i, \varrho_i)$ on the number of Cornell undergraduate partners $(i = f, m)$, with $\varrho_f = T_m/N_m$ and $\varrho_m = T_f/N_f$, derive a Bayes estimator of the population in sexual contact with the Cornell undergraduates, N_m and N_f. The prior distributions are

$$\lambda_i \sim \mathcal{E}xp(\lambda_0), \qquad \varrho_i \sim \mathcal{B}e(\gamma, \delta), \qquad \pi(\gamma, \delta) = 1/\gamma\delta.$$

(*Numerical application:* $y_m = 54$, $x_m = 31$, $y_f = 135$, $x_f = 67$.)

c. Compare your results with the maximum likelihood estimators obtained in the study: $\hat{N}_f = 4186$, $\hat{N}_m = 1473$, $\hat{T}_f = 2323$ and $\hat{T}_m = 615$.

d. Repeat the estimation for the hyperpriors

$$\pi(\alpha, \beta) = e^{-(\alpha+\beta)}, \qquad \pi(\gamma, \delta) = e^{-(\gamma+\delta)},$$

and

$$\pi(\alpha, \beta) = 1/(\alpha + \beta)^2, \qquad \pi(\gamma, \delta) = 1/(\gamma + \delta)^2.$$

6.37 In the setting of logistic regression (see Exercise 6.20), a missing data structure can be exhibited and exploited by a Gibbs sampling algorithm.

a. Derive the distribution of z_i such that the observation y_i is $\mathbb{I}_{z_i \leq x_i^t \theta}$.

b. Give the likelihood of the completed model and examine whether a Gibbs algorithm similar to those of Section 6.4 can be constructed in the special case $\theta \sim \mathcal{N}_p(\mu, \Sigma)$.

c. Compare the performance of this algorithm with a more straightforward Metropolis–Hastings algorithm of your choice.

6.38 A *probit* model is a qualitative regression model where the dependence on the auxiliary variables is given by

$$P_\theta(y_i = 1) = 1 - P_\theta(y_i = 0) = \Phi(\theta^t x_i).$$

a. Show that, as in Exercise 6.37, it is possible to complete the model by exhibiting a continuous latent variable z_i.

b. Propose a Gibbs sampling algorithm based upon the completed data when $\theta \sim \mathcal{N}_p(\mu, \Sigma)$.

Section 6.4

6.39 (Casella et al. (2000)) Gibbs sampling and other MCMC techniques unlocked the difficulties with Bayesian inference on mixtures. It is, however, possible to produce efficient importance sampling estimators in this setting. We assume that a sample $(x_1, \ldots, x_n)$ from

$$\sum_{j=1}^{k} p_j f(x|\theta_j)$$

is available.

a. Given the allocation variables $z_1, \ldots, z_n$, where $x_i | z_i \sim f(x|\theta_{z_i})$, show that the posterior distribution of $\mathbf{z} = (z_1, \ldots, z_n)$ is given by

(6.5.2) $\qquad P(\mathbf{z}|\mathbf{x}) = \dfrac{\displaystyle\prod_{j=1}^{k} \int_\Theta \prod_{\{i:z_i=j\}} f(x_i|\theta_j)\pi_j(\theta_j)d\theta_j}{\displaystyle\sum_{\mathbf{z}\in\mathcal{Z}} \prod_{j=1}^{k} \int_\Theta \prod_{\{i:z_i=j\}} f(x_i|\theta_j)\pi_j(\theta_j)d\theta_j}$,

where $\mathcal{Z}$ is the set of all k^n allocation vectors $\mathbf{z}$.

b. Show that

(6.5.3)
$$P(Z_i = j|x_i) = \frac{p_j m_j(x_i)}{\sum_{j=1}^{k} p_j m_j(x_i)},$$

where $m_j(x) = \int f(x|\theta_j)\pi(\theta_j)d\theta_j$, $(j = 1, \ldots, m)$ is the univariate marginal distribution of x_i.

c. Deduce that, if (6.5.2) and (6.5.3) are both available up to a normalizing constant, the Bayes estimator $\mathbb{E}[h(\theta)|\mathbf{x}]$ can be approximated by importance sampling, the z_i's $(i = 1, \ldots, n)$ being generated from the marginal distributions of b., if $\mathbb{E}[h(\theta)|(x_1, z_1), \ldots, (x_n, z_n)]$ is also known in closed form.

d. Apply to the case of a mixture of exponential distributions,

$$\sum_{j=1}^{k} p_j \lambda_j \exp(-\lambda_j x), \qquad x > 0,$$

under the prior

$$\lambda_j \sim \mathcal{G}(\alpha_j, \beta_j), \qquad j = 1, \ldots, k,$$

when the weights p_j and the hyperparameters α_j, β_j are known. In particular, determine transforms $h(\lambda_1, \ldots, \lambda_k)$ such that the conditional expectations $\mathbb{E}[h(\theta)|(x_1, z_1), \ldots, (x_n, z_n)]$ are known.

6.40 For a normal mixture, detail the reasoning leading to the conditional distributions (6.4.2) and (6.4.3) and give an explicit expression of $\mathbb{E}^\pi[\mu_i|x, z]$.

6.41 For a small sample size, run several simulations to compare Bayesian sampling with a direct computation of the Bayes estimator for a mixture of two normal distributions.

6.42 Show that conjugate priors cannot lead to a noninformative answer in the case of a two-component normal mixture when the variances of the prior distributions go to ∞.

6.43 (Robert and Soubiran (1993)) Derive the formulas equivalent to (6.4.2) and (6.4.3) for a mixture of multidimensional normal distributions. (*Hint:* Use Section 4.4.1 for the choice of conjugate prior distributions and detail the simulation of Wishart distributions.)

6.44 (Binder (1978)) Consider a sample $x_1, \ldots, x_n$ from the mixture

$$x \sim f(x|\theta) = \sum_{i=1}^{k} p_i f_i(x),$$

where the densities f_i and the weights p_i are known. The problem is to identify the origins of the observations, $g = (g_1, \ldots, g_n)$, with

$$g_j = \sum_{i=1}^{k} i \mathbb{I}_{z_{ij}=1} \qquad (1 \le j \le n).$$

a. Show that calculation difficulties also occur in this setting for the computation of the Bayes estimators.

b. Give the Bayes estimator of g when $p \sim \mathcal{D}(1/2, \ldots, 1/2)$ and $f_i(x) = \varphi(x; \mu_i, 1)$ with $\mu_i \sim \mathcal{N}(\xi_i, 1)$.

c. How can Gibbs sampling be implemented for this problem?

6.45 Adapt the Gibbs sampling techniques developed in Section 6.4 in the case of a mixture of distributions to the case of a *censored model*, that is, for observations y_i^* such that

$$y_i^* = \begin{cases} y_i & \text{if } y_i \leq c, \\ c & \text{otherwise,} \end{cases}$$

and $y_i \sim f(y|\theta)$, with $f(\cdot|\theta)$ belonging to an exponential family.

6.46 (Robert et al. (1993))   A *hidden Markov model* generalizes the mixture model studied in Example 6.1.5 and in Section 6.4 by introducing some dependence between the observations $x_1, \ldots, x_t$. When completing these observations by (unknown) missing state indicators z_i, the model becomes hierarchical $(1 \leq i \leq t)$:

$$x_i|z_i, \theta \sim f(x|\theta_{z_i})$$

and (z_i) constitutes a Markov chain on $\{1, \ldots, K\}$ with transition matrix $\mathbb{P} = (p_{jk})$, where

$$p_{jk} = P(z_i = k|z_{i-1} = j) \qquad (2 \leq i \leq t)$$

(taking $z_1 = 1$ for identifiability reasons). We also assume that $f(\cdot|\theta)$ belongs to an exponential family.

a. Give the likelihood of this model and deduce that neither maximum likelihood, nor Bayesian estimation with conjugate priors on θ and $\mathbb{P}$, can be derived explicitly in this case.

b. Considering the particular case when $f(\cdot|\theta)$ is $\mathcal{N}(\xi, \sigma^2)$ with $\theta = (\xi, \sigma^2)$, show that a Gibbs sampling implementation with iterative simulations from $\pi(\theta|\mathbf{x}, \mathbf{z})$ and $\pi(\mathbf{z}|\mathbf{x}, \theta)$ is quite time-consuming because of $\pi(\mathbf{z}|\mathbf{x}, \theta)$.

c. Show that the fully conditional distributions $\pi(z_i|\mathbf{x}, \theta, z_{j \neq i})$ only depend on z_{i-1} and z_{i+1} and are much easier to simulate.

d. Propose a Gibbs sampling algorithm for this model. Show that the condition $p_{kj} > 0$ for all $1 \leq j, k \leq K$ is sufficient to ensure geometric convergence of the chains $(\theta^{(m)})$ and $(\mathbf{P}^{(m)})$ to the true posterior distributions. (*Hint:* Arguments similar to those of Exercise 6.25 can be used.)

6.47 (Robert et al. (1999a))   In the setting of Exercise 6.46, there exists a way to simulate the whole chain $\mathbf{z} = (z_2, \ldots, z_n)$ conditional on the parameters θ and thus to implement a data-augmentation scheme. The representation of the conditional distribution of $\mathbf{z}$ is called *forward–backward* and has been known for a while in the signal processing literature (Baum and Petrie (1966)).

a. Establish the so-called *backward recurrence relation* $(1 \leq i \leq n - 1)$

$$f(x_i, \ldots, x_n|\theta, z_i = j) =$$

(6.5.4) $$\sum_{k=1}^{K} p_{jk} f(x_i|\theta_j) f(x_{i+1}, \ldots, x_n|\theta, z_{i+1} = k),$$

with $f(x_n|z_n = j) = f(x_n|\theta_j)$.

b. Derive from the backward formulas the probability $P(z_1 = j|x_1, \ldots, x_n, \theta)$ under the assumption that z_1 is marginally distributed from the stationary distribution associated with the transition matrix $\mathbb{P}$.

c. Compute the probabilities $P(z_i = j|x_1, \ldots, x_n, \theta, z_1, \ldots, z_{i-1})$ $(i = 2, \ldots, n)$.

Table 6.5.1. *Frequencies of car passages for a sequence of one-minute intervals.*

Number of cars	0	1	2	3	4 or more
Number of occurrences	139	128	55	25	13

d. Conclude that the vector $(z_1, \ldots, z_n)$ can be simulated conditional upon the observations and on θ, thus that the data-augmentation scheme can be implemented in some hidden Markov models.

6.48 In a mixture setting, compare the performance (in terms of computing time) of a Gibbs sampling with that of a more straightforward Metropolis–Hastings algorithm.

Note 6.6.3

6.49 Does the decomposition of the noncentral chi-squared distribution proposed in Example 6.6.1 allow for implementation of the Gibbs sampling? Give an approximation by the Metropolis–Hastings algorithm.

6.50 (Heitjan and Rubin (1991)) *Coarse data* are defined as an aggregation of the observations in classes. Given a "complete" random variable $y_i \sim f(y|\theta)$, taking values in $\mathcal{Y}$, and a partition A_j $(j \in I)$ of $\mathcal{Y}$, the observations are $x_i = j$ if $y_i \in A_j$.

a. Give a real-life justification for this model.

b. Propose a Gibbs sampling algorithm in the case when $f(\cdot|\theta)$ is a normal distribution $\mathcal{N}(\xi, \sigma^2)$ with $\theta = (\xi, \sigma^2)$ and $A_j = [j, j+1)$ $(j \in \mathbb{Z})$.

Frequencies of car passages during a one-minute period have been observed for 360 consecutive minutes and the resulting observations are given in Table 6.50.

c. Assuming a Poisson $\mathcal{P}(\theta)$ distribution for this model, apply Gibbs sampling to estimate the parameter θ for this data set and the prior $\pi(\theta) = 1/\theta$.

Note 6.6.4

6.51 In the setting of Example 6.3.8,

a. Show that the marginal distributions associated with the full conditionals $\pi(\theta|\lambda)$ and $\pi(\lambda|\theta)$ satisfy

$$\frac{\pi(\theta)}{\pi(\lambda)} = \frac{\theta}{\lambda}, \qquad \theta, \lambda < B.$$

b. Deduce that the joint distribution corresponding to these two conditionals is not defined when B goes to infinity.

Note 6.6.6

6.52 For the sequence $(\hat{\theta}_{(j)})_j$ produced by the EM algorithm (see Note 6.6.6 and Robert and Casella (2004, Section 5.3.2)),

a. Show that

$$Q(\hat{\theta}_{(j+1)}|\hat{\theta}_{(j)}, \mathbf{x}) \geq Q(\hat{\theta}_{(j)}|\hat{\theta}_{(j)}, \mathbf{x}).$$

b. If $k(\mathbf{z}|\theta, \mathbf{x})$ denotes the conditional distribution of $\mathbf{z}$ given $\mathbf{x}$, show that

$$\mathbb{E}_{\hat{\theta}_{(j)}}\left[\log\left(\frac{k(\mathbf{z}|\hat{\theta}_{(j+1)}, \mathbf{x})}{k(\mathbf{z}|\hat{\theta}_j, \mathbf{x})}\right)\Big|\hat{\theta}_{(j)}, \mathbf{x}\right] \leq 0.$$

(*Hint:* Use Jensen's inequality.)

c. Conclude that

$$L(\hat{\theta}_{(j+1)}|\mathbf{x}) \geq L(\hat{\theta}_{(j)}|\mathbf{x}),$$

with equality holding if and only if $Q(\hat{\theta}_{(j+1)}|\hat{\theta}_{(j)}, \mathbf{x}) = Q(\hat{\theta}_{(j)}|\hat{\theta}_{(j)}, \mathbf{x})$.

6.6 Notes

6.6.1 Pseudo-random uniform generators.

The algorithms for generating random variables from any distribution all rely on the generation of *uniform random variables* on $[0, 1]$. Since the production of an exact i.i.d. sequence of uniform $\mathcal{U}([0, 1])$ variables is impossible, there are methods that use a fully deterministic process to produce a sequence imitating an i.i.d. $\mathcal{U}([0, 1])$ sequence in the sense that the deterministic sequence is accepted as an i.i.d. $\mathcal{U}([0, 1])$ sequence for all statistical tests. For instance, the generator proposed in Ripley (1987) is a *congruencial* generator, defined as follows.

---

1. Start with an initial arbitrary seed x_0
2. Iterate

$$x_i = (69069 x_{i-1} + 1) \bmod 2^{32},$$
$$u_i = 2^{-32} x_i.$$

The corresponding sequence of u_i's can then be considered as an i.i.d. $\mathcal{U}_{[0,1]}$ sequence, although its actual support is finite.

Pseudo-random uniform generators are available on most machines and in most languages, and can be used as such, even though some of these generators are not thoroughly tested and may possess undesirable features (see Robert and Casella (2004, Exercise 2.44)).

Marsaglia and Zaman (1993) have developed a simple uniform generator with multiple seeds whose period is larger than 2^{95}. See Robert and Casella (2004, Section 2.6.1) for details.

6.6.2 The BUGS *and* CODA *softwares*

An MCMC software has been developed by Spiegelhalter et al. (1995a,b,c) at the MRC Biostatistics Unit in Cambridge, England. This software offers some possibilities to run a partly automated Gibbs sampler (BUGS stands for *Bayesian inference using Gibbs sampling*). Further, it comes as a computer language, which is C or R like, involves declarations about the model, the data, and the prior specifications, including hierarchical modeling, and allows for a large range of transforms of most standard distributions. BUGS produces the Gibbs sample made of simulated values of the parameters, after an open number of warmup iterations, the batch size being also open.

A major restriction on the prior modeling is that conjugate priors or log-concave distributions must be used, for either standard simulation methods or the ARMS algorithm of Gilks et al. (1995) to apply, but more complex distributions can be handled by discretization of their support. The other restriction is that improper priors cannot be used and must be replaced by vague proper priors, that is, priors with large variances.

The BUGS software is completed with a convergence diagnosis software,[9] CODA, which contains some of the most common MCMC convergence assessment techniques. This S-Plus package has been developed by Best et al. (1995) and can be used independently from BUGS. The techniques selected in CODA are described in Robert and Casella (2004, Chapter 11): they include the convergence diagnostics of Gelman and Rubin (1992), Geweke (1992), Heidelberger and Welch (1983), Raftery and Lewis (1992), plus plots of autocorrelation for each variable and of cross-correlations between variables.

6.6.3 Hidden mixtures

The hierarchical decomposition (6.3.3), which is instrumental for the Gibbs sampler, is also interesting for prior selection when the sampling distribution does not belong to an exponential family and there is no conjugate family of priors. For instance, this is the case for Student's t- and noncentral chi-squared distributions. Decomposition of $f(x|\theta)$ of the form

$$f(x|\theta) = \int f(x|\theta, z) g(z|\theta) \, dz$$

may then allow for a prior modeling on θ through conjugate priors (for $f(x|\theta, z)$ or $g(z|\theta)$). As in Section 3.3.3, we call this representation a *hidden mixture*, in contrast with standard mixture problems where the mixture structure itself is of interest. (See also Note 3.8.3.)

Example 6.6.1   Consider $x \sim \chi_p^2(\theta)$, an observation from a noncentral chi-squared distribution. It can be written as a mixture

$$
\begin{aligned}
x|\theta, z &\sim \chi_{p+2z}^2, \\
z|\theta &\sim \mathcal{P}(\theta/2).
\end{aligned}
$$

Therefore, only $g(z|\theta)$ depends on θ and a possible prior distribution on θ is $\mathcal{G}(\alpha, \beta)$, since it is conjugate for the Poisson distribution.   ∥

Example 6.6.2   Consider $x|\mu, \sigma \sim \mathcal{T}(m, \mu, \sigma^2)$, with $\theta = (\mu, \sigma)$ unknown. Following Dickey's representation (1968),

$$x|\theta, z \sim \mathcal{N}(\mu, z), \qquad z|\sigma^2 \sim \mathcal{IG}(m/2, m\sigma^2/2),$$

---

[9] Both software programs are currently available on the web site of the MRC Biostatistics Unit at www.mrc-bsu.cam.ac.uk.

we can propose

$$\mu \sim \mathcal{N}(\xi, \tau^2), \qquad \sigma^2 \sim \mathcal{G}(\alpha, \beta),$$

as a prior distribution and we derive

$$z|x, \theta \quad \sim \quad \mathcal{IG}\left(\frac{m+1}{2}, \frac{m\sigma^2 + (x-\mu)^2}{2}\right),$$

(6.6.1)
$$\sigma^2|x, z \quad \sim \quad \mathcal{G}(\alpha + (m/2), \beta + (m/2z)),$$

$$\mu|x, z \quad \sim \quad \mathcal{N}\left(\frac{z\mu + \tau^2 x}{z + \tau^2}, \frac{z\tau^2}{z + \tau^2}\right).$$

The conditional distributions (6.6.1) directly allow for simulation through Gibbs sampling. Notice the difference with the classical normal example (see Section 4.4). In this case, σ^2 has a gamma prior instead of an inverse gamma distribution and, more importantly, μ and σ are a priori independent. The conditional decomposition thus leads to a modeling that is more satisfactory than in the normal case.   ‖

Resorting to a hidden mixture, either for $f(x|\theta)$ or for $\pi(\theta)$, is obviously helpful to simulate $\pi(\theta|x)$ through Gibbs sampling when the posterior distribution is not available.

Example 6.6.3 (Example 6.6.2 continued) If, for robustness purposes, the prior distribution is actually

$$\mu \sim \mathcal{T}(\nu, \xi, \tau^2), \qquad \sigma^2 \sim \mathcal{G}(\alpha, \beta),$$

the corresponding hidden mixture representation is

$$\mu|\delta \sim \mathcal{N}(\xi, \delta), \qquad \delta \sim \mathcal{IG}(\nu/2, \nu\tau^2/2),$$

and the simulation of $\pi(\mu, \sigma|x)$ can be done by Gibbs sampling according to the following conditional distributions:

$$z|x, \theta \quad \sim \quad \mathcal{IG}\left(\frac{m+\nu}{2}, \frac{m\sigma^2 + (x-\mu)^2}{2}\right),$$

$$\sigma^2|x, z \quad \sim \quad \mathcal{G}(\alpha + (m/2), \beta + (m/2z)),$$

$$\mu|x, z, \delta \quad \sim \quad \mathcal{N}\left(\frac{\delta\mu + \tau^2 x}{\delta + \tau^2}, \frac{\delta\tau^2}{\delta + \tau^2}\right),$$

$$\delta|\theta \quad \sim \quad \mathcal{IG}\left(\frac{\nu+1}{2}, \frac{\nu\tau^2 + (x-\mu)^2}{2}\right).$$

‖

6.6.4 Improper posteriors

As stressed in Note 1.8.3, prior distributions π such that

$$\int_\Theta \pi(\theta) f(x|\theta) d\theta = \infty$$

cannot be used. The difficulties in satisfying this condition for complex models are that (a) an analytic check is often impossible; and (b) the conditional distributions derived from $\pi(\theta) f(x|\theta)$ may well be proper. Take, for instance, the case of Example 6.3.8: when B goes to infinity, the joint distribution on (θ, λ)

is not defined; the conditional distributions are, however, standard exponential $\mathcal{E}xp(\lambda)$ and $\mathcal{E}xp(\theta)$ distributions (Exercise 6.51). The additional difficulty is that a Gibbs sampler implemented with these conditional distributions may fail to expose the impropriety problem (see Hobert and Casella (1996)).

Example 6.6.4 Consider the usual random-effect model ($1 \leq i \leq I,\ 1 \leq j \leq J$)

$$y_{ij} = \theta + u_i + \epsilon_{ij}, \qquad u_i \sim \mathcal{N}(0, \sigma^2),\ \epsilon_{ij} \sim \mathcal{N}(0, \tau^2).$$

The corresponding Jeffreys prior is $\pi(\theta, \tau^2, \sigma^2) = 1/\sigma^2\tau^2$. Then (see Robert and Casella (1999, Example 7.4.3 and Problem 7.38)), the joint posterior distribution on $(\theta, \tau^2, \sigma^2)$ is not defined while the conditional distributions are well defined and can be used in a Gibbs sampler.   ‖

Despite the fundamental impossibility of using improper posteriors, which are truly measures $f(x|\theta)\pi(\theta)$ with infinite mass, for Bayesian inference, there exist settings where such measures can be of use. Indeed, it is possible to augment the parameter θ artificially with an auxiliary parameter α, and to introduce an improper prior $\pi(\alpha)$ such that the joint posterior $\pi(\alpha, \theta|x) = \pi(\alpha)\pi(\theta)f(x|\theta)$ is also improper, while preserving the properness of the well defined $\pi(\theta|x)$ within the Markov chain.

Example 6.6.5 (Meng and van Dyk (1999)) A Student's t-distribution with parameter $\theta = (\mu, \sigma)$, $\mathcal{T}(\nu, \mu, \sigma^2)$, can be written as

$$x = \mu + \sigma y_1/(\nu y_2)^{1/2}, \quad \text{with} \quad y_1 \sim \mathcal{N}(0, 1),\ y_2 \sim \chi_\nu^2.$$

(See Exercise 1.1 and Example 3.3.11.) If we introduce $\alpha > 0$ such that

$$x|y_2 \sim \mathcal{N}(\mu, \alpha\sigma^2/(\nu y_2)), \qquad y_2 \sim \alpha\chi_\nu^2,$$

this does not change the model under study because the quantity α/y_2 does not depend on α. The parameter α is thus nonidentifiable and, with an improper prior on α, $\pi(\alpha) = \alpha^{-1}\exp(-\beta/\alpha)$ say, the marginal posterior distribution on α is equal to this prior: the joint posterior distribution on (θ, α) is not defined. It is nonetheless possible to create a Markov chain $(y_2^{(t)}, \theta^{(t)}, \alpha^{(t)})$ via a simple data-augmentation scheme applied to the full conditionals derived from

$$\pi(\alpha)\pi(\mu, \sigma)f(x|\mu, \alpha, \sigma, y_2)f(y_2|\alpha)$$

such that (a) this σ-finite measure is stationary for this chain; and (b) the subchain $(\theta^{(t)})$ converges to the well-defined posterior distribution $\pi(\theta|x)$. ‖

Improper posteriors then appear as tools to speed up the exploration of the parameter space Θ by way of null-recurrent or even transient Markov chains in larger spaces. See Casella (1996), Meng and van Dyk (1999), Hobert (2000), and Liu and Wu (1999) for more details.

6.6.5 MCMC algorithms for dynamic models

In Section 4.5, we introduced several dynamic models and pointed out that the complex parameter space induced by the stationarity constraints as well as the lack of closed-form likelihoods necessitates the use MCMC algorithms.

The state-space representations of Sections 4.5.3 and 4.5.4 and the reparameterization of Lemma 4.5.4 are instrumental in designing Gibbs samplers in these models.

For instance, in the AR(p) model, the ϱ_j's ($1 \le j \le p$) are linear functions of the partial autocorrelations ψ_k ($1 \le k \le p$), when the ψ_ℓ ($\ell \ne k$) are fixed:

$$\varrho_j = a_{kj} + b_{kj}\psi_k \,,$$

where ($1 \le \ell \le i - 1$)

$$
\begin{aligned}
a^{ii} &= \psi_i,\ b^{ii} = 0,\ a^{i\ell} = a^{(i-1)\ell} - \psi_i a^{(i-1)(i-\ell)},\ b^{i\ell} = 0, && \text{if}\ \ i < k \\
a^{ii} &= 0,\ b^{ii} = 1,\ a^{i\ell} = a^{(i-1)\ell},\ b^{i\ell} = -a^{(i-1)(i-\ell)} && \text{if}\ \ i = k \\
a^{ii} &= \psi_i,\ b^{ii} = 0,\ a^{i\ell} = a^{(i-1)\ell} - \psi_i a^{(i-1)(i-\ell)},\ b^{i\ell} = b^{(i-1)\ell} - \psi_i b^{(i-1)(i-\ell)} \\
& \qquad\qquad \text{if}\ \ i > k
\end{aligned}
$$

and

$$a_{ik} = a^{pi} \,,\quad b_{ik} = b^{pi} \ \ (1 \le i \le p)\,.$$

Therefore, if the ψ_i's are simulated one by one, the likelihood (4.5.8) has a normal structure

$$\prod_{t=1}^{T} \exp\left\{ -\frac{1}{2\sigma^2}\left(x_t - \mu - \sum_{j=1}^{p}(a_{ij} + b_{ij}\psi_i)(x_{t-j} - \mu) \right)^2 \right\}.$$

A similar conditional decomposition can be used for the MA(q) and ARMA(p, q) models of Sections 4.5.3 and 4.5.4, taking advantage of the linear structure of the state-space representation that preserves the normal structure. Alternatives based on the recursive representation (4.5.12) and on Metropolis–Hastings moves have been studied in Billio et al. (1999).

6.6.6 More on mixture estimation

The importance of mixtures of standard distributions cannot be discounted from a modeling perspective: mixture models are located at the boundary between parametric and nonparametric modeling and they allow for the description of more complex phenomena (compared with standard distributions), while preserving the *parsimony principle* (that is, using a reasonably small number of parameters to describe a phenomenon). This was illustrated for the construction of priors in Notes 3.8.3 and 6.6.3. Mixture structures appear in Bayesian nonparametric analysis, as, for instance, with Dirichlet process priors (see Notes 1.8.2 and 6.6.7). They are equally instrumental in classification problems (see, e.g., Bensmail et al. (1999)) and in outlier detection (Verdinelli and Wasserman (1992)).

The classical treatment of estimation of finite mixtures of distributions is presented in Titterington et al. (1985) and MacLachlan and Basford (1987). It can be traced to Pearson (1894), who proposed an estimation method based on moments involving the resolution of a 9th-degree polynomial equation.

For maximum likelihood estimation, Dempster et al. (1977) and Redner and Walker (1984) have developed a special algorithm called EM (for Estimation–Maximization) that has been incredibly popular (see Meng and van Dyk (1997) and MacLachlan and Krishnan (1997)). This algorithm is based on the same completion as the Gibbs sampler. Given a completed likelihood $L^c(\theta|\mathbf{x}, \mathbf{z})$, the

EM algorithm runs as follows.

At iteration m,

1. Compute

$$Q(\theta|\hat{\theta}_{(m)}, \mathbf{x}) = \mathbb{E}_{\hat{\theta}_{(m)}}\left[\log L^c(\theta|\mathbf{x}, \mathbf{z})|\mathbf{x}\right],$$

where the expectation is with respect to $k(\mathbf{z}|\hat{\theta}_m, \mathbf{x})$ (*E-step*) .

2. Maximize $Q(\theta|\hat{\theta}_{(m)}, \mathbf{x})$ in θ and take (*M-step*)

$$\theta_{(m+1)} = \arg\max_{\theta} \ Q(\theta|\hat{\theta}_{(m)}, \mathbf{x}).$$

It is validated by the fact that the *observed* likelihood increases at every step (Exercise 6.52). The sequence $(\hat{\theta}_{(m)})_m$ thus converges to a stationary point of the observed likelihood (which may be a local maximum or a saddlepoint). See Robert and Casella (1999, Section 5.3.3) for more details.

Since the convergence of the EM algorithm depends on the starting point $\hat{\theta}_{(0)}$ and since this algorithm requires the computation of the expectation in the *E-step*, some authors, including Broniatowski et al. (1983), Celeux and Diebolt (1990), Qian and Titterington (1991), and Lavielle and Moulines (1997), have proposed stochastic extensions of the EM algorithm.

From a Bayesian point of view, a more detailed study of MCMC techniques for mixtures is proposed in Robert (1996a), Roeder and Wasserman (1997), Robert and Mengersen (1999), Celeux et al. (2000), or Stephens (2000). In particular, Celeux et al. (2000) show that the ordering of the parameters used to ensure identifiability of the parameters may have disastrous effects on the resulting inference, and they devise specific loss functions to overcome the nonidentifiability problem.

Gibbs sampling and other MCMC schemes have thus brought considerable improvement to the Bayesian approach to mixture models, not only for estimation, as shown above, but also for testing and modeling because tests on the number of components of a mixture have been proposed (Mengersen and Robert (1996)). Moreover, these studies have also exhibited interesting noninformative extensions. As mentioned in Exercise 1.57, the peculiar properties of mixtures prohibit the use of improper priors of the form

$$\prod_{i=1}^{k} \pi_1(\mu_i, \sigma_i).$$

In fact, in the decomposition (6.1.6) of the posterior distribution as a sum over all possible partitions, some of these partitions do not attribute any observation to a given component i^* of the mixture. The prior distribution on the corresponding parameters $(\mu_{i^*}, \sigma_{i^*})$ must therefore be *proper*.

However, as shown in Mengersen and Robert (1996), an improper prior can still be used if the component parameters are a priori dependent. For instance, the mixture model can be reparametrized in terms of a global location-scale parameter (μ, τ), with a corresponding prior $\pi(\mu, \tau) = 1/\tau$. In this case, the prior input can be reduced to the choice of a single hyperparameter $\xi > 0$.

Indeed, if (6.4.1) is written as

$$p_1\mathcal{N}(\mu,\tau^2) + (1-p_1)\left\{p_2\mathcal{N}(\mu+\tau\theta_1,\tau^2\sigma_1^2)\right.$$
$$\left. +(1-p_2)\left\{p_3\mathcal{N}(\mu+\tau\theta_1+\tau\sigma_1\theta_2,\tau^2\sigma_1^2\sigma_2^2)+\ldots\right\}\right\},$$

an acceptable prior distribution is of the form $p_i \sim \mathcal{B}e(1/2,1/2)$, $\theta_i \sim \mathcal{N}(0,\xi^2)$, and $\sigma_i \sim (1/2)\mathcal{U}_{[0,1]} + (1/2)\mathcal{P}a(2,1)$, the later distribution being justified by a uniform distribution on either σ_i or $1/\sigma_i$. (See Roeder and Wasserman (1997) and Robert and Titterington (1998) for similar proposals.)

6.6.7 A Gibbs Sampler for Dirichlet Processes

The interest of using Dirichlet processes for Bayesian nonparametric estimation has been mentioned in Note 1.8.2. We indicate here how a Gibbs sampling implementation proceeds in a normal setting.

Consider $x_i \sim \mathcal{N}(\theta_i,\sigma_i^2)$ $(1 \le i \le n)$ with $(\theta_i,\sigma_i^2) \sim \pi$ and π distributed as a Dirichlet process $\mathcal{D}(\alpha,\pi_0)$. As already mentioned in Note 1.8.2, π_0 is the prior expectation of π and α is a degree of concentration around π_0. The corresponding marginal distribution is a mixture of normal distributions, with a random number of components ranging from 1 up to n. That the number of components can be as high as the sample size is a reflection of the lack of constraints on the model, and this can be related to the fact that the usual *kernel* estimator always uses n components. Another important consequence of this modeling is that the prior conditional distributions of the (θ_i,σ_i^2)'s can be expressed as

$$\pi[(\theta_i,\sigma_i^2)|(\theta_j,\sigma_j^2)_{j\neq i}] = \alpha(\alpha+n-1)^{-1}\pi_0(\theta_i,\sigma_i^2)$$
$$(6.6.2) \qquad\qquad +(\alpha+n-1)^{-1}\sum_{j\neq i}\mathbb{I}((\theta_i,\sigma_i^2)=(\theta_j,\sigma_j^2)).$$

The decomposition (6.6.2) exhibits the moderating effect of the Dirichlet prior: new values of (θ,σ^2) only occur with probability $\alpha/(\alpha+n-1)$.

A similar conditional distribution can be obtained a posteriori, namely, that for observations $x_1,\ldots,x_n$,

$$\pi[(\theta_i,\sigma_i^2)|(\theta_j,\sigma_j^2)_{j\neq i},x_i] = q_{i0}\pi_0(\theta_i,\sigma_i^2|x_i)$$
$$(6.6.3) \qquad\qquad +\sum_{j\neq i}q_{ij}\mathbb{I}((\theta_i,\sigma_i^2)=(\theta_j,\sigma_j^2)),$$

where $q_{i0} + \sum_{j\neq i}q_{ij} = 1$ and $(i \neq j)$

$$q_{i0} \propto \alpha\int e^{-(x_i-\theta_i)^2/2\sigma_i^2}\sigma_i^{-1}\pi_0(\theta_i,\sigma_i^2)d\theta_i d\sigma_i^2, \quad q_{ij} \propto e^{-(x_i-\theta_j)^2/2\sigma_j^2}\sigma_j^{-1}.$$

For the conditional distributions (6.6.3), (θ_i,σ_i^2) is a new parameter with probability q_{i0} and is equal to another parameter with probability $1 - q_{i0}$. Therefore, a Gibbs sampling implementation can proceed by simulating successively from those conditionals for all i's, and propose as a marginal distribution for $(x_1,\ldots,x_n)$ a mixture of k normal distributions, where k is the number of different values of simulated (θ_i,σ_i^2). (Notice that the number k will vary at each iteration.)

Another consequence of this representation is that, if we are interested in the predictive density f, we can simulate a sample of size T from $\pi(\theta,\sigma^2|x_1,\ldots,x_n)$,

$(\theta^{(t)}, \sigma^{(t)^2})$ $(t = 1, \ldots, T)$, by simulating successively (θ_i, σ_i^2) $(1 \leq i \leq n)$ according to (6.6.3) and $(\theta_{n+1}, \sigma_{n+1}^2)$ according to

$$
\begin{aligned}
\pi(\theta_{n+1}, \sigma_{n+1}^2) &= \pi[(\theta_{n+1}, \sigma_{n+1}^2) | (\theta_{i \neq n+1}, \sigma_{i \neq n+1}^2)] \\
&= \alpha(\alpha + n)^{-1} \pi_0(\theta_{n+1}, \sigma_{n+1}^2) \\
&\quad + (\alpha + n)^{-1} \sum_{j=1}^{n} \mathbb{I}((\theta_{n+1}, \sigma_{n+1}^2) = (\theta_j, \sigma_j^2)).
\end{aligned}
$$

The predictive density can then be estimated by

(6.6.4)
$$
\frac{1}{T} \sum_{t=1}^{T} f(x | \theta^{(t)}, \sigma^{(t)^2}),
$$

and is therefore of the same order of complexity as a kernel density estimator, since it involves formally T terms. In fact, the sample of the $(\theta^{(t)}, \sigma^{(t)^2})$ will include a few values simulated according to $\pi_0(\theta_{n+1}, \sigma_{n+1}^2)$ and most of the values of the (θ_i, σ_i^2) $(1 \leq i \leq n)$ that are also simulated according to π_0, but with replications. Improvements upon this direct implementation of the Dirichlet process prior are suggested in Escobar and West (1995), such as a derivation of the average number of components in the distribution of the (θ_i, σ_i^2) $(1 \leq i \leq n)$. However, the choice of the hyperparameters is quite crucial to the performances of the resulting estimator.

CHAPTER 7

Model Choice

"Right this minute, wherever he is, Galad is puzzling over something he may never have faced before. Two things that are right, but opposite."

Robert Jordan, *The Fires of Heaven, Book V of the Wheel of Time.*

7.1 Introduction

As pointed out in Chapter 5, *model choice* can be considered a special case of testing and, still, we feel this problem deserves special treatment (and, hence, a separate chapter). Why is this so? Before embarking on a more precise definition of what model choice is, we present below several arguments in favor of this separate treatment, and which, we hope, can be understood with only the vague idea that model choice deals with the comparison, and maybe the selection, of models.

From a conceptual point of view, the inferential action takes place on a wider scale than in Chapter 5: we are now dealing with models, rather than with parameters. For instance, there seems to be more at stake when comparing an exponential model with a Weibull model than when deciding whether a parameter θ is equal to 1, say. In other words, in contrast with the other chapters of this book, the sampling distribution $f(x)$ is unknown to a larger extent than simply depending on an unknown (finite dimensional) parameter.

From a modeling point of view, model choice often appears to be closer to estimation than to regular testing. While we saw in Chapter 5 that testing $H_0 : \theta \in \Theta_0$ is also equivalent to estimating the indicator function $\mathbb{I}_{\Theta_0}$, model choice may simultaneously involve many possibilities, $\mathcal{M}_1, \ldots, \mathcal{M}_p$ say, and the decision about "the" model is equivalent to estimating the index $\mu \in \{1, \ldots, p\}$ of this model (or, more exactly, getting the posterior distribution of the index). Obviously, many settings require a firm and precise decision about which model is right (meaning, which model is the most appropriate to the data at hand), but this does not sound as definitive as deciding whether H_0 is true.

From a computational point of view, model choice involves more complex structures that, almost systematically, require advanced tools such as those presented in Chapter 6. Hence the break between Chapter 5 and the current chapter, which also allows us to come back to the computation of Bayes factors and pseudo-Bayes factors using Monte Carlo and MCMC methods (Section 7.3). In fact, the comparison between models implies recourse to even more advanced tools than those of Chapter 6, and we shall introduce in Section 7.3.4 simulation methods which handle collections of parameter spaces (also called *spaces of varying dimensions*), specially designed for model comparison.

At last, the larger inferential scope mentioned in the first point means that we are leaving for a while the well charted domain of solid parametric models: we will see repeatedly in this chapter settings where the "true" distribution f is not known, and where we are trying to assess the distance between f and one (or several) family of distributions $\{f_\theta;\ \theta \in \Theta\}$. For instance, in the goodness-of-fit tests of Section 7.6, we need to call for a nonparametric estimate of f. Similar issues will be raised in variable selection (Section 7.5), where an *embedding* model, different from the true model, may be introduced.

Obviously, there is also much in common with Chapter 5, since the tools used here are mostly the same, namely posterior probabilities and Bayes factors. Many authors actually minimize the difference between regular testings and model choice on this basis. See, for instance, Berger and Pericchi (2001), whose survey on model choice mainly proposes examples with tests of null hypotheses like $H_0 : \theta = 0$.

Model choice, and the related topics of variable selection and goodness-of-fit tests, have been the subject of considerable effort in the past years, and this partly owes to the introduction of new computational methods, but we can only give here a partial indication of this effort. We thus refer the reader to Racugno (1999) for deeper treatments on this topic.

7.1.1 Choice between models

Model choice seems to elude the Bayesian paradigm in that the sampling distribution f is itself uncertain, making it difficult to condition on the observation x. We will feel more acutely this shift of paradigm in Section 7.6 when the question is: *Does f belong to the family $\{f_\theta;\ \theta \in \Theta\}$?* the alternative being completely open. First, consider the more restricted setting where several (parametric) models are in competition,

$$\mathcal{M}_i : x \sim f_i(x|\theta_i), \qquad \theta_i \in \Theta_i, \quad i \in I,$$

the index set I being possibly infinite. This reduced perspective is less puzzling from a Bayesian point of view, in the sense that a prior distribution can be constructed for each model $\mathcal{M}_i$ as if it were the only and true model under consideration.

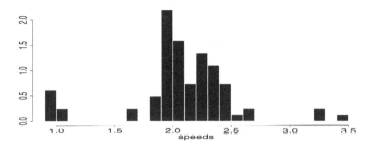

Figure 7.1.1. *Histogram of the galaxy dataset of Roeder (1992).*

In the simplest case, the choice is between a small number of models that have been chosen for convenience, historical or more motivated reasons.

Example 7.1.1 We introduced in Example 1.1.5 a dataset of Lenk (1999) relating the monthly unemployment rate with the monthly number of accidents in Michigan from 1978 to 1987. At a cursory reading, before looking at the connection between these two variates, one may oppose

$$\mathcal{M}_1 : N \sim \mathcal{P}oi(\lambda), \qquad \lambda > 0$$

to

$$\mathcal{M}_2 : N \sim \mathcal{N}eg(m, p), \qquad m \in \mathbb{N}^*, p \in [0, 1]$$

as possible models for the number of accidents N in a given month. ‖

In more complex cases, the number of potential models may be large because the available information is too limited to eliminate most of them. We are then closer to the nonparametric perspective.

Example 7.1.2 A dataset used in most papers on mixture estimation is the *galaxy dataset*. First treated by Roeder (1992), it has been analyzed in Chib (1995), Escobar and West (1995), Phillips and Smith (1996), Richardson and Green (1997), Roeder and and Wasserman (1997) and Robert and Mengersen (1999), among others. It consists in 82 observations of galaxy velocities, described in Figure 7.1.1. For astrophysical reasons, the distribution of this dataset can be represented as a mixture of normal distributions whose number of components k is unknown. (A component of the mixture is to be interpreted as a *cluster of galaxies*.) The models in contention are thus

$$(7.1.1) \qquad \mathcal{M}_i : n_j \sim \sum_{\ell=1}^{i} p_{\ell i} \mathcal{N}(\mu_{\ell i}, \sigma_{\ell i}^2),$$

where i varies between 1 and some arbitrary upper bound. ‖

In other settings, including variable selection (Section 7.5), the variety of models stems from the large number of combinations of *covariates* (or *explanatory variables*) that could be included in the model.

Table 7.1.1. *Orange tree circumferences (in millimeters) against time (in days) for 5 trees. (Source: Gelfand (1996)).*

	tree	number			
time	1	2	3	4	5
118	30	33	30	32	30
484	58	69	51	62	49
664	87	111	75	112	81
1004	115	156	108	167	125
1231	120	172	115	179	142
1372	142	203	139	209	174
1582	145	203	140	214	177

Example 7.1.3 (Gelfand (1996)) For 5 orange trees, the growth of tree i is measured through the circumferences y_{it} at different times T_t, resulting in the data of Table 7.1.1. The models under scrutiny are ($i = 1, \cdots, 5$, $t = 1, \ldots, 7$)

$$\mathcal{M}_1 \quad : \quad y_{it} \sim \mathcal{N}(\beta_{10} + b_{1i}, \sigma_1^2)$$

$$\mathcal{M}_2 \quad : \quad y_{it} \sim \mathcal{N}(\beta_{20} + \beta_{21}T_t + b_{2i}, \sigma_2^2)$$

$$\mathcal{M}_3 \quad : \quad y_{it} \sim \mathcal{N}\left(\frac{\beta_{30}}{1 + \beta_{31}\exp(\beta_{32}T_t)}, \sigma_3^2\right)$$

$$\mathcal{M}_4 \quad : \quad y_{it} \sim \mathcal{N}\left(\frac{\beta_{40} + b_{4i}}{1 + \beta_{41}\exp(\beta_{42}T_t)}, \sigma_4^2\right),$$

where the b_{ji}'s are random-effects, distributed as $\mathcal{N}(0, \tau^2)$. Such models start from the plain individual effect—no time effect in $\mathcal{M}_1$, to the linear time effect in $\mathcal{M}_2$, to a non-linear time effect in $\mathcal{M}_3$, with individual effects in $\mathcal{M}_4$. ‖

As should be obvious from Example 7.1.3, there often is a high degree of arbitrariness involved in the selection of the models to choose from. Similarly, in Example 7.1.2, the normality assumption corresponds to a convenience choice, rather than being motivated by astrophysical reasons.

Examples 7.1.1 to 7.1.3 expose a fundamental difficulty with model choice issues, namely, that while no model is true, *several* models may be appropriate. To be forced to choose one and only one model thus reproduces the dilemma encountered in Chapter 5, where a test procedure taking values in $\{0, 1\}$ did not sound like the right answer. (This problem will find a radical solution in Section 7.4 by refusing the choice of a particular model.) In both Examples 7.1.2 and 7.1.3, we also face an additional *embedding* problem, namely, that some models are submodels of others. For instance, in Example 7.1.2, a k component mixture is a submodel of a $(k+p)$ component mixture, the p remaining components being associated with weights 0. From a modeling point of view, the larger model should be preferred, while, from a statistical point of view, this is not so clear, given that more parameters need to be estimated from the same sample! The model choice criterion must thus include parts that weight the fit, as well as parts that incorporate the estimation error.

7.1.2 Model choice: motives and uses

As motivated by the previous examples, model choice is not a monolithic estimation procedure, but can be undertaken for various reasons, which are not always obvious for (or made explicit by) the experimenter (or the "client"), and which therefore make the construction of a strict decision-theoretic framework very nearly impossible. Among these possible reasons, we can identify the choice of a model as

  (i) a first step in *model construction*, as in Example 7.1.1, where a few models come to mind and the experimenter wants to decide which one fits "best" the data at hand. This perspective is just one step beyond nonparametric Statistics, in the sense that there is no reason to believe that one of these models is correct.

 (ii) conversely, a last step of *model checking*, as in Example 7.1.3. A model or a family of models has been selected for various theoretical and practical reasons, and one wants to know whether the data agrees with this type of model. This is also the domain of goodness-of-fit tests, where the model is not clearly defined outside the null hypothesis (as detailed in Section 7.6).

(iii) a call for *model improvement*, as in the move from $\mathcal{M}_1$ to $\mathcal{M}_2$ or from $\mathcal{M}_3$ to $\mathcal{M}_4$ in Example 7.1.3. Given a model, possibly validated by a goodness-of-fit test, the goal is to introduce possible refinements in the model to improve the fit, or, in other words, to create an *embedding* of the existing model in a class of models to check whether the current model is good enough.

 (iv) the reverse need of *model pruning*,[1] where the current model is deemed to be too complicated to be of practical use, as for instance in Example 7.1.2 with $k = 50$, and where, for parsimony reasons, simpler submodels are examined to see whether they fit the data well enough. This is in particular the setting for variable selection techniques, where a whole range of covariates is proposed, and the aim is to reduce those to a few important covariates.

  (v) a simpler *model comparison*, when a few models are proposed because they fitted correctly other samples and one wonders which of these models best fits the current sample, as in Example 7.1.1.

 (vi) a more ambitious purpose of *hypothesis testing*, as in scientific settings, where several models are built from theoretical considerations and then

---

[1] This expression makes literal sense when considering the tree of possible models in variable selection, which needs to be pruned of most of its branches!

tested through specially designed experiments. (This is the case, for instance, of Einstein's theory of gravitation versus Newton's, or of the cosmological theories of expansion versus contraction of the Universe.)

(vii) a more limited requirement of *prediction efficiency*, as, for instance, in finance. Contrary to (vi), the models are not considered per se since the experimenter is only interested in the prediction performances of different models. In the setting of Example 7.1.2, one could think for instance of the probability to allocate a new galaxy to the correct group of galaxies.

The applications of model choice are obviously as wide as those of Statistics, since there are very few cases when a given model or a parametric family is accepted by one and all! Let us mention here a few settings where model choice is particularly necessary: image analysis, when comparing different neighborhood structures (Cressie (1993)); graphical models and expert systems, when considering removing some links between variables (Spiegelhalter et al. (1993), Cowell et al. (1999)); variable dimension models, as in ARMA(p, q) models where the lags p and q are unknown; *causal inference*, where the question is to decide whether A has an effect on B, given a set of variables $C_1, \ldots, C_p$ (Shafer (1997), Wasserman and Robins (2000)).

7.2 Standard framework

7.2.1 Prior modeling for model choice

As in other settings, the standard Bayesian solution is to put a prior distribution on the unknown, which means extending the prior modeling from parameters to models. The parameter space associated with the set of models (7.1.1) can be written as

$$(7.2.1) \qquad \Theta = \bigcup_{i \in I} \{i\} \times \Theta_i \,,$$

the model indicator $\mu \in I$ being now part of the parameters. So, if one can allocate probabilities p_i to the indicator values, that is, to the models $\mathcal{M}_i$ $(i \in I)$, and then define priors $\pi_i(\theta_i)$ on the parameter subspaces Θ_i, things fold over by virtue of Bayes's theorem, as usual, since we can compute

$$(7.2.2) \qquad p(\mathcal{M}_i | x) = P(\mu = i | x) = \frac{p_i \displaystyle\int_{\Theta_i} f_i(x|\theta_i)\pi_i(\theta_i)d\theta_i}{\displaystyle\sum_j p_j \int_{\Theta_j} f_j(x|\theta_j)\pi_j(\theta_j)d\theta_j} \,.$$

While a common solution based on this prior modeling is simply to take the (marginal) MAP estimator of μ, that is, to determine the model with the largest $p(\mathcal{M}_i|x)$, or even to use directly the average

$$(7.2.3) \qquad \sum_j p_j \int_{\Theta_j} f_j(y|\theta_j)\pi_j(\theta_j|x)d\theta_j = \sum_j p(\mathcal{M}_j|x)\, m_j(y)$$

as a predictive density in y, a deeper-decision theoretic evaluation is often necessary.

There are difficulties with the usual Bayesian formalism, or at least with the prior modeling, in this setting: the solution based on the representation (7.2.1) of the collection of models requires the construction of a prior distribution (π_i, p_i) for each $i \in I$, which is delicate when I is infinite. Moreover, these priors π_i must all be proper because there is no unique scaling for improper priors, as already seen in Chapter 5. In addition, if some models are embedded into others, that is, if $\mathcal{M}_{i_0} \subset \mathcal{M}_{i_1}$, there should be some coherence in the choice of π_{i_0} given π_{i_1}, and maybe also in the choice of p_{i_0} given p_{i_1}. For instance, if $\mathcal{M}_1 = \mathcal{M}_2 \cup \mathcal{M}_3$, one could argue that $p(\mathcal{M}_1) = p(\mathcal{M}_2) + p(\mathcal{M}_3)$, or at least $p(\mathcal{M}_1) \geq p(\mathcal{M}_2) + p(\mathcal{M}_3)$, should hold. Similarly, if two models $\mathcal{M}_{i_0}$ and $\mathcal{M}_{i_1}$ are not embedded in one another, the prior modeling should account for the possibility of a third model $\mathcal{M}_{i_2}$ embedding both $\mathcal{M}_{i_0}$ and $\mathcal{M}_{i_1}$. (In econometrics, this technique of creating a super-model is called *encompassing*.) A last important point, which is specific to model choice, is that *parameters common to several models must be treated as separate entities*. This point is often neglected in the literature, including Jeffreys (1961), because common parameters can be integrated out using the *same* prior, even when the prior is improper. A milder evasion of the above recommendation is to argue, as in Berger and Pericchi (1998), that, for common parameters, the *same* improper prior should be used, thus removing the issue of the normalizing constant (Exercise 7.4), but we cannot advise such an ad-hoc solution on a general basis.

Example 7.2.1 (Example 7.1.3 continued) Consider models $\mathcal{M}_1$ and $\mathcal{M}_2$: while β_{10} and β_{20} are both intercepts, and σ_1^2 and σ_2^2 are both variances, they must be distinguished because of the additional term $\beta_{21}T_t$ in model $\mathcal{M}_2$. In particular, if $\mathcal{M}_2$ is the true model, β_{10} corresponds to β_{20} shifted by the average of the $\beta_{21}T_t$'s, while σ_1^2 is larger than σ_2^2 to account for the poorer fit (see Exercise 7.5).   ‖

From a decision-theoretic perspective, the inferential problem is also hard to formalize because, as pointed out in Section 7.1.2, there are many conflicting potential uses of model choice. Overall, model choice appears as one part of a global *decision process*, where a model is constructed, improved by reduction or extension (as in points (iii) and (iv) above), then selected for future uses as the true model. To build a loss function that incorporates all these stages is clearly impossible, but the stress can be placed on the selection part. For instance, *model averaging* as in (7.2.3) is not acceptable in this sense because, by preserving all models which are compatible with the data, one adopts an estimation perspective which amounts to a lack of decision! If no (or not enough) information can be gathered about the cost of choosing the wrong model, and thus if a decision-based loss function $L((\mu, \theta_\mu), (d, \vartheta))$ cannot be constructed, a solution supported at the end of Section 7.1.1 is to fight overfitting by introducing in the loss function *penalty terms* on the number of parameters (that is, the dimension) of the model, as detailed in Section 7.2.3. See also Carota, Parmigiani and Polson (1996) for a decision-theoretic attempt at model criticism, which pertains more to point (iii) above, and which uses Kullback–Leibler divergences as in Section 7.5.

A third type of difficulty is associated with the computation of predictives, marginals and other quantities related to the model choice procedures. This is not an issue restricted to model choice (see Chapter 6), obviously, but there are specificities that call for tailor-made solutions:

- the parameter spaces are often infinite-dimensional, as in (7.1.1), which may cause measure-theoretic complications.

- The computation of posterior or predictive quantities involves integration over different parameter spaces and thus increases the computational burden, since there is no time savings from one subspace to the next.

- The representation of the parameter space as a direct sum of different subspaces requires more advanced Markovian techniques to implement MCMC algorithms.

- In some settings, such as variable selection, the size of the collection of models is so large that some models cannot be explored.

In all but the simplest models, there is thus a need for computational (or approximation) techniques because no analytic representation is available. Section 7.3 will detail these techniques.

7.2.2 Bayes factors

Once the modeling representation (7.1.1) is accepted, and the corresponding priors are selected, the inferential issues are reduced to a generic testing problem. The solution proposed by Kass and Raftery (1995), as well as Berger and Pericchi (2001), is then to call for Bayes factors, e.g.,

$$B_{12} = \frac{P(\mathcal{M}_1|x)}{P(\mathcal{M}_2|x)} \bigg/ \frac{P(\mathcal{M}_1)}{P(\mathcal{M}_2)}$$

$$= \frac{\displaystyle\int_{\Theta_1} f_1(x|\theta_1)\pi_1(\theta_1)d\theta_1}{\displaystyle\int_{\Theta_2} f_2(x|\theta_2)\pi_2(\theta_2)d\theta_2}$$

for the comparison of models $\mathcal{M}_1$ and $\mathcal{M}_2$. The setting is therefore similar to Section 5.2 and, consequently, the difficulties are also the same, only exacerbated here because we are dealing with more models (possibly an infinite number of them!) and are much more likely to call for noninformative priors. Notice that the comparison of models based on Bayes factors can proceed one pair $(\mathcal{M}_i, \mathcal{M}_j)$ at a time because of the coherence property of Bayes factors, that is $B^\pi_{ij} = B^\pi_{ik} B^\pi_{kj}$, which ensures that the model ordering is transitive. (Recall, though, that this is not the case for the pseudo-Bayes factors of Section 5.2.6.)

For exactly the same reason as in Section 5.2.5, *improper priors* cannot be used (unless, as mentioned above, they bear on parameters common to all models). Moreover, vague priors, that is, proper priors with a large variance—which is the representation adopted in BUGS, see Note 6.6.2—do not solve the difficulty, as already shown with the *Jeffreys–Lindley paradox* (Section 5.2.5).

Example 7.2.2 (Example 7.1.1 continued) Consider the priors

$$\pi_1(\lambda) = \mathcal{G}a(\alpha, \beta), \qquad \pi_2(m, p) = \frac{1}{M} \, \mathbb{I}_{\{1,\cdots,M\}}(m) \mathbb{I}_{[0,1]}(p),$$

where the second prior is uniform over the parameter space Θ_2. Then the Bayes factor

$$
\begin{aligned}
B^\pi_{12} &= \frac{\dfrac{\beta^\alpha}{\Gamma(\alpha)} \displaystyle\int_0^\infty \frac{\lambda^{\alpha+x-1}}{x!} e^{-\lambda\beta} d\lambda}{\dfrac{1}{M} \displaystyle\sum_{m=1}^M \int_0^1 \binom{m}{x-1} p^x (1-p)^{m-x} dp} \\[2em]
&= \frac{\Gamma(\alpha+x)}{x! \, \Gamma(\alpha)} \beta^{-x} \Bigg/ \frac{1}{M} \sum_{m=1}^M \frac{x}{(m-x+1)(m+1)} \\[2em]
&= M(m+1) \frac{(x+\alpha-1)\cdots\alpha}{x(x-1)\cdots 1} \beta^{-x} \Bigg/ \sum_{m=1}^M \frac{x}{m-x+1}
\end{aligned}
$$

depends on the choice of α, β when both go to 0 (Exercise 7.10).   ‖

The solution to this fundamental difficulty with improper priors is to use approximative Bayesian solutions, calling for minimal training samples or virtual observations, as in Section 5.2.6. (Notice that one of the early proposals of pseudo-Bayes factors appeared in Spiegelhalter and Smith (1980) for model choice in linear and log-linear models.) Intrinsic and fractional Bayes factors can then be proposed (with the same provisos as those in Section 5.2.6) as evaluation of the models under improper priors.

Example 7.2.3 (Example 7.1.2 continued) As detailed in Example 5.2.19, there is no minimal training sample for a mixture model, whatever the number of observations is. Therefore, intrinsic and fractional Bayes factors cannot be used there.

A first solution, already used in Diebolt and Robert (1994) for simulation purposes and further validated by Wasserman (1999), is to impose that the sample $(x_1, \ldots, x_n)$ is such that enough observations (in the sense of the training samples) come from each component (see also Green and Richardson (1998)). While being reasonable if all components are clearly distinguishable, this solution creates a dependence between the observations (which remain exchangeable), and the computation of the pseudo-Bayes factors under this assumption becomes very expensive.

The second solution, adopted in Mengersen and Robert (1996) when testing $k = 1$ versus $k = 2$, is to put a non-informative prior $\pi(\mu, \tau)$ on the global location-scale parameter of the model, and to express the parameters of all components as perturbations of this location-scale parameter, using proper priors. Since (μ, τ) is common to all components, the improper prior does not raise so keenly the normalization issue. ‖

7.2.3 Schwartz's criterion

In order to comment on penalty terms and (crude) approximations to Bayesian solutions, we need to briefly consider asymptotic approximations to Bayes factors.[2]

Consider the *Laplace expansion* discussed in Section 6.2.3,

$$\int_{\Theta} \exp\{n\,h(\theta)\}d\theta = \exp\{n\,h(\hat{\theta})\}(2\pi)^{p/2}n^{-p/2}|H^{-1}(\hat{\theta})| + O(n^{-1}),$$

where p is the dimension of Θ, $\hat{\theta}$ is the maximum of h and H is the Hessian of h. If we apply this approximation to the Bayes factor by expanding both numerator and denominator, we get an approximation of the Bayes factor,

$$B_{12}^{\pi} \simeq \frac{L_{1,n}(\hat{\theta}_{1,n})}{L_{2,n}(\hat{\theta}_{2,n})} \left| \frac{H_1^{-1}(\hat{\theta}_{1,n})}{H_2^{-1}(\hat{\theta}_{2,n})} \right|^{1/2} \left(\frac{n}{2\pi}\right)^{(p_2-p_1)/2},$$

where p_1 and p_2 are the dimensions of Θ_1 and Θ_2, $L_{1,n}$ and $L_{2,n}$ are the likelihood functions based on n observations, and $\hat{\theta}_{1,n}$ and $\hat{\theta}_{1,n}$ are the maxima of L_1 and L_2, respectively. Therefore,

$$(7.2.4) \qquad \log(B_{12}^{\pi}) \simeq \log \lambda_n + \frac{p_2 - p_1}{2} \log(n) + K(\hat{\theta}_{1,n}, \hat{\theta}_{2,n}),$$

where λ_n is the standard likelihood ratio for the comparison of $\mathcal{M}_1$ with $\mathcal{M}_2$,

$$\lambda_n = L_{1,n}(\hat{\theta}_{1,n})/L_{2,n}(\hat{\theta}_{2,n}),$$

---

[2] This section aims at illustrating the link between Bayesian approximation and usual penalization criteria, not at promoting such criteria. It can thus be skipped on a first reading.

and $K(\hat{\theta}_{1,n}, \hat{\theta}_{2,n})$ denotes the remainder term.

This approximation leads to *Schwartz's criterion* (1978):

$$S = -\log \lambda_n - \frac{p_2 - p_1}{2} \log(n)$$

when $\mathcal{M}_1 \subset \mathcal{M}_2$, if the remainder term $K(\hat{\theta}_{1,n}, \hat{\theta}_{2,n})$ is negligible compared with both other terms, that is, is a $O(1)$. (See Gelfand and Dey (1994, §8) for an example where this term is not negligible.)

Recall that, for regular models, when $\mathcal{M}_1 \subset \mathcal{M}_2$, the likelihood ratio is approximately distributed as a $\chi^2_{p_2-p_1}$ distribution,

$$-2\log \lambda_n \approx \chi^2_{p_2-p_1}$$

if $\mathcal{M}_1$ is the true model (Lehmann and Casella (1998), Gouriéroux and Monfort (1996)). Since

$$
\begin{aligned}
P(\mathcal{M}_2 \text{ chosen}|\mathcal{M}_1) &= P(\lambda_n < c|\mathcal{M}_1) \\
&\simeq P(\chi^2_{p_2-p_1} > -2\log(c)) > 0,
\end{aligned}
$$

it follows, from a frequentist point of view, that a criterion based solely on the likelihood ratio does not converge to a sure answer under $\mathcal{M}_1$. This is why penalization factors have been added to the (log) likelihood ratio to account for this bias, starting with Akaike's (1983) criterion,

(7.2.5) $$-2\log \lambda_n - \alpha(p_2 - p_1),$$

where $\alpha = \log 2$ also corresponds to an approximation of Aitkin's (1991) procedure, where the author uses the data twice, firstly to build a proper (pseudo-) prior by taking the posterior, and secondly to derive the Bayes factor as if this construct were a genuine prior. (See Exercise 5.15.)

Schwartz's criterion, also called *BIC* (for *Bayes Information Criterion*), thus provides a cursory first-order approximation to the Bayes factor, as defended in Kass and Raftery (1995). However, we do not see the relevance of this criterion in a Bayesian setting, since (a) the dependence on the prior assumption disappears; and (b) the approximation only works for regular models. For instance, in Example 7.1.2, the asymptotics of the log-likelihood ratio $-2\log \lambda_n$ are much more complex than the $\chi^2_{p_2-p_1}$ approximation (see, e.g., Dacunha-Castelle and Gassiat (2000)) and Schwartz's criterion does not work. See Berger and Pericchi (2001) for other examples of irregular likelihoods. Moreover, in non-i.i.d. structures, the definition of both n and p may be ambiguous, as stressed by Spiegelhalter et al. (1998). At a computational level, notice that Schwartz's criterion requires the derivation of the maximum likelihood estimates for all models.

Example 7.2.4 (Example 7.1.2 continued) If we decompose Schwartz's criterion as

$$
\begin{aligned}
S &= \log\left\{L_{2,n}(\hat{\theta}_{2,n})/L_{1,n}(\hat{\theta}_{1,n})\right\} - \frac{p_2 - p_1}{2}\log(n) \\
&= \log L_{2,n}(\hat{\theta}_{2,n}) - \frac{p_2}{2}\log(n) - \log L_{1,n}(\hat{\theta}_{1,n}) + \frac{p_1}{2}\log(n),
\end{aligned}
$$

the part owing to model $\mathcal{M}_i$ can be identified as

$$S_i = \log L_{i,n}(\hat{\theta}_{i,n}) - \frac{p_i}{2} \log(n) \,.$$

For $\mathcal{M}_k$ associated with a k component model, $p_k = 3k - 1$. In the case of the galaxy data, Raftery (1996) gets

$$S_1 = -271.8, \quad S_2 = -249.7, \quad S_3 = -256.7, \quad S_4 = -263.6,$$

using the EM algorithm (see Note 6.6.6) to obtain approximations of the maximum likelihood estimates $\hat{\theta}_{i,n}$ when $k > 1$. This means that, for Schwartz's criterion, the model with two components is to be preferred to the others.   ‖

7.2.4 Bayesian deviance

Spiegelhalter, Best and Carlin (1998) have developed a Bayesian alternative to both AIC (Akaike's Information Criterion) and BIC, based on the *deviance* and called DIC (for Deviance Information Criterion). This criterion is more satisfactory than the two former alternatives because it takes into account the prior information and gives a natural penalization factor to the log-likelihood. Besides, it also allows for improper priors, since each model is considered separately.

As mentioned earlier in Section 7.2.3, for a model $f(x|\theta)$ associated with a prior distribution $\pi(\theta)$, the deviance[3] $D(\theta) = -2\log(f(x|\theta))$ is not a good discriminating measure, given its bias toward higher dimensional models. The same obviously applies to its posterior distribution. Spiegelhalter et al. (1998) introduce a penalized deviance,

$$(7.2.6) \qquad \mathrm{DIC} = \mathbb{E}[D(\theta)|x] + p_D$$
$$= \mathbb{E}[D(\theta)|x] + \{\mathbb{E}[D(\theta)|x] - D(\mathbb{E}[\theta|x])\} \,.$$

The criterion is then implemented for model evaluation as *the smaller the value of DIC, the better the model.*

The factor $\mathbb{E}[D(\theta)|x]$ in (7.2.6) can be interpreted as a measure of *fit*, while p_D is a measure of *complexity*, also called the *effective number of parameters*. Since $\mathrm{DIC} = D(\mathbb{E}[\theta|x]) + 2p_D$, the analogy with Akaike's Information Criterion (7.2.5) is clear. As shown in Spiegelhalter et al. (1998), in a non-hierarchical framework where the posterior distribution of θ is approximately normal, DIC and AIC are in fact equivalent. Notice also that DIC reproduces the usual decomposition of the squared error as the

---

[3] In generalized linear models (McCullagh and Nelder (1989)), the *deviance* is usually calibrated by an additional term $f(y)$ like $f(y|\hat{\theta}(y))$ where $\hat{\theta}(y)$ is an arbitrary estimator of θ. When this term does not depend on the model, or is chosen for a particular model such as a "full" model, the comparison of models based on $D(\theta)$ and $D(\theta) + 2\log f(y|\hat{\theta}(y))$ are obviously the same.

squared bias plus the variance

$$\mathbb{E}_\theta[(\delta - \theta)^2] = (\mathbb{E}_\theta[\delta] - \theta)^2 + \mathbb{E}_\theta[(\delta - \mathbb{E}_\theta[\delta])^2]$$

to a parameter-free framework (except for $\mathbb{E}[\theta|x]$, which depends on the parameterization).

Example 7.2.5 (Spiegelhalter et al. (1998) For a one-way analysis of variance $(i = 1, \ldots, p)$

$$y_i = \theta_i + \sigma_i \epsilon_i, \qquad \epsilon_i \sim \mathcal{N}(0, 1),$$

the divergence is $D(\theta) = \sum_i \sigma_i^{-1}(\theta_i - y_i)^2$. Therefore, if $\theta_i = \theta$ $(i = 1, \ldots, p)$ and $\pi(\theta) = 1$,

$$(7.2.7) \quad \mathbb{E}[D(\theta)|y_1, \ldots, y_p] = \sum_{i=1}^k \sigma_i^{-1}(y_i - \mathbb{E}[\theta|y_1, \ldots, y_p])^2 + 1$$

where $\mathbb{E}[\theta|y_1, \ldots, y_p] = \sum_i \sigma_i^{-1} y_i / \sum_i \sigma_i^{-1}$. And $p_D = 1$ in this case.

If, instead, we consider the model $\theta_i \sim \mathcal{N}(\mu, \tau^2)$ with known hyperparameters μ and τ,

$$(7.2.8) \quad \mathbb{E}[D(\theta)|y_1, \ldots, y_p] = \sum_{i=1}^k \sigma_i^{-1}(1 - \varrho_i)^2(y_i - \mu)^2 + \sum_{i=1}^k \varrho_i,$$

where $\varrho_i = \sigma_i^2 \tau^2 / (\sigma_i^2 + \tau^2)$. ‖

The practical computation of the Bayesian deviance most often requires the call to an MCMC algorithm, since settings such as those of Example 7.2.5 and others developed in Spiegelhalter et al. (1998) are fairly rare. This computation is quite straightforward to implement, however, once an MCMC sample $(\theta^{(1)}, \ldots, \theta^{(T)})$ has been generated, since $\mathbb{E}[D(\theta)|y_1, \ldots, y_p]$ is simply a posterior expectation of an explicit function of θ.

Example 7.2.6 (Spiegelhalter et al. (1998)) A study on lip cancer in 56 regions of Scotland relates the observed numbers of cases y_i with the expected national numbers E_i as

$$y_i \sim \mathcal{P}(\lambda_i E_i),$$

$\lambda_i = \exp(\theta_i)$ being the area-specific risk of lip cancer. Possible covariates are x_i, the percentage of the population working outdoors, and the geographic location of the region, represented by a list $\mathcal{A}_i$ of adjacent regions. Some models under consideration are then

$$\begin{aligned} \mathcal{M}_1: \quad \theta_i &= \alpha + \beta x_i, \\ \mathcal{M}_2: \quad \theta_i &= \varphi_i, \\ \mathcal{M}_3: \quad \theta_i &= \varphi_i + \beta x_i, \end{aligned}$$

where the φ_i's are spatially correlated, that is

$$\varphi_i | \varphi_j, j \neq i \sim \mathcal{N}\left(\sum_{j \in \mathcal{A}_i} \varphi_j / n_i, \tau^2 / n_i\right),$$

where n_i denotes the number of adjacent regions. (This spatial model is called the *autoregressive spatial model* and is often used in Spatial Statistics. See Besag (1974) or Cressie (1992).)

Under noninformative priors for the hyperparameters (except for τ^2 which is distributed as a $\mathcal{IG}(1,1)$ random variable), the MCMC sampler produces approximate DIC's of 242.8, 88.5 and 89.0 for the three models, with effective numbers p_D equal to 2.1, 31.6 and 29.4, respectively. Models $\mathcal{M}_2$ and $\mathcal{M}_3$ are thus similar, while being better than model $\mathcal{M}_1$. Notice that, while the true number of parameters in model $\mathcal{M}_1$ is 2, the number of parameters in models $\mathcal{M}_2$ and $\mathcal{M}_3$ is respectively 57 and 58. $\parallel$

Spiegelhalter et al. (1998) suggest additional uses of the Bayesian deviance like *deviance residuals*. They also notice the lack of parameterization-invariance of $D(\mathbb{E}[\theta|x])$ and suggest using the canonical parameterization in generalized linear models.[4]

7.3 Monte Carlo and MCMC approximations

As in other settings, the customary difficulty with the Bayesian approach is the computation of integrals of the type

(7.3.1) $$m_i(x) = \int f_i(x|\theta_i)\pi_i(\theta_i)d\theta_i$$

and, more crucially, of ratios of integrals

$$\frac{\displaystyle\int f_1(x|\theta_1)\pi_1(\theta_1)d\theta_1}{\displaystyle\int f_2(x|\theta_2)\pi_2(\theta_2)d\theta_2},$$

not to mention the additional difficulties induced by the intrinsic and fractional Bayes factors. The solutions presented in Chapter 6, namely the use of asymptotic approximations, and of Monte Carlo and MCMC simulation methods, obviously apply in this case. But, as mentioned earlier in this chapter and illustrated in Chen et al. (2000), special methods have also been devised for the computation of Bayes factors and related quantities.

7.3.1 Importance sampling

This technique, introduced in Section 6.2.2, is particularly well adapted to the computation of predictive distributions like (7.3.1). Given a proposal (or *importance*) distribution, with density proportional to g, and a sample $\theta^{(1)}, \ldots, \theta^{(T)}$, the marginal density for model $\mathcal{M}_i$, $m_i(x)$, is approximated

---

[4] Notice that replacing $\mathbb{E}[\theta|x]$ with the MAP estimate does produce a truly parameterization-invariant criterion, with the drawback that this estimate is more difficult to derive than the posterior mean.

by

$$m_i^{IS}(x) = \frac{\sum\limits_{t=1}^{T} f_i(x|\theta^{(t)}) \frac{\pi_i(\theta^{(t)})}{g(\theta^{(t)})}}{\sum\limits_{t=1}^{T} \frac{\pi_i(\theta^{(t)})}{g(\theta^{(t)})}},$$

where the denominator takes care of the missing normalizing constant. (Notice that, if q is a density, the expectation of $\pi(\theta^{(t)})/g(\theta^{(t)})$ is 1.)

A compelling incentive, among others, for using importance sampling in the setting of model choice is that the sample $(\theta^{(1)}, \ldots, \theta^{(T)})$ can be recycled for several models $\mathcal{M}_i$ if they all involve the same (type of) parameters. (For instance, this is not possible for Examples 7.1.1 and 7.1.2.) See Chen and Shao (1997) for an illustration on Bayes factors.

The variance of $m^{IS}(x)$ may be infinite, as already discussed in Section 6.2.2. Raftery (1996) examines the choices of importance functions in the present setting for a given model with sampling density $f(x|\theta)$ and prior distribution $\pi(\theta)$. The first natural choice is $g(\theta) = \pi(\theta)$, which leads to the estimator of the marginal

$$m^{IS}(x) = \frac{1}{T} \sum_t f(x|\theta^{(t)}),$$

but is often inefficient if the data is informative because most of the simulated values $\theta^{(t)}$ fall outside the modal region of the likelihood. (In the extreme case when π is improper, this choice is, of course, impossible.) Obviously, given that the tails of π are usually wider than those of $\pi(\theta|x)$, infinite variance problems are rather rare with this importance function.

A second possible choice is $g(\theta) = f(x|\theta)\pi(\theta)$. Since the associated estimator is then

(7.3.2) $$m^{IS}(x) = 1 \left/ \frac{1}{T} \sum_{t=1}^{T} \frac{1}{f(x|\theta^{(t)})} \right.,$$

that is, the *harmonic mean* of the likelihoods, $m^{IS}(x)$ provides an approximation of the normalization constant of g. While this solution allows for improper priors, as long as the posteriors are defined, the corresponding variance is often infinite. A solution to this problem is to call for the so-called *defensive importance sampling* by taking a mixture of g (or, rather, of $\pi(\theta|x)$) and of a distribution with fat tails, $\varpi(\theta)$:

$$(1 - \varrho)\pi(\theta|x) + \varrho\varpi(\theta), \qquad \varrho \ll 1.$$

(See Hesterberg (1998) and Owen and Zhou (2000) for detailed treatments of this technique.) For instance, $\varpi(\theta) = \pi(\theta)$, as proposed by Newton and Raftery (1994).

Another solution, proposed by Gelfand and Dey (1994), is to generate a sample of $\theta^{(t)}$'s from the posterior and to use

$$(7.3.3) \qquad m^{IS}(x) = 1 \bigg/ \frac{1}{T} \sum_{t=1}^{T} \frac{h(\theta^{(t)})}{f(x|\theta^{(t)})\pi(\theta^{(t)})} \,,$$

rather than (7.3.2), where h is an arbitrary density (Exercise 7.18). The estimator (7.3.3) has furthermore a finite variance if

$$\int \frac{h^2(\theta)}{f(x|\theta)\pi(\theta)} d\theta < \infty \,.$$

Since h is a free parameter, it can (in principle) be chosen so that this condition is satisfied.

7.3.2 Bridge sampling

Monte Carlo methods adapted to the estimation of ratios of normalizing constants, or, equivalently, of Bayes factors, have been developed in the past five years. We refer the reader to Chen et al. (2000, Chapter 5) for a complete exposition on such methods and introduce here one solution which pertains from importance sampling.

Bridge sampling was introduced in Meng and Wong (1996), and is based on identities used in the physics literature: if both models cover the same parameter space Θ, if $\pi_1(\theta|x) = c_1 \tilde{\pi}_1(\theta|x)$ and $\pi_2(\theta|x) = c_2 \tilde{\pi}_2(\theta|x)$, then the equality

$$(7.3.4) \qquad \frac{c_2}{c_1} = \frac{\mathbb{E}^{\pi_2}[\tilde{\pi}_1(\theta|x)\,h(\theta)]}{\mathbb{E}^{\pi_1}[\tilde{\pi}_2(\theta|x)\,h(\theta)]}$$

holds for any *bridge function* $h(\theta)$ such that both expectations are finite (Exercise 7.20). The bridge sampling estimator is then

$$(7.3.5) \qquad B_{12}^S = \frac{\dfrac{1}{n_1} \displaystyle\sum_{i=1}^{n_1} \tilde{\pi}_2(\theta_{1i}|x)\,h(\theta_{1i})}{\dfrac{1}{n_2} \displaystyle\sum_{i=1}^{n_2} \tilde{\pi}_1(\theta_{2i}|x)\,h(\theta_{2i})} \,,$$

where the θ_{ji}'s are simulated from $\pi_j(\theta|x)$ ($j = 1, 2$, $i = 1, \ldots, n_j$).

For instance, if

$$h(\theta) = 1/\left[\tilde{\pi}_1(\theta|x)\tilde{\pi}_2(\theta_{1i}|x)\right] \,,$$

then B_{12}^S is a ratio of harmonic means, generalizing (7.3.2). Meng and Wong (1996) have derived an (asymptotically) optimal bridge function

$$h^*(\theta) = \frac{n_1 + n_2}{n_1 \pi_1(\theta|x) + n_2 \pi_2(\theta|x)} \,.$$

This choice is not of direct use, since the normalizing constants of $\pi_1(\theta|x)$ and $\pi_2(\theta|x)$ are unknown (otherwise, we should not need to resort to such techniques!). Nonetheless, it shows that a good bridge function should cover the support of both posteriors, with equal weights if $n_1 = n_2$.

Example 7.3.1 For *generalized linear models*, that is, models from exponential families,

$$f(y|\theta) = h(y) \, e^{\theta \cdot y - \psi(\theta)}$$

where the mean $\mathbb{E}[y|\theta] = \nabla\psi(\theta)$ is a function of covariates, x, of the form $\nabla\psi(\theta) = \Psi(x^t\beta)$, the choice of the *link function* Ψ is never easy. For instance, when the regressed variable y takes values in $\{0,1\}$ and $\mathbb{E}[y|x] = P(y = 1|x)$, three common choices of Ψ are (McCullagh and Nelder (1989)

– the *logit* link function, $\Psi(t) = \exp(t)/(1 + \exp(t))$;

– the *probit* link function, $\Psi(t) = \Phi(t)$, that is, the c.d.f. of the $\mathcal{N}(0,1)$ distribution; and

– the *log–log* link function, $\Psi(t) = 1 - \exp(-\exp(t))$.

The three contending models are then

$$\mathcal{M}_1 \quad : \quad y|x \quad \sim \quad \frac{e^{yx^t\beta_1}}{1 + e^{yx^t\beta_1}}$$

$$\mathcal{M}_2 \quad : \quad y|x \quad \sim \quad \Phi(x^t\beta_2)^y \, [1 - \Phi(x^t\beta_2)]^{1-y}$$

$$\mathcal{M}_3 \quad : \quad y|x \quad \sim \quad \exp\{-(1-y)\exp(x^t\beta_3)\} \, [1 - \exp\{-\exp(x^t\beta_3)\}]^y \, .$$

If the prior distribution π on the β_i's is a normal $\mathcal{N}_p(\xi, \tau^2 I_p)$, and if the bridge function is $h(\beta) = 1/\pi(\beta)$, the bridge sampling estimate is then $(1 \le i < j \le 3)$

$$B_{ij}^S = \frac{\dfrac{1}{n}\sum_{t=1}^{n} L_j(\beta_{it}|x)}{\dfrac{1}{n}\sum_{t=1}^{n} L_i(\beta_{jt}|x)} \, ,$$

where the β_{it} are generated from $\pi_i(\beta_i|x) \propto L_i(\beta_i|x)\pi(\beta_i)$.   ‖

   In a special case when both priors are equal, except for a hyperparameter, Gelman and Meng (1998) have constructed an improvement upon bridge sampling, called *path sampling* and presented in Note 7.8.1.

7.3.3 MCMC methods

While importance sampling seems quite appropriate in this setting, MCMC methods can also be used to generate samples from complex distributions. For instance, the bridge sampling estimate can be based on MCMC samples rather than on i.i.d. samples if the distributions $\pi_j(\theta|x)$ are too involved.

Example 7.3.2 (Example 7.3.1 continued) For model $\mathcal{M}_j$ $(j = 1, 2, 3)$, the likelihood part of the posterior is

$$\prod_{i=1}^{n} \Psi(x_i^t \beta_j)^{y_i} [1 - \Psi(x_i^t \beta_j)]^{1-y_i} .$$

In the case of the probit link function $(j = 2)$, $\Psi(t) = \Phi(t)$, which is not available in closed form. A natural Gibbs sampling solution is then to create auxiliary variables $z_i \sim \mathcal{N}(0, 1)$ such that $\Psi(x_i^t \beta_2) = \mathbb{E}\left[\mathbb{I}_{z_i \leq x_i^t \beta_2}\right]$, that is, to generate from the joint distribution

$$\pi(\beta_2, z_1, \ldots, z_n) \propto \pi(\beta_2) \prod_{i=1}^{n} \mathbb{I}_{z_i \leq x_i^t \beta_2}^{y_i} \mathbb{I}_{z_i \geq x_i^t \beta_2}^{1-y_i} .$$

For both other link functions, a standard slice sampler (see Section 6.3.6) can be used: for the logit model, $u_i \leq \Psi(x_i^t \beta_1)$ can be inverted into

$$x_i^t \beta_1 \geq \log(u_i/(1 - u_i))i ,$$

while, for the log-log model, $u_i \leq \Psi(x_i^t \beta_3)$ is equivalent to

$$x_i^t \beta_3 \geq \log(- \log(1 - u_i)) .$$

For the three models, the components of the β_j's are then simulated from multidimensional truncated normal distributions. ‖

Therefore, the approximation (7.3.3) of the marginal distribution can be based on an MCMC sample $(\theta^{(t)})$ from $\pi(\theta|x)$.

Example 7.3.3 (Example 7.2.1 continued) If the priors for the four models are of the form $(j = 1, \ldots, 4)$

$$\pi_j(\beta_{\cdot j}, \sigma_j^2, \tau_j^2) \propto \sigma_j^2 \tau_j^2 e^{-2(\sigma_j^{-2} + \tau_j^{-2})} ,$$

where $\beta_{\cdot j}$ denotes the vector of β_{ij}'s for model j, Gelfand (1996) suggest to evaluate the four models by generating a sample of $\theta_j^{(t)}$'s from the corresponding posteriors, by approximating the predictive distributions as

$$\hat{f}_j(y|y_1, \ldots, y_n) = \frac{1}{T} \sum_{t=1}^{T} f_j(y|\theta_j^{(t)}) ,$$

then by drawing samples from these predictives to see whether they agree with the sample $y_1, \ldots, y_n$. Table 7.3.3 shows the result of this experiment: these figures exclude models $\mathcal{M}_1$ and $\mathcal{M}_2$, while both models $\mathcal{M}_3$ and $\mathcal{M}_4$ agree with the predictive intervals. Obviously, this is only a first indicator of fit, which should be completed with a computation of the genuine Bayes factors, but this empirical evaluation may allow for the elimination of inadequate models. ‖

Table 7.3.1. *Adequacy results for the four models of orange tree growths, as percentage of observations within the 50% and 90% predictive intervals. (Source: Gelfand (1996).)*

Model	50%	95%
$\mathcal{M}_1$	89	100
$\mathcal{M}_2$	29	51
$\mathcal{M}_3$	46	100
$\mathcal{M}_4$	60	86

Chib (1995) proposes to use the Gibbs sampler to approximate marginal densities, based on the Bayes representation

$$\log m(x) = \log f(x|\theta) + \log \pi(\theta) - \log \pi(\theta|x),$$

which holds for any value of θ. When $\theta = (\theta_1, \theta_2)$, and when both $\pi(\theta_1|\theta_2, x)$ and $\pi(\theta_2|\theta_1, x)$ are available in closed form, *including the normalization constant*, the Rao–Blackwellization argument of Section 6.3.4 gives an approximation to the marginal posterior $\pi(\theta_1|x)$ as

$$\hat{\pi}(\theta_1|x) = \frac{1}{T} \sum_{t=1}^{T} \pi(\theta_1|\theta_2^{(t)}, x),$$

where the $\theta_2^{(t)}$'s are generated from by a Gibbs sampling algorithm. (Notice that the choice of the partition of θ in (θ_1, θ_2) is dictated by the availability of both $\pi(\theta_1|\theta_2, x)$ and $\pi(\theta_2|\theta_1, x)$.) Chib's (1995) approximation of $\log m(x)$ is then

$$\log f(x|\theta) + \log \pi(\hat{\theta}) - \log \pi(\hat{\theta}_2|\hat{\theta}_1, x) - \log \hat{\pi}(\hat{\theta}_1|x),$$

where $\hat{\theta} = (\hat{\theta}_1, \hat{\theta}_2)$ is an approximation of the MAP estimator of θ, for instance. When both conditional densities are not available in closed form, or when one of them misses its normalizing constant, Chib (1995) suggests an extension to more blocks in the partition, but this is much more costly (Exercise 7.23).

The major improvement brought by MCMC techniques to model selection is, however, their ability to deal with *variable dimension models*, that is, with models $\mathcal{M}_k$ on different parameter sets, with no intersection and possibly different dimensions.

Example 7.3.4 (Example 7.1.2 continued) The dimension of the parameter space for a k component normal mixture is $3k - 1$, accounting for the constraint

$$\sum_{\ell=1}^{k} p_{k\ell} = 1.$$

If the prior distribution on k is a Poisson $\mathcal{P}(\lambda)$ distribution, the parameter space is infinite dimensional, since k is unbounded. ∥

While, from a model-choice point of view, the difficulty is mainly in computing the posterior probability of model $\mathcal{M}_k$, $\pi(\mu = k|x)$, there are more fundamental problems related to this representation, the first one being the very notion of model parameter, which can be written as a sequence $(\theta_1, \ldots, \theta_k, \ldots)$ or as a couple (k, θ_k). There are also measure-theoretic difficulties in the representation of the prior density for a direct sum of spaces. Obviously, the corresponding MCMC samplers are much harder to construct.

A first solution, due to Carlin and Chib (1995), consists in *saturating the model*, that is, in considering all models at once: for a finite range of models $\mathcal{M}_k$ $(k = 1, \cdots, K)$ with corresponding priors $\pi_k(\theta_k)$, and prior weights ϱ_k, the parameter space is

$$\Theta = \{1, \ldots, K\} \times \prod_{k=1}^{K} \Theta_k$$

and, if μ denotes the model indicator, the posterior distribution is

$$\pi(\mu, \theta_1, \ldots, \theta_K|x) \propto \varrho_\mu f_\mu(x|\theta_\mu) \prod_{k=1}^{K} \pi_k(\theta_k).$$

Since

$$m(x|\mu = j) = \int f_j(x|\theta_j)\pi(\theta_1, \ldots, \theta_K|\mu = j) \, d\theta = \int f_j(x|\theta_j)\pi_j(\theta_j) \, d\theta_j$$

does not depend on the $\pi_k(\theta_k)$'s for $k \neq i$, Carlin and Chib (1995) propose to use *pseudo-priors* $\tilde{\pi}_k(\theta_k|\mu = j)$ to simulate the parameters θ_k on steps when $k \neq j$. Their method is then implemented as a Gibbs sampler on $(\mu, (\theta_1, \ldots, \theta_K))$, where μ is generated from

$$P(\mu = j|x, \theta_1, \ldots, \theta_K) \propto \varrho_j f_j(x|\theta_j)\pi_j(\theta) \prod_{k \neq j} \tilde{\pi}_k(\theta_k|\mu = j).$$

The authors point out that the method works better when these pseudo-priors are close to the true posteriors, but there is always some danger in the calibration of the pseudo-priors that some important parts of the parameter spaces Θ_k may be omitted. The major drawback of Carlin and Chib's (1995) method is that it requires simulation for all models at every stage, which is costly when K is large. In addition, it cannot be implemented when K is infinite.

Example 7.3.5 (Carlin and Chib (1995)) On a dataset of 42 pine trees, the grain (strength) y_i is regressed against either the wood density x_i, or a modified (resin adapted) density, z_i. Two contending models are

$$\mathcal{M}_1 : y_i = \alpha + \beta x_i + \sigma \varepsilon_i$$

and
$$\mathcal{M}_2 : y_i = \gamma + \delta z_i + \tau \varepsilon_i \,,$$
where both $(\alpha, \beta, \sigma^2)$ and (γ, δ, τ^2) are associated with (empirical Bayes) conjugate priors:

$$\binom{\alpha}{\beta} , \binom{\gamma}{\delta} \sim \mathcal{N} \left(\binom{3000}{185} , \begin{bmatrix} 10^6 & 0 \\ 0 & 10^4 \end{bmatrix} \right) , \quad \sigma^2, \tau^2 \sim \mathcal{IG}(a, b) \,,$$

with (a, b) such that the mean and standard deviation of σ^2 and τ^2 are 300^2. (Notice that, in a realistic Bayesian analysis, the effect of this prior modeling should be evaluated by a robustness analysis. See Section 3.6.)

The pseudo-priors are chosen in this case as the priors for σ^2 and τ^2, and some vague conjugate priors on (α, β) and (γ, δ):

$$\begin{array}{llll}
\alpha | \mu = 2 & \sim & \mathcal{N}(3000, 5^2) , & \beta | \mu = 2 & \sim & \mathcal{N}(185, 12^2) , \\
\gamma | \mu = 1 & \sim & \mathcal{N}(3000, 43^2) , & \delta | \mu = 1 & \sim & \mathcal{N}(185, 9^2) .
\end{array}$$

In order to force visits to the model $\mathcal{M}_1$, the authors used disproportionate weights, $\varrho_1 = .9995$ and $\varrho_2 = .0005$. (This appears to be quite a common feature in the pseudo-prior approach, in order to compensate for the possibly poor choice of the pseudo-priors.)

The result is that B_{21} is estimated by 4420 (after correction for the weights), with a (simulation) confidence interval of $(4353, 4487)$. (The confidence interval is simply deduced from the binomial variance on the posterior probability $P(\mu = 1|x)$.) The model $\mathcal{M}_2$ can therefore be safely chosen as the appropriate model.   ‖

Example 7.3.6 (Example 7.1.2 continued) For the galaxy mixture model, if we only consider the case of 3 (model $\mathcal{M}_1$) versus 4 (model $\mathcal{M}_2$) components, Carlin and Chib (1995) use a complete data model as in Section 6.4, by creating allocations z_i^k $(i = 1, \cdots, n, \ k = 1, 2)$. As in Example 7.3.5, using preliminary runs on both models, the pseudo-priors on the parameters are conjugate distributions matching the posterior estimates for both models, while the pseudo-priors for the z_i^k's, when $\mu \neq k$, are derived from the observed frequencies. The authors evaluate the Bayes factor to be 0.5153, with a standard error of 0.0146, thus concluding against the four-component model. (The authors also mention that the result may be altered *in favor* of the four-component model simply by modifying the prior on the weights.)   ‖

7.3.4 Reversible jump MCMC

For models with variable dimensions, Green (1995) proposes another type of *saturation* technique, at a more local level than Carlin and Chib's (1995). The idea at the core of this technique, given two models $\mathcal{M}_1$ and $\mathcal{M}_2$ of possibly different dimensions, is to remove the difference in the dimensions of models $\mathcal{M}_1$ and $\mathcal{M}_2$ by supplementing the corresponding parameters θ_1 and θ_2 by auxiliary variables $u_{1\to2}$ and $u_{2\to1}$ such that

$$(\theta_1, u_{1\to2}) \text{ and } (\theta_2, u_{2\to1})$$

are in bijection:

(7.3.6) $$(\theta_2, u_{2\to1}) = \Psi_{1\to2}(\theta_1, u_{1\to2}) \,.$$

If θ_1 is distributed from $\pi_1(\theta_1)$ and $u_{1\to2}$ is generated from $g_{1\to2}(u)$, the distribution of (7.3.6) is given by

$$\pi_1(\theta_1) g_{1\to2}(u_{1\to2}) \left| \frac{\partial \Psi_{1\to2}(\theta_1, u_{1\to2})}{\partial(\theta_1, u_{1\to2})} \right|^{-1}$$

by the Jacobian formula. Now, if we want (7.3.6) to be distributed from $\pi_2(\theta_2) g_{2\to1}(u_{2\to1})$, the Metropolis–Hastings acceptance probability is

$$\min\left(\frac{\pi_2(\theta_2) g_{2\to1}(u_{2\to1})}{\pi_1(\theta_1) g_{1\to2}(u_{1\to2})} \left| \frac{\partial \Psi_{1\to2}(\theta_1, u_{1\to2})}{\partial(\theta_1, u_{1\to2})} \right|, 1 \right) \,.$$

In opposition to Carlin and Chib's (1995) technique, this approach only considers local moves between two models: the other θ_j's and the auxiliary variables $u_{i\to j}$ are not explicitly used outside moves from $\mathcal{M}_i$ to $\mathcal{M}_j$.

Obviously, the theory behind *reversible jump MCMC* is more advanced than the justification given above, if only because it requires a deeper understanding of the joint measure on $(\theta_2, u_{2\to1})$ and $(\theta_1, u_{1\to2})$, which must satisfy a *detailed balance condition* as in (6.3.1). We refer the reader to Green (1995) for details. The main issue, however, is that, given a probability $\varrho_{i\to j}$ of choosing model $\mathcal{M}_j$ when in model $\mathcal{M}_i$, the acceptance probability of a move is indeed given by

(7.3.7) $$\min\left(\frac{\varrho_j \varrho_{j\to i} \pi_j(\theta_j) g_{j\to i}(u_{j\to i})}{\varrho_i \varrho_{i\to j} \pi_i(\theta_i) g_{i\to j}(u_{i\to j})} \left| \frac{\partial \Psi_{i\to j}(\theta_i, u_{i\to j})}{\partial(\theta_i, u_{i\to j})} \right|, 1 \right) \,,$$

where $(\theta_j, u_{j\to i}) = \Psi_{i\to j}(\theta_i, u_{i\to j})$. The algorithm can then be completed with additional steps within a given model $\mathcal{M}_i$, or about hyperparameters that are not model-dependent.

As discussed in Robert and Casella (2004, Section 11.2), the freedom allowed by the reversible jump algorithm to move between many models of different dimensions has generated many applications, which are by no means restricted to the model-choice setting. In connection with Example 7.1.2, Richardson and Green (1997) devised a reversible jump algorithm for normal mixtures, with the conclusion that the number of components for the galaxy dataset should be four. We present below the corresponding algorithm for a mixture of exponential distributions, developed in Gruet, Philippe and Robert (1999). (See also Robert, Rydén and Titterington (1999b) for a generalization to hidden Markov models.)

Example 7.3.7 For a mixture of exponential distributions

$$\sum_{j=1}^{k} p_{jk}\mathcal{E}xp(\lambda_{jk}),$$

the reversible jump algorithm can be restricted to moves between adjacent models, that is, between model $\mathcal{M}_k$ and models $\mathcal{M}_{k+1}$ and $\mathcal{M}_{k-1}$. These moves are quite open, in the sense that a component can be added (or removed) at random, as long as there is symmetry in the *up* and *down* moves. For instance, a birth of component $k + 1$ may be proposed by generating $(p_{(k+1)(k+1)}, \lambda_{(k+1)(k+1)})$ from a prior $\varpi_{k+1}(p, \lambda)$, assuming independent identical priors on all components. The transformation is then

$$(p_{1(k+1)}, \ldots, p_{k(k+1)}) = ((1 - p_{(k+1)(k+1)})p_{1k}, \ldots, (1 - p_{(k+1)(k+1)})p_{kk})$$
$$(\lambda_{1(k+1)}, \ldots, \lambda_{(k+1)(k+1)}) = (\lambda_{1k}, \ldots, \lambda_{kk}, \lambda_{(k+1)(k+1)}).$$

The Jacobian of this transform is therefore $(1 - p_{(k+1)(k+1)})^k$ and the probability to accept the birth is

$$\min\left(\frac{\varrho_{k+1}(1 - p_{(k+1)(k+1)})^k \pi_{k+1}(p_{1(k+1)}, \ldots, p_{(k+1)(k+1)},}{\varrho_k \pi_k(p_{1k}, \ldots, p_{kk}, \lambda_{1k}, \ldots, \lambda_{kk})}\right.$$
$$\left.\frac{\lambda_{1(k+1)}, \ldots, \lambda_{(k+1)(k+1)})}{\varpi_{k+1}(p_{(k+1)(k+1)}, \lambda_{(k+1)(k+1)})}, 1\right),$$

if the probabilities of choosing a birth (move to $\mathcal{M}_{k+1}$) or a death (move to $\mathcal{M}_{k-1}$) are the same.

The move from $\mathcal{M}_k$ to $\mathcal{M}_{k+1}$ considered in Gruet et al. (1999) is a *split* of a randomly chosen component j in such a way that the new component parameters $(p_{j(k+1)}, p_{(j+1)(k+1)}, \lambda_{j(k+1)}, \lambda_{(j+1)(k+1)})$ satisfy the moment condition

(7.3.8) $p_{jk} = p_{j(k+1)} + p_{(j+1)(k+1)}$
$$p_{jk}\lambda_{jk} = p_{j(k+1)}\lambda_{j(k+1)} + p_{(j+1)(k+1)}\lambda_{(j+1)(k+1)},$$

the opposite move being a *merge* of two components j and $j + 1$ according to (7.3.8). This proposal can be equivalently represented as a generation of $u_1, u_2 \sim \mathcal{U}([0, 1])$, with $p_{j(k+1)} = u_1 p_{jk}$ and $\lambda_{j(k+1)} = u_2 \lambda_{jk}$, and the Jacobian is then

$$\frac{\partial \Psi_{k \to k+1}(p_{jk}, \lambda_{jk}, u_1, u_2)}{\partial(p_{jk}, \lambda_{jk}, u_1, u_2)} = p_{jk}/(1 - u_1).$$

Figure 7.3.4 presents some condensed analysis of the performances of the reversible jump algorithm, when analyzing a dataset on hospital stays, with a posterior mode for k at 4. The allocation map on the lower right graph represents the successive allocations of the observations by grey levels: it shows that the mixing properties of the chain are quite good, since no pattern emerges. (See Gruet et al. (1999) for more details.)   ‖

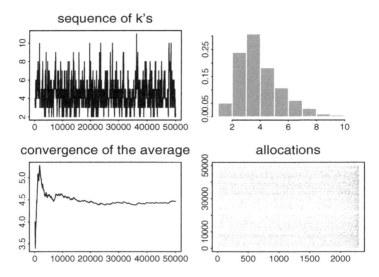

Figure 7.3.1. *Sequence of values $k^{(t)}$ simulated by reversible jump, with corresponding histogram (upper right), convergence of the empirical mean (lower left) and sequence of allocations to the components (lower right) for 50,000 iterations (Source: Gruet, Philippe and Robert (1999).)*

Notice that, in both schemes described in Example 7.3.7, there is no auxiliary variable $u_{k \to (k-1)}$ for the downward moves. This setting often occurs when one model is a completion of the other, although additional auxiliary variables can be introduced for potential improvements in speed.

A related technique has been proposed in Ripley (1987), Grenander and Miller (1994), Phillips and Smith (1996) and Stephens (2000), based on birth and death continuous time processes. See Note 7.8.2.

7.4 Model averaging

A natural approach to model uncertainty, from the Bayesian point of view, is to include all models $\mathcal{M}_k$ under consideration for future decisions, thus bypassing the model-choice step. As proposed in Raftery, Madigan and Volinsky (1996), this solution escapes the usual underestimation of uncertainty resulting from choosing model $\mathcal{M}_{k_0}$, say, at the model-selection stage, and thereafter ignoring the uncertainty about that choice in the subsequent steps.

Obviously, this perspective is not appropriate to all settings: sometimes, the decision maker or the statistician must select a model, as for instance in scientific inference, or they must eliminate some superfluous covariates in variable selection because of prohibitive sampling costs (Section 7.5). In addition, model averaging seems to infringe the parsimony requirement, in

the sense that, since all models are included into the (super) model, this
generates an inflation of parameters and, given that most cases involve
Monte Carlo or MCMC algorithms, implies the generation and storage of
a vast number of MCMC samples. This is the case with Example 7.1.2 for
instance.

In this approach, given a sample $\mathbf{x} = (x_1, \ldots, x_n)$, the predictive distribution is obtained by averaging over all models

$$
\begin{aligned}
f(y|x) &= \int_\Theta f(y|\theta)\pi(\theta|\mathbf{x})d\theta \\
&= \sum_k \int_{\Theta_k} f_k(y|\theta_k)\pi(k, \theta_k|\mathbf{x})d\theta_k \\
&= \sum_k p(\mathcal{M}_k|\mathbf{x}) \int f_k(y|\theta_k)\pi_k(\theta_k|\mathbf{x})\, d\theta_k \,,
\end{aligned}
$$

where Θ denotes the overall parameter space defined in (7.2.1).

Although this approach faces most of the difficulties encountered in Section 7.2, including the computation of many integrals and the simulation on a parameter space Θ, which is a sum of spaces with different dimensions, the removal of the decision step on the model label μ alleviates some of these difficulties. For instance, the fact that the collection of models is possibly infinite (or simply too large, as in variable selection) is not an impediment, in the sense that an MCMC algorithm that explores the space Θ will bypass models with very small probabilities $P(\mathcal{M}_i|\mathbf{x})$.

The issue here is rather at the modeling level, as already discussed in Section 7.2.1: when facing many models, the choice of the prior probabilities $\pi(k)$ is paramount, but difficult to formalize and justify. For instance, in the setting of variable selection (Section 7.5), the models in contention can be represented by vectors of indicator variables,

$$
\mathcal{M}_k : (\delta_{k1}, \cdots, \delta_{kd}), \qquad \delta_{kj} \in \{0, 1\} \,,
$$

where d is the number of potential covariates. Madigan and Raftery (1991) suggest using

$$
\pi(k) \propto \prod_{j=1}^d \left\{ \varrho_j^{\delta_{kj}} (1 - \varrho_j)^{1 - \delta_{kj}} \right\} \,,
$$

where ϱ_j denotes the prior probability that *variable j has an effect*. A drawback to this choice is that covariates are included in a model independently from one another, which only makes sense if they are independent, a risky assumption in most cases. The standard alternative of putting equal weights on all models is also open to criticism: besides the obvious drawback that it does not work if the number of models is infinite, it is awkward to use for nested models, that is, when some models are special cases of others, as in variable selection.

As mentioned above, MCMC techniques such as reversible jump or jump diffusions (Note 7.8.2) naturally solve the difficulty of exploring a large number of models by avoiding those with very small probabilities, assuming that the corresponding algorithms have a proper convergence behavior. Madigan and Raftery (1994) devise another technique, called *Occam's Window*.[5] They suggest that only models such that

$$\frac{\max_k P(\mathcal{M}_k|x)}{P(\mathcal{M}_\ell|x)} \leq C$$

should be considered, that is, models sufficiently likely compared with the most likely model. They propose in addition to exclude models $\mathcal{M}_\ell$, such that there exists a submodel $\mathcal{M}_h \subset \mathcal{M}_\ell$ such that

$$\frac{P(\mathcal{M}_h|x)}{P(\mathcal{M}_\ell|x)} \geq 1.$$

This reduction in the number of models can only be implemented if the number of models is sufficiently small, though, and Clyde (1999) points out the possible bias in the resulting probabilities resulting from this pruning.

In the case of variable selection in normal regression, $y \sim \mathcal{N}(X\beta, \sigma^2 I)$, that is,

$$y_t = \sum_{j=1}^{J} \beta_j x_{jt} + \sigma\varepsilon_t \qquad t = 1, \cdots, T,$$

with orthogonal regressors

$$X^t X = \text{diag}(x_j' x_j),$$

Clyde (1999) proposes to use priors of the form

$$\beta_j \sim \mathcal{N}(0, c_j^2 \gamma_j), \qquad \gamma_j \sim \mathcal{B}(p_j),$$

where the γ_j's act like 0-1 indicators for the jth regressor to be present in the model. Then, under Madigan and Raftery's (1991) prior,

$$(7.4.1) \qquad \pi(\gamma_1, \ldots, \gamma_J | y, \sigma) = \prod_{j=1}^{J} \varrho_j^{\gamma_j} (1 - \varrho_j)^{\gamma_j},$$

[5] William d'Occam or d'Ockham (*circa* 1290–*circa* 1349) was a English theologian (and a Franciscan monk) from Oxford who worked on the bases of empirical induction and, in particular, posed the principle later called *Occam's razor*, which excludes a plurality of reasons for a phenomenon if they are not supported by some experimentation (see Adams (1987)). This principle, *Pluralitas non est ponenda sine neccesitate* (or *Entities are not to be multiplied without necessity*), is often invoked as a *parsimony principle* to choose the simplest among two equally possible explanations, and its use is recurrent in the Bayesian literature (see, for instance, Jeffreys (1961, §6.12) or Jefferys and Berger (1992)). However, we refrain from using this notion because it does not provide a working principle, and is therefore open to misappropriation. In other words, the call to *Occam's razor* does not provide further justification for a given method. (At a more anecdotal level, Umberto Eco's *The Name of the Rose* borrows from Occam to create the character William of Baskerville.)

with

$$\varrho_j = \frac{O_j(y,\sigma)}{1 + O_j(y,\sigma)}$$

and

$$O_j(y,\sigma) = \frac{p_j}{1 - p_j}\left(\frac{x_j'x_j + \sigma^2/c_j^2}{\sigma^2/c_j^2}\right)^{-1/2}$$
$$\exp\left\{\frac{(\hat{\beta}_j x_j'/\sigma^2)^2}{2(x_j'x_j/\sigma^2 + 1/c_j^2)}\right\},$$

which means that the γ_j's are a posteriori independent, and also that the probability of a given submodel can be easily computed, including the most likely submodel (Exercise 7.25). This is not the case with George and McCulloch's (1997) alternative modeling

$$\beta_j \sim \mathcal{N}\left(0, c_j^2\gamma_j + [c_j^2/100](1 - \gamma_j)\right).$$

If σ^2 is unknown, Clyde (1999) uses a single prior $\sigma^2 \sim \mathcal{IG}(\alpha,\beta)$ for all models

$$\pi(\sigma^2|\gamma, y) \sim \mathcal{IG}(\hat{\alpha}, \hat{\beta})$$

and approximates the posterior weights of the different models with either a "plug-in" estimate, that is, replacing σ^2 with an estimate $\hat{\sigma}^2$ in (7.4.1), or with a Rao–Blackwell average.

While such results are interesting, they are quite difficult to extend to other settings, such as generalized linear models, where further approximations are necessary. Besides, the orthogonality assumption is quite restrictive because usual regressors are never orthogonal, and using a orthogonalizing transform such as principal components is not satisfactory because decision-makers are often interested in the values of the coefficients β_j. Furthermore, it violates the principle that common parameters should be treated as different entities in different models, since the β_j's are the same in every model where they appear.

7.5 Model projections

We now present a different approach[6] to model selection, that has been developed by Goutis and Robert (1998) and applied to variable selection in Dupuis and Robert (2001). The approach is based on *projections* of a full model $f(y|\theta)$ on submodels, represented as restrictions on θ, and on the assessment of the approximation error owing to these restrictions. It applies in particular to *variable selection*, that is, to the determination of a set of covariates, among a larger set of covariates.

---

[6] This section contains more specialized material which, while being at the same level of difficulty as the remainder of this chapter, can be skipped on a first reading.

Example 7.5.1 In a study into the influence of dietary factors on breast cancer, the following covariates are considered by Raftery and Richardson (1996):

age	age of first pregnancy
menopausal age	age at the end of study
age at menarche	body mass index
parity	fat intake (total)
alcohol consumption	fat intake (saturated)
family history of breast cancer	history of benign BC

Since the observation takes values in $\{0,1\}$, which corresponds to the no cancer–cancer dichotomy, the models under consideration are *logistic regressions* involving some or all of the covariates ($i = 1, \cdots, 2^{12}$):

$$\mathcal{M}_i : P(y_j = 1 | x_j) = \frac{\exp[\alpha_i + \beta_i^t x_j^{(i)}]}{1 + \exp[\alpha_i + \beta_i^t x_j^{(i)}]},$$

where $x^{(i)}$ denotes the coordinates of x corresponding to the binary decomposition of i. For instance, model $\mathcal{M}_5$ is associated with $i = 5 = 0 \cdots 0101$ and therefore $x^{(5)} = (x_{10}, x_{12})$. ‖

One of the main differences of the projection approach, compared with the usual model-choice axioms of Section 7.2, is that it only requires the construction of a prior $\pi(\theta)$ on the full model $f(x|\theta)$, instead of a prior for every submodel, and that it accommodates improper priors. In fact, as we will see below, it derives the submodel weights and priors from the original prior distribution π, thus escaping marginalization and projection paradoxes with subspaces of different dimensions.

Given a restriction $\theta \in \Theta_0$, Goutis and Robert's (1998) perspective is to consider that this restriction is acceptable if

(7.5.1) $$d(f(\cdot | \theta), \Theta_0) < \epsilon,$$

where d is a divergence measure and

$$\begin{aligned} d(f(\cdot | \theta), \Theta_0) &= d(f(\cdot | \theta), f(\cdot | \theta^\perp)) \\ &= \inf_{\theta_0 \in \Theta_0} d(f(\cdot | \theta), f(\cdot | \theta_0)). \end{aligned}$$

The parameter $\theta^\perp$ is then the projection of the parameter θ onto the submodel. From this perspective, model choice is seen as an assessment of the difference between the true model and a more parsimonious model. It thus provides a realistic modeling of experimental purposes where exact nullity is rarely at stake and escapes parameterization issues, since the representation (7.5.1) is parameter-free. In addition, this formalism only requires a prior distribution on the full parameter θ, since the projection parameter $\theta^\perp$ is a transform of θ. The posterior probability that (7.5.1) holds can therefore be computed with this prior distribution only. Notice that this is

Table 7.5.1. *Values of the parameters for Kullback–Leibler divergences of ϵ in the case of Bernoulli, Poisson and normal distributions (Source: Goutis and Robert (1998).)*

ϵ	0	0.01	0.05	0.1	0.25	0.5	1	2	∞
$\mathcal{B}(p)$	0.5	0.57	0.65	0.71	0.81	0.9	0.96	0.99	1
$\mathcal{P}(\lambda)$	1	1.15	1.35	1.52	1.88	2.36	3.15	4.5	∞
$\mathcal{N}(\mu, 1)$	0	0.14	0.32	0.45	0.71	1	1.41	2	∞

different from deriving the prior distribution on $\theta^\perp$ by projection of $\pi(\theta)$ and then using a standard Bayes factor, as in McCulloch and Rossi (1992) (Exercise 7.32).

There are many choices for the divergence measure d, but a natural choice is the *Kullback–Leibler* pseudo-distance

$$d(f, g) = \int \log\left(\frac{f(z)}{g(z)}\right) f(z)\, dz,$$

already used in (2.5.6). As argued in Bernardo and Smith (1994), there are many good reasons for using this measure, including information theory, scoring rule, transitivity and additivity properties, as well as connection with exponential families and generalized linear models.

Similarly, the factor ϵ in (7.5.1) can be chosen in many ways. For instance, it can be calibrated on simple distributions to evaluate the proper scale of ϵ, as in Table 7.5 (Exercise 7.28). In the case of a single restriction, ϵ can be derived from the (proper) prior π, so that

$$P^\pi(d(f(\cdot\,|\theta), f(\cdot\,|\theta^\perp)) \le \epsilon) = 1/2.$$

This was done in the setting of mixtures by Mengersen and Robert (1996), but the $1/2$ value can be criticized as giving a false impression of objectivity (since the resulting value depends on π). At last, in variable selection settings and related embedded models, there exists a minimal or most rudimentary model, f_0, which corresponds to a regression with only one intercept, and which may scale ϵ as $\epsilon = \varrho d(f, f_0)$. (Dupuis and Robert (2001) call $d(f, f_0)$ the *maximal loss of explanatory power*.)

Once d and ϵ have been chosen, the method can be implemented by either computing the posterior probability $P^\pi(d(f(\cdot\,|\theta), f(\cdot\,|\theta^\perp)) \le \epsilon)$, or by deriving the posterior expectation of $d(f(\cdot\,|\theta), f(\cdot\,|\theta^\perp))$. For the variable-selection problem, when y is regressed on a vector x of p covariates, the issue is complicated by the fact that this distance is computed by integrating over the joint distribution of (x, y), leading to the term (Exercise 7.30)

$$\mathbb{E}_x[d(f(\cdot|x, \theta), f_\mathcal{A}(\cdot|x_\mathcal{A}, \theta^\perp))],$$

where $\mathcal{A} \subset \{1, \dots, p\}$ and $x_\mathcal{A}$ is the corresponding subset of covariates. Because the distribution of the covariate vector x is rarely known, this quantity can be approximated by the empirical average

$$\frac{1}{n}\sum_{i=1}^{n} \mathbb{E}_y \left[\log \left(\frac{f(y|x_i, \theta)}{g(y|x_{i\mathcal{A}}, \theta^{\perp})} \right) \bigg| x_i \right].$$

Besides the usual computational difficulties of approximating posterior expectations or posterior probabilities, an additional issue with variable selection is that, given p potential covariates, there exist 2^p (or $2^p - 1$) models in competition. For large values of p, a complete exploration of all the models is thus impossible. As detailed in Note 7.8.3, some transitivity and additivity properties of the Kullback–Leibler distance allow for a faster pruning of the submodel tree: when selecting among all subsets $\mathcal{A}$ of covariates such that

$$d(\mathrm{M}_g, \mathcal{M}_{\mathcal{A}}) = \mathbb{E}_x[d(f(y|x, \alpha), g(y|x_{\mathcal{A}}, \alpha^{\perp}))] < \epsilon,$$

the submodel with the smallest cardinal, that is, with the smallest number of covariates, it is possible to evaluate this cardinal by a *downward step*—starting from the full model and descending in the submodel tree by eliminating one covariate at a time, the farthest away from M_g, till the distance is too large—and an *upward step*—starting from the constant model and adding one covariate at a time, the closest from M_g, till the distance is smaller than ϵ—and to check afterwards that no other model with the same cardinality p_0 is closer to the full model. This last step may be very costly, though, being of order $\binom{p}{p_0}$ (Exercise 7.29).

Example 7.5.2 (Example 7.5.1 continued) For a flat prior on the regression parameters (α, β), Dupuis and Robert (2001) obtained the results listed in Table 7.5 via this variable-selection procedure (with $\varrho = 0.9$ in the scaling of ϵ). The same submodel 100111111001 was selected at the three stages of this procedure. From the list of explanatory variables given in Example 7.5.1, this means that the selected submodel excludes fat intakes from the important explanatory variables. Notice the agreement between the approach based on the posterior expected distance *(column 3)* and the posterior probability that the distance is less than ϵ *(column 4)*. ‖

Although this approach has the appeal of relying on a loss function to select among submodels, and of eliminating the improper prior problem, it also has some defects. Firstly, in cases such as variable selection when the number of submodels under consideration is large, it may require enormous computing time to implement. Secondly, the determination of the bound ϵ is open to criticism: for instance, why should a fixed proportion of the distance $d(f, f_0)$ be pertinent for the decision-maker? How should it depend on the number of observations? Another drawback to this approach is that it requires a full (or reference) model, thus only works with nested models. Following an approach routinely used in econometrics (see, e.g., Gouriéroux and Monfort (1996)), Goutis and Robert (1998) propose to extend the method to the general case by creating an *encompassing model*, but this is a

Table 7.5.2. *Submodels examined by the variable selection procedure for the breast-cancer dataset. The result of each step is shown in bold, $d(M_g, \mathcal{M}_\mathcal{A})$ denotes the expected Kullback–Leibler divergence between the full model and the projection on the subset of covariates $\mathcal{A}$, and $P(M_\mathcal{A})$ is the posterior probability that the distance $d(M_g, \mathcal{M}_\mathcal{A})$ is less than ϵ. (Source: Dupuis and Robert (2001).)*

step	subset $\mathcal{A}$	$d(M_g, \mathcal{M}_\mathcal{A})$ ($\times 740$)	$P(M_\mathcal{A})$
1	101111111111	0.508	0.98
	101111111011	1.146	0.96
	100111111011	1.800	0.94
	100111111001	**2.726**	**0.91**
2.	000000010000	21.78	0.29
	000010010000	16.97	0.45
	100010010000	13.81	0.55
	100010011000	10.61	0.66
	100010011001	7.601	0.75
	100011011001	5.224	0.83
	100111011001	3.736	0.88
	100111111001	**2.726**	**0.91**
3.	111111110000	8.170	0.73
	111111001010	13.72	0.55
	111100111010	8.349	0.73
	110011111010	5.988	0.81
	001111111010	9.215	0.70
	111110011001	4.542	0.85
	111101011001	4.761	0.85
	111011011001	3.91	0.87
	110111011001	3.265	0.89
	101111011001	3.017	0.90
	011111011001	5.895	0.81
	100111111001	**2.726**	**0.91**
	100111011101	3.109	0.899
	100011111101	3.826	0.88
	111011010011	5.284	0.83
	110110110011	6.04	0.80
	101101110011	5.9	0.81
	101011011011	3.576	0.88
	100111011011	2.77	0.91
	101010111011	5.08	0.84
	011001111011	9.346	0.70
	100110011111	4.151	0.87
	100101011111	4.224	0.86
	100011011111	3.787	0.88

difficult issue given that this encompassing model is not the true model, and is thus of little interest for the decision-maker. Besides, the encompassing model can be defined in many ways, which lead to different answers. For instance, one can oppose arithmetic and geometric averages (Exercise 7.34).

Example 7.5.3 (Example 7.1.1 continued) Given that both Poisson and negative binomial models involve the terms

$$\frac{\lambda^y}{y!},$$

where $\lambda = p/(1-p)$ in the negative binomial case, a possible encompassing model is

$$f(y|\lambda, m, \alpha) \propto \frac{1}{y!}\lambda^y e^{-\alpha\lambda} \left[\frac{m!}{(m-y)!}\frac{1}{(1+e^\lambda)^m}\right]^{1-\alpha} \qquad 0 \le \alpha \le 1,$$

since the Poisson model corresponds to $\alpha = 1$ and the negative binomial model to $\alpha = 0$. This density is in fact the geometric average of both densities, but the normalizing constant, which depends on (λ, m, α), is unknown. A more manageable alternative is to use an arithmetic average, that is, the mixture

$$p\,\mathcal{P}(\lambda) + (1-p)\,\mathcal{N}eg\left(m, \frac{e^\lambda}{1+e^\lambda}\right) \qquad 0 \le p \le 1.$$

$\|$

7.6 Goodness-of-fit

We conclude this chapter with a short introduction to the Bayesian approach to goodness-of-fit, which is, in a way, the most challenging of all model choice problems. Indeed, when considering the question *Is $\mathcal{M}_0$ compatible with x?* or *Does f belong to the family $\{f_\theta; \theta \in \Theta\}$?* there is no alternative to $\mathcal{M}_0$. For instance, in Example 7.1.1, if we only consider the Poisson model, to assess whether this model is compatible with the dataset is difficult, given that, when it is *not* compatible, the model is left undefined![7]

The difficulty here with the Bayesian paradigm seems to be that it cannot provide an answer about the validity of the model without moving "outside" the model, that is, without working in a larger structure (*a meta-model*) that includes the model under consideration as a special case. In fact, this is not a shortcoming of the Bayesian approach, but rather a result of the poor formulation of the question. That the Bayesian paradigm fails to produce an answer to an poorly posed problem does not imply that other systems

---

[7] The frequentist approach to this problem bypasses the difficulty by working only under the null hypothesis. The standard χ^2 test, for instance, is based on the χ^2 approximation, which only works when the model is the true model. *If it is not*, the χ^2 statistic diverges to infinity, but nothing can be said about its distribution for a given sample size.

that produce such an answer, such as the χ^2 test, are in any way validated! Rather, by requiring the construction of an alternative model, the Bayesian paradigm clarifies the issue and formalizes the definition of the meta-model that includes the model under study.

This issue being clarified, there are many ways of defining this alternative model $\mathcal{M}_1$, unless strong prior information is available. For instance, $\mathcal{M}_1$ can be an *embedding model* of $\mathcal{M}_0$. But, as discussed in Section 7.5, there is no single choice of an embedding model, that is, there is no such thing as the smallest (or the most natural) embedding model, besides the trivial $\mathcal{M}_0$ itself! Neyman (1937) proposed the exponential family extension

$$f_1(x|\theta, \varphi) \propto f(x|\theta) \exp\left\{-\varphi \log \frac{f(x|\theta)}{f(x|\hat{\theta}(x))}\right\}, \qquad \varphi \geq 0,$$

where $\hat{\theta}(x)$ is the maximum likelihood estimator (provided it is defined), but other hierarchical extensions are possible. Moreover, embedding models give very restricted representations of the alternative, given that the alternative in a goodness-of-fit problem should be f *is not in* $\mathcal{M}_0$.

Another approach to the definition of the alternative model avoids such restrictions, since it uses a *nonparametric* representation of the alternative. We mentioned in Note 1.8.2 some standard techniques in Bayesian non-parametrics, including the Dirichlet process prior and their generalizations, mixtures, or wavelets. For illustration, we consider here the *orthogonal polynomial* representation of Verdinelli and Wasserman (1998). See Castro et al. (1999) for an alternative in the discrete case (Exercise 7.37).

The model under consideration $\mathcal{M}_0 : x \sim f(x|\theta)$, $\theta \in \Theta$, can be expressed as

$$\mathcal{M}_0 : x = F^-(u|\theta), \qquad \theta \in \Theta, \qquad u \sim \mathcal{U}([0,1]),$$

where $F^-(\cdot|\theta)$ is the generalized inverse of the c.d.f. of $f(\cdot|\theta)$ (Exercise 7.38). We can thus write $\mathcal{M}_0$ as a special case of

$$\mathcal{M}_1 : x = F^-(u|\theta), \qquad \theta \in \Theta, \qquad u \sim g(u|\psi), \qquad \psi \in \mathcal{S},$$

where $g(\cdot|\psi)$ is a distribution on $[0,1]$, which contains as a special case, the uniform distribution, $g(u|\psi_0) = 1$, and $\mathcal{S}$ is an infinite dimensional space. This reparameterization of the model means that we can work on distributions on $[0,1]$, rather than on a general space, and that we want to devise a test of uniformity (conditional upon θ).

There are many possible choices for the infinite dimensional family of distributions $g(\cdot|\psi)$. While a possibility is to consider mixtures of beta densities,

$$g(u|\psi) = \varrho_0 + (1 - \varrho_0) \sum_{j=1}^{+\infty} \varrho_j \frac{u^{\alpha_j}(1-u)^{\beta_j}}{K(\alpha_j, \beta_j)},$$

as in Petrone and Wasserman (2000), which can be estimated by reversible jump techniques, Verdinelli and Wasserman (1998) propose instead using

Legendre polynomials on $[0, 1]$,

$$\phi_j(x) = \frac{1}{2^j j!} \frac{d^j}{dx^j} (x^2 - 1)^j$$

to write the density as

$$g(u|\psi) \propto \exp \left\{ \sum_{j=1}^{+\infty} \psi_j \phi_j(u) \right\}.$$

(See Barron (1988, 1998) and Lenk (1991) for details.) The null model $\mathcal{M}_0$ then corresponds to $\psi_1 = \ldots = \psi_p = \ldots = 0$.

The prior distribution on (θ, ψ) is chosen such that θ and ψ are independent, with a reference prior on θ. This independence assumption is not innocuous, given that θ has the same prior under $\mathcal{M}_0$ and $\mathcal{M}_1$, but is not identifiable under $\mathcal{M}_1$ (Exercise 7.39). The ψ_j's are then modeled as independent normal rv's,

$$\psi_j \sim \mathcal{N}(0, \tau_j^2),$$

with $\tau_j = \tau/2^j$ for consistency reasons (Barron (1988)), and τ is associated with a vague proper prior, $\pi(\tau)$.

The posterior distribution is then given by

$$(7.6.1) \qquad \pi(\theta, \psi, \tau | x_1, \cdots, x_n) \propto \prod_{i=1}^{n} f(x_i|\theta) g(u_i|\psi) \pi(\theta) \, \pi(\psi|\tau) \pi(\tau) \,,$$

where $u_i = F(x_i|\theta)$ (Exercise 7.40). It is obviously intractable, if only because of the dependence of the u_i's on θ. Simulation from $\pi(\theta, \psi, \tau | x_1, \cdots, x_n)$ can nonetheless be implemented via an MCMC algorithm using for instance the Gibbs steps

$$\theta | \psi, x_1, \cdots, x_n \sim \prod_{i=1}^{n} f(x_i|\theta) g(u_i|\psi) \pi(\theta) \,,$$

$$\psi | \tau, \theta, x_1, \cdots, x_n \sim \prod_{i=1}^{n} g(u_i|\psi) \pi(\psi|\tau) \,,$$

$$\tau | \psi \sim \pi(\psi|\tau) \pi(\tau) \,.$$

Some additional Metropolis–Hastings steps are necessary for the generation of θ and ψ.

Once the posterior distribution is approximated, Verdinelli and Wasserman (1998) suggest using a Bayes factor

$$B_{01} = \frac{\displaystyle \int \prod_{i=1}^{n} f(x_i|\theta) \pi(\theta) d\theta}{\displaystyle \int \prod_{i=1}^{n} f(x_i|\theta) g(F(x_i|\theta)|\psi) \pi(\theta, \psi, \tau) d\theta d\psi d\tau}$$

to assess whether $\mathcal{M}_0$ provides a good fit to the data. (They show in addition that the procedure is *consistent*, that is, that B_{01} goes to 0 almost surely if $\mathcal{M}_0$ is the wrong model, and to ∞ in probability if it is the right model.) An alternative evaluation is to notice that $\mathcal{M}_0$ also corresponds to $\tau = 0$, and to use a standard point-null test based on the MCMC sample.

7.7 Exercises

Section 7.1.1

7.1 The *deviance* associated with a model is simply the log-likelihood taken at the maximum likelihood estimator (McCullagh and Nelder (1989)). In the setting of Example 7.1.1, compute the maximum likelihood estimator's $\hat{\lambda}$ and $(\hat{m}, \hat{p})$ and compare both deviances.

7.2 In the setting of Example 7.1.2, show that a mixture of k components can be represented as a mixture of $k + 1$ components by either putting a component weight to 0, or by equating the mean and variance parameters of the $(k+1)$-th component to the same parameters for one of the first kth components. Relate this multiplicity with the non-identifiability property of mixtures raised in Note 6.6.6.

7.3 For the setting of Example 7.1.3, write down the marginal distributions of $y_i = (y_{i1}, \ldots, y_{i7})$ when integrating out the random-effects. Is it possible to obtain closed-form estimates with conjugate priors?

Section 7.2.1

7.4 Given two models $\mathcal{M}_1 : x \sim f_1(x|\theta_1, \gamma)$ and $\mathcal{M}_2 : x \sim f_2(x|\theta_2, \gamma)$, show that, if the prior distribution is of the form

$$\pi(\theta_1, \theta_2, \gamma) = \pi_1(\theta_1|\gamma)\pi_2(\theta_2|\gamma)\pi_0(\gamma),$$

with both π_1 and π_2 proper, the Bayes factor B_{12}^π does not depend on the normalizing constant of π_0 if π_0 is improper.

7.5 In the setting of Example 7.2.1, assume T_t is distributed from a uniform $\mathcal{U}_{[0,\bar{T}]}$ distribution, and that $\beta_{21} \sim \mathcal{N}(0, \tau^2)$.

    a. Compute the marginal model of y_{it} by integrating out the term $\beta_{21}T_t$ in $\mathcal{M}_2$.

    b. Deduce the prior distribution on the parameters of $\mathcal{M}_1$ if $\mathcal{M}_2$ is the true model and $(\beta_{20}, b_{2i}, \sigma_2) \sim \pi(\beta_{20}, b_{2i}, \sigma_2)$.

7.6 *(Barbieri, Liseo and Petrella (1999))* Consider a model $f(x|\varphi, \psi)$, $(\varphi, \psi) \in \Phi \times \Psi$, such that there exists $\psi^* \in \overline{\Psi}$ that satisfies

$$\lim_{\psi \to \psi^*} f(x|\varphi, \psi) = f^*(x|\psi^*),$$

that is, the limiting distribution does not depend on φ.

    a. Show that this setting occurs for the *linear calibration model*,

$$z_1 \sim \mathcal{N}(\psi, 1), \qquad z_2 \sim \mathcal{N}(\phi\psi, 1).$$

    b. If $\pi(\varphi, \psi)$ is a proper prior with a point mass in ψ^*, show that

$$\pi(\varphi|x) = \pi(\varphi|\psi^*)\pi(\psi^*|x) + \pi(\varphi|\psi \neq \psi^*, x) \int_{\psi \neq \psi^*} \pi(\psi|x)\, d\psi.$$

c. If $H_0 : \varphi = \varphi^0$ is to be tested against $H_1 : \varphi \neq \varphi^0$, show that

$$B_{01} = \pi(\psi^*|x) + \frac{m(x|\psi \neq \psi^*)}{\pi(\varphi^0)} \int_{\psi \neq \psi^*} \pi(\psi|x)\, d\psi$$

assuming that π also has a point mass at φ^0.

d. Deduce that the Bayes factor is heavily influenced by the prior modeling on ψ^*, no matter what φ^0 is.

[*Note:* Gleser and Hwang (1987) study such models from a frequentist point of view and show that a confidence interval at a level α on an unbounded function of φ has infinite volume with positive probability.]

Section 7.2.2

7.7 (Berger and Pericchi (2001)) Consider the normal linear model $\mathcal{M}_2$

$$y = \alpha\mathbf{1} + z_1\beta_1 + z_2\beta_2 + \epsilon, \qquad \epsilon \sim \mathcal{N}_n(0, \sigma^2\mathbf{I}_n),$$

where $\beta_1 \in \mathbb{R}^k$ and $\beta_2 \in \mathbb{R}^p$, and where the z_i's are centered and orthogonal, that is, $z_1^t z_2 = 0$. The submodel $\mathcal{M}_1$ is associated with $\beta_2 = 0$.

a. Show that, under the priors

$$\pi_1(\alpha, \beta_1, \sigma) = 1/\sigma \qquad \text{and} \qquad \pi_2(\alpha, \beta_1, \sigma, \beta_2) = h(\beta_2|\sigma)/\sigma,$$

where $h(\beta_2|\sigma)$ is the Cauchy $\mathcal{C}_p(0, z_2^t z_2/n\sigma^2)$ density, the Bayes factor B_{12} cannot be computed in closed form.

b. Show that, under the G-prior of Zellner (1986), which associates with model $\mathcal{M}_1 : y = X\beta + \epsilon, \epsilon \sim \mathcal{N}_n(0, \sigma^2\mathbf{I}_n)$, the prior

$$\pi(\sigma) = 1/\sigma, \qquad \pi(\beta|\sigma) \propto \exp\{-\beta^t X^t X\beta/2g\sigma^2\},$$

the associate marginal is available in closed form.

c. Show that, if $\mathcal{M}_0$ is associated with $\beta = 0$ and if the maximum likelihood estimator $\hat{\beta}$ goes to infinity, the Bayes factor B_{01} goes to $(1 + g)^{(k-n)/2}$ (where k is the dimension of β under model $\mathcal{M}_1$). Conclude about the appeal of the G-prior in this setting.

7.8 *(Exercise 7.6 cont.) In the case of the linear calibration model,

a. Show that the Jeffreys prior is

$$\pi^J(\varphi, \psi) \propto |\psi|.$$

b. When testing $H_0 : \varphi = \varphi^0$, with $\pi_0(\psi) \propto 1$, show that the *fractional Bayes factor* with fraction $0 < b < 1$ (see (5.2.10)) is

$$B_{01}^F = b^{-1/2} \exp\left\{-\frac{1-b}{2} \frac{(z_1 - z_2\varphi^0)^2}{1 + \varphi_0^2}\right\}.$$

c. Show that the *arithmetic intrinsic Bayes factor* (see (5.2.7)) is

$$B_{01}^A = \sqrt{2} \exp\left\{-\frac{1 - 0.5}{2} \frac{(z_1 - z_2\varphi^0)^2}{1 + \varphi_0^2}\right\}.$$

d. Extend to the case of n observations.

7.9 *(Exercise 7.8 cont.)

a. Show that the reference prior is

$$\pi^R(\varphi, \psi) \propto \frac{1}{\sqrt{1 + \varphi^2}}.$$

b. Show that the fractional Bayes factor is

$$B_{01}^F = b^{-1/2} \exp\left\{1 - b2 \frac{[(z_1^2 - z_2^2)(1 - \varphi^0)^2 - 4z_1 z_2 \varphi_0]}{1 + \psi_0^2}\right\}$$

$$\frac{I_0(b(z_1^2 + z_2^2)/4)}{I_0((z_1^2 + z_2^2)/4)},$$

where I_0 is the modified Bessel function (Exercise 4.35).

7.10 *In the setting of Example 7.2.2,

a. Show that the final expression for B_{12}^π is correct.

b. Show that the limit of B_{12}^π varies if α/β goes to 0 as both α and β go to 0, or if α/β^N goes to $c > 0$, when $x < N$.

7.11 Give the marginal distribution of x_1 if $x_1, \ldots, x_n$ is a sample from a two-component normal mixture such that there are at least two observations in each component.

Section 7.2.3

7.12 Show that, for the comparison of two linear models $\mathcal{M}_1$ and $\mathcal{M}_2$ with k_1 and k_2 regressors, respectively, and n observations, under the prior $\pi_j(\beta_j) = \sigma_j^{-1-q_j}$ $(j = 1, 2)$, the BIC writes down as

$$B_{12} = \left(\frac{R_2}{R_1}\right)^{n/2} n^{(k_2 - k_1)/2},$$

where the R_j's are the residual sums of squares.

7.13 (Exercise 7.8 cont.) In the case of the linear calibration model, under the Jeffreys prior, show that Schwartz's criterion gives almost the same result as the fractional Bayes factor with fraction $b = 0$ in the exponent.

Section 7.2.4

7.14 If $f(\cdot|\theta)$ belongs to an exponential family, show that the effective number of parameters p_D is always positive.

7.15 *(Spiegelhalter et al. (1998)) In the setting of Example 7.2.5,

a. Show that, for the saturated model where the θ_i's are independent with flat priors, $p_D = p$ and that the Bayesian deviance is equal to $2p$.

b. Show that the Bayesian deviance associated with the pooled model, $\theta_i = \theta$ for all i's, is given by (7.2.7).

c. Show that (7.2.8) holds.

d. Assume that $\theta_i \sim \mathcal{N}(\mu, \tau^2)$ with τ known and $\pi(\mu) = 1$. Show that

$$p_D = \sum_{i=1}^p \varrho_i + \sum_{i=1}^p \varrho_i(1 - \varrho_i) \bigg/ \sum_{i=1}^p \varrho_i$$

and that the Bayesian deviance is equal to

$$\text{DIC} = \tau^{-2} \sum_{i=1}^{p} \varrho_i (1 - \varrho_i)(y_i - \bar{y})^2 + p_D \,,$$

where $\varrho_i = \sigma_i^2 \tau^2 / (\sigma_i^2 + \tau^2)$ and $\bar{y} = \sum_i \varrho_i y_i / \sum_i \varrho_i$.

7.16 Detail the MCMC implementation for the three models of Example 7.2.6. (*Hint:* The simulation can be handled by BUGS.)

7.17 *(Spiegelhalter et al. (1998)) Consider a general linear model

$$y \sim \mathcal{N}(A\theta_1, \Sigma_1) \,, \qquad \theta_1 \sim \mathcal{N}(B\theta_2, \Sigma_2) \,.$$

a. Show that the posterior distribution of θ_1 is of the form $\mathcal{N}(\bar{\theta}_1, \Psi)$ and specify $\bar{\theta}_1$ and Ψ.

b. Show that $\mathbb{E}[D(\theta)|y] = D(\bar{\theta}_1) + \text{tr}(A'\Sigma_1^{-1}A\Psi)$ and deduce that $p_D = \text{tr}(A'\Sigma_1^{-1}A\Psi)$.

c. Extend to the case when θ_2 is random and $\pi(\theta_2) = 1$.

Section 7.3.1

7.18 Check whether the expectation of

$$\frac{1}{T} \sum_{t=1}^{T} \frac{h(\theta^{(t)})}{f(x|\theta^{(t)})\pi(\theta^{(t)})} \,,$$

when the $\theta^{(t)}$'s are distributed from $\pi(\theta|x)$, is equal to $m(x)$, whatever the probability density h.

7.19 *(Chen and Shao (1997)) Consider two densities, $\pi_1(\theta) = c_1 \tilde{\pi}_1(\theta)$ and $\pi_2(\theta) = c_2 \tilde{\pi}_2(\theta)$, on the same parameter space Θ.

a. If π is a density on Θ, give sufficient conditions on the support of π for

$$\varrho = \frac{c_2}{c_1} = \frac{\mathbb{E}^\pi[\tilde{\pi}_1(\theta)/\pi(\theta)]}{\mathbb{E}^\pi[\tilde{\pi}_2(\theta)/\pi(\theta)]} \,.$$

b. Show that the asymptotic variance of the estimator of

$$\varrho^{US} = \frac{\sum_{i=1}^{n} \tilde{\pi}_1(\theta_i)/\pi(\theta_i)}{\sum_{i=1}^{n} \tilde{\pi}_2(\theta_i)/\pi(\theta_i)} \,,$$

where the θ_i's are i.i.d. from π, is

$$\varrho^2 \mathbb{E}^\pi \left\{ \frac{\pi_1(\theta)}{\pi(\theta)} - \frac{\pi_2(\theta)}{\pi(\theta)} \right\}^2 \,.$$

c. Assuming that

$$\varrho^{-2} \mathbb{E}^\pi[(\varrho^{US} - \varrho)^2] = \frac{1}{n} \mathbb{E}^\pi \left[\frac{\{\pi_1(\theta) - \pi_2(\theta)\}^2}{\pi^2(\theta)} \right] + \text{o}\left(n^{-1}\right) \,,$$

show that the best importance density π is

$$\pi_0(\theta) \propto |\pi_1(\theta) - \pi_2(\theta)| \,,$$

if

$$\int |\pi_1(\theta) - \pi_2(\theta)| \, d\theta < \infty \,.$$

[*Note:* Torrie and Valleau (1977) call this method *umbrella sampling.*]

Section 7.3.2

7.20 Given two densities $\pi_1(\theta) = c_1\tilde{\pi}_1(\theta)$ and $\pi_2(\theta) = c_2\tilde{\pi}_2(\theta)$ on the same parameter space Θ,

a. For an arbitrary function h, express $\mathbb{E}^{\pi_2}[h(\theta)\tilde{\pi}_1(\theta|x)]$ as an integral in terms of π_1 and π_2.

b. Deduce the equality (7.3.4).

7.21 *Chen et. al. (2000) introduce the *relative mean square error*

$$\mathcal{E}(r,\hat{r}) = \frac{\mathbb{E}[\hat{r} - r]}{r}$$

as a measure of the performances of an estimator $\hat{r}$ of the constant ratio r.

a. Show that, if $n = n_1 + n_2$ and if n_1/n_2 goes to ϱ as n goes to infinity,

$$\mathcal{E}(r, B_{12}^S) = \frac{1}{n\varrho(1-\varrho)} \left[\frac{\int \pi_1(\theta)\pi_2(\theta)\{\varrho\pi_1(\theta) + (1-\varrho)\pi_2(\theta)\}h^2(\theta)\,d\theta}{\left(\int \pi_1(\theta)\pi_2(\theta)\,d\theta\right)^2} \right]$$

for the estimator (7.3.5), where the dependence on x is omitted to simplify the expression. (*Hint:* Use the δ-method.)

b. Deduce that the optimal choice of h is

$$h^*(\theta) \propto \frac{1}{\varrho\pi_1(\theta) + (1-\varrho)\pi_2(\theta)}.$$

7.22 For the three link functions of Example 7.3.1, identify the latent variable structure z identifying y as the indicator $\mathbb{I}_{z \leq x^t\beta}$.

Section 7.3.3

7.23 Consider a posterior distribution $\pi(\theta_1, \theta_2, \theta_3|x)$ such that the three full conditional distributions $\pi(\theta_1|\theta_2, \theta_3, x)$, ... and $\pi(\theta_3|\theta_1, \theta_2, x)$ are available.

a. Show that

$$\begin{aligned} \log m(x) \;=\; & \log f(x|\hat{\theta}) + \log \pi(\hat{\theta}) - \log \pi(\hat{\theta}_3|\hat{\theta}_1, \hat{\theta}_2, x) \\ & - \log \pi(\hat{\theta}_2|\hat{\theta}_1, x) - \log \pi(\hat{\theta}_1|x). \end{aligned}$$

b. Show that $\pi(\theta_1|x)$ can be approximated by

$$\hat{\pi}(\theta_1|x) = \frac{1}{T} \sum_{t=1}^{T} \pi(\theta_1, \theta_2^{(t)}, \theta_3^{(t)}|x),$$

where the $(\theta_1^{(t)}, \theta_2^{(t)}, \theta_3^{(t)})$'s are generated by Gibbs sampling.

c. Show that $\pi(\theta_2|\hat{\theta}_1, x)$ can be approximated by

$$\hat{\pi}(\theta_2|\hat{\theta}_1, x) = \frac{1}{T} \sum_{t=1}^{T} \pi(\theta_2|\hat{\theta}_1, \theta_3^{(t)}, x)$$

where the $(\theta_2^{(t)}, \theta_3^{(t)})$'s are generated by Gibbs sampling from the conditional distributions $\pi(\theta_2|\hat{\theta}_1, \theta_3^{(t-1)}, x)$ and $\pi(\theta_3|\hat{\theta}_1, \theta_2^{(t)}, x)$, that is, with θ_1 being kept equal to $\hat{\theta}_1$.

d. Extend to the case where p full conditional distributions are available and evaluate the computing cost of this approximation method.

Section 7.3.4

7.24 In the setting of Example 7.3.7, show that the Jacobians of both the birth and the split moves are given by

$$\left(1 - p_{(k+1)(k+1)}\right)^k \qquad \text{and} \qquad p_{jk}/(1 - u_1).$$

Section 7.4

7.25 For the prior distributions proposed by Clyde (1999),

a. Show that the posterior distribution of $(\gamma_1, \ldots, \gamma_J)$ conditional on σ is given by (7.4.1).

b. Deduce that the most likely submodel corresponds to the regressors X_j with weights ϱ_j larger than $1/2$.

7.26 *(George and Foster (1999)) For a normal regression model

$$y = \beta_1 x_1 + \ldots + \beta_p x_p + \sigma \epsilon, \qquad \epsilon \sim \mathcal{N}(0, I),$$

we denote γ the index of one of the 2^p submodels, q_γ the corresponding number of covariates, X_γ the corresponding matrix of regressors, $\hat{\beta}_\gamma$ the least-squares estimate and s_γ^2 the sum of squares $\hat{\beta}_\gamma' X_\gamma' X_\gamma \hat{\beta}_\gamma$.

a. Consider the prior distributions

$$\beta_\gamma | \sigma, \gamma, c \sim \mathcal{N}_{q_\gamma}\left(0, c\sigma^2 \left(X_\gamma' X_\gamma\right)^{-1}\right), \qquad \pi(\gamma | \omega) = \omega^{q_\gamma} (1 - \omega)^{p - q_\gamma}.$$

Identify this prior with Raftery and Madigan (1994) prior.

b. Show that

$$\pi(\gamma | y, \sigma, c, \omega) \propto \exp\left[\frac{c}{2(1 + c)} \{s_\gamma^2 / \sigma^2 - F(c, \omega) q_\omega\}\right]$$

where

$$F(c, \omega) = \frac{1 + c}{c} \left(2 \log \frac{1 + w}{w} + \log(1 + c)\right).$$

c. Deduce that the integrated posterior $\pi(\gamma | y, \sigma, c, \omega)$ is an increasing function of $s_\gamma^2 / \sigma^2 - F(c, \omega) q_\omega$.

d. Conclude that, with an appropriate choice of (c, ω), the log-posterior can be any standard model-choice criterion, ranging from AIC (with $F(c, \omega) = 2$), to BIC (with $F(c, \omega) = \log n$), to Foster and George's (1994) RIC (with $F(c, \omega) = 2 \log p$).

Section 7.5

7.27 Show that the Kullback–Leibler divergence between a normal $\mathcal{N}(0, 1)$ and a normal $\mathcal{N}(\mu, \sigma^2)$ distribution is

$$\log \sigma + \frac{\mu^2 + 1}{2\sigma^2} - \frac{1}{2}.$$

Extend to the Kullback–Leibler divergence between a normal $\mathcal{N}(\mu_0, \sigma_0^2)$ and a normal $\mathcal{N}(\mu, \sigma^2)$ distribution by the appropriate change of scale.

7.28 For each of the following distributions, show that the equality holds:

(i) Bernoulli $B(p)$

$$d(f(\cdot\,|p_0), f(\cdot\,|p)) = p_0 \log \frac{p_0}{p} + (1 - p_0) \log \frac{1 - p_0}{1 - p} \, ;$$

(ii) Poisson $\mathcal{P}(\lambda)$

$$d(f(\cdot\,|\lambda_0), f(\cdot\,|\lambda)) = \lambda - \lambda_0 + \lambda_0 \log \frac{\lambda_0}{\lambda} \, ; \quad \text{and}$$

(iii) Normal $\mathcal{N}(\mu, 1)$

$$d(f(\cdot\,|\mu_0), f(\cdot\,|\mu)) = (\mu - \mu_0)^2/2 \, .$$

7.29 When considering a variable-selection problem with p covariates,

  a. Show that the number of submodels is $2^p - 1$ if all models have a constant term and $2^p - 2$ otherwise.

  b. Show that the number of submodels with p_0 covariates is $\binom{p}{p_0}$.

  c. Using Stirling's approximation, show that this number is also of order 2^p when $p_0 = p/2$.

7.30 In a model-choice problem where $(x, y) \sim g(x|\alpha) f(y|x, \theta)$,

  a. Show that, for the Kullback–Leibler divergence,

$$d\left(g(\cdot|\alpha)f(\cdot|\cdot, \theta), g(\cdot|\alpha')f(\cdot|\cdot, \theta')\right) = d(g(\cdot|\alpha), g(\cdot|\alpha')) + \mathbb{E}_\alpha \left[d(f(\cdot|x, \theta), f(\cdot|x, \theta')\right] \, ,$$

where the expectation is taken for $x \sim g(x|\alpha)$.

  b. Deduce that, if the submodel puts restrictions on θ only, for example $\varphi(\theta) = 0$, the projection of (α, θ) is $(\alpha, \theta^\perp)$ where $\theta^\perp$ is the solution of

$$\arg \min_{\theta'; \, \varphi(\theta')=0} \mathbb{E}_\alpha \left[d(f(\cdot|x, \theta), f(\cdot|x, \theta')\right] \, .$$

7.31 In the case of a normal linear regression model, $y \sim \mathcal{N}(x'\beta, \sigma^2)$,

  a. Show that, if z is a subvector of x, the Kullback–Leibler divergence between $\mathcal{N}(x'\beta, \sigma^2)$ and $\mathcal{N}(z'\gamma, \sigma^2)$ is $||x'\beta - z'\gamma||^2/2\sigma^2$, conditional upon x.

  b. Deduce that the projection $\beta^\perp$ is given by $\beta^\perp = (zz')x'\beta$.

7.32 (Exercise 7.31 cont.) Assume β is distributed from a conjugate prior $\mathcal{N}(\beta_0, \Sigma)$. Give the induced prior distribution of $\beta^\perp$. What happens in the case of a flat prior on β?

7.33 In the setting of Example 7.5.3, examine whether the normalizing constant of the geometric average of the Poisson and negative binomial distributions, $f(y|\lambda, m, \alpha)$, can be computed.

7.34 When comparing two models $\mathcal{M}_1$ and $\mathcal{M}_2$, with densities $f_1(\cdot|\theta_1)$ and $f_2(\cdot|\theta_2)$ both from an exponential family,

  a. Show that the geometric average

$$f_1(\cdot|\theta_1)^\alpha \, f_2(\cdot|\theta_2)^{1-\alpha}$$

still belongs to an exponential family.

  b. Show that, if $(i = 1, 2)$

$$f_i(y|\theta_i) = h_i(y) \exp\{\theta_i \cdot \varphi_i(y) - \psi_i(\theta_i)\} \, ,$$

$(\varphi_1(y), \varphi_2(y))$ is a sufficient statistic for the geometric average.

c. Deduce that, if $(\varphi_1(y), \varphi_2(y))$ is of full rank, the dimension of this family (see Definition 3.3.2) is the sum of the dimensions of f_1 and f_2.

d. In the special case when $\mathcal{M}_1$ is the exponential $\mathcal{E}xp(\theta_1)$ model and $\mathcal{M}_2$ is the half-normal $\mathcal{N}^+(0, 1/\theta_2)$ model, show that the geometric average model is the half-normal distribution

$$\mathcal{N}^+ \left(-\frac{\alpha\theta_1}{(1-\alpha)\theta_2}, \frac{1}{(1-\alpha)\theta_2} \right),$$

and give its normalizing constant.

Section 7.6

7.35 Examine the Neyman (1937) exponential family extension when $f(x|\theta)$ is the density of (i) a Poisson $\mathcal{P}(\theta)$, (ii) an exponential $\mathcal{E}xp(\theta)$ and (iii) a normal $\mathcal{N}(\theta, 1)$ distribution. In the three cases, determine whether the normalizing constant can be computed.

7.36 Given a density

$$f(y|\theta) = h(y) \exp\{\theta \cdot \varphi(y) - \psi(\theta)\}$$

from an exponential family of dimension d (see Definition 3.3.2), show that its Neyman extension still belongs to an exponential family of dimension $d + 1$.

7.37 *(Castro, Conigliani and O'Hagan (1999)) Consider a multinomial model

$$\mathbf{r} = (r_0, \ldots, r_k) \sim \mathcal{M}_{k+1}(n; \alpha_0, \ldots, \alpha_k),$$

with $\alpha = (\alpha_1, \ldots, \alpha_k)$.

a. If we denote $(0 \leq b \leq 1)$

$$q_2(\mathbf{r}; b) = \frac{\int f(\mathbf{r}|\alpha)\pi_2(\alpha)\, d\alpha}{\int f^b(\mathbf{r}|\alpha)\pi_2(\alpha)\, d\alpha},$$

show that, under the improper prior $\pi_2(\alpha) = 1/\alpha_1 \ldots \alpha_k$,

$$q_2(\mathbf{r}; b) = \frac{\Gamma(bn)}{\Gamma(n)} \prod_{j=0}^{k} \frac{\Gamma(r_j)}{\Gamma(br_j)},$$

if all the r_j's are positive. (Recall that, if one r_j is 0, the posterior is not defined.)

b. If the constraint on the α_j's is that

$$\alpha_j = \binom{k}{j} \mu^j (1 - \mu)^{k-j}, \qquad 0 < \mu < 1,$$

that is, if one wants to test whether the underlying model is truly binomial, show that, under the prior $\pi_1(\mu) = 1/\mu(1 - \mu)$,

$$
\begin{aligned}
q_1(\mathbf{r}; b) &= \frac{\int f(\mathbf{r}|\alpha(\mu))\pi_1(\mu)\, d\mu}{\int f^b(\mathbf{r}|\alpha(\mu))\pi_1(\mu)\, d\mu} \\
&= \frac{B(r, kn - s_r)}{B(br, b(kn - s_r))} \left[\prod_{j=0}^{k} \binom{k}{j}^{r_j} \right]^{1-b},
\end{aligned}
$$

where $s_r = r_1 + \ldots + kr_k$ and $B(a, b)$ is the beta $\mathcal{B}e(a, b)$ normalizing constant (see Appendix A).

Table 7.7.1. *Number of women in a queue of 10 persons in the London Underground (Source: Hoaglin, Mosteller and Tukey (1985).)*

Women	0	1	2	3	4	5	6	7	8	9	10
Occurrences	1	3	4	23	25	19	18	5	1	1	0

    c. Show that the fractional Bayes factor associated with the constraint in h is $B_{12}^F = q_1(\mathbf{r}; b)/q_2(\mathbf{r}; b)$.

    d. Apply b. to the data in Table d..

    e. If the constraint on the α_j's is that the model is a Poisson model, $\alpha_j = e^{-\lambda}\lambda^j/j!$ $(j = 0, \ldots, k)$, show that, under the prior $\pi_1(\lambda) = \lambda^{-t}$,

$$
\begin{aligned}
q_1(\mathbf{r}; b) &= \frac{\int f(\mathbf{r}|\alpha(\lambda))\pi_1(\lambda)\,d\lambda}{\int f^b(\mathbf{r}|\alpha(\lambda))\pi_1(\lambda)\,d\lambda} \\
&= \frac{\Gamma(s_r - t + 1)b^{bs_r - t + 1}n^{s_r(b-1)}}{\Gamma(bs_r - t + 1)}\prod_{j=0}^{k}[j!]^{(b-1)r_j},
\end{aligned}
$$

where s_r is defined as in b..

    f. Show that, in the setting of e., intrinsic Bayes factors do not apply, unless cells are grouped together to obtain positive r_j's.

    g. Show that, for a continuous model, this strategy is the Bayesian equivalent of the χ^2 test and, therefore, that it suffers from the same drawback, the arbitrary grouping of observations in k cells.

7.38 Given F a cumulative distribution function in $\mathbb{R}$, the *generalized inverse* of F is defined as

$$F^-(u) = \inf\{x; F(x) \geq u\}$$

    a. Show that, if $u \sim \mathcal{U}([0,1])$, $F^-(u) \sim F$.

    b. Deduce a simulation technique for both the Cauchy and exponential distributions.

    c. How can you generalize this result for a multidimensional distribution?

7.39 In the setting of Verdinelli and Wasserman (1998), show that the parameter θ is not identifiable under the alternative model $\mathcal{M}_1$. (*Hint:* Show that, for every c.d.f. $F(x)$ and every θ, there exists ψ such that $F_\theta^- \circ G_\psi = F^-$.)

7.40 (Verdinelli and Wasserman (1998)) Show (7.6.1) by establishing that, under model $\mathcal{M}_1$,

$$
\begin{aligned}
x \sim h(x|\theta, \psi) &= g(F(x|\theta)|\psi)\frac{dF(x|\theta)}{dx} \\
&= g(F(x|\theta)|\psi)f(x|\theta).
\end{aligned}
$$

Note 7.8.1

7.41 Establish (7.8.1) by showing that

$$\int\int \frac{d}{d\lambda}\log\tilde{\pi}(\theta|\lambda)\pi(\theta|\lambda)\,d\lambda\,d\theta = -\int_{\lambda_1}^{\lambda_2}\frac{d}{d\lambda}c(\lambda)\,d\lambda.$$

7.42 *(**Exercise 7.41 cont.**) Show that the generalization of (7.8.1) to the multidimensional case is

$$\log(c(\lambda_2)/c(\lambda_1)) = \int_0^1 \mathbb{E}_{\lambda(t)} \left[\sum_{j=1}^k \frac{d\lambda_j(t)}{dt} \frac{\partial}{\partial \lambda_j} \log \tilde{\pi}(\theta|\lambda) \right] dt \,,$$

where $\lambda(t)$ is a continuous function from $[0,1]$ in Λ such that $\lambda(0) = \lambda_1$ and $\lambda(1) = \lambda_2$. Deduce the corresponding path sampler. (*Hint:* See Gelman and Meng (1998) for a detailed resolution.)

Note 7.8.2

7.43 In the setting of Example 7.8.1, give the reversible jump steps which correspond to the birth and death moves.

7.8 Notes

7.8.1 Path sampling

Gelman and Meng (1998) generalize bridge sampling to *path sampling* by considering the special case when both posteriors depend on hyperparameters λ_1 and λ_2,

$$\begin{aligned} \pi_1(\theta|x) &= \pi(\theta|\lambda_1) &= \tilde{\pi}(\theta|\lambda_1)/c(\lambda_1)\,, \\ \pi_2(\theta|x) &= \pi(\theta|\lambda_2) &= \tilde{\pi}(\theta|\lambda_2)/c(\lambda_2)\,. \end{aligned}$$

If those hyperparameters are real, with $\lambda_1 < \lambda_2$, the identity

$$(7.8.1) \qquad \log(c(\lambda_2)/c(\lambda_1)) = \mathbb{E}\left[\frac{1}{\pi_0(\lambda)} \frac{d}{d\lambda} \log \tilde{\pi}(\theta|\lambda) \right]\,,$$

when integrated against the density $\pi(\theta|\lambda)\pi_0(\lambda)$, holds for every density π_0 with support $[\lambda_1, \lambda_2]$ (Exercise 7.41).

The corresponding path sampling estimate of the logarithm of the Bayes factor is then

$$B_{12}^{PS} = \frac{1}{n} \sum_{i=1}^n \frac{\frac{d}{d\lambda} \log \tilde{\pi}(\theta_i|\lambda_i)}{\pi_0(\lambda_i)}\,,$$

with a formal optimal choice for π_0,

$$\pi_0(\lambda) \propto \sqrt{\mathbb{E}\left[\left(\frac{d}{d\lambda} \log \tilde{\pi}(\theta|\lambda) \right)^2 \Big| \lambda \right]}\,.$$

See Exercise 7.42 for an extension to the multidimensional case.

7.8.2 Jump processes

A technique similar to reversible jump has been proposed in the literature (see, e.g., Ripley (1987), Grenander and Miller (1994), or Phillips and Smith (1996)). It does apply in a very general framework, but has only been used, so far, for variable dimension problems, such as Stephens's (2000) approach to the setting of Example 7.1.2.

This method is based on *jump processes*: it generates a continuous-time jump process on the space (7.2.1), that is, a stochastic process $(\xi_t)_{t \in \mathbb{R}^+}$ that remains in a given state (i, θ_i) according to an exponential schedule, $T \sim \mathcal{E}xp(\varphi_i(\theta_i))$, where φ is called the *intensity* of the process, and then jumps to a new state j

with probability $q_{i\to j}$ and generates θ_j from a density $h_{i\to j}(\theta_j|\theta_i)$. Then, similar to the discrete-time setting (see (6.3.1)), if the parameters of the process, φ, q and h, satisfy a *detailed balance condition*

$$\pi(i,\theta_i)\varphi_i(\theta_i)q_{i\to j}h_{i\to j}(\theta_j|\theta_i) = \pi(j,\theta_j)\varphi_j(\theta_j)q_{j\to i}h_{j\to i}(\theta_i|\theta_j),$$

then $\pi(i,\theta_i)$ is the stationary distribution of this Markovian process. For instance, if $h_{i\to j}(\theta_j|\theta_i) = g_j(\theta_j)$ and $q_{i\to j} = 1/k$, where k is the number of states, the balance condition is

$$\pi(i,\theta_i)\varphi_i(\theta_i)g_j(\theta_j) = \pi(j,\theta_j)\varphi_j(\theta_j)g_i(\theta_i)$$

and a choice of intensity is $\varphi_i(\theta_i) \propto g_i(\theta_i)/\pi(i,\theta_i)$. (Because the intensity $\varphi_i(\theta_i)$ is the inverse of the average stay in (i,θ_i), this average stay is logically proportional to $\pi(i,\theta_i)$.)

In the particular case when the moves are restricted to neighboring states, that is, when $q_{i\to i+1} + q_{i\to i-1} = 1$ (with appropriate modifications at the endpoints), the process is called a *birth and death jump process*. It is then customary to write $\varphi_i(\theta_i) = \beta(\theta_i) + \delta(\theta_i)$, $\beta(\theta_i)$ being the *birth rate* and $\delta(\theta_i)$ the *death rate*, and to remove the parameters $q_{i\to j}$. The process stays in state (i,θ_i) an exponential $\mathcal{E}xp[\beta(\theta_i) + \delta(\theta_i)]$ time, then moves to state $(i+1,\theta_{i+1})$ with probability $\beta(\theta_i)/(\beta(\theta_i)+\delta(\theta_i))$, θ_{i+1} being generated from $K_i^+(\theta_{i+1}|\theta_i)$, and to state $(i-1,\theta_{i-1})$ otherwise, θ_{i-1} being generated from $K_i^-(\theta_{i-1}|\theta_i)$.

Example 7.8.1 (Example 7.1.2 continued) For the mixture example, the state labels i correspond to the numbers of components, a birth to the addition of one component and a death to the removal of one component. Then $\theta_i = (p_{1i},\ldots,p_{ii},\mu_{1i},\ldots,\mu_{ii},\sigma_{1i},\ldots,\sigma_{ii})$. In his implementation of the birth and death jump algorithm, Stephens (2000) generates new components from the prior distribution (where all components are i.i.d.) and chooses a fixed birth rate $\beta(\theta_i) = b$. The balance condition then reads as

$$(i+1)\beta(\theta_{i+1})L[(i+1,\theta_{i+1})|x_1,\ldots,x_n]\pi(i+1) = bL[(i,\theta_i)|x_1,\ldots,x_n]\pi(i),$$

where $L(\theta|x_1,\ldots,x_n)$ denotes the likelihood. (The coefficient $(i+1)$ owes to the fact that there are $(i+1)$ components, and thus $(i+1)$ possible removals.) If we denote by $\theta_i/(p_{\ell i},\mu_{\ell i},\sigma_{\ell i})$ the parameter of the $(i-1)$ component model where the component $(p_{\ell i},\mu_{\ell i},\sigma_{\ell i})$ has been removed, the birth and death algorithm can thus be written as follows.

When in state (i,θ_i),

1. Compute component death rates $(\ell = 1,\ldots,i)$
$$\beta_\ell(\theta_i) = \frac{L[(i-1,\theta_i/(p_{\ell i},\mu_{\ell i},\sigma_{\ell i}))|x_1,\ldots,x_n]}{L[(i,\theta_i)|x_1,\ldots,x_n]}$$
and take $\beta(\theta_i) = \sum_{\ell=1}^i \beta_\ell(\theta_i)$
2. Generate jump time as $T \sim \mathcal{E}xp(\beta(\theta_i) + b)$
3. At time T, choose removal of
$$(p_{\ell i},\mu_{\ell i},\sigma_{\ell i})|x_1,\ldots,x_n$$
with probability
$$\frac{\beta_\ell(\theta_i)}{\beta(\theta_i) + b}$$

Otherwise, create

$$\left(p_{(i+1)(i+1)}, \mu_{(i+1)(i+1)}, \sigma_{(i+1)(i+1)}\right)$$

from the prior distribution.

Notice that in Step 3, the new weight is generated from the marginal prior distribution of $p_{(i+1)(i+1)}$, which is a $\mathcal{Be}(i, 1)$ distribution if the prior on $(p_{1(i+1)}, \ldots, p_{(i+1)(i+1)})$ is a Dirichlet $\mathcal{D}_{i+1}(1, \ldots, 1)$ distribution.   ‖

Cappé, Robert and Rydén (2003) gives a reassessment of this technique, with the conclusion that there are little theoretical and practical differences with reversible jump.

7.8.3 Variable selection for generalized linear models

Here we present further developments on the variable selection technique presented in Section 7.5. Consider, thus, a general exponential family ($i = 1, \ldots, n$)

$$y_i | \theta_i \sim \exp\left[\varphi_i \{\theta_i y_i - \psi(\theta_i)\} + c(\varphi_i, y_i)\right]$$

with a *generalized linear model* structure, which relates the mean to a vector of covariates,

$$g(\psi'(\theta_i)) = x_i^t \beta.$$

In this setting, the Kullback–Leibler divergence is available in closed form, since

$$d(f(\cdot\,|\theta), f(\cdot\,|\theta_0)) = \sum_{i=1}^{n} \varphi_i \{\psi'(\theta_i)(\theta_i - \theta_i^0) - \psi(\theta_i) + \psi(\theta_i^0)\}$$

and the projection equations ($j = 1, \ldots, p$)

$$(7.8.2) \qquad \sum_{i=1}^{n} \varphi_i\, \psi'(\theta_i)\, \frac{\partial \theta_i^0}{\partial \beta_j} = \sum_{i=1}^{n} \varphi_i\, \psi'(\theta_i^0)\, \frac{\partial \theta_i^0}{\partial \beta_j},$$

are equivalent to the system of likelihood equations, a fact that facilitates their derivation.

In the case of a logit model,

$$P(y_i = 1 | x_i, \alpha) = 1 - P(y_i = 0 | x_i, \alpha) = \frac{\exp(\alpha^t x_i)}{1 + \exp(\alpha^t x_i)},$$

the projection $\alpha^\perp$ of α for the covariates z_i (which are subvectors of the x_i's) is, for instance, associated with β solution of

$$\sum_{i=1}^{n} \frac{\exp \beta^t z_i}{1 + \exp \beta^t z_i} z_i = \sum_{i=1}^{n} \frac{\exp \alpha^t x_i}{1 + \exp \alpha^t x_i} z_i,$$

which indeed provides a formal equivalence with the maximum likelihood estimator equations

$$\sum_{i=1}^{n} \frac{\exp \beta^t z_i}{1 + \exp \beta^t z_i} z_i = \sum_{i=1}^{n} y_i z_i.$$

A consequence of (7.8.2) is that the Kullback–Leibler projections are transitive in the sense that, if ω is a subvector of z, itself a subvector of x, we get

$$
\sum_{i=1}^{n} \frac{\exp \gamma^t \omega_i}{1 + \exp \gamma^t \omega_i} \omega_i \;=\; \sum_{i=1}^{n} \frac{\exp \beta^t z_i}{1 + \exp \beta^t z_i} \omega_i
$$

$$
\;=\; \sum_{i=1}^{n} \frac{\exp \alpha^t x_i}{1 + \exp \alpha^t x_i} \omega_i
$$

in the logit example. This means, in other words, that the projection γ of the projection β of α is the projection of α on the smaller subspace, a model-choice version of the *two projection theorem*. Notice also that the distance between these projections is additive:

$$
d(f(\cdot \,|\alpha), f(\cdot \,|\gamma)) = d(f(\cdot \,|\alpha), f(\cdot \,|\beta)) + d(f(\cdot \,|\beta), f(\cdot \,|\gamma)) \,.
$$

For the variable selection scheme exposed in Section 7.5, this means that, once a submodel has been rejected as being too far from the full model, all its submodels are also rejected. See Dupuis and Robert (2001) for more details.

CHAPTER 8

Admissibility and Complete Classes

You can turn the worse that comes to your advantage if you only think, his father has always said, and certainly Abell Cauthon was the best horse trader in the Two Rivers. (...) All because he thought about things from every side that there was.

Robert Jordan, *The Dragon Reborn, Book III of the Wheel of Time.*

8.1 Introduction

Chapters 1 through 3 repeatedly mentioned that the Bayes estimators are instrumental for the frequentist notions of optimality, in particular, for admissibility. This chapter provides a more detailed description of this phenomenon. In Section 8.2, we consider the performances of the Bayes and generalized Bayes estimators in terms of admissibility. Then, Section 8.3 studies Stein's sufficient condition in order to relate the admissibility of a given estimator with a sequence of prior distributions. The notion of *complete class* introduced in Section 8.4 is also fundamental because it provides a characterization of admissible estimators, or at least a substantial reduction in the class of acceptable estimators. We show that, in many cases, the set of the Bayes estimators constitutes a complete class and that, in other cases, it is necessary to include generalized Bayes estimators. From a more general, although non-Bayesian, perspective, Section 8.5 presents a method introduced by Brown (1971), and developed by Hwang (1982b), that provides necessary admissibility conditions. For a more technical survey of these topics, see Rukhin (1995).

8.2 Admissibility of Bayes estimators

8.2.1 General characterizations

Recall the two following results about the admissibility of (proper) Bayes estimators, already stated in Chapter 2 (Propositions 2.4.22 and 2.4.23):

Proposition 8.2.1 *If a Bayes estimator is unique, it is admissible.*

Proposition 8.2.2 *When the risk function is continuous in θ for every estimator δ, if π is equivalent to the Lebesgue measure on Θ, that is, is absolutely continuous with a positive density on Θ, then a Bayes estimator associated with π is admissible.*

On the contrary, if the support of π is different from the whole space, it is possible that an associated Bayes estimator is inadmissible. Similarly, the Bayes estimators will often be inadmissible when the Bayes risk is infinite.

Example 8.2.3 Consider a normal setting $x \sim \mathcal{N}(\theta, 1)$ with a conjugate prior $\theta \sim \mathcal{N}(0, \sigma^2)$. The posterior distribution is then $\mathcal{N}(\frac{\sigma^2}{\sigma^2+1}x, \frac{\sigma^2}{\sigma^2+1})$ and the Bayes estimator under quadratic loss is $\delta^\pi(x) = \frac{\sigma^2}{\sigma^2+1}x$, which is admissible, as shown in Corollary 8.2.14. On the contrary, if the quadratic loss is modified into

$$L_\alpha(\theta, \delta) = e^{\theta^2/2\alpha}(\theta - \delta)^2,$$

the corresponding Bayes estimator is inadmissible for α small enough. In fact, the formal generalized Bayes estimator associated with L_α is

$$\delta_\alpha^\pi(x) = \frac{\int_{-\infty}^\infty \theta e^{\theta^2/2\alpha} e^{-(\theta - \delta^\pi(x))^2(\sigma^2+1)/2\sigma^2} d\theta}{\int_{-\infty}^\infty e^{\theta^2/2\alpha} e^{-(\theta - \delta^\pi(x))^2(\sigma^2+1)/2\sigma^2} d\theta},$$

provided both integrals are finite. Since

$$\exp\left\{\frac{\theta^2}{2\alpha} - (\theta - \delta^\pi(x))^2 \frac{\sigma^2+1}{2\sigma^2}\right\}$$

$$= \exp\left\{-\frac{\theta^2}{2}(\frac{\sigma^2+1}{\sigma^2} - \frac{1}{\alpha}) + \delta^\pi(x)\theta\frac{\sigma^2+1}{\sigma^2} - \delta^\pi(x)^2\frac{\sigma^2+1}{2\sigma^2}\right\},$$

δ_α^π is defined for $\alpha > \frac{\sigma^2}{\sigma^2+1}$ and

$$\begin{aligned}\delta_\alpha^\pi(x) &= \frac{\sigma^2+1}{\sigma^2}\left(\frac{\sigma^2+1}{\sigma^2} - \alpha^{-1}\right)^{-1}\delta^\pi(x)\\ &= \frac{\alpha}{\alpha - \frac{\sigma^2}{\sigma^2+1}}\delta^\pi(x).\end{aligned}$$

The corresponding Bayes estimator is

$$r(\pi) = \int_{-\infty}^{+\infty} e^{\theta^2/2\alpha}e^{-\theta^2/2\sigma^2} d\theta,$$

that is, is infinite for $\alpha \leq \sigma^2$. Moreover, since

$$\begin{aligned}\frac{\alpha}{\alpha - \frac{\sigma^2}{\sigma^2+1}}\delta^\pi(x) &= \frac{\alpha}{\alpha - \frac{\sigma^2}{\sigma^2+1}}\frac{\sigma^2}{\sigma^2+1}x\\ &= \frac{\alpha}{\alpha\frac{\sigma^2+1}{\sigma^2} - 1}x,\end{aligned}$$

the Bayes estimator $\delta_\alpha^\pi(x)$ is of the form cx with $c > 1$ when

$$\alpha > \alpha\frac{\sigma^2 + 1}{\sigma^2} - 1,$$

that is, when $\alpha < \sigma^2$. And, in this case,

$$\begin{aligned} R(\theta, \delta_\alpha^\pi) &= \mathbb{E}_\theta[(cx - \theta)^2]e^{\theta^2/2\alpha} \\ &= \{(c-1)^2\theta^2 + c^2\}e^{\theta^2/2\alpha} > e^{\theta^2/2\alpha} \end{aligned}$$

implies that δ_α^π is inadmissible, since it is dominated by $\delta_0(x) = x$. However, δ_0 is also a Bayes estimator under L_α when $\alpha < \sigma^2$, since the Bayes risk is infinite. It is interesting to recognize that the limiting case $\alpha = \sigma^2$ leads to the admissible estimator $\delta_{\sigma^2}^\pi(x) = x$ with an infinite Bayes risk.   ‖

Example 8.2.4 Consider $y \sim \sigma^2\chi_p^2$. The conjugate prior distribution for σ^2 is the inverse gamma distribution $\mathcal{IG}(\nu/2, \alpha/2)$ (see Chapter 3) and $\pi(\sigma^2|y)$ is the distribution $\mathcal{IG}((\nu+p)/2, (\alpha+y)/2)$, leading to the following posterior expectation:

$$\delta_{\nu,\alpha}^\pi(y) = \mathbb{E}^\pi[\sigma^2|y] = \frac{\alpha + y}{\nu + p - 2}.$$

Consider $\nu = 2$. In this case, $\delta^\pi(y) = (y/p) + (\alpha/p)$. Since y/p is an unbiased estimator of σ^2, the estimators $\delta_{2,\alpha}^\pi$ are not admissible under square error (as $\alpha > 0$). The same result holds when $\nu < 2$. It is easy to check that the Bayes risk of δ^π is infinite in this case (see Lehmann (1983, p. 270)).   ‖

Example 8.2.5 The constant estimators $\delta_0(x) = \theta_0$ are the Bayes estimators corresponding to a Dirac mass prior in θ_0, and are almost always admissible under quadratic losses. In fact,

$$\mathbb{E}_{\theta_0}(\delta(x) - \theta_0)^2 = (\mathbb{E}_{\theta_0}[\delta(x)] - \theta_0)^2 + \text{var}_{\theta_0}(\delta(x)) = 0$$

implies that $\text{var}_{\theta_0}(\delta(x)) = 0$ and therefore that $\delta(x) = \theta_0$ uniformly, unless the distribution is degenerated in θ_0 (see Exercise 8.4).   ‖

A result similar to Proposition 8.2.2 can be established in the discrete case (the proof is straightforward and left as an exercise).

Proposition 8.2.6 *If Θ is a discrete set and $\pi(\theta) > 0$ for every $\theta \in \Theta$, then a Bayes estimator associated with π is admissible.*

8.2.2 Boundary conditions

We saw in Section 3.3 that, if x has a distribution from an exponential family

$$f(x|\theta) = h(x)e^{\theta \cdot T(x) - \psi(\theta)},$$

the conjugate distributions are also in exponential families and the posterior expectation of the mean of $T(x)$ is then affine in $T(x)$, that is,

$$(8.2.1) \qquad \mathbb{E}^\pi [\nabla \psi(\theta)|x] = \frac{T(x) + t_0}{\lambda + 1} = \frac{1}{\lambda + 1} T(x) + \frac{\gamma_0 \lambda}{\lambda + 1},$$

when

$$\pi(\theta|t_0, \lambda) = e^{\theta \cdot t_0 - \lambda \psi(\theta)}$$

and $\gamma_0 = t_0/\lambda$. In the case when $\theta \in \mathbb{R}$ and the natural parameter space is $N = [\underline{\theta}, \bar{\theta}]$, Karlin (1958) exhibits a sufficient admissibility condition for these estimators of the mean (see also Exercises 8.1 and 8.2).

Theorem 8.2.7 *If $\lambda > 0$, a sufficient condition for the estimator* (8.2.1) *to be admissible under a quadratic loss is that, for every $\underline{\theta} < \theta_0 < \bar{\theta}$,*

$$\int_{\theta_0}^{\bar{\theta}} e^{-\gamma_0 \lambda \theta + \lambda \psi(\theta)} \, d\theta = \int_{\underline{\theta}}^{\theta_0} e^{-\gamma_0 \lambda \theta + \lambda \psi(\theta)} \, d\theta = +\infty.$$

This theorem is derived from the *Cramér–Rao inequality* (Lehmann and Casella (1998)). It also appears as a corollary to Stein's necessary and sufficient condition (see Section 8.3.3). Berger (1982b) considers the reciprocal to Theorem 8.2.7, that is, shows that, under a few additional assumptions, this condition is also necessary (see Exercise 8.12).

Example 8.2.8 (Example 8.2.4 continued) For the chi-squared distribution, the natural parameterization is

$$\theta = \frac{1}{\sigma^2}, \quad T(y) = -\frac{1}{2}y, \quad \psi(\theta) = -\frac{p}{2} \log(\theta),$$

and

$$\int_0^c e^{-\gamma_0 \lambda \theta} \theta^{-\lambda p/2} \, d\theta$$

is infinite if $\lambda p \geq 2$. Similarly,

$$\int_c^{+\infty} e^{-\gamma_0 \lambda \theta} \theta^{-\lambda p/2} \, d\theta = +\infty$$

if $\gamma_0 \lambda < 0$ or $\gamma_0 \lambda = 0$ and $\lambda p \leq 2$. Therefore, the Bayes estimator

$$\delta^\pi(y) = \frac{\gamma_0 \lambda}{1 + \lambda} - \frac{1}{1 + \lambda} \frac{y}{2}$$

is admissible if $\gamma_0 = 0$ and $\lambda = 2/p$ or $\gamma_0 < 0$ and $\lambda \geq 2/p$; these conditions lead to the estimators

$$\varphi_1(y) = \frac{p}{p + 2} \left(\frac{-y}{2} \right) \quad \text{and} \quad \varphi_2(y) = \frac{\gamma_0 \lambda}{1 + \lambda} + \frac{1}{1 + \lambda} \left(\frac{-y}{2} \right),$$

for the estimation of $\mathbb{E}_\sigma(-y/2) = -\frac{p}{2}\sigma^2$, that is, to the following admissible Bayes estimators of σ^2:

$$\delta_1(y) = \frac{y}{p + 2} \quad \text{and} \quad \delta_2(y) = ay + b, \quad b > 0, \ 0 \leq a \leq \frac{1}{p + 2}. \qquad \|$$

Example 8.2.9 Consider $x \sim \mathcal{B}(n,p)$. The natural parameterization is given by $\theta = n \log(p/q)$ since

$$f(x|\theta) = \binom{n}{x} e^{(x/n)\theta} \left(1 + e^{\theta/n}\right)^{-n}.$$

Then the two integrals

$$\int_{-\infty}^{\theta_0} e^{-\gamma_0 \lambda \theta} \left(1 + e^{\theta/n}\right)^{\lambda n} d\theta \quad \text{and} \quad \int_{\theta_0}^{+\infty} e^{-\gamma_0 \lambda \theta} \left(1 + e^{\theta/n}\right)^{\lambda n} d\theta$$

cannot diverge simultaneously if $\lambda < 0$. Consider thus $\lambda > 0$. The second integral is divergent at $+\infty$ if $\lambda(1 - \gamma_0) > 0$, that is, if $\gamma_0 < 1$. And the first integral is divergent at $-\infty$ if $\gamma_0 \lambda \geq 0$. We then derive from Theorem 8.2.7 that a class of admissible Bayes estimators of p is

$$\delta^\pi(x) = a\frac{x}{n} + b, \qquad 0 \leq a \leq 1, \quad b \geq 0, \quad a + b \leq 1. \qquad \|$$

8.2.3 Inadmissible generalized Bayes estimators

Given that the Bayes estimators are not necessarily admissible, inadmissibility occurs more frequently for generalized Bayes estimators. The particular case when the Bayes risk of a generalized Bayes estimator is finite (and thus when this estimator is admissible—see Proposition 2.4.25) does not occur very often, except in testing and other bounded loss settings (see Example 2.4.26), and it is then necessary to use more advanced results to establish admissibility, such as Stein's condition (Section 8.3.3).

Example 8.2.10 Consider $x \sim \mathcal{N}_p(\theta, I_p)$ and $\delta_0(x) = x$; δ_0 is a generalized Bayes estimator for the prior distribution $\pi(\theta) = 1$. The Stein effect (Note 2.8.2) states that δ_0 is admissible under quadratic loss if $p \leq 2$ (see Corollary 8.2.14) and inadmissible otherwise. $\qquad \|$

Example 8.2.11 The prior distribution used in Example 8.2.10 can also produce more extreme cases of inadmissibility. For instance, if $\pi(\theta) = 1$ and if the parameter of interest is $\eta = ||\theta||^2$, Example 3.5.7 has shown that the posterior distribution of η is a $\chi_p^2(||x||^2)$ distribution, leading to the following generalized Bayes estimator:

$$\delta^\pi(x) = ||x||^2 + p.$$

As already stated, this estimator is inadmissible and dominated by $\tilde{\delta}(x) = (||x||^2 - p)^+$. Example 3.5.7 proposes an alternative prior distribution that is more appropriate in this setting. $\qquad \|$

Example 8.2.12 Consider $x \sim \mathcal{G}(\alpha, \theta)$ when α is known. Because θ is a scale parameter, $\pi(\theta) = 1/\theta$ is an appropriate noninformative distribution (see Chapter 9). The corresponding posterior distribution is $\mathcal{G}(\alpha, x)$ and thus

$$\delta^\pi(x) = \frac{\alpha}{x}$$

is the generalized Bayes estimator of θ under quadratic loss. For an estimator of the form $\delta_c(x) = c/x$, the quadratic risk is

$$R(\theta, \delta_c) = \mathbb{E}_\theta \left(\frac{c}{x} - \theta \right)^2 = c^2 \mathbb{E}_\theta(x^{-2}) - 2c\theta \mathbb{E}_\theta(x^{-1}) + \theta^2.$$

For $\alpha > 2$, we have

$$
\begin{aligned}
\mathbb{E}_\theta(x^{-2}) &= \frac{1}{\Gamma(\alpha)} \int_0^{+\infty} x^{-2} x^{\alpha-1} \theta^\alpha e^{-\theta x} \, dx \\
&= \frac{1}{\Gamma(\alpha)} \int_0^{+\infty} \theta^\alpha x^{\alpha-3} e^{-\theta x} \, dx \\
&= \theta^2 \frac{\Gamma(\alpha-2)}{\Gamma(\alpha)} = \frac{\theta^2}{(\alpha-1)(\alpha-2)}
\end{aligned}
$$

and

$$
\begin{aligned}
\mathbb{E}_\theta(x^{-1}) &= \frac{1}{\Gamma(\alpha)} \int_0^{+\infty} \theta^\alpha x^{\alpha-2} e^{-\theta x} \, dx \\
&= \theta \frac{\Gamma(\alpha-1)}{\Gamma(\alpha)} = \frac{\theta}{\alpha-1}.
\end{aligned}
$$

This implies that the best estimator of the form δ_c is associated with

$$c^* = \frac{\theta \mathbb{E}_\theta(x^{-1})}{\mathbb{E}_\theta(x^{-2})} = \frac{\theta^2/(\alpha-1)}{\theta^2/(\alpha-1)(\alpha-2)} = \alpha - 2,$$

and thus that δ^π is dominated by δ_{c^*}.   ‖

The three previous examples show that all behaviors are indeed possible for generalized Bayes estimators, from the admissibility of x for $p = 1, 2$ (Example 8.2.10) to the strong inadmissibility of the estimators in Examples 8.2.11 and 8.2.12, including the weak[1] inadmissibility of x for $p \geq 3$ (Example 8.2.10).

8.2.4 Differential representations

For multidimensional exponential families, Brown and Hwang (1982) have extended Theorem 8.2.7 to arbitrary generalized prior distributions. Consider a random variable

$$x \sim f(x|\theta) = h(x) e^{\theta \cdot x - \psi(\theta)},$$

---

[1] In fact, $\delta_0(x) = x$ is still a minimax estimator for every dimension and the estimators that dominate δ_0 only improve significantly upon δ_0 (in terms of risk) in a relatively small region of the sample space (see, for instance, Bondar (1987)). The practical implication of this property is that, without prior information on θ, the domination of δ_0 is mostly formal.

where θ and x belong to $\mathbb{R}^p$ and recall that the mean of this distribution is $\nabla\psi(\theta)$. Given a measure π with density g on Θ, we assume that

$$(8.2.2) \qquad I_x(\nabla g) = \int ||\nabla g(\theta)||e^{\theta.x-\psi(\theta)}\,d\theta < +\infty.$$

When estimating $\nabla\psi(\theta)$ under quadratic loss, the Bayes estimator associated with g can be written as the differential representation

$$(8.2.3) \qquad \delta_g(x) = x + \frac{I_x(\nabla g)}{I_x(g)}.$$

The following conditions on g:

$$(8.2.4) \qquad \int_{\{||\theta||>1\}} \frac{g(\theta)}{||\theta||^2 \log^2(||\theta|| \vee 2)}\,d\theta < \infty,$$

$$(8.2.5) \qquad \int \frac{||\nabla g(\theta)||^2}{g(\theta)}\,d\theta < \infty,$$

and

$$(8.2.6) \qquad \forall \theta \in \Theta, \qquad R(\theta,\delta_g) < \infty,$$

are sufficient to establish the admissibility of δ_g.

Theorem 8.2.13 *Under the conditions (8.2.4), (8.2.5), and (8.2.6), the estimator (8.2.3) is admissible.*

The proof of this result is deferred until Example 8.3.5 because it relies on Blyth's condition given in Section 8.3.2. Notice that this result has important consequences, since it covers the estimation of the expectation parameter for all continuous exponential families on $\mathbb{R}^p$. For instance, it gives, as a particular application, Stein's (1955b) admissibility result for all exponential families. It also generalizes Zidek (1970), who was dealing with the one-dimensional case (see Exercise 8.8).

Corollary 8.2.14 *If $\Theta = \mathbb{R}^p$ and $p \le 2$, the estimator $\delta_0(x) = x$ is admissible.*

Proof. Consider $g \equiv 1$, then $\nabla g \equiv 0$ and $\delta_g(x) = x$. Conditions (8.2.4), (8.2.5) and (8.2.6) being satisfied, δ_g is admissible. □□

Example 8.2.15 (Example 8.2.10 continued) If $x \sim \mathcal{N}_p(\theta, I_p)$, θ is the natural parameter of the distribution and Corollary 8.2.14 actually reproduces Stein's (1955a) original result. Notice that Theorem 8.2.13 also provides a means to check the admissibility of other generalized Bayes estimators of θ, including those considered by Strawderman (1971) (see Exercise 10.5) and Berger (1980a). ‖

Example 8.2.16 Consider x_1, x_2, two random variables from $\mathcal{P}(\lambda_i)$ ($i = 1, 2$). If $\theta_i = \log(\lambda_i)$, $\delta_0(x) = (x_1, x_2)$ is an admissible estimator of $(\lambda_1, \lambda_2) = (e_1^\theta, e_2^\theta)$. This result does not extend to larger dimensions, as shown by Hwang (1982a) and Johnstone (1984). ‖

Brown and Hwang (1982) present various generalizations of Theorem 8.2.13 allowing us to include cases where $\Theta \neq \mathbb{R}^p$, like the gamma and geometric distributions. They also show that, in the special case of p observations x_i from independent Poisson distributions, $\mathcal{P}(\lambda_i)$, the generalized Bayes estimator

$$\delta_{CZ}(x) = \left[1 - \frac{\beta + p - 1}{\beta + p - 1 + S}\right] x,$$

with $S = \sum_i x_i$, which was proposed by Clevenson and Zidek (1975) to improve upon $x = (x_1, \ldots, x_p)$, is admissible for $\beta > 0$ and $p \geq 2$ under the loss

$$L(\theta, \delta) = \sum_{i=1}^{p} \frac{1}{\lambda_i}(\delta - \lambda_i)^2.$$

Das Gupta and Sinha (1986) also provide sufficient admissibility conditions for the estimation of independent gamma means.

8.2.5 Recurrence conditions

In the particular case of a normal multidimensional distribution, $\mathcal{N}_p(\theta, \Sigma)$, when Σ is known, Brown (1971) characterizes more thoroughly the admissible Bayes estimators under quadratic loss by providing a necessary and sufficient condition, based on a Markovian representation of the estimation problem. (Notice that Shinozaki (1975) implies that the choice $\Sigma = I_p$ does not restrict the generality of the treatment (see Section 2.5.1 and Exercise 2.39).)

Theorem 8.2.17 *Consider* $x \sim \mathcal{N}_p(\theta, I_p)$. *A generalized Bayes estimator of the form*

$$\delta(x) = (1 - h(||x||))x$$

is

(i) *inadmissible if there exist $\epsilon > 0$ and $K < +\infty$ such that, for $||x|| > K$,*

$$||x||^2 h(||x||) < p - 2 - \epsilon;$$

and

(ii) *admissible if there exist K_1 and K_2 such that $h(||x||)||x|| \leq K_1$ for every x and, for $||x|| > K_2$,*

$$||x||^2 h(||x||) \geq p - 2.$$

The proof of this result is quite advanced and the derivation of (i) and (ii) involves the recurrence or the transience of a random process[2] associated with δ. (See Srinivasan (1981) for a simpler description.) Part (i) also

---

[2] Random walks are always recurrent in dimensions 1 or 2 and may be transient for larger dimensions (see Feller (1971) or Meyn and Tweedie (1993)). The connection exhibited by Brown (1971) then points out that the similar role of $p = 3$ as a limiting case in both problems is not coincidental.

appears as a consequence of Lemma 8.5.1. Notice the factor $(p-2)$, which delineates the boundary between admissibility and inadmissibility of the usual estimator $\delta_0(x) = x$. The relation between this result and the Stein phenomenon is explained in Section 8.5.

Johnstone (1984) provides an equivalent to Theorem 8.2.17 in the case of a Poisson model. If $x_i \sim \mathcal{P}(\lambda_i)$ $(i = 1, \ldots, p)$, the parameter $\lambda = (\lambda_1, \ldots, \lambda_p)$ is estimated under the loss

$$\sum_{i=1}^{p} \frac{1}{\lambda_i} (\delta_i - \lambda_i)^2.$$

Then:

Theorem 8.2.18 *A generalized Bayes estimator of the form*

$$\delta(x) = (1 - h(s))x,$$

where $s = \sum_i x_i$, is

(i) *inadmissible if there exist $\epsilon > 0$ and $K < +\infty$ such that, for $s > K$,*

$$sh(s) < (p - 1 - \epsilon);$$

and

(ii) *admissible if there exist K_1 and K_2 such that $\sqrt{s}\, h(s) \leq K_1$ for every s and, for $s > K_2$,*

$$sh(s) \geq (p - 1).$$

Eaton (1992) exhibits connections similar to those of Brown (1971) between the admissibility of an estimator and the recurrence of an associated Markov chain. We mention the main results of this paper below but urge the reader to investigate the paper not only for the proofs, but also for its deeper implications. The problem considered by Eaton (1992) is to determine whether, for a *bounded* function $g(\theta)$, a generalized Bayes estimator associated with a prior measure π is admissible under quadratic loss. Assuming that the posterior distribution $\pi(\theta|x)$ is well defined, we consider the transition kernel

$$(8.2.7) \qquad K(\theta|\eta) = \int_{\mathcal{X}} \pi(\theta|x) f(x|\eta)\, dx,$$

which is associated with a Markov chain $(\theta^{(n)})$ generated as follows. The transition from $\theta^{(n)}$ to $\theta^{(n+1)}$ is done by generating first $x \sim f(x|\theta^{(n)})$, and then $\theta^{(n+1)} \sim \pi(\theta|x)$. (For the use of this kernel in Markov Chain Monte Carlo methods, and more details about Markov chain theory, see Chapter 6.) For every measurable set C such that $\pi(C) < +\infty$, we define

$$V(C) = \left\{ h \in \mathcal{L}^2(\pi); h(\theta) \geq 0 \text{ and } h(\theta) \geq 1 \text{ when } \theta \in C \right\}$$

and

$$\Delta(h) = \int\int \left\{ h(\theta) - h(\eta) \right\}^2 K(\theta|\eta)\pi(\eta)\, d\theta\, d\eta.$$

The following result then characterizes admissibility for *all bounded functions* in terms of Δ and $V(C)$, that is, independently of the estimated functions g:

Theorem 8.2.19 *If, for every C such that $\pi(C) < +\infty$,*

$$(8.2.8) \qquad \inf_{h \in V(C)} \Delta(h) = 0,$$

then the Bayes estimator $\mathbb{E}^{\pi}[g(\theta)|x]$ is admissible under quadratic loss for every bounded function g.

This result is obviously quite general, but only mildly helpful in the sense that the practical verification of (8.2.8) for every set C can be overwhelming. Notice also that (8.2.8) always holds when π is a proper prior distribution, since $h \equiv 1$ belongs to $\mathcal{L}^2(\pi)$ and $\Delta(1) = 0$ in this case. The extension to improper priors then considers approximations of 1 by functions in $V(C)$. (See Chapter 9 for a similar relation between amenability and minimaxity.) Eaton (1992) exhibits a connection with the Markov chain $(\theta^{(n)})$, which gives a condition equivalent to Theorem 8.2.19. First, for a given set C, a stopping rule σ_C is defined as the first integer $n > 0$ such that $(\theta^{(n)})$ belongs to C (and $+\infty$ otherwise). The chain $(\theta^{(n)})$ is said to be π-*recurrent* if the probability that σ_C is finite is 1 for π-almost every starting point $\theta^{(0)}$.

Theorem 8.2.20 *For every set C such that $\pi(C) < +\infty$,*

$$\inf_{h \in V(C)} \Delta(h) = \int_C \left\{ 1 - P(\sigma_C < +\infty | \theta^{(0)} = \eta) \right\} \pi(\eta) \, d\eta.$$

Therefore, the generalized Bayes estimators of bounded functions of θ are admissible if the associated Markov chain $(\theta^{(n)})$ is π-recurrent.

We refer to Note 8.7.1 and to Eaton (1992, 1999) for extensions, examples, and comments on this result. Its principal appeal is that, besides its mathematical elegance, the verification of the recurrence of the Markov chain $(\theta^{(n)})$ is much easier to operate than the determination of the lower bound of $\Delta(h)$. Moreover, it suggests a possible numerical verification of admissibility based on the generation of the chain $(\theta^{(n)})$, which is in a way similar to the numerical minimaxity verification proposed in Berger and Robert (1990).

8.3 Necessary and sufficient admissibility conditions

The results obtained in the previous section only apply to generalized Bayes estimators. Moreover, some conditions are rather arduous to verify—see, e.g., (8.2.4) or (8.2.5). We present in this section a general necessary and sufficient admissibility condition that does not require estimators to be generalized Bayes estimators. It somehow formalizes our repeated assertion that *"admissible estimators are limits of Bayes estimators...."* A first version of Stein's condition only allows for the comparison of continuous risk estimators, but Section 8.3.1 shows why it is usually sufficient to consider continuous risk estimators.

8.3.1 Continuous risks

It is often necessary to restrict the scope of the study to continuous risk function estimators in order to produce a sufficient admissibility condition. However, in some settings, all estimators have continuous risks. In other cases, the admissible estimators necessarily have continuous risks.

Lemma 8.3.1 *Consider $\Theta \subset \mathbb{R}^m$. The loss function $L(\theta, \delta)$ is assumed to be bounded and continuous as a function of θ for every $\delta \in \mathcal{D}$. If $f(x|\theta)$ is continuous in θ for every x, the risk function of every estimator is continuous.*

Proof. Given an estimator δ, the difference of the risks in θ and $\theta' \in \Theta$ is

$$|R(\theta, \delta) - R(\theta', \delta)| = \left| \int L(\theta, \delta(x)) f(x|\theta) \, dx - \int L(\theta', \delta(x)) f(x|\theta') \, dx \right|$$

$$\leq \int |L(\theta, \delta(x)) - L(\theta', \delta(x))| f(x|\theta) \, dx$$

$$+ \left| \int L(\theta, \delta(x)) (f(x|\theta) - f(x|\theta')) \, dx \right|.$$

Since L is continuous and bounded by C, there exist $\eta_0 > 0$ and a compact set K_0 such that

$$\int_{K_0^c} f(x|\theta) \, dx < \frac{\epsilon}{8C} \quad \text{and} \quad \int_{K_0} |L(\theta, \delta(x)) - L(\theta', \delta(x))| f(x|\theta) \, dx < \frac{\epsilon}{4}$$

when $||\theta - \theta'|| < \eta_0$. Thus,

$$\int |L(\theta, \delta(x)) - L(\theta', \delta(x))| f(x|\theta) \, dx < \frac{\epsilon}{2}.$$

Moreover, $f(x|\theta)$ being a continuous function of θ, a similar argument can be applied: there exist $\eta_1 > 0$ and a compact set K_1 such that

$$\left| \int L(\theta, \delta(x)) (f(x|\theta) - f(x|\theta')) \, dx \right| \leq C \int_{K_1} |f(x|\theta) - f(x|\theta')| \, dx$$

$$+ C \int_{K_1^c} [f(x|\theta) + f(x|\theta')] \, dx < \frac{\epsilon}{2}$$

and

$$\int_{K_1^c} f(x|\theta) \, dx < \frac{\epsilon}{8C},$$

when $||\theta - \theta'|| < \eta_1$. Therefore, $R(\theta, \delta)$ is continuous. $\qquad\square\square$

Lemma 8.3.1 is somehow of limited interest, since the most delicate admissibility problems occur when L is unbounded. Some settings still allow for a reduction in the class of estimators to consider to the class of continuous risk estimators, that is, for a *complete class* characterization.

Definition 8.3.2 *A class $\mathcal{C}$ of estimators is said to be complete if, for every $\delta' \notin \mathcal{C}$, there exists $\delta \in \mathcal{C}$ that dominates δ'. The class is essentially complete if, for every $\delta' \notin \mathcal{C}$, there exists $\delta \in \mathcal{C}$ that is at least as good as δ'.*

Apart from trivial cases such as the class of all estimators, the determination of useful complete classes is not always possible. For instance, there are cases when the class of admissible estimators *is not* a complete class although such settings seldom occur (see Blackwell and Girshick (1954, Theorem 5.7.1) or Brown (1976)). Section 8.4 analyzes the relations between Bayes estimators, generalized Bayes estimators, and complete classes. The following result is a complete class lemma giving sufficient conditions for considering continuous risk estimators only.

Lemma 8.3.3 *Consider a statistical decision model $\mathcal{X}, \Theta \subset \mathbb{R}$ with a closed decision space $\mathcal{D} \subset \mathbb{R}$. Assume that $f(x|\theta)$ has the monotone likelihood ratio property and is continuous in θ. If*

(i) *$\mathrm{L}(\theta, d)$ is a continuous function of θ for every $d \in \mathcal{D}$;*

(ii) *L is decreasing in d for $d < \theta$ and increasing for $d > \theta$; and*

(iii) *there exist two functions K_1 and K_2 bounded on the compact subsets of Θ, such that*

$$\mathrm{L}(\theta_1, d) \leq K_1(\theta_1, \theta_2)\mathrm{L}(\theta_2, d) + K_2(\theta_1, \theta_2),$$

then the estimators with finite and continuous risks form a complete class.

See Ferguson (1967) and Brown (1976) for additional results. For instance, it is possible to show that, if the problem is *monotone*, then *monotone estimators* constitute a complete class (see Exercise 8.23 and Theorem 5.4.4).

8.3.2 Blyth's sufficient condition

Prior to Stein's (1955b) derivation of his necessary and sufficient condition (Section 8.3.3), Blyth (1951) proposed a sufficient admissibility condition, relating admissibility of an estimator with the existence of a sequence of prior distributions approximating this estimator.

Theorem 8.3.4 *Consider a nonempty open set $\Theta \subset \mathbb{R}^p$. Assume that the estimators with continuous risk constitute a complete class. If, for a continuous risk estimator δ_0, there exists a sequence (π_n) of generalized prior distributions such that*

(i) *$r(\pi_n, \delta_0)$ is finite for every n;*

(ii) *for every nonempty open set $C \subset \Theta$, there exist $K > 0$ and N such that, for every $n \geq N$, $\pi_n(C) \geq K$; and*

(iii) $\lim\limits_{n \to +\infty} r(\pi_n, \delta_0) - r(\pi_n) = 0$;

then the estimator δ_0 is admissible.

Proof. If δ_0 is not admissible, there exists an estimator δ' dominating δ_0, that is, such that $R(\theta, \delta) - R(\theta, \delta') \geq 0$ and

$$R(\theta, \delta) \quad R(\theta, \delta') \gg \epsilon$$

on an open set $C \subset \Theta$ (for ϵ small enough). It then follows from assumptions (i) and (ii), that for $n \geq N$,

$$
\begin{aligned}
r(\pi_n, \delta_0) - r(\pi_n) &\geq r(\pi_n, \delta_0) - r(\pi_n, \delta') \\
&= \mathbb{E}^\pi[R(\theta, \delta_0) - R(\theta, \delta')] \\
&\geq \int_C (R(\theta, \delta_0) - R(\theta, \delta'))\pi_n(\theta)\, d\theta \\
&\geq \epsilon \int_C \pi_n(\theta)\, d\theta \geq \epsilon K.
\end{aligned}
$$
□□

This result can be used to establish the admissibility of generalized Bayes estimators, since the measures π associated with these estimators can be written as limits of sequences of proper distributions π_n. However, the choice of these sequences is not necessarily straightforward, as shown by Berger (1982b) or Brown and Hwang (1982). Theorem 8.3.4 also applies to other estimators, in settings where there exist admissible estimators that are not generalized Bayes (see Section 8.4).

Example 8.3.5 The proof of Theorem 8.2.13 is a first illustration of Blyth's condition. Consider h_n taking values in $[0, 1]$, differentiable, satisfying $h_n(\theta) = 0$ if $||\theta|| > n$ and $h_n(\theta) = 1$ on a set S such that

$$\int_S g(\theta)\, d\theta > 0.$$

We then define a sequence of associated measures with densities $g_n(\theta) = h_n^2(\theta)g(\theta)$ and the corresponding Bayes estimator δ_n. Reverting to the notation $I_x(.)$ introduced in (8.2.2), the difference of the integrated Bayes risks is

$$
\begin{aligned}
r(\pi_n, \delta_g) - r(\pi_n) &= \int ||\delta_g(x) - \delta_n(x)||^2 I_x(g_n)\, dx \\
&= \int \left\| \frac{I_x(\nabla g)}{I_x(g)} - \frac{I_x(h_n^2 \nabla g)}{I_x(g_n)} - \frac{I_x(g \nabla h_n)}{I_x(g_n)} \right\|^2 I_x(g_n)\, dx,
\end{aligned}
$$

using the representation (8.2.3). Therefore,

$$r(\pi_n, \delta_g) - r(\pi_n) \leq 2 \int \left\| \frac{I_x(\nabla g)}{I_x(g)} - \frac{I_x(h_n^2 \nabla g)}{I_x(g_n)} \right\|^2 I_x(g_n)\, dx$$

$$+2 \int \left\| \frac{I_x(g\nabla h_n)}{I_x(g_n)} \right\|^2 I_x(g_n)\, dx$$

$$= \quad B_n + A_n.$$

The second term, A_n, is bounded from above by

$$4 \int \|\nabla h_n(\theta)\|^2 g(\theta)\, d\theta.$$

In the particular case where h_n is

$$h_n(\theta) = \begin{cases} 1 & \text{for } \|\theta\| < 1, \\ 1 - \dfrac{\log(\|\theta\|)}{\log(n)} & \text{for } 1 < \|\theta\| < n, \\ 0 & \text{otherwise,} \end{cases}$$

we have actually

$$\|\nabla h_n(\theta)\|^2 \leq \frac{1}{\|\theta\|^2 \log^2(\max(\|\theta\|, 2))} \mathbb{I}_{\|\theta\| > 1}(\theta),$$

and condition (8.2.4) implies that A_n converges to 0 as n goes to infinity. The first term satisfies

$$\begin{aligned} B_n &= \int \left\| I_x \left(g_n \frac{I_x(\nabla g)}{I_x(g)} - h_n^2 \nabla g \right) \right\|^2 \Big/ (I_x(g_n))\, dx \\ &= \int \left\| I_x \left(g_n \left[\frac{I_x(\nabla g)}{I_x(g)} - \frac{\nabla g}{g} \right] \right) \right\|^2 \Big/ (I_x(g_n))\, dx \\ &\leq \int I_x \left(g \left\| \frac{I_x(\nabla g)}{I_x(g)} - \frac{\nabla g}{g} \right\|^2 \right) dx. \end{aligned}$$

Using (8.2.5), we can then derive from the dominated convergence theorem that B_n goes to 0, since g_n converges to g. This completes the proof of Theorem 8.2.13.   ∥

In practice, a usual way to apply Blyth's condition to a generalized Bayes estimator, δ_0, is to exhibit a sequence of proper Bayes estimators that converge to δ_0, and then to de-normalize the sequence of the associated prior distributions by a suitable weight.

Example 8.3.6 Consider $x \sim \mathcal{N}(\theta, 1)$ and $\delta_0(x) = x$, an estimator of θ. Because δ_0 corresponds to $\pi(\theta) = 1$ under quadratic loss, we choose π_n as the measure with density

$$g_n(x) = e^{-\theta^2/2n},$$

that is, the density of the normal distribution $\mathcal{N}(0, n)$ without the normalizing factor $1/\sqrt{2\pi n}$. Because the densities g_n increase with n, condition (ii) of Theorem 8.3.4 is satisfied, as well as (i): the Bayes estimator for π_n is still

$$\delta_n(x) = \frac{nx}{n+1},$$

since the absence of the normalizing factor is of no importance in this case, and

$$r(\pi_n) = \int_{\mathbb{R}} \left[\frac{\theta^2}{(n+1)^2} + \frac{n^2}{(n+1)^2} \right] g_n(\theta) \, d\theta$$

$$= \sqrt{2\pi n} \, \frac{n}{n+1},$$

while

$$r(\pi_n, \delta_0) = \int_{\mathbb{R}} 1 \, g_n(\theta) \, d\theta = \sqrt{2\pi n}.$$

The two risks are then finite. Moreover,

$$r(\pi_n, \delta_0) - r(\pi_n) = \sqrt{2\pi n}/(n+1)$$

converges to 0. The Blyth condition thus provides another proof of admissibility for $\delta_0(x) = x$ in the normal case. On the contrary, the proof of the admissibility of δ_0 in dimension two necessitates a more complex sequence (see Stein (1955a)).   ‖

Example 8.3.7 Consider $x \sim \mathcal{B}(m, \theta)$. The inferential problem is to test the null hypothesis $H_0 : \theta \leq \theta_0$ under the quadratic loss introduced in Section 5.4,

$$\left(\mathbb{I}_{[0,\theta_0]}(\theta) - \gamma(x) \right)^2.$$

The *p-value* is then

$$\varphi(x) = P_{\theta_0}(X \geq x) = \sum_{k=x}^{m} \binom{m}{k} \theta_0^k (1 - \theta_0)^{m-k}.$$

In this case, the natural conjugate distributions are beta distributions. This suggests approximating $\varphi(x)$ by a sequence of estimators associated with an appropriate sequence of beta distributions. In fact, $\varphi(x)$ can be written (for $x \neq 0$)

$$\varphi(x) = \frac{1}{B(x, m - x + 1)} \int_0^{\theta_0} t^{x-1}(1 - t)^{m-x} \, dt = P(T \leq \theta_0 | x)$$

when $T \sim \mathcal{B}e(x, m - x + 1)$, which corresponds to the generalized prior distribution

$$\pi(\theta) = \theta^{-1} \qquad (0 < \theta < 1).$$

Consider then π_n with density

$$g_n(\theta) = \theta^{\alpha_n - 1}$$

on $[0, 1]$ and let the sequence (α_n) go to 0. In this case, the Bayes procedure is

$$\gamma^{\pi_n}(x) = P^{\pi_n}(\theta \leq \theta_0 | x) = \frac{1}{B(x + \alpha_n, m - x + 1)} \int_0^{\theta_0} t^{x+\alpha_n - 1}(1-t)^{m-x} \, dt$$

and

$$r(\pi_n) \;\; = \;\; \sum_{k=0}^{m} B(k+\alpha_n, m-k+1)\gamma^{\pi_n}(k)(1-\gamma^{\pi_n}(k)),$$

$$r(\pi_n,\varphi) \;\; = \;\; \sum_{k=0}^{m} B(k+\alpha_n, m-k+1)(\gamma^{\pi_n}(k)-2\gamma^{\pi_n}\varphi(k)+\varphi^2(k)).$$

Therefore,

$$r(\pi_n,\varphi) - r(\pi_n) = \sum_{k=0}^{m} B(k+\alpha_n, m-k+1)(\gamma^{\pi_n}(k)-\varphi(k))^2.$$

If $k \neq 0$, it is straightforward to verify that

$$\lim_{\alpha_n \to 0}(\varphi(k)-\gamma^{\pi_n}(k))^2 = 0.$$

Similarly, we get

$$\lim_{\alpha \to 0}\frac{\int_0^{\theta_0} t^{\alpha-1}(1-t)^{m-1}dt}{\int_0^1 t^{\alpha-1}(1-t)^{m-1}dt} = 1,$$

which takes care of the case $k = 0$. Moreover, condition (ii) is also satisfied. The *p-value* φ is then admissible in this setting. Example 5.4.6 provides a more direct proof of this result because the Bayes risk is finite. ∥

Examples 8.3.5 and 8.3.7 take advantage of a general result, namely, the fact that, under quadratic loss, condition (iii) of Theorem 8.3.4 implies the quadratic convergence of the Bayes estimators to δ_0 in the sense of the marginal measures.

Proposition 8.3.8 *If L is a quadratic loss and if there exists a sequence (π_n) satisfying conditions* (i), (ii) *and* (iii) *of Theorem 8.3.4, then the Bayes estimators δ^{π_n} converge quadratically to δ_0 for the marginal measures*

$$m_n(x) = \int_\Theta f(x|\theta)\pi_n(\theta)\,d\theta.$$

Proof. It is enough to write the difference of the risks as

$$r(\pi_n,\delta_0) - r(\pi_n)$$
$$= \int_{\mathcal{X}}\int_\Theta (||\delta_0(x)-\theta||^2 - ||\delta^{\pi_n}(x)-\theta||^2)\pi_n(\theta|x)\,d\theta\,m_n(x)\,dx$$
$$= \int_{\mathcal{X}}\Big[||\delta_0(x)-\delta^{\pi_n}(x)||^2$$
$$+ 2(\delta_0(x)-\delta^{\pi_n}(x))\cdot\int_\Theta(\delta^{\pi_n}(x)-\theta)\pi_n(\theta|x)\,d\theta\Big]m_n(x)\,dx$$
$$= \int_{\mathcal{X}}||\delta_0(x)-\delta^{\pi_n}(x)||^2 m_n(x)\,dx,$$

since

$$\int_\Theta(\delta^{\pi_n}(x)-\theta)\pi_n(\theta|x)\,d\theta = 0. \qquad\qquad \square\square$$

Unfortunately, this convergence result depends on (m_n), except when it is possible to establish a uniform equivalence with the Lebesgue measure, or another fixed measure, in which case there is quadratic convergence in the usual sense. This is, for instance, what occurs when the sequence (m_n) is increasing, as in Examples 8.3.5, 8.3.6, and 8.3.7. Section 8.3.4 provides a more fundamental result of Brown (1986), which shows that pointwise convergence of the δ^{π_n} to δ_0, independently of the measures m_n, is actually necessary.

8.3.3 Stein's necessary and sufficient condition

The completion of the previous condition by Stein (1955b) and Farrell (1968b) is even more important than Theorem 8.3.4, since it shows that *all admissible estimators are limits of sequences of Bayes estimators* (in the sense of the Bayes risk). The assumptions in Farrell (1968b) are that

(i) $f(x|\theta)$ is continuous in θ and strictly positive on Θ; and

(ii) the loss L is strictly convex, continuous and, if $E \subset \Theta$ is compact,

$$\lim_{\|\delta\| \to +\infty} \inf_{\theta \in E} L(\theta, \delta) = +\infty.$$

Notice that this second assumption necessarily eliminates bounded losses.

Theorem 8.3.9 *Under the hypotheses* (i) *and* (ii), *an estimator δ is admissible if, and only if, there exist a sequence (F_n) of increasing compact sets such that $\Theta = \bigcup_n F_n$, a sequence (π_n) of finite measures with support F_n, and a sequence (δ_n) of Bayes estimators associated with π_n such that*

(i) *there exists a compact set $E_0 \subset \Theta$ such that $\inf_n \pi_n(E_0) \geq 1$;*

(ii) *if $E \subset \Theta$ is compact, $\sup_n \pi_n(E) < +\infty$;*

(iii) *$\lim_n r(\pi_n, \delta) - r(\pi_n) = 0$; and*

(iv) *$\lim_n R(\theta, \delta_n) = R(\theta, \delta)$.*

This fundamental theorem underlies most of the admissibility and complete class results presented in Section 8.4. A proof of Theorem 8.3.9 is beyond our reach; see Farrell (1968b). The *sufficient* part of this result is close to Blyth's condition, but the *necessary* part allows for the exclusion of many inadmissible estimators.

8.3.4 Another limit theorem

Brown (1986) provides an alternative, and quite general, characterization of admissible estimators. Consider $x \sim f(x|\theta)$, with $f(x|\theta) > 0$, and assume that $\mathcal{D}$ is a closed convex set. Moreover, the loss function L is supposed to be lower semicontinuous and such that

$$\lim_{\|\delta\| \to +\infty} L(\theta, \delta) = +\infty.$$

(Notice that this is roughly assumption (ii) of Farrell (1968b).) The main result of Brown (1986) is to show that, under these conditions, the *closure* (for the pointwise convergence) of the set of the Bayes estimators is a complete class. The following convergence result rephrases this completeness (see Brown (1986, pp. 254–267)).

Proposition 8.3.10 *If* L *is strictly convex, every admissible estimator of* θ *is a pointwise limit of Bayes estimators for a sequence of prior distributions with finite supports.*

This result can be compared with the results of Dalal and Hall (1983) and Diaconis and Ylvisaker (1985), presented in Section 3.4, which show that, in an exponential family setting, every prior distribution is a limit of mixtures of conjugate prior distributions. Therefore, for exponential families, an admissible estimator is also a limit of Bayes estimators associated with a mixture of conjugate prior distributions. When the model is invariant under spherical transformations, the distributions with finite support can be replaced by distributions supported on embedded spheres, since they preserve symmetry. In this case, if π_c is the uniform distribution on the sphere with radius c,

$$\mathcal{S}_c = \{\theta; \ ||\theta|| = c\},$$

and δ_c is the associated Bayes estimator under quadratic loss, that is, the posterior mean, Robert (1990a) derives the following limit theorem.

Proposition 8.3.11 *If* $x \sim \mathcal{N}_p(\theta, I_p)$ *and* π *is a prior distribution which is spherically symmetric around* 0, *then there exist two sequences,* (q_n^i) *and* (c_n^i), *such that* $\sum_{i=1}^n q_n^i = 1$ *and*

$$m^\pi(x) = \int_{\mathbb{R}^p} f(x|\theta)\pi(\theta)\,d\theta = \lim_{n \to +\infty} \sum_{i=1}^n q_n^i m_{c_n^i}(x),$$

where

$$m_{c_n^i} = \int_{\mathbb{R}^p} f(x|\theta)\pi_{c_n^i}(\theta)\,d\theta.$$

Moreover, under quadratic loss,

$$(8.3.1) \qquad \delta^\pi(x) = \lim_{n \to +\infty} \sum_{i=1}^n \frac{q_n^i m_{c_n^i}(x)}{\sum_j q_n^j m_{c_n^j}(x)} \delta_{c_n^i}(x).$$

Therefore, in the normal case, every Bayes estimator associated with a spherically symmetric prior distribution is a pointwise limit of Bayes estimators associated with uniform distributions on spheres. Recall that the estimators δ_c can be written as

$$(8.3.2) \qquad \delta_c(x) = c\,\frac{I_{p/2}(||x||c)}{I_{p/2-1}(||x||c)}\,\frac{x}{||x||},$$

where I_ν is the *modified Bessel function* (Exercises 4.35 and 4.36). It actually follows from Kempthorne (1988) that *every admissible estimator* $\delta(x)$

can be written under the form (8.3.2) *or that there exists an estimator* δ' *of the form* (8.3.2) *equivalent to* δ (*in risk*).

8.4 Complete classes

The previous section established in a general setting that admissible estimators are limits of Bayes estimators from several points of view. In some particular cases, it is possible to improve upon this description of admissible estimators, and to show that they are necessarily generalized Bayes estimators. Such results are interesting because, on one hand, they restrict further the class of estimators to be considered and, on the other hand, they point out the advantage of using solely Bayes and generalized Bayes estimators from a frequentist point of view. This is, for instance, the case when testing procedures are evaluated under a quadratic loss, as seen in Section 5.4 (Theorems 5.4.3 and 5.4.4). This section provides similar results for point estimation. Additional references are given by Brown (1986) and Rukhin (1995).

As an introductory example, consider the extremely simple case when $\Theta = \{\theta_1, \theta_2\}$, since it allows for a graphical representation of the *risk set*,

$$\mathcal{R} = \{r = (R(\theta_1, \delta), R(\theta_2, \delta)), \ \delta \in \mathcal{D}^*\},$$

where $\mathcal{D}^*$ is the set of randomized estimators. Assume the risk set $\mathcal{R}$ is *bounded and closed from below*, that is, such that all risk points on the lower boundary of $\mathcal{R}$ are in $\mathcal{R}$ and have finite components. This is the case when the loss is positive. This lower boundary, denoted $\Gamma(\mathcal{R})$, is important because it actually provides the *admissible* points of $\mathcal{R}$. Indeed, if $r \in \Gamma(\mathcal{R})$, there cannot exist $r' \in \mathcal{R}$ such that $r'_1 \leq r_1$ and $r'_2 \leq r_2$ with strict inequality on one of the two axes. Moreover, for every $r \in \Gamma(\mathcal{R})$, there exists a tangent line to $\mathcal{R}$ going through r, with positive slope and equation

$$p_1 r_1 + p_2 r_2 = k,$$

that is, such that every $r' \in \mathcal{R}$ satisfies $p_1 r'_1 + p_2 r'_2 \geq k$, as shown by Figure 8.4. (In fact, this is a consequence of the convexity of $\mathcal{R}$.) This property implies that r is a Bayes estimator for the prior distribution $\pi(\theta_i) = p_i$ ($i = 1, 2$), since it minimizes the Bayes risk $p_1 r_1 + p_2 r_2$. We derive from this argument the following general result.

Proposition 8.4.1 *If Θ is finite and if the risk set $\mathcal{R}$ is bounded and closed from below, then the set of Bayes estimators constitutes a complete class.*

This characterization relies on a *separating hyperplane* theorem since, under the assumptions of the theorem, there exists a hyperplane tangent to the risk set for each point of the lower boundary, and that this hyperplane defines a prior distribution on Θ by duality. The extension of this complete class result to denumerable and non-denumerable parameter spaces

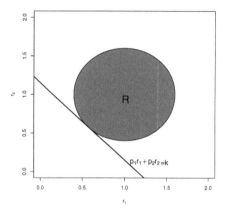

Figure 8.4.1. *Risk set and admissible estimators for* $\Theta = \{\theta_1, \theta_2\}$.

Θ calls for an equivalent generalization of separating hyperplane theorems to spaces of functions on Θ. For instance, Brown (1976) mentions the following result, where $\overset{\circ}{S}$ denotes the interior set of S.

Lemma 8.4.2 *Consider* S *a convex subset of a topological vector space* $\mathcal{E}$. *If* $\overset{\circ}{S} \neq \emptyset$ *and* $y_0 \notin \overset{\circ}{S}$, *there exists* $f \in \mathcal{E}^*$ *such that* S *is included in* $\{y;\ f(y) \geq f(y_0)\}$.

We derive from this lemma the following complete class result, due to Wald (1950), which generalizes Proposition 8.4.1.

Theorem 8.4.3 *Consider the case when* Θ *is compact and the risk set* $\mathcal{R}$ *is convex. If all estimators have a continuous risk function, the Bayes estimators constitute a complete class.*

Proof. This result is indeed a consequence of Lemma 8.4.2 since, if δ_0 is admissible, the risk function $R(\theta, \delta_0)$ belongs to the lower boundary of the risk set. Therefore, there exists a linear functional on $\mathcal{R}$, ψ^*, such that, for every estimator δ,

$$\psi^*(R(\cdot, \delta)) \geq \psi^*(R(\cdot, \delta_0)).$$

We can then derive from the *Riesz representation theorem* that there exists a finite measure π on Θ such that

$$\psi^*(R(\cdot, \delta)) = \int_\Theta R(\theta, \delta)\pi(\theta)d\theta,$$

and that this measure can be renormalized into $\tilde{\pi}(\theta) = \pi(\theta)/\pi(\Theta)$, thus defining a prior distribution. The above inequality can be rewritten as

$$\int R(\theta, \delta)\tilde{\pi}(\theta)\,d\theta \geq \int R(\theta, \delta_0)\tilde{\pi}(\theta)\,d\theta$$

and implies that δ_0 is a Bayes estimator for $\tilde{\pi}$. □□

If Θ is not a compact space, we have already seen examples where the Bayes estimators cannot constitute a complete class. For instance, when estimating the mean θ of a normal random variable $x \sim \mathcal{N}(\theta, 1)$, the estimator $\delta(x) = x$ is admissible but is not a Bayes estimator. However, in many cases, complete classes are made of generalized Bayes estimators (meaning, obviously, *Bayes and generalized Bayes* estimators). For instance, Berger and Srinivasan (1978) have established that, when estimating the natural parameter θ of an exponential family

$$x \sim f(x|\theta) = e^{\theta \cdot x - \psi(\theta)} h(x), \quad x, \theta \in \mathbb{R}^k,$$

under quadratic loss, every admissible estimator is a generalized Bayes estimator. They thus extend Brown (1971), who considered the normal case.

Example 8.4.4 In the normal case, $x \sim \mathcal{N}_p(\theta, I_p)$, we repeatedly mentioned the truncated James–Stein estimator,

$$(8.4.1) \qquad \qquad \delta^{JS}(x) = \left(1 - \frac{p-2}{||x||^2}\right)^+ x.$$

Although well performing, this estimator is not admissible because, otherwise, it would be a generalized Bayes estimator. Since δ^{JS} is not analytical, this is impossible (see Exercise 8.26). ‖

Chow (1987) establishes a similar result for families with noncentrality parameters, $\chi_p^2(\lambda)$ and $\mathcal{F}_{p,q}(\lambda)$, thus giving an illustration of the completeness of the generalized Bayes rules outside the framework of exponential families. This complete class theorem leads, in particular, to the inadmissibility of the classical estimator, $(x - p)^+$, in the case of the distribution $\chi_p^2(\lambda)$, although Saxena and Alam (1982) have shown that this estimator is already efficient, since it dominates the maximum likelihood estimator (see also Exercise 3.25).

Fraisse et al. (1990) derive a result similar to Berger and Srinivasan (1978) in the presence of a *nuisance parameter*. Consider $x = (u, z)$ with $u \in \mathbb{R}^k$ and $z \in \mathbb{R}$. The density of x with respect to ν is

$$f(x|\theta, \delta) = h(u, z) e^{\theta \cdot u + \delta z - \psi(\theta, \delta)},$$

with $\theta \in \Theta \subset \mathbb{R}^k$ and $\delta \in \Delta$, a compact interval of $\mathbb{R}_+^*$. As in the normal setting, the problem at hand is to estimate θ/δ under a quadratic loss. For this model, the complete class theorem is given by:

Proposition 8.4.5 *If φ is an admissible estimator of θ/δ, there exists a measure π on $\Theta \times \Delta$ such that, for ν-almost every (u, z),*

$$(8.4.2) \qquad \qquad \varphi(u, z) = \frac{\int_{\Theta \times \Delta} \theta e^{\theta \cdot u + \delta z - \psi(\theta, \delta)} \pi(d\theta, d\delta)}{\int_{\Theta \times \Delta} \delta e^{\theta \cdot u + \delta z - \psi(\theta, \delta)} \pi(d\theta, d\delta)}.$$

Therefore, the complete class theorem of Berger and Srinivasan (1978) is still valid in the presence of nuisance parameters. The proof of Proposition 8.4.5 actually relies on Proposition 8.3.10 (see Exercise 8.27).

Example 8.4.6 Consider $x \sim \mathcal{N}_p(\theta, \sigma^2 I_p)$ and $s^2 \sim \sigma^2 \chi_q^2$, independent of x. In this setting, $\delta_0(x, s^2) = x$ is also inadmissible for $p \geq 3$. We consider extensions of the James–Stein estimator (8.4.1) of the form

$$\varphi(x, s^2) = (I_p - h(\|x\|_C^2, s^2)B)x,$$

where B and C are $(p \times p)$ matrices, h is a.e. differentiable, and $\|x\|_C^2 = x^t C x$. These estimators are called *matricial shrinkage estimators* (see Judge and Bock (1978)). Proposition 8.4.5 implies that a necessary condition for φ to be admissible is that h be infinitely differentiable and also that B and C be proportional (see Exercise 8.28). ‖

Brown (1988) considers the estimation of the mean of an exponential family, $\xi(\theta)$. In dimension one, he establishes that admissible estimators have an integral expression close to (8.4.2). In fact, admissible estimators are then equal to generalized Bayes estimators on intervals.

In the case of distributions with *discrete support*, the completeness of generalized Bayes estimators does not always hold and complete classes involve piecewise-Bayesian procedures (see Berger and Srinivasan (1978), Brown (1981), and Brown and Farrell (1985)). The complete class results obtained in Section 5.4 for testing under quadratic loss are of this type, since we saw that admissible estimators are identical to generalized Bayes estimators on *truncation* intervals.

8.5 Necessary admissibility conditions

When a complete class theorem restricting the choice of estimators to generalized Bayes estimators is not available, it is necessary to find some other way to exclude inadmissible estimators as much as possible. While being necessary, Stein's condition does not usually provide an efficient tool for the elimination of inadmissible estimators because its main practical interest is Blyth's sufficient condition. Moreover, the results of Section 8.3 cannot be applied in this general context, since they deal with generalized Bayes estimators only. For quadratic losses, Hwang (1982b) developed a technique introduced in Brown (1971) called STUB (for *semi-tail upper bounds*), which gives an effective necessary admissibility condition. It is based upon the following lemma.

Lemma 8.5.1 *Consider two estimators δ_1 and δ_2 with finite risks, such that*

$$R(\theta, \delta_1) = \mathbb{E}_\theta \left[(\delta_1(x) - \theta)^t Q(\delta_1(x) - \theta) \right] < R(\theta, \delta_2)$$

for every $\theta \in \Theta$ and a given positive-definite matrix Q. Then, every *estimator δ satisfying a.e. the inequality*

$$\delta(x)^t Q(\delta_1(x) - \delta_2(x)) < \delta_2(x)^t Q(\delta_1(x) - \delta_2(x))$$

is inadmissible under every quadratic loss.

Proof. Consider the new estimator $\delta'(x) = \delta(x) + \delta_1(x) - \delta_2(x)$. Then

$$
\begin{aligned}
R(\theta, \delta') &= \mathbb{E}_\theta \left[(\delta'(x) - \theta)^t Q(\delta'(x) - \theta) \right] \\
&= R(\theta, \delta) + 2\mathbb{E}_\theta \left[(\delta_1(x) - \delta_2(x))^t Q(\delta(x) - \theta) \right] \\
&\quad + \mathbb{E}_\theta \left[(\delta_1(x) - \delta_2(x))^t Q(\delta_1(x) - \delta_2(x)) \right] \\
&\leq R(\theta, \delta) + 2\mathbb{E}_\theta \left[(\delta_1(x) - \delta_2(x))^t Q(\delta_2(x) - \theta) \right] \\
&\quad + \mathbb{E}_\theta \left[(\delta_1(x) - \delta_2(x))^t Q(\delta_1(x) - \delta_2(x)) \right] \\
&= R(\theta, \delta) + R(\theta, \delta_1) - R(\theta, \delta_2) < R(\theta, \delta)
\end{aligned}
$$

and δ' dominates δ.   □□

This lemma may appear quite rudimentary at first sight, but it is actually quite powerful because it introduces a new necessary admissibility condition for *every* (risk) ordered couple (δ_1, δ_2). Moreover, since admissibility does not depend on the matrix Q, it provides an extended battery of inadmissibility criteria. In particular, it recovers the necessary admissibility condition (i) of Theorem 8.2.17.

Example 8.5.2 Consider $x \sim \mathcal{N}_p(\theta, I_p)$. It follows from James and Stein (1961) (see Note 2.8.2) that, among the estimators

$$\delta_a(x) = \left(1 - \frac{a}{||x||^2} \right) x,$$

δ_{p-2} is optimal for the usual quadratic loss. Therefore, Lemma 8.5.1 implies that every estimator δ satisfying

(8.5.1) $$\delta(x)^t x \frac{a - (p - 2)}{||x||^2} \leq \left(1 - \frac{a}{||x||^2} \right) (a - (p - 2))$$

is inadmissible. Consider δ of the form

$$\delta(x) = \left(1 - \frac{h(x)}{||x||^2} \right) x.$$

Then (8.5.1) implies that δ is inadmissible if

$$h(x) \leq a < p - 2 \qquad \text{or} \qquad h(x) \geq a > p - 2.$$

Therefore, every estimator such that h is uniformly greater or smaller than $(p - 2)$ is inadmissible. The necessary part of Theorem 8.2.17 follows by considering instead the truncated estimators ($a \leq p - 2$),

$$\varphi_a(x) = \left(1 - \frac{a}{||x||^2} \mathbb{I}_{[K, +\infty[}(||x||^2) \right) x,$$

and showing that $a^* = p - 2$ corresponds to the optimal estimator of this class (see Exercise 8.20). Lemma 8.5.1 then implies that, if

$$h(x) \leq a < p - 2$$

for $||x||^2 > K$, then the estimator δ is also inadmissible. $\|$

In fact, notice also that in Lemma 8.5.1 the *strict* inequality $R(\theta, \delta_1) < R(\theta, \delta_2)$ does not need to hold for every θ, but only for some θ's, as long as $R(\theta, \delta_1) \leq R(\theta, \delta_2)$ is satisfied for every $\theta \in \Theta$.

Example 8.5.3 Das Gupta (1984) derives from Lemma 8.5.1 a necessary admissibility condition for exponential distributions. If $x_1, \ldots, x_p$ are random variables from $\mathcal{E}xp(\theta_i)$, every estimator δ of $(\theta_1^{-1}, \ldots, \theta_p^{-1})$ satisfying

$$\sum_{i=1}^{p} x_i^{-3} \delta_i(x) \leq \sum_{i=1}^{p} x_i^{-3} \delta_{c,i}^{B}(x)$$

for $x_i \leq M$, $x = (x_1, \ldots, x_n)$, and

$$\delta_{c,i}^{B}(x) = \frac{x_i}{2} \left[1 + \frac{cx_i^{-4}}{2 \left(\sum_{j=1}^{p} x_j^{-2} \right)^2} \right],$$

$0 < c < 2(p - 1)$, is inadmissible. The estimator $\delta_{c,i}^{B}$ was introduced by Berger (1980b) to improve upon the usual estimator, $x/2$, for $p \geq 2$. Notice that $x/2$ actually dominates the maximum likelihood estimator, x (see Exercise 8.32). $\|$

It is also possible to derive from Lemma 8.5.1 a necessary admissibility condition for the estimation of a normal mean vector when the variance is known up to a multiplicative factor, σ^2, as in Example 8.4.6. Consider $x \sim \mathcal{N}_p(\theta, \sigma^2 I_p)$ and $s^2 \sim \sigma^2 \chi_q^2$ an independent observation of σ^2. The following result provides a necessary admissibility condition (Robert (1987)).

Proposition 8.5.4 *If, for the estimator*

$$\delta(x) = (1 - h(||x||^2, s^2))x,$$

there exist α, M_1, and M_2 such that

(i) *for $t \geq M_1$ and $u \leq M_2$,*

$$\frac{t}{u} h(t, u) \leq \alpha < \frac{p - 2}{q + 2};$$

    or

(ii) *for $t \leq M_1$ and $u \geq M_2$,*

$$\frac{t}{u} h(t, u) \geq \alpha > \frac{p - 2}{q + 2};$$

δ is inadmissible under quadratic loss.

The proof of this result is based on the existence of an optimal estimator in the class

$$\varphi_c(x, s^2) = x - \frac{cs^2}{||x||^2} \mathbb{I}_A(||x||^2, s^2)x,$$

where $A = [K_1, +\infty) \times [0, M_2]$ or $A = [0, K_1] \times [M_2, +\infty)$. To support Proposition 8.5.4, recall that, in this setting, the James–Stein estimators are of the form

$$\delta_a^{JS}(x, s^2) = \left(1 - \frac{as^2}{||x||^2}\right)x$$

and that

$$a^* = \frac{p - 2}{q + 2}$$

gives an optimal estimator in the class δ_a^{JS}. Therefore, $\delta_{a^*}^{JS}$ gives the minimal shrinkage factor for $||x||^2/s^2$ large and the maximal shrinkage for $||x||^2/s^2$ small. The estimator $\delta_{a^*}^{JS}$ is nonetheless inadmissible (Example 8.4.4). Fraisse et al. (1998) extend this result to exponential families with a nuisance parameter in the same setting as in Proposition 8.4.5.

Example 8.5.5 (Example 8.4.6 continued) Among matricial shrinkage estimators, the only interesting estimators are of the form

$$(8.5.2) \qquad \varphi(x, s^2) = (I_p - h(x^t Bx, s^2)B)x,$$

since the other estimators are inadmissible. Proposition 8.5.4 implies that, if, for every (t, u),

$$\frac{t}{u}h(t, u) \le \alpha < \frac{p - 2}{q + 2},$$

these estimators are also inadmissible. In addition, notice that a necessary minimaxity condition under quadratic loss is

$$\frac{t}{u}h(t, u) \le 2 \frac{\text{tr}(B) - 2\lambda_{\max}(B)}{\lambda_{\max}(B)} \frac{1}{q + 2},$$

where $\text{tr}(B)$ denotes the trace, and $\lambda_{\max}(B)$ the largest eigenvalue, of B (see Brown (1975) and Cellier et al. (1989)). Therefore, a necessary condition for the existence of an estimator satisfying both admissibility and minimaxity requirements is

$$\text{tr}(B) > \lambda_{\max}(B) \frac{p + 2}{2},$$

which excludes estimators shrinking toward subspaces of small dimensions.

This result also points out that admissibility and minimaxity are not totally compatible. In fact, an estimator admissible under a quadratic loss is so under all quadratic losses. On the contrary, Brown (1975) shows that the unique estimator of the form (8.5.2), which is minimax for all quadratic losses, is $\delta_0(x) = x$. This result also relates to the U-admissibility of the estimator δ_0 established in Brown and Hwang (1989) (see Section 2.6). ‖

8.6 Exercises

Section 8.2.1

8.1 (Lehmann (1986))   Consider a random variable x with mean μ and variance σ^2.

  a. Show that $\delta(x) = ax + b$ is an inadmissible estimator of μ under quadratic loss if

  (a) $a > 1$;
  (b) $a < 0$; and
  (c) $a = 1$ and $b \neq 0$.

  b. Generalize to the case where $\delta(x) = (1 + h(x))x$ with $h(x) > 0$.

8.2 (Exercise 8.1 cont.) Deduce that it is sufficient to consider $\lambda \geq 0$ for the estimators (8.1) used in Theorem 8.2.7.

8.3 Consider $x \sim \mathcal{U}_{[-\theta,\theta]}$ when $\pi(\theta)$ is the uniform distribution $\mathcal{U}_{[0,1]}$.

  a. Show that

  $$\delta_1^\pi(x) = \begin{cases} \dfrac{1 - |x|}{\log(1/|x|)} & \text{if } |x| \leq 1, \\ 0 & \text{otherwise}, \end{cases}$$

  is a Bayes estimator that is inadmissible and dominated under the usual quadratic loss by

  $$\delta_2^\pi(x) = \begin{cases} \delta_1^\pi(x) & \text{if } |x| \leq 1, \\ |x| & \text{otherwise}. \end{cases}$$

  b. Show that δ_2^π is also a Bayes estimator for π.

8.4 Consider $x \sim \mathcal{B}(n, p)$ and determine whether $\delta_0 \equiv 0$ is an admissible estimator of p under quadratic loss.

8.5 (Johnson (1971)) Consider $x \sim \mathcal{B}(n, \theta)$.

  a. Show that $\delta_0(x) = x$ is the maximum likelihood estimator of θ and also a Bayes estimator under quadratic loss for $\pi(\theta) = 1/\theta(1 - \theta)$.

  b. Show that $(\delta_0, 1 - \delta_0)$ is admissible under the loss

  (8.6.1) $L(\theta, \delta) = (\theta - \delta_1)^2 + (1 - \theta - \delta_2)^2$.

  (*Hint:* Use the Bayes representation of δ_0 to show that

  $$\int [R(\theta, \delta) - R(\theta, (\delta_0, 1 - \delta_0))] \frac{d\theta}{\theta(1 - \theta)} \geq 0$$

  and to deduce that equality only occurs for $\delta_1 = \delta_0$, $\delta_2 = 1 - \delta_0$.)

  c. Show that a complete class for the loss (8.6.1) is made of the estimators such that $\delta_1 = 1 - \delta_2$.

  d. Generalize the result of b. to the multinomial setting $x \sim \mathcal{M}_k(n, p_1, \ldots, p_k)$. (*Hint:* Use induction.)

Section 8.2.2

8.6 Determine the beta priors $\mathcal{B}e(\alpha, \beta)$ which correspond to the admissible estimators of Example 8.2.9.

Section 8.2.3

8.7 In the setting of Example 8.2.12, show that the Bayes risk of δ^π is infinite and determine whether δ_{c*} is a Bayes estimator.

8.8 *(Zidek (1970)) For $x \sim f(x|\theta)$, $\theta \in \mathbb{R}$, such that $\{\theta; f(x|\theta) > 0\}$ is an interval, consider the estimation of $g(\theta)$ under quadratic loss. We want to study a sufficient condition for the generalized Bayes estimator

$$\delta^\pi(x) = \frac{\int g(\theta)f(x|\theta)\pi(\theta)\,d\theta}{\int f(x|\theta)\pi(\theta)\,d\theta}$$

to be admissible when π is a measure and

$$\int R(\theta, \delta^\pi)\pi(\theta)\,d\theta = +\infty.$$

a. Let us define

$$M(x, \theta) = \int_\theta^{+\infty} [g(t) - \delta^\pi(x)]^2 f(x|t)\pi(t)\,dt$$

and

$$h(\theta) = \int \left[\frac{M(x, \theta)}{f(x|\theta)\pi(\theta)}\right]^2 f(x|\theta)\,dx.$$

Show that there exists a function $q(\theta)$ such that $\tilde{\pi}(\theta) = q(\theta)\pi(\theta)$ is a probability density and that

$$\int R(\theta, \delta^\pi)\tilde{\pi}(\theta)\,d\theta < +\infty.$$

b. Let $\tilde{\delta}$ be the Bayes estimator associated with $\tilde{\pi}$. Show that

$$r = \int [R(\theta, \delta^\pi) - R(\theta, \delta)]\tilde{\pi}(\theta)\,d\theta = \int \frac{[\int q'(\theta)M(x, \theta)\,d\theta]^2}{\int f(x|\theta)\pi(\theta)\,d\theta}\,dx.$$

c. Denoting $q(\theta)$ by $f^2(\theta)$, derive from the Cauchy–Schwarz inequality that

$$r \leq 4\int [f'(\theta)]^2 h(\theta)\pi(\theta)\,d\theta.$$

d. Show that if, for every (θ_0, θ_1) and $\epsilon > 0$, there exist a function q such that $q(t) = 1$ on (θ_0, θ_1) and a real number $r < \epsilon$, then the estimator δ^π is admissible.

e. Consider the condition (E): If

$$\int_t^{+\infty} R(\theta, \delta^\pi)\pi(\theta)\,d\theta = +\infty, \qquad \text{then} \qquad \int_t^{+\infty} \frac{1}{h(\theta)\pi(\theta)}\,d\theta = +\infty.$$

Let

$$y(\theta) = \int_{\theta_1}^\theta \frac{1}{h(t)\pi(t)}\,dt$$

and

$$f(t) = \left(1 - \frac{y(t)}{F}\right)\mathbb{I}_{0 \leq y(t) \leq F}.$$

Show that

$$f'(t) = -\frac{1}{Fh(t)\pi(t)} \qquad (0 \leq y(t) \leq F),$$

and that

$$\int_{\theta_1}^{+\infty} [f'(t)]^2 h(t)\pi(t)\, dt = \frac{1}{F}.$$

Deduce from (E) that it is possible to choose F such that $r < \epsilon$. Conclude by deriving a sufficient admissibility condition.

f. Repeat question e. under the symmetric assumption, that is, if

$$\int_{-\infty}^{t} R(\theta, \delta^\pi)\pi(\theta)\, d\theta = +\infty, \quad \text{then} \quad \int_{-\infty}^{t} \frac{1}{h(\theta)\pi(\theta)}\, d\theta = +\infty.$$

8.9 Consider the bounded loss

$$L(\theta, \delta) = 1 - e^{-a(\theta-\delta)^2} \qquad (a > 0),$$

for the estimation of θ when $x \sim \mathcal{N}(\theta, 1)$.

a. Determine the Bayes estimators associated with the conjugate priors $\theta \sim \mathcal{N}(\mu, \tau^2)$.

b. Determine the Bayes estimators associated with the prior distributions $\pi(\theta) \propto \exp(-\lambda|\theta - \mu|)$.

c. Examine the admissibility of the generalized Bayes estimator associated with the Jeffreys prior $\pi(\theta) = 1$ when a varies. (*Hint:* Determine whether the Bayes risk is finite and apply Blyth's method if necessary.)

Section 8.2.4

8.10 Establish the representation formula (8.2.3) and check the equalities in Example 8.3.5.

8.11 Show that the estimators δ_{CZ} proposed in Section 8.2 in a Poisson framework are indeed generalized Bayes estimators by exhibiting the corresponding prior distributions.

8.12 *(Berger (1982b)) Consider x distributed according to

$$x \sim f(x|\theta) = h(x)e^{\theta x - \psi(\theta)}$$

for $x \in [a, b]$. Given two positive differentiable functions, m_0 and d, define

$$\delta_0(x) = \frac{m'_0(x)}{m_0(x)} - \frac{h'(x)}{h(x)}, \quad \gamma(x) = 2\frac{d'(x)}{d(x)}, \quad \text{and} \quad \delta(x) = \delta_0(x) + \gamma(x).$$

a. Show that, under quadratic loss,

$$R(\theta, \delta) - R(\theta, \delta_0) = \mathbb{E}_\theta\left(\frac{4}{d(x)}\left[d''(x) + d'(x)\frac{m'_0(x)}{m_0(x)}\right]\right),$$

under some regularity conditions including

$$\lim_{x \to a} h(x)\gamma(x)e^{\theta x} = \lim_{x \to b} h(x)\gamma(x)e^{\theta x} = 0.$$

b. Assume that one of the two functions

$$g_1(x) = \int_a^x \frac{1}{m_0(y)}\, dy \quad \text{or} \quad g_2(x) = \int_x^b \frac{1}{m_0(y)}\, dy$$

is finite on $[a, b]$. Denote this function by g_i. Show that if, in addition,

$$\mathbb{E}_\theta\left|\frac{d}{dx}\log g_i\right|^2 < +\infty$$

and

$$\lim_{x \to a} h(x)e^{\theta x}\frac{g_i'(x)}{g_i(x)} = \lim_{x \to b} h(x)e^{\theta x}\frac{g_i'(x)}{g_i(x)} = 0,$$

then δ_0 is inadmissible and dominated by δ for $\gamma(x) = 2\alpha g_i'(x)/g_i(x)$ if $0 \le \alpha \le 1$.

c. Apply to the case where $x \sim \mathcal{G}(\nu, \theta)$ and

$$\pi(\theta) = \frac{1}{\pi}\frac{1}{1 + \theta^2}.$$

Section 8.2.5

8.13 Show that the transition kernel (8.2.7) is associated with the stationary measure π. (*Hint:* Establish that the detailed balance condition holds.) Deduce that the corresponding chain is either null-recurrent or transient when the prior π is improper.

8.14 Consider $x \sim \mathcal{N}(\theta, 1)$ and $\pi(\theta) \propto \exp\{-b\theta^2/2 + ab\theta\}$.

a. Give necessary and sufficient conditions on (a, b) for the posterior distribution to be defined. Show that, in this case, the posterior distribution is normal with mean $(x + ab)/(1 + b)$ and variance $1/(b + 1)$.

b. Show that the transition kernel (8.2.7) is then given by

$$\eta|\theta \sim \mathcal{N}\left(\frac{\theta + ab}{1 + b}, \frac{b + 2}{(1 + b)^2}\right).$$

c. Deduce that the Markov chain is an AR(1) model (Section 4.5.2)

$$\theta^{(t+1)} = \frac{1}{1 + b}\theta^{(t)} + \frac{ab}{1 + b} + \frac{\sqrt{b + 2}}{1 + b}\epsilon_t.$$

Conclude that it is transient when $b < 0$ and recurrent when $b = 0$.

Section 8.3.1

8.15 Verify that the three conditions of Lemma 8.3.3 are actually satisfied in the case of a quadratic loss,

$$L(\theta, \delta) = (\delta - \theta)^t Q(\delta - \theta),$$

for every positive-definite matrix Q.

Section 8.3.2

8.16 *(Clevenson and Zidek (1975)) Consider $(x_1, \ldots, x_n)$ distributed as independent Poisson random variables, $x_i \sim \mathcal{P}(\lambda_i)$.

a. Use a sequence of conjugate priors and Blyth's method to show that $\delta_0(x_i) = x_i$ is an admissible estimator of λ_i under quadratic loss.

b. For $n \ge 2$, show that

$$\mathbb{E}_\lambda\left[\sum_{i=1}^n \frac{1}{\lambda_i}\left\{x_i\left(1 + \frac{n-1}{\sum_{i=1}^n x_i}\right)^{-1} - \lambda_i\right\}^2\right] \le \mathbb{E}_\lambda\left[\sum_{i=1}^n \frac{1}{\lambda_i}(x_i - \lambda_i)^2\right]$$

and deduce that $\delta_0(x_1, \ldots, x_n) = (x_1, \ldots, x_n)$ is an inadmissible estimator of $\lambda = (\lambda_1, \ldots, \lambda_n)$. (*Hint:* Minimize (in λ) $\mathbb{E}_\lambda[\sum_i \lambda_i^{-1}(ax_i - \lambda_i)^2]$ and replace the solution a by $\sum_i x_i/\sum_i x_i + n - 1$.)

8.17 Establish the equivalent of Example 8.3.7 for the Poisson distribution, that is, show that, if $H_0 : \lambda \leq \lambda_0$ and $\varphi(x) = P_{\lambda_0}(X \geq x)$, with $X \sim \mathcal{P}(\lambda_0)$, then φ is admissible under quadratic loss. (*Hint:* Use Blyth's condition.)

8.18 Solve Exercise 8.17 for the gamma distribution, $\mathcal{G}(\nu, \theta)$ and $H_0 : \theta \leq \theta_0$.

8.19 Consider $x \sim \mathcal{N}_2(\theta, I_2)$. Check whether Blyth's condition for the admissibility of $\delta_0(x) = x$ is satisfied by the sequence $\pi_n(\theta)$ equal to

$$e^{-||\theta||^2/2n}.$$

If this sequence cannot be used, propose another.

8.20 *(Hwang and Brown (1991)) Consider $x \sim \mathcal{N}_p(\theta, I_p)$. The usual confidence region is

$$C_x = \{\theta; \ ||\theta - x|| < c\},$$

with $P_\theta(\theta \in C_x) = 1 - \alpha$. Using Blyth's method, show that the evaluation $\gamma_0(x) = 1 - \alpha$ is admissible under the quadratic loss

$$L(\theta, \gamma) = (\gamma - \mathbb{I}_{C_x}(\theta))^2,$$

for $p \leq 4$. [*Note:* Robert and Casella (1993) complete this result by showing that this constant estimator is inadmissible for $p \geq 5$. On the contrary, Hwang and Brown (1991) establish that, under frequentist validity, γ_0 is admissible for every p (see Section 5.5).]

8.21 In the setting of Example 8.3.7, show that, for $x \neq 0$,

$$\varphi(x) = \frac{1}{B(x, m - x + 1)} \int_0^{\theta_0} t^{x-1}(1 - t)^{m-x} \, dt$$

and derive the corresponding representation for $x = 0$.

Section 8.4

8.22 A class $\mathcal{C}$ is said to be *minimal complete* if $\mathcal{C}$ is complete and every proper subset of $\mathcal{C}$ is not complete.

a. Show that every complete class contains every admissible estimator.

b. Show that, if there exists a minimal complete class, it is exactly made of the admissible estimators.

8.23 *(Karlin and Rubin (1956)) Consider $f(x|\theta)$ satisfying the monotone likelihood ratio property (in $x \in \mathbb{R}$), with $\theta \in \Theta$. The estimation problem is said to be *monotone* if $L(\theta, \delta)$ is minimum for $\delta = q(\theta)$, with q increasing in θ, and if $L(\theta, \delta)$ is an increasing function of $|\delta - q(\theta)|$.

a. Show that, if L is convex, the estimators which are increasing functions of x constitute a complete class.

b. Show that, if δ_0 is not monotone, the monotone estimator δ_M, defined by

$$P_{q^{-1}(a)}(\delta_M(X) \leq a) = P_{q^{-1}(a)}(\delta_0(X) \leq a), \quad \forall a$$

dominates δ_0.

c. If δ_M is strictly increasing, show that the above relation implies that $\delta_M(x)$ is a number a such that

$$F(x|q^{-1}(a)) = P_{q^{-1}(a)}(\delta_0(X) \leq a).$$

8.24 Apply Exercise 8.23 to the case where $x \sim \mathcal{N}(\theta, 1)$, $L(\theta, \delta) = (\theta - \delta)^2$ and $\delta_0(x) = -cx + b$, with $c > 0$.

8.25 (Berger (1985a)) Consider Θ, a finite set of cardinality p, and assume that the risk set $\mathcal{R}$ is bounded and closed from below. Let $\Gamma(\mathcal{R})$ be the lower boundary of $\mathcal{R}$, that is,

$$\Gamma(\mathcal{R}) = \{r \in \mathcal{R}; \not\exists r' \in \mathcal{R}, \ r' \neq r \text{ and } r'_i \leq r_i, \ 1 \leq i \leq p\} \subset \mathcal{R}.$$

The loss L is assumed to be convex.

a. Show that the set of the estimators with risk vector in $\Gamma(\mathcal{R})$ is a minimal complete class.

b. Show that the set of the Bayes estimators is a complete class and that the set of the admissible Bayes estimators is a minimal complete class.

c. Generalize to the case where L is not convex.

8.26 *(Berger and Srinivasan (1978)) Consider $x \sim \mathcal{N}_p(\theta, \Sigma)$ with Σ known. The mean θ is estimated under quadratic loss. Show that an estimator δ_0 is a generalized Bayes estimator if, and only if,

(i) $g(x) = \Sigma^{-1}\delta_0(x)$ is continuously differentiable, with symmetric Jacobian matrix $J_g(x) = \nabla\nabla^t g(x)$; and

(ii) for $g(x) = \nabla r(x)$, $\exp\{r(x)\}$ can be expressed as a Laplace transform.

8.27 *(Fraisse et al. (1990)) Consider $x = (u, z)$ with $u \in \mathbb{R}^k$ and $z \in \mathbb{R}$, with density

$$f(x|\theta, \delta) = \exp\{\theta \cdot u + \delta z - K(\theta, \delta)\}$$

with respect to a σ-finite ν measure, with $\theta \in \Theta \subset \mathbb{R}^k$ and $\delta \in \Delta$, compact subset of $\mathbb{R}_+^*$.

a. Show (or accept) the following lemma: *If (μ_n) is a sequence of measures with finite support such that, for almost every (u, z),*

$$\sup_n \|\nabla\psi_{\mu_n}(z)\| < +\infty,$$

then there exist a measure μ and a subsequence (n_k) such that

$$\lim_{k \to +\infty} \psi_{\mu_{n_k}}(u, z) = \psi_\mu(u, z) \quad and \quad \lim_{k \to +\infty} \nabla\psi_{\mu_{n_k}}(u, z) = \nabla\psi_\mu(u, z),$$

with

$$\psi_\mu(u, z) = \int_{\Theta \times \Delta} e^{\theta \cdot u + \delta z}\mu(d\theta, d\delta).$$

b. Deduce from Proposition 8.3.10 that, for every admissible estimator φ of θ/δ under the squared error loss $\delta^2\|\varphi - \theta/\delta\|^2$, there exists a sequence of measures (ϱ_n) with finite supports on $\Theta \times \Delta$ such that

$$\varphi(u, z) = \lim_{n \to +\infty} \frac{\int \theta e^{\theta \cdot u + \delta z}\mu_n(d\theta, d\delta)}{\int \delta e^{\theta \cdot u + \delta z}\mu_n(d\theta, d\delta)},$$

with $\mu_n(d\theta, d\delta) = e^{-K(\theta, \delta)}\varrho_n(\theta, \delta)$.

c. Show that the condition of the above lemma is satisfied and that, for every admissible estimator φ, there exists μ_0 such that, a.e., it satisfies

$$\varphi(u, z) = \frac{\int \theta e^{\theta \cdot u + \delta z}\mu_0(d\theta, d\delta)}{\int \delta e^{\theta \cdot u + \delta z}\mu_0(d\theta, d\delta)},$$

that is, φ is a generalized Bayes estimator associated with μ_0.

8.28 (Fraisse et al. (1990)) Consider $x \sim \mathcal{N}_p(\theta, \sigma^2 I_p)$ and $s^2 \sim \sigma^2 \chi_q^2$. The mean θ is estimated under a quadratic loss, with $\sigma \in [a, b]$.

   a. Show that this model fits into the framework of Exercise 8.26.

   b. Consider the estimator

$$\varphi(x, s^2) = (I_p - h(x^t Bx, s^2)C)x.$$

   Show that, if φ is admissible, there exists $\varrho \in \mathbb{R}_+^*$ such that $B = \varrho C$.

   c. Compare with the results of Exercise 8.26.

8.29 *(Moors (1981)) Consider $x \sim \mathcal{Be}(p)$, with $0.2 \le p \le 0.8$, and the parameter p is to be estimated under quadratic loss.

   a. Show that $\delta^\pi(1) = 1 - \delta^\pi(0)$ when the prior $\pi(p)$ is symmetric around $1/2$.

   b. Show that $\delta^\pi(1) \le \max_p[1 - 2p(1-p)] = 0.68$.

   c. Deduce that, if an estimator satisfies $\delta(1) = 1 - \delta(0)$ and $\delta(1) > 0.68$, it is inadmissible.

8.30 *(Johnson (1971)) Consider $x \sim \mathcal{B}(n, \theta)$, with θ to be estimated under quadratic loss.

   a. Recall why all admissible estimators are necessarily Bayes estimators.

   b. Show that the reverse is false, that is, exhibit some inadmissible Bayes estimators.

   c. Show that the set of admissible Bayes estimators is constituted of the estimators

$$\delta_\tau(x) = \begin{cases} 0 & \text{if } 0 \le x \le \underline{n}, \\[2mm] \dfrac{\displaystyle\int_0^1 \theta^{x-\underline{n}}(1-\theta)^{n-x-\overline{n}-1}\, d\tau(\theta)}{\displaystyle\int_0^1 \theta^{x-\underline{n}-1}(1-\theta)^{n-x-\overline{n}-1}\, d\tau(\theta)} & \text{if } \underline{n} < x < \overline{n}, \\[4mm] 1 & \text{if } \overline{n} \le x \le n. \end{cases}$$

   d. Explain why δ_τ is a Bayes estimator for a whole class of prior distributions τ.

8.31 (Hwang, et al. (1992)) Consider $H_0 : \theta \in \Theta_0$. Testing procedures γ are compared under a strictly convex loss, $L(\mathbb{I}_{\Theta_0}(\theta), \gamma)$.

   a. Show that $\gamma_0(x) = 1/2$ is the only minimax estimator.

   b. Deduce that γ_0 is admissible.

   c. Is it possible to write γ_0 as a generalized Bayes estimator? Does this phenomenon contradict the complete class result of Section 5.4 (Theorems 5.4.3 and 5.4.4)?

8.32 Given $x \sim \mathcal{E}xp(\theta)$ and $\delta_c(x) = cx$, determine the best estimator δ_c of θ^{-1} under quadratic loss. Show that this estimator is a generalized Bayes estimator and discuss its admissibility.

Section 8.5

8.33 (Robert and Casella (1994)) Consider $x \sim \mathcal{N}(\theta, 1)$ and the usual confidence set $C_x = [x - c, x + c]$. Instead of using the fixed confidence report $\alpha = P_\theta(\theta \in C_x)$, a procedure φ is proposed and evaluated under the loss

$$L(\theta, \varphi) = d(\theta, C_x)(\mathbb{I}_{C_x}(\theta) - \varphi)^2,$$

the weight $d(\theta, C_x)$ being a measure of the distance between θ and the border of C_x.

   a. Justify the use of a data-dependent assessment of a confidence interval and of the distance weight.

   b. In the particular case

$$d(\theta, C_x) = 1 - e^{-\omega(\theta - x)^2} \left(1 - e^{-\omega(\theta - x)^2} \right),$$

      show that this distance is minimal for $|\theta - x| = c$ if $\omega = \log(2)/c^2$.

   c. Give the general form of the Bayes estimators under this loss and show that the complete class result obtained in Theorem 5.4.3 still applies.

   d. For the class of symmetric distances of the form $h(|\theta - x|)$ and $\pi(\theta) = 1$, show that the Bayes estimators are constant and not necessarily equal to α. Are these estimators admissible?

8.34 Derive Proposition 8.5.4 from Lemma 8.5.1.

8.35 Show that, if $x \sim \mathcal{N}_p(\theta, I_p)$, $\delta_0(x) = x$ is admissible when considered under the class of losses

$$L(\theta, \delta) = (\theta - \delta)^t Q(\theta - \delta),$$

where Q varies in the set of symmetric positive-definite matrices.

8.36 (Hwang (1982b)) Consider $x \sim \mathcal{N}_p(\theta, I_p)$. A class of estimators of θ is given by

$$\phi_a(x) = x - \frac{a}{||x||^2} \mathbb{I}_{[K, +\infty[}(||x||^2)x,$$

for $0 \leq a \leq (p - 2)$.

   a. Show that φ_{a^*} associated with $a^* = p - 2$ is optimal among the estimators φ_a under the usual quadratic loss.

   b. Derive the result of Example 8.5.2.

   c. Reproduce this technique with

$$\varphi_b(x) = x - \frac{b}{||x||^2} \mathbb{I}_{[0, K]}(||x||^2)x$$

     and $b \geq (p - 2)$. Derive a STUB condition.

Note 8.7.1

8.37 *Show that K in (8.2.7) and K^* in (8.7.1) are either both recurrent, or both transient. (*Hint:* Use the indicator variable $\sum_{t=1}^{\infty} \mathbb{I}_B(x_t)$ for an arbitrary set B.)

8.38 *In the setting of Example 8.7.1,

   a. Show that (8.7.2) holds.

b. Show that the Markov chain associated with K^* can be written as $x_{t+1} = (x_t + b)z_{t+1}$, where the z_t's are independent with density

$$f(z) = \frac{\Gamma(2\alpha + a)}{\Gamma(\alpha + a)\Gamma(\alpha)} \frac{z^{\alpha-1}}{(z+1)^{2\alpha+a}}.$$

c. Show that z_t has an infinite mean when $a + \alpha \leq 1$ and that $\mathbb{E}[\log(z_t)]$ is negative when $a < 0$, zero when $a = 0$, and positive when $a < 0$.

d. In the case $b = 0$, show that (x_t) is recurrent if, and only if, $a = 0$.

8.39 *(Hobert and Robert (1999)) Consider $x \sim \mathcal{P}(\theta)$ with $\pi(\theta) \propto \theta^{a-1} \exp\{-b\theta\}$.

a. Give necessary and sufficient conditions on (a, b) for the posterior distribution to be defined and the prior distribution to be improper.

b. Give the transition kernel (8.2.7) in this case. In the special case $b = -1/2$ and $a = k/2$, show that the transition is a noncentral chi-squared distribution.

c. Show that the transition kernel (8.7.1) is

$$K^*(x, y) = \frac{\Gamma(y + x + a)}{y!\Gamma(x + a)} p^{x+a}(1 - p)^y,$$

where $p = (b+1)/(b+2)$.

d. Show that this distribution corresponds to the usual negative binomial distribution when a is a positive integer.

e. In the general case, the distribution with mass function

$$P(Z = z) = \frac{\Gamma(z + c)}{z!\Gamma(c)} p^c(1 - p)^z$$

is called a *generalized negative binomial* distribution $\mathcal{N}eg(c, p)$. Deduce from the probability generating function that, if $z_1, \ldots, z_n$ are independent with $z_i \sim \mathcal{NB}(c_i, p)$, $z_1 + \ldots + z_n \sim \mathcal{N}eg(c_1 + \ldots + c_n, p)$.

f. Deduce that (x_t) associated with the kernel K^* is a *branching process*,

$$x_{t+1} = \sum_{i=1}^{x_t} \eta_{i,t} + \omega_{t+1}$$

where $\eta_{i,t} \sim \mathcal{N}eg(1, p)$ and $\omega_{t+1} \sim \mathcal{N}eg(a, p)$.

g. Conclude that the chain (x_t) is recurrent when $b = 0$ and $0 < a < 1$, and transient otherwise.

8.40 *(Hobert and Robert (1999)) Consider $x \sim \mathcal{N}eg(k, \theta)$ with $\pi(\theta) \propto \theta^{a-1}(1 - \theta)^{b-1}$.

a. Give necessary and sufficient conditions on (a, b) for the posterior distribution to be defined and the prior distribution to be improper.

b. Give the transition kernel (8.2.7) when $b = k$.

c. Show that the corresponding Markov chain $(\theta^{(t)})$ can be written as $\theta^{(t+1)} = \theta^{(t)}/(\omega_t + \theta^{(t)})$, where the ω_t's are i.i.d.

d. Show that $\mathbb{E}[\log \omega_t] = 0$ if, and only if, $a = 0$ and deduce that the chain $(\theta^{(t)})$ is recurrent when $a = 0$ and transient otherwise;

8.41 For the transition kernel (8.7.3),

a. Show that the stationary measure is $\pi(\theta)\Psi(\theta)$, where π is the prior distribution and $\Psi(\theta)$ is the normalizing constant in $T(\theta, \eta)$. (*Hint:* Establish that the detailed balance condition holds and use the equality $\pi(\theta)f(x|\theta)\pi(\eta|x) = \pi(\eta)f(x|\eta)\pi(\theta|x)$.)

b. Deduce that the chain is positive recurrent when the Bayes risk

$$\int\int (\varphi(\theta) - \mathbb{E}[\varphi(\theta)|x])^2 \, f(x|\theta)\pi(\theta)dxd\theta$$

is finite.

8.7 Notes

8.7.1 Extensions on Eaton's sufficient admissibility condition

Hobert and Robert (1999) show that Eaton's sufficient condition also applies to a *dual chain*, with transition kernel

$$(8.7.1) \qquad K^*(x,y) = \int_\Theta f(x|\theta)\pi(\theta|x)\,d\theta,$$

since both kernels K in (8.2.7) and K^* are of the same nature, that is, they are both recurrent, or both transient. This duality result is of interest when the sampling space is simpler than the parameter space, for instance when K^* is on a finite state space. (This property was used in a completely different setting by Diebolt and Robert (1994) to establish finer convergence properties for the Gibbs sampler in latent variable models. See Robert and Casella (1999, Section 7.2.4).) Hobert and Robert (1999) demonstrate the interest of this condition in some standard settings (Exercises 8.39 and 8.40).

Example 8.7.1 Consider a gamma model, $x \sim \mathcal{G}a(\alpha, \theta)$, with $\pi(\theta) \propto \theta^{\alpha-1} \exp\{-b\theta\}$. The posterior is defined when $b \geq 0$ and $a > -\alpha$. While the transition kernel K cannot be written in closed form, K^* can be expressed as

$$(8.7.2) \qquad K^*(x,y) = \frac{\Gamma(2\alpha + a)\Gamma(b+x)^{\alpha+a}}{\Gamma(\alpha+a)\Gamma(\alpha)} \frac{y^{\alpha-1}}{(x+y+b)^{2\alpha+a}}.$$

Hobert and Robert (1999) then show that this kernel is recurrent if, and only if, $a = 0$ (Exercise 8.38). ‖

Eaton (1999) generalizes Eaton's (1992) result to arbitrary functions of θ, $\varphi(\theta)$, estimated under squared-error loss, through another Markov representation. Instead of $K(\theta|\eta)$ defined by (8.2.7), Eaton (1999) proposes to use the transition kernel

$$(8.7.3) \qquad T(\theta|\eta) = \Psi(\eta)^{-1}(\varphi(\theta) - \varphi(\eta))^2 K(\theta|\eta),$$

where $\Psi(\theta)$ is the normalizing factor of this density. The equivalent of Theorem 8.2.19 in this case is then that the Bayes estimator $\mathbb{E}^\pi[\varphi(\theta)|x]$ is admissible when the Markov chain associated with T is recurrent. Although this result is case-dependent, that is, requires a study of the Markov chain for each function φ of interest, the extension to unbounded functions φ makes it quite valuable. (As noted in Eaton (1999), the result extends to the estimation of vector-valued functions $\varphi(\theta)$ under squared-error loss.)

Example 8.7.2 In the special case of a location family, that is, when $f(x|\theta) = g(x - \theta)$, if θ is estimated under the loss function $L(\theta, d) = (\theta - d)^2$, then the Markov kernel K associated with the flat prior $\pi(\theta) = c$ is

$$K(\theta|\eta) = \int g(x - \theta)g(x - \eta)dx = r(\theta - \eta),$$

by an appropriate change of integrand in the integral. Then,

$$T(\theta, \eta) \propto (\theta - \eta)^2 r(\theta - \eta) = t(\theta - \eta),$$

and the proportionality factor is independent of θ. Therefore, the Markov chain associated with T is a random walk, which is recurrent in dimension one if the first moment of t exists. As shown in Eaton (1999), this is equivalent to the existence of the third moment of g.   ‖

CHAPTER 9

Invariance, Haar Measures, and Equivariant Estimators

The ring certainly looked like stone, but it felt harder than steel and heavier than lead. And the circle of it was twisted. If she ran a finger along one edge, it would go around twice, inside as well as out; it only had one edge. She ran her finger along that edge twice, just to convince herself.

Robert Jordan, *The Dragon Reborn, Book III of the Wheel of Time.*

9.1 Invariance principles

Invariance can be seen as a notion introduced in frequentist settings to restrict the range of acceptable estimators sufficiently so that an optimal estimator can be derived. From this point of view, it appears as an alternative to unbiasedness, and is thus similarly at odds with the Bayesian paradigm. However, invariance can also be justified on a non-decision-theoretic heuristic, namely, that estimators should meet some consistency requirements under a group of transformations, and it is thus of interest to consider this notion. Moreover, optimal (equivariant) estimators are always Bayes or generalized Bayes estimators. The corresponding measures can then be considered as noninformative priors induced by the invariance structure. Therefore, a Bayesian study of invariance is appealing, not because classical optimality once more relies on Bayesian estimators, but mainly because of the connection between invariance structures and the derivation of noninformative distributions.

A first version of the *invariance principle* is to consider that the properties of a statistical procedure should not depend on the *unit of measurement*. If x and θ are measured in unit u_1, and if y and η are the transforms of x and θ for the new unit u_2, then an estimator $\delta_2(y)$ of η should then correspond to the estimator $\delta_1(x)$ of θ by the same change of unit. Of course, the notion of *unit of measurement* is to be understood in a general sense: for instance, it can indicate the choice of a particular scale (*cm* vs. *m*)—and estimators should be *scale-invariant*—the choice of a particular origin—in which case estimators should be *translation-equivariant*—or even the ordering of the observations of the sample x—in which case estimators should be symmetric.

Example 9.1.1 Consider the problem of estimating the speed of light, θ, given an observation x, with distribution $\mathcal{U}_{[\theta-\epsilon,\theta+\epsilon]}$, measured in meters per second. A typical change of unit in this setting is the *scale modification*, $y = \tau x$, where, for instance, $\tau = 10^{-3}$ for a conversion from meters to kilometers. In this case, $y \sim \mathcal{U}_{[\eta-\epsilon',\eta+\epsilon']}$ with $\eta = \tau\theta$, $\epsilon' = \tau\epsilon$, but the quantity η still represents the same intrinsic quantity, that is, the speed of light. If δ_0 is an estimator of θ in the initial unit, it seems legitimate to require that the estimator in the transformed problem, δ^*, satisfy the *scale equivariance* property

$$\delta^*(y) = \tau\delta_0(y/\tau).$$

Moreover, assume that the loss is the scaled quadratic loss

$$L(\theta, d) = \left(1 - \frac{d}{\theta}\right)^2.$$

It satisfies

$$L(\theta, d) = \left(1 - \frac{\tau d}{\tau\theta}\right)^2 = \left(1 - \frac{d^*}{\eta}\right)^2 = L(\eta, d^*).$$

Therefore, the loss is invariant under this change of unit and the two estimation problems are *formally identical*. It is then natural to select the same estimator for both problems, namely, $\delta^*(y) = \delta_0(y)$. Considering both equations simultaneously, it follows that the estimator should satisfy

$$\delta_0(\tau y) = \tau\delta_0(y)$$

for all τ and y. Therefore, the decision rules complying with the invariance requirements are necessarily of the form $\delta_0(x) = ax$, where a is a positive constant.   ‖

This principle is often extended to a *formal invariance principle*, according to which two problems with identical formal *structures*, $(\mathcal{X}, f(x|\theta), L)$, should lead to the same decision rule, even though they are not related physically, as in the restricted invariance principle. In the previous example, the speed of light *is always the same object*. This extension does not seem so natural in a Bayesian perspective because the prior information is not necessarily the same in both problems. It is thus only in a noninformative setting that both approaches can be compatible.

A Bayesian approach to invariance is thus justified for the following reasons:

(i) The best invariant (or *equivariant*) estimator is a generalized Bayes estimator with respect to a particular measure called a *Haar measure*.

(ii) This measure can be defended in a noninformative setting because invariance provides an alternative method for constructing noninformative prior distributions.

(iii) The most efficient method for deriving the best equivariant estimators is to use a Bayesian approach.

Therefore, invariance considerations reinforce the Bayesian paradigm, since it underlies once more a frequentist optimality criterion. In this chapter, we justify the connection between invariance and the Bayesian approach in the case of location parameters in Section 9.2, then present the general invariant framework in Section 9.3, the Haar measure as a potential noninformative prior in Section 9.4, and the Hunt–Stein theorem, which links invariance with minimaxity, in Section 9.5. For deeper coverages of invariance, from both general and Bayesian perspectives, see Berger (1985a, Chapter 8), Eaton (1989), and Wijsman (1990).

9.2 The particular case of location parameters

Consider $(x_1, \ldots, x_n)$ with density $f(x_1 - \theta, \ldots, x_n - \theta)$, with an unknown *location parameter* $\theta \in \mathbb{R}$. The natural invariance structure of this problem is *invariance by translation*. In fact, if $(x_1, \ldots, x_n)$ is transformed into $(y_1, \ldots, y_n) = (x_1 + a, \ldots, x_n + a)$, this new random variable $(y_1, \ldots, y_n)$ is distributed according to $f(y_1 - \theta - a, \ldots, y_n - \theta - a)$ and $\eta = \theta + a$ is the corresponding location parameter. Therefore, the transformed vector has the same type of density, and the problem is invariant under translation transformations. It seems logical to reproduce this structure by imposing the following invariance restriction on the estimators δ of θ:

$$(9.2.1) \qquad \delta(x_1 + a, \ldots, x_n + a) = \delta(x_1, \ldots, x_n) + a.$$

This condition is satisfied, for instance, by $\delta_0(x_1, \ldots, x_n) = \bar{x}$. Moreover, it also seems natural to impose a similar invariance restriction on the loss function, namely, that $L(\theta + a, d + a) = L(\theta, d)$ for every a. A loss compatible with the invariance structure should then be of the form

$$(9.2.2) \qquad L(\theta, d) = L(0, d - \theta) = \varrho(d - \theta).$$

The estimators satisfying (9.2.1) are called *equivariant* and the loss functions satisfying (9.2.2) *invariant*, both under the action of the translation group. The main purpose of these restrictions is to reduce sufficiently the class of "acceptable" estimators so that there might be a single *best equivariant estimator* under the loss (9.2.2), since it is not possible without this restriction (see Exercise 2.36). The following lemma shows why there can be a single optimal estimator.

Lemma 9.2.1 *Under losses of the form (9.2.2), equivariant estimators have a constant risk.*

Proof. In fact,

$$
\begin{aligned}
R(\delta, \theta) &= \mathbb{E}_\theta[\varrho(\delta(x) - \theta)] \\
&= \mathbb{E}_\theta[\varrho(\delta(x_1 - \theta, \ldots, x_n - \theta))] \\
&= \mathbb{E}_0[\varrho(\delta(x_1, \ldots, x_n))] = R(\delta, 0). \qquad \square\square
\end{aligned}
$$

Therefore, the particular setting of equivariant estimators under an equivariant loss is similar to the general case of all estimators under the Bayes risk: there exists a *total ordering* on this restricted class, since the comparison of estimators is equivalent to a comparison of real numbers. This is why a best equivariant estimator can exist.

The classical determination of this best estimator proceeds by conditioning on a maximal ancillary statistic, such as $y = (x_1 - x_n, \ldots, x_{n-1} - x_n)$. It is then straightforward to check that every equivariant estimator can be written as $\delta_0(x) + v(y)$, where δ_0 is a particular equivariant estimator, $\delta_0(x) = x_n$ say.

Lemma 9.2.2 *If there exists a function $v^*(y)$ that minimizes*

$$
\mathbb{E}_0[\varrho(\delta_0(x) + v(y))|y],
$$

the best equivariant estimator under the loss (9.2.2) is given by

$$
\delta^*(x) = \delta_0(x) + v^*(y).
$$

Proof. By definition, the best equivariant estimator minimizes (in v) the constant risk

$$
R(\delta, \theta) = \mathbb{E}_0[\varrho(\delta(x))] = \mathbb{E}_0[\varrho(\delta_0(x) + v(y))].
$$

Since y is an ancillary statistic, it is possible to condition on y, decomposing the risk as

$$
\mathbb{E}_0[\varrho(\delta_0(x) + v(y))] = \mathbb{E}\left[\mathbb{E}_0[\varrho(\delta_0(x) + v(y))|y]\right].
$$

If $v^*(y)$ minimizes the integrand for each y, δ^* minimizes the risk over the class of the equivariant estimators. $\qquad \square\square$

In the particular case where $\varrho(\delta - \theta) = (\delta - \theta)^2$, the optimal factor v^* is given by

$$
v^*(y) = -\mathbb{E}_0[\delta_0(x)|y].
$$

We have then derived the best equivariant estimator of Pitman (1939).

Corollary 9.2.3 *Under the quadratic loss $L(\theta, d) = (\theta - d)^2$, the best equivariant estimator of θ is*

$$
\delta^*(x_1, \ldots, x_n) = \frac{\int_{-\infty}^{+\infty} \theta f(x_1 - \theta, \ldots, x_n - \theta)\, d\theta}{\int_{-\infty}^{+\infty} f(x_1 - \theta, \ldots, x_n - \theta)\, d\theta}.
$$

Proof. Consider $\delta_0(x) = x_n$ as the particular equivariant estimator used in Lemma 9.2.2 and let us denote $y_n = x_n$ to complete y. The density of

$(y_1, \ldots, y_n)$ is then (for $\theta = 0$)

$$g_Y(y_1, \ldots, y_n) = f(y_1 + y_n, \ldots, y_{n-1} + y_n, y_n),$$

since $y_i = x_i - x_n$ $(i \neq n)$ and the Jacobian determinant is equal to 1. Moreover,

$$
\begin{aligned}
\mathbb{F}_0[y_n | y_1, \ldots, y_{n-1}] &= \frac{\int_{-\infty}^{+\infty} t f(y_1 + t, \ldots, y_{n-1} + t, t) \, dt}{\int_{-\infty}^{+\infty} f(y_1 + t, \ldots, y_{n-1} + t, t) \, dt} \\
&= \frac{\int_{-\infty}^{+\infty} t f(x_1 - x_n + t, \ldots, x_{n-1} - x_n + t, t) \, dt}{\int_{-\infty}^{+\infty} f(x_1 - x_n + t, \ldots, x_{n-1} - x_n + t, t) \, dt} \\
&= x_n - \frac{\int_{-\infty}^{+\infty} \theta f(x_1 - \theta, \ldots, x_{n-1} - \theta, x_n - \theta) \, d\theta}{\int_{-\infty}^{+\infty} f(x_1 - \theta, \ldots, x_n - \theta) \, d\theta},
\end{aligned}
$$

by the change of variable $\theta = x_n - t$. Since

$$\delta^*(x_1, \ldots, x_n) = x_n - \mathbb{E}_0[y_n | y_1, \ldots, y_{n-1}],$$

the above expression of δ^* follows.   □□

The main appeal of Corollary 9.2.3, besides providing the best equivariant estimator, is to exhibit this estimator as a Bayes estimator, although the whole derivation does not involve any Bayesian input. The Pitman estimator is indeed a Bayes estimator associated with the prior distribution $\pi(\theta) = 1$, that is, the usual noninformative distribution for location parameters (see Chapter 3). This result actually holds for the other invariant losses (9.2.2), as shown in Section 9.4. Therefore, the best equivariant estimator can be derived as the estimator δ that minimizes the posterior loss

$$\mathbb{E}^\pi[L(\theta, \delta)|x] = \frac{\int_{-\infty}^{+\infty} \varrho(\theta - \delta) f(x_1 - \theta, \ldots, x_n - \theta) \, d\theta}{\int_{-\infty}^{+\infty} f(x_1 - \theta, \ldots, x_n - \theta) \, d\theta},$$

or, equivalently,

$$\int_{-\infty}^{+\infty} \varrho(\theta - \delta) f(x_1 - \theta, \ldots, x_n - \theta) \, d\theta,$$

and this representation drastically simplifies the derivation of the best equivariant estimators. Section 9.4 shows that the connection between best equivariant estimators and a particular class of Bayes estimators holds in much greater generality than for the estimation of location parameters.

9.3 Invariant decision problems

A general description of the relations between invariant problems and Bayesian analysis involves an abstract description of invariance through the action of *invariance groups*. Consider a statistical model, $(\mathcal{X}, \Theta, f(x|\theta))$, and an inferential problem on θ represented by a *decision space*, $\mathcal{D}$. Assume in addition that a *group $\mathcal{G}$ of transformations on $\mathcal{X}$* is provided with the problem. (It can also be perceived as a particular case of prior information.) The existence of such a group is important in the following case.

Definition 9.3.1 *The statistical model is said to be invariant (or closed) under the action of the group $\mathcal{G}$ if, for every $g \in \mathcal{G}$, there exists a unique $\theta^* \in \Theta$ such that $y = g(x)$ is distributed according to the density $f(y|\theta^*)$. We denote $\theta^* = \bar{g}(\theta)$.*

Example 9.3.2 Consider $x \sim f(x - \theta)$ and the *translation group*

$$\mathcal{G} = \{g_c; \ g_c(x) = x + c, c \in \mathbb{R}\}.$$

The statistical model is then actually invariant under the action of $\mathcal{G}$. This is not the case with the multiplicative group

$$\mathcal{G}' = \{g_c; \ g_c(x) = cx, c > 0\},$$

since

$$y = cx \sim \frac{1}{c} f\left(\frac{y - c\theta}{c}\right).$$ ‖

When the group $\mathcal{G}$ has a globally invariant action on the model, it naturally induces a set $\bar{\mathcal{G}}$ of transformations on Θ and the fact that $\bar{\mathcal{G}}$ is also a *group* is left as an exercise for the reader. To simplify notations, we will write $g(x)$ as gx and $\bar{g}(\theta)$ as $\bar{g}\theta$ in the sequel.

We assume in addition that the loss function associated with the model, L from $\Theta \times \mathcal{D}$ in $\mathbb{R}^+$, is discriminant in the sense that two different decisions are associated with different losses and, moreover, that the loss is compatible with the invariance structure in the following sense.

Definition 9.3.3 *If the model is invariant under the action of $\mathcal{G}$, the loss L is said to be invariant under $\mathcal{G}$ if, for every $g \in \mathcal{G}$ and $d \in \mathcal{D}$, there exists a unique decision $d^* \in \mathcal{D}$ such that $\mathrm{L}(\theta, d) = \mathrm{L}(\bar{g}\theta, d^*)$ for every $\theta \in \Theta$. This decision is denoted $d^* = \tilde{g}(d)$ and the decisional problem is said to be invariant under $\mathcal{G}$.*

In this case, the group $\mathcal{G}$ induces a second group $\tilde{\mathcal{G}}$, acting on $\mathcal{D}$. Given these three groups $\mathcal{G}$, $\bar{\mathcal{G}}$, and $\tilde{\mathcal{G}}$, and the above assumptions on the decision problem, it seems logical to restrict the class of estimators to the *equivariant estimators*, that is, to those satisfying

$$\delta(gx) = \tilde{g}\delta(x).$$

These estimators were also called *invariant* in the past. A particular case of interest is when θ is estimated, that is, when $\mathcal{D} = \Theta$, since $\mathcal{G} = \tilde{\mathcal{G}}$ in such settings.

Example 9.3.4 (Example 9.3.2 continued) Consider the estimation of θ under the quadratic loss $(\theta - d)^2$. The decisional problem is then invariant and $\bar{\mathcal{G}} = \tilde{\mathcal{G}} = \mathcal{G}$. ‖

Example 9.3.5 Consider $x \sim \mathcal{N}(0, \sigma^2)$. The variance σ^2 is estimated under the *entropy* loss,

$$L(\sigma, \delta) = \frac{\delta}{\sigma^2} - \log(\delta/\sigma^2) - 1,$$

introduced in Chapter 2. If the group considered in this case is the group of *scale transformations*,

$$\mathcal{G} = \{g_c;\ g_c(x) = cx, c > 0\},$$

the associated groups are

$$\bar{\mathcal{G}} = \tilde{\mathcal{G}} = \{\bar{g}_c(\sigma^2) = c^2\sigma^2, c > 0\}$$

and the loss, thus the decisional problem, is also invariant under the action of $\mathcal{G}$.   ‖

Example 9.3.6 Consider $x \sim \mathcal{T}_p(\nu, \theta, I_p)$ and let $||\theta||^2$ be the parameter of interest. A natural invariance structure is then invariance under *orthogonal transformations*,

$$\mathcal{G} = \bar{\mathcal{G}} = \{g_A;\ g_A(x) = Ax,\ A^tA = I_p\},$$

and the problem is invariant if the loss can be written as

$$L(\theta, \delta) = \tilde{L}(||\theta||^2, \delta),$$

since there is always an orthogonal matrix A such that $A\theta = ||\theta||(1, 0, \ldots, 0)^t$ and $\tilde{\mathcal{G}}$ reduces to the identity transformation. In this case, the equivariant estimators only depend on $||x||^2$.   ‖

Definition 9.3.7 *When $\bar{\mathcal{G}}$ is a group operating on Θ, θ_1 and θ_2 are said to be equivalent if there exists $\bar{g} \in \bar{\mathcal{G}}$ with $\theta_2 = \bar{g}\theta_1$. An orbit of Θ is an equivalence class for this relation and the group $\bar{\mathcal{G}}$ is said to be transitive if Θ has a single orbit.*

    If the group $\mathcal{G}$ is small enough, there may be many orbits. For instance, when $x \sim \mathcal{B}(n, p)$ and $\mathcal{G}$ is restricted to $g_0(x) = x$ and $g_1(x) = n - x$, $\bar{\mathcal{G}} = \{\bar{g}_0, \bar{g}_1\}$ with $\bar{g}_1(p) = 1 - p$. Then there is an orbit associated with every $p \in [0, 0.5]$. However, when $\mathcal{G}$ is larger, this notion usually allows for the generalization of the phenomenon observed in the case of the location parameters.

Theorem 9.3.8 *The risk of an equivariant estimator is constant within each orbit of Θ, that is,*

$$R(\delta, \theta) = R(\delta, \bar{g}\theta)$$

for every $g \in \mathcal{G}$.

Proof. As for the estimators of location parameters, we derive that

$$\begin{aligned} R(\delta, \theta) = \mathbb{E}_\theta \left[L(\theta, \delta(x)) \right] &= \mathbb{E}_\theta \left[L(\bar{g}\theta, \tilde{g}\delta(x)) \right] \\ &= \mathbb{E}_\theta \left[L(\bar{g}\theta, \delta(gx)) \right] \\ &= \mathbb{E}_{\bar{g}\theta} \left[L(\bar{g}\theta, \delta(x)) \right] \\ &= R(\delta, \bar{g}\theta) \end{aligned}$$

for every $g \in \mathcal{G}$. ☐☐

An immediate consequence of Theorem 9.3.8 is the following result.

Corollary 9.3.9 *If $\bar{\mathcal{G}}$ is transitive, every equivariant estimator has a constant risk.*

For transitive groups, it is thus legitimate to look for the *best equivariant estimator* by minimizing the *constant* risk $R(\delta, \theta_0)$ on the class of equivariant estimators. However, the call to *ancillary statistics*, as in Section 9.2, is not always straightforward. A classical approach is to consider the *maximal invariant* statistic in order to reduce the dimension of the problem, similar to the call to minimal sufficient statistics under convex losses.

Definition 9.3.10 *For a group of transformations $\mathcal{G}$, a statistic $T(x)$ is said to be invariant if $T(gx) = T(x)$ for every $x \in \mathcal{X}$ and every $g \in \mathcal{G}$. It is said to be maximal invariant if it is invariant and $T(x_1) = T(x_2)$ implies that x_1 and x_2 are equivalent.*

In other words, a maximal invariant statistic indexes the orbits of $\bar{\mathcal{G}}$. In particular, if $\bar{\mathcal{G}}$ is transitive, the only maximal invariant statistics are constant. Moreover, it is straightforward to see that *every invariant statistic is a function of a maximal invariant statistic.* Notice also that, if $\bar{\mathcal{G}}$ is transitive, $T(x)$ is necessarily ancillary.

Example 9.3.11 Consider a distribution with a scale parameter σ,

$$x = (x_1, \ldots, x_n) \sim \frac{1}{\sigma^n} f\left(\frac{x_1}{\sigma}, \ldots, \frac{x_n}{\sigma} \right),$$

when $\mathcal{G}$ is the multiplicative group, made of the transformations

$$g_c(x_1, \ldots, x_n) = (cx_1, \ldots, cx_n) \qquad (c > 0).$$

Then, if $z = ||x||$,

$$T(x) = \begin{cases} 0 & \text{if } z = 0, \\ \frac{x}{z} & \text{otherwise,} \end{cases}$$

is a maximal invariant statistic. ‖

Example 9.3.12 (Example 9.3.6 continued) Similarly, if $z = ||x||$, the statistic

$$T(x) = \begin{cases} 0 & \text{if } z = 0, \\ \frac{x}{z} & \text{otherwise,} \end{cases}$$

is also maximal invariant for this problem. ‖

When determining the best equivariant estimator, the maximal invariant statistic can be used by *conditioning*. (Notice that the choice of a particular maximal invariant statistic does not matter, since each generates the same σ-algebra, being in one-to-one correspondence with each other.) In fact, if $\bar{\mathcal{G}}$ is transitive and T is a maximal invariant statistic, every equivariant estimator δ satisfies

$$
\begin{aligned}
R(\delta, \theta) &= R(\delta, \theta_0) \\
&= \mathbb{E}_{\theta_0}[\mathrm{L}(\theta_0, \delta(x))] \\
&= \mathbb{E}_{\theta_0}^T\{\mathbb{E}_{\theta_0}[\mathrm{L}(\theta_0, \delta(x))|T(x) = t]\}
\end{aligned}
$$

for an arbitrary value θ_0 (because the risk is constant). Since T is maximal invariant, every x such that $T(x) = t$ can be written gx_t, where x_t is a selected member of the orbit of x (assuming the axiom of choice holds). Therefore, for an equivariant estimator, $\delta(x) = \bar{g}\delta(x_t)$. It is then sufficient to minimize the above quantity in $\delta(x_t)$, conditional upon T, to obtain the best equivariant estimator. Although straightforward, the above conditioning is actually instrumental in the determination of the best equivariant estimators, and will be used in the sequel.

Example 9.3.13 (Example 9.3.11 continued)   Notice that, in this case, T is also an ancillary statistic. For the entropy loss, the minimization problem (in δ) is over

$$
\begin{aligned}
\mathbb{E}_1[\delta(x) - \log \delta(x)|T(x) = t] &= \mathbb{E}_1[\delta(zt) - \log \delta(zt)|T(x) = t] \\
&= \mathbb{E}_1[z\delta(t) - \log \delta(t) - \log(z)|T(x) = t]
\end{aligned}
$$

(where $z = ||x||$). By linearity of the expectation, $\delta(t)$ is minimizing

$$
\mathbb{E}_1[z|T = t]\delta(t) - \log \delta(t),
$$

and therefore satisfies

$$
\delta^*(t) = \frac{1}{\mathbb{E}_1[z|T = t]}.
$$

The best equivariant estimator of σ is thus

$$
\delta^*(x) = \frac{||x||}{\mathbb{E}_1[z|T = x/||x||]}.
$$

In the particular case when $x_i \sim \mathcal{N}(0, \sigma^2)$, we derive that the best equivariant estimator of σ is

$$
\delta^*(x) = \frac{||x||}{\mathbb{E}_1(||x||)} = \frac{\Gamma(p/2)}{\sqrt{2}\Gamma(p + 1/2)} ||x||.
$$

‖

Further details on this technique are given in Berger (1985a, §6.5) and Eaton (1989, §2.3).

9.4 Best equivariant estimators and noninformative distributions

In this section, we generalize the result obtained in Section 9.2 in the particular case of location parameters. We show that it is indeed possible to relate the best equivariant estimator to a σ-finite measure on Θ called a *right-invariant Haar measure*. For a more detailed and rigorous treatment, see Eaton (1989) and Wijsman (1990).

Let us assume first that, for a statistical problem invariant under the action of $\mathcal{G}$, there exists a probability distribution π^* on Θ that is also *invariant under the action of $\bar{\mathcal{G}}$*, that is, such that

$$\pi^*(\bar{g}A) = \pi^*(A)$$

for every measurable set in Θ, that is, every $A \in \mathcal{B}(\Theta)$, and for every $g \in \mathcal{G}$. In this case, the Bayes estimator associated with π^*, δ^*, minimizes

$$
\begin{aligned}
\int_\Theta R(\delta^*, \theta) \, d\pi^*(\theta) &= \int_\Theta R(\delta^*, \bar{g}\theta) \, d\pi^*(\theta) \\
&= \int_\Theta \mathbb{E}_\theta \left[\mathrm{L}(\theta, \tilde{g}^{-1}\delta^*(gx)) \right] d\pi^*(\theta)
\end{aligned}
$$

and, if the Bayes estimator is unique, it satisfies

$$\delta^*(x) = \tilde{g}^{-1}\delta^*(gx)$$

π-almost everywhere, the set of measure zero where the equality does not hold depending on g. Therefore, *a Bayes estimator associated with an invariant prior and a strictly convex invariant loss is almost equivariant.* When $\mathcal{G}$ is not countable, the collection of the above sets of measure zero over all g's is not necessarily of measure zero, but it is possible to show that, under additional conditions (see Lehmann (1986, Chapter 6, Theorem 4)), there exists *an equivariant estimator which is a Bayes estimator with respect to π^** (see also Strasser (1985)).

Example 9.4.1 Consider $\delta^\pi(x) = \mathbb{E}^\pi[\theta|x]$ under an invariant proper loss. If π^* is an invariant probability distribution, the Bayes estimator associated with π^* satisfies

$$
\begin{aligned}
\delta^\pi(gx) &= \frac{\int_\Theta \theta f(gx|\theta) \, d\pi^*(\theta)}{\int_\Theta f(gx|\theta) \, d\pi^*(\theta)} \\
&= \frac{\int_\Theta \theta f(x|\bar{g}^{-1}\theta) \, d\pi^*(\theta)}{\int_\Theta f(x|\bar{g}^{-1}\theta) \, d\pi^*(\theta)} \\
&= \frac{\int_\Theta \bar{g}\eta f(x|\eta) \, d\pi^*(\eta)}{\int_\Theta f(x|\eta) \, d\pi^*(\eta)}.
\end{aligned}
$$

Therefore, if

$$\int_\Theta \bar{g}\eta f(x|\eta) \, d\pi^*(\eta) = \bar{g} \int_\Theta \eta f(x|\eta) \, d\pi^*(\eta),$$

for every $g \in \mathcal{G}$, δ^* is indeed equivariant.   ‖

Actually, invariant probability distributions are rare, since they can only exist for compact groups[1] $\bar{\mathcal{G}}$ (see Lehmann (1983, Chapter 4, Example 4.2) for an illustration in a noncountable group). In other settings, it is necessary to consider *invariant measures*, for which the above results do not always hold (because formal Bayes estimators are not always defined).

Example 9.4.2 (Example 9.3.2 continued) If π is invariant under the action of the translation group, it satisfies $\pi(\theta) = \pi(\theta + c)$ for every θ and for every c, which implies that $\pi(\theta) = \pi(0)$ uniformly on $\mathbb{R}$ and thus leads to the Lebesgue measure as an invariant measure.   ‖

Example 9.4.3 Consider $x_1, \ldots, x_n$ a sample from $\mathcal{N}(\theta, \sigma^2)$, with unknown θ and σ^2. For sufficiency reasons, we can consider the couple $(\bar{x}, s)$, where $\bar{x}$ is the empirical mean and s^2 is the sum of the squared errors. In this setting, the group to consider is the *affine group*

$$\mathcal{G} = \{g_{a,b};\ g_{a,b}(\bar{x}, s) = (a\bar{x} + b, as),\ a > 0, b \in \mathbb{R}\}\,,$$

and $\bar{\mathcal{G}} = \tilde{\mathcal{G}} = \mathcal{G}$ if the parameter to be estimated is (θ, σ). If π is an invariant measure, its density satisfies

$$a^2 \pi(a\theta + b, a\sigma) = \pi(\theta, \sigma), \qquad \forall a > 0,\ \forall b \in \mathbb{R}\,,$$

which implies

$$\pi(\theta, \sigma) = \pi(0, 1)/\sigma^2.$$

Therefore, an invariant measure is proportional to $\pi(\theta, \sigma) = 1/\sigma^2$ and corresponds to the Jeffreys measure obtained in Chapter 3.   ‖

In general, given a locally compact topological group $\mathcal{G}$, and defining $K(\mathcal{G})$ as the set of continuous real functions on $\mathcal{G}$ with compact support, we introduce for $g \in \mathcal{G}$ the transformation L_g on $K(\mathcal{G})$ as

$$(L_g f)(x) = f(gx) \qquad \text{for } f \in K(\mathcal{G}),\ x \in \mathcal{G}.$$

An integral J on $K(\mathcal{G})$ is said to be *left-invariant* if

$$J(L_g f) = J(f)$$

for every $f \in K(\mathcal{G})$ and for every $g \in \mathcal{G}$. The Radon measure ν_ℓ associated with J is said to be a *left Haar measure*, and it can be shown (see Nachbin (1965)) that this measure is unique up to a multiplicative factor. Defining R_g on $K(\mathcal{G})$ by

$$(R_g f)(x) = f(xg), \qquad \text{for } f \in K(\mathcal{G}),\ x \in \mathcal{G},$$

---

[1] When $\bar{\mathcal{G}}$ is not a subset of $\mathbb{R}^p$, the topological structure induced by $\bar{\mathcal{G}}$ is the topology induced by the group composition and inversion, that is, the smallest collection of open sets such that the group composition and inversion are continuous (see Rudin (1976)).

we derive similarly *right-invariant integrals* and a *right Haar measure* ν_r, also defined up to a multiplicative constant. As mentioned above, the *finiteness of the Haar measure*, that is, the existence of an invariant probability distribution, is in fact *equivalent to the compactness of* $\mathcal{G}$. See Eaton (1989, Chapter 1) for examples of Haar measures; Berger (1985a) details the case where $\mathcal{G} \subset \mathbb{R}^k$.

The *modulus of* $\mathcal{G}$ is defined as the *multiplier* Δ—that is, a real-valued function satisfying $\Delta(g_1 g_2) = \Delta(g_1)\Delta(g_2)$—that relates left and right Haar measures by

$$\nu_r(dx) = \Delta(x^{-1})\nu_\ell(dx)$$

(see Exercises 9.13 and 9.15). We assume the existence of a Radon measure μ on $\mathcal{X}$ such that, for every f,

$$\int_{\mathcal{X}} f(g^{-1}x)\mu(dx) = \Delta^{-1}(g) \int_{\mathcal{X}} f(x)\mu(dx).$$

This relation shows the connection between the modulus of $\mathcal{G}$ and the Jacobian of the transformation of x in gx. Consider the distributions P_θ, $\theta \in \Theta$, with density $f(x|\theta)$ with respect to μ. Then, for every $g \in \mathcal{G}$,

$$f(x|\theta) = f(gx|\bar{g}\theta)\Delta^{-1}(g).$$

We also assume that $\bar{\mathcal{G}}$ acts *transitively* on Θ. As shown in Eaton (1989, p. 84), some additional assumptions then ensure the validity of a theorem à la Fubini: *If ν_r is the right Haar measure on $\mathcal{G}$, Q is the projection of $\mathcal{X}$ on $\mathcal{X}/\mathcal{G}$, and (Tf) is defined on $\mathcal{X}/\mathcal{G}$ by*

$$(Tf)(Q(x)) = \int_{\mathcal{G}} f(gx)\nu_r(dg),$$

then there exists an integral J_1 defined on $K(\mathcal{X}/\mathcal{G})$ such that

$$J_1(Tf) = \int_{\mathcal{X}} f(x)\mu(dx).$$

This can be rewritten as the fact that the integral of f with respect to μ is the integral over all the orbits of $\mathcal{X}$ (that is, on $\mathcal{X}/\mathcal{G}$) of the average of f with respect to the right Haar measure on each orbit, Tf.

Consider a nonrandomized estimator δ and, for a fixed $\theta \in \Theta$, let us define

$$f_0(x) = L(\theta, \delta(x))f(x|\theta),$$

then

$$R(\delta, \theta) = \int_{\mathcal{X}} f_0(x)\mu(dx).$$

It follows from the above theorem à la Fubini that there exists an integral J_1 on $K(\mathcal{X}/\mathcal{G})$ such that

$$R(\delta, \theta) = J_1(Tf_0),$$

with

$$(Tf_0)(Q(x)) = \int_{\mathcal{G}} L(\theta, \delta(gx)) f(gx|\theta) \nu_r(dg)$$

$$= \int_{\mathcal{G}} L(\bar{g}\theta, \delta(x)) f(x|\bar{g}\theta) \nu_r(dg)$$

(see Eaton (1989, p. 85)). We also define

$$H(a, x) = \int_{\mathcal{G}} L(\bar{g}\theta, a) f(x|\bar{g}\theta) \nu_r(dg),$$

which does not depend on θ (since $\bar{\mathcal{G}}$ acts transitively on Θ). Notice that $H(\delta(x), x)$ gives the risk of δ conditional upon the orbit of x. It is instrumental in the derivation of the best equivariant estimator.

Theorem 9.4.4 *If there exists $a_0(x)$ such that*

(i) $H(a, x) \geq H(a_0(x), x)$ *for every $a \in \mathcal{D}$, $x \in \mathcal{X}$; and*

(ii) $a_0(gx) = \tilde{g}a_0(x)$ *for every $g \in \mathcal{G}$, $x \in \mathcal{X}$,*

then $\delta_0(x) = a_0(x)$ is a best equivariant estimator.

Proof. Consider an equivariant estimator δ. Then

$$\int_{\mathcal{G}} L(\bar{g}\theta, \delta(x)) f(x|\bar{g}\theta) \nu_r(dg) \geq \int_{\mathcal{G}} L(\bar{g}\theta, a_0(x)) f(x|\bar{g}\theta) \nu_r(dg).$$

Integrating with respect to J_1, it follows that $R(\delta, \theta) \geq R(\delta_0, \theta)$. The estimator δ_0 then dominates δ. □□

This theorem points out the relation existing between the best equivariant estimator and a particular Bayes estimator, since $H(a, x)$ can also be interpreted as a posterior Bayes risk. In fact, if $\theta_0 \in \Theta$ is arbitrarily selected, the function $\tau(g) = \bar{g}\theta_0$ defines a surjection from $\mathcal{G}$ to Θ because of the transitivity of $\bar{\mathcal{G}}$. It therefore induces a measure on Θ, called a *right Haar measure on* Θ, which is defined by $\pi^*(B) = \nu_r(\tau^{-1}(B))$ for every $B \in \mathcal{B}(\Theta)$, and is obviously invariant under the action of $\bar{\mathcal{G}}$. Moreover,

$$H(a, x) = \int_{\Theta} L(\theta, a) f(x|\theta) \, d\pi^*(\theta).$$

This extension of the right Haar measure to Θ implies the following result, which expresses the best equivariant estimator as a Bayes estimator for every transitive group acting on a statistical model.

Corollary 9.4.5 *The best equivariant estimator of θ is the Bayes estimator associated with the right Haar measure on Θ, π^*, and the corresponding invariant loss.*

Therefore, we have obtained a method that derives the best equivariant estimators directly from the right Haar measure. (See Stein (1965) and Zidek (1969) for similar results.)

In the above development, the dominating measure is μ, which is *relatively invariant with multiplier the modulus* Δ^{-1}. In fact, if the measure μ is relatively invariant with an arbitrary multiplier χ, that is, such that for every $f \in K(\mathcal{G})$,

$$\int_{\mathcal{X}} f(gx)\mu(dx) = \chi(g) \int_{\mathcal{X}} f(x)\mu(dx),$$

Corollary 9.4.5 still holds (see Eaton (1989, p. 87)).

Example 9.4.6 (Example 9.4.3 continued) We obtained the following *left Haar measure* on Θ:

$$\pi^\ell(\theta, \sigma) = 1/\sigma^2.$$

The *right Haar measure* can be derived by inversion: if $g = (a, b)$ and $g_0 = (a_0, b_0)$, $gg_0 = (aa_0, ab_0 + b)$ for the group composition. Taking the Jacobian into account, we want the right Haar measure to satisfy

$$a_0 \pi^r (b_0 \sigma + \theta, a_0 \sigma) = \pi^r(\theta, \sigma)$$

for every (θ, σ) and uniformly in a_0, b_0; this implies

$$\pi^r(\theta, \sigma) = 1/\sigma,$$

up to a multiplicative factor. Therefore, the right Haar measure is different from the left Haar measure and gives the noninformative alternative to the Jeffreys prior (see Section 3.6). Under the invariant quadratic loss,

$$(9.4.1) \qquad L((\theta, \sigma), \delta) = \frac{(\theta - \delta_1)^2}{\sigma^2} + \left(\frac{\delta_2}{\sigma} - 1\right)^2,$$

the best equivariant estimator is the Bayes estimator associated with the prior distribution π^r, that is,

$$\delta_1^*(\bar{x}, s) = \frac{\mathbb{E}^{\pi^r}[\theta/\sigma^2|\bar{x}, s]}{\mathbb{E}^{\pi^r}[1/\sigma^2|\bar{x}, s]}, \qquad \delta_2^*(\bar{x}, s) = \frac{\mathbb{E}^{\pi^r}[1/\sigma|\bar{x}, s]}{\mathbb{E}^{\pi^r}[1/\sigma^2|\bar{x}, s]}.$$

Since

$$\pi^r(\theta, \sigma|\bar{x}, s) \propto \sigma^{-(n+1)} e^{-n(\bar{x}-\theta)^2/2\sigma^2} e^{-s^2/2\sigma^2},$$

this is a special case of conjugate distribution on (θ, σ) and

$$\delta_1^*(\bar{x}, s) = \bar{x}, \qquad \delta_2^*(\bar{x}, s) = \frac{\Gamma(n/2)}{\sqrt{2}\Gamma((n+1)/2)} s.$$

Notice that δ_2 is also the estimator obtained in Example 9.3.11.   ‖

Example 9.4.7 (Eaton (1989)) Consider a *multiplicative model* $\mathcal{N}(\theta, \theta^2)$, with n observations $x_1, \ldots, x_n$. This model appears in settings where the difficulty of measuring an object increases with its magnitude (particle physics, astronomy, etc.). If we estimate θ under the loss

$$L(\theta, d) = \frac{(\theta - d)^2}{\theta^2},$$

the problem is invariant under the action of the multiplicative group. The right Haar measure is then $\pi(\theta) = 1/|\theta|$. (It is also the left Haar measure, since the group is commutative.)

Therefore, the best equivariant estimator of θ is

$$\delta^*(x_1, \ldots, x_n) = \frac{\mathbb{E}^\pi[1/\theta|x_1, \ldots, x_n]}{\mathbb{E}^\pi[1/\theta^2|x_1, \ldots, x_n]}$$

and

$$\pi(\theta|x) \quad \propto \quad \frac{1}{\theta^2} \exp\left\{ -\sum_{i=1}^n (x_i - \theta)^2/2\theta^2 \right\}$$

$$\propto \quad \frac{1}{\theta^2} \exp\left\{ -\frac{1}{2}\left(\frac{n\bar{x}}{s^2} - \frac{1}{\theta} \right)^2 s^2 \right\},$$

for $s^2 = \sum_{i=1}^n x_i^2$. The posterior distribution is then a *generalized inverse normal* distribution $\mathcal{IN}(2, n\bar{x}/s^2, 1/s^2)$ (Robert (1991)) and

$$\mathbb{E}^\pi[1/\theta|\bar{x}, s^2] = \sqrt{2}s \, \frac{{}_1F_1(1; 1/2; n^2\bar{x}^2/2s^2)}{\Gamma(1/2){}_1F_1(1/2; 1/2; n^2\bar{x}^2/2s^2)}.$$

Therefore,

$$\delta^*(x_1, \ldots, x_n) = \sqrt{2} \, \Gamma(3/2)\frac{{}_1F_1(3/2; 1/2; n^2\bar{x}^2/2s^2)}{\Gamma(1/2){}_1F_1(1; 1/2; n^2\bar{x}^2/2s^2)} \, s.$$

In this case, the best equivariant estimator dominates the maximum likelihood estimator

$$\hat{\delta}(\bar{x}, s) = \frac{-\bar{x} + (\bar{x}^2 + 4s^2)^{1/2}}{2},$$

which is also equivariant. For additional results on the multiplicative models, see Gleser and Healy (1976), Kariya (1984), Kariya et al. (1988), and Perron and Giri (1990).   ‖

The reader is referred to Eaton (1989), Lehmann (1986), and Berger (1985a) for other examples of the use of Haar measures in the derivation of best equivariant estimators in the case of tests and confidence regions. A general mathematical treatise on Haar measures is Nachbin (1965).

9.5 The Hunt–Stein theorem

If we consider the particular case discussed at the beginning of the previous section, namely the case when $\mathcal{G}$ is compact and where there exists an invariant probability distribution on Θ, the best equivariant estimator is a (proper) Bayes estimator, and therefore admissible in most cases. Because its risk is constant when $\bar{\mathcal{G}}$ is transitive, the best equivariant estimator is also *minimax*. When $\mathcal{G}$ is not compact, the best equivariant estimator is a generalized Bayes estimator associated with the right Haar measure, and therefore not necessarily admissible. The Stein effect (see Note 2.8.2) is an illustration of this possible suboptimality because the best equivariant estimator of a location parameter, x, is inadmissible for the quadratic loss in dimension 3 and above. Therefore, the question of the admissibility of the best equivariant estimator cannot be considered in general for noncompact groups.

On the contrary, it is possible to extend the minimaxity property beyond than for the compact case, through the *Hunt–Stein Theorem*[2]. This result is intuitively sound because, when a problem is invariant, there exists an equivariant estimator with a constant risk that attains the lower bound of the maximal risks,

$$\inf_\delta \sup_\theta R(\delta, \theta).$$

Furthermore, using the natural invariant structures of the model, it seems legitimate to improve on an estimator δ, by averaging it, that is, by integrating over $\mathcal{G}$

$$\delta^*(x) = \int_{\mathcal{G}} \delta(gx)\nu_r(dg),$$

if $L(\theta, d)$ is convex in d and if the theorem à la Fubini given in Section 9.4 applies (assuming δ^* is well defined). In fact, we would then get (in an informal way)

$$
\begin{aligned}
R(\delta, \theta) &= \mathbb{E}_\theta[L(\theta, \delta(x))] \\
&= \mathbb{E}^T(\mathbb{E}_\theta[L(\theta, \delta(x))|Q(x) = T]) \\
&\geq \mathbb{E}^T[L(\theta, \delta^*(t))] \\
&= R(\delta^*, \theta).
\end{aligned}
$$

This improvement is similar to the domination result of the Rao–Blackwell Theorem, when conditioning on a sufficient statistic.

We formalize this sketch of proof a bit further by introducing the notion of *amenable group* presented in detail in Bondar and Milnes (1981). Firstly, the following counter-example shows that intuition is not always satisfactory, in particular when the invariance structures are too strong, that is, when $\mathcal{G}$ is too large.

Example 9.5.1 (Stein (1965)) Consider $x \sim \mathcal{N}_p(0, \Sigma)$ and $y \sim \mathcal{N}_p(0, \varrho\Sigma)$ with $p \geq 2$. The parameter ϱ is estimated under the loss function

$$L((\varrho, \Sigma), d) = \mathbb{I}_{[1/2, +\infty)}\left(\left|1 - \frac{d}{\varrho}\right|\right).$$

The problem is then invariant under the action of the linear group GL_p because, if B is a nonsingular matrix, $Bx \sim \mathcal{N}_p(0, B\Sigma B^t)$ and $By \sim \mathcal{N}_p(0, \varrho B\Sigma B^t)$. As $\bar{g}_B(\varrho, \Sigma) = (\varrho, B\Sigma B^t)$, the equivariant estimators are actually invariant

$$\delta(Bx, By) = \delta(x, y)$$

---

[2] This theorem is also famous for remaining without published proof for a long time, although Kiefer (1957) established this result in a particular case.

for every x, y, and B. If x and y are linearly independent (an event that occurs with probability 1), we can find B such that

$$Bx = (1, 0, \ldots, 0)^t \qquad \text{and} \qquad By = (0, 1, 0, \ldots, 0)^t,$$

which implies that the equivariant estimators are *almost everywhere constant*. Since

$$R(\delta_0, (\varrho, \Sigma)) = 1 \qquad \text{if} \qquad \left| 1 - \frac{\delta_0}{\varrho} \right| > 1/2$$

for a given constant δ_0, the minimax risk of the equivariant estimators is 1.

Defining

$$\delta_1(x, y) = \left| \frac{y_1}{x_1} \right|,$$

the risk of δ_1 is

$$\begin{aligned}
R(\delta_1, \theta) &= P_{\varrho, \Sigma} \left(\left| 1 - \frac{y_1}{x_1 \varrho} \right| \geq 1/2 \right) \\
&= P \left(\left| 1 - \frac{z_1}{z_2} \right| \geq 1/2 \right),
\end{aligned}$$

where z_1, z_2 are i.i.d. $\mathcal{N}(0, 1)$. Therefore, this risk is also *constant*, but strictly smaller than 1. Notice that δ_1 is also an equivariant estimator for the multiplicative group, which then appears as a more appropriate invariance structure. ‖

For a general approach to this problem, consider a locally compact group of transformations $\mathcal{G}$, with a right Haar measure ν_r. Let $\mathcal{V}$ be an algebra of real-valued essentially bounded measurable functions on $\mathcal{G}$, such that the constant function $\mathbf{1}$ is in $\mathcal{V}$.

Definition 9.5.2 *A mean on $\mathcal{V}$ is a linear and continuous functional m on $\mathcal{V}$ such that*

(i) $m(\mathbf{1}) = 1$; and

(ii) $m(f) \geq 0$ if $f \in \mathcal{V}$ and $f \geq 0$ (a.s.).

That such a functional m exists is actually a necessary and sufficient condition for the Hunt–Stein Theorem to hold. In this case, it is then possible to average on the orbits of $\mathcal{X}$ with respect to $\mathcal{G}$, as suggested at the beginning of this section.

Example 9.5.3 (Bondar and Milnes (1981)) For $\mathcal{G} = \mathbb{R}$ and $n \in \mathbb{N}$, consider

$$m_n(f) = \frac{1}{2n} \int_{-n}^{n} f(x) \, dx;$$

then m_n defines a mean on $\mathcal{L}_\infty(\mathbb{R})$. Moreover, the sequence (m_n) has an accumulation point m in the sense of the *weak topology* on $\mathcal{L}_\infty$: for every $f \in \mathcal{L}_\infty$, $\epsilon > 0$, and $n_0 \in \mathbb{N}$, there exists $n \geq n_0$ such that

$$|m_n(f) - m(f)| < \epsilon.$$

In particular, this accumulation point satisfies $m(f) = 0$ for every f such that $f(x)$ goes to 0 when x goes to $\pm\infty$. Notice also that m is not σ-additive, and that the sequence (m_n) does not converge to m in the sense of the weak topology. ‖

Definition 9.5.4 *The mean m is said to be right-invariant if, for every $f \in \mathcal{V}$ and $g \in \mathcal{G}$, $m(f_g) = m(f)$, where $f_g(x) = f(xg)$. The group $\mathcal{G}$ is said to be amenable if there exists a right-invariant mean on $\mathcal{L}_\infty(\mathcal{G})$ or, equivalently, on $\mathcal{C}_B(\mathcal{G})$, the space of the continuous bounded functions on $\mathcal{G}$.*

As shown in Bondar and Milnes (1981), the existence of an amenable group is equivalent to *the existence of a sequence of almost right-invariant probability measures*: in such a case, there exists a sequence (P_n) of probability measures on $\mathcal{G}$ such that, for every $B \in \mathcal{B}(\mathcal{G})$ and every $g \in \mathcal{G}$,

$$\lim_{n \to +\infty} |P_n(Bg) - P_n(B)| = 0.$$

Moreover, there exists a sequence (G_n) of nested compact sets such that the density of P_n is $\nu_r(G_n)^{-1}\mathbb{I}_{G_n}(g)$ (with respect to ν_r). Therefore, the existence of the sequence (G_n) allows for the approximation of the Haar measure ν_r by a sequence of probability distributions, and these probabilities are almost invariant in the sense that if $B \cap G_n = Bg \cap G_n$, $P_n(B) = P_n(Bg)$ (see also Strasser (1985) and Lehmann (1986)). Example 9.5.3 provides a direct illustration of this result.

Examples of amenable groups are the additive and multiplicative groups, the group of location-scale transformations (see Example 9.4.1), and the group T_p of invertible upper triangular matrices. On the contrary, the linear group GL_p and the group SL_p of matrices with determinant 1 are not amenable. Bondar and Milnes (1981) provide many other structural examples of amenable and nonamenable groups.

The Hunt–Stein Theorem then states the minimaxity of the best equivariant estimator.

Theorem 9.5.5 *If the group $\mathcal{G}$ is amenable and the statistical problem $(\mathcal{X}, f(x|\theta), \mathcal{D}, L)$ is invariant under the action of $\mathcal{G}$, the existence of a minimax estimator implies the existence of a minimax equivariant estimator. Moreover, an equivariant estimator that is minimax among the equivariant estimators is minimax.*

A proof of this theorem is provided by Berger (1985a, §6.7) in the case where $\mathcal{G}$ is finite, by Lehmann (1983, §9.5) for tests, and by Le Cam (1986,

§8.6) in more general settings, as a consequence of the fixed-point theorem of Markov and Kakutani. As mentioned at the beginning of this section, the Hunt–Stein Theorem relies on an adapted version of the Fubini Theorem. To give a quick sketch of a proof, let us assume L is convex. For a real-valued estimator δ, define

$$\delta^*(x) = m(\tilde{\delta}_x),$$

where m is the right-invariant mean and $\tilde{\delta}_x(g) = \delta(gx)$. The estimator δ^* is then equivariant since, if $g_0 \in \mathcal{G}$,

$$
\begin{aligned}
\delta^*(g_0 x) &= \int_{\mathcal{G}} \tilde{g}^{-1} \delta(g g_0 x)\, dm(g) \\
&= \int_{\mathcal{G}} \tilde{g}_0 \tilde{g}_0^{-1} \tilde{g}^{-1} \delta(g g_0 x)\, dm(g) \\
&= \tilde{g}_0 \int_{\mathcal{G}} \tilde{g}^{-1} \delta(g x)\, dm(g) \\
&= \tilde{g}_0 \delta^*(x),
\end{aligned}
$$

because of the right-invariance of m. Moreover,

$$(9.5.1) \quad \sup_\theta R(\delta^*, \theta) \le \sup_\theta \int_{\mathcal{G}} \int_{\mathcal{X}} L\left(\theta, \tilde{g}^{-1} \delta(g x)\right) f(x|\theta)\, dx\, dm(g)$$

from the convexity of L. Therefore

$$
\begin{aligned}
\sup_\theta R(\delta^*, \theta) &\le \sup_\theta \int_{\mathcal{G}} \int_{\mathcal{X}} L\left(\bar{g}\theta, \delta(g x)\right) f(x|\theta)\, dx\, dm(g) \\
&= \sup_\theta \int_{\mathcal{G}} R(\bar{g}\theta, \delta)\, dm(g) \\
&\le \sup_\theta R(\delta, \theta),
\end{aligned}
$$

which implies[3] the domination of δ by δ^*.

A consequence of the Hunt–Stein Theorem is that, in the normal case, the maximum likelihood estimator, $x \sim \mathcal{N}_p(\theta, I_p)$, is minimax for every value of p, although inadmissible for $p \ge 3$. The same result holds if $x \sim \mathcal{N}_p(\theta, \sigma^2 I_p)$ and if the unknown variance σ^2 is estimated by s^2/q, when $s^2 \sim \sigma^2 \chi_q^2$.

9.6 The role of invariance in Bayesian Statistics

To conclude this chapter, let us reiterate the reservations expressed in Berger (1985a) over the implications of the invariance requirements on the Bayesian approach, in particular, as a determination technique for non-

---

[3] Notice that these indications do not constitute a rigorous proof, since the application of the Fubini Theorem in (9.3.8) is not always justified. In fact, this "averaging" can only be used under particular conditions. Otherwise, it would also lead to a general admissibility result for the best equivariant estimator under convex losses, a result negated by the Stein effect.

informative distributions, even though it provides a justification for the choice of an alternative to the Jeffreys prior (see Example 9.4.1).

One criticism of the notion of invariance is that, although intuitively attractive, it is not devoid of ambiguity and that, since it is sometimes possible to consider several globally invariant groups, the resulting best equivariant estimators can lead to distinct inferences, which contradicts the Likelihood Principle.

Another criticism is that the natural invariance structures of a statistical model can be either too weak, and thus of no use to determine an estimator, or too strong, and therefore too constraining. An extreme example of the first setting is the Poisson distribution, where there is actually no invariance structure at all. The following example illustrates the opposite case (see also Example 9.5.1).

Example 9.6.1 Consider a distribution family symmetric around a location parameter θ, that is, such that $x \sim f(|x - \theta|)$. The loss function is $\varrho(|d - \theta|)$. If the invariance by symmetry, that is, the fact that the distribution of $y = -x$ belongs to the same family, is taken into account, the estimators that correspond to $\pi(\theta) = 1$ and satisfy

$$\delta(x + c) = \delta(x) + c \qquad \text{and} \qquad \delta(-x) = -\delta(x)$$

reduce to $\delta(x) = x$, which is not necessarily a good estimator.   ‖

An excess of invariance can obviously be reduced by taking into account only a part of the invariance structures, that is, by considering a subgroup $\mathcal{G}_0$ of $\mathcal{G}$ that induces a transitive action on Θ, while being as small as possible. Nonetheless, the choice of this subgroup, when possible, can be crucial in the resulting inference.

As a last but important criticism, notice that a modeling of statistical problems based on invariant structures can be damaging from a *subjective* point of view, since it constrains the decision structures to be compatible with invariance—therefore, in particular, to choose an invariant loss—and can conflict with the prior information—the only compatible prior distribution being the Haar measure. Such a modeling can also be damaging from an *efficiency* point of view, since the equivariant estimators can be strongly inadmissible, as shown by the Stein effect and Example 9.5.1 (see also Examples 4.1.4–4.2.3 in Lehmann (1983, §4.4)). Moreover, invariance does not necessarily lead to a good noninformative distribution, as shown by Example 9.6.1. And, in practice, the computation and derivation of right Haar measures can be quite involved.

9.7 Exercises

Section 9.2

9.1 (Blackwell and Girshick (1954)) Consider the distribution f with weights $f(k) = 1/k(k + 1)$ for $k = 1, 2, \ldots$ and $x \sim f(x - \theta)$, with $\theta \in \mathbb{R}$. Under the loss function,

$$L(\theta, d) = \begin{cases} d - \theta & \text{if } d > \theta, \\ 0 & \text{otherwise,} \end{cases}$$

show that the equivariant estimators are of the form $x - c$, and that every equivariant estimator has an infinite risk. Compare with the constant estimator $\delta_0(x) = c$.

9.2 Consider x an observation from a $\mathcal{C}(\theta, 1)$ distribution. Under quadratic loss, show that all equivariant estimators have infinite risk. Propose a finite risk estimator other than the constant estimator.

9.3 (Berger (1985a)) Consider

$$x = (x_1, \dots, x_n) \sim f(x_1 - \theta, \dots, x_n - \theta),$$

where θ is unknown. The hypothesis $H_0 : f = f_0$ is to be tested against $H_1 : f = f_1$ under the $0 - 1$ loss.

a. Show that $T(x) = (x_1 - x_n, \dots, x_{n-1} - x_n)$ is a maximal invariant statistic for the group of transformations

$$\mathcal{G} = \{g_c; \ g_c(x_1, \dots, x_n) = (x_1 + c, \dots, x_n + c), \ c \in \mathbb{R}\}.$$

b. Deduce that every invariant test only depends on $y = T(x)$, and that the optimal tests have the following rejection region:

$$W = \{f_1^*(y) \geq K f_0^*(y)\},$$

where f_i^* is the density of y under H_i.

9.4 (Berger (1985a)) Consider x distributed according to

$$P_\theta(x = \theta - 1) = P_\theta(x = \theta + 1) = 1/2.$$

The associated loss function is

$$L(\theta, d) = \begin{cases} |\theta - d| & \text{if } |\theta - d| \leq 1, \\ 1 & \text{otherwise.} \end{cases}$$

Give the best equivariant estimators for the translation group and show that they are dominated by

$$\delta^*(x) = \begin{cases} x + 1 & \text{if } x \leq 0, \\ x - 1 & \text{otherwise.} \end{cases}$$

9.5 (Berger (1985a)) Consider $x_1, \dots, x_n$ a sample from the truncated normal distribution, with density

$$f(x|\theta) = \left(\frac{2}{\pi}\right)^{1/2} e^{-(x-\theta)^2/2} \mathbb{I}_{[\theta, +\infty)}(x).$$

Show that the best equivariant estimator of θ under quadratic loss is

$$\delta^*(x) = \bar{x} - \frac{\exp\{-n(x_{(1)} - \bar{x})^2/2\}}{\sqrt{2n\pi} \Phi(\sqrt{n}(x_{(1)} - \bar{x}))}.$$

Section 9.3

9.6 Consider $x \sim \mathcal{N}(\theta, a\theta^2)$, with $\theta \in \mathbb{R}$ and known $a > 0$. The parameter θ is estimated under the loss $L(\theta, d) = (\frac{d}{\theta} - 1)^2$.

a. Show that the problem is invariant under the group of transformations
$$\mathcal{G} = \{g_c; \ g_c(x) = cx, \ c > 0\}.$$

Is the action of the group transitive?

b. Give the best equivariant and the maximum likelihood estimators of θ.

c. Compare with the estimators obtained in Exercise 3.33 and in Example 9.4.7.

d. Use Exercise 3.33 to show that the best equivariant estimator δ_0 is a generalized Bayes estimator.

9.7 (Lehmann (1983)) Consider the estimation of a scale parameter σ, under the loss

(9.7.1) $$L(\sigma, \delta) = \left(\frac{\delta}{\sigma} - 1\right)^2,$$

for n observations
$$x_1, \ldots, x_n \sim \frac{1}{\sigma^n} f\left(\frac{x_1}{\sigma}, \ldots, \frac{x_n}{\sigma}\right).$$

a. If $z = (x_1/x_n, \ldots, x_{n-1}/x_n, x_n/|x_n|)$, show that every estimator of σ equivariant under scale transformations can be written
$$\delta(x) = \delta_0(x)/\omega(z),$$

where δ_0 is a particular equivariant estimator, and that z is a maximal invariant statistic.

b. Determine the function ω^* that minimizes
$$\mathbb{E}[L(\sigma, \delta(x))|z]$$

under (9.7.1), and deduce the best equivariant estimator.

c. Write this estimator in an appropriate form in order to find the result of Section 9.4 with the corresponding Haar measure.

d. Consider the previous questions for the estimation of σ^r $(r \in \mathbb{R}_+^*)$ under the loss
$$L(\sigma, \delta) = \left(\frac{\delta}{\sigma^r} - 1\right)^2.$$

9.8 Apply the results of Exercise 9.7 to the following cases:

(i) $x_1, \ldots, x_n$ are i.i.d. $\mathcal{N}(0, \sigma^2)$;

(ii) $x_1, \ldots, x_n$ are i.i.d. $\mathcal{G}(\alpha, \sigma)$; and

(iii) $x_1, \ldots, x_n$ are i.i.d. $\mathcal{U}[0, \sigma]$.

9.9 Examine Exercise 9.7 under the following losses:
$$L(\sigma, \delta) = \frac{|\delta - \sigma|}{\sigma}, \quad L(\sigma, \delta) = \frac{\delta}{\sigma} - \log(\delta/\sigma) - 1, \quad L(\sigma, \delta) = \left(\frac{\sigma}{\delta} - 1\right)^2.$$

9.10 (Lehmann (1983)) Consider the estimation of σ in the case when
$$x = (x_1, \ldots, x_n) \sim \frac{1}{\sigma^n} f\left(\frac{x_1 - \theta}{\sigma}, \ldots, \frac{x_n - \theta}{\sigma}\right),$$

under the action of the *affine group*
$$\mathcal{G}_a = \{g_{a,b}; \ g_{a,b}(x) = ax + b\mathbf{1}, \ a > 0, b \in \mathbb{R}\},$$

where $\mathbf{1} = (1, \ldots, 1) \in \mathbb{R}^n$.

a. Determine the best equivariant estimator under the loss (9.7.1) similarly to Exercise 9.7. (*Hint:* Use the transformations $y_i = x_i - x_n$ and define $z_i = y_i/y_{n-1}$ $(i \neq n-1)$, $z_{n-1} = y_{n-1}/|y_{n-1}|$.)

b. Compare with a Bayesian formulation using the right Haar measure.

c. Consider the previous questions when estimating θ under the loss

$$L(\theta, \delta) = \frac{(\theta - \delta)^2}{\sigma^2}.$$

d. Apply to the case where $x_i - \theta \sim \mathcal{E}xp(\sigma)$, and show that the best equivariant estimator of θ is

$$\delta^*(x) = x_{(1)} - \frac{1}{n^2} \sum_{i=1}^{n} (x_i - x_{(1)}).$$

9.11 *(Eaton (1989)) Consider $\mathcal{G} \subset \mathbb{R}_+^* \times \mathbb{R}$, with the group operation

$$(a_1, b_1)(a_2, b_2) = (a_1 a_2, a_1 b_2 + b_1).$$

If $D = \{x \in \mathbb{R}^n; \ x_1 = \cdots = x_n\}$, consider $\mathcal{X} = \mathbb{R}^n - D$. Assume that $\mathcal{G}$ acts on $\mathcal{X}$ by

$$(a, b)x = ax + be_n,$$

where $e_n = (1, \ldots, 1)^t$. Show that the maximal invariant statistic is

$$f(x) = \frac{x - \bar{x} e_n}{s(x)},$$

with $\bar{x} = \sum x_i/n$, $s^2(x) = \sum (x_i - \bar{x})^2$.

9.12 *(Eaton (1989)) Verify that if there exists a *multiplier* ξ on $\mathcal{G}$, that is, a real-valued function such that $\xi(g_1 g_2) = \xi(g_1)\xi(g_2)$, that satisfies

$$f(x|\theta) = f(gx|\bar{g}\theta)\xi(g)$$

uniformly on $\mathcal{X}$, Θ, $\mathcal{G}$, the family

$$\mathcal{P} = \{f(x|\theta); \theta \in \Theta\}$$

is $\mathcal{G}$-invariant. Deduce that, in this case, the maximum likelihood estimator is equivariant, as is every Bayes estimator associated with a *relatively invariant* prior measure, that is, such that there exists a multiplier ξ_1 with $\pi(gB) = \xi_1(g)\pi(B)$ uniformly in B and g.

9.13 *(Delampady (1989b)) Consider $x \sim \mathcal{N}_p(\theta, I_p)$. The hypothesis $H_0 : \theta = \theta_0$ is tested against $H_1 : \theta \neq \theta_0$. This problem is invariant under the action of the orthogonal group $\mathcal{G}_o$ and we only consider prior distributions in the invariant class

$$I = \{\pi; \ \pi(gA) = \pi(A), \forall A \in \mathcal{B}(\mathbb{R}^p), \forall g \in \mathcal{G}_o\}.$$

a. Show that $t(x) = ||x||^2$ is a maximal invariant statistic, distributed as a noncentral χ_p^2, with the noncentrality parameter, $\eta(\theta) = ||\theta||^2$ (the corresponding maximal invariant statistic on $\bar{\mathcal{G}}_0$), and that its density can be written

$$q(t(x)|\eta(\theta)) = \int_{\mathcal{G}_o} f(gx|\theta) \, d\mu(g),$$

where μ is the Haar measure on $\mathcal{G}_0$.

b. Deduce that if B^π is the Bayes factor, it satisfies

$$\inf_{\pi \in I} B^\pi(x) = \frac{q(t(x)|\theta_0)}{q(t(x)|\hat{\eta})},$$

where $\hat{\eta}$ is the maximum likelihood estimator of η.

c. Compare with the *p-value* for different values of $t(x)$.

9.14 Show that the intrinsic losses defined in Section 2.5.4 are naturally invariant.

9.15 Consider $x \sim \mathcal{N}(\theta, \sigma^2)$ and the parameter of interest is e^θ, when σ^2 is known.

a. Show that

$$\mathbb{E}_\theta[e^{ax}] = e^{a\theta + a^2 \sigma^2/2}.$$

b. Among the estimators of the form $\delta_c(x) = e^{x + c\sigma^2}$, determine the best estimator for the quadratic loss L_2, δ^*. Show that δ^* is a Bayes estimator and determine the corresponding prior π^*. (Consider first the Lebesgue measure and the weighted quadratic loss

$$L_0(\theta, \delta) = e^{-2\theta}(e^\theta - \delta)^2.$$

What is the Bayes estimator for the Lebesgue prior under L_2?)

c. Consider the previous question for the absolute error loss,

$$L_1(\theta, \delta) = |e^\theta - \delta|.$$

Show that the best estimator is associated with $\pi(\theta) = e^{-\theta}$. Is this answer surprising from an invariance point of view?

d. Given the estimator δ^*, we want to evaluate the performances of δ^* under L_0 and L_2, that is, to estimate $L_0(\theta, \delta^*(x))$ and $L_2(\theta, \delta^*(x))$ under the quadratic loss

(9.7.2) $(L_0(\theta, \delta^*(x)) - \gamma)^2$.

Show that, for $\pi(\theta) = 1$, the posterior loss $\mathbb{E}^\pi[L_0(\theta, \delta^*)|x]$ is constant and equal to the constant risk of δ^*.

e. Show that, for $\pi^*(\theta) = \exp(-2\theta)$, the posterior variance of δ^* is

$$\gamma^\pi(x) = e^{2x - 2\sigma^2}\left(1 - e^{-\sigma^2}\right).$$

Show that γ^π is an unbiased estimator of the risk, $\mathbb{E}_\theta[L_2(\theta, \delta^*(x))]$, and that it is dominated by the Bayes estimator of $L_2(\theta, \delta^*(x))]$ under $\pi(\theta) = e^{-4\theta}$. Can you justify the use of this prior on invariance grounds?

Section 9.4

9.16 *(Eaton (1989)) Show that, for a topological group $\mathcal{G}$, two left-invariant integrals, that is, functionals such that

$$\int_\mathcal{G} f(gx)\mu(dx) = \int_\mathcal{G} f(x)\mu(dx)$$

for every $f \in \mathcal{L}_1(\mu)$ and $g \in \mathcal{G}$, are necessarily proportional.

9.17 *(Eaton (1989)) Consider ν_ℓ a left Haar measure, $f \in K(\mathcal{G})$ and

$$J_1(f) = \int_\mathcal{G} f(xg^{-1})\nu_\ell(dx).$$

a. Show that J_1 is left-invariant. Deduce that there exists a function Δ on $\mathcal{G}$ such that

$$J_1(f) = \Delta(g) \int_{\mathcal{G}} f(x)\nu_\ell(dx) = \Delta(g)J(f).$$

The function Δ is called the *modulus of* $\mathcal{G}$.

b. Show that Δ does not depend on the choice of J_1 and that $\Delta(g_1 g_2) = \Delta(g_1)\Delta(g_2)$ (that is, Δ is a *multiplier*).

c. Consider J_2 such that

$$J_2(f) = \int_{\mathcal{G}} f(x)\Delta(x^{-1})\nu_\ell(dx).$$

Show that J_2 is right-invariant and satisfies

$$J_2(f) = \int_{\mathcal{G}} f(x^{-1})\nu_\ell(dx).$$

Deduce that, if ν_ℓ is a left Haar measure,

$$\nu_r(dx) = \Delta(x^{-1})\nu_\ell(dx)$$

is a right Haar measure.

d. If $\mathcal{G}$ is compact, show that Δ is identically equal to 1. (*Hint:* Use the continuity of Δ and the fact that $\Delta(\mathcal{G})$ is compact.)

e. Consider $\mathcal{G} = GL_n$, the linear group of $\mathbb{R}^n$. We denote by dx the Lebesgue measure on $\mathcal{L}_{n,n}$, the vector space of $n \times n$ matrices. Show that

$$J(f) = \int_{\mathcal{G}} f(x)\frac{dx}{|\det(x)|^n}$$

is simultaneously right- and left-invariant. Deduce that $\Delta = 1$. Is $\mathcal{G}$ compact?

9.18 *(Eaton (1989)) Consider $\mathcal{G}$ a compact group acting on $\mathcal{X}$, and ν the unique Haar probability distribution on $\mathcal{G}$. Define U, a uniform random variable on $\mathcal{G}$, by

$$P(U \in B) = \nu(B).$$

a. Consider $x \in \mathcal{X}$. Show that μ_x, defined by

$$\mu_x(B) = P(Ux \in B)$$

is the unique $\mathcal{G}$-invariant probability on the orbit of x, O_x.

b. If P is a $\mathcal{G}$-invariant distribution on $\mathcal{X}$, show that

$$P = \int_{\mathcal{X}} \mu_x P(dx).$$

c. A *measurable section* $\mathcal{Y} \subset \mathcal{X}$ is defined by
(i) $\mathcal{Y}$ is measurable;
(ii) $\forall x \in \mathcal{X}$, $\mathcal{Y} \cap O_x = \{y(x)\}$; and
(iii) the function $t(x) = y(x)$ is measurable for the σ-algebra induced on $\mathcal{Y}$ by $\mathcal{X}$.
Show that, for every probability distribution Q on $\mathcal{Y}$,

$$P = \int_{\mathcal{Y}} \mu_y Q(dy)$$

is $\mathcal{G}$-invariant on $\mathcal{X}$ and that, reciprocally, every $\mathcal{G}$-invariant probability can be written this way.

d. Consider U a uniform random variable on $\mathcal{G}$, $\mathcal{Y}$ a measurable section of $\mathcal{X}$, and X a random variable on $\mathcal{X}$. Deduce from c. the equivalence between the following properties:
(i) the distribution of gX is independent of $g \in \mathcal{G}$; and
(ii) there exists Y, random variable on $\mathcal{Y}$, independent of U, such that UY has the same distribution as X.

e. Apply to the case $\mathcal{X} = \{0, 1\}^n$.

9.19 Consider $x \sim \mathcal{N}(\theta, 1)$ when the quantity of interest is $h(\theta) = e^{c\theta}$.

a. Give the risk of the Bayes estimator of $h(\theta)$ associated with $\pi(\theta) = 1$ and the quadratic loss, $R(\theta, \delta^\pi)$, and show that the Bayes estimator of $h(\theta)$ associated with $\pi'(\theta) = R(\theta, \delta^\pi)^{-1}$ dominates δ^π.

b. Notice that $R(\theta, \delta^\pi)^{-1}(e^{c\theta} - \delta)^2$ is an invariant loss and derive the following result: *For every invariant loss* $L(\theta, \delta)$, *if* δ^π *is the estimator associated with* L *and the Haar measure* π, *and if* $w(\theta) = \mathbb{E}_\theta[L(\theta, \delta^\pi(x))]$, *the estimator associated with* L *and* $\pi'(\theta) = \pi(\theta)/w(\theta)$ *is the best equivariant estimator.*

Section 9.5

9.20 *(Berger (1985a)) Consider the particular case where the group $\mathcal{G}$ is finite, that is,
$$\mathcal{G} = \{g_1, \ldots, g_m\}.$$
Let us assume that the loss $L(\theta, a)$ is invariant, convex in a, and, in addition, that the action induced by the group $\mathcal{G}$ on $\mathcal{D}$ satisfies
$$\tilde{g}\left(\frac{1}{m}\sum_{i=1}^m a_i\right) = \frac{1}{m}\sum_{i=1}^m \tilde{g}(a_i).$$
Establish the Hunt–Stein Theorem under the additional assumption that $\mathcal{D}$ is convex. (*Hint:* Show that, for every estimator δ, there exists an associated invariant estimator δ^I which dominates δ.)

9.21 In the setting of Example 9.5.1, derive the exact risk of the estimator δ_1 (*Hint:* Notice that z_1/z_2 is distributed as a Cauchy random variable.)

9.22 Examine the estimation of ϱ in Example 9.5.1 for the invariance structure induced by the multiplicative group.

9.23 Consider $(x_1, \ldots, x_p)$ and $(y_1, \ldots, y_p)$ with normal distributions $\mathcal{N}_p(0, \Sigma)$ and $\mathcal{N}_p(0, \Delta\Sigma)$. The hypothesis $H_0 : \Delta \leq \Delta_0$ is to be tested versus $H_1 : \Delta > \Delta_0$.

a. Show that the problem is invariant under $\mathcal{GL}_p$, the group of nonsingular linear transformations.

b. Show that $\mathcal{GL}_p$ is transitive on the sample space, up to a set of measure 0. Deduce that the equivariant estimators are constant, that is, that invariant tests at level α are $\varphi_\alpha(x, y) = 1 - \alpha$.

c. Consider $\varphi_c(x, y) = \mathbb{I}_{y_1^2 \leq cx_1^2}$ and show that it dominates φ_α under the $0-1$ loss for $\alpha = P_{\Delta_0}(y_1^2 > cx_1^2)$.

d. Is $\mathcal{GL}_p$ amenable?

9.24 Consider $x_1, \ldots, x_n \sim \mathcal{C}(\mu, \sigma^2)$.

   a. Show that the Haar measure is $\pi^H(\mu, \sigma) \propto 1/\sigma$.

   b. Consider the reparameterization $y_i = 1/x_i$. Show that $y_i \sim \mathcal{C}(\nu, \tau^2)$ and derive ν and τ in terms of μ and σ.

   c. Show that $\pi^H(\nu, \tau) \propto 1/\tau$ is not the transform of $\pi^H(\mu, \sigma)$ and conclude about the limitations of invariance in terms of reparameterization.

Section 9.6

9.25 *(Villegas (1990)) Consider a family of probability distributions P_θ on $\mathcal{X}$, with $\theta \in \Theta$ and $T(x)$ taking values in a Euclidean affine space E, such that the likelihood function is

$$\ell(\theta|x) = c_1(x)c_2(\theta) \exp\{-||T(x) - \theta||^2/2\}.$$

This model is called *Euclidean Bayesian* if $\pi(\theta) = 1$.

   a. Deduce that the corresponding Euclidean prior distribution in the case of a Poisson model $\mathcal{P}(\lambda)$ is $\pi(\lambda) = 1/\lambda$.

   b. Show that the p-value $p(x) = P_{\lambda_0}(X \geq x)$ when testing $H_0 : \lambda \leq \lambda_0$ against $H_1 : \lambda > \lambda_0$ is related to this prior distribution, but that this relation does not hold for the alternative test of $H_0 : \lambda \geq \lambda_0$ against $H_1 : \lambda < \lambda_0$.

   c. Show that the Haldane prior distribution

(9.7.3)
$$\pi(p) = \frac{1}{p(1-p)}$$

   also appears as a Euclidean model when $x \sim \mathcal{B}(n, p)$. Is (9.7.3) still the Euclidean prior for the negative binomial distribution, $\mathcal{N}eg(n, p)$?

   d. If $0 < x < n$, show that, in the binomial case, the p-values $P_{p_0}(X \leq x)$ and $P_{p_0}(X \geq x)$ associated with the hypotheses $H_0 : p \geq p_0$ and $H_0 : p \leq p_0$ do not correspond to the Euclidean distribution (9.7.3).

   e. In the normal case $\mathcal{N}(\mu, \sigma^2)$, show that the Euclidean prior distributions are the following ones:
    (i) $\pi(\theta) = 1$ if $\theta = \mu$;
    (ii) $\pi(\theta) = 1$ if $\theta = \sigma^{-2}$; and
    (iii) $\pi(\theta) = \theta_2$ if $(\theta_1, \theta_2) = (\mu, \sigma^{-2})$.

9.26 Examine the issue of compatibility between the invariance requirements and the Likelihood Principle. In particular, determine whether the maximum likelihood estimator is always an invariant estimator.

9.27 *For an arbitrary loss function $L(\theta, \delta)$ and a given prior distribution π, assume that the Bayes estimator δ^π is such that $0 < R(\theta, \delta^\pi) < \infty$ for every θ.

   a. If we define $L^\pi(\theta, \delta) = L(\theta, \delta)/R(\theta, \delta^\pi)$, show that δ^π has constant risk 1. Does this imply that δ^π is minimax? (*Hint:* Notice that δ^π is not necessarily the Bayes estimator under L^π.)

   b. Consider the special case when $x \sim \mathcal{N}(\theta, 1)$ and π is $\mathcal{N}(\theta_0, \tau^2)$. Compute $R(\theta, \delta^\pi)$ and study the behavior of the Bayes estimator associated with π and L^π, compared with δ^π (numerically, if necessary).

c. In the event δ_1^π is associated with π and $L_1^\pi = L^\pi$ is different from δ^π, a sequence of estimators δ_n^π can be defined sequentially by $L_n^\pi = L/R(\theta, \delta_{n-1}^\pi)$. What can be said about the limit of the sequence (δ_n^π)?

d. Study the above sequence when $x \sim \mathcal{P}(\lambda)$, $\pi(\lambda) = 1$, and $L(\lambda, \delta) = (1 - \lambda/\delta)^2$.

9.8 Notes

9.8.1 Invariance and marginalization paradoxes

Besides bringing a new perspective on equivariant estimation, Helland (1999) advocates the use of right Haar measures as means of avoiding the *marginalization paradoxes*, as already pointed out in Dawid et al. (1973). (See Exercises 3.44—3.50.)

More precisely, given a model $(\mathcal{X}, \Theta, f(x|\theta))$ with a group $\mathcal{G}$ acting on $\mathcal{X}$ and the corresponding group $\bar{\mathcal{G}}$ acting on Θ, a function $h(\theta)$ is said to be *invariantly estimable* if $h(\theta_1) = h(\theta_2)$ implies $h(\bar{g}\theta_1) = h(\bar{g}\theta_2)$ for all $\bar{g} \in \bar{\mathcal{G}}$ (Hora and Buehler (1966)). Helland (1999) supports the view that the invariance perspective and the use of the corresponding Haar measure should be restricted to the estimation of invariantly estimable functions of the parameters. For instance, although the Lebesgue measure is the right Haar measure for the translation group, and thus applies to the normal distribution $\mathcal{N}_p(\theta, I_p)$, it should not be used for the estimation of $||\theta||^2$, thus avoiding the subefficiency pointed out in Section 3.5.4. This point of view somehow relates to the construction of reference priors: given a parameter of interest, the relevant invariance structure should be determined first, and the corresponding right Haar measure could then be derived as the appropriate noninformative prior. Some drawbacks to this approach are that there are functions for which non-trivial invariance groups cannot be found, and that there is always arbitrariness involved in the choice of these groups, when they exist.

Example 9.8.1 (Helland (1999)) When $x \sim \mathcal{N}_p(\theta, \sigma^2 I_p)$ and θ is on the sphere of radius c, the best equivariant estimator of θ under the group of rotations is associated with the uniform measure on the sphere and is given by (8.3.2). (See Exercise 4.36.) If $c = ||\theta||^2$ is unknown, it can be estimated based on $||x||$ rather than x (see Examples 3.5.9 and 3.5.11), using for instance the maximum likelihood estimator, or Saxena and Alam's (1982) improvement $(||x||^2 - p)^+$. If one plugs this estimator $\kappa(x)$ of c in δ_c, the resulting estimator

$$\tilde{\delta} = \kappa(x) \frac{I_{p/2}(\kappa(x)||x||)}{I_{p/2-1}(\kappa(x)||x||)} \frac{x}{||x||},$$

where I_ν is the *modified Bessel function*, behaves similarly to a shrinkage estimator (2.8.4) for large values of $||x||$, as shown by Bock and Robert (1990) and Beran (1996). (See Exercise 10.37.) ‖

For an invariantly estimable function $h(\theta)$, Helland (1999) considers the subgroup of $\bar{\mathcal{G}}$,

$$\bar{K} = \{\bar{g} \in \bar{\mathcal{G}}; h(\bar{g}\theta) = h(\theta) \text{ for all } \theta \in \Theta\},$$

since $h(\theta)$ is a maximal invariant for $\bar{K}$. Given the corresponding subgroup of $\mathcal{G}$,

$$K = \{g \in \mathcal{G}; \bar{g} \in \bar{K}\},$$

let z be a maximally invariant variable for K. If $\eta = h(\theta)$, Helland (2000) then shows that the marginalization paradox does not occur for (η, z) when using the right Haar measure associated with $\bar{\mathcal{G}}$. This means that, under this measure, if $\theta = (\eta, \xi)$ and $x = (z, y)$, and if the marginal posterior distribution of η depends only on z, it can also be obtained as a posterior distribution based on x alone.

CHAPTER 10

Hierarchical and Empirical Bayes Extensions

Books and papers and scrolls covered nearly every flat surface, with all sorts of odd things interspeded among the piles, and sometimes on top of them. Strange shapes of glass or metal, spheres and tubes interlinked, and circles held inside circles, stood among bones and skulls of every shape and description.

Robert Jordan, *The Dragon Reborn, Book III of the Wheel of Time.*

10.1 Incompletely Specified Priors

In the previous chapters, we have noticed (and even stressed!) the ambivalent aspect of Bayesian analysis: it is sufficiently reductive to produce an effective decision, but this efficiency can also be misused. For instance, the subjective aspects of a Bayesian analysis can always be modified to produce conclusions established in advance. Of course, the same misappropriation is possible in a frequentist framework through the choice of the loss or estimation criteria, while the classical approach does not distinguish between the subjective and objective inputs of an analysis. The main point here is that, as mentioned in Chapter 3, the choice of a prior distribution should always be *justifiable* by the statistician, that is, it must be based on sound (or "repeatable") arguments. So the fact that *Bayesian tools* may lead to illegitimate inferences cannot be taken as a flaw of the *Bayesian paradigm*.

A more pertinent criticism is that the prior information is rarely rich enough to define a prior distribution exactly. In such cases, it seems necessary to include this uncertainty in the Bayesian model, although the notion of prior distribution may seem insufficient to represent such a degree of ignorance. In fact, the residual uncertainty prompted some extensions of the Bayesian paradigm, such as the *upper and lower probabilities* of Dempster (1968), or the *imprecise probabilities* of Walley (1991).

The *hierarchical Bayes analysis* considers, nonetheless, that this is can be done within the Bayesian paradigm. This particular modeling of the prior information decomposes the prior distribution into several conditional levels of distributions, and thus allows for a distinction between structural and subjective items of information. According to the Bayesian paradigm, uncertainty at any of these levels is incorporated into additional prior distributions. In the simplest cases, the hierarchical structure is reduced to two levels, the parameters of the first being associated with a prior distribution defined on the second. The first-level distribution is generally a conjugate prior, owing to the computational tractability of these distributions, and also because the upper level somehow compensates for modeling errors at the lower levels. (Another justification for a conjugate modeling is provided by Dalal and Hall (1983) and Diaconis and Ylvisaker (1985); see Section 3.4.) We have already encountered examples of such modelings in Chapter 6, such as in Example 6.3.10.

A general characteristic of the hierarchical modeling is that it improves the robustness of the resulting Bayes estimators: while still incorporating prior information, the estimators are also well performing from a frequentist point of view (minimaxity and admissibility), although these two requirements are usually difficult to reconcile.

Additional justifications of hierarchical Bayes modeling stem from real life, since there are settings in medicine, biology, animal breeding, economics, and so on, where the population of interest can be perceived as a subpopulation of a meta-population, or even as a subpopulation of a subpopulation of a global population. This is, for instance, the case with *meta-analysis*, according to which several experiments about the same phenomenon undertaken at different places with different subjects and different protocols are pooled together (see, e.g., Mosteller and Chalmers (1992), Mengersen and Tweedie (1993), or Givens et al. (1997)).

Example 10.1.1 (Guihenneuc-Jouyaux et al. (1998)) The *Human Immunodeficiency Virus* (HIV) is the virus that leads to AIDS. For a given patient, the progression of the HIV infection towards AIDS can be represented as a transition between seven stages of increasing severity, the final stage corresponding to AIDS. The moves between the stages are directed by a continuous time Markov model with *infinitesimal generator* Λ. (This means that the distribution of the state at time T, given a distribution $\omega_0 = (\omega_{01}, \ldots, \omega_{07})$ at time 0, is given by the matrix product $\omega_0 \cdot \exp\{T\Lambda\}$.)[1]

For a given patient infected with HIV, the first six stages are not observed directly, but only through random variables ($1 \leq i \leq n$, $1 \leq j \leq n_i$),

$$x_{ij} \sim \mathcal{N}\left(\mu_{S_{ij}}, \sigma^2\right),$$

where i denotes the individual and j the follow-up point, and where S_{ij} denotes the HIV stage. The x_{ij}'s represent blood markers (CD4 counts)

---

[1] The extension of the exponential function to multivariate settings such as this one is obtained by using the series representation of $\exp(x)$. See Exercise 10.2.

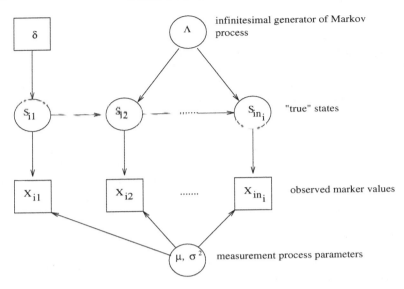

Figure 10.1.1. *Directed acyclic graph of the hierarchical model.*
(Source: Guihenneuc–Jouyaux et al. (1998).)

subject to great variability, and to measurement error. This is thus a special case of *hidden Markov model* (see Exercise 6.46), with the additional difficulty that the hidden Markov chain is operating in continuous time. But the forward–backward formulas developed in Exercise 6.47 still apply (Exercise 10.2). A similar model has been proposed by Kirby and Spiegelhalter (1994).

The S_{ij}'s therefore constitute the first level of a hierarchical model, with hyperparameters such as the generator matrix Λ corresponding to the second level and common to all individuals. A second hyperparameter is δ, the prior distribution on the HIV stage at the first observation. It is usually quite helpful to represent such hierarchical models as graphs (more exactly, *directed acyclic graphs* or DAG's). Figure 10.1 provides this representation for the HIV model, the convention being that *boxes* correspond to known or observed quantities, and *circles* to unknown quantities; *arrows* indicate probabilistic dependence. (See Note 10.7.1 and Lauritzen (1996) for more details on graphical models.) ‖

The empirical Bayes analysis is based on the same perception of imprecision over the prior information, but at a more pragmatic level. In fact, this approach considers that it is illusory to try to model this imprecision by several levels of conditional distributions when the first level is already partly or totally unknown. Rather paradoxically, the empirical Bayes analysis still relies on a conjugate prior modeling, where the hyperparameters are estimated from the observations and this "estimated prior" is then used as a regular prior in the subsequent inference. Needless to say, the substitution of hyperparameters by estimated hyperparameters that is the core of the empirical Bayes analysis definitely excludes this technique from the Bayesian paradigm, but allows the statistician to take advantage of diffuse prior information in a simplified way. Moreover, it is often the case that the resulting estimators have good frequentist properties, although there is too much arbitrariness in the determination of the estimated hyperparameters to make a general rule of this fact. A related interest in the empirical Bayes modeling is to provide Bayesian support for the Stein effect (see Note 2.8.2). The empirical Bayes analysis may also appear as a convenient alternative in settings where a hierarchical Bayes analysis is too involved, even though increasingly efficient computation techniques progressively suppress the need for such approximations, as we saw in Chapter 6.

10.2 Hierarchical Bayes analysis

This section provides only a short introduction to hierarchical Bayes analysis, and focuses on a few features of interest. For a more thorough treatment of this topic, see Berger (1985a), in relation to the robustness aspects, and Deely and Lindley (1981), Dumouchel and Harris (1983), George (1986a), Angers and MacGibbon (1990), Gelman et al. (1996), Hobert (2000) and Draper (2001). For applications to animal breeding, see, e.g., Fouley et al. (1992).

10.2.1 Hierarchical models

For reasons owing to the modeling of the observations or to the decomposition of the prior information, it may happen that the Bayesian statistical model is *hierarchical*, that is, involves several levels of conditional prior distributions.

Definition 10.2.1 *A hierarchical Bayes model is a Bayesian statistical model, $(f(x|\theta),\ \pi(\theta))$, where the prior distribution $\pi(\theta)$ is decomposed in conditional distributions*

$$\pi_1(\theta|\theta_1),\ \pi_2(\theta_1|\theta_2),\ldots,\ \pi_n(\theta_{n-1}|\theta_n)$$

and a marginal distribution $\pi_{n+1}(\theta_n)$ such that

$$(10.2.1)\quad \pi(\theta) = \int_{\Theta_1\times\ldots\times\Theta_n} \pi_1(\theta|\theta_1)\pi_2(\theta_1|\theta_2)\cdots\pi_{n+1}(\theta_n)\,d\theta_1\cdots d\theta_{n+1}.$$

The parameters θ_i are called hyperparameters of level i $(1 \leq i \leq n)$.

Before we justify this decomposition by its usefulness, notice that hierarchical structures can also occur in classical statistical models.

Example 10.2.2 A classical occurrence of hierarchical models is the inclusion of *random-effects* in a linear model. This extension can be written under the form

$$y|\theta \sim \mathcal{N}_p(\theta, \Sigma_1),$$
$$\theta|\beta \sim \mathcal{N}_p(X\beta, \Sigma_2),$$

with no reference to a Bayesian modeling. The mean of y, θ, is decomposed in *fixed effects*, $X\beta$, and in *random-effects*, $Z\eta$, where η is normal with mean 0 (the covariance matrix Σ_2 can then be singular). These models are often used in biometry, in particular in *animal breeding*, to distinguish between the influence of the fixed effect (e.g., sires, breed, year, etc.) and the influence of random factors (e.g., dams within sires). ‖

Another classical class of non-Bayesian hierarchical structures comprises *latent variable models*, as in mixtures (Section 6.4) and other hidden mixture representations (Note 6.6.3). The vector of latent variables z then constitutes the first level of a hierarchical Bayesian model, the prior modeling per se occurring at higher levels.

Such examples also point out that the boundary between classical and Bayesian hierarchical models is sometimes fuzzy, and mainly depends on the interpretation of the parameters. For instance, in the random-effect model, we operate according to a classical perspective if the inference is about the fixed effects (β), and according to a Bayesian perspective when considering the global effect (θ). Similarly, if the latent variables z_t are of interest, as in the stochastic volatility model (4.7.2) for the y_t^*'s, they are treated as missing *data*, whereas, if they are introduced for a more convenient representation of the model, as in the case of mixtures used for non-parametric modeling, the z_t's can be treated as parts of the prior modeling.

Let us stress here that a hierarchical Bayes model is simply a special type of Bayesian model. In fact, if

$$(10.2.2) \qquad x \sim f(x|\theta), \ \theta \sim \pi_1(\theta|\theta_1), \ldots, \theta_n \sim \pi_{n+1}(\theta_n),$$

we recover the usual Bayesian model

$$x \sim f(x|\theta), \quad \theta \sim \pi(\theta),$$

for the prior

$$\pi(\theta) = \int_{\Theta_1 \times \ldots \times \Theta_n} \pi_1(\theta|\theta_1) \ldots \pi_n(\theta_{n-1}|\theta_n)\pi_{n+1}(\theta_n) \, d\theta_1 \cdots d\theta_n.$$

This reduction shows that hierarchical modelings are indeed included in the Bayesian paradigm, and therefore that this approach enjoys the general optimality properties of the Bayesian approach with some additional advantages related to the decomposition of the prior distribution (see Section 10.3). It also shows why it is seldom necessary to go beyond two conditional levels in the hierarchical decomposition. In fact, if the hyperparameters $\theta_1, \ldots, \theta_n$ are of no interest for the inference (about θ), it is equivalent to consider the simpler hierarchical model

$$x|\theta \sim f(x|\theta), \qquad \theta|\theta_1 \sim \pi_1(\theta|\theta_1),$$

with

$$\theta_1 \sim \pi_2(\theta_1) = \int_{\Theta_2 \times ... \times \Theta_n} \pi_1(\theta_1|\theta_2) \cdots \pi_{n+1}(\theta_n) \, d\theta_2 \cdots d\theta_n,$$

which eliminates intermediary steps and additional hyperparameters. Nonetheless, a more elaborate decomposition may still be of interest for the construction and the practical computation of the Bayes estimators, as already shown in Chapter 6.

Example 10.2.3 Robert and Reber (1998) consider an experiment under which rats are intoxicated by a substance, then treated by either a placebo or a drug. The model associated with this experiment is a *linear additive model effect*: given x_{ij}, y_{ij} and z_{ij}, jth responses of the ith rat at the control, intoxication and treatment stages, respectively, we assume that $(1 \leq i \leq I)$

$$\begin{aligned} x_{ij} &\sim \mathcal{N}(\theta_i, \sigma_c^2), & 1 \leq j \leq J_i^c, \\ y_{ij} &\sim \mathcal{N}(\theta_i + \delta_i, \sigma_a^2), & 1 \leq j \leq J_i^a, \\ z_{ij} &\sim \mathcal{N}(\theta_i + \delta_i + \xi_i, \sigma_t^2), & 1 \leq j \leq J_i^t, \end{aligned}$$

where θ_i is the average control measurement, δ_i the average intoxication effect and ξ_i the average treatment effect for the ith rat, the variances of these measurements being constant for the control, the intoxication and the treatment effects. An additional (observed) variable is w_i, which is equal to 1 if the rat is treated with the drug, and 0 otherwise.

Given that the purpose of the experiment is to assess the overall effect of the tested drug, the different individual averages are related through a common (conjugate) prior distribution $(1 \leq i \leq I)$,

$$\theta_i \sim \mathcal{N}(\mu_\theta, \sigma_\theta^2), \qquad \delta_i \sim \mathcal{N}(\mu_\delta, \sigma_\delta^2),$$

and

$$\xi_i \sim \mathcal{N}(\mu_P, \sigma_P^2) \qquad \text{or} \qquad \xi_i \sim \mathcal{N}(\mu_D, \sigma_D^2),$$

depending on whether the ith rat is treated with a placebo or a drug. The hyperparameters of the model,

$$\mu_\theta, \mu_\delta, \mu_P, \mu_D, \sigma_c, \sigma_a, \sigma_t, \sigma_\theta, \sigma_\delta, \sigma_P, \sigma_D,$$

are then associated with Jeffreys' noninformative priors. This prior leads to a well defined posterior distribution if there are at least two observations for each stage of the experiment.   ‖

10.2.2 Justifications

The hierarchical Bayes analysis is partly based on the work of Good (see Good (1980, 1983) for references) and is exposed in Lindley and Smith

(1972) for the particular case of the linear model, where the authors use the duality between the usual Bayesian analysis of a random-effect model and the hierarchical Bayes analysis of a regular regression model. Despite the reduction (10.2.1), which shows that a hierarchical Bayes model is actually a particular case of a Bayesian model, the decomposition

$$\pi(\theta) = \int_{\Theta_1} \pi_1(\theta|\theta_1)\pi_2(\theta_1)\, d\theta_1$$

or its generalization (10.2.1) may be preferred for several reasons:

(i) *Objective* reasons based on the modeling of the observed phenomenon as a special case of a meta-population may lead to the first two levels, the Bayesian approach being justified by a prior knowledge of the meta-population. This is the case of Examples 10.2.2 and 10.2.3. More generally, as mentioned in Section 10.1, hierarchical Bayes models naturally appear in meta-analysis where several studies have to be pooled together.

Example 10.2.4 (Berger (1985a)) Consider $x_i \sim \mathcal{N}(\beta_i, 10)$ $(i = 1, \ldots, 7)$, which represent yearly and independent measures of the intelligence quotient (IQ) of a child, for seven consecutive years. Since IQ tests are supposed to account for an age effect, it is reasonable to consider that the β_i's have the same mean θ, the "true" value of the IQ. A corresponding first-level prior distribution is then

$$\beta_i|\theta \sim \mathcal{N}(\theta, \sigma_\pi^2) \qquad (i = 1, \ldots, 7).$$

Moreover, if the child belongs to a thoroughly-studied population of children, it may be the case that, for this population,

$$\theta \sim \mathcal{N}(\xi, \tau^2),$$

where ξ and τ are known and this provides the second level of the analysis. Otherwise, a noninformative alternative is to take $\pi_2(\theta) = 1$.   ‖

(ii) In connection with the above justification, a researcher may wish to separate the prior modeling into two parts, the first corresponding to the *structural* information about the model, the second to the more *subjective* information. For instance, the information may consist of uncertain linear restrictions on the parameters of a regression model, whereas the distribution on the hyperparameters $\pi_2(\theta_1)$ accounts for the imprecision of these restrictions.

Example 10.2.5 Albert (1988) represents uncertainties about generalized linear models (McCullagh and Nelder (1989)) $(i = 1, \ldots, n)$

$$y_i|x_i \sim \exp\{\theta_i \cdot y_i - \psi(\theta_i)\}, \qquad \nabla\psi(\theta_i) = \mathbb{E}[y_i|x_i] = h(x_i^t\beta),$$

where h is the *link* function and $x_i \in \mathbb{R}^q$ a vector of covariates, by removing the linear constraint $\nabla\psi(\theta_i) = h(x_i^t\beta)$ to an higher level of the hierarchy, that is, by introducing a conjugate prior

$$\theta_i \sim \exp\left\{\lambda\left[\theta_i \cdot \xi_i - \psi(\theta_i)\right]\right\}$$

such that $\mathbb{E}[\nabla\psi(\theta_i)] = h(x_i^t\beta)$. The regression parameter β is thus transfered to a second level with, possibly, a normal prior $\beta \sim \mathcal{N}_q(0, \tau^2 I_q)$,
including the flat prior for $\tau = \infty$. The posterior variance of $\psi(\theta_i)$ then
reveals how accurate the generalized linear model is, that is, allows assessment of the departure from the linearity assumption.   ∥

Example 10.2.6 (Example 10.2.3 continued) An alternative also
considered in Robert and Reber (1998) is to assume the prior distribution

(10.2.3) $$\delta_i \sim p\mathcal{N}(\mu_{\delta 1}, \sigma_{\delta 1}^2) + (1-p)\mathcal{N}(\mu_{\delta 2}, \sigma_{\delta 2}^2),$$

which allows for two possible levels of intoxication, that is, for two different reactions to the intoxication in the rat population. As detailed in
Robert and Reber (1998), there are metabolic reasons for this modification of the prior distribution. While this mixture structure also transfers
to the marginal distribution of the y_{ij}'s, it differs from a regular mixture
model, since it requires that the y_{ij}'s belong to the same component of
the mixture for $1 \leq j \leq J_i^a$.   ∥

(iii) On the contrary, in a *noninformative setting*, a hierarchical Bayes
model suggests a compromise between the Jeffreys noninformative distributions, which are diffuse but sometimes difficult to use or justify,
and the conjugate distributions, which have limited subjective justification but are analytically tractable. When the hyperparameters have
a prior *hyperdistribution* (or *hyperprior*), the noninformative perspective is reinforced, while generally providing a defined posterior distribution on θ. A possibility is to iterate this argument by introducing a
conjugate distribution on θ_1, $\pi_2(\theta_1|\theta_2)$, and a noninformative distribu-
tion on θ_2. However, the choice of a conjugate distribution on θ_1 does
not guarantee closed-form expressions for the Bayes estimators, and,
more fundamentally, does not seem to improve the robustness of the
model. Whatever the number of levels in the distribution, the averaging on the unknown hyperparameters can only reinforce the robustness of the prior distribution when compared with a classical conjugate
approach. See Berger (1985a) for an explanation of why hierarchical
modeling is interesting from the robustness point of view.

Example 10.2.7 Consider the usual regression model, $y = X\beta + \epsilon$,
that is, $y \sim \mathcal{N}_n(X\beta, \sigma^2 I_n)$, where $\beta \in \mathbb{R}^p$. For structural reasons, the
coefficients of the regression are similar. For instance, the β_i's can de-
scribe the investment rates of different European automobile companies,
for which the rates are quite similar. We then assume that $\beta_i \sim \mathcal{N}(\xi, \sigma_\pi^2)$,
where ξ represents the common value. Such a model is called *exchange-
able* (see Note 3.8.2, Bernardo and Smith (1994) and Gelman et al.

(1996)). If additional information is available about the common value, we can take $\xi = \xi_0$ or $\xi \sim \mathcal{N}(\xi_0, \tau^2)$. Otherwise, the second level can be noninformative, that is, $\pi_2(\xi) = 1$.   ‖

Example 10.2.8 (Example 10.2.2 continued) In the setting of the random-effect linear model,

$$y|\theta \;\; \sim \;\; \mathcal{N}_p(\theta, \Sigma_1),$$
$$\theta|\beta \;\; \sim \;\; \mathcal{N}_p(X\beta, \Sigma_2),$$

Lindley and Smith (1972) and Smith (1973) assume that β also satisfies a linear relation and use the following prior:

$$\beta \sim \mathcal{N}_n(Z\xi, \Sigma_3).$$

An alternative prior which robustifies the model is

$$\beta \sim \mathcal{T}_n(\alpha, Z\xi, \Sigma_3),$$

but this distribution simply involves an additional level in the hierarchical model compared with the original normal distribution, as shown by Dickey (1968). In fact,

$$\beta|z \sim \mathcal{N}_p(Z\xi, \Sigma_3/z), \qquad z \sim \mathcal{G}(\alpha/2, \alpha/2),$$

in this case. If we consider $\beta|z \sim \mathcal{N}_p(\mu, z\Sigma_3)$ (conjugate distribution) and $\pi(z) = 1/z$ (noninformative distribution) the marginal distribution

$$\beta \sim \mathcal{T}_p(p/2, \mu, \Sigma_3),$$

is thus a proper distribution, contrary to the noninformative distribution $\pi(\beta) = 1$.   ‖

(iv) Another beneficial aspect of the hierarchical Bayes analysis is that it also robustifies the usual Bayesian analysis from a frequentist point of view, since it reduces the arbitrariness of the hyperparameter choice (sometimes transferred to a higher level), and averages the conjugate Bayesian answers. Section 10.3 demonstrates that, in the normal case, many prior distributions on the hyperparameters lead to minimax generalized Bayes estimators.

(v) A last justification of the hierarchical Bayes approach is that it can often simplify *Bayesian calculations*. In fact, the decomposition of a prior distribution π in its components $\pi_1, \ldots, \pi_n$ (which can be, for instance, conjugate distributions) may allow for an easier approximation of some posterior quantities by simulation, as already stressed in Section 6.3.5 for Gibbs sampling.

10.2.3 Conditional decompositions

A particularly appealing aspect of hierarchical models is that they allow for conditioning on all levels, and this easy decomposition of the posterior distribution may compensate for the apparent complexity occasioned by these successive levels. For instance, if

$$\theta|\theta_1 \sim \pi_1(\theta|\theta_1), \qquad \theta_1 \sim \pi_2(\theta_1),$$

we have the following result.

Lemma 10.2.9 *The posterior distribution of θ is*

$$\pi(\theta|x) = \int_{\Theta_1} \pi(\theta|\theta_1, x)\pi(\theta_1|x)\, d\theta_1,$$

where

$$\pi(\theta|\theta_1, x) = \frac{f(x|\theta)\pi_1(\theta|\theta_1)}{m_1(x|\theta_1)},$$

$$m_1(x|\theta_1) = \int_{\Theta} f(x|\theta)\pi_1(\theta|\theta_1)\, d\theta,$$

$$\pi(\theta_1|x) = \frac{m_1(x|\theta_1)\pi_2(\theta_1)}{m(x)},$$

$$m(x) = \int_{\Theta_1} m_1(x|\theta_1)\pi_2(\theta_1)\, d\theta_1.$$

Moreover, this decomposition works for the posterior moments, that is, for every function h,

$$\mathbb{E}^{\pi}[h(\theta)|x] = \mathbb{E}^{\pi(\theta_1|x)}\left[\mathbb{E}^{\pi_1}[h(\theta)|\theta_1, x]\right],$$

where

$$\mathbb{E}^{\pi_1}[h(\theta)|\theta_1, x] = \int_{\Theta} h(\theta)\pi(\theta|\theta_1, x)\, d\theta.$$

This result is a straightforward consequence of Bayes's Theorem, the last equality being easily established by the Fubini Theorem. It has important consequences in terms of the computation of Bayes estimators, though, since it shows that $\pi(\theta|x)$ can be simulated by generating, first, θ_1 from $\pi(\theta_1|x)$ and then θ from $\pi(\theta|\theta_1, x)$, if these two conditional distributions are easier to work with.

Example 10.2.10 (Example 10.2.3 continued) The posterior distribution of the complete parameter vector is given by

$$\pi((\theta_i, \delta_i, \xi_i)_i, \mu_\theta, \ldots, \sigma_c, \ldots |\mathcal{D}) \propto$$

$$\prod_{i=1}^{I} \left\{ \exp -\{(\theta_i - \mu_\theta)^2/2\sigma_\theta^2 + (\delta_i - \mu_\delta)^2/2\sigma_\delta^2\} \right.$$

$$\prod_{j=1}^{J_i^c} \exp -\{(x_{ij} - \theta_i)^2/2\sigma_c^2\} \prod_{j=1}^{J_i^a} \exp -\{(y_{ij} - \theta_i - \delta_i)^2/2\sigma_a^2\}$$

$$\left. \prod_{j=1}^{J_i^t} \exp -\{(z_{ij} - \theta_i - \delta_i - \xi_i)^2/2\sigma_t^2\} \right\}$$

$$(10.2.4) \quad \prod_{\ell_i=0} \exp-\{(\xi_i - \mu_P)^2/2\sigma_P^2\} \prod_{\ell_i=1} \exp-\{(\xi_i - \mu_D)^2/2\sigma_D^2\}$$

$$\sigma_c^{-\sum_i J_i^c - 1} \sigma_a^{-\sum_i J_i^a - 1} \sigma_t^{-\sum_i J_i^t - 1} (\sigma_\theta \sigma_\delta)^{-I-1} \sigma_D^{-I_D - 1} \sigma_P^{-I_P - 1},$$

where $\mathcal{D}$ denotes the sample. The posterior distribution thus fails to integrate into explicit formulas for the marginal distributions of the parameters of interest and does not lead to closed-form expressions for the posterior expectations of these parameters. However, the full conditional distributions are available, as detailed in Exercise 10.14.   ‖

Obviously, Lemma 10.2.9 only holds when the various integrals are well defined. In fact, because the second-level distributions are generally improper, this is not always the case. The following lemma gives a sufficient condition for the posterior moments to exist when $x|\theta \sim \mathcal{N}_p(\theta, \Sigma)$ (see Berger and Robert (1990) for a proof):

Lemma 10.2.11 *If the marginal distribution*

$$m(x) = \int_\Theta f(x|\theta)\pi(\theta)\, d\theta$$

is finite for every $x \in \mathbb{R}^k$, then the mean and the variance of the posterior distribution $\pi(\theta|x)$ always exist.

Another important feature of hierarchical models that influences the computation of hierarchical Bayesian estimators is given by the following result:

Lemma 10.2.12 *For the hierarchical model (10.2.2), the full conditional distribution of θ_i given x and the θ_j's $(j \neq i)$ satisfies*

$$\pi(\theta_i|x, \theta, \theta_1, \ldots, \theta_n) = \pi(\theta_i|\theta_{i-1}, \theta_{i+1})$$

with the convention $\theta_0 = \theta$ and $\theta_{n+1} = 0$.

Proof. Since

$$\begin{aligned} \pi(\theta_i|x, \theta, \theta_1, \ldots, \theta_n) &\propto f(x|\theta)\pi_1(\theta|\theta_1)\cdots\pi_{n+1}(\theta_{n+1}) \\ &\propto \pi_{i-1}(\theta_{i-1}|\theta_i)\pi_i(\theta_i|\theta_{i+1}), \end{aligned}$$

the posterior distribution only depends on the two adjacent levels in the hierarchy.   □□

This simple result is indeed of importance because it means that the conditional distributions in a hierarchical model only involve local hyperparameters. In settings such as graphical or spatial models, where the joint distribution is defined locally on a group of hyperparameters (called *cliques* in graphical models), Lemma 10.2.12 thus shows that computational techniques such as the Gibbs sampler (Section 6.3.3) are the only way to process complex models of this type.

10.2.4 Computational issues

A drawback to hierarchical models is that they usually prevent an explicit derivation of the corresponding Bayes estimators, even when the successive levels are conjugate, and therefore that they call for numerical approximations.

Example 10.2.13 Consider $x \sim \mathcal{B}(n,p)$ and $p|m \sim \mathcal{B}e(m,m)$ with $m \in \mathbb{N}^*$. Therefore,

$$
\begin{aligned}
\pi_1(p|m) &= \frac{\Gamma(2m)}{\Gamma(m)^2} [p(1-p)]^{m-1} \\
&= (2m-1)\binom{2m-2}{m-1} [p(1-p)]^{m-1}.
\end{aligned}
$$

If the second-level prior distribution is $\pi_2(m) = 1/(2m-1)$, the prior distribution on p is then

$$
\begin{aligned}
\pi(p) &= \int_{\mathbb{N}^*} \pi_1(p|m)\pi_2(m)\, dm \\
&= \sum_{n=0}^{+\infty} \binom{2n}{n} [p(1-p)]^n.
\end{aligned}
$$

The posterior distribution

$$
\pi(p|x) = \int \pi_1(p|m,x)\pi_2(m|x)\, dm
$$

cannot be obtained in a closed form, since, although $\pi(p|m,x)$ is the beta distribution $\mathcal{B}e(m+x, m+n-x)$, $\pi(m|x)$ is the beta-binomial distribution

$$
\frac{(m+x-1)\ldots m\,(m+n-x-1)\ldots m}{(2m+n-1)\ldots(2m)(2m-1)}
$$

up to a normalization factor. Posterior quantities such as $\mathbb{E}^\pi[p|x]$ cannot be obtained in closed form.   ‖

    The most natural solution in hierarchical settings is to use a simulation-based approach. Indeed, as mentioned above, the decomposition provided by Lemmas 10.2.9 and 10.2.12 is most useful in this respect, since it allows for simulations via the Gibbs sampler or alternative MCMC techniques (Sections 6.3.2 and 6.3.3). This was already clear in the examples of Section 6.3.5 and the following examples reinforce the point that MCMC methods are ideally suited for hierarchical models (see also Gelman et al. (1996) and Robert and Casella (2004, §10.2.3)).

Example 10.2.14 Consider

$$
x \sim \mathcal{N}_p(\theta, \Sigma) \qquad \text{and} \qquad \theta|\mu, \xi \sim \mathcal{N}_p(\mu, B(\xi)),
$$

where $B(\xi) = \xi C - \Sigma$. The positive definite matrix C is fixed and ξ varies on the half-line $[\lambda_{\max}(C^{-1}\Sigma), +\infty)$, where $\lambda_{\max}(A)$ denotes the largest eigenvalue of A. This representation of the posterior covariance matrix simplifies

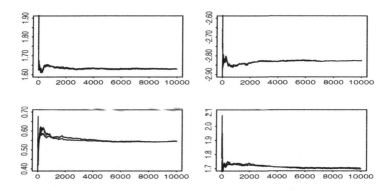

Figure 10.2.1. *Convergence curves for* μ_θ *(top left),* μ_δ *(top right),* μ_P *(bott. left) and* μ_D *in the experiment of Example 10.2.3. The dotted curves represent the partial Rao–Blackwellized averages and are almost indistinguishable from the standard averages. (Source: Robert and Reber (1999).)*

the computations while providing robust estimators. For instance, a second level modeling on (μ, ξ) may involve a noninformative distribution. However, a usual assumption is that $\mu = Y\beta$ for $\beta \in \mathbb{R}^k$, with a given regressor Y such that $Y^t C Y$ is full rank, with a noninformative distribution on β. It can then be shown that $m(x) < +\infty$ if $p > 2 + k$ (see Exercise 10.12).   ‖

Example 10.2.15 (Example 10.2.3 continued) Since the full conditional distributions correspond to standard distributions (see Exercise 10.14), the Gibbs sampler applies in this setting. Figure 10.2.4 provides the convergence graphs of the posterior expectations of the four means, against the number of iterations k, for both the partial average and the *Rao-Blackwellized* estimator (see Section 6.3.4). Since both quantities converge to the same Bayes estimator, the strong similarity of both curves is a partial assessment of convergence, suggesting that 10,000 iterations of the Gibbs sampler could be sufficient to ensure stability.

Because we wish to assess the effects of the intoxication, and of both treatments, we focus here on the comparisons of μ_δ, μ_D, μ_P, and $\mu_D - \mu_P$ with 0. Table 10.2.4 provides posterior probabilities that the effects are significant, that is, whether $0 > \mu_\delta$, $\mu_D > 0$, etc., as well as the confidence intervals, both approximated from the Gibbs samples represented in Figure 10.2.15. This allows us to conclude that intoxication, drug and placebo effects are significant, although to a lesser degree for the placebo,. Also, we see that the drug effect significantly differs from the placebo effect.   ‖

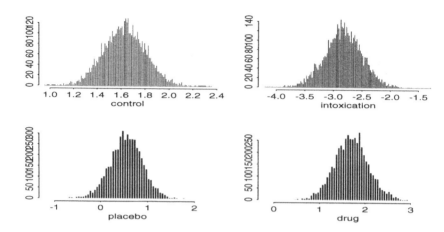

Figure 10.2.2. *Histograms of the Gibbs samples for* μ_θ, μ_δ, μ_P *and* μ_D *in the experiment of Example 10.2.3. (Source: Robert and Reber (1999).)*

Table 10.2.1. *Posterior probabilities of significance and 95% confidence intervals for the mean effects.*

	μ_δ	μ_D	μ_P	$\mu_D - \mu_P$
Probability	1.00	0.9998	0.94	0.985
Confidence	[-3.48,-2.17]	[0.94,2.50]	[-0.17,1.24]	[0.14,2.20]

10.2.5 Hierarchical extensions for the normal model

In this section, as well as in Section 10.3, we consider the special case of the normal distribution,

$$x \sim \mathcal{N}_p(\theta, \Sigma)$$

because it leads to (partly) closed-form expressions and analytic results. As in Lindley and Smith (1972), Smith (1973) and Berger (1985a), we also use a first-level conjugate distribution $\theta \sim \mathcal{N}_p(\mu, \Sigma_\pi)$, since this choice allows for an easier decomposition of the estimators.

Lemma 10.2.16 *In the conjugate normal model, the hierarchical Bayes estimator is*

$$\delta^\pi(x) = \mathbb{E}^{\pi_2(\mu, \Sigma_\pi | x)}[\delta(x | \mu, \Sigma_\pi)],$$

with

$$
\begin{aligned}
\delta(x | \mu, \Sigma_\pi) &= x - \Sigma W(x - \mu), \\
W &= (\Sigma + \Sigma_\pi)^{-1}, \\
\pi_2(\mu, \Sigma_\pi | x) &\propto (\det W)^{1/2} \exp\{-(x-\mu)^t W(x-\mu)/2\} \pi_2(\mu, \Sigma_\pi).
\end{aligned}
$$

The proof is a direct consequence of Lemma 10.2.9, and of the fact that the marginal $m_1(x|\mu, \Sigma_\pi)$ is a normal distribution $\mathcal{N}_p(\mu, W^{-1})$.

Example 10.2.17 (Example 10.2.14 continued) The choice of a flat prior distribution on β leads to a closed-form expression for $\delta^\pi(x)$. In fact, there exists a function h_k (see Exercise 10.19) such that

$$\delta^\pi(x) = x - h_p \ _{k-2}(||x||_*^2)\Sigma C^{-1}(x - Px),$$

with

$$\begin{aligned} P &= Y(Y^t C^{-1} Y)^{-1} Y^t C^{-1}, \\ ||x||_*^2 &= xC^{-1}(I_p - P)x. \end{aligned}$$

Notice that Px is the orthogonal projection of x on the subspace $H = \{\mu = Y\beta, \ \beta \in \mathbb{R}^k\}$ according to the metrics defined by C^{-1}. The estimator δ^π is thus a weighted sum of x and of this projection. Therefore, δ^π takes the prior information into account in an adaptive way, depending on the distance $||x||_*$ of x to H. $\parallel$

Example 10.2.18 Consider the *exchangeable* hierarchical model:

$$\begin{aligned} x|\theta &\sim \mathcal{N}_p(\theta, \sigma_1^2 I_p), \\ \theta|\xi &\sim \mathcal{N}_p(\xi\mathbf{1}, \sigma_\pi^2 I_p), \\ \xi &\sim \mathcal{N}(\xi_0, \tau^2), \end{aligned}$$

where $\mathbf{1} = (1, \ldots, 1)^t \in \mathbb{R}^p$. In this case,

$$\delta(x|\xi, \sigma_\pi) = x - \frac{\sigma_1^2}{\sigma_1^2 + \sigma_\pi^2}(x - \xi\mathbf{1}),$$

$$\begin{aligned} \pi_2(\xi, \sigma_\pi^2|x) &\propto (\sigma_1^2 + \sigma_\pi^2)^{-p/2} \exp\{-\frac{||x - \xi\mathbf{1}||^2}{2(\sigma_1^2 + \sigma_\pi^2)}\} e^{-(\xi-\xi_0)^2/2\tau^2} \pi_2(\sigma_\pi^2) \\ &\propto \frac{\pi_2(\sigma_\pi^2)}{(\sigma_1^2 + \sigma_\pi^2)^{p/2}} \exp\left\{ -\frac{p(\bar{x} - \xi)^2}{2(\sigma_1^2 + \sigma_\pi^2)} - \frac{s^2}{2(\sigma_1^2 + \sigma_\pi^2)} - \frac{(\xi - \xi_0)^2}{2\tau^2} \right\} \end{aligned}$$

with $s^2 = \sum_i (x_i - \bar{x})^2$. Therefore, $\pi_2(\xi|\sigma_\pi^2, x)$ is the normal distribution $\mathcal{N}(\mu(x, \sigma_\pi^2), V_\pi(\sigma_\pi^2))$, where

$$\mu(x, \sigma_\pi^2) = \bar{x} - \frac{\sigma_1^2 + \sigma_\pi^2}{\sigma_1^2 + \sigma_\pi^2 + p\tau^2}(\bar{x} - \xi_0), \qquad V_\pi(\sigma_\pi^2) = \frac{\tau^2(\sigma_1^2 + \sigma_\pi^2)}{\sigma_1^2 + \sigma_\pi^2 + p\tau^2}.$$

Then

$$\delta^\pi(x) = \mathbb{E}^{\pi_2(\sigma_\pi^2|x)}\left[x - \frac{\sigma_1^2}{\sigma_1^2 + \sigma_\pi^2}(x - \bar{x}\mathbf{1}) - \frac{\sigma_1^2 + \sigma_\pi^2}{\sigma_1^2 + \sigma_\pi^2 + p\tau^2}(\bar{x} - \xi_0)\mathbf{1} \right]$$

and

$$(10.2.5) \quad \pi_2(\sigma_\pi^2|x) \propto \frac{\tau \exp -\frac{1}{2}\left[\dfrac{s^2}{\sigma_1^2 + \sigma_\pi^2} + \dfrac{p(\bar{x} - \xi_0)^2}{p\tau^2 + \sigma_1^2 + \sigma_\pi^2} \right]}{(\sigma_1^2 + \sigma_\pi^2)^{(p-1)/2}(\sigma_1^2 + \sigma_\pi^2 + p\tau^2)^{1/2}} \pi_2(\sigma_\pi^2).$$

Berger (1985a, pp. 184-185) provides a detailed proof of this result, as well as the corresponding expression for the posterior variance of θ.

Notice the particular form of the hierarchical Bayes estimator

$$\delta^\pi(x) \;=\; x - \mathbb{E}^{\pi_2(\sigma_\pi^2|x)}\left[\frac{\sigma_1^2}{\sigma_1^2 + \sigma_\pi^2}\right](x - \bar{x}\mathbf{1})$$

(10.2.6)
$$\qquad\qquad - \mathbb{E}^{\pi_2(\sigma_\pi^2|x)}\left[\frac{\sigma_1^2 + \sigma_\pi^2}{\sigma_1^2 + \sigma_\pi^2 + p\tau^2}\right](\bar{x} - \xi_0)\mathbf{1}.$$

This means that the two hierarchical levels induce two different types of shrinkage in the Bayes estimator. Firstly, the exchangeability assumption justifies the second term, $(x - \bar{x}\mathbf{1})$, which is shrinking the observation toward the common mean $\bar{x}$; this would be the estimator to use in the case of an *exact* relation between the parameters of the model. Similarly, the third term originates from the assumption that the common mean varies around ξ_0.

In the event that the information about ξ_0 is unreliable, a noninformative distribution can be proposed for the second level, that is, $\pi_2(\sigma_\pi^2) = 1$ and $\tau^2 = +\infty$. Then, for $p \geq 4$,

$$\delta^\pi(x) \;=\; x - \mathbb{E}^{\pi_2(\sigma_\pi^2|x)}\left[\frac{\sigma_1^2}{\sigma_1^2 + \sigma_\pi^2}\right](x - \bar{x}\mathbf{1})$$

(10.2.7)
$$\qquad\qquad =\; x - h_{p-2}(\|x - \bar{x}\mathbf{1}\|^2)(x - \bar{x}\mathbf{1})$$

and

(10.2.8) $$\pi_2(\sigma_\pi^2|x) \propto (\sigma_1^2 + \sigma_\pi^2)^{-(p-1)/2}\exp\left\{-\frac{s^2}{2(\sigma_1^2 + \sigma_\pi^2)}\right\},$$

the function h_k being introduced in Example 10.2.7 (see also Exercise 10.19). It can be verified that (10.2.7) and (10.2.8) are derived from (10.2.5) and (10.2.6) when τ^2 goes to $+\infty$, and that (10.2.8) only corresponds to a proper distribution when $p \geq 4$. The usefulness of the exchangeability assumption in dimension 3 relies on an additional amount of information, that is, a prior information on the location of the common mean ξ. This constraint agrees with frequentist results on the minimaxity of (10.2.7), which only holds for $p \geq 4$ (Brown (1988)).

Notice that, if σ_1 is also unknown, with (possibly noninformative) prior distribution π_0, the quantities (10.2.6) and (10.2.7) are still valid provided the expectations are considered with respect to the posterior distribution $\pi(\sigma_1^2, \sigma_\pi^2|x)$. Similarly, if ξ is distributed according to a Student's t-distribution $\mathcal{T}(\alpha, \xi_0, \tau^2)$ instead of a normal distribution, we showed in Example 10.2.2 that this distribution can be decomposed as a mixture of a normal distribution $\mathcal{N}(\xi_0, \tau^2/z)$ by a gamma distribution $\mathcal{G}(\alpha/2, \alpha/2)$ on z. Therefore, δ^π can be derived from the expressions (10.2.6) and (10.2.7) by integrating with respect to z. See Angers (1987, 1992) for a more detailed treatment of a prior modeling by Student's t-distributions.   ‖

Example 10.2.19 (Example 10.2.7 continued) In the setting of the usual regression model, an exchangeability assumption on the parameters β_i $(1 \leq i \leq p)$ leads to estimators similar to the ones derived above. When

$$\beta_i \sim \mathcal{N}(\xi, \sigma_\pi^2) \qquad \text{and} \qquad \pi(\xi) = 1,$$

an analysis similar to Example 10.2.18 was conducted by Lindley and Smith (1972) and provided the estimator

$$\delta^\pi(y) = \left\{ I_p + \frac{\sigma^2}{\sigma_\pi^2}(X^t X)^{-1}(I_p - p^{-1}J_p) \right\}^{-1} \hat{\beta},$$

where $\hat{\beta}$ is the least-squares estimator $\hat{\beta} = (X^t X)^{-1} X^t y$ and J_p is the $(p \times p)$ matrix made of 1. The analogy with the above example is more striking when δ^π is written in the form

$$\delta^\pi(y) = \bar{\beta}\mathbf{1} + \left\{ I_p + \frac{\sigma^2}{\sigma_\pi^2}(X^t X)^{-1}(I_p - p^{-1}J_p) \right\}^{-1} (\hat{\beta} - \bar{\beta}\mathbf{1})$$

(since $(I_p - p^{-1}J_p)\bar{\beta}\mathbf{1} = 0$) because the Bayes estimator is shrinking toward the common mean $\bar{\beta}$ (in a matricial sense). Notice that it can also be written as

$$\delta^\pi(y) = \left\{ X^t X + \frac{\sigma^2}{\sigma_\pi^2}(I_p - p^{-1}J_p) \right\}^{-1} X^t y.$$

This expression points out how the exchangeability assumption alleviates the numerical and statistical problems caused by near-collinearities in the columns of X. Indeed, the matrix

$$\frac{\sigma^2}{\sigma_\pi^2}(I_p - p^{-1}J_p)$$

plays the role of stabilizer in the estimator. If, in the second-level prior, we consider instead $\xi = 0$, the Bayes estimator is then (see Exercise 10.23)

$$\begin{aligned} \delta^\pi(y) &= \left\{ I_p + \frac{\sigma^2}{\sigma_\pi^2}(X^t X)^{-1} \right\}^{-1} \hat{\beta} \\ &= \left(X^t X + \frac{\sigma^2}{\sigma_\pi^2}I_p \right)^{-1} X^t y. \end{aligned}$$

These estimators are called *ridge estimators* and have been introduced by Hoerl and Kennard (1970) as a remedy to *multicollinearity* problems in the matrix $X^t X$, that is, when two (or more) of the regressors are almost collinear. The matricial factor

$$\left[I_p + k(X^t X)^{-1} \right]^{-1}$$

stabilizes the least-squares estimator when some eigenvalues of $X^t X$ are close to 0 (see also Lindley and Smith (1972) and Goldstein and Smith (1974)). These estimators have been generalized later by considering a matricial factor of the form

$$\left[I_p + h(y)(X^t X)^{-1} \right]^{-1},$$

which may correspond to the case when σ_π^2 is unknown, with prior distribution $\pi_2(\sigma_\pi^2)$, since the Bayes estimator is then

$$\delta^\pi(y) = \mathbb{E}^{\pi_2(\sigma_\pi^2|y)} \left[I_p + \frac{\sigma^2}{\sigma_\pi^2}(X^tX)^{-1} \right]^{-1} \hat\beta.$$

From a classical point of view, it appears that the imperatives of a reduction of multicollinearity and of minimaxity are contradictory, since Casella (1980, 1985b) has shown that necessary minimaxity conditions for the ridge estimators cannot accord with a stabilizing influence of these estimators. Robert (1988) exhibits the same phenomenon for other classes of shrinkage estimators and points out that the antagonism owes to the *unidimensionality* of the multicollinearity problem, which explains why a uniform improvement over $\hat\beta$ is impossible.   ‖

10.3 Optimality of hierarchical Bayes estimators

From[2] a general point of view, since hierarchical Bayes estimators cannot really be distinguished from the usual Bayes estimators, these estimators are not more, and not less, admissible than the Bayes estimators derived in the previous chapters. For instance, the necessary and sufficient conditions obtained in Chapter 8 also apply in the case of hierarchical Bayes estimators. Similarly, the invariance aspects of Chapter 9 ignore the possibly hierarchical structure of prior distributions.

On the other hand, we will see in a particular case that it is possible to derive a general minimaxity condition taking advantage of the specificity of hierarchical Bayes estimators, since this condition involves the second-level prior distributions. Such results point out the robustifying aspect of the hierarchical Bayes approach, which assigns the more subjective aspects of prior modeling to higher levels, and thus provides an intermediary position between a straightforward Bayesian analysis and frequentist imperatives.

Consider again the normal model, $x \sim \mathcal{N}_p(\theta, \Sigma)$ where Σ is known. As in Section 10.2.5, the first-level prior distribution on θ is conjugate, $\theta \sim \mathcal{N}_p(\mu, \Sigma_\pi)$. The prior distribution π_2 of the hyperparameters μ, Σ_π can be decomposed as follows:

$$\pi_2(\mu, \Sigma_\pi) = \pi_2^1(\Sigma_\pi|\mu)\pi_2^2(\mu).$$

In this case,

$$m(x) = \int_{\mathbb{R}^p} m(x|\mu)\pi_2^2(\mu)\,d\mu,$$

with

$$m(x|\mu) = \int f(x|\theta)\pi_1(\theta|\mu, \Sigma_\pi)\pi_2^1(\Sigma_\pi|\mu)\,d\theta\,d\Sigma_\pi.$$

---

[2] This Section can be omitted in a first reading because it deals with the minimaxity of a particular class of hierarchical Bayes estimators in the normal case. Its purpose is to illustrate the increased robustness brought about by hierarchical modeling.

Moreover, the Bayes estimator

(10.3.1) $$\delta^\pi(x) = x + \Sigma \nabla \log m(x)$$

can be written

$$\delta^\pi(x) = \int \delta(x|\mu) \pi_2^2(\mu|x)\, d\mu,$$

with

$$\delta(x|\mu) = x + \Sigma \nabla \log m(x|\mu),$$
$$\pi_2^2(\mu|x) = \frac{m(x|\mu)\pi_2^2(\mu)}{m(x)}.$$

These conditional decompositions will be used below.

Consider Q, a $(p \times p)$ symmetric positive-definite matrix associated with the quadratic loss

(10.3.2) $$L_Q(\theta, \delta) = (\theta - \delta)^t Q(\theta - \delta).$$

An estimator δ is minimax for the loss (10.3.2) if it satisfies

$$R(\theta, \delta) = \mathbb{E}_\theta[L_Q(\theta, \delta(x))] \le \text{tr}(\Sigma Q),$$

since $\text{tr}(\Sigma Q)$ is the minimax risk of $\delta_0(x) = x$. The *method of the unbiased estimator of the risk* has been developed by Stein (1973, 1981) to derive sufficient minimaxity conditions. (See Brown (1988) and Rukhin (1995) for detailed reviews of this method.) It consists of obtaining a differential operator $\mathcal{D}$, independent of θ, such that

$$R(\theta, \delta) = \mathbb{E}_\theta[\mathcal{D}\delta(x)],$$

for every parameter θ and every estimator δ. This technique indeed gives a sufficient minimaxity condition of the form $\mathcal{D}\delta(x) \le \text{tr}(Q\Sigma)$ (see Exercise 2.56). In the particular case of (10.3.1), the differential operator is provided by the following result (Berger and Robert (1990)).

Lemma 10.3.1 *If $m(x)$ satisfies the three conditions*

(1) $\mathbb{E}_\theta \|\nabla \log m(x)\|^2 < +\infty;$ (2) $\mathbb{E}_\theta \left| \dfrac{\partial^2 m(x)}{\partial x_i \partial x_j} \middle/ m(x) \right| < +\infty;$

and $(1 \le i \le p)$

(3) $\displaystyle \lim_{|x_i| \to +\infty} |\nabla \log m(x)| \exp\{-(1/2)(x - \theta)^t \Sigma^{-1}(x - \theta)\} = 0,$

the unbiased estimator of the risk of δ^π is given by

$$\mathcal{D}\delta^\pi(x) = \text{tr}(Q\Sigma)$$
$$+ \frac{2}{m(x)}\text{tr}(H_m(x)\tilde{Q}) - (\nabla \log m(x))^t \tilde{Q}(\nabla \log m(x)),$$

where

$$\tilde{Q} = \Sigma Q \Sigma, \qquad H_m(x) = \left(\frac{\partial^2 m(x)}{\partial x_i \partial x_j} \right).$$

This unbiased estimator of the risk then leads to a sufficient minimaxity condition,

$$\frac{2}{m(x)}\mathrm{tr}(H_m(x)\tilde{Q}) - (\nabla \log m(x))^t \tilde{Q}(\nabla \log m(x)) \leq 0.$$

We denote by div the *divergence* operator, that is,

$$\mathrm{div}f(x) = \sum_{i=1}^{n}\frac{\partial f_i}{\partial x_i}(x),$$

for a differentiable function f from $\mathbb{R}^n$ to $\mathbb{R}^n$.

Corollary 10.3.2 *If m satisfies the conditions of Lemma 10.3.1 and if*

$$(10.3.3) \qquad\qquad \mathrm{div}\left(\tilde{Q}\nabla\sqrt{m(x)}\right) \leq 0,$$

δ^π is minimax.

Proof. It is sufficient to consider the development of $\mathrm{div}(\tilde{Q}\nabla\sqrt{m(x)})$ to obtain

$$\mathrm{div}(\tilde{Q}\nabla\sqrt{m(x)}) = \frac{1}{2}\mathrm{div}\left(\tilde{Q}\frac{\nabla m(x)}{\sqrt{m(x)}}\right)$$

$$= \frac{1}{2\sqrt{m(x)}}\mathrm{div}(\tilde{Q}\nabla m(x)) - \frac{1}{4}\left(\frac{\nabla m(x)}{m(x)\sqrt{m(x)}}\right)^t \tilde{Q}\nabla m(x)$$

$$= \frac{\sqrt{m(x)}}{4}\left[\frac{2}{m(x)}\mathrm{tr}(H_m(x)\tilde{Q}) - \nabla \log m(x)^t \tilde{Q}\nabla \log m(x)\right]$$

and derive the additional term in $\mathcal{D}\delta^\pi(x)$. $\qquad\qquad\qquad\square\square$

In the particular case when $\Sigma = Q = I_p$, the condition of Corollary 10.3.2 can be written more simply as a condition on the *Laplacian* of $m(x)^{1/2}$ because it is

$$\Delta\sqrt{m(x)} = \sum_{i=1}^{n}\frac{\partial^2}{\partial x_i^2}(\sqrt{m(x)}) \leq 0$$

($\sqrt{m(x)}$ is then said to be *superharmonic*). The verification of this condition is generally quite complicated. A more explicit minimaxity condition can be derived from Corollary 10.3.2 by conditioning on μ.

Lemma 10.3.3 *The estimator δ^π is minimax if*

$$(10.3.4) \qquad\qquad \mathrm{div}\left(\tilde{Q}\nabla m(x|\mu)\right) \leq 0.$$

Proof. In fact,

$$\mathrm{div}(\tilde{Q}\nabla m(x)) = \int \mathrm{div}\left(\tilde{Q}\nabla m(x|\mu)\right)\pi_2^2(\mu)\,d\mu$$

and (10.3.4) implies (10.3.3). $\qquad\qquad\qquad\qquad\qquad\qquad\square\square$

Therefore, if $\tilde{Q} = I_p$ and $m(x|\mu)$ is superharmonic, the corresponding hierarchical Bayes estimator is minimax. This result may appear to be trivial in its proof and statement, but it is actually quite general. In fact, it provides a *necessary and sufficient condition of minimaxity that does not depend on $\pi_2^2(\mu)$*, and thus allows for every possible modeling on the hyperparameter μ. From a subjective point of view, to have complete freedom of choice on the prior distribution of μ is more much important than the alternative choice on Σ_π, since it is usually easier to get information on μ than on Σ_π. The following example shows, moreover, that the condition (10.3.4) is satisfied by a large class of distributions π_2^1.

Example 10.3.4 (Example 10.2.14 continued) Consider again the case when $\Sigma_\pi = \xi C - \Sigma$ and $Q = \Sigma^{-1}C\Sigma^{-1}$ (therefore $\tilde{Q} = C$). It follows from Lemma 10.2.11 that

$$m(x|\mu) \propto \int_0^\infty \xi^{-p/2} \exp\left\{ -\frac{(x-\mu)^t C^{-1}(x-\mu)}{2\xi} \right\} \pi_2^1(\xi|\mu)\, d\xi.$$

Therefore,

$$\mathrm{div}\left(\tilde{Q}\nabla m(x|\mu) \right) \propto \int_0^\infty \left(-\frac{p}{\xi} + \frac{(x-\mu)^t C^{-1}(x-\mu)}{\xi^2} \right)$$
$$\times e^{-(x-\mu)^t C^{-1}(x-\mu)/2\xi} \xi^{-p/2} \pi_2^1(\xi|\mu)\, d\xi\,,$$

and (10.3.4) is equivalent to

$$\psi(a) = \int_0^\infty (2a - p\xi)\xi^{-(p+4)/2} e^{-a/\xi} \pi_2^1(\xi|\mu)\, d\xi \leq 0, \qquad \forall a \geq 0.$$

If π_2^1 is a.e. differentiable, an integration by parts gives

$$\psi(a) = -2e^{-a/\xi_0} \xi_0^{-p/2} \pi_2'(\xi_0|\mu) - \int_{\xi_0}^{+\infty} \xi^{-p/2} e^{-a/\xi} \pi_2^1(\xi|\mu)\, d\xi,$$

where $\xi_0 = \inf(\mathrm{supp}(\pi_2^1))$ and π_2' is the derivative of π_2^1. This expression implies:

Proposition 10.3.5 *If, for every $\mu \in \mathbb{R}^p$, $\pi_2^1(\xi|\mu)$ is nondecreasing, δ^π is minimax for every prior distribution π_2^2.*

Therefore, if $\pi_2^1(\xi|\mu) = 1$ for $\xi_0 \leq \xi$ when $\lambda_{\max}(C^{-1}\Sigma) \leq \xi_0$, the corresponding Bayes estimator is minimax.   ‖

The above example can be extended to the case when $\theta \sim \mathcal{N}_p(\mu, \sigma_\pi^2\Sigma)$ and when $\pi_2^1(\sigma_\pi^2|\mu)$ is increasing ($C = \Sigma$ and $\xi = \sigma_\pi^2 - 1$). This class obviously fails to include all hierarchical estimators or all minimax estimators, but it is large enough to contain the minimax estimators proposed by Strawderman (1971) and Berger (1976, 1980a), some of them being, moreover, admissible (see also Kubokawa (1991) and Exercise 10.36).

Notice that Proposition 10.3.5 suggests the use of unnatural prior distributions: actually, it seems difficult to argue in favor of an increasing distribution on the variance, on a subjective or noninformative basis. On the contrary, the prior distributions are, in general, decreasing for large σ_π^2's. This is, for instance, the case for the Jeffreys noninformative distribution, $\pi(\sigma_\pi^2) = 1/\sigma_\pi^2$. Therefore, this result stresses indirectly the *artificial aspect of the notion of minimaxity*: to similarly weight *a posteriori* all the possible values of the parameter is equivalent to favoring *a priori* the more unlikely (or the *least favorable*) values.

The example below illustrates the advantage of the hierarchical Bayes modeling from a minimax viewpoint, even when the first-level distribution is more rudimentary. It also exhibits a minimaxity robustness property, in the sense that minimaxity does not depend so much on the normality of the prior distribution as on its spherical symmetry. This result thus appears as a Bayesian counterpart to the frequentist results of Cellier et al. (1989).

Example 10.3.6 Consider $x \sim \mathcal{N}_p(\theta, I_p)$. The mean θ is estimated under quadratic loss. Instead of assuming a conjugate first-level distribution, we propose the uniform distribution on the sphere of radius c,

$$\pi_1(\theta|c) \propto \mathbb{I}_{\{||\theta||^2=c\}},$$

thus assuming only spherical symmetry for the overall prior distribution. The second-level prior distribution $\pi_2(c)$ is a gamma distribution, $\mathcal{G}(\alpha, \beta)$. The Bayes estimator is then (see Robert et al. (1990))

$$\delta^\pi(x) = \frac{2\alpha}{p} \frac{1}{1+2\beta} \frac{{}_1F_1(\alpha+1; (p+2)/2; ||x||^2/(2+4\beta))}{{}_1F_1(\alpha; p/2; ||x||^2/(2+4\beta))} x,$$

where ${}_1F_1$ is the confluent hypergeometric function. When $\alpha < 1$ and $\beta = 0$, we get

$$\delta^\pi(x) = \frac{2\alpha}{p} \frac{{}_1F_1(\alpha+1; (p+2)/2; ||x||^2/2)}{{}_1F_1(\alpha; p/2; ||x||^2/2)} x,$$

which is a minimax and admissible estimator (see Alam (1973)).   ‖

10.4 The empirical Bayes alternative

The method we examine in the remainder of this chapter does not follow from the Bayesian principles,[3] since it approximates the prior distribution by *frequentist* methods when the prior information is too vague. We still consider it in this book for several reasons

(i) it can be perceived as a dual method of the hierarchical Bayes analysis presented above;

(ii) it is *asymptotically* equivalent to the Bayesian approach;

(iii) it is usually classified as Bayesian by frequentists and practitioners; and

---

[3] The appellation *empirical Bayes* is doubly defective because firstly, the method is not Bayesian and, secondly, genuine Bayesian methods are empirical, since they are based on the data!

(iv) it may be an acceptable approximation in problems for which a genuine Bayes modeling is too complicated or too costly.

We will see how the empirical Bayes analysis occupies an intermediate position between the classical and Bayesian methods, and also that the hierarchical alternative is often preferable. Notice that this section is only a short introduction to the empirical Bayes approach. See Morris (1983b), Berger (1985a), Maritz and Lwin (1989) or Carlin and Louis (2000a,b) for more extensive treatments of the topic.

10.4.1 Nonparametric empirical Bayes

Introduced by Robbins (1951, 1955, 1964, 1983), the *empirical Bayes perspective* can be stated as follows. Given $(n + 1)$ independent observations $x_1, \ldots, x_{n+1}$ with densities $f(x_i|\theta_i)$, the problem is to draw an inference on θ_{n+1}, under the additional assumption that the θ_i's have all been generated according to the same *unknown* prior distribution g. From a Bayesian point of view, this means that the sampling distribution is known, but the prior distribution is not. The marginal distribution,

$$(10.4.1) \qquad f_g(x) = \int f(x|\theta)g(\theta)\, d\theta,$$

can then be used to recover the distribution g from the observations, since $x_1, \ldots, x_n$ can be considered an i.i.d. sample from f_g. Deriving an approximation $\hat{g}_n$ in this manner, we can use it as a substitute for the true prior distribution, and propose the plug-in approximation to the posterior distribution

$$(10.4.2) \qquad \tilde{\pi}(\theta_{n+1}|x_{n+1}) \propto f(x_{n+1}|\theta_{n+1})\hat{g}_n(\theta_{n+1}).$$

Obviously, this derivation is not Bayesian, although it relies on the Bayes formula (10.4.2), and can also correspond to a classical modeling. A Bayesian approach, arguing from the ignorance on g, would index this distribution by a hyperparameter λ and would thus represent ignorance by a second-level prior distribution, $\pi_2(\lambda)$.[4] Deely and Lindley (1981) compare the two approaches in the case of a Poisson distribution.

The initial approach of Robbins (1955) is essentially *nonparametric* and uses the observations $x_1, \ldots, x_{n+1}$ to estimate f_g. (In the general case, the marginal density f_g can be estimated by the kernel method; see, e.g., Devroye and Györfi (1985).)

Example 10.4.1 Consider x_i distributed according to $\mathcal{P}(\theta_i)$ $(i = 1, \ldots, n)$. If $p_k(x_1, \ldots, x_n)$ is the number of observations equal to k, $k \in \mathbb{N}$, $p_k(x_1, \ldots, x_n)$ gives an estimation of the marginal distribution,

$$f_g(k) = \int_0^{+\infty} e^{-\theta}\frac{\theta^k}{k!}g(\theta)\, d\theta.$$

---

[4] Indexing by λ is not formally restrictive, as shown in Exercise 1.2.

If $x_{n+1} \sim \mathcal{P}(\theta_{n+1})$ and θ_{n+1} is estimated under quadratic loss, the Bayes estimator is

$$\delta^g(x_{n+1}) \;=\; \mathbb{E}^g[\theta|x_{n+1}] = \frac{\int_0^{+\infty} e^{-\theta}\theta^{x_{n+1}+1}g(\theta)\,d\theta}{\int_0^{+\infty} e^{-\theta}\theta^{x_{n+1}}g(\theta)\,d\theta}$$

$$=\; \frac{f_g(x_{n+1}+1)}{f_g(x_{n+1})}(x_{n+1}+1).$$

Therefore, the empirical Bayes approximation of δ^g is

(10.4.3) $$\delta^{\mathrm{EB}}(x_{n+1}) = \frac{p_{x_{n+1}+1}(x_1,\ldots,x_n)}{p_{x_{n+1}}(x_1,\ldots,x_n)+1}(x_{n+1}+1),$$

where f_g is replaced by its approximation.   ‖

Several problems can be identified in this method.

(a) To use nonparametric estimation, for instance of the prior density, to initiate a parametric estimation procedure seems to be suboptimal because the errors made in the nonparametric estimation step are always more difficult to assess. For instance, in the above example, if the numerator of (10.4.3) is null, the estimator is null.

(b) More generally, nonparametric estimates of mixing densities g in (10.4.1) by maximum likelihood techniques usually are rudimentary, since they correspond to distributions with finite support (see Laird (1978), Böhning (1999), or Carlin and Louis (2000a)).[5] Such priors are hardly acceptable from a Bayesian point of view.

(c) Functional relations between the mean (or any other quantity of interest) and the marginal distribution, as in Example 10.4.1, are quite rare. When such a relation does not exist, the derivation of an estimator of g is generally too complicated to guarantee the resulting estimators be good approximations of the true Bayes estimators.

(d) The approximation is actually justified for large sample sizes only, that is, when the estimator of the marginal distribution, $\hat{f}_g^n$, provides an acceptable approximation of the marginal distribution. Otherwise, as shown by Example 10.4.1, $\hat{f}_g^n$ varies too widely and needs to be smoothed to be of any use (see Maritz and Lwin (1989)).

(e) The assumption that many identical and independent problems are available about the same prior distribution is a strong one, and can fail to be satisfied in practice. Therefore, a single sample, even when very large, cannot lead to the estimator of f_g because it corresponds to an unique observation of θ. This criticism remains valid for the parametric approach (see, for instance, Proposition 10.4.5).

For these reasons, we shall not proceed any further in the study of the nonparametric empirical Bayes analysis; we consider only a restricted version called *parametric empirical Bayes* by Morris (1983b).

---

[5] This is also the case with kernel or Dirichlet process prior approaches.

10.4.2 Parametric empirical Bayes

One appeal of the empirical Bayes techniques is to provide approximations in noninformative settings. We showed in the previous chapters that the Bayesian approach produces an efficient tool to obtain frequentist optimal procedures, besides providing a unified technique of statistical inference. The empirical Bayes analysis can then be perceived as a practical approximation of this tool.

In exponential family settings, the prior distribution being unavailable, a simple choice is to take a conjugate prior associated with $f(x|\theta)$, $\pi(\theta|\lambda)$. While the hierarchical approach introduces an additional distribution on the hyperparameters λ, the empirical Bayes analysis proposes to estimate these hyperparameters from the marginal distribution

$$m(x|\lambda) = \int_\Theta f(x|\theta)\pi(\theta|\lambda)\, d\theta$$

by $\hat\lambda(x)$ and to use $\pi(\theta|\hat\lambda(x), x)$ as a pseudo-posterior distribution. This method then appears as the parametric version of the original approach of Robbins (1955).

A defect with the empirical Bayes perspective is that it relies on frequentist methods to estimate the hyperparameters of $m(x|\lambda)$, although Bayesian techniques could be used as well, as shown in Note 10.7.2. Therefore, a wide range of options is available: for instance, the estimator of λ can be derived by the moment method or the maximum likelihood method. The corresponding arbitrariness of empirical Bayes analysis is the major flaw of this theory, since it prohibits a decision-theoretic treatment of the empirical Bayes estimators, and often appears as a posterior justification of *existing* estimators, as shown in Section 10.5. The most common approach is to use maximum likelihood estimators, for practical and theoretical reasons, in particular because of the proximity of the maximum likelihood estimation to the Bayesian paradigm. An additional justification of this choice is given below in the particular case of the estimation of the natural parameter of an exponential family under quadratic loss.

Lemma 10.4.2 *Consider*

$$x \sim f(x|\theta) = e^{\theta \cdot x - \psi(\theta)} h(x), \qquad x \in \mathbb{R}^k.$$

If θ is distributed according to $\pi(\theta|\lambda)$, $\lambda \in \mathbb{R}^p$, and $\hat\lambda(x)$ is the solution of the likelihood equations associated with $m(x|\lambda)$, the empirical Bayes estimator of θ satisfies

$$
\begin{aligned}
\delta^{\mathrm{EB}}(x) &= \left. (\nabla \log m(x|\lambda)) \right|_{\lambda=\hat\lambda(x)} - \nabla \log h(x) \\
&= \nabla[\log m(x|\hat\lambda(x))] - \nabla \log h(x).
\end{aligned}
$$

Proof. In fact,

$$\nabla \log m(x|\hat\lambda(x)) = \left. (\nabla \log m(x|\lambda)) \right|_{\lambda=\hat\lambda(x)} + \nabla_x \hat\lambda(x) \nabla_\lambda m(x|\lambda) \big|_{\lambda=\hat\lambda(x)},$$

where $\nabla_\lambda m(x|\lambda)$ is the vector with components

$$\frac{\partial m(x|\lambda)}{\partial \lambda_i} \qquad (1 \le i \le p),$$

and $\nabla_x \hat\lambda(x)$ is the $(k \times p)$ matrix with components

$$\frac{\partial \hat\lambda_i(x)}{\partial x_j} \qquad (1 \le i \le p, 1 \le j \le k).$$

By definition of $\hat\lambda(x)$, the second term is null.   □□

Therefore, a regular Bayesian derivation using the approximate posterior distribution $\pi(\theta|\hat\lambda(x))$ gives the same result as the genuine empirical Bayes approach, where λ is replaced by $\hat\lambda(x)$. This justification is obviously quite limited, since it only works for the posterior mean of the natural parameter in exponential families.

Example 10.4.3 (Example 10.4.1 continued) Consider the case when $\pi(\theta|\lambda)$ is an exponential distribution $\mathcal{E}xp(\lambda)$. Then

$$\begin{aligned} m(x_i|\lambda) &= \int_0^{+\infty} e^{-\theta}\frac{\theta^{x_i}}{x_i!}\lambda e^{-\theta\lambda} d\theta \\ &= \frac{\lambda}{(\lambda+1)^{x_i+1}} = \left(\frac{1}{\lambda+1}\right)^{x_i}\frac{\lambda}{\lambda+1}, \end{aligned}$$

and $x_i|\lambda \sim \mathcal{G}eo(\lambda/\lambda+1)$. The maximum likelihood estimator of λ is $\hat\lambda(x) = 1/\bar{x}$ and the empirical Bayes estimator of θ_{n+1} is

$$\delta^{\text{EB}}(x_{n+1}) = \frac{x_{n+1}+1}{\hat\lambda+1} = \frac{\bar{x}}{\bar{x}+1}(x_{n+1}+1),$$

the average $\bar{x}$ being taken on the n first observations. ||

Example 10.4.4 Consider $x_1,\ldots,x_n$, n independent observations from $\mathcal{B}(m, p_i)$. Casella (1985a) (see also Morisson (1979)) applies this model to the intentions of buying a new car in the coming year. The assumption is that the parameters p_i $(1 \le i \le n)$ are distributed according to the same conjugate prior distribution

$$p_i \sim \mathcal{B}(\alpha, \beta).$$

The corresponding Bayes estimator of p_i is

$$\delta_i^\pi(x_i) = \frac{\alpha+\beta}{\alpha+\beta+1}\frac{\alpha}{\alpha+\beta} + \left(1 - \frac{\alpha+\beta}{\alpha+\beta+1}\right)\frac{x_i}{m}$$

and the marginal distribution of x_i is called *beta-binomial*,

$$P(x_i = k|\alpha, \beta) = \frac{B(k+\alpha, m-k+\beta)}{B(\alpha, \beta)}.$$

as in Example 10.2.13. It is shown in Kendall and Stuart (1979) that, for this marginal distribution,

$$\mathbb{E}(x_i|m) = \frac{\alpha}{\alpha+\beta}, \qquad \mathrm{var}(x_i|m) = \frac{1}{m}\frac{\alpha\beta}{(\alpha+\beta)^2}\frac{\alpha+\beta+m}{\alpha+\beta+1}.$$

When α and β are estimated by the method of moments, the resulting empirical Bayes estimator of p_i is

$$\gamma_i^{\mathrm{EB}}(x_1,\ldots,x_n) = \frac{\hat\alpha + (x_i/m)}{\hat\alpha + \hat\beta + 1}.$$

(Exercise 10.23 provides the data of Morisson (1979).)   ‖

Section 10.5 indicates how the Stein effect is strongly related to the empirical Bayes approach and how the latter can provide estimators that perform well for point estimation, as well as for tests and confidence regions. The following result shows, on the contrary, why empirical Bayes tests are of limited interest for a single sample.

Proposition 10.4.5 *Consider the test of $H_0 : \theta = \theta_0$ against $H_1 : \theta = \theta_1$ based on a sample $x_1,\ldots,x_n$, i.i.d. $f(x|\theta)$. An empirical Bayes approach gives the likelihood ratio test procedure*

$$(10.4.4) \qquad \varphi(x) = \begin{cases} 1 & \text{if } \prod_{i=1}^n f(x_i|\theta_0) > \prod_{i=1}^n f(x_i|\theta_1), \\ 0 & \text{otherwise,} \end{cases}$$

for every confidence level.

Proof. In this setting, the unknown parameters are reduced to π_0, the prior probability of H_0. The marginal distribution of x is then

$$m(x|\pi_0) = \pi_0 \prod_{i=1}^n f(x_i|\theta_0) + (1-\pi_0)\prod_{i=1}^n f(x_i|\theta_1)$$

and gives the following maximum likelihood estimator of π_0:

$$\hat\pi_0(x_1,\ldots,x_n) = \begin{cases} 1 & \text{if } \prod_{i=1}^n f(x_i|\theta_0) > \prod_{i=1}^n f(x_i|\theta_1), \\ 0 & \text{otherwise.} \end{cases}$$

The Bayesian answer being

$$\varphi^\pi(x_1,\ldots,x_n) = \begin{cases} 1 & \text{if } P(\theta = \theta_0|x_1,\ldots,x_n,\pi_0) > \alpha, \\ 0 & \text{otherwise,} \end{cases}$$

the posterior probability of H_0 is

$$P(\theta = \theta_0|x_1,\ldots,x_n,\hat\pi_0) = \frac{\hat\pi_0 \prod_{i=1}^n f(x_i|\theta_0)}{\hat\pi_0 \prod_{i=1}^n f(x_i|\theta_0) + (1-\hat\pi_0)\prod_{i=1}^n f(x_i|\theta_1)}$$

and (10.4.4) follows.   □□

When several testing problems are considered simultaneously, this extreme behavior of the empirical Bayes tests disappears (see Maritz and

Lwin (1989)). However, it is rather rare to have to test *simultaneously* hypotheses on parameters from the *same distribution*, and the practical interest of the empirical Bayes approach for tests is thus quite limited. We consider the estimation of the confidence regions in Section 10.5, in relation to the Stein effect. For an alternative review, see Laird and Louis (1987) or Carlin and Gelfand (1990).

Notice in conclusion that a refinement of the empirical Bayes approach is to consider instead *mixtures* of conjugate distributions, since they also constitute a conjugate family (see Lemma 3.4.2). If $x_i \sim f(x_i|\theta_i)$ and

$$\theta_i \sim \sum_{j=1}^{n} p_j \pi(\theta_i|\lambda_j),$$

the marginal distribution of x_i is

$$x_i|p, \lambda \sim \sum_{j=1}^{n} p_j \int_{\Theta} f(x_i|\theta)\pi(\theta|\lambda_j) \, d\theta.$$

See Section 6.4 and Note 6.6.6 for details on the Bayesian analysis of this problem. Maritz and Lwin (1989) consider more particularly the application to the empirical Bayes analysis. A drawback to this extension is obviously that it calls for a larger number of hyperparameters, and thus for a larger number of independent samples, while suffering from some of the defects mentioned before.

Let us stress again that the legitimacy of the empirical Bayes methods is *asymptotic* (see Deely and Lindley (1981)). Their popularity owes to the good frequentist properties of some resulting estimators, and also to the simplification they brought to the treatment of complex problems, compared with a hierarchical Bayes analysis. (See, for instance, Carter and Rolph (1974) or Hui and Berger (1983).) For finite sample size problems, the empirical Bayes methods are only approximations of the exact Bayesian methods and cannot claim the same coherence. In particular, it is not possible to draw a full Bayesian inference using $\pi(\theta|x, \lambda(x))$, because it is not a posterior distribution. Moreover, with increasing computational power (see Chapter 6), the need for empirical approximations to more complex hierarchical analyses diminishes (see Berger (1985a), Berger and Berliner (1986), and Berger and Robert (1990)).

10.5 Empirical Bayes justifications of the Stein effect

The empirical Bayes analysis of the Stein effect described in Note 2.8.2 unifies the different occurrences of this paradox, according to which the simultaneous estimation of independent parameters can lead to a global improvement in estimation performances, although each component cannot be improved uniformly. Moreover, this analysis explains the form of the original James–Stein estimators, and points out that they correspond to the vague prior information that θ is close to 0.

10.5.1 Point estimation

Example 10.5.1 Consider $x \sim \mathcal{N}_p(\theta, I_p)$ and $\theta_i \sim \mathcal{N}(0, \tau^2)$. The marginal distribution of x is then

$$x|\tau^2 \sim \mathcal{N}_p(0, (1 + \tau^2)I_p)$$

and leads to the following maximum likelihood estimator of τ^2,

$$\hat{\tau}^2 = \begin{cases} (||x||^2/p) - 1 & \text{if } ||x||^2 > p, \\ 0 & \text{otherwise.} \end{cases}$$

The corresponding empirical Bayes estimator of θ_i under quadratic loss is derived by replacing τ^2 by $\hat{\tau}^2$ in the Bayes estimator,

$$\delta^{\text{EB}}(x) = \frac{\hat{\tau}^2 x}{1 + \hat{\tau}^2}$$

$$(10.5.1) \qquad\qquad = \left(1 - \frac{p}{||x||^2}\right)^+ x.$$

The estimator (10.5.1) is actually a truncated James–Stein estimator. Therefore, these estimators can be interpreted as empirical Bayes estimators related to the vague information that the expectations of the observations are close to 0. The original James–Stein estimator can also be expressed as an empirical Bayes estimator, using an alternative frequentist estimation method. In fact, given the marginal distribution of x, the best unbiased estimator of $1/(1 + \tau^2)$ is $(p - 2)/||x||^2$, which leads to

$$(10.5.2) \qquad\qquad \delta^{\text{EB}}(x) = \left(1 - \frac{p - 2}{||x||^2}\right) x.$$

$\parallel$

This example illustrates the weakness in the justifications of the empirical Bayes approach, which cannot compare the different methods used for estimating the hyperparameters. This lack of ordering is actually characteristic of the whole frequentist approach. The comparison between the estimators (10.5.1) and (10.5.2) must thus rely on external considerations.

Example 10.5.2 Consider two independent vectors, $x \sim \mathcal{N}_p(\theta, \sigma^2 I_p)$ and $y \sim \mathcal{N}_q(0, \sigma^2 I_q)$, as for a linear regression. The parameter of interest is the variance factor σ^2, evaluated under the entropic loss,

$$L(\sigma^2, d) = \frac{d}{\sigma^2} - \log(d/\sigma^2) - 1.$$

Apart from intrinsic considerations (see Section 2.5.4), this loss is often preferred to the quadratic loss, since it gives the maximum likelihood estimator

$||y||^2/p + q$ as the best equivariant estimator[6] of σ^2. Under this loss, the Bayes estimator of σ^2 is

$$(10.5.3) \qquad \delta^\pi(x) = \left(\mathbb{E}^\pi[\sigma^{-2}|x] \right)^{-1} .$$

For the gamma-normal conjugate distribution on (θ, σ^2),

$$\theta|\sigma^2 \sim \mathcal{N}_p(0, \tau\sigma^2 I_p), \qquad \sigma^{-2} \sim \mathcal{G}(\nu/2, \beta/2) ,$$

the estimator (10.5.3) is then

$$\delta^\pi(x, y) = \frac{1}{p+q+\nu} \left(\frac{||x||^2}{1+\tau} + ||y||^2 + \beta \right)$$

and maximization of the marginal likelihood (in (τ, ν, β)) leads to the following empirical Bayes estimator (see Kubokawa et al. (1992)):

$$(10.5.4) \qquad \delta^{EB}(x, y) = \min \left(\frac{||y||^2}{q}, \frac{||x||^2 + ||y||^2}{p+q} \right) .$$

Notice the intuitive aspect of this estimator, which uses the additional information about σ^2 contained in x only if $||x||^2$ is not too large, that is, if θ is close to 0, as

$$\frac{||x||^2 + ||y||^2}{p+q}$$

is the best scale equivariant estimator of σ^2 when $\theta = 0$.

The real interest in this result is that (10.5.4) has been obtained in Brewster and Zidek (1974) as a uniform improvement on the best equivariant estimator $\delta^\star(x, y) = ||y||^2/q$ under entropy loss. (See Maatta and Casella (1990) for an exhaustive review of the different perspectives in variance estimation.) ∥

Morris (1983b) considers the Stein effect more generally than in Example 10.5.1. In fact, he studies the Bayesian model

$$\begin{aligned} x|\theta &\sim \mathcal{N}_p(\theta, \Lambda), \\ \theta|\beta, \sigma_\pi^2 &\sim \mathcal{N}_p(Z\beta, \sigma_\pi^2 I_p), \end{aligned}$$

with $\Lambda = \mathrm{diag}(\lambda_1, \dots, \lambda_p)$ and Z a $(p \times q)$ full rank matrix. The marginal distribution of x is then

$$x_i|\beta, \sigma_\pi^2 \sim \mathcal{N}(z_i'\beta, \sigma_\pi^2 + \lambda_i)$$

and the posterior distribution of θ is

$$\theta_i|x_i, \beta, \sigma_\pi^2 \sim \mathcal{N}\left((1 - b_i)x_i + b_i z_i'\beta, \lambda_i(1 - b_i) \right) ,$$

---

[6] This argument does not justify the use of the entropic loss, since it legitimizes a posteriori a given estimator, instead of being based on utility considerations and leading to the determination of an estimator.

with $b_i = \lambda_i / (\lambda_i + \sigma_\pi^2)$. If all the variances λ_i are identical and equal to σ^2, the best equivariant estimators of β and b are given by

$$\hat\beta = (Z^t Z)^{-1} Z^t x \qquad \text{and} \qquad \hat b = \frac{(p - q - 2)\sigma^2}{s^2},$$

with $s^2 = \sum_{i=1}^p (x_i - z_i'\hat\beta)^2$. We deduce from these estimators of the hyperparameters the corresponding empirical Bayes estimator of θ

(10.5.5) $$\delta^{EB}(x) = Z\hat\beta + \left(1 - \frac{(p - q - 2)\sigma^2}{\|x - Z\hat\beta\|^2}\right)(x - Z\hat\beta),$$

which is of the form of the general Stein estimators.

In the particular case where the means are assumed to be identical (exchangeability), the matrix Z reduces to the vector $\mathbf{1}$ and β is a real number; the empirical Bayes estimator is then

$$\delta^{EB}(x) = \bar x \mathbf{1} + \left(1 - \frac{(p - 3)\sigma^2}{\|x - \bar x \mathbf{1}\|^2}\right)(x - \bar x \mathbf{1}).$$

It thus provides the Stein estimator that shrinks toward the common mean, as in Efron and Morris (1975). See Morris (1983b) for an extension to the case when the variances λ_i are not identical.

10.5.2 Variance evaluation

As mentioned above, the estimation of the hyperparameters β and σ_π^2 considerably modifies the behavior of the resulting procedures. If the resulting point estimators are generally efficient, as shown in the above examples, the estimation of the posterior variance of $\pi(\theta | x, \beta, b)$ by the empirical variance, $\mathrm{var}(\theta_i | x, \hat\beta, \hat b)$, induces an underestimation of this variance. Thus, using empirical Bayes analysis to assess the performances of δ^{EB} by estimating its quadratic loss $(\theta_i - \delta_i^{EB})^2$ as $\mathrm{var}(\theta_i | x, \hat\beta, \hat b)$ is misleading because it underrates the error resulting from using δ^{EB}.

Morris (1983b) takes into account the additional variability resulting from the estimation of the hyperparameters through a modification of the estimators. In the exchangeable case, the resulting procedures are

$$\delta^{EB}(x) = x - \tilde B(x - \bar x \mathbf{1}),$$
$$V_i^{EB}(x) = \left(\sigma^2 - \frac{p-1}{p}\tilde B\right) + \frac{2}{p-3}\hat b(x_i - \bar x)^2,$$

with

$$\hat b = \frac{p-3}{p-1}\frac{\sigma^2}{\sigma^2 + \hat\sigma_\pi^2}, \qquad \hat\sigma_\pi^2 = \max\left(0, \frac{\|x - \bar x \mathbf{1}\|^2}{p-1} - \sigma_\pi^2\right)$$

and

$$\tilde B = \frac{p-3}{p-1}\min\left(1, \frac{\sigma^2(p-1)}{\|x - \bar x \mathbf{1}\|^2}\right).$$

This last quantity estimates the ratio $\sigma^2/(\sigma^2 + \sigma_\pi^2)$. However, this modification, although more satisfactory, suffers from the general drawback of empirical Bayes inference, namely, that the procedures are usually justified by ad-hoc reasons that cannot be extended to a general principle (although Kass and Steffey (1989) provide a partial generalization).

Notice the analogy between the modified empirical variance V_i^{EB} and the hierarchical variance for the same model,

$$V_i^{\text{HB}}(x) = \sigma^2 \left(1 - \frac{p-1}{p} \ \mathbb{E}^\pi \left[\frac{\sigma^2}{\sigma^2 + \sigma_\pi^2}\bigg|x\right]\right) + \text{var}\left[\frac{\sigma^2}{\sigma^2 + \sigma_\pi^2}\bigg|x\right](x_i - \bar{x})^2$$

(see Berger (1985a)). This resemblance is not coincidental, since this modification brings an improvement in the original empirical Bayes approach by taking advantage of the true Bayesian approach one step further. Ghosh et al. (1989) and Blattberg and George (1991) provide econometric illustrations of the empirical Bayes analysis and the connection with Stein estimators in regression models.

10.5.3 Confidence regions

Another aspect of the Stein effect can be interpreted in an empirical Bayes manner. In the case of *recentered confidence regions* (see Section 5.5), Hwang and Casella (1982) have shown that some of these regions allow for a larger coverage probability than does the usual confidence set for an identical volume. These sets can also be expressed as empirical HPD regions.

Example 10.5.3 In Hwang and Casella (1982), the usual confidence region

$$C_0(x) = \{\theta; \ \|\theta - x\|^2 \leq c_\alpha\},$$

with $x \sim \mathcal{N}_p(\theta, I_p)$, is compared with

$$C_a(x) = \{\theta; \ \|\theta - \delta_a(x)\|^2 \leq c_\alpha\},$$

where $\delta_a(x) = [1 - (a/\|x\|^2)]^+ x$. Hwang and Casella (1982) show that, for a small enough and $p \geq 4$, the set C_a satisfies, for every θ,

$$P_\theta(\theta \in C_a(x)) > P_\theta(\theta \in C_0(x)) = 1 - \alpha.$$

Casella and Hwang (1983) also consider recentered regions with a *variable volume*

$$C_\delta^v(x) = \{\theta; \ \|\theta - \delta(x)\|^2 \leq v(x)\},$$

and they determine δ and v by an empirical Bayes analysis based on an α-credible HPD region. The center of the region is the James–Stein estimator

$$\delta(x) = \left(1 - \frac{p-2}{\|x\|^2}\right)^+ x$$

and the radius is provided by

$$v(x) = \begin{cases} \left(1 - \frac{p-2}{c_\alpha}\right)\left[c_\alpha - p\log\left(1 - \frac{p-2}{c_\alpha}\right)\right] & \text{if } ||x||^2 < c_\alpha, \\ \left(1 - \frac{p-2}{||x||^2}\right)\left[c_\alpha - p\log\left(1 - \frac{p-2}{||x||^2}\right)\right] & \text{otherwise.} \end{cases}$$

The shape of the variable radius is justified in terms of a linear loss

$$L(\theta, C) = k\,\text{vol}(C) - \mathbb{I}_C(\theta),$$

already presented in Section 5.5 (see Exercise 10.29). This empirical Bayes confidence region has then at least a confidence level of $1 - \alpha$ (in the frequentist sense), except for the smallest values of p. ‖

Example 10.5.4 A usual rejection of recentered confidence regions is based on the fact that they are useless in practice, since the reported confidence level is still

$$\inf_\theta P_\theta(\theta \in C_a(x)) = 1 - \alpha = P_\theta(\theta \in C_0(x)).$$

In this sense, the usual regions can be argued to be more accurate because they coincide exactly with the reported confidence level. The actual value of such confidence levels has already been discussed in Section 5.5 and the reader is referred to Chapter 5 for criticisms of the artificial aspect of the notion of confidence levels. An alternative answer is also offered at the end of Chapter 5. It is to propose a conditional confidence level, $\gamma(x)$, which is more adapted to the recentered region $C_a(x)$, and to evaluate it under the quadratic loss

(10.5.6) $$(\gamma(x) - \mathbb{I}_{C_a(x)})^2.$$

For the model presented in Example 10.5.2, George and Casella (1994) propose an empirical Bayes solution to this evaluation problem for a recentered region of the form

$$C^{EB}(x) = \{\theta;\ ||\theta - (1 - \hat{b})x||^2 \le c\}$$

and a confidence report

$$\gamma^{EB}(x) = P(\chi_p^2 \le c/(1 - \hat{b})).$$

In fact, if $\theta \sim \mathcal{N}_p(0, \tau^2 I_p)$, the Bayesian answer would be

$$\begin{aligned} \gamma^\pi(x) &= P^\pi(\theta \in C_B(x)|x) \\ &= P^\pi(||\theta - (1 - b)x||^2 \le c|x) \\ &= P(\chi_p^2 \le c/(1 - b)), \end{aligned}$$

since $\theta|x \sim \mathcal{N}_p((1 - b)x, (1 - b))$ with $1 - b = \tau^2/(\sigma^2 + \tau^2)$. The empirical Bayes estimators derived by George and Casella (1994) in γ^{EB} are

$$1 - \hat{b}(x) = \max\left(d, 1 - \frac{a}{||x||^2}\right) = u_{a,d}(||x||^2),$$

while C^{EB} is centered in the truncated Stein estimator associated with a and $d \leq 1$. Actually, George and Casella (1994) show that the empirical Bayes estimator obtained this way,

$$\gamma^{EB}(x) = P\left[\chi_p^2 \leq \frac{c}{\max\{d, (||x||^2 - a)/||x||^2\}}\right],$$

dominates the constant report $1 - \alpha$ under the quadratic loss (10.5.6), for $d \leq 1$ and a small enough. A suggested value of d is

$$d = \frac{2c}{c + 2a + \sqrt{c(c + 4a)}}.$$

See Lu and Berger (1989b) for a different solution. ||

10.5.4 Comments

To conclude this overview of empirical Bayes methods, let us stress once more their dual nature: they draw strength from both frequentist and Bayesian methods to derive inferential procedures. It can be argued that the improvements these estimators bring on classical frequentist estimators owes to their mimicking of the Bayesian approach, whereas their suboptimality (in terms of admissibility for instance) can be attributed to the refusal to adopt a fully Bayesian perspective, and to the subsequent arbitrariness in the choice of the resulting method. Fundamentally, it is quite logical that a method that relies on classical, but suboptimal, estimators (such as the maximum likelihood estimator of the mean in the multidimensional normal case) and ad-hoc concepts lacking a decision-theoretic basis (such as unbiased estimation or moment methods) cannot provide optimal procedures. The domination of these estimators by genuine Bayes estimators (see Brown (1988)) is another argument in favor of a complete adoption of the Bayesian paradigm, even if it requires a hierarchical modeling. As shown in Chapter 6, the development of new numerical tools to deal with far more complex models than before comes as a last blow to these methods that had earlier alleviated the computational difficulties of fully Bayesian analyses.

10.6 Exercises

Section 10.1

10.1 For a model represented by a directed acyclic graph, as in Figure 10.1, show that the full conditional distribution of a variable (or *node*), given the other variables of the model, is identical to the distribution of this node, given only the nodes it is connected with.

10.2 In the setting of Example 10.1.1,

    a. Show that, if the generator Λ can be decomposed as $P\tilde{\Lambda}P^t$, where P is the orthogonal matrix of the eigenvectors of Λ and $\tilde{\Lambda}$ is the diagonal matrix of the eigenvalues of Λ, λ_i $(i = 1, \ldots, 7)$, then

$$\exp\{\Lambda\} = P \begin{pmatrix} e^{\lambda_1} & 0 & \cdots & 0 \\ 0 & e^{\lambda_2} & \cdots & 0 \\ & & \ddots & \\ 0 & 0 & \cdots & e^{\lambda_7} \end{pmatrix} P^t,$$

and that $\exp\{T\Lambda\} = \exp\{\Lambda\}^T$.

b. Deduce that the forward–backward formulas (6.5.4) of Exercise 6.47 also apply in this setting when p_{ij} is replaced with $p_{ij}^{(T)}$, (i,j)-th element of the matrix $\exp\{\Lambda\}^T$.

c. Determine the computational time of these formulas and compare it with the following alternative representation: introduce missing values x_{ij}^* so that individuals are observed at regular intervals with inter-observation time η, complete the sample with these missing values, and compute the forward–backward formulas on the completed sample. (Notice that this completion implies that $\exp\{\Lambda\}^\eta$ is computed only once.)

Section 10.2.1

10.3 Show that the hyperprior chosen in Example 10.2.3 leads to a well defined posterior distribution provided there are at least two observations for each stage of the experiment.

10.4 Represent the hierarchical model of Example 10.2.3 as a directed acyclic graph, as in Figure 10.1.

10.5 Consider $J \sim \mathcal{M}_k(N; p_1, \ldots, p_k)$, a multinomial random variable. Assume that N is generated according to a Poisson distribution with parameter λ. Determine the marginal distribution of J. Give, in particular, the covariance matrix. Extend to the case when $p = (p_1, \ldots, p_k) \sim \mathcal{D}(\alpha_1, \ldots, \alpha_k)$, a Dirichlet distribution.

10.6 A fly lays N eggs according to a Poisson distribution $\mathcal{P}(\lambda)$ and each egg survives with probability p.

a. Show that the distribution of the number of surviving eggs x is then hierarchical

$$x|N \sim \mathcal{B}(N, p), \qquad N \sim \mathcal{P}(\lambda).$$

b. Compute the marginal distribution of x and the posterior distribution of N.

10.7 In the setting of Example 10.6, when p is known, give the posterior distribution of N if $\pi_2(\lambda) = 1/\lambda$. Examine the generalization to the case when p is unknown and $\pi_1(p) = 1$.

Section 10.2.2

10.8 If $y|\theta \sim \mathcal{N}_p(\theta, \Sigma_1)$, $\theta|\beta \sim \mathcal{N}_p(X\beta, \Sigma_2)$, and $\beta \sim \mathcal{N}_q(\mu, \Sigma_3)$, compute the prior and posterior distributions of θ.

10.9 Consider the setting of a *logistic regression*, that is, of observations (x_1, y_1), $\ldots, (x_n, y_n)$ such that $x_i \in \mathbb{R}^k$ and $y_i \in \{0, 1\}$ with

$$P(y_i = 1|x_i) = \exp(x_i^t \beta)/(1 + \exp(x_i^t \beta))$$

and derive a sufficient condition on $\pi(\tau)$ for the posterior distribution of β to be defined when $\beta|\tau \sim \mathcal{N}_q(0, \tau^2 I_p)$. (The x_i's are deemed to be fixed.)

10.10 Reproduce Exercise 10.9 in the setting of a *probit* model, that is, when

$$P(y_i = 1|x_i) = \Phi(x_i^t\beta)$$

and Φ is the c.d.f. of the standard normal distribution.

Section 10.2.3

10.11 Establish Lemmas 10.2.9 and 10.2.11.

10.12 (Berger and Robert (1990)) In the setting of Example 10.2.14, assume that $\mu \in H = \{\mu = Y\beta; \; \beta \in \mathbb{R}^\ell\}$ and $\pi_2(\beta, \sigma_\pi^2) = 1$. Show that $m(x) < +\infty$ if $p > 2 + \ell$.

10.13 Spiegelhalter et al. (1990) consider the following model: the weights y_{ij} of 60 rats are measured weekly, the first 30 observations being associated with the control group ($1 \le i \le 60, 1 \le j \le 5$). The corresponding model is

$$y_{ij} \sim \mathcal{N}(\alpha_i + \beta_i j, \sigma_i),$$

with $\sigma_i = \sigma_c$ for $i \le 30$ and $\sigma_i = \sigma_t$ for $31 \le i \le 60$ and

$$
\begin{aligned}
(\alpha_i, \beta_i) &\sim \mathcal{N}_2\left((\alpha_c, \beta_c), \Sigma_c\right) & (i = 1, \dots, 30),\\
(\alpha_i, \beta_i) &\sim \mathcal{N}_2\left((\alpha_t, \beta_t), \Sigma_t\right) & (i = 31, \dots, 60).
\end{aligned}
$$

Complete the model with noninformative priors on the hyperparameters and examine whether the posterior distribution is well defined.

10.14 In the setting of Example 10.2.10, define the averages

$$\bar{x}_i = \frac{1}{J_i^c}\sum_{i=1}^{J_i^c} x_{ij}, \quad \bar{y}_i = \frac{1}{J_i^a}\sum_{i=1}^{J_i^a} y_{ij}, \quad \bar{z}_i = \frac{1}{J_i^t}\sum_{i=1}^{J_i^t} z_{ij}$$

and

$$\bar{\theta} = \frac{1}{I}\sum_{i=1}^{I} \theta_i, \quad \bar{\delta} = \frac{1}{I}\sum_{i=1}^{I} \delta_i.$$

Show that the full conditional distributions are given by ($1 \le i \le I$)

$$\mu_\theta \sim \mathcal{N}(\bar{\theta}, \sigma_\theta^2/I), \qquad \mu_\delta \sim \mathcal{N}(\bar{\delta}, \sigma_\delta^2/I),$$

$$\mu_P \sim \mathcal{N}\left(\sum_{\ell_i=0} \xi_i/I_P, \sigma_P^2/I_P\right), \qquad \mu_D \sim \mathcal{N}\left(\sum_{\ell_i=1} \xi_i/I_D, \sigma_D^2/I_D\right),$$

$$\theta_i \sim \mathcal{N}\left(\frac{\sigma_\theta^{-2}\mu_\theta + J_i^c\sigma_c^{-2}\bar{x}_i + J_i^a\sigma_a^{-2}(\bar{y}_i - \delta_i) + J_i^t\sigma_t^{-2}(\bar{z}_i - \delta_i - \xi_i)}{\sigma_\theta^{-2} + J_i^c\sigma_c^{-2} + J_i^a\sigma_a^{-2} + J_i^t\sigma_t^{-2}},\right.$$

$$\left.(\sigma_\theta^{-2} + J_i^c\sigma_c^{-2} + J_i^a\sigma_a^{-2} + J_i^t\sigma_t^{-2})^{-1}\right)$$

$$\delta_i \sim \mathcal{N}\left(\frac{\sigma_\delta^{-2}\mu_\delta + J_i^a\sigma_a^{-2}(\bar{y}_i - \theta_i) + J_i^t\sigma_t^{-2}(\bar{z}_i - \theta_i - \xi_i)}{\sigma_\delta^{-2} + J_i^a\sigma_a^{-2} + J_i^t\sigma_t^{-2}},\right.$$

$$\xi_i \sim \mathcal{N}\left(\frac{\sigma_D^{-2\ell_i}\sigma_P^{-2(1-\ell_i)}\mu_D^{\ell_i}\mu_P^{1-\ell_i} + J_i^t\sigma_t^{-2}(\bar{z}_i - \theta_i - \delta_i)}{\sigma_D^{-2\ell_i}\sigma_P^{-2(1-\ell_i)} + J_i^t\sigma_t^{-2}}, \right.$$
$$\left. (\sigma_\delta^{-2} + J_i^a\sigma_a^{-2} + J_i^t\sigma_t^{-2})^{-1} \right)$$
$$(\sigma_D^{-2\ell_i}\sigma_P^{-2(1-\ell_i)} + J_i^t\sigma_t^{-2})^{-1} \Big)$$

$$\sigma_c^{-2} \sim \mathcal{G}a\left(\sum_i \frac{J_i^c}{2}, \sum_{i,j} \frac{(x_{ij} - \theta_i)^2}{2}\right),$$

$$\sigma_a^{-2} \sim \mathcal{G}a\left(\sum_i \frac{J_i^a}{2}, \sum_{i,j} \frac{(y_{ij} - \theta_i - \delta_i)^2}{2}\right),$$

$$\sigma_t^{-2} \sim \mathcal{G}a\left(\sum_i \frac{J_i^t}{2}, \sum_{i,j} \frac{(z_{ij} - \theta_i - \delta_i - \xi_i)^2}{2}\right),$$

$$\sigma_\theta^{-2} \sim \mathcal{G}a\left(\frac{I}{2}, \sum_i \frac{(\theta_i - \mu_\theta)^2}{2}\right), \quad \sigma_\delta^{-2} \sim \mathcal{G}a\left(\frac{I}{2}, \sum_i \frac{(\delta_i - \mu_\delta)^2}{2}\right),$$

$$\sigma_P^{-2} \sim \mathcal{G}a\left(\frac{I_P}{2}, \sum_{\ell_i=0} \frac{(\xi_i - \mu_P)^2}{2}\right), \quad \sigma_D^{-2} \sim \mathcal{G}a\left(\frac{I_D}{2}, \sum_{\ell_i=1} \frac{(\xi_i - \mu_D)^2}{2}\right).$$

10.15 (Exercise 10.14 cont.) When the δ_i's are distributed from (10.2.3), give the corresponding full conditionals.

Section 10.2.4

10.16 (Berger and Robert (1990)) Consider $x \sim \mathcal{N}_p(\theta, \Sigma)$, $\theta \sim \mathcal{N}_p(y\beta, \sigma_\pi^2 I_p)$, and $\beta \sim \mathcal{N}_\ell(\beta_0, A)$, with $\text{rank}(A) = m$.

a. Show that if, for $K > 0$, the two integrals

$$\int_0^K \pi_2(\sigma_\pi^2)d\sigma_\pi^2 \quad \text{and} \quad \int_K^{+\infty} \frac{1}{(\sigma_\pi^2)^{(p-\ell+m)/2}}\pi_2(\sigma_\pi^2)d\sigma_\pi^2$$

are finite, then $m(x) < +\infty$ for every $x \in \mathbb{R}^p$.

b. Show that condition a. is satisfied if, for $\epsilon > 0$, $K_1 > 0$, $K_2 > 0$,

$$\pi_2(\sigma_\pi^2) < \frac{K_1}{K_2 + (\sigma_\pi^2)^{(2+\epsilon-p+\ell-m)/2}},$$

thus if $\pi_2(\sigma_\pi^2) = 1$ and $p - l + m > 2$.

10.17 (Berger (1985a)) In the setting of Example 10.2.18, compute the posterior variance. Consider also the noninformative case.

10.18 (Lindley and Smith (1972)) Extend Example 10.2.18 to the general model

$$x|\theta \sim \mathcal{N}_p(A_1\theta, \Sigma_1), \qquad \theta|\beta \sim \mathcal{N}_\ell(A_2\beta, \Sigma_2), \qquad \beta|\xi \sim \mathcal{N}_q(A_3\xi, \Sigma_3),$$

and check the results of Example 10.2.7.

10.19 (Berger (1985a)) Show that, for the model of Example 10.2.18 and a noninformative distribution on ξ and σ_π^2, the hierarchical Bayes estimator is

$$\delta^\pi(x) = x - h_{p-2}(||x - \bar{x}\mathbf{1}||^2)(x - \bar{x}\mathbf{1})$$

with

$$h_p(t) \;=\; \frac{p}{2t}(1 - H_p(t)),$$

$$H_p(t) \;=\; \begin{cases} \dfrac{t^{p/2}}{(p/2)!\left\{e^t - \sum_{i=1}^{(p-2)/2} t^i/i!\right\}} & \text{if } p \text{ is even,} \\[4mm] \dfrac{t^{p/2}}{\Gamma(p/2)\left\{e^t[2\Phi(\sqrt{2t}) - 1] - \sum_{i=1}^{(p-3)/2} \frac{t^{(i+3)/2}}{\Gamma(i+3/2)}\right\}} & \text{if } p \text{ is odd.} \end{cases}$$

10.20 In the setting of Example 10.2.13, compute the posterior mean of p when $x = 3$, $n = 5$.

Section 10.2.5

10.21 Compare the models

$$x \sim \mathcal{N}_p(\theta, I_p), \quad \theta|\mu \sim \mathcal{N}_p(\mu, \tau^2 I_p), \quad \pi_2(\mu, \tau^2) = 1/\tau^2,$$

and

$$x \sim \mathcal{N}_p(\theta, I_p), \quad \theta|\mu \sim \mathcal{N}_p(\mu, I_p), \quad \mu|\xi \sim \mathcal{N}_p(\xi, \tau^2 I_p), \quad \pi_2(\xi, \tau^2) = 1/\tau^2,$$

in terms of estimators of θ.

10.22 Consider $x_i \sim \mathcal{N}(\mu_i, \sigma^2)$ and $\mu_i|\mu, \tau \sim \mathcal{N}(\mu, \tau^2)$ $(i = 1, \dots, n)$.

a. Show that $\pi(\mu, \tau) = 1/\tau$ leads to an undefined posterior distribution.

b. Show that $\pi(\mu, \tau) = 1$ avoids the above problem.

10.23 In the setting of Example 10.2.7, show that the Bayes estimator

$$\delta^\pi(y) = \mathbb{E}^{\pi_2(\sigma_\pi^2|y)}\left[I_p + \frac{\sigma^2}{\sigma_\pi^2}(X^t X)^{-1}\right]^{-1}\hat{\beta}$$

can be written in the form

$$\left[I_p + h(y)(X^t X)^{-1}\right]^{-1}\hat{\beta}.$$

(*Hint:* Use a simultaneous diagonalization of I_p and $X^t X$.) Explain how this estimator can help to reduce multicollinearity.

Section 10.3

10.24 (Stein (1981)) Establish Lemma 10.3.1 using an integration by parts and relate this result to Exercise 2.56.

10.25 If H is the Hessian matrix defined in Lemma 10.3.1, show that the equivalent of (10.3.2) for the covariance matrix is

$$V^{\mathrm{EB}}(x) = \Sigma + \Sigma\frac{H(x)}{m(x)}\Sigma - \Sigma(\nabla \log m(x))(\nabla \log m(x))^t\Sigma.$$

Using a technique as in Exercise 10.24, show that an unbiased estimator of the average matricial error

$$\mathbb{E}_\theta[(\theta - \delta(x))(\theta - \delta(x))^t]$$

can be written in the differential form

$$\hat{V}_{\delta\mathrm{HB}}(x) = \Sigma + 2\Sigma\frac{H(x)}{m(x)}\Sigma - \Sigma(\nabla \log m(x))(\nabla \log m(x))^t\Sigma.$$

Derive from this expression the unbiased estimator of the quadratic risk.

Table 10.6.1. *Car-buying intentions of households.*

Intentions	0.0	0.1	0.2	0.3	0.4	0.5	0.6	0.7	0.8	0.9	1.0
Answers	293	26	21	21	10	9	12	13	11	10	21

10.26 Use the following approximation of $_1F_1(a; b; z)$:

$$_1F_1(u; b; z) \simeq \frac{\Gamma(b)}{\Gamma(a)} e^{z/2} (z/2)^a \ ^b \left(1 + \frac{(1-a)(b \quad a)}{(z/2)} \right)$$

to provide an approximation of the estimator δ^π given in Example 10.3.6 and compare with the James–Stein estimator.

10.27 Consider $x \sim \mathcal{N}_p(\theta, I_p)$, $\theta \sim \mathcal{N}_p(0, \tau^2 I_p)$, and, if $\eta = 1/(1+\tau^2)$, assume $\pi_2(\eta) = \eta^{2-(p/2)}$. Show that the corresponding hierarchical Bayes estimator can be written explicitly as

$$\delta^{HB}(x) = \left(\frac{1}{1 - e^{-||x||^2/2}} - \frac{2}{||y||^2} \right) x,$$

and determine whether it is minimax and admissible.

10.28 *(Hartigan (1983)) Consider an observation $x \sim \mathcal{N}_p(\theta, I_p)$.

a. If f is a positive nondecreasing function bounded above by $2(p-2)$, show that

$$\delta_f(x) = \left(1 - \frac{f(||x||^2)}{||x||^2} \right) x$$

dominates $\delta_0(x) = x$ for the usual quadratic loss. (*Hint:* Use the unbiased estimator of the risk obtained in Exercise 2.56.)

b. Let π be a prior on θ such that, conditional upon τ^2, $\theta \sim \mathcal{N}_p(0, \tau^2)$ and $\tau^2 \sim \pi_1$. The hyperprior π_1 is assumed to be a log-concave function of $\log(\tau^2+1)$ and $(\tau^2 + 1)^{1-\alpha}\pi_1(\tau^2)$ is increasing in τ^2. Using the general result of a., show that the hierarchical Bayes estimator associated with π dominates δ_0 if $4 - 2\alpha \le p$. (*Hint:* Show that $\delta^\pi(x) = (1 - \mathbb{E}[(\tau^2 + 1)^{-1}|x])x$ and that $\mathbb{E}[(\tau^2 + 1)^{-1}|x])$ is increasing in $||x||^2$ while being obviously bounded by $2(p-2)$.)

c. Show that such priors can only be proper for $\alpha < 0$, and therefore that these minimax Bayes estimators are guaranteed to be admissible only for $p \ge 5$.

d. Show that the Bayes risk is actually finite for $\alpha < 2$ and deduce that the resulting hierarchical Bayes estimators are admissible for every p. [*Note:* Strawderman (1971) considered the particular case $\pi_1(\tau^2) = (1+\tau^2)^{\alpha-1}$ to show that the limiting dimension for the existence of proper Bayes minimax estimators is exactly $p = 5$.]

Section 10.4.2

10.29 (Casella (1985a)) In a survey of car-buying intentions, 447 households provide their evaluation of the probability of buying a new car in the coming year. The result of the survey is given in Table 10.28.

Table 10.6.2. *Proportions of car acquisitions depending on the intention.*

Intentions	0	0.1—0.3	0.4—0.6	0.7—0.9	1
Declared	0	0.19	0.51	0.79	1
Realized	0.07	0.19	0.41	0.48	0.583

The answers x_i $(1 \le i \le 447)$ are modeled as issued from a renormalized binomial $\mathcal{B}(10, p_i)$ distribution, that is, $10x_i \sim \mathcal{B}(10, p_i)$, and the p_i are distributed according to $\mathcal{B}e(\alpha, \beta)$.

a. Use the marginal distribution to provide estimators of α and β by the method of moments.

b. Derive an empirical Bayes estimator of the p_i's under quadratic loss.

c. The true intentions p_i have actually been observed at the end of the year, and Table 10.29 gives the difference with the declared intentions. Compare the quadratic losses of the classical estimator (that is, $\hat{p}_i = x_i$), the empirical Bayes estimator, and a Bayes estimator of your choice.

10.30 Establish the equivalent of Proposition 10.4.5 if the test is about H_0 : $\theta = \theta_0$ versus H_1 : $\theta = \theta_1$ for two independent problems with samples $x_1, \ldots, x_n \sim f(x|\theta)$, $y_1, \ldots, y_m \sim f(y|\theta')$, and $P(\theta = \theta_0) = P(\theta' = \theta_0) = \pi_0$. Generalize to p samples and apply in the case of the test of $\theta_i = 0$ versus $\theta_i = 1$ for $x_i \sim \mathcal{N}(\theta_i, 1)$ $(1 \le i \le p)$.

10.31 *(Hartigan (1983)) Consider $x \sim \mathcal{N}_p(\theta, \sigma^2 I_p)$ and $\theta \sim \mathcal{N}_p(0, \tau^2 I_p)$, with σ^2 unknown and $s^2 \sim \sigma^2 \chi_k^2$.

a. Give an empirical Bayes estimator of θ based on the maximum likelihood estimators of τ^2 and σ^2 and determine whether the resulting estimator is minimax. (*Hint:* Use Exercise 10.28.)

b. Compare with the empirical Bayes estimators based on the moment estimators of σ^2 and τ^2.

c. If $\pi(\sigma^2, \tau^2) \propto (\sigma^2 + \sigma_0^2)^{\alpha-1}(\sigma^2)^{\beta-1}$, show that the posterior distribution of $(\sigma^{-2}, (\sigma^2 + \tau^2)^{-1})$ is

$$\chi_{k-2\beta}^2/s \times \chi_{p-2\alpha}^2/||x||^2 \mathbb{I}_{\sigma^2 \le \sigma^2 + \tau^2}.$$

Show that the resulting estimator is minimax if

$$\frac{p-\alpha}{k-\beta-2} \le \frac{2(p-2)}{k+1}.$$

(*Hint:* Use Theorem 2.8.1.)

10.32 (Hartigan (1983)) Consider a multinomial model $\mathcal{M}_k(n; p_1, \ldots, p_k)$ and the observation $(n_1, \ldots, n_k)$. A possible conjugate prior is the Dirichlet distribution, $\mathcal{D}(\alpha_1, \ldots, \alpha_k)$.

a. Show that

$$\mathbb{E}\left[\sum_{i=1}^{k} n_i^2\right] = n + (n-1)\frac{\alpha+1}{k\alpha+1},$$

and determine when the moment equation derived from this equality has a positive solution. Derive an empirical Bayes estimator of $(p_1, \ldots, p_k)$ in this case.

b. Compute an alternative empirical Bayes estimator by using maximum likelihood estimators of the α_i's. [*Note:* See Good (1975) for details on this model.]

10.33 (Morris (1983a)) An exponential family with density

$$f(x|\theta) = h(x)e^{\theta x - \psi(\theta)}$$

has a *quadratic variance* if the variance can be written

$$V(\mu) = \psi''(\theta) = v_0 + v_1\mu + v_2\mu^2,$$

where $\mu = \psi'(\theta)$ is the expectation of $f(x|\theta)$. Morris (1982) has characterized the six families with quadratic variance (see Exercise 3.9). These distributions are denoted $NEF(\mu, V(\mu))$.

a. Show that the conjugate distribution in μ can be written

(10.6.1) $\qquad g(\mu) = K(m, \mu_0)e^{m\mu_0\theta(\mu) - m\psi(\theta(\mu))}V^{-1}(\mu)$

and that

$$\mathbb{E}^\pi[\mu] = \mu_0, \qquad V^\pi(\mu) = \tau_0^2 = \frac{V(\mu_0)}{m - v_2},$$

Therefore, the conjugate distribution is also an exponential family with quadratic variance. Derive a table of the correspondence between sample and conjugate prior distributions for the six families obtained in Exercise 3.24.

b. Show that the Bayes estimator associated with (10.6.1) for n independent observations $x_1, \ldots, x_n$ and quadratic loss is

$$\delta^\pi(x_1, \ldots, x_n) = (1 - B)\bar{x} + B\mu_0,$$

where

$$B = \frac{V(\mu_0) + v_2\tau_0^2}{V(\mu_0) + (n + v_2)\tau_0^2}.$$

c. Show that, for the conjugate distribution (10.6.1), the marginal moments of $\bar{x}$ are

$$\mathbb{E}[\bar{x}] = \mu_0, \qquad \text{var}(\bar{x}) = \frac{V(\mu_0)}{n}\frac{m + n}{m - v_2}.$$

d. Consider k independent observations

$$x_i|\mu_i \sim NEF(\mu_i, V(\mu_i)/n) \qquad (1 \le i \le k),$$

with independent parameters μ_i from the conjugate distribution (10.6.1). If $\bar{x} = \sum_i x_i/k$ and $s = \sum_i(x_i - \bar{x})^2$ and if

$$\frac{\mathbb{E}[V(\bar{x})(k - 1)]}{\mathbb{E}[s]} = \mathbb{E}\left[\frac{V(\bar{x})(k - 3)}{s}\right]$$

(the expectations being taken under the marginal distribution), show that an empirical Bayes estimator for μ_i is

$$\delta_i^{EB}(x_1, \ldots, x_k) = (1 - \hat{B})x_i + \hat{B}\bar{x},$$

with

$$\hat{B} = \min\left(\frac{v_2}{n + v_2}\frac{k-1}{k} + \frac{n}{n+v_2}\frac{(k-3)V(\bar{x})}{ns}, 1\right).$$

Section 10.5.1

10.34 Show that, for the marginal distribution of Example 10.5.1, $(p-2)/||x||^2$ is indeed an unbiased estimator of $1/(1+\tau^2)$.

10.35 Derive formula (10.5.4) of Example 10.5.2.

10.36 *(Kubokawa (1991)) Consider $\delta^{JS}(x) = [1 - (p-2)/||x||^2]x$, the original James–Stein estimator and $x \sim \mathcal{N}_p(\theta, I_p)$. Define $\lambda = ||\theta||^2/2$; $f_p(t; \lambda)$ is the noncentral chi-squared density with noncentrality parameter λ.

a. For the truncation of δ^{JS}

$$\delta_1(x; c, r) = \begin{cases} \left(1 - \dfrac{c}{||x||^2}\right)x & \text{if } ||x||^2 < r, \\ \delta^{JS}(x) & \text{otherwise}, \end{cases}$$

show that the quadratic risk of $\delta_1(x; c, r)$ is minimized for

$$c_1(r, \lambda) = p - 2 - \frac{2f_p(r; \lambda)}{\int_0^r (1/t)f_p(t; \lambda)\, dt}.$$

b. Let us define

$$c_1(r) = p - 2 - \frac{2}{\int_0^1 t^{p/2-2} e^{(1-t)r/2}\, dt}.$$

Show that $\delta_1(x; c_1(r), r)$ dominates δ^{JS} for every r.

c. Using a limiting argument, show that

$$\delta_1^*(x) = \left(1 - \frac{c_1(||x||^2)}{||x||^2}\right)x$$

dominates δ^{JS}. [*Note:* This estimator is proposed in Strawderman (1971) and Berger (1976). See Exercise 10.28.]

d. Show that δ_1^* is admissible. (*Hint:* The sufficient condition of Theorem 8.2.13 can be used.)

10.37 *(Bock and Robert (1991)) Consider $x \sim \mathcal{N}_p(\theta, I_p)$ and $\theta \sim \mathcal{U}_{\{||\theta||^2 = c\}}$, the uniform distribution on the sphere with radius c. Propose an empirical Bayes estimator of θ based on $||x||^2$ and show that, if this estimator is derived from the maximum likelihood estimator of c, then $\delta^{EB}(x) = h(x)x$ with

$$\left(1 - \frac{p}{||x||^2}\right)^+ \le h(x) \le \left(1 - \frac{p-1}{||x||^2}\right)^+.$$

Discuss the robustness of the Stein effect in terms of spherical symmetry.

10.38 *(George (1986a)) Consider $y \sim \mathcal{N}_p(\theta, I_p)$. This exercise derives an estimator that selects among several partitions of y into subvectors before shrinking the observation toward each of these subvectors. For $k = 1, \ldots, K$, let us denote

$$y = (y_{k1}, \ldots, y_{kJ_k})C_k \qquad \text{and} \qquad \theta = (\theta_{k1}, \ldots, \theta_{kJ_k})C_k$$

as the partitions of y and θ in subvectors y_{kj} and θ_{kj} of dimension p_{kj} ($1 \leq j \leq J_k$), where C_k is a permutation matrix comprising 0's and 1's with a single 1 per row and per column. For $k = 1, \ldots, K$, consider $\delta_k = (\delta_{k1}, \ldots, \delta_{kJ_k}) C_k$ an estimator with components

$$\delta_{kj}(y_{kj}) = y_{kj} + \nabla \log m_{kj}(y_{kj}),$$

where the functions m_{kj} from $\mathbb{R}^{p_{kj}}$ in $\mathbb{R}$ are twice differentiable. We also denote

$$m_k(y) = \prod_{j=1}^{J_k} m_{kj}(y_{kj}) \qquad \text{and} \qquad m_*(y) = \sum_{k=1}^{K} \omega_k m_k(y),$$

for $\omega_k \geq 0$ ($1 \leq k \leq K$) and $\sum_k \omega_k = 1$.

a. If π_{kj} is a prior distribution on θ_{kj} and m_{kj} is the corresponding marginal distribution on y_{kj} ($1 \leq k \leq K$, $1 \leq j \leq J_k$), show that m_k is the marginal distribution of y for the prior distribution

$$\pi_k(\theta) = \prod_{j=1}^{J_k} \pi_{kj}(\theta_{kj}),$$

and that δ_k is the posterior mean for this posterior distribution.

b. Deduce that

$$\delta^*(y) = y + \nabla \log m_*(y)$$

is the Bayes estimator for the prior distribution

$$\pi^*(\theta) = \sum_{k=1}^{K} \omega_k \pi_k(\theta).$$

c. Show that δ^* can also be written under the form

$$\delta^*(y) = \sum_{k=1}^{K} \varrho_k(y) \delta_k(y),$$

with $\varrho_k(y) = \omega_k m_k(y)/m_*(y)$, and interpret this result.

d. Show that if, for $k = 1, \ldots, K$,

$$\mathbb{E}_\theta \left| \frac{\partial^2 m_k(y)}{\partial y_i^2} \middle/ m_k(y) \right| < +\infty,$$

$$\mathbb{E}_\theta \|\nabla \log m_k(y)\|^2 < +\infty,$$

then the unbiased estimator of the risk of δ^* can be written

$$\mathcal{D}\delta^*(y) = p - \sum_{k=1}^{K} \varrho_k(y) \left[\mathcal{D}\delta_k(y) - (1/2) \sum_{\ell=1}^{K} \varrho_\ell(y) \|\delta_k(y) - \delta_\ell(y)\|^2 \right],$$

with

$$\mathcal{D}\delta_k(y) = \|\nabla \log m_k(y)\|^2 - 2\Delta m_k(y)/m_k(y).$$

(Hint: Use Lemma 10.3.1 with $Q = \Sigma = I_p$.)

e. Deduce that, if m_{kj} is superharmonic, that is, such that $\Delta m_{kj}(y_{kj}) \leq 0$ for $1 \leq k \leq K$, $1 \leq j \leq J_k$, δ^* is minimax. [Note: This result can be described as the fact that a "proper" convex combination of minimax estimators is still minimax.]

f. For $1 \leq k \leq K$, $1 \leq j \leq J_k$, we denote by V_{kj} a subspace of $\mathbb{R}^{p_{kj}}$, with dim $V_{kj} = p_{kj} - q_{kj}$ and $q_{kj} \geq 3$; P_{kj} is the associated orthogonal projector from $\mathbb{R}^{p_{jk}}$ on V_{kj} and $s_{kj} = ||y_{kj} - P_{kj}y_{kj}||^2$. Give the multiple shrinkage estimator δ^* associated with

$$m_{kj}(y_{kj}) = \begin{cases} \left(\dfrac{q_{kj} - 2}{es_{kj}} \right)^{(q_{kj}-2)/2} & \text{if } s_{kj} \geq q_{kj} - 2, \\ \exp(-s_{kj}/2) & \text{otherwise.} \end{cases}$$

(*Hint:* The solution is the truncated James–Stein estimator.)

Section 10.5.2

10.39 *(Kubokawa et al. (1993)) Consider $x \sim N_p(\theta, \sigma^2 I_p)$, $y \sim N_q(\xi, \sigma^2 I_q)$, and $s \sim \sigma^2 \chi_n^2$, with unknown θ, ξ, and σ. An empirical Bayes estimator of θ is the James–Stein estimator

$$\delta^{JS}(x, s) = \left(1 - \frac{(p - 2)s}{(n + 2)||x||^2} \right) x.$$

The purpose of this exercise is to show that the replacement of s by a more efficient estimator of σ^2 can lead to an improvement in the estimation of θ.

a. Show that, if $\gamma_h(y, s) = sh(||y||^2/s)$ dominates $\gamma_0(s) = s/(n+2)$ under the invariant quadratic loss

$$L(\sigma^2, \gamma) = \left(\frac{\gamma}{\sigma^2} - 1 \right)^2,$$

δ^{JS} is dominated by

$$\hat{\delta}(x, y, s) = \left(1 - \frac{(p - 2)\gamma(y, s)}{||x||^2} \right) x$$

under quadratic loss. (*Hint:* Recall that γ_0 is the best equivariant estimator of σ^2.)

b. Consider

$$\delta_g(x, y, s) = \left(1 - \frac{(p - 2)s}{||x||^2} g(||y||^2/s, ||x||^2/s) \right) x.$$

Define

$$g^*(u, v) = \min \left(g(u, v), \frac{1 + u + v}{p + q + n} \right),$$

and assume g and g^* are absolutely continuous functions of v. Show that, if

$$\mathbb{E} \left[\frac{\partial g^*(U, V)}{\partial v} - \frac{\partial g(U, V)}{\partial v} \right] \geq 0,$$

when $U = ||y||^2/s$ and $V = ||x||^2/s$, δ_{g^*} dominates δ_g.

c. Deduce that

$$\delta_2(x, y, s) = x - \frac{p - 2}{||x||^2} \min \left\{ \frac{s}{n + 2}, \frac{s + ||y||^2}{n + q + 2}, \frac{s + ||x||^2 + ||y||^2}{n + p + q + 2} \right\} x$$

dominates δ^{JS}.

Section 10.5.3

10.40 *(Casella and Hwang (1983)) Consider $x \sim \mathcal{N}_p(\theta, I_p)$. Under the linear loss,

$$L(\theta, C) = k \, \text{vol}(C) - \mathbb{I}_C(\theta),$$

recall that the Bayes estimators are HPD regions of the form $\{\theta; \, \pi(\theta|x) \geq k\}$ when $\pi(\{\theta; \pi(\theta|x) = k\}) = 0$. Moreover, if

$$k = k_0 = e^{-c^2/2}/(2\pi)^{p/2},$$

Joshi (1969) has established that the usual region

$$C_x^0 = \{\theta; \, \|\theta - x\| \leq c\},$$

is minimax.

a. Show that, if $\theta \sim \mathcal{N}_p(0, \tau^2 I_p)$, the Bayes set is

$$C_x^\pi = \left\{ \theta; \, \|\theta - \delta^\pi(x)\|^2 \leq -\frac{2\tau^2}{\tau^2 + 1} \log \left[k \left(\frac{2\pi\tau^2}{\tau^2 + 1} \right)^{p/2} \right] \right\},$$

where $\delta^\pi(x) = (\tau^2/\tau^2 + 1)x$ is the Bayes estimator of θ. For $k = k_0$, show that this set can be written

$$C_x^\pi = \left\{ \theta; \, \|\theta - \delta^\pi(x)\|^2 \leq \frac{\tau^2}{\tau^2 + 1} \left[c^2 - p \log \left(\frac{\tau^2}{\tau^2 + 1} \right) \right] \right\}.$$

b. Deduce that a simple empirical Bayes set is

$$C_x^{\text{EB}} = \left\{ \theta; \, \|\theta - \delta^{\text{EB}}(x)\|^2 \leq v^{\text{EB}}(x) \right\},$$

with $\delta^{\text{EB}}(x) = (1 - [(p - 2)/\|x\|^2])x$ and

$$v^{\text{EB}}(x) = \left(1 - \frac{p - 2}{\|x\|^2} \right) \left(c^2 - p \log \left| 1 - \frac{p - 2}{\|x\|^2} \right| \right).$$

c. Explain why it is preferable to consider

$$\delta^+(x) = \left(1 - \frac{p - 2}{\|x\|^2} \right)^+ x$$

and

$$v^e(x) = \begin{cases} \left(1 - \frac{p-2}{c^2} \right) \left(c^2 - p \log \left[1 - \frac{p-2}{c^2} \right] \right) & \text{if } \|x\|^2 < c, \\ \left(1 - \frac{p-2}{\|x\|^2} \right) \left(c^2 - p \log \left[1 - \frac{p-2}{\|x\|^2} \right] \right) & \text{otherwise.} \end{cases}$$

d. Extend to the case when $x \sim \mathcal{N}_p(\theta, \sigma^2 I_p)$ and $s^2 \sim \sigma^2 \chi_q^2$.

Note 10.7.2

10.41 In the setting of Example 10.7.3,

a. Show that

$$L(\eta, \hat{\eta}) = \frac{p}{2} \log \left(\frac{\eta}{\hat{\eta}} \right) + \frac{1}{2} \left(\frac{1}{\hat{\eta}} - \frac{1}{\eta} \right) p\eta,$$

and deduce (10.7.3).

b. Show that the posterior distribution associated with π_d is well defined for $d > (4 - p)/2$.

c. Show that (10.7.4) and (10.7.5) hold.

d. Derive from the approximation

$$
{}_1F_1(a, b, z) = \Gamma(b) \, z^{-a} \left\{ 1 - \frac{a(1 + a - b)}{z} + O(z^2) \right\}
$$

the equivalence (10.7.6).

10.42 *(Exercise 10.41 cont.) Using STUB conditions (see Section 8.5) and the approximation (10.7.6), show that the choice $d = 1$ is optimal.

10.43 *(Exercise 10.41 cont.) Derive from Alam (1973) the minimaxity of δ_1^{EB}. (*Hint:* See Example 10.3.6.)

10.44 In the setting of Example 10.7.4,

a. Show that the entropy loss is given by (10.7.7) and deduce the general form of Bayes estimators under that loss.

b. Show that, when $\pi(\lambda) = \lambda^{-d}$, the posterior distribution is given by

$$
\pi_d(\lambda|x) \propto \lambda^{n-d}(\lambda + 1)^{-\sum_i x_i - n}
$$

and deduce that the Bayes estimator of λ is given by (10.7.8).

c. Show that the conditional distribution $\pi(\theta|x, \lambda) \propto \theta^x e^{-(\lambda+1)\theta}$ and deduce that the Bayes estimator of θ_i conditional on λ is

$$
\mathbb{E}[\theta_i|x_i, \lambda] = \frac{x_i + 1}{\lambda + 1} \, .
$$

10.45 (Exercise 10.44 cont.) Show that the usual estimator of θ, that is, $\gamma_0(x) = x$, has an infinite risk under entropy loss.

10.46 (Exercise 10.44 cont.) Show that the Bayes estimator associated with the integrated prior

$$
\pi(\theta) = \int \pi(\theta, \lambda) d\lambda
$$

is given by

$$
\delta^\pi(x) = \left(1 - \frac{n - d + 1}{\sum_{i=1}^n x_i + n} \right) (x + 1) \, .
$$

Deduce from the difference of the entropy risks that the Bayes estimator dominates its empirical Bayes counterpart under entropy loss.

10.7 Notes

10.7.1 Graphical models[7]

Graphical models analyze statistical models by graphs. They have been developed mainly to represent conditional independence relations, primarily in the field of *expert systems* (Whittaker (1990), Spiegelhalter et al. (1993)). The Bayesian approach to these models, as a way to incorporate model uncertainty, has been aided by the advent of MCMC techniques, as stressed by Madigan and York (1995) in an expository paper on which this note is based.

---

[7] This note closely follows Note 7.6.6 in Robert and Casella (1999).

The construction of a graphical model is based on a collection of independence assumptions represented by a graph. We briefly recall here the essentials of graph theory and refer to Lauritzen (1996) for details. A *graph* is defined by a set of *vertices* or *nodes*, $\alpha \in \mathcal{V}$, which represents the random variables or factors under study, and by a set of *edges*, $(\alpha, \beta) \in \mathcal{V}^2$, which can be ordered (the graph is then said to be *directed*) or not (the graph is *undirected*) and represent the dependence connections between the variables. For an undirected graph, the variables α and β are connected by an edge if, conditional on all the other variables, they are *not* independent. For a directed graph, α is a *parent* of β if (α, β) is an edge (and β is then a *child* of α).[8] Graphs are also often assumed to be *acyclic*, that is, without directed paths linking a node α with itself. This leads to the notion of *directed acyclic graphs* introduced by Kiiveri and Speed (1982), often represented by the acronym DAG.

For the construction of probabilistic models on graphs, an important notion is that of a *clique*.[9] A *clique* C is a maximal subset of nodes that are all joined by an edge (in the sense that there is no subset containing C and satisfying this condition). An ordering of the cliques of an undirected graph $(C_1, \ldots, C_n)$ is *perfect* if the nodes of each clique C_i contained in a previous clique are all members of one previous clique (these nodes are called the *separators*, $\alpha \in S_i$). In this case, the joint distribution of the random variable V taking values in $\mathcal{V}$ is

$$p(V) = \prod_{v \in V} p(v | \mathcal{P}(v)) \, ,$$

where $\mathcal{P}(v)$ denotes the parents of v. This can also be written as

$$(10.7.1) \qquad\qquad p(V) = \frac{\displaystyle\prod_{i=1}^{n} p(C_i)}{\displaystyle\prod_{i=1}^{n} p(S_i)} \, ,$$

and the model is then called *decomposable*; see Spiegelhalter and Lauritzen (1990), Dawid and Lauritzen (1993) or Lauritzen (1996). As stressed by Spiegelhalter et al. (1993), the representation (10.7.1) leads to a *principle of local computation*, which enables the building of a prior distribution, or the simulation from a conditional distribution on a single clique. (In other words, the distribution is *Markov with respect to the undirected graph*, as shown by Dawid and Lauritzen (1993).) The appeal of this property for a Gibbs sampling implementation is then obvious.

When the densities or probabilities are parametrized, the parameters are denoted by θ_A for the marginal distribution of $V \in A$, $A \subset \mathcal{V}$. (In the case of discrete models, $\theta = \theta_V$ may coincide with p itself; see Example 10.7.1.) The prior distribution $\pi(\theta)$ must then be compatible with the graph structure:

---

[8] Directed graphs can be turned into undirected graphs by adding edges between nodes sharing a child, and dropping the directions.

[9] *Clique* is a French word meaning faction, gang or group. Its usual connotation is rather pejorative.

Dawid and Lauritzen (1993) show that a solution is of the form

$$(10.7.2) \qquad \pi(\theta) = \frac{\displaystyle\prod_{i=1}^{n} \pi_i(\theta_{C_i})}{\displaystyle\prod_{i=1}^{n} \tilde{\pi}_i(\theta_{S_i})} \,,$$

thus reproducing the clique decomposition (10.7.1).

Example 10.7.1 Consider a decomposable graph such that the random variables corresponding to all the nodes of $\mathcal{V}$ are discrete. Let $w \in W$ be a possible value for the vector of these random variables and $\theta(w)$ be the associated probability. For the perfect clique decomposition $(C_1, \ldots, C_n)$, $\theta(w_i)$ denotes the marginal probability that the subvector $(v, v \in C_i)$ takes the value w_i ($\in W_i$) and, similarly, $\theta(w_i^s)$ is the probability that the subvector $(v, v \in S_i)$ takes the value w_i^s when $(S_1, \ldots, S_n)$ is the associated sequence of separators. In this case,

$$\theta(w) = \frac{\displaystyle\prod_{i=1}^{n} \theta(w_i)}{\displaystyle\prod_{i=1}^{n} \theta(w_i^s)} \,.$$

As illustrated by Madigan and York (1995), a Dirichlet prior can be constructed on $\theta_W = (\theta(w), w \in W)$. It leads to genuine Dirichlet priors on the $\theta_{W_i} = (\theta(w_i), w_i \in W_i)$, under the constraint that the Dirichlet weights are identical over the intersection of two cliques. Dawid and Lauritzen (1993) establish that this prior is unique, given the marginal priors on the cliques. ‖

Example 10.7.2 Giudici and Green (1998) provide another illustration of prior specification in the case of a *graphical Gaussian model*, $\mathbf{X} \sim \mathcal{N}_p(0, \Sigma)$, where the precision matrix $K = \{k_{ij}\} = \Sigma^{-1}$ must comply with the conditional independence relations on the graph. For instance, if $\mathbf{X}_v$ and $\mathbf{X}_w$ are independent given the rest of the graph, then $k_{vw} = 0$. The likelihood can then be factorized as

$$f(\mathbf{x}|\Sigma) = \frac{\displaystyle\prod_{i=1}^{n} f(\mathbf{x}_{C_i}|\Sigma^{C_i})}{\displaystyle\prod_{i=1}^{n} f(\mathbf{x}_{S_i}|\Sigma^{S_i})} \,,$$

with the same clique and separator notations as above, where $f(\mathbf{x}_C|\Sigma^C)$ is the Normal $\mathcal{N}_{p_C}(0, \Sigma^C)$ density, following (10.7.1). The prior on Σ can be chosen as the conjugate inverse Wishart priors on the Σ^{C_i}'s, under some compatibility conditions. ‖

Madigan and York (1995) discuss an MCMC approach to model choice and model averaging in this setting, whereas Dellaportas and Forster (1996) and Giudici and Green (1998) implement reversible jump algorithms for determining the probable graph structures associated with a given dataset, the latter under a Gaussian assumption.

10.7.2 Bayes empirical Bayes

As noticed in Section 10.4, the difficulty with the empirical Bayes approach is that it is a two–stage estimation procedure, where, firstly, the hyperparameter is estimated from the marginal distribution, and, secondly, the parameter is estimated based on a pseudo-prior where the hyperparameter is replaced by the estimate of the first stage. Although the inefficiency of this procedure—when compared with a genuine Bayesian approach—cannot be fully remedied, it seems natural to aim at a maximal efficiency by using the most efficient estimation method at the first stage, that is, a noninformative Bayesian approach. Because the first stage estimation is not induced by a decision problem, a loss function is most likely unavailable and the intrinsic losses presented in Section 2.5.4 have been argued as natural default losses in such cases. This solution, which removes the arbitrariness attached to the estimation of the hyperparameters in the empirical Bayes approach, is surprisingly ignored in the literature. We present below two examples treated in Fourdrinier and Robert (1995).

Example 10.7.3 (Example 10.5.1 continued) The marginal distribution of x, $m(x|\eta)$, is $\mathcal{N}_p(0, \eta I_p)$, where $\eta = 1 + \tau^2$, and the corresponding entropy loss for the estimation of η is

$$
\begin{aligned}
L(\eta, \hat{\eta}) &= \int \log\left(\frac{m(x|\eta)}{m(x|\hat{\eta})}\right) m(x|\eta) dx \\
(10.7.3) \qquad &= \frac{p}{2}\left(\frac{\eta}{\hat{\eta}} - \log\left(\frac{\eta}{\hat{\eta}}\right) - 1\right).
\end{aligned}
$$

Since η is a scale parameter for the marginal distribution, a natural family of noninformative priors is $\pi_d(\eta) = \eta^{-d}$ on $[1, \infty[$ and the corresponding estimator of η is

$$
(10.7.4) \qquad \hat{\eta}_d = \frac{\int_0^1 \nu^{(p/2)+d-3} e^{-||x||^2 \nu/2} d\nu}{\int_0^1 \nu^{(p/2)+d-2} e^{-||x||^2 \nu/2} d\nu}.
$$

The empirical Bayes estimator is thus

$$
\begin{aligned}
\delta_d^{EB}(x) &= (1 - \hat{\eta}^{-1})x \\
&= \frac{\int_0^1 \nu^{(p/2)+d-3}(1-\nu)e^{-||x||^2 \nu/2} d\nu}{\int_0^1 \nu^{(p/2)+d-2} e^{-||x||^2 \nu/2} d\nu} x \\
(10.7.5) \qquad &= \frac{2}{p+2d-2} \frac{{}_1F_1(2, d+p/2, ||x||^2/2)}{{}_1F_1(2, d+p/2-1, ||x||^2/2)} x,
\end{aligned}
$$

where ${}_1F_1$ is the confluent hypergeometric function (Abramowitz and Stegun (1964, Chapter 13)). Since $\delta_d^{EB}(x)$ is asymptotically equivalent to

$$
(10.7.6) \qquad \left(1 - \frac{p+2(d-2)}{||x||^2}\right) x
$$

the choice $d = 1$, that is, $\pi(\eta) = 1/\eta$ provides the optimal choice of d (Exercise 10.42). ‖

Example 10.7.4 (Example 10.4.3 continued) The entropy loss associated with $m(x|\lambda)$ is

$$(10.7.7) \qquad L(\lambda, \hat{\lambda}) = \log\left(\frac{\lambda}{\hat{\lambda}}\right) + \left(1 + \frac{1}{\lambda}\right) \log\left(\frac{\hat{\lambda}+1}{\lambda+1}\right)$$

and, for $\pi(\lambda) = \lambda^{-d}$, the corresponding Bayes estimator of λ is

$$(10.7.8) \qquad \hat{\lambda} = \frac{n-d}{\sum_{i=1}^{n} x_i + n - 1}.$$

If we also use an entropy loss for the estimation of λ, $L(\theta, \hat{\theta}) = \hat{\theta} - \theta - \log(\hat{\theta}/\theta)$, the empirical Bayes estimator of $\theta = (\theta_1, \ldots, \theta_n)$ is

$$(10.7.9) \qquad \theta^{EB}(x) = \left(1 - \frac{n-d}{\sum_{i=1}^{n} x_i + n - 1}\right)(x + 1),$$

where $\mathbf{1} = (1, \ldots, 1) \in \mathbb{R}^n$. Fourdrinier and Robert (1995) show in addition that there is an optimal choice d^* of d for the entropy loss, with $d^* \leq 2$, and that there is a range of values of d for which θ^{EB} dominates $\hat{\theta}_0 = x + \mathbf{1}$. ‖

CHAPTER 11

A Defense of the Bayesian Choice

A series of steps, each taken for good cause or pure necessity, each seeming so reasonable at the time, and each leading to things he had never imagined. He always seemed to find himself caught in that sort of dance.

Robert Jordan, *Lord of Chaos, Book VI of the Wheel of Time.*

This book has presented the main aspects of Bayesian inference in Statistics from a decision-theoretic point of view. Our coverage has scarcely been exhaustive: on one hand, the topics we consider are often treated in more detail. On the other hand, Bayesian analysis can be applied to many fields, and is increasingly done, thanks to the computational advances described in Chapter 6. Among these fields, we can mention *biostatistics* (see, e.g., Berry and Stangl (1996)); *econometrics* (Zellner (1971, 1984), Box and Tiao (1973), Poirier (1995), Bauwens et al. (1999), or Geweke (1999)); *environmetrics* (Parent et al. (1998)); *expert systems* (Gilks et al. (1993); Cowell et al. (1999)); *finance* (Jacquier, Polson and Rossi (1994), Pitt and Shephard (1996)); *image processing and pattern recognition* (Geman and Geman (1984), Besag (1986), or Fitzgerald et al. (1999)); *neural networks* (Ripley (1992), Neal (1996)); *signal processing* (Andrieu and Doucet (1999), Andrieu, Doucet and Fitzgerald (2001)); *Bayesian networks* (Chickering and Heckerman (2000), Kontkanen et al. (2000)). (See the recent survey by Berger (2000), as well as Gatsonis et al. (1993, 1995, 1997, 1999), Gilks et al. (1996), and Carlin and Louis (2000a) for additional references.)

Nonetheless, we found it important to consider Bayesian statistical analysis primarily from this theoretical and decisional perspective, before paying more attention to its potential for applications. Firstly, this study exhibits the inherent *coherence* of the Bayesian approach in comparison with alternative classical theories. Secondly, to develop an efficient approach for the processing of applications requires a solid background in theoretical issues.

We present, in this concluding chapter, a justification of the Bayesian approach that summarizes the various arguments advanced so far.[1]

(1) **Opting for a probabilistic representation**

Proposing a distribution on the unknown parameters of a statistical model can be characterized as a *probabilization of uncertainty*, that is, as an axiomatic reduction from the notion of unknown to the notion of random. This reduction being acceptable—and it is usually accepted by most statisticians—for sampling models, it should be acceptable as well for the parameters directing those models. For one thing, the distinction between sample and parameters is not always clear-cut. Consider, for instance, random-effects models (Chapter 10) or allocation vectors in a mixture model (Chapter 6).

More fundamentally, a probabilistic model is most often nothing but an *interpretation* of a given phenomenon—as opposed to an *explanation* of it. If we consider, for instance, *econometric* models, where the differences between the realizations of *endogenous* variables and their linear prediction on *exogenous* variables are explained by a random perturbation, it is clear that the random nature of this difference is of little importance because the experiment cannot be replicated.[2] Therefore, the representation of unknown phenomena by a probabilistic model, at the observational level as well as at the parameter level, does not need to correspond effectively—or physically—to a generation from a probability distribution, nor does it compel us to enter a supradeterministic scheme, fundamentally because of the nonrepeatability of most experiments.

The probabilistic representation of partly explained phenomena should be perceived mainly as a simplifying but efficient *tool* conveying and quantitatively analyzing these phenomena (see point (4) below). This perspective is really similar to the way physics can be seen as an interpretation of the world and is a *tool*, efficient enough to allow for a better understanding of this world (and incidentally for technical progress), while not needing to correspond to a truth definitely unattainable.[3]

---

[1] The presentation of this chapter is thus quite different from the other chapters, with no theorem or even example, but a sequence of points (from (1) to (10)) arguing in favor of the Bayesian approach, followed by a rebuttal of the most common criticisms of it. The tone is thus mildly philosophical, rather than methodological (or mathematical), and the reader can judge whether her or his impression, after going through the book, coincides with the points set forth below.

[2] To put it bluntly, an arbitrary number can always be perceived as a single realization from an infinity of distributions!

[3] See also Popper (1983) for his alternative justification of scientific modeling through the *metaphysical realism* that he sets in opposition to this *instrumental* approach.

(2) **Conditioning on the data**

The basis of statistical inference is fundamentally an *inversion process*, since it aims at deriving effects from causes by taking into account the probabilistic nature of the model and the influence of totally random (that is, unexplained) factors. In both its discrete and continuous versions, *Bayes's Theorem* formalizes this inversion, as does the notion of the likelihood function $\ell(\theta|x)$, substituted for the density $f(x|\theta)$. The failure of *Fiducial Statistics* to provide a satisfactory inferential system (see Note 1.8.1) can be attributed to a refusal to pursue this inversion to its logical consequences and, relatedly, to a certain confusion between observations and random variables.

From a probabilistic point of view, a quantitative analysis on the parameters θ that is operated *conditional upon* x strictly requires a corresponding distribution on the parameter θ, $\pi(\theta)$, in order to invert the probabilistic model. Taking this requirement into account, the Bayesian approach is thus the unique *coherent* paradigm which respects the inversion perspective. The practical problem of the determination of the prior distribution π does not take place in the same conceptual space (see point (ii), below).

(3) **Exhibiting the true likelihood**

In relation to points (1) and (2), notice also that the prior modeling on the parameters of the model authorizes a complete *quantitative inference* on these parameters, therefore the effective determination of the *likelihood* of θ conditional on the observation x. On the contrary, classical Statistics fails to attain this completeness. In particular, as long as θ is taken to be an unknown but *fixed* quantity, the likelihood function $\ell(\theta|x)$ cannot be treated as a density conditional upon x, despite the formal resemblance.

This impossibility of the classical approach to provide quantitative conclusions is particularly well illustrated in the case of confidence regions and tests, where it proposes an inappropriate problematic (and, consequently, an inappropriate answer). As we saw in Chapter 5, the classical procedure, whether a 95% interval or a p-value, derives its probabilistic nature from a frequentist interpretation. It is not the parameter θ that belongs to an interval with probability 95% conditional upon x, but the interval derived from x that contains the fixed value θ with probability 0.95. Again, the nonrepeatability of most practical experiments impeaches this frequentist point of view (see also point (9) below).

(4) **Using priors as tools and summaries**

The choice of a prior distribution π does not require any kind of *belief* in this distribution. It is actually rare to have a completely specified prior distribution, the original example of Thomas Bayes being, paradoxically, an exceptional counter-example where a physical knowledge of the experiment leads to the construction of the prior distribution. In general, π should rather be considered either a *tool* that provides a single inferential procedure with acceptable frequentist properties (see points (6) and (8)), or a way to *summarize* the available prior information and the uncertainty surrounding this information. That Bayesian analysis can be extended to *noninformative* settings—with a few exceptions, such as some testing situations—is actually an indicator of this polyvalence. Moreover, that many usual estimators can be recovered through a noninformative modeling means that the use of a prior distribution does not necessarily introduce a bias in the statistical process, but, on the contrary, authorizes in addition the quantitative treatment already mentioned in point (3). These coincidences actually enhance the superiority of the Bayesian approach, since it provides at once a full inferential treatment that supersedes these classical estimators.

(5) **Accepting the subjective basis of knowledge**

From a philosophical point of view, it is generally agreed that knowledge stems from a confrontation between a prioris and experiments. For instance, according to Kant, *although knowledge starts with experimenting, it does not follow that knowledge is entirely derived from experimenting.* In fact, without *a prioris*, that is, without a pre-established structure of the world, observation is meaningless because it does not come as a support of or as a confrontation to a referential model. Therefore, the building of knowledge through experimentation implies the existence of a prior representation system, which is very primitive at the beginning, but gets progressively actualized via these experiments. From this perspective, *learning* can be expressed as the critical examination of pre-existent external systems subjected to experiments with respect to this overall referential representation of the world.

This point of view is also found in Poincaré (1902):

> It is often stated that one should experiment without preconceived ideas. This is simply impossible; not only would it make every experiment sterile, but even if we were ready to do so, we could not implement this principle. Everyone stands by his own conception of the world, which he cannot get rid of so easily.

The Bayesian approach is obviously in accordance with this perspective, since prior distributions are most often based on the results of previous experiments. Actually, even the *subjective* aspect of the choice of the prior distribution can be integrated in this theory of knowledge, since it implies that every acquisition of knowledge is essentially *subjective*, resulting from an interaction between individual perceptions and exterior reality.[4]

---

[4] In fact, Bayesian Statistics could answer this wish of Kant's in the Introduction to his *Critique of Pure Reason*: *"Philosophy needs a science able to determine the possibility, the principles and the scope of our whole prior knowledge."*

In his epistemological theory, Feyerabend (1975) also stresses that individualism (that is, *subjectivity*) is an important, but blatantly ignored, factor in scientific discoveries. Although opposed to this subjectivist approach to knowledge, Popper (1983) also recognizes the role of prior intuitions (or systems), not always grounded on experiments, in the history of science—the most striking example being, from his point of view, *atomism*, that is, the representation of matter as being constituted by atoms, which took more than twenty centuries to ascertain experimentally.

(6) **Choosing a coherent and unique system of inference**

The ultimate goal of Statistics is, arguably, to provide an *inference* about a parameter θ given some observations x related to θ through a probability distribution $f(x|\theta)$. Moreover, it seems only natural to seek *efficiency* (or *optimality*) in this inference, the notion of optimality being defined *explicitly* by the statistician (or the decision-maker). To force inference into a decision-theoretic mold through the choice of a *loss function* allows for a clarification of the way inferential tools should be evaluated, and therefore implies a conscious (although subjective) choice of the *retained optimality*. In addition, when the decision-theoretic framework is completed by the choice of a prior distribution, the above inferential goals are automatically satisfied, since the Bayesian approach usually leads to a *unique procedure*, depending on the requested properties and the prior knowledge. Obviously, *uniqueness* of the decision procedure is not a sufficient validation per se, since many meaningless, although unique, procedures can be proposed instantly.

The important feature of a Bayesian approach is, thus, that Bayes estimators are derived by an eminently *logical process*: starting from requested properties, summarized in the loss function and the prior distribution, the Bayesian approach derives the best solution satisfying these properties. On the contrary, classical procedures are ad hoc in the sense that they start from an "arbitrary" estimator (maximum likelihood estimator, least-squares estimator, etc.) and then examine its frequentist properties, not necessarily in a decision-theoretic setting and with no pretension to global optimality, as shown by the Stein effect. In other cases, classical approaches also establish a criterion for the choice of an estimator (best unbiased estimator, best equivariant estimator, uniformly most powerful test, etc.), but they cannot provide a universal method, that is, an algorithm, for the derivation of optimal estimators (see also point (10) below), and it is even sometimes necessary to restrict further the class of considered estimators as, for instance, in the case of the uniformly most powerful unbiased tests.

This opposition in the logical foundations of the two theories reinforces the *coherence* of the Bayesian approach, since it is the *only one*—when incorporating the case of the best equivariant estimators as a Bayesian estimation method under the invariant Haar measure—to provide a *universal and implementable process stemming from inferential requirements.*

(7) **Implementing the Likelihood Principle**

The *Likelihood Principle*, as shown in Chapter 1, is based on the quite logical *Sufficiency* and *Conditionality Principles*. Therefore, it should always direct the choice of estimation procedures, adding a desirable property to those already discussed in point (6). The Bayesian paradigm provides an implementation technique for this principle, since it allows for the derivation of decisions compatible with these different requirements.

Moreover, while formally incorporating the maximum likelihood estimation method as a particular case (for $\pi(\theta) = 1$), the Bayesian approach can also avoid some likelihood paradoxes such as those presented in Section 4.1, by using the Jeffreys noninformative distributions, even though these priors are not entirely compatible with the Likelihood Principle. An additional important advantage of the Bayesian approach, compared with the maximum likelihood method, is that it can also incorporate the requirements of a loss function, and thus enter into the framework of Decision Theory, while being acceptable for the Likelihood Principle.

(8) **Looking for optimal frequentist procedures**

From the point of view of frequentist theory, the most convincing argument in favor of the Bayesian approach is that *it intersects widely with the three notions of classical optimality*, namely, minimaxity, admissibility and equivariance. Indeed, we saw in Chapters 2, 8 and 9 that most estimators that are optimal according to one of these criteria are Bayes estimators or limits of Bayes estimators (the notion of *limit* depending on the context). Thus, not only is it possible to produce Bayes estimators that satisfy one, two, or three of the optimality criteria, but, more importantly, the Bayes estimators are essentially the only ones to achieve this aim. Therefore, a frequentist statistician may be opposed to any *subjective* input in his inferential treatment and still remain in agreement with his principles by using only Bayes or generalized Bayes estimators, since most of them behave satisfactorily.[5] Moreover, these estimators are easy to derive, compared with the choice of an ad-hoc estimator and the subsequent verification that it is actually a limit of Bayes estimators. From this point of view, prior distributions are again considered an *inferential tool*, not an exhaustive summary of the prior information, but their shape and their posterior uses obviously remain the same. The optimality of the Bayes procedures also holds for asymptotic criteria because, under the conditions ensuring efficiency of the maximum likelihood estimator, most Bayes estimators are asymptotically efficient and become

---

[5] In this regard, notice that the frequentist pretensions to *objectivity* (as opposed to the Bayesian inherent subjectivity) are actually quite limited. In fact, it is necessary to select the estimators to be compared, and the choice of these estimators is partly subjective.

equivalent to the maximum likelihood estimator as the sample size increases (see Lehmann (1983) and Ibragimov and Has'minskii (1982)), even though this type of optimality is less important in our opinion.

(9) **Solving the actual problem**

It is also necessary to provide an alternative to the frequentist approach from a *practical* point of view. In fact, frequentist methods are justified on a *long-term* basis. For instance, a confidence interval at level 95% used for many independent problems will have a success rate close to 95%, which is satisfactory for the statistician. On the contrary, for a decision-maker ("the client"), these long-term properties have little appeal because she is interested by the performances of the proposed procedure *for the problem at hand*. For instance, the fact that a drug is successful 99% of the time is not reassuring for a particular patient, compared with that patient's chances of recovery! This request obviously calls for an inference that is *conditional* on x, and thus brings us back to the Bayesian approach (see point (2)).

This argument does not seem to apply to statistical settings involving repeated experiments where the decision is taken by the same individual, as in *quality control*. But such settings may also justify a Bayesian implementation, since they are most likely to allow the researcher to borrow strength from previous studies through a prior distribution.

(10) **Computing procedures as a minimization problem**

An important point in favor of the Bayesian choice is that the Bayesian procedures are *easier to compute* than procedures of alternative theories. This assertion may appear paradoxical when considering the developments of Chapters 6 and 10 and, for instance, the difficulties encountered in the treatment of finite mixtures; more generally, we saw that the Bayes estimators are seldom derived in closed form, except in the rather special case of conjugate distributions. However, an additional appeal of the Bayesian approach is to provide a *universal method* for the computation of Bayes estimators, whatever the loss and the distribution of the observations are, which is to minimize the posterior loss, even if such minimization requires a call to numerical or Monte Carlo algorithms.

On the contrary, the frequentist theory does not provide any indication of the derivation of minimax or admissible estimators, except to use a Bayesian approach through least-favorable priors or proper distributions.[6] Similarly, the only procedure providing a general derivation of *best equivariant estimators* entails using Haar measures and the corresponding Bayesian representation, as shown in Chapter 9.

---

[6] Although they both minimize losses, the main difference between the two approaches is that, for the frequentist approach, the minimization is done on a functional space—the space of estimators—whereas the Bayesian approach carries out the minimization on a decision space—the space of estimates. The respective complexities of these two spaces are, generally, considerably different.

From another point of view, maximum likelihood estimators also proceed from a general optimization program, but their derivation can get quite complicated and, more importantly, this method does not provide a complete inferential scope. Moreover, the Bayes estimators allow for integral representations under usual losses, whereas the maximum likelihood method does not necessarily lead to an estimator. For instance, this is the case for normal mixtures, where the likelihood is not bounded, or in settings where there are several maxima of the likelihood function.

From a pragmatic and computational point of view, it can be argued that the *effective* calculation of the Bayes estimators is often more delicate since it usually involves multiple integrations. Although a concern, this type of drawback takes place at a material (or software) level, in the sense that it should progressively disappear as computational methods evolve and improve. This defect is indeed of another magnitude, though the opening provided by Markov chain Monte Carlo methods, and the subsequent derivation of Bayesian solutions in many new domains, shows that computational issues may be an important slowing factor. Nevertheless, what really matters is the existence of a unique process leading to the Bayes estimator, whatever the inferential problem, the loss and the prior distribution. This perspective definitely singles out Bayesian analysis.

The reader still skeptical about the advantages of a Bayesian approach is referred to the books mentioned in the previous chapters, in particular to Jeffreys (1961), Lindley (1971), Berger (1985a, §4.1 and §4.12), Berger and Wolpert (1988), Bernardo and Smith (1994), Carlin and Louis (2000a) and Gelman et al. (2000). Smith (1984) also provides a similar list of justifications for the Bayesian choice.

For the sake of objectivity, we should also present a corresponding list of criticisms of the Bayesian approach by other statistical approaches. However, we believe none of these criticisms bring out strong incoherences of the Bayesian paradigm.[7] Therefore, we will only consider below three issues about the *prior distributions*, which are the basis of the Bayesian paradigm and the focus of most criticisms.

(i) *The passage from prior information, which can be vague or poorly defined, to the prior distribution is not explained by the Bayesian paradigm.*

A partial, although superficial, answer to this point would be that a similar criticism applies to the sampling distribution, which is always assumed to be known exactly. In many cases, and under most theories, modeling strongly influences the further developments of the analysis, but it cannot be formalized as precisely as these subsequent steps. The diversity of information sources, the various degrees of precision of this

---

[7] Notice, however, that there are non-consistent Bayes estimators, as discussed in Note 1.8.4.

information and the assessment of the consequences of the prior distribution selection keep modeling closer to an art than to science. Furthermore, we saw in Note 3.8.1 that some coherence axioms on the prior likelihood ordering justify the existence of a prior distribution, albeit on a cruder σ-algebra than expected.

From a practical point of view, the development of a prior distribution relies on the ability of individuals to represent their knowledge, and the limitations of this knowledge, in terms of probabilities. That individuals are not presently able to do so does not imply that they will never be able to assess probability distributions, nor that they should not be trained toward this goal. In fact, a proper training could get us closer to this aim, in the same sense that social evolution has allowed the majority of individuals to deal constantly with figures. (See also Smith (1988).)

Another point worth mentioning is that Bayesian analysis provides, in addition, some tools to deal with imprecisions on the prior distribution, through robustness and hierarchical analyses. The important aspect of the partial arbitrariness associated with the choice of a prior distribution is the influence of the prior information modeling on the resulting inference. In fact, different modelings can provide similar inferences. When discrepancies occur, it is necessary to assess more thoroughly the influence of the choice of the prior by a *sensitivity analysis*, in order to expose the influential factors of the prior modeling, instead of rejecting the available information. This very exposition of the influential factors is actually an additional advantage of a Bayesian analysis (see below).

(ii) *Subjectivity is nothing but a pretext for all kinds of abductions, including the choice of the most advantageous procedures.*

Again, a similar criticism could be addressed to frequentist methods over the choice of the loss or of the estimator to be studied; for instance, Brown (1980) shows that, for every dimension p_0, there exists a loss function such that the Stein effect occurs only when the dimension of the problem is larger than p_0 (see Note 2.8.2).

Nonetheless, the above remark is justified, since the recourse to an additional factor in the inferential model can always be diverted and misappropriated. Dirac masses as prior distributions are a straightforward illustration of such a danger, but there are also subtler devices to produce inference at will. This is unfortunately the price we have to pay for greater freedom and superior power of adaptivity, but an implicit requirement of the Bayesian paradigm is, however, that the choice of the prior distribution be justifiable (or *testable*), in the sense that the statistician must be able to account for the passage from prior information to prior distribution—even if the justification is partly based on computational simplicity, or personal feelings.

This possibility of verification is quite similar to the imperative of the *repeatability* of the experiments in other fields, and is not directly present in alternative statistical approaches, which reflect an inherent ambiguity about the choice of an estimation procedure, for instance, between the maximum likelihood estimator and the least-squares estimator. It can actually be argued to the contrary, namely, that the Bayesian approach is essentially *more objective* than other inferential methods because, firstly, it separates the different subjective inputs of the inferential process (sample distribution, prior, loss function), thus leaving ground for possible modifications, and, secondly, it develops objective tools to assess the influence of the prior distribution (noninformative distributions, sensitivity analysis, etc.). In this regard, Poincaré (1902) brings an additional argument following the quotation provided in point (5):

> We have, for one thing, to use a language and our language is entirely made of preconceived ideas and has to be so. However, these are unconscious preconceived ideas, which are a million times more dangerous than the other ones. Were we to assert that if we are including other preconceived ideas, consciously stated, we would aggravate the evil! I do not believe so: I rather maintain that they would balance one another.

The *Stopping Rule Principle* illustrates this objectivity, since the Bayesian decision is independent of the stopping criterion, therefore is not influenced by the subjective motivations that led to the resulting sample size. Again, in a frequentist framework, the choice of the statistical model and the loss function are equally determining factors that are usually overlooked, or "swept under the carpet" (Good (1973)).

(iii) *In a completely noninformative setting, the choice of the so-called non-informative prior has no justification whatsoever, and only stands as a pretext for an extension of the Bayesian scope.*

Although we see no harm in extending the Bayesian scope, points (1), (2), (3), (4), and (6) partly address this issue. In fact, in a noninformative setting, while the prior distribution cannot correspond to a modeling of the prior information, it can still be perceived as an efficient inferential tool.[8] In this sense, noninformative Bayesian methods are no more ad hoc than the maximum likelihood method, since they all stem from the distribution of the observations representing the only available information. If a loss function is also provided by the decision-maker, it constitutes an additional piece of information and the Bayesian approach can make good use of it; this is not so with the maximum likelihood method. Furthermore, since these methods often provide the usual estimators, they cannot be rejected solely on the grounds that they are Bayesian. A Bayesian rebuttal would be that the acceptable properties of these estimators owes to their Bayesian justification (see Jaynes (1980)). Lastly, we insist in most of the above points on the necessity

---

[8] The above criticism proceeds from an argument that rejects the use of the prior information—except when it is unavailable!

of *conditioning on x*. This conditioning necessarily *implies* a probabilistic modeling of θ through a prior distribution $\pi(\theta)$, since the maximum likelihood approach cannot provide a complete statistical inference and does not usually function as an "objective" distribution on θ.

Jeffreys's approach thus appears as a *technique* that takes advantage of the information of the model (that is, of the information brought by x about θ), while retaining the richness of the Bayesian approach and the compatibility with intuitive requirements, such as invariance, and including most of the usual statistical procedures. The necessity of this approach is made quite clear in testing theory, where the Neyman–Pearson approach has been seen to be suboptimal from several perspectives.

Although there are some *technical* difficulties in the treatment of the *nuisance parameters* (as in the marginalization paradoxes in Section 3.5), the reference prior generalization of Bernardo (1979) brings a partial solution to this problem. The other difficulty mentioned in this book, namely, *mixture estimation* as presented in Chapter 6, is more delicate but fundamentally linked to an *identifiability* problem, the maximum likelihood estimator being also undetermined in this case.

A benefit of such criticisms is that they point out the necessity for Bayesians to build up rigorously the prior distribution, and to strengthen noninformative techniques, for instance, by taking advantage of the information contained in the loss function. They also push toward a faster development of "automatic" (or semiautomatic) techniques of determination of the prior distribution for a more widespread use of Bayesian methods in applied Statistics. Bayesian software is already available in this area (see Note 6.6.2 and Berger (2000)). In connection with the approximation techniques presented in Chapter 6, and developing robustness methods, such techniques should encourage the diffusion of Bayesian methods to a wider audience. The current explosion of applied Bayesian studies is a sure indicator that this diffusion is under way (see Berger (2000)).

Let us conclude this book on a moderating note. External observers may get perplexed, and even weary, by the continual bickering between Bayesians and frequentists. Recent developments in Decision Theory have reinforced the Bayesian foundations of the frequentist optimality notions (see point (6)), whereas the latest works in the Bayesian robustness area have been aiming to reduce prior misspecification errors by taking into account these frequentist criteria (such as minimaxity or *Bayes minimaxity* as in Kempthorne (1988)). Leading figures such as James O. Berger are actively working towards a reformed decision-theoretic framework that would result in procedures acceptable to both schools, as illustrated by Note 5.7.4.

In practice, it is also often necessary to call for frequentist approximations when the *complete* elicitation of a prior distribution gets too complicated, for instance, when the Fisher information is not available in closed form, or when the number of parameters is too large. The various developments of the empirical Bayes techniques provide a rather persuasive illustration[9] of the need for an interface between the Bayesian and frequentist approaches.

The Bayesian choice is thus based on the reconciliation of most classical procedures with a Bayes, or generalized Bayes, analysis, on the strong appeal of its completeness and global coherence, and also on its ability to push the inferential process further. It is not based on a categorical rejection of all classical procedures. This Bayesian choice really stems from the growing realization that the Bayesian approach is indeed more appropriate for inference, as well as being more attractive intellectually.

---

[9] However, let us warn the reader of the dangers of an empirical Bayes analysis where unavoidable resorts to ad-hoc manipulations mar its credibility. The approximation of a genuine Bayesian analysis provided by the empirical Bayes methods is only partial, yet gives the illusion of providing a true alternative prior distribution. The suboptimality of the resulting empirical Bayes estimators (see Chapter 10) emphasizes the fundamental differences between the two approaches.

Probability Distributions

We recall here the density and the first two moments of most of the distributions used in this book. An exhaustive review of probability distributions is provided by Johnson and Kotz (1969–1972), or the more recent Johnson et al. (1992, 1994, 1995). The densities are given with respect to the Lebesgue or the counting measure, depending on the context.

A.1 Normal distribution, $\mathcal{N}_p(\theta, \Sigma)$

$\theta \in \mathbb{R}^p$ and Σ is a $(p \times p)$ symmetric positive-definite matrix,

$$f(\mathbf{x}|\theta, \Sigma) = (\det \Sigma)^{-1/2} (2\pi)^{-p/2} e^{-(\mathbf{x}-\theta)^t \Sigma^{-1}(\mathbf{x}-\theta)/2}.$$

$\mathbb{E}_{\theta, \Sigma}[\mathbf{X}] = \theta$ and $\mathbb{E}_{\theta, \Sigma}[(\mathbf{X} - \theta)(\mathbf{X} - \theta)^t] = \Sigma$.

When Σ is not definite, the $\mathcal{N}_p(\theta, \Sigma)$ distribution has no density with respect to Lebesgue measure on $\mathbb{R}^p$. For $p = 1$, the *log-normal* distribution is defined as the distribution of e^X when $X \sim \mathcal{N}(\theta, \sigma^2)$.

A.2 Gamma distribution, $\mathcal{G}(\alpha, \beta)$

$\alpha, \beta > 0$,

$$f(x|\alpha, \beta) = \frac{\beta^\alpha}{\Gamma(\alpha)} x^{\alpha-1} e^{-\beta x} \mathbb{I}_{[0,+\infty)}(x).$$

$\mathbb{E}_{\alpha, \beta}[X] = \alpha/\beta$ and $\operatorname{var}_{\alpha, \beta}(X) = \alpha/\beta^2$.

Particular cases of the gamma distribution are the *Erlang distribution*, $\mathcal{G}(\alpha, 1)$, the *exponential distribution* $\mathcal{G}(1, \beta)$ (denoted by $\mathcal{E}xp(\beta)$), and the *chi-squared distribution*, $\mathcal{G}a(\nu/2, 1/2)$ (denoted by χ_ν^2). Notice also that the opposite convention is sometimes adopted for the parameter, namely that $\mathcal{G}(\alpha, \beta)$ may also be noted as $\mathcal{G}(\alpha, 1/\beta)$. See, e.g., Berger (1985).

A.3 Beta distribution, $\mathcal{B}e(\alpha, \beta)$

$\alpha, \beta > 0$,

$$f(x|\alpha, \beta) = \frac{x^{\alpha-1}(1 - x)^{\beta-1}}{B(\alpha, \beta)} \mathbb{I}_{[0,1]}(x),$$

where

$$B(\alpha, \beta) = \frac{\Gamma(\alpha)\Gamma(\beta)}{\Gamma(\alpha + \beta)}.$$

$\mathbb{E}_{\alpha,\beta}[X] = \alpha/(\alpha + \beta)$ and $\text{var}_{\alpha,\beta}(X) = \alpha\beta/[(\alpha + \beta)^2(\alpha + \beta + 1)]$.

The beta distribution can be obtained as the distribution of $Y_1/(Y_1 + Y_2)$ when $Y_1 \sim \mathcal{G}(\alpha, 1)$ and $Y_2 \sim \mathcal{G}(\beta, 1)$.

A.4 Student's t-distribution, $\mathcal{T}_p(\nu, \theta, \Sigma)$

$\nu > 0$, $\theta \in \mathbb{R}^p$, and Σ is a $(p \times p)$ symmetric positive-definite matrix,

$$f(\mathbf{x}|\nu, \theta, \Sigma) = \frac{\Gamma((\nu + p)/2)/\Gamma(\nu/2)}{(\det \Sigma)^{1/2}(\nu\pi)^{p/2}} \left[1 + \frac{(\mathbf{x} - \theta)^t\Sigma^{-1}(\mathbf{x} - \theta)}{\nu}\right]^{-(\nu+p)/2}.$$

$\mathbb{E}_{\nu,\theta,\Sigma}[\mathbf{X}] = \theta$ $(\nu > 1)$ and $\mathbb{E}_{\theta,\Sigma}[(\mathbf{X} - \theta)(\mathbf{X} - \theta)^t] = \nu\Sigma/(\nu - 2)$ $(\nu > 2)$.

When $p = 1$, a particular case of Student's t-distribution is the *Cauchy distribution*, $\mathcal{C}(\theta, \sigma^2)$, which corresponds to $\nu = 1$. Student's t-distribution $\mathcal{T}_p(\nu, 0, I_p)$ can be derived as the distribution of $\mathbf{X}/Z$ when $\mathbf{X} \sim \mathcal{N}_p(0, I_p)$ and $\nu Z^2 \sim \chi_\nu^2$.

A.5 Fisher's F-distribution, $\mathcal{F}(\nu, \varrho)$

$\nu, \varrho > 0$,

$$f(x|\nu, \varrho) = \frac{\Gamma((\nu + \varrho)/2)\nu^{\varrho/2}\varrho^{\nu/2}}{\Gamma(\nu/2)\Gamma(\varrho/2)} \frac{x^{(\nu-2)/2}}{(\nu + \varrho x)^{(\nu+\varrho)/2}} \mathbb{I}_{[0,+\infty)}(x).$$

$\mathbb{E}_{\nu,\varrho}[X] = \varrho/(\varrho-2)$ $(\varrho > 2)$ and $\text{var}_{\nu,\varrho}(X) = 2\varrho^2(\nu+\varrho-2)/[\nu(\varrho-4)(\varrho-2)^2]$ $(\varrho > 4)$.

The distribution $\mathcal{F}(p, q)$ is also the distribution of $(\mathbf{X} - \theta)^t\Sigma^{-1}(\mathbf{X} - \theta)/p$ when $\mathbf{X} \sim \mathcal{T}_p(q, \theta, \Sigma)$. Moreover, if $X \sim \mathcal{F}(\nu, \varrho)$, $\varrho X/(\nu + \varrho X) \sim \mathcal{B}e(\nu, \varrho)$.

A.6 Inverse gamma distribution, $\mathcal{IG}(\alpha, \beta)$

$\alpha, \beta > 0$,

$$f(x|\alpha, \beta) = \frac{\beta^\alpha}{\Gamma(\alpha)} \frac{e^{-\beta/x}}{x^{\alpha+1}} \mathbb{I}_{[0,+\infty[}(x).$$

$\mathbb{E}_{\alpha,\beta}[X] = \beta/(\alpha-1)$ $(\alpha > 1)$ and $\text{var}_{\alpha,\beta}(X) = \beta^2/((\alpha-1)^2(\alpha-2))$ $(\alpha > 2)$.

This distribution is the distribution of X^{-1} when $X \sim \mathcal{G}(\alpha, \beta)$.

A.7 Noncentral chi-squared distribution, $\chi_\nu^2(\lambda)$

$\lambda \geq 0$,

$$f(x|\lambda) = \frac{1}{2}(x/\lambda)^{(p-2)/4} I_{(p-2)/2}(\sqrt{\lambda x}) e^{-(\lambda+x)/2}.$$

$\mathbb{E}_\lambda[X] = p + \lambda$ and $\text{var}_\lambda(X) = 3p + 4\lambda$.

This distribution can be derived as the distribution of $X_1^2 + \cdots + X_p^2$ when $X_i \sim \mathcal{N}(\theta_i, 1)$ and $\theta_1^2 + \ldots + \theta_p^2 = \lambda$.

A.8 Dirichlet distribution, $\mathcal{D}_k(\alpha_1, \ldots, \alpha_k)$

$\alpha_1, \ldots, \alpha_k > 0$ and $\alpha_0 = \alpha_1 + \cdots + \alpha_k$,

$$f(x|\alpha_1, \ldots, \alpha_k) = \frac{\Gamma(\alpha_0)}{\Gamma(\alpha_1) \ldots \Gamma(\alpha_k)} x_1^{\alpha_1 - 1} \ldots x_k^{\alpha_k - 1} \mathbb{I}_{\{\sum x_i = 1\}}.$$

$\mathbb{E}_\alpha[X_i] = \alpha_i/\alpha_0$, $\mathrm{var}(X_i) = (\alpha_0 - \alpha_i)\alpha_i/[\alpha_0^2(\alpha_0 + 1)]$ and $\mathrm{cov}(X_i, X_j) = -\alpha_i\alpha_j/[\alpha_0^2(\alpha_0 + 1)]$ $(i \neq j)$.

As a particular case, notice that $(X, 1 - X) \sim \mathcal{D}_2(\alpha_1, \alpha_2)$ is equivalent to $X \sim \mathcal{B}e(\alpha_1, \alpha_2)$.

A.9 Pareto distribution, $\mathcal{P}a(\alpha, x_0)$

$\alpha > 0$ and $x_0 > 0$,

$$f(x|\alpha, x_0) = \alpha \frac{x_0^\alpha}{x^{\alpha+1}} \mathbb{I}_{[x_0, +\infty[}(x).$$

$\mathbb{E}_{\alpha, x_0}[X] = \alpha x_0/(\alpha - 1)$ $(\alpha > 1)$ and $\mathrm{var}_{\alpha, x_0}(X) = \alpha x_0^2/[(\alpha - 1)^2(\alpha - 2)]$ $(\alpha > 2)$.

A.10 Binomial distribution, $\mathcal{B}(n, p)$.

$0 \leq p \leq 1$,

$$f(x|p) = \binom{n}{x} p^x (1 - p)^{n-x} \mathbb{I}_{\{0, \ldots, n\}}(x).$$

$\mathbb{E}_p(X) = np$ and $\mathrm{var}(X) = np(1 - p)$.

A.11 Multinomial distribution, $\mathcal{M}_k(n; p_1, \ldots, p_k)$

$p_i \geq 0$ $(1 \leq i \leq k)$ and $\sum_i p_i = 1$,

$$f(x_1, \ldots, x_k|p_1, \ldots, p_k) = \binom{n}{x_1 \ \ldots \ x_k} \prod_{i=1}^k p_i^{x_i} \mathbb{I}_{\sum x_i = n}.$$

$\mathbb{E}_p(X_i) = np_i$, $\mathrm{var}(X_i) = np_i(1 - p_i)$, and $\mathrm{cov}(X_i, X_j) = -np_i p_j$ $(i \neq j)$.

Notice that, if $X \sim \mathcal{M}_k(n; p_1, \ldots, p_k)$, $X_i \sim \mathcal{B}(n, p_i)$, and that the binomial distribution $X \sim \mathcal{B}(n, p)$ corresponds to $(X, n - X) \sim \mathcal{M}_2(n; p, 1 - p)$.

A.12 Poisson distribution, $\mathcal{P}(\lambda)$

$\lambda > 0$,

$$f(x|\lambda) = e^{-\lambda} \frac{\lambda^x}{x!} \mathbb{I}_{\mathbb{N}}(x).$$

$\mathbb{E}_\lambda[X] = \lambda$ and $\mathrm{var}_\lambda(X) = \lambda$.

A.13 Negative Binomial distribution, $\mathcal{N}eg(n,p)$

$0 \le p \le 1$,

$$f(x|p) = \binom{n+x+1}{x} p^n (1-p)^x \mathbb{I}_{\mathbb{N}}(x).$$

$\mathbb{E}_p[X] = n(1-p)/p$ and $\mathrm{var}_p(X) = n(1-p)/p^2$.

A.14 Hypergeometric distribution, $\mathcal{H}yp(N;n;p)$

$0 \le p \le 1$, $n < N$ and $pN \in \mathbb{N}$,

$$f(x|p) = \frac{\binom{pn}{x}\binom{(1-p)N}{n-x}}{\binom{N}{n}} \mathbb{I}_{\{n-(1-p)N,\ldots,pN\}}(x) \mathbb{I}_{\{0,1,\ldots,n\}}(x).$$

$\mathbb{E}_{N,n,p}[X] = np$ and $\mathrm{var}_{N,n,p}(X) = (N-n)np(1-p)/(N-1)$.

APPENDIX B

Usual Pseudo-random Generators

This appendix provides some pseudo-random generators for the usual probability distribution. They can be of use in the implementation of the Monte Carlo techniques described in Chapter 6. Additional details about their performances, their limitations, and their justification are given in Devroye (1985), Fishman (1996), Gentle (1998), and Robert and Casella (1999. 2004, Chapter 2). Notice that these algorithms should not be used blindly: for extreme values of the parameters or extreme simulation needs, their efficiency decreases rapidly. In fact, when pseudo-random generators are already available on the machine (e.g., in `Gauss`, `R`, or `Mathematica`), those are reliable enough to be used directly. The algorithms below all depend on the generation of *uniform random variables* on $[0, 1]$ (see Note 6.6.1).

B.1 Normal distribution, $\mathcal{N}(0,1)$

The Box–Muller method (1958) provides two independent normal observations out of two uniform random variables.

---

1. Generate U_1, U_2.

2. Take

$$
\begin{aligned}
x_1 &= \sqrt{-2 \log U_1} \cos(2\pi U_2), \\
x_2 &= \sqrt{-2 \log U_1} \sin(2\pi U_2).
\end{aligned}
$$

B.2 Exponential distribution, $\mathcal{E}xp(\lambda)$

Given that the c.d.f. of the exponential distribution is $1 - e^{-\lambda x}$ on $\mathbb{R}_+$, it can be inverted as follows.

---

1. Generate U.

2. Take $x = -\log(U)/\lambda$.

This generator can also be used for the *geometric* distribution $\mathcal{G}eo(p)$ because, if $x \sim \mathcal{G}eo(p)$, $P(x = r) = P(r \leq E < r + 1)$, with $E \sim \mathcal{E}xp(-\log(1 - p))$.

B.3 Student's t-distribution, $\mathcal{T}(\nu, 0, 1)$

Kinderman et al. (1977) provide an alternative to the generation of a normal random variable and a chi-square random variable.

---

1. Generate U_1, U_2.
2. If $U_1 < 0.5$, $x = 1/(4U_1 - 1)$ and $v = x^{-2}U_2$; otherwise, $x = 4U_1 - 3$ and $v = U_2$.
3. If $v < 1 - (|x|/2)$ or $v < (1 + (x^2/\nu))^{-(\nu+1)/2}$, take x; otherwise, repeat.

B.4 Gamma distribution, $\mathcal{G}(\alpha, 1)$

The simulation methods differ according to the value of α (notice that the scale factor β can be assumed to be 1). When $\alpha > 1$, the Cheng and Feast algorithm (1979) is:

---

0. Define $c_1 = \alpha - 1$, $c_2 = (\alpha - (1/6\alpha))/c_1$, $c_3 = 2/c_1$, $c_4 = 1 + c_3$ and $c_5 = 1/\sqrt{\alpha}$.
1. Repeat
     generate U_1, U_2 and take $U_1 = U_2 + c_5(1 - 1.86U_1)$ if $\alpha > 2.5$
   until $0 < U_1 < 1$.
2. $W = c_2 U_2/U_1$.
3. If $c_3 U_1 + W + W^{-1} \le c_4$ or $c_3 \log U_1 - \log W + W \le 1$, take $c_1 W$; otherwise, repeat.

If α is very large ($\alpha > 50$), it is better to use a normal approximation based on the Central Limit Theorem.

When $\alpha < 1$, a possible algorithm is:

---

1. Generate U and $y \sim \mathcal{G}(\alpha + 1, 1)$.
2. Take $y U^{1/\alpha}$.

Ahrens and Dieter (1974) propose the following alternative.

---

1. Generate U_0, U_1.
2. If $U_0 > e/(e + \alpha)$, $x = -\log\{(\alpha + e)(1 - U_0)/\alpha e\}$ and $y = x^{\alpha - 1}$; otherwise, $x = \{(\alpha + e)U_0/e\}^{1/\alpha}$ and $y = e^{-x}$.
3. If $U_1 < y$, take x; otherwise, repeat.

The beta, Fisher, and chi-squared distributions can also be simulated using these algorithms, since they can be derived from the gamma distribution by elementary transformations (see Appendix A). Ahrens and Dieter (1974) and Schmeiser and Shalaby (1980) provide alternative algorithms.

B.5 Binomial distribution, $\mathcal{B}(n, p)$

When n is reasonably small ($n \leq 30$), an elementary algorithm is to generate n uniform random variables and to count those less than p. For large n's, Knuth (1981) provides an alternative algorithm.

---

0. Define $k = n$, $\theta = p$ and $x = 0$.

1. Repeat
   $i = [1 + k\theta]$
   $v \sim \mathcal{B}e(i, k + 1 - i)$
   if $\theta > v$, $\theta = \theta/v$ and $k = i - 1$;
        otherwise, $x = x + i$, $\theta = (\theta - v)/(1 - v)$ and $k = k - i$
   until $k \leq K$.

2. For $i = 1, 2, \ldots, k$,
   generate U_i
   if $U_i < p$, $x = x + 1$.

3. Take x.

The constant K can be chosen as a function of n in order to increase the efficiency of the algorithm.

B.6 Poisson distribution, $\mathcal{P}(\lambda)$

Again, if λ is reasonably small ($\lambda < 30$), a simple algorithm is to generate uniform variables, in relation to the Poisson process.

---

0. Take $p = 1$, $N = 0$, $c = e^{-\lambda}$.

1. Repeat
   generate U_i
   $p = pU_i$, $N = N + 1$
   until $p < c$.

3. Take $x = N - 1$.

For large λ's, Atkinson (1979) proposes a more efficient alternative.

---

0. Define $c = 0.767 - (3.36/\lambda)$, $\beta = \pi(3\lambda)^{-1/2}$, $\alpha = \beta\lambda$, $k = \log c - \lambda - \log \beta$.

1. Repeat
   generate U_1
   $x = [\alpha - \log((1 - U_1)/U_1)]/\beta$
   until $x > -1/2$.

2. Generate U_2.

3. Take $N = [x + 0.5]$.

4. If $\alpha - \beta x + \log\{U_2/[1 + \exp(\alpha - \beta x)^2]\} \leq k + N\log\lambda - \log N!$, take N; otherwise, repeat.

For large n's, the *negative binomial* distribution $\mathcal{N}eg(n,p)$ can also be generated from this algorithm since, if $y \sim \mathcal{G}(n, (1-p)/p)$ and $x|y \sim \mathcal{P}(y)$, then $x \sim \mathcal{N}eg(n,p)$ (see Devroye (1985)).

Notations

C.1 Mathematical

$A \prec B$	$(B - A)$ is a positive definite matrix
$\|A\|$	determinant of the matrix A
a^+	$\max(a, 0)$
C_n^p, $\binom{n}{p}$	binomial coefficient
$\Delta f(z)$	Laplacian of $f(z)$, $\sum(\partial^2/\partial z_i^2) f(z)$
$\Delta(g)$	multiplier acting on a group
$f(t) \propto g(t)$	the functions f and g are proportional
$_1F_1(a; b; z)$	confluent hypergeometric function
F^-	generalized inverse of F
$\Gamma(x)$	gamma function $(x > 0)$
$\mathbf{h} = (h_1, \ldots, h_n) = \{h_i\}$	boldface signifies a vector
$H = \{h_{ij}\}$	uppercase signifies matrices
$I, \mathbf{1}, J = \mathbf{11}'$	Identity matrix, vector of ones, and matrix of ones
$\mathbb{I}_A(t)$	indicator function (1 if $t \in A$, 0 otherwise)
$I_\nu(z)$	modified Bessel function $(z > 0)$
$\lambda_{\max}(A)$	largest eigenvalue of the matrix A
$\binom{n}{p_1 \ldots p_n}$	multinomial coefficient
$\nabla f(z)$	gradient of $f(z)$, the vector with coefficients $(\partial/\partial z_i) f(z)$ $(f(z) \in \mathbb{R}$ and $z \in \mathbb{R}^p)$
$\nabla^t f(z)$	divergence of $f(z)$, $\sum(\partial/\partial z_i) f(z)$ $(f(z) \in \mathbb{R}^p$ and $z \in \mathbb{R})$
$\|\cdot\|_{TV}$	total variation norm
$O(n), o(n)$ or $O_p(n), o_p(n)$	big "Oh", little "oh." As $n \to \infty$, $\frac{O(n)}{n} \to$ constant, $\frac{o(n)}{n} \to 0$, and the subscript p denotes *in probability*
$\Psi(x)$	digamma function, $(d/dx)\Gamma(x)$ $(x > 0)$
$\mathrm{supp}(f)$	support of f
$\mathrm{tr}(A)$	trace of the matrix A
$\|\mathbf{x}\| = (\sum_i x_i^2)^{1/2}$	Euclidean norm
$[x]$ or $\lfloor x \rfloor$	greatest integer less than x
$\lceil x \rceil$	smallest integer larger than x
$< x, y >$	scalar product of x and y in $\mathbb{R}^p$
$x \vee y$	maximum of x and y
$x \wedge y$	minimum of x and y

C.2 Probabilistic

β_n	β-mixing coefficient
$\delta_{\theta_0}(\theta)$	Dirac mass at θ_0
$E(\theta)$	energy function of a Gibbs distribution
$\mathcal{E}(\pi)$	entropy of the distribution π
$\mathbb{E}_\theta[g(X)]$	expectation of $g(x)$ under the distribution $X \sim f(x|\theta)$
$\mathbb{E}^V[h(V)]$	expectation of $h(v)$ under the distribution of V
$\mathbb{E}^\pi[h(\theta)|x]$	expectation of $h(\theta)$ under the distribution of θ conditional on x, $\pi(\theta|x)$
i.i.d.	independent and identically distributed
$F(x|\theta)$	cumulative distribution function of X, conditional on the parameter θ
$f(x|\theta)$	density of X, conditional on the parameter θ, with respect to Lebesgue or counting measure
$\lambda(dx)$	Lebesgue measure, also denoted by $d\lambda(x)$
ν_r, ν_ℓ	right and left Haar measures
P_θ	probability distribution, indexed by the parameter θ
$\Phi(t)$	cumulative distribution function of the normal distribution
$\varphi(t)$	density of the normal distribution $\mathcal{N}(0,1)$
X, Y	random variable (uppercase)
$X \sim f(x|\theta)$	X is distributed with density $f(x|\theta)$
$(\mathcal{X}, \mathcal{P}, \mathcal{B})$	Probability triple: sample space, probability distribution, and σ-algebra of sets

C.3 Distributional

$\mathcal{B}(n,p)$	binomial distribution
$\mathcal{B}e(\alpha, \beta)$	beta distribution
$\mathcal{C}(\theta, \sigma^2)$	Cauchy distribution
$\mathcal{D}_k(\alpha_1, \ldots, \alpha_k)$	Dirichlet distribution
$\mathcal{E}xp(\lambda)$	exponential distribution
$\mathcal{F}(p,q)$	Fisher's F-distribution
$\mathcal{G}a(\alpha, \beta)$	gamma distribution
$\mathcal{IG}(\alpha, \beta)$	inverse gamma distribution
χ_p^2	chi-squared distribution
$\chi_p^2(\lambda)$	noncentral chi-squared distribution with noncentrality parameter λ
$\mathcal{M}_k(n; p_1, .., p_k)$	multinomial distribution
$\mathcal{N}(\theta, \sigma^2)$	univariate normal distribution
$\mathcal{N}_p(\theta, \Sigma)$	multivariate normal distribution
$\mathcal{N}eg(n,p)$	negative binomial distribution
$\mathcal{P}(\lambda)$	Poisson distribution
$\mathcal{P}a(x_0, \alpha)$	Pareto distribution
$\mathcal{T}_p(\nu, \theta, \Sigma)$	multivariate Student's t-distribution
$\mathcal{U}_{[a,b]}$	continuous uniform distribution

| $\mathcal{W}e(\alpha, c)$ | Weibull distribution |
| $\mathcal{W}_k(p, \Sigma)$ | Wishart distribution |

C.4 Decisional

$\mathcal{D}$	decision space
$\mathcal{G}$	group acting on $\mathcal{X}$
$\bar{g}$	element of $\bar{\mathcal{G}}$ associated to $g \in \mathcal{G}$
$\bar{\mathcal{G}}$	group induced by $\mathcal{G}$ acting on Θ
$\tilde{g}$	element of $\tilde{\mathcal{G}}$ associated to $g \in \mathcal{G}$
$\tilde{\mathcal{G}}$	group induced by $\mathcal{G}$ acting on $\mathcal{D}$
$L(\theta, \delta)$	loss function of δ in θ
$\mathcal{M}_0$	model under consideration
$R(\theta, \delta)$	frequentist risk of δ in θ
$r(\pi, \delta)$	the Bayes risk of δ for the prior distribution π
$\varrho(\pi, \delta|x)$	posterior risk of δ for the prior distribution π
Θ	parameter space
$\mathcal{X}$	observation space

C.5 Statistical

$B_{12}^{\pi}(x)$, B_{12}	Bayes factor
$B_{12}^{A}(x)$, B_{12}^{G}, B_{12}^{M}	pseudo-Bayes factor
$\underline{B}$	lower bound on Bayes factor
C_α	confidence region
$\delta^{JS}(x)$	James–Stein estimator
$\delta^{\pi}(x)$	Bayes estimator
$\delta^{+}(x)$	positive-part James–Stein estimator
$\delta^{\star}(x)$	randomized estimator
H_0	null hypothesis
H_1	alternative hypothesis
$I(\theta)$	Fisher information
$L(\theta, \delta)$	loss function, loss of estimating θ with δ
$L(\theta|x)$	likelihood function, a function of θ for fixed x, mathematically identical to $f(x|\theta)$
$\ell(\theta|x)$	the logarithm of the likelihood function
$L^P(\theta|x)$, $\ell^P(\theta|x)$	profile likelihood
$m(x)$	marginal density
$\overset{P}{\succ}$	Pitman closeness domination
$\pi(\theta)$	generic prior density for θ
$\pi^{J}(\theta)$	Jeffreys prior density for θ
$\pi(\theta|x)$	generic posterior density θ
s^2	sample variance
θ, λ	parameters (lowercase Greek letters)

Θ, Ω	parameter space (uppercase script Greek letters)
$\bar{x}$	sample mean
$\mathcal{X}, \mathcal{Y}$	sample space (uppercase script Roman letters)
X^*, Y^*, x^*, y^*	latent or missing variables (data)

C.6 Markov chains

$\mathrm{AR}(p)$	autoregressive process of order p
$\mathrm{ARMA}(p, q)$	autoregressive moving average process of order (p, q)
K_ϵ	kernel of the resolvant
$\mathrm{MA}(q)$	moving average process of order q
$P(x, A)$, $K(x, y)$	transition kernel
$P^m(x, A)$	transition kernel of the chain $(X_{mn})_n$
$P_\mu(\cdot)$	probability distribution of the chain (X_n) with initial state $X_0 \sim \mu$
$P_{x_0}(\cdot)$	probability distribution of the chain (X_n) with initial state $X_0 = x_0$
$X_t, X^{(t)}$	generic element of a Markov chain

References

Abramovich, F., Spatinas, T. and Silverman, B.W. (1998) Wavelet thresholding via a Bayesian approach. *J. Roy. Statist. Soc., Ser. B* **60**, 725–749.

Abramowitz, M. and Stegun, I. (1964) *Handbook of Mathematical Functions.* Dover, New York.

Adams, M. (1987) *William Ockham.* University of Notre Dame Press, Notre Dame, Indiana.

Ahrens, J. and Dieter, U. (1974) Computer methods for sampling from gamma, beta, Poisson and binomial distributions. *Computing* **12**, 223–246.

Aitkin, M. (1991) Posterior Bayes factors (with discussion). *J. Roy. Statist. Soc., Ser. B* **53**, 111–142.

Akaike, H. (1978) A new look at the Bayes procedure. *Biometrika* **65**, 53–59.

Akaike, H. (1983) Information measure and model selection. *Bull. Int. Statist. Inst.* **50**, 277–290.

Alam, K. (1973) A family of admissible minimax estimators of the mean of a multivariate normal distribution. *Ann. Statist.* **1**, 517–525.

Albert, J.H. (1981) Simultaneous estimation of Poisson means. *J. Multivariate Analysis* **11**, 400–417.

Albert, J.H. (1988) Computational methods using a Bayesian hierarchical generalized linear model. *J. Amer. Statist. Assoc.* **83**, 1037–1044.

Anderson, T.W. (1984) *An Introduction to Multivariate Statistical Analysis* (2nd edition). J. Wiley, New York.

Andrieu, C. and Doucet, A. (1999) Joint Bayesian Detection and Estimation of Noisy Sinusoids via Reversible Jump MCMC. *IEEE Trans. Signal Proc.* **47**(10), 2667–2676.

Andrieux, C., Doucet, A. and Fitzgerald, W.J. (2001) On Monte Carlo Methods for Bayesian Data Analysis. In *Nonlinear Dynamics and Statistics*, A. Mees and R.L. Smith (eds.). Birkhauser, Boston.

Angers, J.F. (1987) Development of robust Bayes estimators for a multivariate normal mean. Ph.D. thesis, Purdue University, West Lafayette, Indiana.

Angers, J.F. (1992) Use of the Student's *t*-prior for the estimation of normal means: A computational approach. In *Bayesian Statistics* 4, J.M. Bernardo, J.O. Berger, A.P. Dawid and A.F.M. Smith (eds.), 567–575. Oxford University Press, Oxford.

Angers, J.F. and MacGibbon, K.B. (1990) Hierarchical Bayes estimation in linear models with robustness against partial prior misspecification. Rapport n°69, Dépt. de Mathématiques et d'Informatique, Université de Sherbrooke.

Arrow, K.S. (1951) *Social Choice and Individual Values.* J. Wiley, New York.

Atkinson, A. (1979) The computer generation of Poisson random variables. *Appl. Statist.* **28**, 29–35.

Baranchick, A.J. (1970) A family of minimax estimators of the mean of a multivariate normal distribution. *Ann. Math. Statist.* **41**, 642–645.

Barbieri, M., Liseo, B. and Petrella, L. (1999) Bayes factor at work in a challenging class of problems. In *Model Choice*, W. Racugno (ed.), 109–132. Collana Atti di Congressi, Pitagora Editrice, Bologna.

Bar-Lev, S., Enis, P. and Letac, G. (1994) Models which admit a given exponential family as an a priori conjugate model. *Ann. Statist.* **22**(3), 1555–1586.

Barnett, G., Kohn, R. and Sheather, S. (1996) Bayesian estimation of an autoregressive model using Markov chain Monte Carlo. *J. Econometrics* **74**, 237–254.

Barron, A. (1988) The exponential convergence of posterior probabilities with implication for Bayes estimators of density functions. Tech. report 7, Dept. of Statistics, University of Illinois.

Barron, A. (1998) Information-theoretic characterization of Bayes performances and the choice of priors in parametric and nonparametric problems (with discussion). In *Bayesian Statistics 6*, J. Bernardo, J. Berger, A.P. Dawid and A.F.M. Smith (eds.), 27–52. Oxford University Press, Oxford.

Barron, A., Schervish, M.J. and Wasserman, L. (1999) The consistency of posterior distributions in nonparametric problems. *Ann. Statist.* **27**(2), 536–561.

Barnard, G.A. (1949) Statistical inference (with discussion). *J. Roy. Statist. Soc., Ser. B* **11**, 115–159.

Bartlett, M.S. (1937) Properties of sufficiency and statistical tests. *Proc. Roy. Soc. London* (Series A) **130**, 268–282.

Basu, D. (1988) *Statistical Information and Likelihood.* J.K. Ghosh (ed.), Springer-Verlag, New York.

Baum, L.E. and Petrie, T. (1966) Statistical inference for probabilistic functions of finite state Markov chains. *Ann. Math. Statist.* **37**, 1554–1563.

Bauwens, L. (1984) *Bayesian Full Information of Simultaneous Equations Models Using Integration by Monte Carlo.* Lecture Notes in Economics and Mathematical Systems **232**. Springer-Verlag, New York.

Bauwens, L. (1991) The "pathology" of the natural conjugate prior density in the regression model. *Ann. d'Eco. et Statist.* **23**, 49–64.

Bauwens, L., Lubrano, M. and Richard, J.F. (1999) Bayesian inference in dynamic econometric models. In *Advanced Texts in Econometrics*, C.W.J. Granger and G.E. Mizon (eds.). Oxford University Press, Oxford.

Bauwens, L. and Richard, J.F. (1985) A 1-1 Poly-*t* random variable generator with application to Monte Carlo integration. *J. Econometrics* **29**, 19–46.

Bayarri, M.J. and DeGroot, M.H. (1988) Gaining weight: a Bayesian approach. In *Bayesian Statistics* **3**, J.M. Bernardo, M.H. DeGroot, D. Lindley and A.F.M. Smith (eds.), 25–44. Oxford University Press, Oxford.

Bayes, T. (1763) An essay towards solving a problem in the doctrine of chances. *Phil. Trans. Roy. Soc.* **53**, 370–418.

Bechofer, R.E. (1954) A single-sample multiple decision procedure for ranking means of normal populations with known variance. *Ann. Math. Statist.* **25**, 16–39.

Bensmail, H., Celeux, G., Raftery, A.E. and Robert, C.P. (1997) Inference in model-based cluster analysis. *Statist. Comput.* **7**(1), 1–10.

Beran, R. (1996) Stein estimation in high dimension: A retrospective. In *Research Developments in Probability and Statistics: Madan L. Puri Festschrift*, 91–110. Universiteit Utrecht.

Bergé, P., Pommeau, Y. and Vidal, C. (1984) *Ordre Within Chaos*. J. Wiley, New York.

Berger, J.O. (1975) Minimax estimation of location vectors for a wide class of densities. *Ann. Statist.* **3**, 1318–1328.

Berger, J.O. (1976) Admissibility results for generalized Bayes estimators of a location vector. *Ann. Statist.* **4**, 334–356.

Berger, J.O. (1980a) A robust generalized Bayes estimators and confidence region for a multivariate normal mean. *Ann. Statist.* **8**, 716–761.

Berger, J.O. (1980b) Improving on inadmissible estimators in continuous exponential families with applications to simultaneous estimation of gamma scale parameters. *Ann. Statist.* **8**, 545–571.

Berger, J.O. (1982a) Selecting a minimax estimator of a multivariate normal mean. *Ann. Statist.* **10**, 81–92.

Berger, J.O. (1982b) Estimation in continuous exponential families: Bayesian estimation subject to risk restrictions and inadmissibility results. In *Statistical Decision Theory and Related Topics* **3**, S.S. Gupta and J.O. Berger (eds.), 109–142. Academic Press, New York.

Berger, J.O. (1984a) The robust Bayesian viewpoint (with discussion). In *Robustness of Bayesian Analysis*, J. Kadane (ed.). North-Holland, Amsterdam.

Berger, J.O. (1984b) The frequentist viewpoint and conditioning. In *Proceedings of the Berkeley Conference in Honor of Kiefer and Neyman*, L. Le Cam and R. Olshen (eds.). Wadsworth, Belmont, California.

Berger, J.O. (1985a) *Statistical Decision Theory and Bayesian Analysis* (2nd edition). Springer-Verlag, New York.

Berger, J.O. (1985b) Discussion of 'Quantifying prior opinion' by Diaconis and Ylvisaker. In *Bayesian Statistics* 2, J.M. Bernardo, M. DeGroot, D.V. Lindley and A.F.M. Smith (eds.). North-Holland, Amsterdam.

Berger, J.O. (1990a) Robust Bayesian analysis: sensitivity to the prior. *J. Statist. Plann. Inference* 25, 303–328.

Berger, J.O. (1990b) On the inadmissibility of unbiased estimators. *Statist. Prob. Letters* 5, 71–75.

Berger, J.O. (2000) Bayesian Analysis: A look at today and thoughts of tomorrow. *J. Amer. Statist. Assoc.* 95, 1269–1277.

Berger, J.O. and Berliner, L.M. (1986) Robust Bayes and empirical Bayes analysis with ε-contamined priors. *Ann. Statist.* 14, 461–486.

Berger, J.O. and Bernardo, J.M. (1989) Estimating a product of means: Bayesian analysis with reference priors. *J. Amer. Statist. Assoc.* 84, 200–207.

Berger, J.O. and Bernardo, J.M. (1990) Reference priors in a variance components problem. Tech. Report # 89–32C, Purdue University, West Lafayette, Indiana.

Berger, J.O. and Bernardo, J.M. (1992a) Ordered group reference priors with application to the multinomial problem. *Biometrika* 25, 25–37.

Berger, J.O. and Bernardo, J.M. (1992b) On the development of the reference prior method. In *Bayesian Statistics 4*. J.O. Berger, J.M. Bernardo, A.P. Dawid and A.F.M. Smith (Eds.). Oxford University Press, London, 35–60. Oxford University Press, Oxford.

Berger, J.O. and Bock, M.E. (1976) Eliminating singularities of Stein-type estimators of location vectors. *J. Roy. Statist. Soc., Ser. B* 39, 166–170.

Berger, J.O., Boukai, B. and Wang, Y. (1997) Unified frequentist and Bayesian testing of a precise hypothesis (with discussion). *Statistical Science* 12, 133–160.

Berger, J.0., Boukai, B. and Wang, Y. (1999) Simultaneous Bayesian–Frequentist Sequential Testing of Nested Hypotheses. *Biometrika* 86, 79–92.

Berger, J.O., Brown, L. and Wolpert, R. (1994) A Unified Conditional Frequentist and Bayesian Test for Fixed and Sequential Hypothesis Testing. *Ann. Statist.* 22, 1787-1807.

Berger, J.O. and Deely, J.J. (1988) A Bayesian approach to ranking and selection of related means with alternatives to Anova methodology. *J. Amer. Statist. Assoc.* 83, 364–373.

Berger, J.O. and Delampady, M. (1987) Testing precise hypotheses (with discussion). *Statist. Science* 2, 317–352.

Berger, J.O. and Mortera, J. (1991) Interpreting the stars in precise hypothesis testing. *Int. Statist. Rev.* 59, 337–353.

Berger, J.O. and Pericchi, L.R. (1996a) The Intrinsic Bayes Factor for Model Selection and Prediction. *J. Amer. Statist. Assoc.* **91**, 109–122.

Berger, J.O. and Pericchi, L.R. (1996b) The Intrinsic Bayes Factor for Linear Models. In *Bayesian Statistics 5*. J.O. Berger, J.M. Bernardo, A.P. Dawid and A.F.M. Smith (Eds)., 23–42. Oxford University Press, Oxford.

Berger, J.O. and Pericchi, L.R. (1998) Accurate and Stable Bayesian Model Selection: the Median Intrinsic Bayes Factor. *Sankhya B* **60**, 1–18.

Berger, J.O. and Pericchi, L.R. (2001) Objective Bayesian methods for model selection: introduction and comparison. *J. Statist. Plan. Inf.* (to appear).

Berger, J.O., Pericchi, L.R., and Varshavsky, J. (1998) Bayes Factors and Marginal Distributions in Invariant Situations. *Sankhya A* **60**, 307–321.

Berger, J.O., Philippe, A. and Robert, C.P. (1998) Estimation of quadratic functions: noninformative priors for non-centrality. *Statistica Sinica* **8**, 359–375.

Berger, J.O. and Robert, C.P. (1990) Subjective hierarchical Bayes estimation of a multivariate normal mean: on the frequentist interface. *Ann. Statist.* **18**, 617–651.

Berger, J.O. and Sellke, T. (1987) Testing a point-null hypothesis: the irreconcilability of significance levels and evidence (with discussion). *J. Amer. Statist. Assoc.* **82**, 112–122.

Berger, J.O. and Srinivasan, C. (1978) Generalized Bayes estimators in multivariate problems. *Ann. Statist.* **6**, 783–801.

Berger, J.O. and Wolpert, R. (1988) *The Likelihood Principle* (2nd edition). IMS lecture notes – Monograph Series, **9**. Hayward, California.

Berger, J.O. and Yang, R. (1994a) Noninformative priors and Bayesian testing for the AR(1) model. *Econometric Theory* **10**, 461–482.

Berger, J.O. and Yang, R. (1994b) Estimation of a covariance matrix using the reference prior. *Ann. Statist.* **22**, 1195–1211.

Bergman, N., Doucet, A. and Gordon, N.J. (2001) Optimal estimation and Cramér-Rao bounds for partial non-Gaussian state-space models. *Annals of the Institute of Statistical Mathematics* **53**(1), 97-112.

Berliner, L.M. (1991) Likelihood and Bayesian prediction of chaotic models. *J. Amer. Statist. Assoc.* **86**, 938–952.

Berliner, L.M. (1992) Statistics, probability and chaos. *Statist. Science* **7**, 69–122.

Bernardo, J.M. (1979) Reference posterior distributions for Bayesian inference (with discussion). *J. Roy. Statist. Soc., Ser. B* **41**, 113–147.

Bernardo, J.M. (1980) A Bayesian analysis of classical hypothesis testing. In *Bayesian Statistics*. J.M. Bernardo, M.H. deGroot, D.V. Lindley and A.F.M. Smith (Eds.). University Press, Valencia.

Bernardo, J.M. and Girón, F.J. (1986) A Bayesian approach to cluster analysis. In *Second Catalan International Symposium on Statistics*, Barcelona,

Spain.

Bernardo, J.M. and Girón, F.J. (1988) A Bayesian analysis of simple mixture problems. In *Bayesian Statistics* **3**, J.M. Bernardo, M.H. DeGroot, D.V. Lindley and A.F.M. Smith (eds.), 67–78. Oxford University Press, Oxford.

Bernardo, J.M. and Smith, A.F.M. (1994) *Bayesian Theory*. J. Wiley, New York.

Berry, D.A. and Stangl, D.K. (1996) *Bayesian Biostatistics*. Marcel Dekker, New York.

Bertrand, J. (1889) *Calcul des Probabilités*. Gauthier-Villars, Paris.

Besag, J. (1974) Spatial interaction and the statistical analysis of lattice systems (with discussion). *J. Roy. Statist. Soc., Ser. B* **36**, 192–326.

Besag, J. (1986) Statistical analysis of dirty pictures (with discussion). *J. Roy. Statist. Soc., Ser. B* **48**, 259–302.

Besag, J. (2000) Markov chain Monte Carlo for statistical inference. Tech. Report no. 9, University of Washington.

Besag, J. and Green, P.J. (1992) Spatial Statistics and Bayesian computation (with discussion). *J. Roy. Statist. Soc., Ser. B* **55**, 25–38.

Best, N.G., Cowles, M.K. and Vines, K. (1995) CODA: Convergence diagnosis and output analysis software for Gibbs sampling output, Version 0.30. Tech. Report, MRC Biostatistics Unit, University of Cambridge.

Bhattacharya, R.N. and Ghosh, J.K. (1978) Validity of formal Edgeworth expansion. *Ann. Statist.* **6**, 434–451.

Bhattacharya, R.N. and Rao, R. (1986) *Normal approximations and asymptotic expansions* (2nd edition). J. Wiley, New York.

Bickel, P.J. (1981) Minimax estimation of the mean of a normal distribution when the parameter space is restricted. *Ann. Math. Statist.* **9**, 1301–1309.

Bickel, P. and Ghosh, J.K. (1990) A decomposition for the likelihood ratio statistic and the Bartlett correction – a Bayesian argument. *Ann. Statist.* **18**, 1070–1090.

Billingsley, P. (1965) *Ergodic Theory and Information*. J. Wiley, New York.

Billingsley, P. (1986) *Probability and Measure* (2nd edition). J. Wiley, New York.

Billio, M., Monfort, A. and Robert, C.P. (1999) Bayesian estimation of switching ARMA models. *J. Econometrics* **93**, 229–255.

Bilodeau, M. (1988) On the simultaneous estimation of scale parameters. *Canad. J. Statist.* **14**, 169–174.

Binder, D. (1978) Bayesian cluster analysis. *Biometrika* **65**, 31–38.

Birnbaum, A. (1962) On the foundations of statistical inference (with discussion). *J. Amer. Statist. Assoc.* **57**, 269–326.

Bjørnstad, J. (1990) Predictive likelihood: a review. *Statist. Science* **5**, 242–265.

Blackwell, D. and Girshick, M.A. (1954) *Theory of Games and Statistical Decisions.* J. Wiley, New York.

Blattberg, R.C. and George, E.I. (1991) Shrinkage estimation of price and promotion elasticities: seemingly unrelated equations. *J. Amer. Statist. Assoc.* **86**, 304–315.

Blyth, C.R. (1951) On minimax statistical decisions procedures and their admissibility. *Ann. Math. Statist.* **22**, 22–42.

Blyth, C.R. (1972) Some probability paradoxes in choice from among random alternatives (with discussion). *J. Amer. Statist. Assoc.* **67**, 366–387.

Blyth, C.R. (1993) Discussion of Robert, Hwang and Strawderman (1993). *J. Amer. Statist. Assoc.* **88**, 72–74.

Blyth, C.R. and Hutchinson, D. (1961) Tables of Neyman-shortest confidence interval for the binomial parameter. *Biometrika* **47**, 381–391.

Blyth, C.R. and Pathak, P.K. (1985) Does an estimator distribution suffice? In *Proc. Berkeley Conf. in Honor of J. Neyman and J. Kiefer* **1**, L. Le Cam and A. Olshen (eds.). Wadsworth, Belmont, California.

Bock, M.E. (1985) Minimax estimators that shift towards a hypersphere for location of spherically symmetric distributions. *J. Multivariate Anal.* **9**, 579–588.

Bock, M.E. (1988) Shrinkage estimators: pseudo-Bayes rules for normal vectors. In *Statistical Decision Theory and Related Topics* **4**, S.S. Gupta and J.O. Berger (eds.), 281–297. Springer-Verlag, New York.

Bock, M.E. and Robert, C.P. (1991) Bayes estimators with respect to uniform distributions on spheres (I): the empirical Bayes approach. Unpublished report, Purdue University, West Lafayette, Indiana.

Böhning, D. (1999) *Computer–Assisted Analysis of Mixtures and Applications.* Chapman & Hall, London.

Bondar, J.V. (1987) How much improvement can a shrinkage estimator give? In *Foundations of Statistical Inference*, I. McNeill and G. Umphreys (eds.). Reidel, Dordrecht.

Bondar, J.V. and Milnes, P. (1981) Amenability: a survey for statistical applications of Hunt-Stein and related conditions on groups. *Z. Wahrsch. verw. Gebiete* **57**, 103–128.

Boole G. (1854) *A Investigation of the Laws of Thought.* Walton and Maberly, London.

Bose, S. (1991) Some properties of posterior Pitman closeness. *Comm. Statist.* (Ser. A), **20**, 3697–3412.

Bosq, D. and Lecoutre, J.P. (1988) *Théorie de l'Estimation Fonctionnelle.* Economica, Paris.

Box, G.E.P. and Jenkins, G.M. (1976) *Time Series Analysis: Forecasting and Control.* Holden-Bay, San Francisco.

Box, G.E.P. and Muller, M. (1958) A note on the generation of random normal variates. *Ann. Math. Statist.* **29**, 610–611.

Box, G.E.P. and Tiao, G.C. (1973) *Bayesian Inference in Statistical Analysis*. Addison-Wesley, Reading, Massachusetts.

Brandwein, A. and Strawderman, W.E. (1980) Minimax estimators of location parameters for spherically symmetric distributions with concave loss. *Ann. Statist.* **8**, 279–284.

Brandwein, A. and Strawderman, W.E. (1990) Stein estimation: the spherically symmetric case. *Statist. Science* **5**, 356–569.

Brandwein, A. and Strawderman, W.E. (1991) Generalizations of James-Stein estimators under spherical symmetry. *Ann. Statist.* **19**, 1639–1650.

Brandwein, A., Strawderman, W.E. and Ralescu, S. (1992) Stein estimation for non-normal spherically symmetric location families in three dimensions. *J. Multivariate Analysis* **42**, 35–50.

Brewster, J.F. and Zidek, J.V. (1974) Improving on equivariant estimators. *Ann. Statist.* **2**, 21–38.

Brockwell and Davis (1998) *Introduction to Time Series and Forecasting*. Springer–Verlag, New York.

Broniatowski, M., Celeux, G. and Diebolt, J. (1983) Reconnaissance de mélanges de densités par un algorithme d'apprentissage probabiliste. In *Data Analysis and Informatics* **3**, E. Diday (ed.). North-Holland, Amsterdam.

Brown, L.D. (1966) On the admissibility of invariant estimators of one or more location parameters. *Ann. Math. Statist.* **37**, 1087–1136.

Brown, L.D. (1967) The conditional level of Student's *t*-test. *Ann. Math. Statist.* **38**, 1068–1071.

Brown, L.D. (1971) Admissible estimators, recurrent diffusions, and insoluble boundary-value problems. *Ann. Math. Statist.* **42**, 855–903.

Brown, L.D. (1975) Estimation with incompletely specified loss functions. *J. Amer. Statist. Assoc.* **70**, 417–426.

Brown, L.D. (1976) *Notes on Statistical Decision Theory*. Unpublished lecture notes, Ithaca, New York.

Brown, L.D. (1978) A contribution to Kiefer's theory of conditional confidence procedures. *Ann. Statist.* **6**, 59–71.

Brown, L.D. (1980) Examples of Berger's phenomenon in the estimation of independent normal means. *Ann. Statist.* **9**, 1289–1300.

Brown, L.D. (1981) A complete class theorem for statistical problems with finite sample spaces. *Ann. Statist.* **9**, 1289–1300.

Brown, L.D. (1986) *Foundations of Exponential Families*. IMS lecture notes – Monograph Series **6**. Hayward, California.

Brown, L.D. (1988) The differential inequality of a statistical estimation problem. In *Statistical Decision Theory and Related Topics* **4**, S.S. Gupta and J.O. Berger (eds.). Springer–Verlag, New York.

Brown, L.D. (1990) An ancilarity paradox which appears in multiple linear regression (with discussion). *Ann. Statist.* **18**, 471–538.

Brown, L.D. (1993) Minimaxity, more or less. In *Statistical Decision Theory and Related Topics* **5**, S.S. Gupta and J.O. Berger (eds.), 1–18. Springer-Verlag, New York.

Brown, L.D. (2000) An essay on Statistical Decision Theory. *J. Amer. Statist. Assoc.* **95**, 1277–1282.

Brown, L.D. and Farrell, R.H. (1985) Complete class theorems for estimation of multivariate Poisson means and related problems. *Ann. Statist.* **8**, 377–398.

Brown, L.D. and Hwang, J.T. (1982) A unified admissibility proof. In *Statistical Decision Theory and Related Topics* **3**, S.S. Gupta and J.O. Berger (eds.), 205–230. Academic Press, New York.

Brown, L.D. and Hwang, J.T. (1989) Universal domination and stochastic domination: U-admissibility and U-inadmissibility of the least-squares estimator. *Ann. Statist.* **17**, 252–267.

Buehler, R.J. (1959) Some validity criteria for statistical inference. *Ann. Math. Statist.* **30**, 845–863.

Cappé, O. and Robert, C.P. (2000) MCMC: Ten years and still running! *J. Amer. Statist. Assoc.* **95**, 1282–1286.

Cappé, O., Robert, C.P. and Rydén, T. (2003) Reversible jump, birth-and-death, and more general continuous time MCMC samplers. *J. Roy. Statist. Soc., Ser. B* **65**(3), 679–700.

Carlin, B.P. and Chib, S. (1995), Bayesian Model Choice via Markov Chain Monte Carlo. *J. Roy. Statist. Soc., Ser. B* **57**, 473–484.

Carlin, B.P. and Gelfand, A. (1990) Approaches for empirical Bayes confidence intervals. *J. Amer. Statist. Assoc.* **85**, 105–114.

Carlin, B.P. and Louis, A. (1996) *Bayes and Empirical Bayes Methods for Data Analysis.* Chapman & Hall, London.

Carlin, B.P. and Louis, A. (2000a) *Bayes and Empirical Bayes Methods for Data Analysis* (2nd edition). Chapman & Hall, London.

Carlin, B.P. and Louis, A. (2000b) Empirical Bayes: Past, present and future. *J. Amer. Statist. Assoc.* **95**, 1286–1290.

Caron, N. (1994) Approches alternatives d'une théorie non-informative des tests bayésiens. Thèse d'université, Dépt. de Mathématique, Université de Rouen.

Carota, C., Parmigiani, G. and Polson, N.G. (1996) Diagnostic measures for model criticism. *J. Amer. Statist. Assoc.* **91**, 753–762.

Carter, G. and Rolph, J. (1974) Empirical Bayes methods applied to estimating fire alarm probabilities. *J. Amer. Statist. Assoc.* **69**, 882–885.

Casella, G. (1980) Minimax ridge regression estimation. *Ann. Statist.* **8**, 1036–1056.

Casella, G. (1985a) An introduction to empirical Bayes data analysis. *Amer. Statist.* **39**, 83–87.

Casella, G. (1985b) Condition number and minimax ridge regression estimation. *J. Amer. Statist. Assoc.* **80**, 753–758.

Casella, G. (1987) Conditionally acceptable recentered set estimators. *Ann. Statist.* **15**, 1364–1371.

Casella, G. (1990) Estimators with nondecreasing risks: application of a chi-squared identity. *Statist. Prob. Lett.* **10**, 107–109.

Casella, G. (1992) Conditional inference for confidence sets. *Current Issues in Statistical Inference: Essays in Honor of D. Basu*, M. Ghosh and P.K. Pathak (eds.), 1–12. IMS lectures notes – Monograph Series **17**. Hayward, California.

Casella, G. (1996) Statistical inference and Monte Carlo algorithms (with discussion). *Test* **5**, 249–344.

Casella, G. and Berger, R. (1987) Reconciling Bayesian and frequentist evidence in the one-sided testing problem. *J. Amer. Statist. Assoc.* **82**, 106–111.

Casella, G. and Berger, R. (1990) *Statistical Inference*. Wadsworth, Belmont, California.

Casella, G. and George, E.I. (1992) An introduction to Gibbs sampling. *Ann. Math. Statist.* **46**, 167-174.

Casella, G. and Hwang, J.T. (1983) Empirical Bayes confidence sets for the mean of a multivariate normal distribution. *J. Amer. Statist. Assoc.* **78**, 688–698.

Casella, G. and Hwang, J.T. (1987) Employing vague prior information in the construction of confidence sets. *J. Multivariate Anal.* **21**, 79–104.

Casella, G., Hwang, J.T. and Robert, C.P. (1993a) A paradox in decision-theoretic set estimation. *Statist. Sinica* **3**, 141–155.

Casella, G., Hwang, J.T.G. and Robert, C.P. (1993b) Loss function for set estimation. In *Statistical Decision Theory and Related Topics V*, J.O. Berger and S.S. Gupta (Eds.) Springer-Verlag, New York., 237–252. Springer-Verlag, New York.

Casella, G. and Robert, C.P. (1988) Non-optimality of randomized confidence sets. Tech. Report # 88-9, Dept. of Statistics, Purdue University, West Lafayette, Indiana.

Casella, G., Robert, C.P. and Wells, M.T. (2000) Mixture models, latent variables and partitioned importance sampling. Tech. report 00-15, CREST, INSEE, Paris.

Casella, G. and Strawderman, W.E. (1981) Estimating a bounded normal mean. *Ann. Statist.* **4**, 283–300.

Casella, G. and Wells, M. (1993) Discussion of Robert, Hwang and Strawderman (1993). *J. Amer. Statist. Assoc.* **88**, 70–71.

Castledine, B. (1981) A Bayesian analysis of multiple-recapture sampling for a closed population. *Biometrika* **67**, 197–210.

Castro, I., Conigliani, C. and O'Hagan, A. (1999) Bayesian assessment of goodness of fit against nonparametric alternatives (with discussion). In *Model Selection*, W. Racugno (ed.). Collana Atti di Congressi, Pitagora Editrice Bologna.

Celeux, G. and Diebolt, J. (1985) The SEM algorithm: a probabilistic teacher algorithm derived from the EM algorithm for the mixture problem. *Comput. Statist. Quater.* **2**, 73–82.

Celeux, G. and Diebolt, J. (1990) Une version de type recuit simulé de l'algorithme EM. *Notes aux Comptes Rendus de l'Académie des Sciences*, **310**, 119–124.

Celeux, G., Hurn, M. and Robert, C.P. (2000) Computational and inferential difficulties with mixture posterior distributions. *J. Amer. Statist. Assoc.* **95**(3), 957–979.

Cellier, D., Fourdrinier, D. and Robert, C.P. (1989) Robust shrinkage estimators of the location parameter for elliptically symmetric distributions. *J. Multivariate Anal.* **29**, 39–52.

Chamberlain, G. (2000) *Econometrics*. Springer–Verlag, New York.

Chen, M.H. and Shao, Q.M. (1997) On Monte Carlo methods for estimating ratios of normalizing constants. *Ann. Statist.* **25**, 1563–1594.

Chen, M.H., Shao, Q.M. and Ibrahim, J.G. (2000) *Monte Carlo Methods in Bayesian Computation*. Springer–Verlag, New York.

Cheng, R. and Feast, G. (1979) Some simple gamma variate generators. *Appl. Statist.* **28**, 290–295.

Chernoff, H. and Yahav, J.A. (1977) A subset selection employing a new criterion. In *Statistical Decision Theory and Related Topics* 2, S.S. Gupta and D. Moore (eds.). Academic Press, New York.

Chib, S. (1995) Marginal likelihood from the Gibbs output. *J. Amer. Statist. Assoc.* **90** 1313–1321.

Chib, S. and Greenberg (1994) Bayes Inference in Regression Models with ARMA (p,q) Errors. *J. Econometrics* **64**, 183–206.

Chickering, D.M. and Heckerman, D. (2000) A comparison of scientific and engineering criteria for Bayesian model selection. *Statist. Comput.* **10**, 55–62.

Chow, G.C. (1983) *Econometrics*. McGraw-Hill, New York.

Chow, M.S. (1987) A complete class theorem for estimating a non-centrality parameter. *Ann. Statist.* **15**, 869–876.

Chow, M.S. and Hwang, J.T. (1990) The comparison of estimators for the noncentrality of a chi-square distribution. Tech. Report, Dept. of Mathematics, Cornell University, Ithaca, New York.

Chow, Y.S. and Teicher, H. (1988) *Probability Theory*. Springer–Verlag, New York.

Chrystal, G. (1891) On some fundamental principles in the theory of probability. *Trans. Actuarial Soc. Edinburgh* **2**, 421–439.

Clarke, B. and Barron, A. (1990) Information-theoretic asymptotics of Bayes methods. *IEEE Trans. Inform. Theory* **36**, 453–471.

Clarke, B. and Barron, A. (1994) Jeffreys prior is asymptotically least favorable under entropy risk. *J. Statist. Plan. Inf.* **41**, 36–60.

Clarke, B. and Wasserman, L. (1995) Information trade-off. *TEST* **4**, 19–38.

Clevenson, M. and Zidek, J.V. (1975) Simultaneous estimation of the mean of independant Poisson laws. *J. Amer. Statist. Assoc.* **70**, 698–705.

Clyde, M. (1999), Bayesian model averaging and model search strategies. In *Bayesian Statistics* **6**, J.M. Bernardo, A.P. Dawid, J.O. Berger, and A.F.M. Smith (eds.), 157–185. Oxford University Press, Oxford.

Cohen, A. (1972) Improved confidence intervals for the variance of a normal distribution. *J. Amer. Statist. Assoc.* **67**, 382–387.

Cohen, A. and Sackrowitz, H. (1984) Decision Theoretic results for vector risks with applications. *Statist. Decisions* Supplement Issue **1**, 159–176.

Cohen, A. and Strawderman, W.E. (1973) Admissible confidence intervals and point estimators for translation or scale parameters. *Ann. Statist.* **1**, 545–550.

Consonni, G. and Veronese, P. (1987) Coherent distributions and Lindley's paradox. In *Probability and Bayesian Statistics*, R. Viertl (ed.), 111–120. Plenum, New York.

Cowell, R.G., Dawid, A.P., Lauritzen, S.L. and Spiegelhalter, D.J. (1999) *Probabilistic Networks and Expert Systems*. Springer–Verlag, New York.

Cox, D.R. (1958) Some problems connected with statistical inference. *Ann. Math. Statist.* **29**, 357–425.

Cox, D.R. (1990) Role of models in statistical analysis. *Statist. Science* **5**, 169–174.

Cox, D.R. and Hinkley, D. (1987) *Theoretical Statistics*. Chapman & Hall, London.

Cox, D.R. and Reid, N. (1987) Orthogonal parameters and approximate conditional inference (with discussion). *J. Roy. Statist. Soc., Ser. B* **49**, 1–18.

Crawford, S.L., DeGroot, M.H., Kadane, J.B. and Small, M.J. (1992) Modeling lake-chemistry distributions: Approximate Bayesian methods for estimating a finite-mixture model. *Technometrics* **34**, 441–453.

Cressie, N. (1993) *Spatial Statistics*. J. Wiley, New York.

Dacunha-Castelle, D. and Gassiat, E. (1999) Testing the order of a model using locally conic parametrization: population mixtures and stationary ARMA processes. *Ann. Statist.* **27**(4), 1178–1209.

Dalal, S.R. and Hall, W.J. (1983) Approximating priors by mixtures of natural conjugate priors. *J. Roy. Statist. Soc., Ser. B* **45**, 278–286.

Dale, A.I. (1991) *A History of Inverse Probability*. Springer–Verlag, New York.

Damien, P., Wakefield, J. and Walker, S. (1999) Gibbs sampling for Bayesian non-conjugate and hierarchical models by using auxiliary variables. *J. Roy. Statist. Soc., Ser. B* **61**(2), 331–344.

Darroch, J. (1958) The multiple-recapture census. I: Estimation of a closed population. *Biometrika* **45**, 343–359.

Das Gupta, A. (1984) Admissibility in the gamma distribution: two examples. *Sankhya* (Ser. A) **46**, 395–407.

Das Gupta, A. and Sinha, B.K. (1986) Estimation in the multiparameter exponential family: admissibility and inadmissibility results. *Statist. Decisions* **4**, 101–130.

Das Gupta, A. and Studden, W. (1988) Frequentist behavior of smallest volume robust Bayes confidence sets. Tech. Report, Dept. of Statistics, Purdue University, West Lafayette, Indiana.

Datta, G.S. and Ghosh, M. (1994) On the invariance of noninformative priors. Technical Report 94-20, Dept. of Statistics, University of Georgia.

Datta, G.S. and Ghosh, M. (1995a) On priors providing frequentist validity for Bayesian inference. *Biometrika* **82**, 37–45.

Datta, G.S. and Ghosh, M. (1995b) Some remarks on noninformative priors. *J. Amer. Statist. Assoc.* **90**, 1357–1363.

Dawid, A.P. (1984) *Probability Forecasts*. Research report, University College London.

Dawid, A.P. (1992) Prequential analysis, stochastic complexity and Bayesian inference. In *Bayesian Statistics* **4**, J.O. Berger, J.M. Bernardo, A.P. Dawid and A.F.M. Smith (eds.), 109–121. Oxford University Press, Oxford.

Dawid, A.P., DeGroot, M.H. and Mortera, J. (1993) Coherent combination of experts' opinions. In *Statistical Decision Theory and Related Topics V*, J.O. Berger and S.S. Gupta (Eds.) Springer-Verlag, New York.

Dawid, A.P. and Lauritzen, S.L. (1993) Hyper Markov laws in the statistical analysis of decomposable graphical models. *Ann. Statist.* **21**, 1272-1317.

Dawid, A.P., Stone, N. and Zidek, J.V. (1973) Marginalization paradoxes in Bayesian and structural inference (with discussion). *J. Roy. Statist. Soc., Ser. B* **35**, 189–233.

Deely, J.J. and Gupta, S.S. (1968) On the property of subset selection per order. *Sankhya* (Ser. A), **30**, 37–50.

Deely, J.J. and Lindley, D.V. (1981) Bayes empirical Bayes. *J. Amer. Statist. Assoc.* **76**, 833–841.

DeGroot, M.H. (1970) *Optimal Statistical Decisions*. McGraw-Hill, New York.

DeGroot, M.H. (1973) Doing what comes naturally: Interpreting a tail area as a posterior probability or as a likelihood ratio. *J. Amer. Statist. Assoc.* **68**, 966–969.

DeGroot, M.H. and Fienberg, S. (1983) The comparison and evaluation of forecasters. *The Statistician* **32**, 12–22.

Delampady, M. (1989a) Lower bounds on Bayes factors for interval null hypotheses. *J. Amer. Statist. Assoc.* **84**, 120–124.

Dellaportas, P. and Forster, J.J. (1996) Markov chain Monte Carlo model determination for hierarchical and graphical log-linear models. Technical Report, University of Southampton.

Delampady, M. (1989b) Lower bounds on Bayes factors for invariant testing situations. *J. Multivariate Anal.* **28**, 227–246.

Delampady, M. and Berger, J.O. (1990) Lower bounds on Bayes factors for multinomial and chi-squared tests of fit. *Ann. Statist.* **18**, 1295–1316.

Dempster, A.P. (1968) A generalization of Bayesian inference (with discussion). *J. Roy. Statist. Soc., Ser. B* **30**, 205–248.

Dempster, A.P., Laird, N.M. and Rubin, D.B. (1977) Maximum likelihood from incomplete data via the EM algorithm (with discussion). *J. Roy. Statist. Soc., Ser. B* **39**, 1–38.

DeRobertis, L. and Hartigan, J.A. (1981) Bayesian inference using intervals of measures. *Ann. Statist.* **9**, 235–244.

De Santis, F., and Spezzaferri, F. (1997) Alternative Bayes Factors for model selection. *Canad. J. Statist.*, **25**, 503–515.

Devroye, L. (1985) *Non-Uniform Random Variate Generation*. Springer–Verlag, New York.

Devroye, L. and Györfi, L. (1985) *Nonparametric Density Estimation: the L_1 View*. J. Wiley, New York.

Dey, D., Müller, P. and Sinha, D. (1998) *Practical Nonparametrics and Semiparametrics in Bayesian Statistical Inference*. Lecture notes in Statistics **133**, Springer–Verlag, New York.

Diaconis, P. (1988) Bayesian numerical analysis. In *Statistical Decision Theory and Related Topics* 4, S. Gupta and J.O. Berger (eds.), 163–176. Springer–Verlag, New York.

Diaconis, P. and Freedman, D.A. (1986) On the consistency of Bayes estimates. *Ann. Statist.* **14**, 1–26.

Diaconis, P. and Kemperman, J. (1996) Some new tools for Dirichlet priors (with discussion). In *Bayesian Statistics 5*. J.O. Berger, J.M. Bernardo, A.P. Dawid and A.F.M. Smith (Eds)., 97–106. Oxford University Press, Oxford.

Diaconis, P. and Mosteller, F. (1989) Methods for studying coincidences. *J. Amer. Statist. Assoc.* **84**, 853–861.

Diaconis, P. and Ylvisaker, D. (1979) Conjugate priors for exponential families. *Ann. Statist.* **7**, 269–281.

Diaconis, P. and Ylvisaker, D. (1985) Quantifying prior opinion. In *Bayesian Statistics 2*, J.M. Bernardo, M.H. DeGroot, D.V. Lindley, A. Smith (eds.), 163–175. North-Holland, Amsterdam.

Diaconis, P. and Zabell, S. (1991) Closed form summation for classical distributions: variations on a theme of De Moivre. *Statist. Science* **6**, 284–302.

DiCiccio, T.J. and Stern, S.E. (1993) On Bartlett adjustments for approximate Bayesian inference, *Biometrika* **80**, 731–740.

DiCiccio, T.J. and Stern, S.E. (1994) Frequentist and Bayesian Bartlett correction of test statistics based on adjusted profile likelihoods. *J. Roy. Statist. Soc., Ser. B* **56**, 397–408.

Dickey, J.M. (1968) Three multidimensional integral identities with Bayesian applications. *Ann. Statist.* **39**, 1615–1627.

Diebolt, J. and Robert, C.P. (1990a) Bayesian estimation of finite mixture distributions, Part I: Theoretical aspects. Rapport Tech. #110, LSTA, Université Paris VI.

Diebolt, J. and Robert, C.P. (1990b) Bayesian estimation of finite mixture distributions, Part II: Sampling implementation. Rapport Tech. #111, LSTA, Université Paris VI.

Diebolt, J. and Robert, C.P. (1994) Estimation of finite mixture distributions by Bayesian sampling. *J. Roy. Statist. Soc., Ser. B* **56**, 363–375.

Doucet, A., deFreitas, N. and Gordon, N. (2001) *Sequential Monte Carlo in Practice*. Springer–Verlag, New York.

Draper, D. (1995) Assessment and Propagation of Model Uncertainty. *J. Roy. Statist. Soc., Ser. B* **57**, 45-98.

Draper, D. (2000) *Bayesian Hierarchical Modelling*. University of Bath (in preparation).

Drèze, J.H. (1978) Bayesian regression analysis using poly-t densities. *J. of Econometrics* **6**, 329–354.

Drèze, J.H. and Morales, J.A. (1976) Bayesian full information analysis of the simultaneous equations. *J. Amer. Statist. Assoc.* **71**, 919–923.

Dudewicz, E.J. and Koo, J.O. (1982) *The Complete Categorized Guide to Statistical Selection and Ranking Procedures*. American Science Press, Columbus, Ohio.

Dumouchel, W.M. and Harris, J.E. (1983) Bayes methods for combining the results of cancer studies in human and other species (with discussion). *J. Amer. Statist. Assoc.* **78**, 293–315.

Dupuis, J.A. (1993) Bayesian estimation of movement probabilities in open populations using hidden Markov chains. Rapport Technique 9341, CREST, INSEE, Paris.

Dupuis, J.A. (1995) Bayesian estimation of movement probabilities in open populations using hidden Markov chains. *Biometrika* / **82**(4), 761–772.

Dupuis, J.A. and Robert, C.P. (2001) Bayesian variable selection in qualitative models by Kullback-Leibler projections. *J. Statist. Plan. Inf.* (to appear).

Durbin, J. and Watson, G.S. (1950) Testing for serial correlation in least-squares regression, *Biometrika* **37**, 409–428.

Dynkin, E.B. (1951) Necessary and sufficient statistics for a family of probability distributions. *Selected Transl. Math. Statist. Prob.* **1**, 23–41.

Eaton, M.L. (1982) *Multivariate Statistics*. J. Wiley, New York.

Eaton, M.L. (1986) A characterization of spherical distributions. *J. Multivariate Anal.* **20**, 272–276.

Eaton, M.L. (1989) *Group Invariance Applications in Statistics*. Regional Conference Series in Probability and Statistics, Vol. 1. Institute of Mathematical Statistics, Hayward, California.

Eaton, M.L. (1992) A statistical dyptich: admissible inferences–recurrence of symmetric Markov chains. *Ann. Statist.* **20**, 1147–1179.

Eaton, M.L. (1999) Markov chain conditions for admissibility in estimation problems with quadratic loss. Report PN1–R9904, Centrum voor Wiskunde en Informatica, Amsterdam.

Eberly, L. and Casella, G. (1999) Comparison of Bayesian credible intervals in hierarchical models. Division of Biostatistics Manuscript 99-004, University of Minnesota.

Efron, B. (1975) Biased versus unbiased estimation. *Adv. in Math.* **16**, 259–277.

Efron, B. (1982) *The Jacknife, the Bootstrap and Other Resampling Plans.* Regional Conference in Applied Mathematics **38**. SIAM, Philadelphia.

Efron, B. (1992) Regression percentile using asymmetric squared error loss. *Statist. Sinica* **1**, 93–125.

Efron, B. and Morris, C. (1973) Stein's estimation rule and its competitors—an empirical Bayes approach. *J. Amer. Statist. Assoc.* **68**, 117–130.

Efron, B. and Morris, C. (1975) Data analysis using Stein's estimator and its generalizations. *J. Amer. Statist. Assoc.* **70**, 311–319.

Efron, B. and Thisted, R.A. (1976) Estimating the number of species: How many words did Shakespeare know? *Biometrika* **63**, 435–447.

Eichenauer, J. and Lehn, J. (1989) Gamma-minimax estimators for a bounded normal mean under squared error–loss. *Statist. Decisions* **7**, 37–62.

Engle, R.F. (1982) Autoregressive conditional heteroscedasticity with estimates of the variance of United Kingdom inflation. *Econometrica* **50**, 987–1008.

Escobar, M.D. (1989) Estimating the means of several normal populations by estimating the distribution of the means. Unpublished Ph.D. thesis, Yale University, New Haven, Connecticut.

Escobar, M.D. and West, M. (1995) Bayesian prediction and density estimation. *J. Amer. Statist. Assoc.* **90**, 577–588.

Evans, M., Fraser, D.A.S., and Monette, G. (1986) On principles and arguments to likelihood (with discussion). *Canad. J. Statist.* **14**, 181-199.

Fabius, J. (1964) Asymptotic behavior of Bayes estimates. *Ann. Math. Statist.* **35**, 846–856.

Fang, K.T. and Anderson, T.W. (1990) *Statistical Inference in Elliptically Contoured and Related Distributions*. Allerton Press, New York.

Farrell, R.H. (1968a) Towards a theory of generalized Bayes tests. *Ann. Math. Statist.* **38**, 1–22.

Farrell, R.H. (1968b) On a necessary and sufficient condition for admissibility of estimators when strictly convex loss is used. *Ann. Math. Statist.* **38**, 23–28.

Farrell, R.H. (1985) *Multivariate Calculation*. Springer–Verlag, New York.

Feller, W. (1970) *An Introduction to Probability Theory and its Applications*, Vol. 1. J. Wiley, New York.

Feller, W. (1971) *An Introduction to Probability Theory and its Applications*, Vol. 2. J. Wiley, New York.

Ferguson, T.S. (1967) *Mathematical Statistics: a Decision-Theoretic Approach*. Academic Press, New York.

Ferguson, T.S. (1973) A Bayesian analysis of some nonparametric problems. *Ann. Statist.* **1**, 209–230.

Ferguson, T.S. (1974) Prior distributions in spaces of probability measures. *Ann. Statist.* **2**, 615–629.

Fernandez, C. and Steel, M. (1999) On the dangers of modelling through continuous distributions: a Bayesian perspective (with discussion). In *Bayesian Statistics 6*. J.O. Berger, J.M. Bernardo, A.P. Dawid and A.F.M. Smith (Eds)., 213–238. Oxford University Press, Oxford.

Feyerabend, P. (1975) *Against Method*. New Left Books, London.

Field, A. and Ronchetti, E. (1990) *Small Sample Asymptotics*. IMS lecture notes – Monograph Series. Hayward, California.

Fieller, E.C. (1954) Some problems in interval estimation. *J. Roy. Statist. Soc., Ser. B* **16**, 175–185.

de Finetti, B. (1972) *Probability, Induction and Statistics*. J. Wiley, New York.

de Finetti, B. (1974) *Theory of Probability*. Vol. 1. J. Wiley, New York.

de Finetti, B. (1975) *Theory of Probability*. Vol. 2. J. Wiley, New York.

Fishburn, P.C. (1988) *Non-Linear Preferences and Utility Theory*. Harvester Wheatsheaf, Brighton, Sussex.

Fisher, R.A. (1912) On an absolute criterion for fitting frequency curves. *Messenger of Mathematics* **41**, 155–160.

Fisher, R.A. (1922) On the mathematical foundations of theoretical statistics. *Philos. Trans. Roy. Soc. London Ser. A* **222**, 309–368.

Fisher, R.A. (1930) Inverse probability. *Proc. Cambridge Philos. Soc.* **26**, 528–535.

Fisher, R.A. (1956) *Statistical Methods and Scientific Inference*. Oliver and Boyd, Edinburgh.

Fisher, R.A. (1959) Mathematical probability in the natural sciences. *Technometrics* **1**, 21–29.

Fishman, G.S. (1996) *Monte Carlo*. Springer–Verlag, New York.

Fitzgerald, W.J., Godsill, S., Kokaram, A.C. and Stark, J.A. (1999) Bayesian methods in signal and image processing. In *Bayesian Statistics 6*. J.O. Berger, J.M. Bernardo, A.P. Dawid and A.F.M. Smith (Eds)., 239–254. Oxford University Press, Oxford

Florens, J.P., Mouchart, M. and Rolin, J.M. (1990) *Elements of Bayesian Statistics*. Marcel Dekker, New York.

Foster, D.P. and George, E.I. (1996) A simple ancillarity paradox. *Scand. J. Statist.* **23**, 233–242.

Fouley, J.L., San Cristobal, M., Gianola, D. and Im, S. (1992) Marginal likelihood and Bayesian approaches to the analysis of heterogeneous residual variances in mixed linear gaussian models. *Comput. Statist. Data Anal.* **13**, 291–305.

Fourdrinier, D., Strawderman, W.E. and Wells, M.T. (1998) On the construction of Bayes minimax estimators. *Ann. Statist.* **26**(2), 660–671.

Fourdrinier, D. and Wells, M. (1993) Risk comparison of variable selection rules. Doc. Travail, Université de Rouen.

Fourdrinier, D. and Wells, M. T. (1994) Règle de sélection de variables : une approche décisionnelle. *C.R. Acad. Sci. Paris* **319**, Série I, 865–870.

Fraisse, A.M., Raoult, J.P., Robert, C.P. and Roy, M. (1990) Une condition nécessaire d'admissibilité et ses conséquences sur les estimateurs à rétrécisseur de la moyenne d'une loi normale. *Canad. J. Statist.* **18**, 213–220.

Fraisse, A.M., Robert, C.P. and Roy, M. (1987) Estimateurs à rétrécisseur matriciel, pour un coût quadratique général. *Ann. d'Eco. Statist.* **8**, 161–175.

Fraisse, A.M., Roy, M. and Robert, C.P. (1998) Semi-tail upper bounds for admissible estimators in exponential families with nuisance parameters. *Statistics & Decisions* **16**(2) 147–162.

Francq, C. and Zakoïan, J.M. (2001) Stationarity of multivariate Markov–switching ARMA models. *J. Econometrics* **102**, 339–364.

Fraser, D.A.S., Monette, G. and Ng, K.W. (1984) Marginalization, likelihood and structural models. In *Multivariate Analysis 6*, P. Krishnaiah (ed.). North-Holland, Amsterdam.

Gassiat, E. and Dacunha-Castelle, D. (1997) Estimation of the number of components in a mixture. *Bernoulli* **3**(3), 279–299.

Gatsonis, C., Hodges, J.S., Kass, R.E. and Singpurwalla, N. (1993) *Case Studies in Bayesian Statistics*. Lecture Notes in Statistics **83**, Springer–Verlag, New York.

Gatsonis, C., Hodges, J.S., Kass, R.E. and Singpurwalla, N. (1995) *Case Studies in Bayesian Statistics*, vol. II. Lecture Notes in Statistics **105**,

Springer–Verlag, New York.

Gatsonis, C., Hodges, J.S., Kass, R.E., McCulloch, R.E. and Singpurwalla, N. (1997) *Case Studies in Bayesian Statistics*, vol. III. Lecture Notes in Statistics **121**, Springer–Verlag, New York.

Gatsonis, C., Kass, R.E., Carlin, B.P., Carriquiry, A., Gelman, A. and Verdinelli, I. (1999) *Case Studies in Bayesian Statistics*, vol. IV. Lecture Notes in Statistics **140**, Springer–Verlag, New York.

Gatsonis, C., MacGibbon, K.B. and Strawderman, W.E. (1987) On the estimation of a truncated normal mean. *Statist. Prob. Lett.* **6**, 21–30.

Gauss, C.F. (1810) *Méthode des Moindres Carrés. Mémoire sur la Combination des Observations*. Transl. J. Bertrand (1955). Mallet-Bachelier, Paris.

Geisser, S. and Cornfield, J. (1963) Posterior distributions for multivariate normal parameters. *J. Roy. Statist. Soc., Ser. B* **25**, 368–376.

Gelfand, A.E. (1996) Model determination using sampling-based methods. In *Markov Chain Monte Carlo in Practice*, W.R. Gilks, S. Richardson and D.J. Spiegelhalter (eds.), 145–162. Chapman & Hall, London.

Gelfand, A.E. (2000) Gibbs sampling. *J. Amer. Statist. Assoc.* **95**, 1300–1304.

Gelfand, A.E. and Dey, D.K. (1994) Bayesian Model Choice: Asymptotics and Exact Calculations. *J. Roy. Statist. Soc., Ser. B* **56**, 501–514.

Gelfand, A.E. Dey, D.K. and Chang, H. (1992) Model determination using predictive distributions with implementation via sampling-based methods. In *Bayesian Statistics 4*. J.O. Berger, J.M. Bernardo, A.P. Dawid and A.F.M. Smith (Eds.). Oxford University Press, London, 147–167. Oxford University Press, Oxford.

Gelfand, A.E., Hills, S., Racine-Poon, A. and Smith, A.F.M. (1990) Illustration of Bayesian inference in normal models using Gibbs sampling. *J. Amer. Statist. Assoc.* **85**, 972–982.

Gelfand, A. and Smith, A.F.M. (1990) Sampling based approaches to calculating marginal densities. *J. Amer. Statist. Assoc.* **85**, 398–409.

Gelfand, A., Smith, A.F.M. and Lee, T.M. (1992) Bayesian analysis of constrained parameters and truncated data problems using Gibbs sampling. *J. Amer. Statist. Assoc.* **87**, 523–532.

Gelman, A., Carlin, J.B., Stern, H.S. and Rubin, D.B. (1995) *Bayesian Data Analysis*. Chapman & Hall, London.

Gelman, A. and Meng, X.L. (1998) Simulating normalizing constants: From importance sampling to bridge sampling to path sampling. *Statist. Sci.* **13** 163-185.

Gelman, A. and Rubin, D.B. (1992) Inference from iterative simulation using multiple sequences (with discussion). *Statist. Sci.* **7**, 457–511.

Geman, S. (1988) Experiments in Bayesian image analysis. In *Bayesian Statistics 3*, J.M. Bernardo, M.H. DeGroot, D.V. Lindley and A.F.M. Smith (eds.). Oxford University Press, Oxford.

Geman, S. and Geman, D. (1984) Stochastic relaxation, Gibbs distributions and the Bayesian restoration of images. *IEEE Trans. Pattern Anal. Mach. Intell.* **6**, 721–740.

Genest, C. and Zidek, J.V. (1986) Combining probability distributions: A critique and an annotated bibliography. *Statist. Science* **1**, 114–135.

Gentle, J.E. (1998) *Random Number Generation and Monte Carlo Methods.* Springer–Verlag, New York.

George, E.I. (1986a) Combining minimax shrinkage estimators. *J. Amer. Statist. Assoc.* **81**, 437–445.

George, E.I. (1986b) Minimax multiple shrinkage estimators. *Ann. Statist.* **14**, 188–205.

George, E.I. and Casella, G. (1994) Empirical Bayes confidence estimation. *Statist. Sinica* **4**(2), 617–638.

George, E.I. and Foster, D. (1999) Empirical Bayes variable selection. In *Model Choice*, W. Racugno (ed.). Collana Atti di Congressi, Pitagora Editrice, Bologna.

George, E.I. and McCulloch, R.E. (1993) Variable Selection Via Gibbs Sampling. *J. Amer. Statist. Assoc.* **88**, 881–889.

George, E.I. and McCulloch, R.E. (1997) Approaches for Bayesian variable selection. *Statistica Sinica* **7**, 339–374.

George, E.I. and Robert, C.P. (1992) Calculating Bayes estimates for capture-recapture models. *Biometrika* **4**, 677–683.

Geweke, J. (1988) Antithetic acceleration of Monte Carlo integration in Bayesian inference. *J. Econometrics* **38**, 73–90.

Geweke, J. (1989) Bayesian inference in econometric models using Monte Carlo integration. *Econometrica* **57**, 1317–1340.

Geweke, J. (1991) Efficient simulation from the multivariate normal and Student *t*-distributions subject to linear constraints. *Computer Sciences and Statistics: Proc. 23d Symp. Interface*, 571–577.

Geweke, J. (1992) Evaluating the accuracy of sampling-based approaches to the calculation of posterior moments (with discussion). In *Bayesian Statistics 4*, J.M. Bernardo, J.O. Berger, A.P. Dawid and A.F.M. Smith (eds.), 169–193. Oxford University Press, Oxford.

Geweke, J. (1999) Using simulation methods for Bayesian Econometric models: inference, development and communication (with discussion). *Econometric Reviews* **18**, 1–73.

Geyer, C.J. (1992) Practical Monte Carlo Markov Chain (with discussion). *Statist. Sci.* **7**, 473–511.

Geyer, C.J. (1995) Conditioning in Markov Chain Monte Carlo. *J. Comput. Graph. Statis.* **4**, 148–154.

Geyer, C.J. (1996) Estimation and optimization of functions. In *Markov chain Monte Carlo in Practice*, W.R. Gilks, S.T. Richardson and D.J.

Spiegelhalter (eds.). 241–258. Chapman & Hall, London.

Geyer, C.J. and Thompson, E.A. (1992) Constrained Monte Carlo maximum likelihood for dependent data (with discussion). *J. Roy. Statist. Soc., Ser. B* **54**, 657–699.

Ghosh, J.K. and Mukerjee, R. (1991) Characterization of priors under which Bayesian and frequentist Bartlett corrections are equivalent in the multiparameter case. *J. Multivariate Anal.* **38**, 385–393.

Ghosh, J.K. and Mukerjee, R. (1992a) Bayesian and frequentist Bartlett corrections for likelihood ratio tests. *J. Roy. Statist. Soc., Ser. B* **56**, 396–408.

Ghosh, J.K. and Mukerjee, R. (1992b) Noninformative priors (with discussion). In *Bayesian Statistics* 4, J.M. Bernardo, J.O. Berger, A.P. Dawid and A.F.M. Smith (eds.). Oxford University Press, Oxford.

Ghosh, J.K. and Mukerjee, R. (1993) Frequentist validity of highest posterior density regions in the multiparameter case. *Annals of the Institute of Statistical Mathematics*, **45**, 293–302.

Ghosh, J.K. and Mukerjee, R. (1994) Adjusted versus conditional likelihood: power properties and Bartlett-type adjustment. *J. Roy. Statist. Soc., Ser. B* **56**, 185–188.

Ghosh, M., Carlin, B. and Srivastava, M.S. (1995) Probability matching priors for linear calibration. *Test* 4(2), 333–358.

Ghosh, M., Hwang, J.T. and Tsui, K. (1983) Construction of improved estimators in multiparameter estimation for discrete exponential families (with discussion). *Ann. Statist.* **11**, 351–376.

Ghosh, M., Keating, J.P. and Sen, P.K. (1993) Discussion of Robert, Hwang and Strawderman (1993). *J. Amer. Statist. Assoc.* **88**, 63–66.

Ghosh, M. and Mukerjee, R. (1992) Hierarchical and empirical Bayes multivariate estimation. In *Current Issues in Statistical Inference: Essays in Honor of D. Basu*, M. Ghosh and P.K. Pathak (eds.), 1–12. IMS Lecture Notes – Monograph Series **17**. Hayward, California.

Ghosh, M. and Sen, P.K. (1989) Median unbiasedness and Pitman closeness. *J. Amer. Statist. Assoc.* **84**, 1089–1091.

Ghosh, M., Sen, P.K. and Saleh, A.K.Md.E. (1989) Empirical Bayes subset estimation in regression models. *Statist. Decisions* **7**, 15–35.

Ghosh, M. and Yang, M.C. (1996) Noninformative priors for the two sample normal problem *Test* **5**, 145–157.

Gibbons, J.D., Olkin, I. and Sobel, M. (1977) *Selecting and Ordering Populations*. J. Wiley, New York.

Gigerenzer, G. (1991) The Superego, the Ego and the Id in statistical reasoning. In *Methodological and Quantitative Issues in the Analysis of Psychological Data*, G. Keren and C. Lewis (eds.). Erlbaum, Hillsdale, New Jersey.

Gilks, W.R., Best, N.G. and Tan, K.K.C. (1995) Adaptive rejection Metropolis sampling within Gibbs sampling. *Applied Statist.* Series C **44**, 455–472.

Gilks, W.R., Clayton, D.G., Spiegelhalter, D.J., Best, N.G., McNeil, A.J., Sharples, L.D., and Kirby, A.J. (1993) Modelling complexity: applications of Gibbs sampling in medicine (with discussion). *J. Roy. Statist. Soc., Ser. B* **55**, 39–52.

Gilks, W., Richardson, S. and Spiegelhalter, D. (1996) *Practical Monte-Carlo Markov Chain.* Chapman & Hall, London.

Gilks, W.R. and Wild, P. (1992) Adaptive rejection sampling for Gibbs sampling. *Appl. Statist.* **41**, 337–348.

Gill, R.D. and Levit, B.Y. (1995) Applications of the Van Trees inequality: a Bayesian Cramér-Rao bound. *Bernouilli* **1**, 59–79.

Giudici, P. and Green, P.J. (1998) Decomposable graphical Gaussian model determination. Technical Report, Università di Pavia.

Givens, G.H., Smith, D.D. and Tweedie, R.L. (1997) Publication bias in meta-analysis: a Bayesian data-augmentation approach to account for issues exemplified in the passive smoking debate. *Statist. Sci.* **12**, 221–250.

Gleick, J. (1987) *Chaos.* Penguin, New York.

Gleser, L.J. and Healy, J.D. (1976) Estimating the mean of a normal distribution with known coefficient of variation. *J. Amer. Statist. Assoc.* **71**, 977–981.

Gleser, L.J. and Hwang, J.T. (1987) The non-existence of $100(1-\alpha)\%$ confidence sets of finite expected diameters in errors-in-variable and related models. *Ann. Statist.* **15**, 1351–1362.

Goel, P.K. (1988) Software for Bayesian analysis. In *Bayesian Statistics 3*, J.M. Bernardo, M.H. DeGroot, D.V. Lindley and A.F.M. Smith (eds.), 173–188. Oxford University Press, London.

Goel, P.K. and Rubin, H. (1977) On selecting a subset containing the best population–a Bayesian approach. *Ann. Statist.* **5**, 969–983.

Goldstein, M. and Smith, A.F.M. (1974) Ridge-type estimators for regression analysis. *J. Roy. Statist. Soc., Ser. B* **36**, 284–219.

Good, I.J. (1952) Rational decisions. *J. Roy. Statist. Soc., Ser. B* **14**, 107–114.

Good, I.J. (1973) The probabilistic explication of evidence, causality, explanation and utility. In *Foundations of Statistical Inference*, V.P. Godambe and D.A. Sprott (eds.). Holt, Rinehart and Winston, Toronto.

Good, I.J. (1975) Bayesian estimation methods for two-way contingency tables. *J. Roy. Statist. Soc., Ser. B* **37**, 23–37.

Good, I.J. (1980) Some history of the hierarchical Bayesian methodology. In *Bayesian Statistics 2*, J.M. Bernardo, M.H. DeGroot, D.V. Lindley, A.F.M. Smith (eds.). North-Holland, Amsterdam.

Good, I.J. (1983) *Good Thinking: The Foundations of Probability and Its Applications.* University of Minnesota Press, Minneapolis.

Gouriéroux, C. (1997) *ARCH Model.* Springer–Verlag, New York.

Gouriéroux, C. and Monfort, A. (1996) *Statistical Methods in Econometrics*, volumes 1 and 2. Cambridge University Press, Cambridge.

Goutis, C. (1990) Ranges of posterior measures for some classes of priors with specified moments. Tech. Report 70, University College London. [Published in a modified version without Table 3.3.4 as Goutis (1994).]

Goutis, C. (1994) Ranges of posterior measures for some classes of priors with specified moments. *International Statistical Review* **62**(2), 245–256.

Goutis, C. and Casella, G. (1991) Improved invariant confidence intervals for a normal variance. *Ann. Statist.* **19**, 2015–2031.

Goutis, C. and Casella, G. (1992) Increasing the confidence in Student's *t*-interval. *Ann. Statist.* **20**(3), 1501–1513.

Goutis, C., Casella, G. and Wells, M.T. (1996) Assessing evidence in multiple hypotheses. *J. Amer. Statist. Assoc.* **91**, 1268–1277.

Goutis, C. and Robert, C.P. (1998) Model Choice in Generalized Linear Models: a Bayesian Approach via Kullback-Leibler Projections. *Biometrika* **85**, 29–37.

Gradshteyn, I. and Ryzhik, I. (1980) *Tables of Integrals, Series and Products.* Academic Press, New York.

Green, P. (1995) Reversible Jump Markov Chain Monte Carlo Computation and Bayesian Model Determination. *Biometrika* **82**, 711–732.

Green, P. and Richardson, S. (1998) Modelling heterogeneity with and without the Dirichlet process. *Biometrika* **82**, 711–732.

Grenander, U. and Miller, M.I. (1994) Representations of knowledge in complex systems (with discussion). *J. Roy. Statist. Soc., Ser. B* **56**, 549–603.

Gruet, M.A., Philippe, A. and Robert, C.P. (1999) MCMC control spreadsheets for exponential mixture estimation. *J. Comp. Graph Statist.* **8**, 298–317.

Guihenneuc–Jouyaux, C., Richardson, S. and Lasserre, V. (1998) Convergence assessment in latent variable models: apllication to longitudinal modelling of a marker of HIV progression. In *Discretization and MCMC Convergence Assessment.* C.P. Robert (ed.). Chapter 7, 147–160. Lecture Notes in Statistics **135**, Springer-Verlag, New York.

Gupta, S.S. (1965) On multiple decision (selection and ranking) rules. *Technometrics*, **7**, 222–245.

Gupta, S.S. and Panchapakesan, S. (1979) *Multiple Decision Procedures.* J. Wiley, New York.

Gutmann, S. (1982) Stein's paradox is impossible in problems with finite sample space. *Ann. Statist.* **10**, 1017–1020.

Hadjicostas, P. and Berry, S.M. (1999) Improper and proper posteriors with improper priors in a Poisson-gamma hierarchical model. *Test* **8**, 147–166.

Haff, L. and Johnstone, R.W. (1986) The superharmonic condition for simultaneous estimation of means in exponential families. *Canad. J. Statist.* **14**, 43–54.

Hajek, J. and Sidàk, Z. (1968) *Theory of Rank Test.* Academic Press, New York.

Hald, A. (1998) *An History of Mathematical Statistics.* J. Wiley, New York.

Haldane, J. (1931) A note on inverse probability. *Proc. Cambridge Philos. Soc.* **28**, 55–61.

Hall, P. (1992) *The Bootstrap and Edgeworth Expansion.* Springer–Verlag, New York.

Hamilton J.D. (1989), A new approach to the economic analysis of nonstationary time series and the business cycle, *Econometrica,* **57**(2), 357-384.

Hammersley, J.M. (1974) Discussion of Besag's paper. *J. Roy. Statist. Soc., Ser. B* **36**, 230–231.

Hansen, M. and Yu, B. (2001) Model selection and minimum description length principle. *J. Amer. Statist. Assoc.* **96**, 746–774.

Hartigan, J.A. (1983) *Bayes Theory.* Springer–Verlag, New York.

Hastings, W.K. (1970) Monte Carlo sampling methods using Markov chains and their application. *Biometrika* **57**, 97–109.

Heath, D. and Sudderth, W. (1989) Coherent inference from improper priors and from finitely additive priors. *Ann. Statist.* **17**, 907–919.

Heidelberger, P. and Welch, P.D. (1983) A spectral method for confidence interval generation and run length control in simulations. *Comm. Assoc. Comput. Machinery* **24**, 233–245.

Heitjan, D.F. and Rubin, D.B. (1991) Ignorability and coarse data. *Ann. Statist.* **19**, 2244–2253.

Helland, I.S. (1999) Statistical inference under a fixed symmetry group. Tech. report, Dept. of Mathematics and Statistics, University of Oslo.

Hesterberg, T. (1998) Weighted average importance sampling and defensive mixture distributions. *Technometrics* **37**, 185–194.

Hinkley, D. (1997) Discussion of "Unified frequentist and Bayesian testing of a precise hypothesis." *Statistical Science* **12**, 155–156.

Hjort, N.L. (1990) Nonparametric Bayes estimates based on beta processes in models for life history data. *Ann. Statist.* **18**, 1501–1555.

Hjort, N.L. (1996) Bayesian approaches to non- and semiparametric density estimation (with discussion). In *Bayesian Statistics 5.* J.O. Berger, J.M. Bernardo, A.P. Dawid and A.F.M. Smith (Eds)., 223–253. Oxford University Press, Oxford.

Hoaglin, D., Mosteller, F. and Tukey, J. (1985) *Exploring Data Tables, Trends, and Shapes.* J. Wiley, New York.

Hobert, J. P (2000a) Stability relationships among the Gibbs sampler and its subchains. Tech. report, University of Florida.

Hobert, J. P (2000b) Hierarchical models: a current computational perspective. *J. Amer. Statist. Assoc.* **95**, 1312–1316.

Hobert, J. P and Casella, G. (1996) The effect of improper priors on Gibbs sampling in hierarchical linear models. *J. Amer. Statist. Assoc.* **91** 1461–1473.

Hobert, J. P. and Casella, G. (1998) Functional Compatibility, Markov Chains, and Gibbs Sampling with Improper Posteriors. *J. Comp. Graph Statist.* **7**, 42–60.

Hobert, J. P. and Robert, C.P. (1999) Eaton's Markov chain, its conjugate partner and *P*-admissibility. *Ann. Statist.* **27**, 361–373.

Hoerl, A. and Kennard, R. (1970) Ridge regression: biased estimators for non-orthogonal problems. *Technometrics* **12**, 55–67.

Hora, R.B. and Buehler, R.J. (1966) Fiducial theory and invariant estimation. *Ann. Math. Stat.* **37**, 361–379.

Huber, P.J. (1964) Robust estimation of a location parameter. *Ann. Math. Statist.* **35**, 73–101.

Huber, P.J. (1972) Robust Statistics: a review. *Ann. Math. Statist.* **47**, 1041–1067.

Huerta, G. and West, M. (2000) Bayesian inference on periodicities and component spectral structure in time series. *J. Time Series Analysis* (to appear).

Hui, S. and Berger, J.O. (1983) Empirical Bayes estimation of rates in longitudinal studies. *J. Amer. Statist. Assoc.* **78**, 753–760.

Huzurbazar, V.S. (1976) *Sufficient Statistics*. Marcel Dekker, New York.

Hwang, J.T. (1982a) Improving upon standard estimators in discrete exponential families with applications to Poisson and negative binomial cases. *Ann. Statist.* **10**, 857–867.

Hwang, J.T. (1982b) Semi-tail upper bounds on the class of admissible estimators in discrete exponential families, with applications to Poisson and negative binomial distributions. *Ann. Statist.* **10**, 1137–1147.

Hwang, J.T. (1985) Universal domination and stochastic domination: decision theory simultaneously under a broad class of loss functions. *Ann. Statist.* **13**, 295–314.

Hwang, J.T. and Brown, L.D. (1991) Estimated confidence under the validity constraint. *Ann. Statist.* **19**, 1964–1977.

Hwang, J.T. and Casella, G. (1982) Minimax confidence sets for the mean of a multivariate normal distribution. *Ann. Stat.* **10**, 868–881.

Hwang, J.T. and Casella, G. (1984) Improved set estimators for a multivariate normal mean. *Statist. Decisions* Supplement Issue **1**, 3–16.

Hwang, J.T., Casella, G., Robert, C.P., Wells, M.T. and Farrel, R. (1992) Estimation of accuracy in testing. *Ann. Statist.* **20**, 490–509.

Hwang, J.T. and Chen, J. (1986) Improved confidence sets for the coefficients of a linear model with spherically symmetric errors. *Ann. Statist.*

14, 444–460.

Hwang, J.T. and Pemantle, R. (1994) Evaluation of estimators of statistical significance under a class of proper loss functions. *Statist. Decisions* **15**, 103–128.

Hwang, J.T. and Ullah, A. (1994) Confidence sets recentered at James–Stein estimators—A surprise concerning the unknown variance case. *J. Econometrics* **60**(1-2), 145–156.

Ibragimov, I. and Has'minskii, R. (1981) *Statistical Estimation. Asymptotic Theory.* Springer-Verlag, New York.

Jacquier, E., Polson, N.G. and Rossi, P.E. (1994) Bayesian analysis of stochastic volatility models. *J. Econ. Business Statist.* **12**, 371–389.

James, W. and Stein, C. (1961) Estimation with quadratic loss. In *Proc. Fourth Berkeley Symp. Math. Statist. Probab.* **1**, 361–380. University of California Press.

Jaynes, E.T. (1980) Marginalization and prior probabilities. In *Bayesian Analysis in Econometrics and Statistics*, A. Zellner (ed.). North-Holland, Amsterdam.

Jaynes, E.T. (1983) *Papers on Probability, Statistics and Statistical Physics*, R.D. Rosencrantz (ed.). Reidel, Dordrecht.

Jefferys, W. and Berger, J.O. (1992) Ockham's razor and Bayesian analysis. *American Scientist* **80**, 64–72.

Jeffreys, H. (1939) *Theory of Probability.* Oxford University Press, London.

Jeffreys, H. (1946) An invariant form for the prior probability in estimation problems. *Proceedings of the Royal Society of London* (Ser. A) **186**, 453–461.

Jeffreys, H. (1961) *Theory of Probability* (third edition). Oxford University Press, London.

Johnson, B.M. (1971) On the admissible estimators for certain fixed sample binomial problems. *Ann. Math. Statist.* **41**, 1579–1587.

Johnson, N.L. and Kotz, S.V. (1969–1972) *Distributions in Statistics* (4 vols.). J. Wiley, New York.

Johnstone, D.J. and Lindley, D.V. (1995) Bayesian inference given data "significant at α": tests of point-null hypotheses. *Theory and Decision* **38**(1), 51–60.

Johnstone, I.M. (1984) Admissibility, difference equations, and recurrence in estimating a Poisson mean. *Ann. Statist.* **12**, 1173–1198.

Johnstone, I.M. (1986) Admissible estimation, Dirichlet principles and recurrence of birth-death chains in $\mathbb{Z}_+^p$. *Z. Wahrsch. Verw. Gebiete* **71**, 231–270.

Johnstone, I.M. (1988) On the inadmissibility of Stein's unbiased estimate of loss. In *Statistical Decision Theory and Related Topics* **4**, S.S. Gupta and J.O. Berger (eds.). Springer-Verlag, New York.

Johnstone, I.M. and MacGibbon, B.K. (1992) Minimax estimation of a constrained Poisson vector. *Ann. Statist.* **20**, 807–831.

Jones, M.C. (1987) Randomly choosing parameters from the stationarity and invertibility region of autoregressive-moving average models. *Applied Statistics* (Series C) **38**, 134–138.

Joshi, V.M. (1967) Admissibility of the usual confidence set for the mean of a multivariate normal population. *Ann. Math. Statist.* **38**, 1868-1875.

Joshi, V.M. (1969) Admissibility of the usual confidence set for the mean of a univariate or bivariate normal population. *Ann. Math. Statist.* **40**, 1042–1067.

Joshi, V.M. (1990) The censoring concept and the likelihood principle. *J. Statist. Plann. Inference* **26**, 109–111.

Judge, G. and Bock, M.E. (1978) *Implications of Pre-Test and Stein Rule Estimators in Econometrics*. North-Holland, Amsterdam.

Kadane, J.B. and Chuang, D. (1978) Stable decision problems. *Ann. Statist.* **6**, 1095–1111.

Kariya, T. (1984) An invariance approach to estimation in a curved model. Tech. Report 88, Hifotsubashi University, Japan.

Kariya, T., Giri, N. and Perron, F. (1988) Invariant estimation of mean vector μ of $\mathcal{N}(\mu, \Sigma)$ with $\mu'\Sigma^{-1}\mu = 1$ or $\Sigma^{-1/2}\mu = C$ or $\Sigma = \delta^2\mu\mu'I$. *J. Multivariate Anal.* **27**, 270–283.

Karlin, S. (1958) Admissibility for estimation with quadratic loss. *Ann. Math. Statist.* **29**, 406–436.

Karlin, S. and Rubin, H. (1956) The theory of decision procedures for distributions with monotone likelihood ratio. *Ann. Math. Statist.* **27**, 272–299.

Kass, R.E. (1989) The geometry of asymptotic inference. *Statist. Science* **4**, 188–234.

Kass, R.E. and Raftery, A.E. (1995) Bayes factor and model uncertainty. *J. Amer. Statist. Assoc.* **90**, 773–795.

Kass, R.E. and Steffey, D. (1989) Approximate Bayesian inference in conditionally independent hierarchical models (parametric empirical Bayes models). *J. Amer. Statist. Assoc.* **87**, 717–726.

Kass, R.E. and Wasserman, L. (1995) A reference Bayesian test for nested hypotheses and its relationship to the Schwarz criterion. *J. Amer. Statist. Assoc.* **90**, 928–934.

Kass, R.E. and Wasserman, L. (1996) Formal rules of selecting prior distributions: a review and annotated bibliography. *J. Amer. Statist. Assoc.* **91**, 343–1370.

Keating, J.P. and Mason, R. (1988) James–Stein estimation from an alternative perspective. *Amer. Statist.* **42**, 160–164.

Keeney, R.L. and Raiffa, H. (1976) *Decisions with Multiple Objectives*. J. Wiley, New York.

Kelker, D. (1970) Distribution theory of spherical distributions and a location-scale parameter generalization. *Sankhya* (Ser. A) **32**, 419–430.

Kempthorne, P.J. (1987) Numerical specification of discrete least favorable prior distributions. *SIAM J. Statist. Comput.* **8**, 178–184.

Kempthorne, P.J. (1988) Controlling risks under different loss functions: the compromise decision problem. *Ann. Statist.* **16**, 1594–1608.

Kendall, M. and Stuart, A. (1979) *The Advanced Theory of Statistics*, Volume II: *Inference and Relationships* (4th edition). Macmillan, New York.

Keynes, J.M. (1921) *A Treatise on Probability*. Macmillan, London.

Kiefer, J. (1957) Invariance, minimax sequential estimation and continuous time–processes. *Ann. Math. Statist.* **28**, 573–601.

Kiefer, J. (1977) Conditional confidence statements and confidence estimators (theory and methods). *J. Amer. Statist. Assoc.* **72**, 789–827.

Kiiveri, H. and Speed, T.P. (1982) Structural analysis of multivariate data: A review. In *Sociological Methodology, 1982*. S. Leinhardt (Ed.). 209–289. Jossey Bass, San Francisco.

Kinderman, A., Monahan, J. and Ramage, J. (1977) Computer methods for sampling from Student's *t*-distribution. *Math. Comput.* **31**, 1009–1018.

Kirby, A. J. and Spiegelhalter, D. J. (1994) Statistical modelling for the precursors of cervical cancer. In *Case Studies in Biometry*, N. Lange (ed.). John Wiley, New York.

Kleibergen, F. and van Dijk, H.K. (1993) Non-stationarity in GARCH models: a Bayesian analysis. *J. of Appl. Econometrics* **8**, 41–61.

Knuth, D. (1981) *The Art of Computer Programing*. Volume 2: *Seminumerical Algorithms* (2nd edition). Addison-Wesley, Reading, Mass.

Kontkanen, P., Myllymäki, P., Silander, T., Tirri, H. and Grünwald, P. (2000) On predictive distributions and Bayesian networks. *Statist. Comput.* **10**, 39–54.

Koopman, B. (1936) On distributions admitting a sufficient statistic. *Trans. Amer. Math. Soc.* **39**, 399–409.

Kubokawa, T. (1987) Admissible minimax estimation of a common mean of two normal populations. *Ann. Statist.* **15**, 1245–1256.

Kubokawa, T. (1991) An approach to improving James–Stein estimator. *J. Multivariate Analysis* **36**, 121–126.

Kubokawa, T., Morita, S., Makita, S. and Nagakura, K. (1993) Estimation of the variance and its applications. *J. Statist. Plann. Inference* **35**, 319–333.

Kubokawa, T. and Robert, C.P. (1994) New perspectives on linear calibration. *J. Multivariate Analysis* **51**, 178-200.

Kubokawa, T., Robert, C.P. and Saleh, A.K.Md.E. (1991) Robust estimation of common regression coefficients under spherical symmetry. *Ann. Inst. Statist. Math.* **43**, 677–688.

Kubokawa, T., Robert, C.P. and Saleh, A.K.Md.E. (1992) Empirical Bayes estimation of the covariance matrix of a normal distribution with unknown mean under an entropy loss. *Sankhya* (Ser. A) **54**, 402–410.

Kubokawa, T., Robert, C.P. and Saleh, A.K.Md.E. (1993) Estimation of noncentrality parameters. *Canad. J. Statist.* **21**, 54–58.

Lad, F. (1996) *Operational Subjective Statistical Methods: a Mathematical, Philosophical and Historical Introduction.* J. Wiley, New York.

Laird, N.M. and Louis, T.A. (1987) Confidence intervals based on bootstrap samples. *J. Amer. Statist. Assoc.* **82**, 739–750.

Laplace, P.S. (1773) Mémoire sur la probabilité des causes par les événements. *Mémoires de l'Académie Royale des Sciences présentés par divers savans* **6**, 621–656. [Reprinted in Laplace (1878) **8**, 27–65.]

Laplace, P.S. (1786) Sur les naissances, les mariages et les morts à Paris depuis 1771 jusqu'à 1784 et dans toute l'étendue de la France, pendant les années 1781 et 1782. *Mémoires de l'Académie Royale des Sciences présentés par divers savans.* [Reprinted in Laplace (1878), **11**, 35–46.]

Laplace, P.S. (1795) *Essai Philosophique sur les Probabilités.* [Reprinted in Christian Bourgeois, coll. Epistémé, 1986.]

Laplace, P.S. (1812) *Théorie Analytique des Probabilités.* Courcier, Paris.

Laplace, P.S. (1878–1912) *Œuvres Complètes de Laplace.* Gauthier-Villars, Paris.

Lauritzen, S.L. (1996) *Graphical Models.* Oxford University Press, London.

Lavielle, M. and Moulines, E. (1997) On a stochastic approximation version of the EM algorithm. *Statist. Comput.* **7**, 229–236.

Lavine, M. (1992) Some aspects of Pólya tree distributions for statistical modeling; *Ann. Statist.* **22**, 1222–1235.

Lawley, D.N. (1956) A general method for approximating to the distribution of the likelihood ratio criteria. *Biometrika* **43**, 295–303.

Le Cam, L. (1986) *Asymptotic Methods in Statistical Decision Theory.* Springer–Verlag, New York.

Le Cam, L. (1990) Maximum likelihood: an introduction. *Int. Statist. Rev.* **58**, 153–172.

Lee, P. (1989) *Bayesian Statistics: an Introduction.* Oxford University Press, London.

Legendre, A. (1805) *Nouvelles Méthodes pour la Détermination des Orbites des Comètes.* Courcier, Paris.

Lehmann, E.L. (1983) *Theory of Point Estimation.* J. Wiley, New York.

Lehmann, E.L. (1986) *Testing Statistical Hypotheses* (2nd edition). J. Wiley, New York.

Lehmann, E.L. (1990) Model specification. *Statist. Science* **5**, 160–168.

Lehmann, E.L. and Casella, G. (1998) *Theory of Point Estimation* (second edition). Springer–Verlag, New York.

Lenk, P. (1999) Bayesian inference for semiparametric regression using a Fourier representation. *J. Roy. Statist. Soc., Ser. B* **61**, 863–879.

Leonard, T. (1982) Comments on Lejeune and Faulkenberry (1982) *J. Amer. Statist. Assoc.* **77**, 657–658.

Letac, G. (1990) Personal communication.

Letac, G. and Mora, M. (1990) Natural real exponential families with cubic variance functions. *Ann. Statist.* **18**, 1–37.

Lindley, D.V. (1957) A statistical paradox. *Biometrika* **44**, 187–192.

Lindley, D.V. (1961) The use of prior probability distributions in statistical inference and decision. In *Proc. Fourth Berkeley Symp. Math. Statist. Probab.* **1**, 453–468. University of California Press.

Lindley, D.V. (1962) Discussion of Professor Stein's paper 'Confidence sets for the mean of a multivariate normal distribution'. *J. Roy. Statist. Soc., Ser. B* **24**, 265–296.

Lindley, D.V. (1965) *Introduction to Probability and Statistics from a Bayesian Viewpoint* (Parts 1 and 2). Cambridge University Press, Cambridge.

Lindley, D.V. (1971) *Bayesian Statistics, A Review*. SIAM, Philadelphia.

Lindley, D.V. (1980) Approximate Bayesian methods. In *Bayesian Statistics* 2, J.M. Bernardo, M. DeGroot, D.V. Lindley and A.F.M. Smith (eds.), North-Holland, Amsterdam.

Lindley, D.V. (1982) Scoring rules and the inevitability of probability. *Int. Statist. Rev.* **50**, 1–26.

Lindley, D.V. (1985) *Making Decisions* (2nd edition). J. Wiley, New York.

Lindley, D.V. (1990) The present position in Bayesian Statistics (with discussion). *Statist. Sci.* **5**(1), 44–89.

Lindley, D.V. and Phillips, L.D. (1976) Inference for a Bernouilli process (a Bayesian view). *Amer. Statist.* **30**, 112–119.

Lindley, D.V. and Smith, A.F.M. (1972) Bayes estimates for the linear model. *J. Roy. Statist. Soc., Ser. B* **34**, 1–41.

Liseo, B. (1993) Elimination of nuisance parameters with reference priors. *Biometrika* **80**(2), 295–304.

Liu, J.S. and Wu, Y.N. (1999) Parameter expansion scheme for data augmentation. *J. Amer. Statist. Assoc.* **94**, 1264-1274.

Liu, J.S., Wong, W.H. and Kong, A. (1994) Covariance structure of the Gibbs sampler with applications to the comparisons of estimators and sampling schemes. *Biometrika* **81**, 27–40.

Liu, J.S., Wong, W.H. and Kong, A. (1995) Correlation structure and convergence rate of the Gibbs sampler with various scans. *J. Roy. Statist. Soc., Ser. B* **57**, 157–169.

Louis, T. (1997) Discussion of "Unified frequentist and Bayesian testing of a precise hypothesis". *Statistical Science* **12**, 152–155.

Lu, K. and Berger, J.O. (1989a) Estimated confidence procedures for multivariate normal means. *J. Statist. Plann. Inference* **23**, 1–19.

Lu, K. and Berger, J.O. (1989b) Estimation of normal means: frequentist estimators of loss. *Ann. Statist.* **17**, 890–907.

Maatta, J. and Casella, G. (1990) Developments in decision theoretic variance estimation (with discussion). *Statist. Science* **5**, 90–120.

Machina, G. (1982) Expected utility analysis without the independence axiom. *Econometrica* **50**, 277–323.

Machina, G. (1987) Choice under uncertainty: problems solved and unsolved. *J. Econom. Perspectives* **1**, 121–154

MacLachlan, G. and Krishnan, T. (1997) *The EM Algorithm and Extensions*. J. Wiley, New York.

MacLachlan, G. and Basford, K. (1987) *Mixture Models*. Marcel Dekker, New York.

Madigan, D. and Raftery, A.E. (1991) Model selection and accounting for model uncertainty in graphical models using Occam's Window. Technical Report 213, University of Washington.

Madigan, D. and Raftery, A.E. (1994) Model selection and accounting for model uncertainty in graphical models using Occam's Window. *J. Amer. Statist. Assoc.* **89**, 1535–1546.

Madigan, D. and Raftery, A.E. (1995) Bayesian graphical models for discrete data. *Int. Statist. Rev.* **63**, 215–232.

Madigan, D. and York, J. (1995) Bayesian graphical models for discrete data. *International Statistical Review* **63**, 215-232.

Maddala, G. (1977) *Econometrics*. McGraw-Hill, New York.

Marin, J.-M. and Robert, C.P. (2007) *Bayesian Core: a Practical Approach to Computational Bayesian Statistics*, Springer–Verlag, New York.

Maritz, J.S. and Lwin, T. (1989) *Empirical Bayes Methods* (2nd edition). Chapman & Hall, London.

Marsaglia, G. and Zaman, A. (1993) The KISS Generator. Tech. Report, Dept. of Statistics, University of Florida.

McCullagh, P. and Nelder, J. (1989) *Generalized Linear Models*. Chapman & Hall, London.

McCullogh, R. and Rossi, P.E. (1992) Bayes Factors for Nonlinear Hypotheses and Likelihood Distributions. *Biometrika* **79**, 663–676.

Meeden, G. and Vardeman, S. (1985) Bayes and admissible set estimation. *J. Amer. Statist. Assoc.* **80**, 465–471.

Meng, X.L. and van Dyk, D.A. (1997) The EM algorithm–an old folk-song sung to a new tune (with discussion). *J. Roy. Statist. Soc., Ser. B* **59**, 511–568.

Meng, X.L. and van Dyk, D.A. (1999) Seeking efficient data augmentation schemes via conditional and marginal augmentation. *Biometrika* **86**,

301–320.

Meng, X.L. and Wong, W.H. (1996) Simulating ratios of normalizing constants via a simple identity: a theoretical exploration. *Statist. Sinica* **6**, 831–860.

Mengersen, K.L. and Robert, C.P. (1996) MCMC Convergence Diagnostics: a "Reviewww" (with discussion). In *Bayesian Statistics 6*. J.O. Berger, J.M. Bernardo, A.P. Dawid and A.F.M. Smith (Eds)., 415–440. Oxford University Press, London.

Mengersen, K.L. and Robert, C.P. (1996) Testing for mixtures: A Bayesian entropic approach (with discussion). In *Bayesian Statistics 5*, J.O. Berger, J.M. Bernardo, A.P. Dawid, D.V. Lindley and A.F.M. Smith (eds.). 255–276. Oxford University Press, London.

Mengersen, K.L. and Tweedie, R.L. (1993) Meta-analysis approaches to dose-response relationships with application in studies of lung cancer and passive smoking. *Statist. Medicine–Proc. NIH Conf. on Meta-Analysis*, D. Williamson (ed.).

Mengersen, K.L. and Tweedie, R.L. (1996) Rates of convergence of the Hastings and Metropolis algorithms. *Ann. Statist.* **24** 101–121.

Metropolis, N., Rosenbluth, A.W., Rosenbluth, M.N., Teller, A.H., Teller, E. (1953) Equations of state calculations by fast computing machines. *J. Chem. Phys.* **21**, 1087–1092.

Metropolis, N. and Ulam, S. (1949) The Monte Carlo method. *J. Amer. Statist. Assoc.* **44**, 335–341.

Meyer, Y. (1990) *Ondelettes*. Hermann, Paris.

Meyn, S.P. and Tweedie, R.L. (1993) *Markov Chains and Stochastic Stability*. Springer-Verlag, London.

Mira, A. and Tierney, L. (2002) On the use of auxiliary variables in Markov chain Monte Carlo methods. *Scand. J. Statist.* **29**(1), 1–12.

Monahan J.F. (1984) A note on enforcing stationarity in autoregressive-moving average models. *Biometrika* **71**, 403–404.

Moors, J.J.A. (1981) Inadmissibility of linearly invariant estimators in truncated parameter spaces. *J. Amer. Statist. Assoc.* **76**, 910–915.

Moreno, E., Bertolino, F., and Racugno, W. (1998a) An intrinsic limiting procedure for model selection and hypothesis testing. *J. Amer. Statist. Assoc.* **93**, 1451–1460.

Morisson, D. (1979) Purchase intentions and purchase behavior. *J. Marketing* **43**, 65–74.

Morris, C. (1982) Natural exponential families with quadratic variance functions. *Ann. Statist.* **10**, 65–80.

Morris, C. (1983a) Natural exponential families with quadratic variance functions: statistical theory. *Ann. Statist.* **11**, 515–529.

Morris, C. (1983b) Parametric empirical Bayes inference: theory and applications. *J. Amer. Statist. Assoc.* **78**, 47–65.

Mosteller, F. and Chalmers, T.C. (1992) Some progress and problems in meta-analysis of clinical trials. *Statist. Science.* **7**, 227–236.

Mosteller, F. and Wallace, D.L. (1984) *Applied Bayesian and Classical Inference.* Springer–Verlag, New York.

Mukerjee, R. and Dey, D.K. (1993) Frequentist validity of posterior quantiles in the presence of a nuisance parameter: higher order asymptotics. *Biometrika* **80**, 499–505.

Müller, P. (1991) A generic approach to posterior integration and Gibbs sampling. Tech. Report # 91–09, Purdue University, West Lafayette, Indiana.

Müller, P. and Vidakovic, B. (1999) *Bayesian Inference in Wavelet-Based Models.* Lecture Notes in Statistics, **141**, Springer–Verlag, New York.

Murphy, A.H. and Winkler, R.L. (1984) Probability forecasting in meteorology. *J. Amer. Statist. Assoc.* **79**, 489–500.

Musio, M. and Racugno, W. (1999) Discussion of Fernandez and Steel's paper. In *Bayesian Statistics 6.* J.O. Berger, J.M. Bernardo, A.P. Dawid and A.F.M. Smith (Eds)., 231–233. Oxford University Press, London.

Mykland, P., Tierney, L. and Yu, B. (1995) Regeneration in Markov chain samplers. *J. Amer. Statist. Assoc.* **90**, 233–241.

Nachbin, L. (1965) *The Haar Integral.* Van Nostrand, New York.

Naylor, J.C. and Smith, A.F.M. (1982) Application of a method for the efficient computation of posterior distributions. *Appl. Statist.* **31**, 214–225.

Nelson, D.B. (1990) Stationarity and persistence in the GARCH(1,1) model. *Econometric Theory* **6**, 318–334.

Newton, M.A. and Raftery, A.E. (1994) Approximate Bayesian inference by the weighted likelihood boostrap (with discussion). *J. Roy. Statist. Soc., Ser. B* **56**, 1–48.

Neyman, J. (1934) On the two different aspects of the representative method: The method of stratified sampling and the method of purposive selection. *J. Roy. Statist. Soc., Ser. A* **97**, 558–625.

Neyman, J. (1937) "Smooth" test for goodness of fit. *Skand. Aktvariebidokr.* **20**, 150–199.

Neyman, J. and Pearson, E.S. (1933a) On the problem of the most efficient tests of statistical hypotheses. *Phil. Trans. Royal Soc. Ser. A* **231**, 289–337.

Neyman, J. and Pearson, E.S. (1933b) The testing of statistical hypotheses in relation to probabilities a priori. *Proc. Cambridge Philos. Soc.* **24**, 492–510.

Neyman, J. and Scott, E.L. (1948) Consistent estimates based on partially consistent observations. *Econometrica* **16**, 1–32.

Novick, M.R. and Hall, W.J. (1965) A Bayesian indifference procedure. *J. Amer. Statist. Assoc.* **60**, 1104–1117.

Nummelin, E. (1984) *General Irreducible Markov Chains and Non-Negative Operators.* Cambridge University Press, Cambridge.

Oh, M.S. (1989) Integration of multimodal functions by Monte Carlo importance sampling, using a mixture as an importance function. Tech. Report, Dept. of Statistics, University of California.

Oh, M.S. and Berger, J.O. (1993) Integration of multimodal functions by Monte-Carlo importance sampling. *J. Amer. Statist. Assoc.* **88**, 450–456.

O'Hagan, A. (1992) Some Bayesian numerical analysis. In *Bayesian Statistics 4*. J.O. Berger, J.M. Bernardo, A.P. Dawid and A.F.M. Smith (Eds.). Oxford University Press, London, 345–355. Oxford University Press, London.

O'Hagan, A. (1994) *Kendall's Advanced Theory of Statistics. Volume 2B: Bayesian Inference.* Chapman & Hall, London.

O'Hagan, A. (1995) Fractional Bayes factors for model comparisons. *J. Roy. Statist. Soc., Ser. B* **57**, 99–138.

O'Hagan, A. (1997) Properties of intrinsic and fractional Bayes factors, *Test* **6**, 101–118.

O'Hagan, A. and Berger, J.O. (1988) Ranges of posterior probabilities for quasi-unimodal priors with specified quantiles. *J. Amer. Statist. Assoc.* **83**, 503–508.

Olkin, I., Petkau, A.J. and Zidek, J.V. (1981) A comparison of n estimators for the binomial distribution. *J. Amer. Statist. Assoc.* **76**, 637–642.

Olver, F.W.J. (1974) *Asymptotics and Special Functions.* Academic Press, New York.

Osborne, C. (1991) Statistical calibration: a review. *Int. Statist. Rev.* **59**, 309–336.

Owen, A. and Zhou, Y. (2000) Safe and effective importance sampling. *J. Amer. Statist. Assoc.* **95**, 135–143.

Parent, E., Bobée, B., Hubert, P. and Miquel, J. (1998) Statistical and Bayesian Methods in Hydrological Sciences. In *Selected Proceedings from the UNESCO conference in honner of Pr. Bernier.* Unesco, IHP-V Technical Documents in Hydrology N20.

Pearl, J. (1988) *Probabilistic Reasoning in Intelligent Systems.* Morgan Kaufman, Palo Alto, California.

Pearson, K. (1894) Contribution to the mathematical theory of evolution. *Proc. Trans. Roy. Soc. A* **185**, 71–110.

Peddada, S. and Khattree, R. (1986) On Pitman nearness and variance of estimators. *Comm. Stat.* **15**, 3005–3018.

Peers, H.W. (1968) Confidence properties of Bayesian interval estimates. *J. Roy. Statist. Soc., Ser. B* **30**, 535–544.

Peers, H.W. (1965) On confidence Points and Bayesian probability points in the case of several parameters. *J. Roy. Statist. Soc., Ser. B* **27**, 9–16.

Perk, W. (1947) Some observations on inverse probability including a new indifference rule. *J. Inst. Actuaries* **73**, 285–312.

Perron, F. and Giri, N. (1990) On the best equivariant estimator of mean of a multivariate normal population. *J. Multivariate Anal.* **32**, 1–16.

Pettit, L.I. (1992) Bayes factors for outlier models using the device of imaginary observations. *J. Amer. Statist. Assoc.* **87**, 541–545.

Pfangzagl, J. (1968) A characterization of the one parameter exponential family by existence of uniformly most powerful tests. *Sankhya* (Ser. A) **30**, 147–156.

Phillips, D.M. and Smith, A.F.M. (1996) Bayesian model comparison via jump diffusions. In *Markov Chain Monte Carlo in Practice*, W.R. Gilks, S. Richardson and D.J. Spiegelhalter (eds.), 215–240. Chapman & Hall, London.

Phillips, P.C.B. (1991) Bayesian routes and unit roots: *de rebus prioribus semper est disputandum*. *J. Appl. Econometrics* **6**, 435–474.

Pierce, D. (1973) On some difficulties in a frequency theory of inference. *Ann. Statist.* **1**, 241–250.

Pilz, J. (1991) *Bayesian Estimation and Experimental Design in Linear Regression Models* (2nd edition). J. Wiley, New York.

Pitman, E.J.G. (1936) Sufficient statistics and intrinsic accuracy. *Proc. Cambridge Philos. Soc.* **32**, 567–579.

Pitman, E.J.G. (1937) The closest estimates of statistical parameters. *Proc. Cambridge Philos. Soc.* **33**, 212–222.

Pitman, E.J.G. (1939) The estimation of location and scale parameters of a continuous population of any given form. *Biometrika* **30**, 391–421.

Pitt, M.K. and Shephard, N. (1999) Filtering via simulation: auxiliary particle filter. *J. Amer. Statist. Assoc.* **94**, 590–599.

Plessis, B. (1989) Context dependent enhancements for digitized radiographs. MSc. thesis, Dept. of Electrical Engineering, University of Ottawa.

Poincaré, H. (1902) *La Science and l'Hypothèse*. Flammarion, Paris. [Reprinted in Champs, 1989.]

Pollock, K. (1991) Modelling capture, recapture and removal statistics for estimation of demographic parameters for fish and wildlife populations: past, present and future. *J. Amer. Statist. Assoc.* **86**, 225–238.

Poirier, D.J. (1995) *Intermediate Statistics and Econometrics: a Comparative Approach*. MIT Press, Cambridge, Mass.

Popper, K. (1983) *Postface to the Logic of Scientific Discovery*. I–*Realism and Science*. Hutchinson, London.

Press, J.S. (1989) *Bayesian Statistics*. J. Wiley, New York.

Qian, W. and Titterington, D.M. (1991) Estimation of parameters in hidden Markov models. *Phil. Trans. Roy. Soc. London* A **337**, 407–428.

Racugno, W. (1999) *Model Selection*. Collana Atti di Congressi, Pitagora Editrice, Bologna.

Raftery, A.E. (1988) Inference for the binomial N parameter hierarchical Bayes approach. *Biometrika* **75**, 355–363.

Raftery, A.E. (1996) Hypothesis Testing and Model Selection Via Posterior Simulation. In *Markov chain Monte Carlo in Practice* W.R. Gilks, S.T. Richardson and D.J. Spiegelhalter (eds.). 115–130. Chapman & Hall, London.

Raftery, A.E. and Lewis, S. (1992) How many iterations in the Gibbs sampler? In *Bayesian Statistics 4*. J.O. Berger, J.M. Bernardo, A.P. Dawid and A.F.M. Smith (Eds.). Oxford University Press, London, 763–773. Oxford University Press, London.

Raftery, A., Madigan, D. and Hoeting, J. (1997) Bayesian Model Averaging for Linear Regression Models. *J. Amer. Statist. Assoc.* **92**, 179–191.

Raftery, A., Madigan, D. and Volinsky, C. (1996) Accounting for model uncertainty in survival analysis improves predictive performance (with discussion). In *Bayesian Statistics 5*, J.O. Berger, J.M. Bernardo, A.P. Dawid, D.V. Lindley and A.F.M. Smith (eds.), 323–349. Oxford University Press, London.

Raftery, A. and Richardson, S. (1995) Model selection for generalized linear models via GLIB, with application to epidemiology. In *Bayesian Biostatistics*, D.A. Berry and D.K. Stangl (eds.). Marcel Dekker, New York.

Raiffa, H. (1968) *Decision Analysis: Introductory Lectures on Choices under Uncertainty*. Addison-Wesley, Reading, Mass.

Raiffa, H. and Schlaifer, R. (1961) *Applied Statistical Decision Theory*. Division of Research, Graduate School of Business Administration, Harvard University.

Rao, C.R. (1980) Discussion of J. Berkson's paper 'Minimum chi-square, not maximum likelihood'. *Ann. Statist.* **8**, 482–485.

Rao, C.R. (1981) Some comments on the minimum mean square error as criterion of estimation. In *Statistics and Related Topics*, M. Csörgo, D. Dawson, J.N.K. Rao, and A. Saleh (eds.), 123–143.

Rao, C.R., Keating, J.P. and Mason, R. (1986) The Pitman nearness criterion and its determination. *Comm. Statist.–Theory Methods* **15**, 3173–3191.

Redner, R. and Walker, H. (1984) Mixture densities, maximum likelihood and the EM algorithm. *SIAM Rev.* **26**, 195–239.

Revuz, D. (1984) *Markov Chains* (2nd edition). North-Holland, Amsterdam.

Richard, J.F. (1973) *Posterior and Predictive Densities for Simultaneous Equation Models*. Springer-Verlag, Berlin.

Richard, J.F. and Tompa, H. (1980) On the evaluation of poly-t density functions. *J. Econometrics* **12**, 335–351.

Richardson, S., and Green, P.J. (1997) On Bayesian analysis of mixtures with an unknown number of components (with discussion). *J. Roy. Statist. Soc., Ser. B* **59**, 731–792.

Ripley, B. (1986) Statistics, images and pattern recognition. *Canad. J. Statist.* **14**, 83–111.

Ripley, B. (1987) *Stochastic Simulation.* J. Wiley, New York.

Ripley, B. (1992) Neural networks. In *Networks and Chaos—Statistical and Probabilistic Aspects*, O. Barnorff-Nielsen et al. (eds.). Monographs in Statistics and Applied Probabilities, Chapman & Hall, London.

Rissanen, J. (1983) A universal prior for integers and estimation by minimum description length. *Ann. Statist.* **11**, 416 431.

Rissanen, J. (1990) Complexity of models. In *Complexity, Entropy, and the Physics of Information* **8**, W. Zurek (ed.), Addison-Wesley, Reading, Mass.

Robbins, H. (1951) Asymptotically subminimax solutions to compound statistical decision problems. In *Proc. Second Berkeley Symp. Math. Statist. Probab.* **1**. University of California Press.

Robbins, H. (1955) An empirical Bayes approach to statistics. In *Proc. Third Berkeley Symp. Math. Statist. Probab.* **1**. University of California Press.

Robbins, H. (1964) The empirical Bayes approach to statistical decision problems. *Ann. Math. Statist.* **35**, 1–20.

Robbins, H. (1983) Some thoughts on empirical Bayes estimation. *Ann. Statist.* **1**, 713–723.

Robert, C.P. (1988) Performances d'estimateurs à rétrécisseur en situation de multicolinéarité. *Ann. d'Eco. Statist.* **10**, 97–119.

Robert, C.P. (1990a) Modified Bessel functions and their applications in Probability and Statistics. *Statist. Prob. Lett.*, **9**, 155–161.

Robert, C.P. (1990b) On some accurate bounds for the quantiles of a non-central chi-squared distribution. *Statist. Prob. Lett.* **10**, 101–106.

Robert, C.P. (1990c) Hidden mixtures and Bayesian sampling. Rapport tech. 115, LSTA, Université Paris VI.

Robert, C.P. (1991) Generalized Inverse Normal distributions. *Statist. Prob. Lett.* **11**, 37–41.

Robert, C.P. (1993a) Prior Feedback: A Bayesian approach to maximum likelihood estimation. *Comput. Statist.* **8**, 279–294.

Robert, C.P. (1993b) A Note on the Jeffreys-Lindley paradox. *Statist. Sinica* **3**, 601–608.

Robert, C.P. (1995a) Simulation of truncated normal variables. *Statist. Comput.* **5**, 121–125.

Robert, C.P. (1995b) Convergence control techniques for Markov chain Monte Carlo algorithms. *Statis. Science* **10**(3), 231–253.

Robert, C.P. (1996a) Inference in mixture models. In *Markov Chain Monte Carlo in Practice*, W.R. Gilks, S. Richardson and D.J. Spiegelhalter (eds.), 441–464. Chapman & Hall, London.

Robert, C.P. (1996b) Intrinsic loss functions. *Theory and Decision* **40** (2), 191–214.

Robert, C.P. (1999) Two techniques of integration by parts and some applications. In *Zeitschrift for Professor Saleh's 65th Birthday*, E. Amed, (ed.). The Nova Science Publishers Inc.

Robert, C.P., Bock, M.E. and Casella, G. (1990) Bayes estimators associated with uniform distributions on spheres (II): the hierarchical Bayes approach. Tech. Report BU-1002-M, Cornell University.

Robert, C.P. and Caron, N. (1996) Noninformative Bayesian testing and neutral Bayes factors. *TEST* **5**, 411–437.

Robert, C.P. and Casella, G. (1990) Improved confidence sets for spherically symmetric distributions. *J. Multivariate Anal.* **32**, 84–94.

Robert, C.P. and Casella, G. (1993) Improved confidence statements for the usual multivariate normal confidence set. In *Statistical Decision Theory and Related Topics V*, J.O. Berger and S.S. Gupta (Eds.) Springer-Verlag, New York., 351–368. Springer–Verlag, New York.

Robert, C.P. and Casella, G. (1994) Distance penalized losses for testing and confidence set evaluation. *Test* **3**(1), 163–182.

Robert, C.P. and Casella, G. (1999) *Monte Carlo Statistical Methods*. Springer-Verlag, New York.

Robert, C.P. and Casella, G. (2004) *Monte Carlo Statistical Methods* (second edition). Springer–Verlag, New York.

Robert, C.P., Celeux, G. and Diebolt, J. (1993) Bayesian estimation of hidden Markov models: A stochastic implementation. *Statistics & Probability Letters* **16**, 77–83.

Robert, C.P. and Hwang, J.T.G. (1996) Maximum likelihood estimation under order constraints. *J. Amer. Statist. Assoc.* **91**, 167–173.

Robert, C.P., Hwang, J.T.G. and Strawderman, W.E. (1993) Is Pitman closeness a reasonable criterion? (with discussion). *J. Amer. Statist. Assoc.* **88**, 57–76.

Robert, C.P. and Mengersen, K.L. (1999) Reparametrisation issues in mixture estimation and their bearings on the Gibbs sampler. *Comput. Statis. Data Ana.* **29**, 325–343.

Robert, C.P. and Reber, A. (1998) Bayesian Modelling of a Biopharmaceutical Experiment with Heterogeneous Responses. *Sankhya B* **60**(1), 145–160.

Robert, C.P., Rydén, T. and Titterington, D.M. (1999a) Convergence controls for MCMC algorithms, with applications to hidden Markov chains. *J. Statist. Computat. Simulat.* **64**, 327–355.

Robert, C.P., Rydén, T. and Titterington, D.M. (1999b) Jump Markov chain Monte Carlo algorithms for Bayesian inference in hidden Markov models. *J. Roy. Statist. Soc., Ser. B* **62**(1), 57–75.

Robert, C.P. and Saleh, A.K.Md.E. (1991) Point estimation and confidence set estimation in a parallelism model: an empirical Bayes approach. *Ann. d'Eco. Statist.* **23**, 65–89.

Robert, C.P. and Soubiran, C. (1993) Estimation of a mixture model through Bayesian sampling and prior feedback. *Test* **2**, 125–146.

Robert, C.P. and Titterington, M. (1998) Reparameterisation strategies for hidden Markov models and Bayesian approaches to maximum likelihood estimation. *Statist. Comput.* **8**(2), 145–158.

Roberts, G.O. and Rosenthal, J.S. (1998) Markov chain Monte Carlo: Some practical implications of theoretical results (with discussion). *Can. J. Statist.* **26**, 5–32.

Roberts, G. and Polson, N. (1990) A note on the geometric convergence of the Gibbs sampler. Tech. Report, Dept. of Statistics, University of Nottingham.

Roberts, G.O. and Sahu, S.K. (1997) Updating schemes, covariance structure, blocking and parametrisation for the Gibbs sampler. *J. Roy. Statist. Soc., Ser. B* **59**, 291–318.

Robertson, T., Wright, F.T. and Dykstra, R.L. (1988) *Order Restricted Statistical Inference.* J. Wiley, New York.

Robins, J. and Ritov, Y. (1997) A curse of dimensionality appropriate (CODA) asymptotic for semiparametric models. *Statist. Medicine* **16**, 285–319.

Robins, J. and Wasserman, L. (2000) Conditioning, likelihood and concepts: A review of some foundational concepts. *J. Amer. Statist. Assoc.* **95**, 1340–1346.

Robinson, G.K. (1976) Properties of Student's t and of the Behrens-Fisher solution to the two means problem. *Ann. Statist.* **4**, 963–971.

Robinson, G.K. (1979) Conditional properties of statistical procedures. *Ann. Statist.* **7**, 742–755.

Robinson, G.K. (1982) Behrens-Fisher problem. In *Encyclopedia of Statistical Science* **1**, S.V. Kotz and N.J. Johnson (eds.), 205–209. J. Wiley, New York.

Roeder, K. (1992) Density estimation with confidence sets exemplified by superclusters and voids in galaxies. *J. Amer. Statist. Assoc.* **85**, 617–624.

Roeder, K. and Wasserman, L. (1997) Practical Bayesian density estimation using mixtures of normals. *J. Amer. Statist. Assoc.* **92**, 894–902.

Romano, J.P. and Siegel, A.F. (1986) *Counterexamples in Probability and Statistics.* Wadsworth, Belmont, California.

Rousseau, J. (1997) *Performances fréquentistes des lois de référence et propriétés asymptotiques des procédures bayésiennes*, Ph.D. thesis, Université Paris VI.

Rousseau, J. (2000) Coverage properties of one-sided intervals in the discrete case and application to matching priors. *Annals of the Institute of*

Statistical Mathematics **52**(1), 28–42.

Rousseau, J. (2002) Coverage properties of HPD regions in the discrete case. *J. Multivariate Analysis* **83**(1), 1–21.

Rousseau, J. (2005) Asymptotic coverage of joint two-sided confidence intervals. *Scan. J. Statist.* **32**, 639–660.

Rubin, D.B. (1984) Bayesianly justifiable and relevant frequency calculations for the applied statistician. *Ann. Statist.* **12**, 1151–1172.

Rubin, G., Umbach, D., Shyu, S.F. and Castillo-Chavez, C. (1992) Using mark-recapture methodology to estimate the size of a population at risk for sexually transmitted diseases. *Statist. Medicine* **11**, 1533–1549.

Rubin, H. (1987) A weak system of axioms for rational behavior and the nonseparability of utility from prior. *Statist. Decision* **5**, 47–58.

Rubinstein, R.Y. (1981) *Simulation and the Monte Carlo Method*. J. Wiley, New York.

Rudin, W. (1976) *Principles of Real Analysis*. McGraw-Hill, New York.

Rue, H. (1995) New loss functions in Bayesian imaging. *J. Amer. Statist. Assoc.* **90**, 900–908.

Rukhin, A.L. (1978) Universal Bayes estimators. *Ann. Statist.* **6**, 345–351.

Rukhin, A.L. (1988a) Estimated loss and admissible loss estimators. In *Statistical Decision Theory and Related Topics* IV, S.S. Gupta and J.O. Berger (eds.), 409–420. Springer–Verlag, New York.

Rukhin, A.L. (1988b) Loss functions for loss estimations. *Ann. Statist.* **16**, 1262–1269.

Rukhin, A.L. (1995) Admissibility: Survey of a concept in progress. *Intern. Statist. Review* **63**, 95–115.

Santner, T.J. and Duffy, D. (1989) *The Statistical Analysis of Discrete Data*. Springer-Verlag, New York.

Savage, L.J. (1954) *The Foundations of Statistical Inference*. J. Wiley, New York.

Saxena, K. and Alam, K. (1982) Estimation of the non-centrality parameter of a chi-squared distribution. *Ann. Statist.* **10**, 1012–1016.

Schaafsma, W., Tolboom, J. and van der Meulen, B. (1989) Discussing truth or falsity by computing a Q-value. In *Statistics, Data Analysis and Informatics*, V. Dodge (ed.). North-Holland, Amsterdam.

Schervish, M.J. (1989) A general method for comparing probability assessors. *Ann. Statist.* **17**, 1856–1879.

Schervish, M.J. (1995) *Theory of Statistics*. Springer–Verlag, New York.

Schervish, M.J. and Carlin, B.P. (1992) On the convergence of successive substitution sampling. *J. Comput. Graphical Statist.* **1**, 111–127.

Schmeiser, B. and Shalaby, M. (1980) Acceptance/rejection methods for beta variate generation. *J. Amer. Statist. Assoc.* **75**, 673–678.

Schwarz, G. (1978) Estimating the dimension of a model. *Annals of Statistics* **6**, 461–464.

Seber, G.A.F. (1983) Capture-recapture methods. In *Encyclopedia of Statistical Science*, S. Kotz and N. Johnson (eds.). J. Wiley, New York.

Seber, G.A.F. (1986) A review of estimation of animal abundance. *Biometrics* **42**, 267–292.

Seidenfeld, T. (1987) Entropy and uncertainty. In *Foundations of Statistical Inference*, I.B. MacNeill and G.J. Umphrey (eds.), 259 287. Reidel, Boston.

Seidenfeld, T. (1992) R.A. Fisher's fiducial argument and Bayes' theorem. *Statist. Sci.* **7**(3), 358–368.

Sen, P.K., Kubokawa, T. and Saleh, A.K.Md.E. (1989) The Stein paradox in the sense of Pitman measure of closeness. *Ann. Statist.* **17**, 1375–1384.

Seneta, E. (1993) Lewis Carroll's pillow problems. *Statist. Science* **8**, 180–186.

Severini, T.A. (1991) On the relationship between Bayesian and non-Bayesian interval estimates. *J. Roy. Statist. Soc., Ser. B* **53**, 611–618.

Severini, T.A. (1993) Bayesian interval estimates which are also confidence intervals. *J. Roy. Statist. Soc., Ser. B* **55**, 533–540.

Shafer, G.R. (1996) *Art of Causal Conjecture*. MIT Press, MIT, Cambridge.

Shannon, C. (1948) A mathematical theory of communication. *Bell System Tech. J.* **27**, 379–423 and 623–656.

Shao, J. (1989) Monte Carlo approximation in Bayesian decision theory. *J. Amer. Statist. Assoc.* **84**, 727–732.

Shao, J. and Strawderman, W.E. (1993) Improving on truncated estimators. In *Statistical Decision Theory and Related Topics V*, J.O. Berger and S.S. Gupta (Eds.) Springer-Verlag, New York., 369–376. Springer–Verlag, New York.

Shao, J. and Strawderman, W.E. (1996) Improving on the James–Stein positive-part estimator. *Statistica Sinica* **6**(1), 259–274.

Shinozaki, N. (1975) Ph.D. thesis, Keio University.

Shinozaki, N. (1980) Estimation of a multivariate normal mean with a class of quadratic loss. *J. Amer. Statist. Assoc.* **75**, 973–976.

Shinozaki, N. (1984) Simultaneous estimation of location parameters under quadratic loss. *Ann. Statist.* **12**, 322–335.

Shinozaki, N. (1990) Improved confidence sets for the mean of a multivariate normal distribution. *Ann. Inst. Statist. Math.* **41**, 331–346.

Shorrock, G. (1990) Improved confidence intervals for a normal variance. *Ann. Statist.* **18**, 972–980.

Silverman, B. (1980) Some asymptotic properties of the probabilistic teacher. *IEEE Trans. Inform. Theory* **26**, 246–249.

Sivaganesan, S. and Berger, J.O. (1989) Ranges of posterior measures for priors with unimodal contaminations. *Ann. Statist.* **17**, 868–889.

Small, C. (1990) A survey of multidimensional medians. *Int. Statist. Rev.* **58**, 263–277.

Smith, A.F.M. (1973) A general Bayesian linear model. *J. Roy. Statist. Soc., Ser. B* **35**, 67–75.

Smith, A.F.M. (1984) Present position and potential developments: some personal view on Bayesian statistics. *J. Roy. Statist. Soc., Ser. A* **147**, 245–259.

Smith, A.F.M. and Hills, S. (1992) Parametrizations issues in Bayesian inference. In *Bayesian Statistics 4*. J.O. Berger, J.M. Bernardo, A.P. Dawid and A.F.M. Smith (Eds.). Oxford University Press, London, 227–238. Oxford University Press, London.

Smith, A.F.M. and Makov, U.E. (1978) A quasi–Bayes sequential procedure for mixtures. *J. Roy. Statist. Soc., Ser. B* **40**, 106–112.

Smith, A.F.M. and Roberts, G.O. (1992) Bayesian computation via Gibbs and related Markov chain Monte Carlo methods (with discussion). *J. Roy. Statist. Soc., Ser. B* **55**, 3–24.

Smith, A.F.M., Sken, A., Shaw, J., Naylor, J.C. and Dransfield, M. (1985) The implementations of the Bayesian paradigm. *Comm. Statist.–Theory Methods* **14**, 1079–1102.

Smith, A.F.M. and Spiegelhalter, D.J. (1982) Bayes factors for linear and log–linear models with vague prior information. *J. Roy. Statist. Soc., Ser. B* **44**, 377–387.

Smith, J.Q. (1988) *Decision Analysis: A Bayesian Approach*. Chapman & Hall, London.

Spiegelhalter, D.J., Best, N.G., and Carlin, B.P. (1998) Bayesian deviance, the effective number of parameters and the comparison of arbitrarily complex models. MRC Biostatistics Unit, Cambridge University.

Spiegelhalter, D.J. and Cowell, R. (1992) Learning in probabilistic expert systems. In *Bayesian Statistics 4*. J.O. Berger, J.M. Bernardo, A.P. Dawid and A.F.M. Smith (Eds.). Oxford University Press, London, 447–460. Oxford University Press, London.

Spiegelhalter, D.J., Dawid, A.P., Lauritzen, S.L. and Cowell, R.G. (1993) Bayesian analysis in expert systems (with discussion). *Statist. Science* **8**, 219–283.

Spiegelhalter, D. and Smith, A.F.M. (1980) Bayes factors and choice criteria for linear models. *J. Roy. Statist. Soc., Ser. B* **42**, 215–220.

Spiegelhalter, D.J. and Lauritzen, S.L. (1990) Sequential updating of conditional probabilities on directed graphical structures. *Networks* **20**, 579-605.

Spiegelhalter, D.J., Thomas, A., Best, N. and Gilks, W.R. (1995a) BUGS: Bayesian Inference Using Gibbs Sampling. Version 0.50. Medical Research Council Biostatistics Unit, Institute of Public Health, Cambridge University.

Spiegelhalter, D.J., Thomas, A., Best, N. and Gilks, W.R. (1995b) BUGS Examples Volume 1, Version 0.50. MRC Biostatistics Unit, Cambridge University.

Spiegelhalter, D.J., Thomas, A., Best, N. and Gilks, W.R. (1995c) BUGS Examples Volume 2, Version 0.50. MRC Biostatistics Unit, Cambridge University.

Srinivasan, C. (1981) Admissible generalized Bayes estimators and exterior boundary value problems. *Sankhya* (Ser. A) **43**, 1–25.

Srivastava, M. and Bilodeau, M. (1988) Estimation of the MSE matrix of the Stein estimator. *Canad. J. Statist.* **16**, 153–159.

Stein, C. (1955a) Inadmissibility of the usual estimator for the mean of a multivariate normal distribution. In *Proc. Third Berkeley Symp. Math. Statist. Probab.* **1**, 197–206. University of California Press.

Stein, C. (1955b) A necessary and sufficient condition for admissibility. *Ann. Math. Statist.* **26**, 518–522.

Stein, C. (1959) An examination of wide discrepancy between fiducial and confidence intervals. *Ann. Math. Statist.* **30**, 877–880.

Stein, C. (1962a) Confidence sets for the mean of a multivariate normal distribution (with discussion). *J. Roy. Statist. Soc., Ser. B* **24**, 573–610.

Stein, C. (1962b) A remark on the likelihood principle. *J. Roy. Statist. Soc., Ser. A* **125**, 565–568.

Stein, C. (1965) Approximation of improper prior measures by prior probability measures. In *Bernoulli, Bayes, Laplace Anniversary Volume.* Springer-Verlag, New York.

Stein, C. (1973) Estimation of the mean of a multivariate distribution. In *Proceedings of the Prague Symposium on Asymptotic Statistics.*

Stein, C. (1981) Estimation of the mean of a multivariate normal distribution. *Ann. Statist.* **9**, 1135–1151.

Stephens, M. (1997) Bayesian methods for mixtures of normal distributions. Ph.D. thesis, Oxford University.

Stephens, M. (2000) Bayesian methods for mixtures of normal distributions. *Ann. Statist.* **28**, 40–74.

Steward, G. (1987) Collinearity and least-squares regression. *Statist. Science* **2**, 68–100.

Steward, L. (1979) Multiparameter univariate Bayesian analysis. *J. Amer. Statist. Assoc.* **74**, 684–693.

Steward, L. (1983) Bayesian analysis using Monte Carlo integration—a powerful methodology for handling some difficult problems. *The Statistician* **32**, 195–200.

Stigler, S. (1986) *The History of Statistics.* Belknap, Harvard.

Stone, M. (1967) Generalized Bayes decision functions, admissibility and the exponential family. *Ann. Math. Statist.* **38**, 818–822.

Stone, M. (1976) Strong inconsistency from uniform priors (with discussion). *J. Amer. Statist. Soc.* **71**, 114–125.

Strasser, H. (1985) *Mathematical Theory of Statistics*. W. de Gruyter, Berlin.

Strawderman, W.E. (1971) Proper Bayes minimax estimators of the multivariate normal mean. *Ann. Math. Statist.* **42**, 385–388.

Strawderman, W.E. (1973) Proper Bayes minimax estimation of the multivariate normal mean. *Ann. Math. Statist.* , **42**, 385–388.

Strawderman, W.E. (1974) Minimax estimation of location parameters for certain spherically symmetric distributions. *J. Multivariate Anal.* **4**, 255–264.

Strawderman, W.E. (2000) Minimaxity. *J. Amer. Statist. Assoc.* **95**, 1364–1368.

Studden, W. (1990) Private communication.

Sweeting, T.J. (1985) Consistent prior distributions for transformed models. In *Bayesian Statistics* **2**, J.M. Bernardo, M.H. DeGroot, D.V. Lindley and A.F.M. Smith (eds.), 755–762. Elsevier Science Publishers, Amsterdam.

Tanner, M. (1991) *Tools for Statistical Inference: Observed Data and Data Augmentation Methods*. Lecture Notes in Statistics **67**, Springer–Verlag, New York.

Tanner, M. and Wong, W.H. (1987) The calculation of posterior distributions by data augmentation. *J. Amer. Statist. Assoc.* **82**, 528–550.

Thatcher, A.R. (1964) Relationships between Bayesian and confidence limits in prediction. *J. Roy. Statist. Soc., Ser. B* **26**, 176–210.

Thisted, R.A. and Efron, B. (1987) Did Shakespeare write a newly-discovered poem? *Biometrika* **74**, 445–468.

Thompson, P.M. (1989) *Admissibility of p-value rules*. Ph.D. thesis, University of Illinois, Urbana.

Tibshirani, R. (1989) Noninformative priors for one parameter of many. *Biometrika* **76**, 604–608.

Tierney, L. (1991) Markov chains for exploring posterior distributions. *Computer Sciences and Statistics: Proc. 23d Symp. Interface*, 563–570.

Tierney, L. (1994) Markov chains for exploring posterior distributions (with discussion). *Ann. Statist.* **22**, 1701–1786.

Tierney, L. and Kadane, J.B. (1986) Accurate approximations for posterior moments and marginal densities. *J. Amer. Statist. Assoc.* **81**, 82–86.

Tierney, L., Kass, R.E. and Kadane, J.B. (1989) Fully exponential Laplace approximations to expectations and variances of non-positive functions. *J. Amer. Statist. Assoc.* **84**, 710–716.

Titterington, D.M., Smith, A.F.M. and Makov, U.E. (1985) *Statistical Analysis of Finite Mixture Distributions*. J. Wiley, New York.

Tong, H. (1991) *Non-linear Time Series: a Dynamical Systems Approach.* Oxford University Press, London.

Torrie, G.M. and Valleau, J.P. (1977) Nonphysical sampling distributions in Monte Carlo free-energy estimation: Umbrella sampling. *J. Chemical Physics* **23**, 187–199.

van der Meulen, B. (1992) Assessing weights of evidence for discussing classical statistical hypotheses. Ph.D. thesis, University of Groningen.

Van Dijk, H.K. and Kloeck, T. (1984) Experiments with some alternatives for simple importance sampling in Monte Carlo integration. In *Bayesian Statistics* II, J.M. Bernardo, M.H. DeGroot, D.V. Lindley and A.F.M. Smith (eds.). North-Holland, Amsterdam.

van Eeden, C. and Zidek, J. (1993) Group Bayes estimation of the exponential mean: a retrospective view of the Wald theory. In *Statistical Decision Theory and Related Topics V*, J.O. Berger and S.S. Gupta (Eds.) Springer–Verlag, New York., 35–50. Springer–Verlag, New York.

Venn, J. (1886) *The Logic of Chance*. Macmillan, London.

Verdinelli, I. and Wasserman, L. (1992) Bayesian analysis of outliers problems using the Gibbs sampler. *Statist. Comput.* **1**, 105–117.

Verdinelli, I. and Wasserman, L. (1995) Computing Bayes Factors Using a Generalization of the Savage–Dickey Density Ratio. *Journal of the American Statistical Association* **90**, 614-618.

Verdinelli, I. and Wasserman, L. (1998) Bayesian goodness-of-fit testing using infinite-dimensional exponential families. *Ann. Statist.* **26**, 1215–1241.

Villegas, C. (1977) On the representation of ignorance. *J. Amer. Statist. Assoc.* **72**, 651–654.

Villegas, C. (1990) Bayesian inference in models with euclidian structure. *J. Amer. Statist. Assoc.* **85**, 1159–1164.

Von Neumann, J. (1951) Various techniques used in connection with random digits. *J. Resources of the National Bureau of Standards – Applied Mathematics Series* **12**, 36–38.

Von Neumann, J. and Morgenstern, O. (1947) *Theory of Games and Economic Behavior* (2nd edition). Princeton University Press, Princeton.

Wakefield, J.C., Gelfand, A.E. and Smith, A.F.M. (1991) Efficient generation of random variates via the ratio-of-uniforms method. *Statist. Comput.* **1**, 129–33.

Wald, A. (1950) *Statistical Decision Functions*. J. Wiley, New York.

Wallace, C.S. and Boulton, D.M. (1975) An invariant Bayes method for point estimation. *Classification Society Bulletin* **3**(3), 11–34.

Walley, P. (1991) *Statistical Reasoning with Imprecise Probability*. Chapman & Hall, London.

Wasserman, L. (1992) Recent methodological advances in robust Bayesian inference. In *Bayesian Statistics 4*. J.O. Berger, J.M. Bernardo, A.P. Dawid and A.F.M. Smith (Eds.). Oxford University Press, London, 483–490. Oxford University Press, London.

Wasserman, L. (1999) Asymptotic inference for mixture models by using data-dependent priors. *J. Roy. Statist. Soc., Ser. B* **61**(1), 159–180.

Welch, B.L. (1965) On comparisons between confidence point procedures in the case of a single parameter. *J. Roy. Statist. Soc., Ser. B* **27**, 1–8.

Welch, B.L. and Peers, H.W. (1963) On formulae for confidence points based on integrals of weighted likelihoods. *J. Roy. Statist. Soc., Ser. B* **25**, 318–329.

Wells, M.T. (1992) Private communication.

West, M. (1992) Modelling with mixtures. In *Bayesian Statistics 4*. J.O. Berger, J.M. Bernardo, A.P. Dawid and A.F.M. Smith (Eds.). Oxford University Press, London, 503–525. Oxford University Press, London.

West, M. and Harrison, J. (1998) *Bayesian Forecasting and Dynamic Models* (2nd edition). Springer–Verlag, New York.

Whittaker, J. (1990) *Graphical Models in Applied Multivariate Statistics*. Wiley, Chichester.

Wijsman, R.A. (1990) *Invariant Measures on Groups and their Use in Statistics*. IMS lecture notes–Monographs Series. Hayward, California.

Wilkinson, G. (1977) On resolving the controversy in statistical inference. *J. Roy. Statist. Soc., Ser. B* **39**, 119–171.

Wolter, W. (1986) Some coverage error models for census data. *J. Amer. Statist. Assoc.* **81**, 338–346.

Zabell, S.L. (1989) R.A. Fisher on the history of inverse probability. *Statist. Science* **4**, 247–263.

Zabell, S.L. (1992) R.A. Fisher and the fiducial argument. *Statist. Science* **7**, 369–387.

Zellner, A. (1971) *An Introduction to Bayesian Inference in Econometrics*. J. Wiley, New York.

Zellner, A. (1976) Bayesian and non-Bayesian analysis of the regression model with multivariate Student-*t* error term. *J. Amer. Statist. Assoc.* **71**, 400–405.

Zellner, A. (1984) *Basic Issues in Econometrics*. University of Chicago Press, Chicago.

Zellner, A. (1986a) Bayesian estimation and prediction using asymmetric loss functions. *J. Amer. Statist. Assoc.* **81**, 446–451.

Zellner, A. (1986b) On assessing prior distributions and Bayesian regression analysis with *g*-priors distributions. In *Bayesian Inference and Decision Techniques*, P. Goel and A. Zellner (eds.), 233–243. Elsevier North-Holland, Amsterdam.

Zidek, J.V. (1969) A representation of Bayes invariant procedures in terms of Haar measure. *Ann. Inst. Statist. Math.* **21**, 291–308.

Zidek, J.V. (1970) Sufficient conditions for the admissibility under squared error loss of formal Bayes estimators. *Ann. Math. Statist.* **41**, 1444-1447.

Zucchini, W. (1999) Frequentist model choice. Summer school lecture, Cagliari, Sardinia, 23rd October 1999.

Yao, J.F. and Attali, J.G. (2000) On stability of nonlinear AR processes with Markov switching. *Applied Probability* **32**, 394–407.

Author Index

Subject Index

Springer Texts in Statistics

Bayesian Core: A Practical Approach to Computational Bayesian Statistics
Jean-Michel Marin and Christian P. Robert

This Bayesian modeling book is intended for practitioners and applied statisticians looking for a self-contained entry to computational Bayesian statistics. Focusing on standard statistical models and backed up by discussed real datasets available from the book website, it provides an operational methodology for conducting Bayesian inference, rather than focusing on its theoretical justifications. Special attention is paid to the derivation of prior distributions in each case and specific reference solutions are given for each of the models.

2007. 270 pp. (Springer Texts in Statistics) Hardcover
ISBN 978-0-387-38979-0

Monte Carlo Statistical Methods
Second Edition
Christian P. Robert and George Casella

The second edition has been revised towards a coherent and flowing coverage of these simulation techniques. This is a textbook intended for a second year graduate course, but someone who either wants to apply simulation techniques for the resolution of practical problems or wishes to grasp the fundamental principles behind those methods can also use it. Chapters 1–5 cover non-Markov Monte Carlo techniques for integration and optimization, while Chapters 7—12 provide a complete coverage of Markov chain Monte Carlo (MCMC) methods. Chapters 13 and 14 provide a path to more recent developments.

2004. 645 pp. (Springer Texts in Statistics) Hardcover
ISBN 978-0-387-21239-5

An Introduction to Bayesian Analysis
Jayanta K. Ghosh, Mohan Delampady, and Tapas Samanta

This is a graduate level textbook on Bayesian analysis blending modern Bayesian theory, methods, and applications. Starting from basic statistics, undergraduate calculus and linear algebra, ideas of both subjective and objective Bayesian analysis are developed to a level where real-life data can be analyzed using the current techniques of statistical computing. Advances in both low-dimensional and high-dimensional problems are covered, as well as important topics such as empirical Bayes and hierarchical Bayes methods and Markov chain Monte Carlo (MCMC) techniques.

2006. 365 pp. (Springer Texts in Statistics) Hardcover
ISBN 978-0-387-40084-6

Easy Ways to Order▶   Call: Toll-Free 1-800-SPRINGER • E-mail: orders-ny@springer.com • Write: Springer, Dept. S8113, PO Box 2485, Secaucus, NJ 07096-2485 • Visit: Your local scientific bookstore or urge your librarian to order.